CHEMICAL ENGINEERING DESIGN AND ANALYSIS

Second Edition

This textbook puts design at the center of introducing students to the course in mass and energy balances in chemical engineering. Employers and accreditations increasingly stress the importance of design in the engineering curriculum, and design-driven analysis will motivate students to dig deeply into the key concepts of the field.

The Second Edition has been completely revised and updated. It introduces the central steps in design, and three methods of analysis: mathematical modeling, graphical methods, and dimensional analysis. Students learn how to apply engineering skills – such as simplification of calculations through assumptions and approximations; verification of calculations; identification of significant figures; application of spreadsheets; graphical analysis (standard, semi-log, and log–log); and the use of data maps – in the contexts of contemporary chemical processes such as the hydrogen economy, petrochemical and biochemical processes, polymers, semiconductors, and pharmaceuticals.

T. Michael Duncan joined the School of Chemical Engineering at Cornell University in 1990, where he holds the Thorpe Chair in Chemical Engineering and has served as Associate Director for the undergraduate program since 1993. Duncan has received many teaching awards: he has been selected four times for the Tau Beta Pi / Cornell Engineering Alumni Association Excellence in Teaching Award and was named Professor of the Year for New York State by the Carnegie Foundation in 2007. He is also a Weiss Scholar at Cornell University, a distinction bestowed on three faculty members each year at Cornell, from the entire campus of over 1600 instructors.

Jeffrey A. Reimer is The C. Judson King Endowed Professor and Warren and Katharine Schlinger Distinguished Professor and Chair of the Chemical and Biomolecular Engineering Department at the University of California at Berkeley. Reimer has received many teaching awards, culminating in the university's Distinguished Teaching Award, the highest award bestowed on faculty for their teaching. He is a Fellow of the American Association for the Advancement of Science, a Fellow of the American Physical Society, and a Fellow of the International Society for Magnetic Resonance, and was the recipient of a Humboldt Research Award in 2014.

CAMBRIDGE SERIES IN CHEMICAL ENGINEERING

Books in the Series

Baldea and Daoutidis, *Dynamics and Nonlinear Control of Integrated Process Systems*

Chamberlin, *Radioactive Aerosols*

Chau, *Process Control: A First Course with MATLAB*

Cussler, *Diffusion: Mass Transfer in Fluid Systems, Third Edition*

Cussler and Moggridge, *Chemical Product Design, Second Edition*

De Pablo and Schieber, *Molecular Engineering Thermodynamics*

Deen, *Introduction to Chemical Engineering Fluid Mechanics*

Denn, *Chemical Engineering: An Introduction*

Denn, *Polymer Melt Processing: Foundations in Fluid Mechanics and Heat Transfer*

Dorfman and Daoutidis, *Numerical Methods with Chemical Engineering Applications*

Duncan and Reimer, *Chemical Engineering Design and Analysis: An Introduction 2E*

Fan, *Chemical Looping Partial Oxidation Gasification, Reforming, and Chemical Syntheses*

Fan and Zhu, *Principles of Gas–Solid Flows*

Fox, *Computational Models for Turbulent Reacting Flows*

Franses, *Thermodynamics with Chemical Engineering Applications*

Leal, *Advanced Transport Phenomena: Fluid Mechanics and Convective Transport Processes*

Lim and Shin, *Fed-Batch Cultures: Principles and Applications of Semi-batch Bioreactors*

Litster, *Design and Processing of Particulate Products*

Marchisio and Fox, *Computational Models for Polydisperse Particulate and Multiphase Systems*

Mewis and Wagner, *Colloidal Suspension Rheology*

Morbidelli, Gavriilidis, and Varma, *Catalyst Design: Optimal Distribution of Catalyst in Pellets, Reactors, and Membranes*

Nicoud, *Chromatographic Processes*

Noble and Terry, *Principles of Chemical Separations with Environmental Applications*

Orbey and Sandler, *Modeling Vapor–Liquid Equilibria: Cubic Equations of State and their Mixing Rules*

Pfister, Nicoud, and Morbidelli, *Continuous Biopharmaceutical Processes: Chromatography, Bioconjugation, and Protein Stability*

Petyluk, *Distillation Theory and its Applications to Optimal Design of Separation Units*

Ramkrishna and Song, *Cybernetic Modeling for Bioreaction Engineering*

Rao and Nott, *An Introduction to Granular Flow*

Russell, Robinson, and Wagner, *Mass and Heat Transfer: Analysis of Mass Contactors and Heat Exchangers*

Schobert, *Chemistry of Fossil Fuels and Biofuels*

Shell, *Thermodynamics and Statistical Mechanics*

Sirkar, *Separation of Molecules, Macromolecules and Particles: Principles, Phenomena and Processes*

Slattery, *Advanced Transport Phenomena*

Varma, Morbidelli, and Wu, *Parametric Sensitivity in Chemical Systems*

Chemical Engineering Design and Analysis

An Introduction

Second Edition

T. Michael Duncan

Raymond G. Thorpe Professor of Chemical Engineering
Cornell University

Jeffrey A. Reimer

The Warren and Katharine Schlinger Distinguished Professor in Chemical Engineering
The C. Judson King Endowed Professor of Chemical and Biomolecular Engineering
University of California at Berkeley

CAMBRIDGE
UNIVERSITY PRESS

University Printing House, Cambridge CB2 8BS, United Kingdom

One Liberty Plaza, 20th Floor, New York, NY 10006, USA

477 Williamstown Road, Port Melbourne, VIC 3207, Australia

314-321, 3rd Floor, Plot 3, Splendor Forum, Jasola District Centre, New Delhi - 110025, India

103 Penang Road, #05-06/07, Visioncrest Commercial, Singapore 238467

Cambridge University Press is part of the University of Cambridge.

It furthers the University's mission by disseminating knowledge in the pursuit of
education, learning and research at the highest international levels of excellence.

www.cambridge.org
Information on this title: www.cambridge.org/9781108421478
DOI: 10.1017/9781108377096

First published 1998
10th printing 2011
Second edition 2019

A catalogue record for this publication is available from the British Library

Library of Congress Cataloging in Publication data
Names: Duncan, T. Michael, author. | Reimer, Jeffrey A. (Jeffrey Allen), author.
Title: Chemical engineering design and analysis : an introduction / T. Michael Duncan, Raymond G. Thorpe Professor of Chemical Engineering, Cornell University, Jeffrey A. Reimer, The Warren and Katharine Schlinger Distinguished Professor in Chemical Engineering, The C. Judson King Professor of Chemical and Biomolecular Engineering, University of California at Berkeley.
Description: 2nd edition. | New York : Cambridge University Press, 2019. | Series: Cambridge series in chemical engineering | Includes bibliographical references and index.
Identifiers: LCCN 2018034173 | ISBN 9781108421478 (hardback : alk. paper)
Subjects: LCSH: Chemical engineering.
Classification: LCC TP155 .D74 2018 | DDC 660–dc23
LC record available at https://lccn.loc.gov/2018034173

ISBN 978-1-108-42147-8 Hardback

Additional resources for this publication at duncan.cbe.cornell.edu/Graphs

To Deborah and Maxwell
T. Michael Duncan

To Karen, Jennifer, Jonathan, Charlotte, and Martin
Jeffrey A. Reimer

Brief Contents

	Preface	*page* xv
	Acknowledgements	xviii
1	**An Overview of Chemical Engineering**	1
2	**Chemical Process Design**	8
3	**Models Derived from Laws and Mathematical Analysis**	89
4	**Models Derived from Graphical Analysis**	243
5	**Dimensional Analysis and Dynamic Similarity**	423
6	**Transient-State Processes**	499
	Appendix A List of Symbols	546
	Appendix B Units, Conversion Factors, and Physical Constants	548
	Appendix C Significant Figures	551
	Appendix D Log–Log Graph Paper	554
	Appendix E Mathematics, Mechanics, and Thermodynamics	559
	Appendix F Glossary of Chemical Engineering	562
	Index	569

Contents

	Preface	*page* xv
	Acknowledgements	xviii
1	**An Overview of Chemical Engineering**	1
	1.1 Achievements of Chemical Engineering	3
	1.2 Opportunities for Chemical Engineering	5
	Reference	7
2	**Chemical Process Design**	8
	2.1 Designing a Chemical Process	8
	2.1.1 Design Evolution by Successive Problem Solving	8
	2.1.2 Analyzing Data to Design Chemical Processes	16
	2.1.3 Conventions for Chemical Process Flowsheets	17
	2.2 Chemical Process Design and Creative Problem Solving	19
	2.2.1 Defining the *Real* Problem in Successive Problem Solving	19
	2.2.2 Conventions for Streams on Chemical Process Flowsheets	23
	2.3 Designing a Chemical Process for the Semiconductor Industry	25
	2.4 Designing Processes to Produce Hydrogen	29
	2.4.1 Hydrogen from Methane	29
	2.4.2 Methane from Natural Gas	32
	2.4.3 Hydrogen from Coal	35
	2.4.4 Hydrogen from Thermal Energy and Water	36
	2.4.5 Tips for Chemical Process Design: Analyzing a Process Flowsheet	40
	2.5 Designing a Process to Store Hydrogen	42
	2.5.1 Options for Storing Hydrogen	42
	2.5.2 Storing Hydrogen as Fuel for Vehicles	44
	Summary	47
	References	48
	Chemical Process Design Bibliography	48
	Exercises	48
	Process Analysis	48
	Process Design	57
	Problem Redefinition	80
	Physical Properties at 1 atm.	86

3 **Models Derived from Laws and Mathematical Analysis** 89
3.1 Mass Balances on Processes with No Chemical Reaction 90
3.1.1 Mass Balances on a Single Unit at Steady State 90
3.1.2 Mass Balances on a Process with Several Units and a Recycle Stream 95
3.2 Mass Balances on Processes with Chemical Reactions 99
3.2.1 A Chemical Reactor with a Separator 99
3.2.2 A Recycle for Minimal Reactant Input and Minimal Waste Output 104
3.2.3 A Purge for Moderate Reactant Input and Moderate Waste Output 106
3.3 Informal Mass Balances for Design Evolution 110
3.4 Mathematical Modeling with Mass Balances: Summary 114
3.4.1 Some Tips on System Borders 114
3.4.2 Mass Balances and Learning Styles 116
3.5 Energy Balances on a Single Unit with No Chemical Reaction 117
3.5.1 Energy Balances for Temperature and Phase Changes 117
3.5.2 Energy Balances for Temperature Changes with Variable Heat Capacity 123
3.5.3 Heat Integration: Matching Energy Needs to Energy Sources 128
3.6 Energy Balances and Chemical Reactions 132
3.6.1 Chemical Reactions with Complete Conversion 132
3.6.2 Chemical Reactions with Incomplete Conversion 135
3.6.3 Chemical Reactions with Conversion Limited by Equilibrium 139
3.7 Chemical Process Economics 145
3.7.1 Economic Analysis of Operating a Chemical Process 145
3.7.2 Economic Analysis of Modifying a Chemical Process 150
3.7.3 Economic Analysis for Evaluating Design Schemes 152
Summary 155
References 156
Exercises 156
Mass Balances 156
Mass Balances with Chemical Reactions 172
Informal Mass Balances 178
Mass Balances on Spreadsheets 182
Energy Balances 187
Energy Balances with Chemical Reactions 195
Process Economics 201
Process Design with Mathematical Modeling 215
Engineering Calculations 241

4 **Models Derived from Graphical Analysis** 243
4.1 Tie Lines, Mixing Lines, and the Lever Rule 244
4.1.1 Graphical Mass Balances 244
4.1.2 Graphical Energy Balances 247
4.1.3 Graphical Mass Balances for Single-Stage Liquid–Vapor Separations 253
4.1.4 Combined Mass and Energy Balances on Two-Component Mixtures 262

4.2 Operating Lines for Two-Phase Systems 267
4.2.1 Single-Stage Absorbers 267
4.2.2 Multistage Absorbers 272
4.2.3 Multistage Liquid–Vapor Separations 277
4.2.4 Multistage Cascading Flash Drums 283
4.3 Trajectories on Pure-Component Phase Diagrams 289
4.3.1 Mapping Solid–Liquid–Gas Phases of a Pure Component 289
4.3.2 Condensation from a Non-condensible Gas 294
Summary 301
References 301
Exercises 302
Linear and Logarithmic Scales 302
Graphical Energy Balances on Pure Substances 303
Graphical Mass Balances on Temperature–Composition and Pressure–Composition Phase Diagrams 306
Combined Mass and Energy Balances on Two-Component Mixtures: Enthalpy–Composition Phase Diagrams 324
Operating Lines for Multistage Countercurrent Separators: Absorbers and Strippers 329
Operating Lines for Multistage Countercurrent Separators: Distillation Columns 350
Phase Diagrams of Pure Substances 376
Process Design with Graphical Modeling 390

5 Dimensional Analysis and Dynamic Similarity 423
5.1 Units and Dimensions 426
5.2 Dimensional Analysis 428
5.2.1 Dimensional Analysis of a Pendulum Swinging 428
5.2.2 Dimensional Analysis of a Person Walking and Running 431
5.2.3 Dimensional Analysis of a Solid Sphere Moving through a Fluid 436
5.3 Dynamic Similarity 447
5.3.1 Dynamic Similarity of Fluid Flow in a Smooth Pipe 447
5.3.2 Dynamic Similarity of Fluid Flow in a Rough Pipe 452
5.3.3 Dynamic Similarity of Heat Transfer from a Fluid Flowing in a Tube 454
5.3.4 Dynamic Similarity of Vapor–Liquid Equilibrium Stages 457
5.4 Applications of Dimensional Analysis 459
5.4.1 Dimensional Analysis of Gases 459
5.4.2 Dimensional Analysis of Biological Systems 463
5.4.3 Dimensional Analysis of Microchemical Systems 465
Summary 466
References 467
Exercises 468
Units and Dimensions 468
Deriving Dimensionless Groups 469
Analyzing Graphical Data 476
Design of Dynamically Similar Models 481
Data Analysis on Spreadsheets 487

6	**Transient-State Processes**	499
6.1	Transient-State Mass Balances: A Surge Tank	500
6.2	Residence Times and Sewage Treatment	507
6.3	Rate Constants: Modeling Atmospheric Chemistry	512
6.4	Optimization: Batch Reactors	517
6.5	Multiple Steady States: Catalytic Converters	523
6.6	Mass Transfer: Citric Acid Production	528
6.7	Heat Transfer: Chemical Reactor Runaway	531
	Summary	534
	References	534
	Exercises	534
	Transient-State Processes	534
	Numerical Integration of Differential Equations	542

Appendix A	List of Symbols	546
Appendix B	Units, Conversion Factors, and Physical Constants	548
Appendix C	Significant Figures	551
Appendix D	Log–Log Graph Paper	554
Appendix E	Mathematics, Mechanics, and Thermodynamics	559
Appendix F	Glossary of Chemical Engineering	562
Index		569

Preface

The teaching of introductory chemical engineering traditionally begins with analysis. Our contention is that design is the essential prerequisite to analysis – and that, in fact, beginning the study of chemical engineering with design is more motivating and engaging to students. We originally developed this textbook to demonstrate that design – the quintessential skill for chemical engineers – can be taught at the first-year or second-year level. This new edition, refined and reorganized following many semesters of classroom use at Cornell University and the University of California at Berkeley, continues to adhere to that paradigm.

Why is the central theme of this textbook process design substantiated by analysis? Design is a key skill in the chemical engineering curriculum. Employers and accreditation boards increasingly stress design training, yet design is typically postponed until a capstone course in the final semester. One design experience at the end of a four-year program is not enough and we, along with other educators, believe that design is an experience that must grow with the student's development. This textbook, we believe, succeeds in providing students with that experience.

THE APPROACH OF THIS TEXT

The choice of textbook for a course, and in particular the first course in the chemical engineering discipline, is a vexing one. Instructors who teach mass and energy balance often feel as though the weight of the whole discipline is resting on their choice; that is, that they must convey a comprehensive set of deep and abiding truths, teach the skill set that will launch a career, and deploy the most modern methods from educational psychology and teaching practice.

Traditional mass and energy balance textbooks represent this approach well. Indeed, these books are so powerful and comprehensive that the independent learner can engage with them in the absence of a professor or classroom; they are a self-study course in print, open to anyone willing to invest the time. Dedicated and brilliant authors prepare traditional comprehensive texts, and many dedicated teachers use them effectively.

We chose to create a different type of textbook – namely a conversational narrative that focuses on design. "To create what never has been," to quote Theodore von Kármán, is a thrill at all stages of preparation in the discipline, and a delight at all ability levels. As such, we focus our narratives less on comprehensive coverage of topics typical of the traditional Mass And Energy Balance course. We believe that the notion of learning being governed by coverage of topics is a legacy of the twentieth century and has not been substantiated by contemporary educational psychology. Comprehensive coverage of topics further complicates instruction in the introductory course because the number of topics encompassed by chemical engineering continues to grow substantially. We offer something special with our textbook – namely that the concepts of algebra and calculus, the visual display of information, and dimensional analysis are *timeless themes* that are the perfect platform for a book that is neither an encyclopedia nor a list of current industrial/academic topics.

FEATURES OF THIS TEXT

- **A story-telling approach.** We intend this textbook to be read, not studied. We are storytellers, and the student will find this book to be a collection of stories crafted to show the reader how to design chemical processes. The end-of-chapter exercises afford practice at the design skills introduced in the stories. These exercises have been time-tested by students at our institutions. As is true for most designs, the exercises are often without a unique solution (as is true for most designs), and indeed are best worked with fellow students engaged in the process of discussing, posing, correcting, and re-posing ideas.
- **The essential tools to create new designs.** If design is to be a full partner in a foundational course, then it is appropriate to ask what tools are necessary for students to create new designs. Here is where we deviate significantly from other "mass and energy balance" textbooks: we demonstrate three different skills for students to employ in design:
  - Like our peer texts, we rely heavily on algebra and macroscopic mathematical models for mass, energy, and asset conservation.
  - Unlike others, we affirm that the visual display of information is in its own right a design tool for systems whose behavior is well described phenomenologically, e.g. systems described by thermodynamic phase diagrams.
  - Finally, we augment algebra and graphs with dimensional analysis for those systems that elude a fundamental basis. This topic is often relegated to later courses in the curriculum and is so heavily contextualized in those specific topics that its general applicability is lost; Chapter 5 rectifies this omission.

The denouement of our textbook engages all three approaches to design and analyze systems that vary with time.

- **"Context, concepts, defining question"** introduction for each section, which establishes for the student a framework for thinking about chemical engineering
- **An abundance of design-oriented exercises.** A key feature of our textbook is the exercises; 359 exercises in total, including 71 open-ended design exercises. Every exercise has been assigned at least once in our courses at Cornell University and the University of California at Berkeley. The wording, the just-in-time delivery of concepts, and the solutions have been crafted so students can practice design skills and then deploy them in situations completely unfamiliar to them.

ORGANIZATION OF THIS TEXT

The coverage of such a vast array of topics suggests that students need to have been exposed to many chemical engineering topics prior to using this book. On the contrary, we chose to eschew an encyclopedic presentation of chemical engineering information: our textbook organizes skills in a "just-in-time" fashion, where each skill is presented to answer a pending design question.

After a brief overview of contemporary chemical engineering in Chapter 1, we introduce concepts and methods of qualitative process design in Chapter 2. We then demonstrate three quantitative methods to analyze design options. Chapter 3 introduces mathematical modeling with mass and energy balances, including techniques for informal mass balances, essential for quickly assessing a design change. We further improve this traditional topic by introducing a third tool essential for process design: process economics, which is mathematical modeling based on the conservation of assets. Chapter 4 demonstrates graphical modeling for quantitative process analysis and design; graphical representations of mass and energy balances. Chapter 5 introduces dimensional analysis and dynamic scaling for process design scale-up from pilot plant to commercial process or scale-down from bench scale to microchemical process. Chapter 6 applies mathematical modeling, graphical modeling, and dynamic scaling to transient-state processes.

TOOLS FOR THE INSTRUCTOR

Our online resources include:

- A solutions manual providing detailed solutions for every exercise in our textbook
- Alternative designs (with relative pros and cons)
- A list of common errors
- Additional exercises, similar to the ones in the textbook, which will be posted regularly, together with detailed solutions
- Design projects
- Grading rubrics, which will be posted for:
  - The design exercises published in the second edition
  - The additional design exercises that we will be regularly adding online
- PowerPoint presentations for exercises suitable for in-class team solutions
- Suggestions for abbreviated syllabi
- Lectures in PowerPoint or in Word, with different contexts to support the concepts of many textbook sections. Please contact us for any of this supplementary material.

We have had success in adopting this text for the Mass and Energy Balance course taught in the 14-week semester system. At various times, and with various stages of preparation, we have chosen to amplify some material. For example, well-prepared students can successfully use Chapter 6 for the organizing topics, with back-reflection to earlier chapters when steady-state design is to be considered. We have also used this text in a course that emphasizes Chapters 2 and 3 for a large fraction of the semester, an especially effective approach when students have minimal chemistry, physics, and mathematics backgrounds. Finally, we have discovered in conversations with colleagues around the world, as well as thoughtful reviews provided of a draft of this text, that the introductory chemical engineering course is often used to introduce chemical thermodynamics. This text does not serve that need well, though we have had considerable success with Chapter 4 as preparation for a chemical engineering thermodynamics course without having to parse exercises into constructs "ideal" and "real." Data drive our description of phase behavior, graphs represent those data well, and graphs can be used to conduct design, including McCabe–Thiele type analyses.

We hope you enjoy reading our book and we welcome your feedback. We are pleased to regularly post new exercises and resources onto our website.

Acknowledgements

We are grateful to the hundreds of Cornell University and University of California at Berkeley students in the past two decades for their patience and critical comments in preparing this textbook. We have been fortunate to work with dozens of dedicated teaching assistants who also improved the text and paid particular attention to the exercises and solutions. During this time we benefited from the steadfast support of Cornell University and the University of California at Berkeley and the freedom to teach and learn alongside our students.

Several textbooks influenced this work. Two fine textbooks on design – *Process Synthesis* by Dale Rudd, Gary Powers, and Jeffrey Siirola, and *Process Modeling* by Morton Denn – inspired us and spawned the material on process design (Chapter 2), mathematical modeling (Chapter 3), and transient process (Chapter 6).

Mass and energy balances, introduced in Chapter 3 as examples of mathematical modeling, is a mature topic in chemical engineering. We are grateful to the authors of three excellent textbooks – Richard Felder and Ronald Rousseau, *Elementary Principles of Chemical Processes*, William Luyben and Leonard Wenzel, *Chemical Process Analysis: Mass and Energy Balances*, and Gintaras Reklaitis, *Introduction to Material and Energy Balances* – for permission to adapt and reprint examples and exercises.

1

An Overview of Chemical Engineering

Chemical engineering is a compelling discipline. It has a rich intellectual history and has provided humankind with technologies that have transformed our lives. Introductory textbooks in chemical engineering, as well as numerous websites, detail our history and contributions to society. The appeal of chemical engineering lies in its balance of two diverse endeavors: to serve society and to advance fundamental science and engineering. Chemical engineering is the academic embodiment of F. Scott Fitzgerald's contention that "... the test of a first rate intelligence is the ability to hold two opposed ideas in the mind at the same time, and still retain the ability to function."[1]

A challenge for chemical engineering derives from its two ostensibly disparate goals: deepening the great intellectual concepts befitting academic engineering, yet enabling the practical "hands-on" skills for professional practice. Chemical engineering is eminently practical because the skills associated with it are integral to people who make products via chemical transformations. It remains one of the highest-paid engineering professions (often second only to petroleum engineering, a specialty within chemical engineering), and students seek training in chemical engineering to join great worldwide enterprises in chemicals, food, energy, electronics, pharmaceuticals, and biotechnology, for example. Academic chemical engineering therefore must enable the professional practice of its graduates, including practices and tools used in the workplace.

Chemical engineering is also a rich intellectual discipline that embraces enduring concepts from science and mathematics. For example, Josiah Willard Gibbs is considered one of the greatest scientists of the twentieth century: "Gibbs energy" is the foundation for undergraduate chemical engineering thermodynamics. The differential control volume constructed for calculus to express conservation of mass, energy, and momentum is an engineering science triumph of the twentieth century and a hallmark of undergraduate chemical engineering. Computational methods as analytical tools (such as non-linear mathematics, and digital data acquisition for process control and analysis) were developed by the twentieth-century intellectual and academic elite, and are now routinely employed in the university classroom and laboratory.

The practical and the theoretical are the yin and yang of chemical engineering, in constant balance. Most chemical engineering curricula achieve this balance by relying on research professors to teach analysis, with the design aspects taught via internships, "co-ops," lab research, and a capstone design course typically taken just before graduation

[1] *The Crack-Up*, by F. Scott Fitzgerald, www.esquire.com/lifestyle/a4310/the-crack-up. Originally published in the February, March, and April 1936 issues of *Esquire*.

and taught by licensed engineers. Yet within each course, and indeed within the definition of chemical engineering itself, this balance is manifest in many ways.

In this textbook we introduce chemical engineering to students as "the design and analysis of systems . . ." because this expression nicely frames the contrasting attributes of the chemical engineer. Design is often associated with the moniker "right brain activity," where creative and practical arts merge to conceive a plan for something new. The great twentieth-century engineer Theodore von Kármán stated, "Science discovers what is, engineers build what never has been."[2] Analysis, on the other hand, is convergent thinking that typically involves mathematics. This "left-brain activity" constitutes a considerable portion of the engineering practice because mathematical modeling is an essential tool in the "creation of what has never been." Chemical engineering is regarded as one of the most challenging majors because it requires creative and practical designs, as well as mathematical analysis in systems highly constrained by physical laws. But with challenge comes pride in achievement – as an introductory student you stand proudly on the bridge spanning these two complementary views.

We complete the phrase above: chemical engineering is "the design and analysis of systems governed by physical and chemical rate processes." At the start of the twentieth century, chemical engineering was born considering the "unit operations" associated with commodity chemical production, sometimes called industrial chemistry. The post-World War II generation of chemical engineers designed and still operate the refineries that fuel modern transportation. The notion of chemical plant design as parsed into "units" where each performs a specific operation (mixing, reactions, separations, etc.) is canon to chemical engineers and remains to the present day. The worldwide volume (or value) of chemicals produced has grown by a factor of 6 over the period 1980–2020, yet the number of BS chemical engineers entering the profession each year has remained roughly constant.[3] This is superb testimony to the increased production efficiency of the systems attributable to our discipline. It is worth noting that early chemical engineers embraced computational modeling and design which amplified the economic value and quantity of chemical goods produced. Emerging economies, where new manufacturing and chemicals production are often co-located, still seek the "unit operations" training in the contemporary chemical engineer. The practical engineer indeed has much to contribute here. Finally, it is interesting to note that the stunning success of the "unit operation" style of design and analysis of chemical plants in the mid twentieth century is a primary motivator for the application of chemical engineering to emerging technologies.

Should growth in large-scale manufacturing return to the USA, no doubt unit operations as deployed by the chemical engineering professional will wax. In the last 20 years or so, however, nations such as the USA have built few new chemical plants, and even fewer new refineries. Specialty chemical, small-scale manufacturing, electronic and "niche" industries, medical diagnostics, health and personal-care products require chemical engineers who will optimize "physical and chemical rates" in devices and situations not previously explored by chemical engineers. It is tempting to conclude that academic chemical engineering, with its emphasis on molecular engineering and science fundamentals with mathematical modeling, is the most valuable skill set for the

[2] www.quotesby.net/Theodore-von-Karman; www.nationalmedals.org/laureates/theodore-von-karman.

[3] Worldwide chemical production is posted at several websites, including www.americanchemistry.com/Jobs/EconomicStatistics/Industry-Profile/Global-Business-of-Chemistry. We have used data from UNEP (2013) (ISBN: 978-92-807-3320, © United Nations Environment Programme). The number of BS graduates per year is typically reported in *Chemical and Engineering News*'s Annual Reports on Professional Training.

twenty-first-century chemical engineer, with "unit operations design" a vestige of history. Von Kármán would be annoyed at this sweeping generalization and correctly argue, we believe, that the natural sciences have moved considerably closer to the historical values of engineering ("to create what never has been . . .") while engineers have moved closer to science ("to discover what is . . .") to innovate in a marketplace suited for nine billion individuals. Chemical engineering is quite functional *and yet* employs both the right and left brain, much to the delight and sometimes frustration of all who participate.

The physical and chemical rate processes that dominate chemical engineering practice have endured for many decades, starting with production rates that tie to economic productivity. How fast can a product be made? What economic rates of return are necessary to justify the interest rate paid to build the manufacturing plant? The chemical engineer looks at the "conservation of assets" and envisions the accumulation of assets for the company, which in turn creates jobs and economic opportunity for others. The movement of fluid from one place to another involves momentum transfer rates, which transcend pumping macroscopic fluids to the motion of complex fluids composed of highly interacting molecules that move through complex geometries in devices with small length scales. How long must you wait before your fuel-cell-powered car will move in winter? Heat-transfer rates often couple strongly with fluid flow, as well as with molecular constitution and structure. Rates of mass transfer mostly concern the relative movement of atoms and molecules, often between different phases, when driven by some external "driving force." Thus mass transfer dominates the design of anything that reacts and separates chemicals, from fuel cells to artificial kidneys to distillation columns. All chemical engineering programs now focus more heavily on rate processes from the point of view of the molecules, yet to some extent must be familiar with empirical (industrial) design correlations for the coupling of mass, momentum, and heat transfer in bulk manufacture.

Finally, the control of chemical rates through the use of elegantly designed apparatuses is where the chemical engineer shines. The twentieth century brought the sub-discipline of "chemical reaction engineering" to extraordinary heights, where chemical rates in conjunction with all the other rates described above are brought to bear to design new equipment and products. The "chemical engineering student as chemist" seeks to extract an understanding of chemical rate phenomena within the sub-disciplines of chemistry. Chemical engineering programs are faced with the application of twentieth-century knowledge of rate processes for traditional roles but must employ twenty-first-century concepts and tools for the next new industries. John Prausnitz (Professor of Chemical Engineering, University of California at Berkeley) captured this theme eloquently when he said that we are now teaching chemical engineering as "the future practical." [4]

1.1 ACHIEVEMENTS OF CHEMICAL ENGINEERING

Chemical engineers have a rich history of transforming society. The design and analysis skills mentioned are manifested in the following list of significant achievements by chemical engineers, as published by the American Institute of Chemical Engineers (AIChE) in 2008.[5]

[4] www.annualreviews.org/userimages/ContentEditor/1315331682902/JohnPrausnitzInterviewTranscript.pdf.

[5] *Chemical Engineering Progress*, November 2008, pages 7–25.

Energy. Chemical engineers created equipment and processes on a massive scale to crack hydrocarbon molecules from crude oil and transform them into fuels. The ability to synthesize high-octane fuel was a crucial factor in the Battle of Britain and World War II. In the latter twentieth century, chemical engineers transformed natural gas to premium blend gasolines. Catalytic cracking also creates the molecular building blocks used to assemble complex chemicals.

Carbon-free energy. Chemical engineers developed massive-scale separation processes to prepare isotopically enriched uranium, thereby launching the nuclear power industry (and other industries associated with more destructive ends). Electrochemical engineers brought analytical rigor to the design and operation of batteries and fuel cells, creating new industries for vehicular transport and portable power.

Environmental technologies. The twentieth century witnessed the discovery by chemical engineers of the hollow-fiber membrane, the basis for reverse-osmosis water purification. Polymer replacements for asbestos, anaerobic wastewater treatment, and removal of toxic and/or dangerous solvents in the chemical industry were all contributions by chemical engineers. Chemical engineers also work to design processes with minimal offensive by-products and devise strategies to remediate polluted sites.

Air pollution. The modeling of air pollution in cities based on principles of chemical reactor engineering provided sound strategies for policymakers to create laws aimed at reducing urban pollution. Chemical engineers developed the three-way automobile catalyst to remove CO, NO_x, and hydrocarbons from the exhausts of cars, thereby reducing their emissions. The solid-state electrochemical oxygen sensor, developed by chemical engineers, affords computer control of automobile combustion, further reducing emissions and greatly improving automobile mileage.

Polymers. Beginning with Bakelite, the first hard, moldable plastic, chemical engineers developed the plastics – such as PVC, nylon, polystyrene, and polyethylene – that are the predominant materials for consumer products. Plastics have replaced wood, metal, and glass in many applications because of superior strength:weight ratio, chemical resistance, and mechanical properties. Chemical engineers built the reactors that produce polymers and designed the processing tools to prepare fibers, films, and macroscopic pieces for a panoply of industries.

Synthetic rubber. Elastic materials, such as automobile tires and drive belts, are an integral part of everyday life. Beginning with rayon in 1933, chemical engineers have brought stretchable polymers to the consumer in a host of technologies. The annual production of rubber typically exceeds a million tons. Remarkably, this industry was developed in only two years, just in time to replace shortages of natural rubber during World War II.

Manufacturing at scale. Chemical engineers have brought their fundamental skills in thermodynamics, kinetics and reactor design, and transport phenomena to many industries, affording mass production of many products. Borosilicate glass, isopropyl alcohol, fertilizers, dyes, laundry detergents, zeolite adsorbents, and diapers are just a few of the many products AIChE has identified as being the result of scale-up by chemical engineers.

Medicine. In 1918 an influenza epidemic killed 20 million people worldwide, half a million in the United States alone. Venereal diseases were incurable. Until the 1950s polio crippled millions. Discovering medicines was only part of the solution. After it was observed that a mold inhibited bacterial growth in a Petri dish, chemical engineering developed the technology to ultimately produce millions of pounds per year of penicillin. Chemical engineering made possible the mass production of medicines and their subsequent availability to people worldwide. Chemical engineering principles have been used to model the processes of the human body as well as to develop artificial organs, such as the kidney, heart, and lungs. Sunscreens, medical oxygen, wafer-delivered chemotherapy, and contact lenses are a few examples of products and processes developed by chemical engineers for health care.

Food and agriculture. Beginning with peanut butter in 1922, chemical engineers have developed food and agricultural processes that provide food for more people than ever before in recorded history. Freeze-drying was enabled by chemical engineers, allowing for juices to be stored and delivered worldwide. Insecticides and pesticides were developed that greatly improved crop yields. While "industrial farming" is coming under criticism in the twenty-first century, no one is advocating a return to the low-yield, low-quality crop harvests of the early 1900s.

Electronic materials. A chemical engineer, Andrew Groves, founded Intel. Intel co-founder Gordon Moore's prediction about transistor density in integrated circuits was made in part by applying the principles of chemical engineering to each of the processing steps in circuit manufacture, from preparing high-quality silicon to plasma etching and deposition of nanometer-thin layers to packaging in ceramics and polymers. Whenever electronic device manufacture requires chemistry (photolithography, etching, heat-shrink tubing, or organic LEDs, for example), or when the length scale of a device approaches the molecular regime (e.g. oxide gate barriers, nanowires) then the toolkit of the chemical engineer is indispensable.

The contributions of chemical engineers influenced the evolution of modern society. Most of the top-ten achievements listed above were achieved during the heyday of engineering – when it seemed that society's needs could be met by technology, with engineers as the purveyors of technology. Around the mid 1950s, however, technology came to be perceived as dangerous. People began to feel that society and the environment were dominated by technology, even victimized by technology. This perception remains today. The chemical engineering curriculum attempts to sensitize students to these issues by encouraging studies in humanities, social sciences, and ethics.

1.2 OPPORTUNITIES FOR CHEMICAL ENGINEERING

And what of the future? What will the chemical engineer bring to the twenty-first century? Contemporary chemical engineers are increasingly involved in services, compared to the historical emphasis on manufacturing. This trend will probably continue as chemical engineers are enlisted to remedy environmental contamination and modify existing processes to meet modern business and manufacturing agendas. We anticipate that a few frontier areas will emerge as new opportunities for chemical engineers, including:

Decarbonizing the atmosphere. Chemical engineers will play an increasingly large role in mitigating climate change as climate science becomes carbon policy. Direct air-capture of CO_2, pre- and post-combustion carbon capture, bioenergy carbon capture, and carbon utilization are ideally addressed with the toolkit of chemical engineering. Low- or zero-carbon energy generation and storage has already emerged within the chemical engineering academic and industrial enterprises, including fuel cells, batteries, capacitors, flow systems, and materials for high-power electrical controls.

Innovation at the nanometer scale. Personalized health care, home maintenance, consumer electronics, and manufacturing efficiency all portend the creation of devices and products that comprise extremely small components. Microfluidic devices will become (and indeed already are) miniature chemical plants, with fluid hydraulics, reaction, separations, and chemical analysis all happening at sub-millimeter length scales. These technologies require engineering design and analysis in the presence of chemical reactions and separations, at surfaces, and in structures that are of nanometer lengths.

Production of novel materials. Chemical engineers will design processes to produce ceramic parts for engines, high-T_c superconductors, polymer composites for structural components, and specialty chemicals produced in small amounts to exacting specifications. Chemical processes will shift from the traditional area of petrochemicals to inorganic compounds, from liquids to solids, and from large scale to small scale.

Biotechnology. Chemical engineers will improve methods of isolating bioproducts, design processes for chemical production from biomass, and capitalize on advances in genetic engineering to produce drugs, foods, and materials. Whereas chemical engineering has traditionally sought new reaction paths to produce established chemical commodities, biotechnology will seek ways to produce new chemicals, such as secondary metabolites and so-called customized proteins. With the emergence of point-of-care and personal-care medical technologies, the systems and rates associated with control of cellular processes ("applied molecular biology") have become important for the chemical engineer to master. "Synthetic biology" and "cellular engineering" are new fields that seek to modify plant, animal, and viral life forms for great societal benefit, though not without considerable ethical debate. For this reason many chemical engineering programs require a course in modern molecular biology and biochemistry.

Product design. The creation of new products and generation of economic growth are not limited to process design. Chemical engineers with knowledge and field experience in the complex process of transforming chemicals can also contribute to transforming technical innovations into commercially successful products. The "Post-It Note" paper products used everywhere to secure notes to various surfaces over and over again is an example of transforming a chemical process – the preparation of a pressure-sensitive adhesive – into a hugely successful commercial product. We see product design as an enormous opportunity for chemical engineers in the twenty-first century.

Solid wastes. Chemical engineers will invent methods to treat landfills as well as to remedy contaminated sites. Chemical engineers will also design alternatives for waste, such as incineration, biological decomposition, and recycling. Whereas the reactants entering traditional chemical processes are well characterized and invariable from day to day, processing wastes requires designs that accept reactants with ill-defined compositions that may change daily.

Pollution control. Chemical engineers will continue to reduce pollution at its sources – for example, by recycling intermediate outputs, redesigning chemical reactors, and re-engineering entire processes. Gone are the days when a public waterway was designated "for industrial use." Wastewater today sometimes exceeds the purity of the public waterway it enters. Chemical engineers will design processes that not only meet current regulations but also anticipate regulations. Chemical engineers will aspire to the ultimate goal of "zero emissions."

Energy. Chemical engineers will continue to improve the efficiency of present energy sources as well as develop new sources and energy storage systems.

Process control. Chemical engineers will develop and implement better sensors for temperature, pressure, and chemical composition. Processes will be designed to integrate artificial intelligence for process control, monitoring, and safety.

This is it, the beginning of your journey through one of the greatest disciplines on your campus.

REFERENCE

UNEP 2013. *Global Chemicals Outlook: Towards Sound Management of Chemicals*, ed. E. Kemf, United Nations Environment Programme, Geneva.

2

Chemical Process Design

In the first chapter we learned that chemical engineers create processes based on physical and chemical changes. In this chapter we develop designs for seven chemical processes. The step-by-step examples will also introduce strategies for design and conventions for depicting a chemical process.

2.1 DESIGNING A CHEMICAL PROCESS

2.1.1 Design Evolution by Successive Problem Solving

Context: A process to synthesize ammonia.

Concepts: Unit operations, process flowsheets, continuous operation, and separation based on boiling points.

Defining Question: How does one design a chemical process? How does one start?

Let's design a chemical process based on the chemical change of N_2 and H_2 into ammonia, NH_3, used primarily as a fertilizer. When crop fertilizing was introduced in the mid 1800s, nitrogen fertilizer – mostly ammonia and urea – was mined as seabird guano in coastal rookeries and bat guano in caves. Clearly, guano was not a sustainable source – the nitrogen-rich deposits from centuries of droppings were soon exhausted. Sustainable production of nitrogen fertilizer was realized by chemist Fritz Haber's 1905 invention of a synthetic route to ammonia by the chemical reaction

$$N_2 + 3H_2 \rightarrow 2NH_3. \qquad \text{rxn (2.1)}$$

This reaction is the basis of an efficient chemical process; the worldwide production of ammonia was almost 310 *billion* pounds in 2016, perennially in the top ten largest chemical productions.

Reaction 2.1 is the key to the ammonia synthesis. So we start with reaction 2.1. The device that conducts this reaction, the *reactor*, is the core of the process. The details of this reactor are key to the viability of the process, but these are not important now. We assume a viable reactor is available and devise a process to deliver the reactants and purify the product.

We need a means of describing our design. We could describe the process with words, such as "a mixture of N_2 and H_2 is piped into the reactor." This would be cumbersome. A chemical process exemplifies the adage "a picture is worth a thousand words." A chemical process is represented by a diagram known as a process *flowsheet*, a key tool in chemical engineering design.

2.1 Designing a Chemical Process

A process flowsheet comprises *units*, represented by simple shapes, such as rectangles or circles:

The pipes that conduct material between units are called *streams*. The streams are represented by arrows:

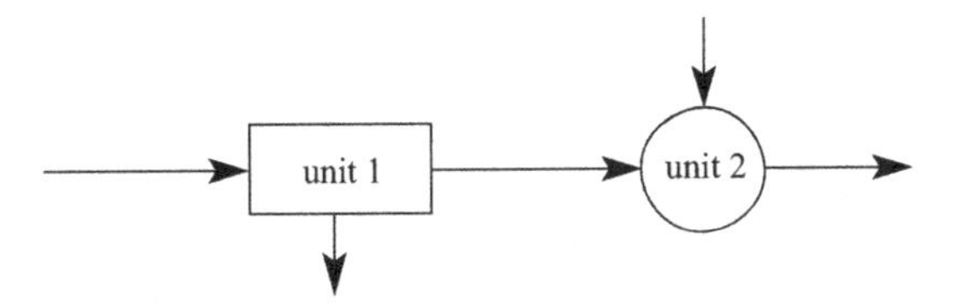

The specifics of a stream, such as its composition, are typically written above the stream. For example, the ammonia reactor can be represented as

N_2
H_2
reactor
$N_2 + 3H_2 \rightarrow 2NH_3$
NH_3

The ammonia reactor is a *continuous* process; material constantly moves through the unit. A *batch* process, in contrast, is characterized by chemical or physical change without material moving in or out, except at the beginning and end of a cycle. A chemical reaction you conduct in a flask in your introductory chemistry laboratory is a batch process. Continuous processes are rarely encountered in undergraduate chemical laboratories, yet they dominate in the chemical industry.

The ammonia reactor above is idealized. Reactants are rarely converted entirely into products. There is almost always some residual reactant in the effluent stream. A realistic description includes N_2 and H_2 in the reactor effluent, as shown below.

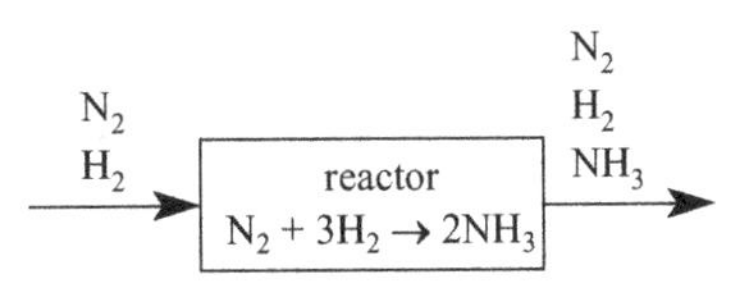

The customer for our NH_3 would probably not be satisfied with N_2 and H_2 in the product. And we would be wasting reactants. We need to purify the NH_3. How might this be done? To separate substances, we need a physical or chemical basis. The most common basis for separations is *states of matter*, or *phase*. A gas phase is easily separated from a liquid phase. N_2, H_2, and NH_3 are gases at room temperature and pressure. Let's explore the possibility of condensing one or more of the compounds without condensing the others. A handbook of chemical data gives the information listed in Table 2.1.

Table 2.1 Boiling points for the ammonia process.

Compound	Boiling point at 1 atm (°C)
NH_3	−33
H_2	−253
N_2	−196

These data show that if we cool the gaseous mixture below −33°C, NH_3 will condense. We can thus separate liquid NH_3 from gaseous H_2 and N_2. We add to our flowsheet a unit to condense NH_3 and then separate NH_3 liquid from the $N_2 + H_2$ gas mixture. We will call this unit a "*liquid–gas separator*." It is conventional to show liquids leaving the bottom of a unit and gases leaving the top.

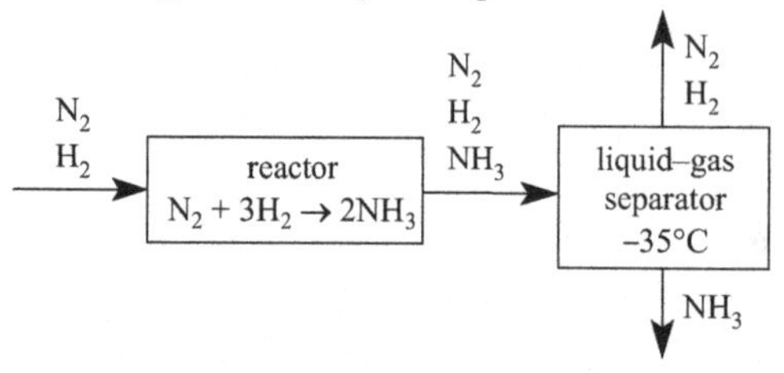

We have assumed the separation of NH_3 from N_2 and H_2 is perfect, which is idealistic. In Chapter 4 we will consider realistic separators and methods to design for specific product purity. We have also ignored the removal of heat from the liquid–gas separator.

What shall we do with the N_2 and H_2 exiting the liquid–gas separator? Both are environmentally benign and could be discharged to the air. But this is wasteful. *Recycling* the reactants to the reactor is more efficient, as shown below.

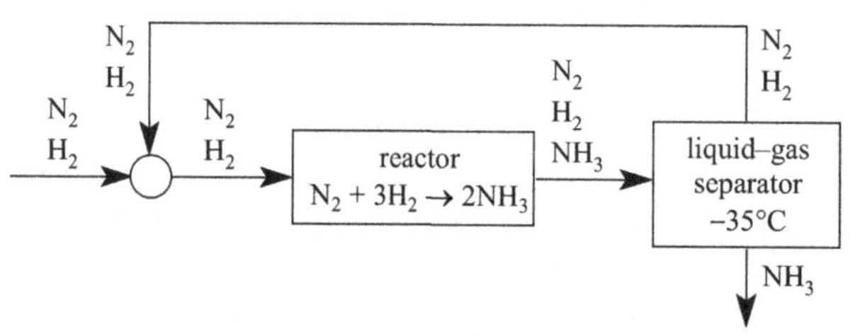

This flowsheet contains a new feature, a *combiner*, which is represented by a circle with two or more streams entering and one stream leaving. A combiner merely combines streams; no mixing is implied. But in this case no mixing is needed because the two streams combined have the same composition.

A process based on this rudimentary design will produce ammonia. One would purchase N_2 and H_2 and sell NH_3. But gases such as N_2 and H_2 are expensive. Furthermore, gases are bulky and must be compressed for efficient transport. H_2 has the added problem of explosion risk. Let's augment the process for producing the reactants N_2 and H_2. If our potential supplier can produce N_2 and H_2, perhaps we can, too. Consider first a source for N_2. Air is an obvious choice. Air is about 79% nitrogen and 21% oxygen and air costs nothing. A subtle benefit is that air contains nitrogen in the chemical form we desire. Thus we need only a separator and not a reactor. (As a rule, reactors are more expensive to build and operate.) It seems reasonable to consider separating nitrogen from air. (Nitrogen gas is another high-production chemical commodity. The US production is typically 90 *billion* pounds per year.) Again we first consider separation based on phase: liquid vs. gas. We add the boiling point of oxygen to our table.

Table 2.2 More boiling points for the ammonia process.

Compound	Boiling point at 1 atm (°C)
NH_3	−33
H_2	−253
N_2	−196
O_2	−183

We thus add a liquid–gas separator to provide N_2 and our process becomes:

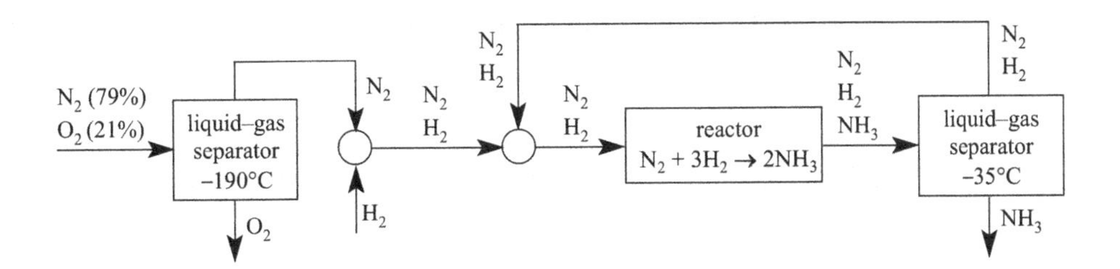

What is a good source for hydrogen? Unlike nitrogen, hydrogen has extremely low abundance in its desired chemical form, H_2; hydrogen is less than 0.00005% of air. We need to produce hydrogen by chemical reaction. What is a practical source of hydrogen? Water is 67% hydrogen and water is inexpensive. One possibility is to decompose, or "crack" water vapor:

$$H_2O(\text{gas}) + 242\,\text{kJ} \rightarrow H_2 + \frac{1}{2}O_2. \qquad \text{rxn (2.2)}$$

Decomposing water vapor requires 242 kJ/mol, a great deal of energy. Indeed, the reverse reaction is a $H_2 + O_2$ torch, which produces considerable energy. We could decompose H_2O with electricity. You have probably seen this demonstrated on a small scale in secondary school. We add an electrolytic reactor to produce H_2.

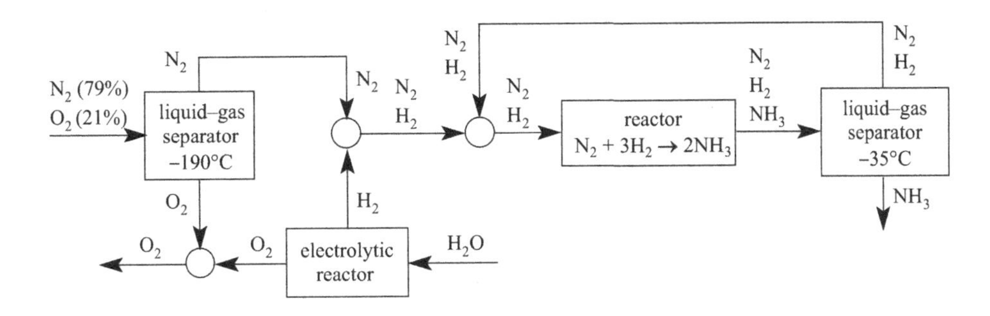

Our process consumes air and water (and energy) and produces ammonia as well as oxygen. Again, the process is workable, but again it can be improved. Let's focus on the source of hydrogen. Electricity is an expensive form of energy. So we look to another means to provide the energy to decompose water into hydrogen and oxygen. Consider the *chemical* energy of fuels such as CH_4, the principal component of natural gas:

$$CH_4 + 2O_2 \rightarrow CO_2 + 2H_2O\ (\text{gas}) + 802\,\text{kJ}. \qquad \text{rxn (2.3)}$$

Each mol of CH_4 provides enough energy to decompose 3.3 mol of H_2O (802 kJ/mol CH_4 divided by 242 kJ/mol H_2O equals 3.3 mol of H_2O/mol CH_4). So if one combines 3.3 times chemical reaction 2.2 with reaction 2.3 the sum is

$$3.3H_2O + 3.3 \times 242\ \text{kJ} \rightarrow 3.3H_2 + \frac{3.3}{2}O_2$$

$$CH_4 + 2O_2 \rightarrow CO_2 + 2H_2O + 802\ \text{kJ} \qquad \text{rxn (2.4)}$$

$$\overline{CH_4 + 0.35O_2 + 1.3H_2O \rightarrow CO_2 + 3.3H_2 + {\sim}0\ \text{kJ}.}$$

Reaction 2.4 is convenient because it supplies the energy needed and it produces H_2, a reactant in our process. Our methane reactor is thus

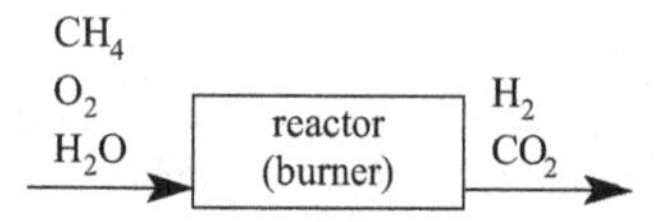

As usual, reaction 2.4 does not convert all the reactants to products. We must include reactants in the effluent:

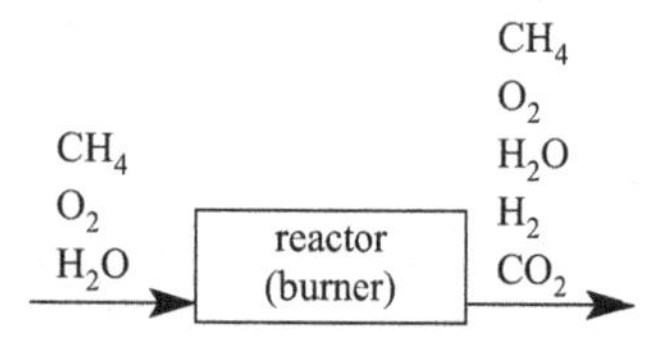

The hydrogen must be separated from the methane, oxygen, water, and carbon dioxide. As before we first consider separation by condensing some of the gases to liquids. We add more boiling points to our table.

Table 2.3 Even more boiling points for the ammonia process.

Compound	Boiling point at 1 atm (°C)
NH_3	−33
H_2	−253
N_2	−196
O_2	−183
CH_4	−164
H_2O	100
CO_2	−79 (sublimation: solid to gas)

Oxygen is the most crucial substance to be removed because it will react with hydrogen (and perhaps nitrogen) in the ammonia synthesis reactor. As luck would have it, the boiling points reveal that oxygen is the most difficult to separate from hydrogen. Therefore, we could design a liquid–gas separator similar to the N_2–O_2 separator that operates below −183°C. Again, such a design would be workable, but there is a better strategy: eliminate oxygen from the effluent by consuming all the oxygen in the reactor. Instead of using stoichiometric proportions indicated in reaction 2.4, use excess methane. Our reactor (burner) thus becomes

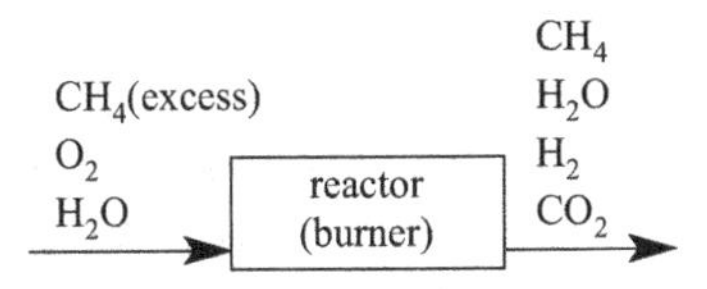

With oxygen depleted from the effluent, the temperature in the liquid–gas separator is set by methane; we must operate the liquid–gas separator below −164°C to condense methane. This is not much of an improvement over the cooling dictated by oxygen. Must methane be separated from hydrogen? If we allow the methane (which is only a small amount) to continue with the hydrogen, we can focus on removing the major by-product, carbon dioxide. We consult with our reactor experts who conduct some experiments and report that a small amount of methane in the ammonia reactor is not a problem. Fine. Our hydrogen production module, then, looks like this:

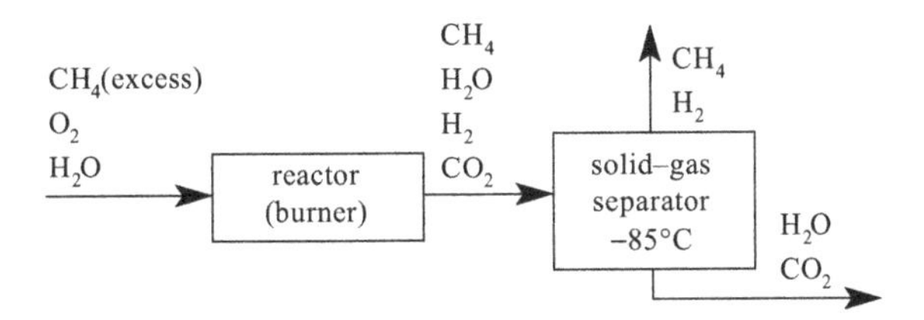

It is not obvious that producing H_2 by burning CH_4 is better than electrolysis of water. Indeed, there are at least three reasons why the contrary seems true: the methane process requires an additional unit, methane impurity enters the ammonia synthesis reactor, and we no longer produce oxygen as a by-product – we instead consume some of the oxygen produced by the air separator. And it remains to be seen whether the air separator produces sufficient oxygen. Also, the original motivation for using methane (cheaper energy) could be undone by the energy requirements of operating the liquid–gas separator. As we shall see when we develop the tools to compare these alternatives, the methane burner is indeed better.

We replace the electrolytic reactor with a methane burner and our ammonia synthesis process becomes:

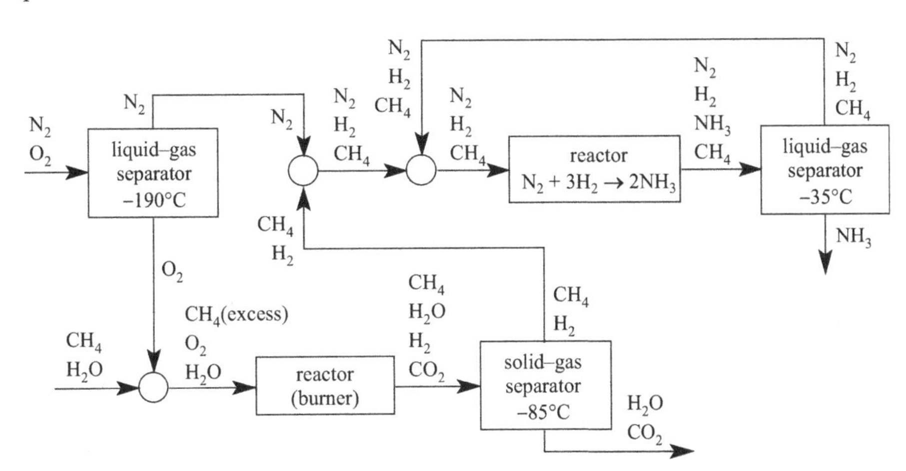

What is the fate of the methane that tags along with the hydrogen? Our reactor experts assure us the methane passes through the N_2–H_2 reactor unchanged. Because the boiling point of methane is lower than −35°C, the liquid–gas separator does not condense methane.

That's good – we need not separate methane from our ammonia product. The methane leaves the liquid–gas separator via the N_2–H_2 recycle stream. The methane recycles to the reactor and accumulates indefinitely in our process. That's bad – there is no outlet for unreacted methane. Similarly, other impurities in air, notably argon, will also accumulate indefinitely in the recycle loop! This is a drawback of any recycling scheme: the accumulation of undesired substances. We could add a unit to separate nitrogen and hydrogen from methane (and argon) but this is expensive. *Purging* a small amount of the recycle stream, perhaps 1% to 10%, is simpler (and cheaper). The revised flowsheet is then:

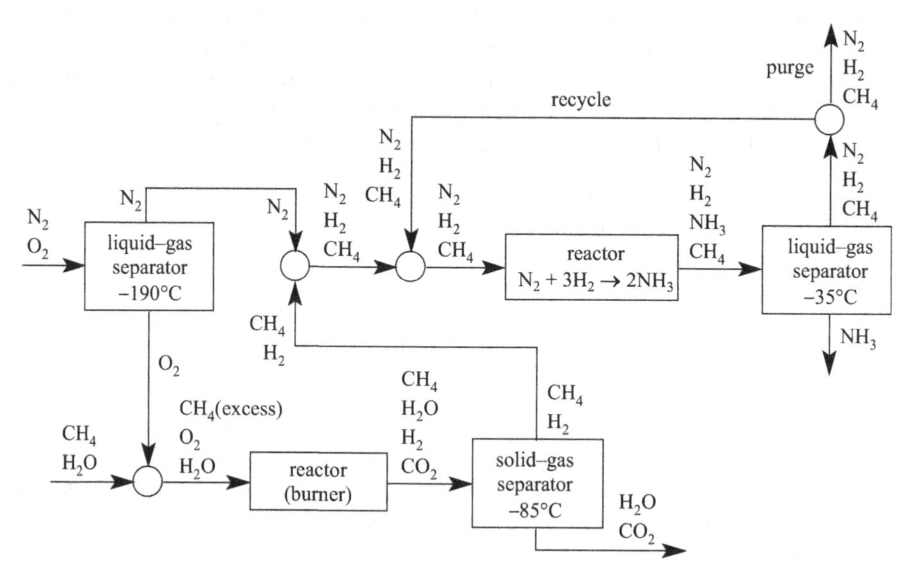

A new type of unit, a splitter, produces the purge stream. As shown in the ammonia process and in the figure below, a circle represents a splitter:

A *B* → ○ → *A* *B*
↓ *A* *B*

A splitter divides a stream into two or more streams of *identical* composition. A splitter only splits streams without separating mixtures, just as a combiner only combines streams without mixing. Although the split streams are the same composition, the streams are not necessarily the same flow rate. For example, the purged stream in the ammonia process may be only 1% of the recycled stream.

We need to check another consequence of adding the methane burner: does the air liquid–gas separator yield enough oxygen to supply the methane burner? Let's estimate the flow rates based on a stoichiometric reaction of 1 mol of N_2 and 3 mol of H_2. One mol of N_2 requires 1/0.79 mol of air into the separator, which yields 0.21/0.79 = 0.27 mol of oxygen. From reaction 2.4, 3.3 mol of hydrogen requires 0.35 mol of oxygen, so 3 mol of hydrogen requires (3/3.3) × 0.35 = 0.32 mol of oxygen. We have a slight oxygen deficit of 0.32 – 0.27 = 0.05 mol. Thus we must either buy extra oxygen or run the air liquid–gas separator at a rate 0.32/0.27 = 1.2 times faster than needed for the stoichiometric amount of nitrogen. We will assume producing oxygen is cheaper than buying oxygen. We will sell the excess nitrogen; we add a splitter to the N_2 stream.

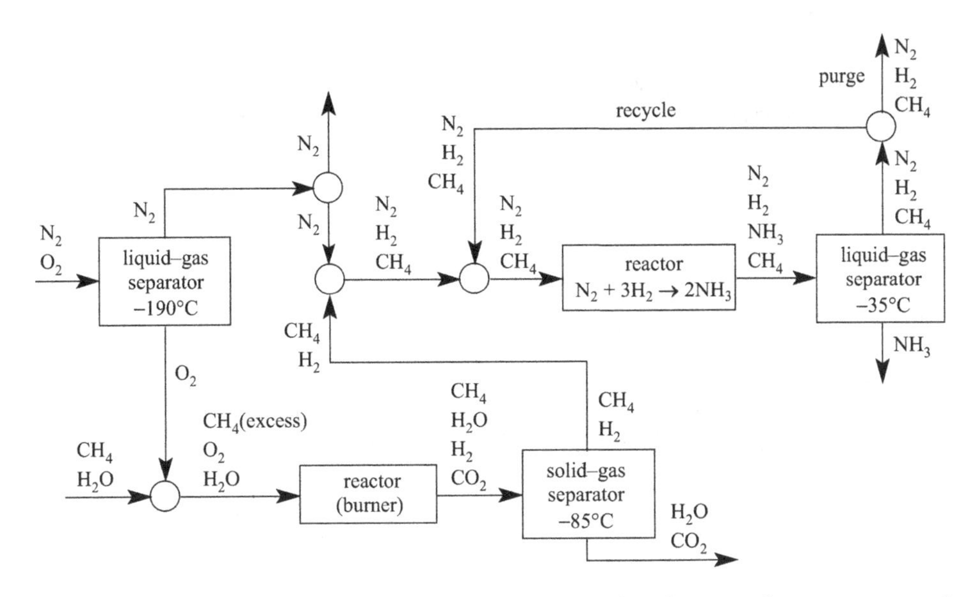

We now have a qualitative process to make ammonia. The mass flows are arranged logically; we have carefully reused by-products such as oxygen, nitrogen, and hydrogen. But what of the energy flow carried inherently by the mass? Are we discarding energy? Let's examine the water–methane reactor. We need to heat the reactants and we need to cool the products (to −85°C in the liquid–gas separator). We need a *heat exchanger.*

A heat exchanger transfers heat from one stream to another stream without mixing the streams. An example of a heat exchanger is shown below:

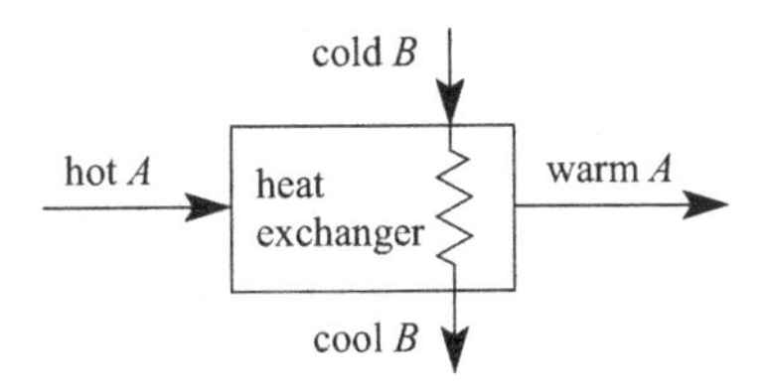

The zigzag line drawn inside the heat exchanger represents tubing that contains *B* and reminds us that the two streams are in thermal contact, but not in physical contact. The radiator on an automobile is a heat exchanger: liquid coolant (usually a mixture of ethylene glycol and water) enters the radiator hot from the engine and flows through tubes. Air also flows through the radiator but outside the tubes. The liquid coolant is cooled and the air is heated. If the radiator is operating properly, the liquid coolant does not mix with the air.

A heat exchanger can be inserted before the methane burner in our ammonia process.

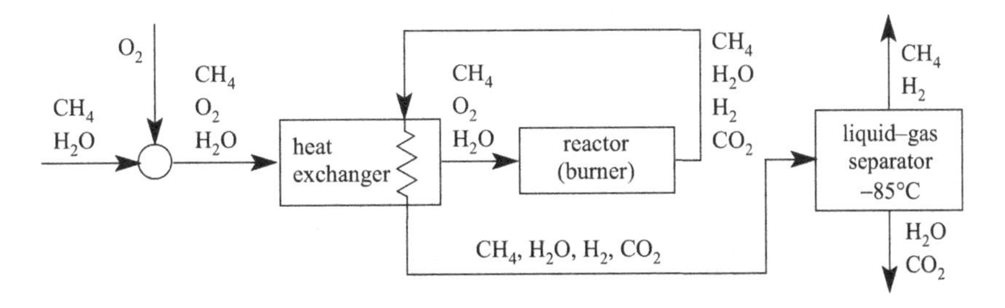

The air liquid–gas separator also wastes energy: the air must be cooled to −190°C but products of the liquid–gas separator – oxygen and nitrogen – must be heated before they enter their respective reactors. Two heat exchangers satisfy both needs.

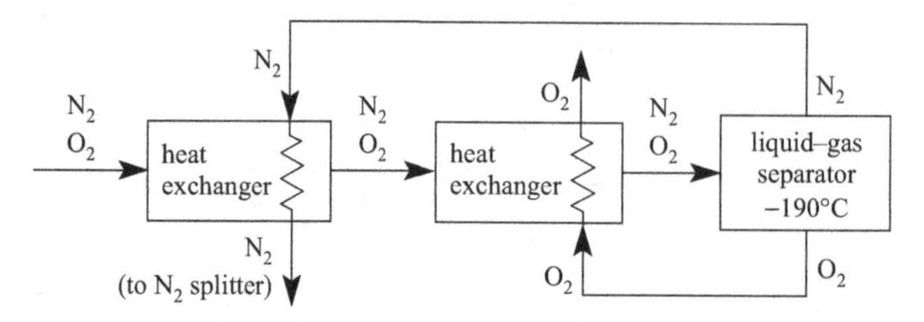

Similarly, the ammonia synthesis reactor's input must be heated and the effluent must be cooled. Again, a heat exchanger could be added.

In summary we have assembled a system of discrete units to synthesize ammonia. Each unit has a specific function. We started with the key unit: the ammonia synthesis reactor. We built the process by successive problem solving, such as "What is a source for nitrogen?"

2.1.2 Analyzing Data to Design Chemical Processes

Context: Phase separation data: melting points and boiling points.

Concepts: Visual vs. verbal presentation of data.

Defining Question: Why is it awkward for a left-handed person to drive a car with a manual transmission?

We used boiling points to specify temperatures in the liquid–gas separators of the ammonia process. Selecting the separator temperature can be tedious, especially if the data are not sorted by boiling point. But there is another reason the data may have been difficult to assimilate: tabulated data are in a verbal format, whereas most people assimilate better from data in a visual format.

The boiling points in Table 2.3 are presented as symbols. Lists of numbers are awkward to absorb and analyze for some engineers. Rather, a graph is more "natural" for some engineers. A graphical alternative to Table 2.3 is boiling points plotted on a temperature line.

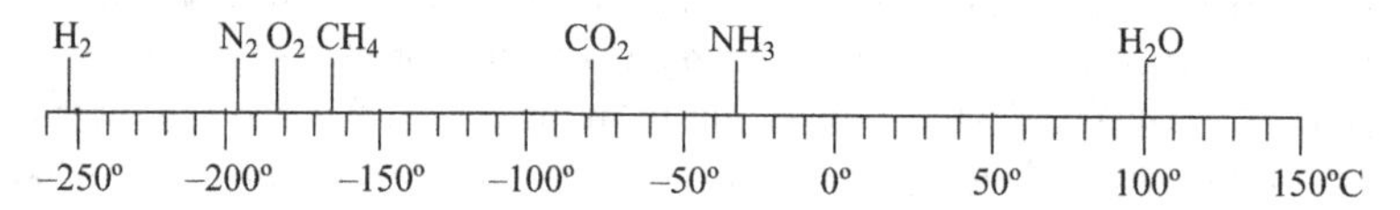

If the goal is to separate CH_4 and CO_2, draw a vertical line at −120°C. All substances to the left of the vertical line are gases and all substances to the right are liquids (or solids).

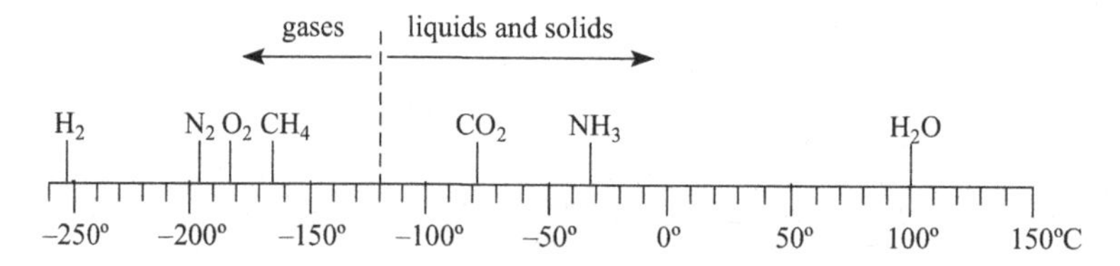

Visual presentation of the data allows one to *see* which boiling points are similar and which boiling points are well separated. Substances with similar boiling points are more expensive to separate by liquid–gas separation. Visual presentation also shows the separator temperature relative to ambient, 20°C. Cryogenic separations – far below ambient – will require expensive refrigeration.

Why is it unnatural for a left-handed person to drive a car with a manual transmission? Cars with manual transmissions usually have the gearshift to the right of the driver. If one is left-handed, performing tasks with one's right hand is unnatural. Similarly, if one is a visual assimilator, using data presented in a verbal format is unnatural.

We will describe other learning stages as appropriate to the context. If you wish to learn more about learning styles, we recommend the comprehensive and comprehensible work of chemical engineer Richard Felder.[1]

2.1.3 Conventions for Chemical Process Flowsheets

Context: Unit operations.

Concepts: Combiners vs. mixers, pumps vs. compressors, and streams crossing vs. streams mixing.

Defining Question: How does one represent specific process units on a flowsheet?

Every chemical process is a collection of units interconnected by streams. This is the concept of *unit operations*, the first paradigm – and most venerable paradigm – adopted by the fledgling discipline of chemical engineering near the beginning of the twentieth century.

The five most common units are reactors, heat exchangers, pumps, mixers, and separators. The design and operation of these five units, which exist in hundreds of incarnations, are the focus of the traditional chemical engineering curriculum. Furthermore, most chemical engineering curricula have a senior-level laboratory course entitled *unit operations*.

As we saw in the ammonia process, the reactor is usually the key unit. The reactor often dictates whether a process is *chemically* possible. Separators are secondary only to the reactor in most processes; the performance of the separator(s) often determines if a process is *economically* possible. Examples of separators include dryers, filters, absorbers, extractors, and centrifuges. The method used to separate a mixture – condensation, evaporation, crystallization, or filtering – is usually indicated by a descriptive symbol on the flowsheet.

Additional units could be added to the ammonia process. Reactants should be mixed before entering a reactor, not just combined. Simply *combining* flour, eggs, water, and baking soda is not enough. The reactants must be *mixed* to bake a cake. The simplest *mixer* combines two or more streams into one homogeneous stream.

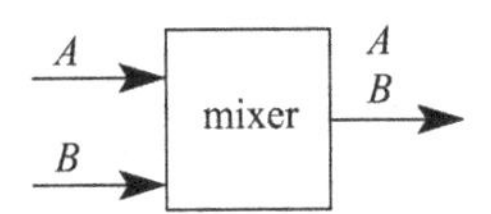

Representations for some process units have shapes and attributes that describe visually the unit operation. For example, a propeller at the end of a shaft indicates mixing.

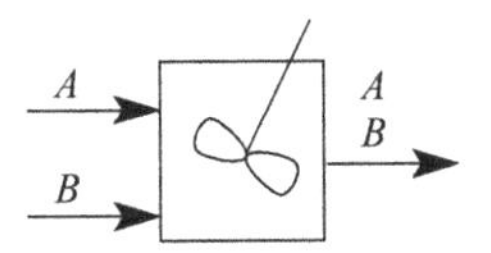

[1] Richard Felder's *Resources in Science and Engineering Education*, www4.ncsu.edu/unity/lockers/users/f/felder/public/RMF.html. See "Learning Styles."

Mixers are important for solids and liquids. Gases, such as N_2 and H_2, mix well without a mixer.

We neglected to add pumps to the ammonia process. Fluids are usually moved about a process by pumps. A pump increases a fluid's pressure and the fluid flows from high pressure to lower pressure. Or, fluid pressure might be increased to condense the fluid (from a gas to a liquid) or speed a chemical reaction. A generic pump may be represented as

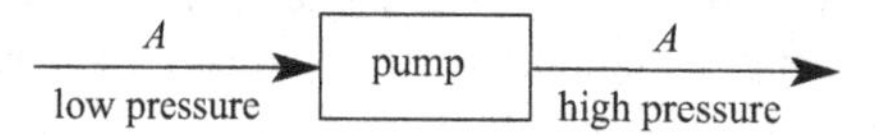

Because liquids and gases respond differently to an increase in pressure (a gas is compressed whereas a liquid is not) different units are used to distinguish this difference. A centrifugal pump for liquids may be represented as

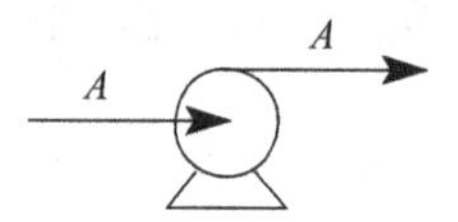

whereas a centrifugal gas compressor may be represented as

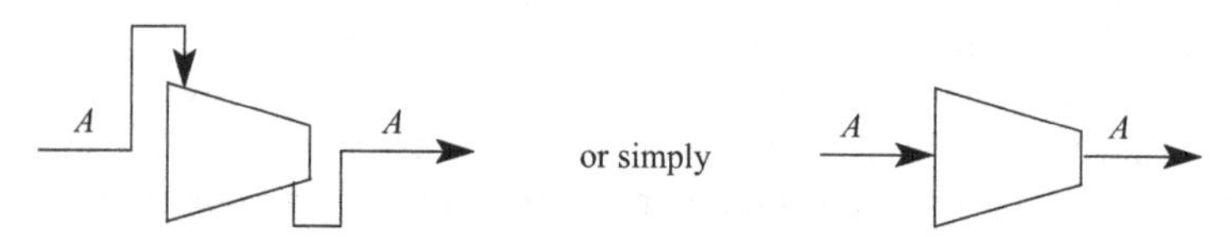

A gas expander, or turbine, is similar to the gas compressor above, but with the flows reversed. We need some valves to control the flows. The simplest representation of a valve in a stream is

A

Finally, we offer stylistic conventions for flowsheets. When convenient, the process flowsheet should begin in the upper left corner and flow left to right, top to bottom. Input streams should start at the flowsheet periphery and output streams should end at the periphery. For separators, gases should depart the top of the unit and liquids (and/or solids) should depart the bottom.

Avoid crossing streams. If streams must cross, indicate explicitly whether the streams mix or don't mix. A flowsheet with two streams as shown below on the left is ambiguous. Do the contents of the streams mix? If the contents of the streams combine (and split), indicate this explicitly with a splitter or combiner, as shown in the middle diagram below. If the streams cross without mixing, this should be indicated as shown on the right below. The primary stream should be unbroken when streams cross.

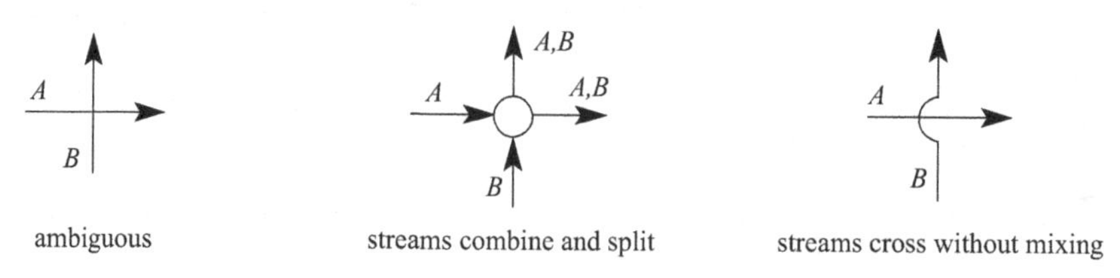

Only horizontal and vertical lines should be used for streams on a flowsheet. Arrowheads should appear only where a stream terminates at a unit, not at a bend in a stream. A process flowsheet should be readable and precise. A representation for a unit

should have a logical shape and show streams entering and leaving at logical locations on the shape. For example, a reactor should be a long rectangle with exactly one stream in and one stream out. A separator should have one stream in and two or more streams out, with the less dense stream leaving the top and the more dense stream leaving the bottom.

2.2 CHEMICAL PROCESS DESIGN AND CREATIVE PROBLEM SOLVING

2.2.1 Defining the *Real* Problem in Successive Problem Solving

Context: A process to purify heptane.

Concepts: Separation based on melting points and separation by chemical reactivity.

Defining Question: How does one get down off an elephant?

Creating a process flowsheet for a chemical or physical process is at the heart of engineering design. These flowsheets, however, are often difficult to conceive and an engineer must use many skills to develop the optimal design. A chief stumbling block for many designs is the failure to *properly define the problem to be solved.* We redefined the problem during our design of the ammonia process. For example, the problem with the effluent from the methane burner was not "separate hydrogen from oxygen" but rather "don't have oxygen in the effluent." Another example was the methane in the hydrogen stream – the goal was not to purify hydrogen but rather to have a hydrogen source suitable for reacting with nitrogen to produce ammonia. Chemically inert substances mixed with the hydrogen are acceptable. In this section we consider other examples of properly defining the problem.

You design a chemical process that uses *n*-heptane, $CH_3(CH_2)_5CH_3$, as a solvent. The *n*-heptane must be 99.999% pure when it enters your process. During the process the *n*-heptane acquires an impurity: 10% *i*-propanol, also known as isopropyl alcohol, $CH_3CH(OH)CH_3$.

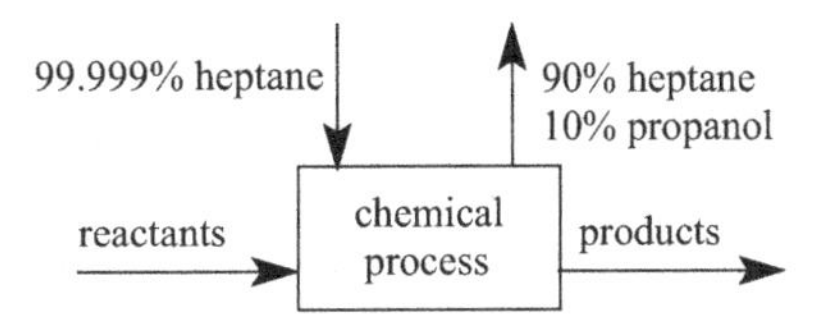

Of course, we should recycle the *n*-heptane. But first we must purify the *n*-heptane. Our challenge is thus:

Problem: "Separate *n*-heptane from *i*-propanol."

This simple statement of the problem carries subtle meaning. Compare this to "Separate *i*-propanol and *n*-heptane," which implies that we wish to produce high-purity *n*-heptane *and* high-purity *i*-propanol. We only wish to produce a stream of high-purity *n*-heptane; we do not care if the *i*-propanol is impure.

As we did several times in our design of the ammonia process, we investigate the possibility of separating the *n*-heptane by phase differences. We begin with the most efficient separation, liquid–gas or solid–gas, for which we need to know boiling points. Again we make a table of the boiling points.

Table 2.4 Boiling points for *n*-heptane purification.

Compound	Boiling point at 1 atm (°C)
n-heptane	98.4
i-propanol	97.8

Evaporating *i*-propanol gas from *n*-heptane liquid is feasible, but the boiling points are very close. When we explore the design of liquid–gas separators in Chapter 4 we will learn the implications of "the boiling points are very close." In theory any difference in boiling points can be exploited for a liquid–gas separation. But in practice the process may be too expensive. It is prudent to consider other methods.

Another method is solid–liquid separation. To explore this possibility, we add a column to our table.

Table 2.5 Boiling points and melting points for *n*-heptane purification.

Compound	Boiling point at 1 atm (°C)	Melting point (°C)
n-heptane	98.4	−90.6
i-propanol	97.8	−126.2

The difference in melting points is large, so solid–liquid separation is feasible. Below −90.6°C solid *n*-heptane will form in liquid *i*-propanol. (The melting point of a substance in a mixture is usually lower than the melting point of the pure substance.) The first unit in our process is thus a solid–liquid separator:

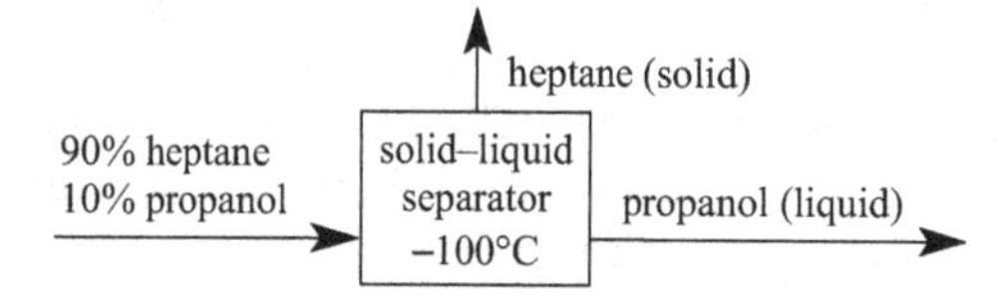

Because *n*-heptane is less dense than *i*-propanol, the solid floats on the liquid. We therefore show the solid effluent leaving the top of the separator.

But solid–liquid separation is more difficult than liquid–gas separation, for at least three reasons. A solid separates from a liquid more slowly than a gas separates from a liquid. Because a gas is a factor of ~1000 less dense than a liquid, gas bubbles rise rapidly through a liquid. But an organic solid is only a factor of ~1.2 or so different in density from an organic liquid. Solid particles settle (or rise) very slowly through a liquid. (In Chapter 5, we will calculate the velocities of rising bubbles and settling solids, which one needs to design a separator.) A second issue is the transport of the product streams. Both outputs from a liquid–gas separator are fluids, which are easily pumped through pipes. The solids from the solid–liquid separator will require special handling, such as an extruder or conveyor. But perhaps the most important issue is how "cleanly" the phases separate. Whereas it may be possible to produce a liquid free of solids (by filtration or settling, for example), it is usually not possible to produce solids free of liquid.

So, given a choice, choose liquid–gas separation over solid–liquid separation. But we do not have a choice. The output of our solid–liquid separator is:

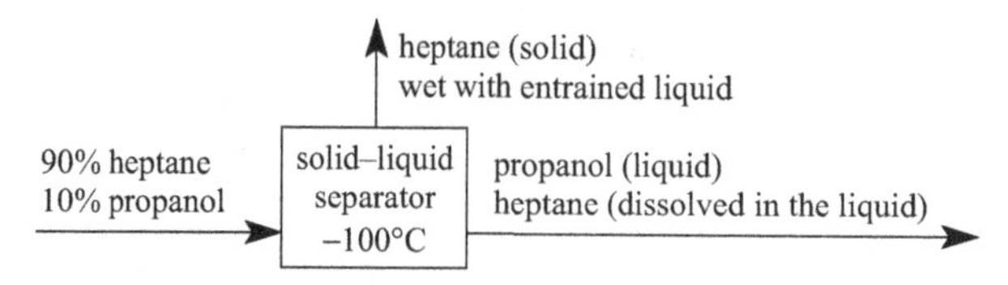

The impurity in the liquid stream is of little consequence. We regret losing some *n*-heptane, but it may not be feasible to recover *all* the *n*-heptane. The liquid entrained with the solid *n*-heptane poses a problem. Our new problem statement becomes:

Problem: "Reduce (or eliminate) the liquid entrained by the solid *n*-heptane."

How do you remove liquid from a solid? What solids do we dry routinely? How do you remove water from clothes after washing? The first step is the spin-dry, also known as a centrifuge. The next step is the air-dryer. Perhaps we could blow air (or some other gas) over the wet solid *n*-heptane at −100°C? When washing dishes, it was once common to dry the dishes with a towel. Perhaps we could design an absorbent conveyor belt to dry the solid *n*-heptane?

Centrifugation, air drying, or toweling might work. However each is probably expensive, especially because the solid *n*-heptane must be dried at −100°C. And it is doubtful that any of the methods could remove all but 0.001% of the liquid, to yield 99.999% *n*-heptane.

A poor statement of the problem unduly restricts our ideas. The problem is not that the *n*-heptane is wet with a liquid. The *real* problem is that the liquid is *i*-propanol; separating *n*-heptane and *i*-propanol is difficult. So we wash the solid *n*-heptane with a liquid that is easier to separate from *n*-heptane, either by distillation or perhaps liquid–liquid immiscibility.

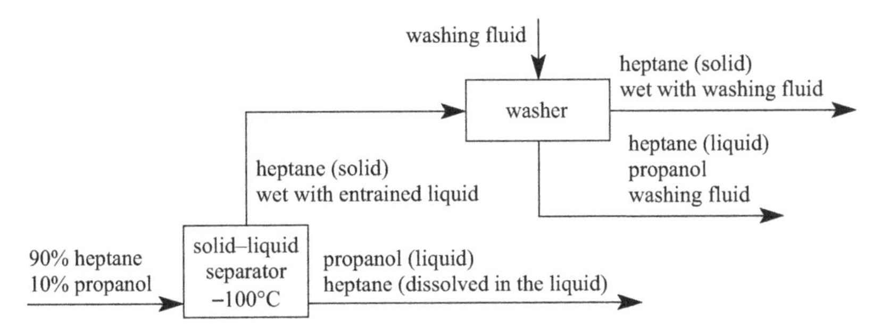

What is a suitable washing fluid? We need a substance that is liquid at −100°C and whose boiling point is different from *n*-heptane. But there are better criteria for the washing fluid. Wash with a liquid that need not be separated from *n*-heptane. We could review the chemistry in our process and perhaps find a compatible liquid. But we already know a compatible liquid: *n*-heptane. Wash solid *n*-heptane with liquid *n*-heptane. The process to purify *n*-heptane becomes

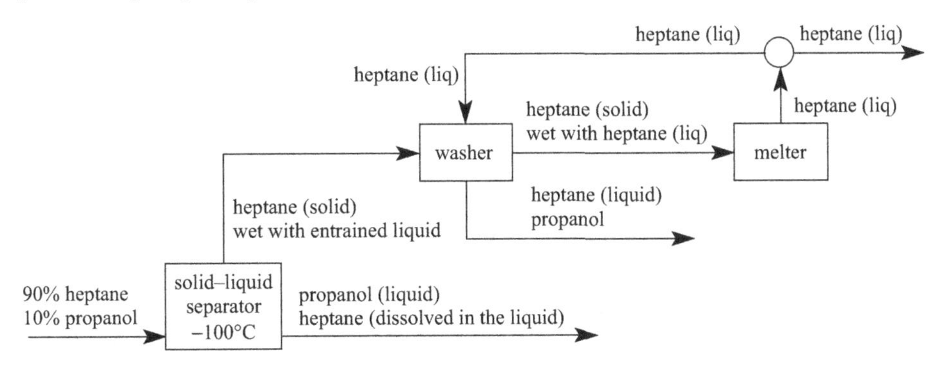

The temperature of liquid *n*-heptane recycled to wash the solid *n*-heptane should be close to its freezing point, to minimize melting the solid *n*-heptane.

Solid–liquid separation followed by washing should work. The process can be improved further by recycling the *n*-heptane + *i*-propanol effluent from the washer to the

90% heptane + 10% propanol stream. The *n*-heptane + *i*-propanol effluent from the washer cannot be recycled to the stream entering the washer. This would not allow the propanol to leave the washer–melter loop, except by a purge stream on the recycle. The purge would have the same flow rate of *i*-propanol as the stream leaving the washer in the above design.

There is a better design. The initial statement of the problem, "Separate *n*-heptane from the *i*-propanol," restricted us because it implied that we needed to separate *n*-heptane from *i*-propanol. We do not care what happens to the *i*-propanol. It can be destroyed in the process of purifying the *n*-heptane. The *real* problem is:

Problem: "Purify the *n*-heptane to 99.999%."

By changing the problem statement, we have more design options. For example, if *n*-heptane and *i*-propanol are difficult to separate, change *i*-propanol into something easy to separate from *n*-heptane. One might suspect that *n*-heptane and *i*-propanol have different reactivity. Why is *n*-heptane an acceptable solvent in this process? Among other properties, we can assume *n*-heptane does not react with any chemical in the process. Why is *i*-propanol unacceptable? We can assume *i*-propanol reacts with something in the process.

Chemical engineering students learn in organic chemistry that an alcohol is generally more reactive than a saturated hydrocarbon, such as *n*-heptane. One example is reactivity toward aluminum:

$$3\left[CH_3-\underset{H}{\overset{CH_3}{\underset{|}{\overset{|}{C}}}}-OH\right] + Al \rightarrow ((CH_3)_2CHO)_3Al + \frac{3}{2}H_2 \qquad \text{rxn (2.5)}$$

$$CH_3(CH_2)_5CH_3 + Al \rightarrow \text{no reaction.} \qquad \text{rxn (2.6)}$$

The products of reaction 2.5 – aluminum propoxide and hydrogen – can be separated from *n*-heptane by distillation, as shown in Table 2.6.

Table 2.6 More boiling points and melting points for *n*-heptane purification.

Compound	Boiling point at 1 atm (°C)	Melting point (°C)
n-heptane	98.4	−90.6
i-propanol	97.8	−126.2
Al propoxide	248	106
H_2	−253	−259

The process below uses reactions 2.5 and 2.6.

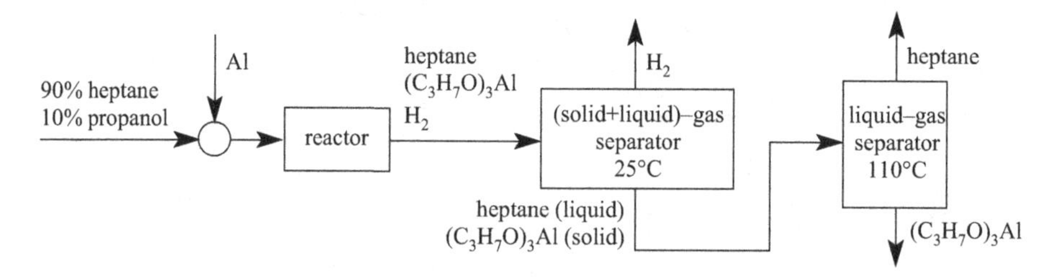

Is reacting *i*-propanol with aluminum a better process than solid–liquid separation of *n*-heptane and *i*-propanol? It is not possible to determine this from the qualitative flowsheets developed here. Reacting *i*-propanol with aluminum has potential problems.

The reaction to aluminum propoxide may not be sufficiently complete. The reactor may be expensive. In Chapters 3, 4, and 5 we will take up methods to analyze designs for product purity and cost, for example. During the design phase, it is essential to generate different possibilities, which can be analyzed later.

Proper definition of the problem statement is key to generating ideas. A quote from Albert Einstein eloquently makes this point:[2]

> The mere formulation of a problem is far more often essential than its solution, which may be merely a matter of mathematical or experimental skill. To raise new questions, new possibilities, to regard old problems from a new angle requires creative imagination and makes real advances in science.

Proper definition of the problem statement also makes real advances in engineering.

Defining the problem is the first step in the McMaster problem-solving heuristic, which is described in two resources by chemical engineers: *Problem-Based Learning* by D. R. Woods[3] and *Strategies for Creative Problem Solving* by H. S. Fogler and S. E. LeBlanc.[4] The McMaster five-step agenda for problem solving is:

Define the problem. Identify the true objective. List the constraints, criteria, and assumptions.
Generate ideas.
Choose a course of action.
Execute a plan.
Evaluate the plan.

This chapter (and especially the exercises at the end of the chapter) provides experience in defining the problem and generating ideas. Chapters 3, 4, and 5 introduce three methods of analysis, the basis for the third step.

The defining question for this section – "How does one get down off an elephant?" – is an example of a poorly defined problem. If one needs down – the soft fluffy feathers used in pillows – a better source is a goose, not an elephant. The *real* problem is better defined as "How does one get down off a goose?"

2.2.2 Conventions for Streams on Chemical Process Flowsheets

Context: Unit operations, continued.
Concepts: Separation strategies and more flowsheet conventions.
Defining Question: What is the prime criterion for process flowsheets?

Separation based on liquid–gas phases is usually the more effective and convenient, but liquid–gas separation is sometimes not feasible, usually because the substances to be separated have similar boiling points or one of the substances is harmed by heat. In the previous section we added two additional separation methods to our repertoire: separation based on solid–liquid phases and separation based on differences in chemical reactivity. Solid–liquid separations produce wet solids, but one can usually melt some of the solid to recycle and rinse the solid. Rinsing works for molecular solids, such as heptane, but not for ionic solids, such as NaCl. We can exploit differences in reactivity to selectively convert an undesired substance to a substance that is more easily separated.

[2] Einstein, A. and Infield, L. 1938. *The Evolution of Physics*, Cambridge University Press, Cambridge, p. 92.
[3] Woods, D. L. Problem-Based Learning, www.eng.mcmaster.ca/chemeng/problem-based-learning-pbl.
[4] See Fogler and LeBlanc (1995) as well as material at www.engin.umich.edu/~problemsolving.

The preceding designs used several conventions for process flowsheets. Each unit should represent only one physical or chemical operation. For example, unit 1 below is unacceptable.

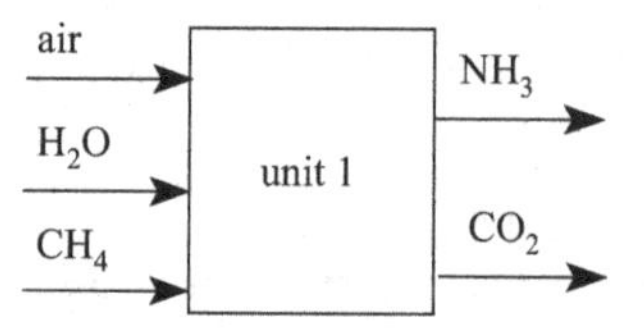

In practice, some units perform more than one operation. Some reactors are also mixers and some reactors are also separators. In this textbook, we use reactors for chemical reactions only; every reactor should have one stream in and one stream out.

Process streams and units should be labeled. The chemical content of the streams and the physical properties of the unit should be included whenever possible. For example, a process stream might be labeled "1240 mol/min methane at 77°C," and a unit might be labeled "reactor, 500°C, 10 atm."

Streams should be drawn to and from units such that matter is neither created nor destroyed. Consider a washer for recycled glass bottles:

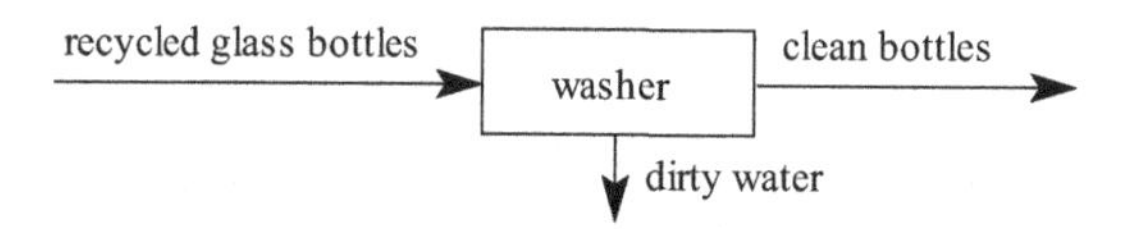

Water appears from nowhere in the diagram above. A proper unit for the washer must include the water that enters.

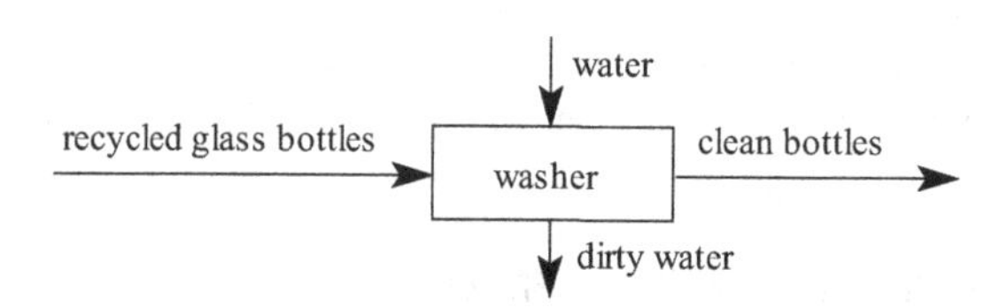

Reactors are sometimes a source of confusion. For example, a reactor may seem to destroy matter, such as nitrogen and hydrogen in the reactor below. A reactor may also seem to create matter. The reactor below seems to create ammonia.

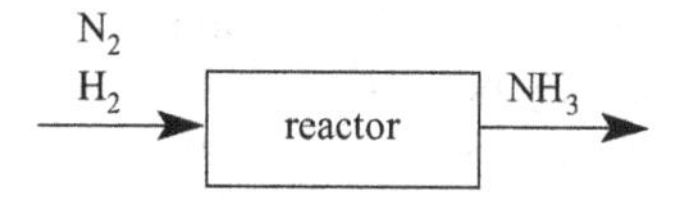

Careful inspection of this unit, however, shows that NH_3 is created from its constituent atoms, H and N, which enter the unit as H_2 and N_2. Matter is not created. Similar reasoning reveals that although nitrogen and hydrogen do not appear in the effluent, matter is not destroyed. There is chemical change in the reactor, but matter is neither created nor destroyed. Labeling the reactor with the reaction decreases ambiguity.

N_2
H_2 → reactor $N_2 + 3H_2 \rightarrow 2NH_3$ → NH_3

Flowsheets are a graphical representation of a chemical process. The overriding consideration is the effectiveness of accurately describing the process. Eliminate ambiguity and improve readability by labeling streams and units, and by drawing descriptive shapes for units.

2.3 DESIGNING A CHEMICAL PROCESS FOR THE SEMICONDUCTOR INDUSTRY

Context: A process for electronic-grade silicon.

Concepts: Separation based on chemical reactivity, continued.

Defining Question: How does one purify a crystalline substance?

Chemical engineering provides the specialty chemicals that enabled the development of solid-state electronic components and semiconductor devices such as integrated circuits (computer chips), light-emitting diodes (LEDs), solid-state lasers, and photovoltaic cells (solar panels). Chemical engineers collaborate to devise sophisticated processes to manufacture these devices, such as lithographic techniques to create circuits on the surface of silicon wafers: chemical etching, ion implantation, chemical vapor deposition, and electrochemical deposition. Chemical engineers are integral members of teams that design fabricating facilities with exacting requirements for air purity (clean rooms) and high standards for protecting workers. Chemical engineers optimize processes for efficient use of materials and energy in manufacturing as well as in recycling electronic devices.

In this section we devise a process to produce silicon used in the manufacture of integrated circuits. This seems easy – silicon is a non-toxic element, is naturally abundant, and is thermally robust. However, the lack of a means to produce *electronic-grade* silicon once inhibited the mass production of solid-state electronic devices, such as transistors and diodes. The complication is the extraordinary purity needed.

Problem: "Produce elemental silicon with less than 1 part per billion impurities."

To put this purity in perspective, 1 part per billion (ppb) corresponds to one grain of salt in a railroad car of sugar.

Is this problem correctly defined? Must the impurity levels be so low? The answers to these questions are not trivial; approximately 40 years of research and development in the field of condensed-matter physics have been devoted to this issue. Without discussing the details of semiconductor device physics, let it suffice to say that the electrical properties of elements such as Si and Ge are extremely sensitive to impurities. Solid-state devices require ultra-pure materials that may be used in their pristine state or intentionally "doped" with impurities to tailor their electrical properties.

What is a good source of silicon? One need not look far. Silicon is 25% of the Earth's crust, the second-most abundant element (oxygen is the most abundant). Virtually all this silicon exists as silicon dioxide (SiO_2) in sand and various silicate minerals. The obvious route to preparing silicon is to decompose sand:

$$SiO_2 + 911\,\text{kJ} \rightarrow Si + O_2. \qquad \text{rxn (2.7)}$$

However, this strongly endothermic reaction proceeds only above 10,000°C. What material could we use to construct the reactor? Most metals are molten at 10,000°C. And how would we heat the vessel? Recall that electricity is an expensive source of energy.

Recall the alternative source of energy we used in the ammonia synthesis. To produce hydrogen we burned water in methane to transfer the oxygen atom from hydrogen to carbon. For example,

$$CH_4 + \frac{1}{2}O_2 + H_2O \rightarrow CO_2 + 3H_2. \qquad \text{rxn (2.8)}$$

The analogous chemical reaction transfers the oxygen atom from silicon to carbon:

$$CH_4 + O_2 + SiO_2 + 350\,kJ \rightarrow Si + CO_2 + 2H_2O. \qquad \text{rxn (2.9)}$$

The additional 350 kJ can be provided by excess CH_4 and O_2, as given in reaction 2.3. Energy from methane is less expensive than energy from electricity, but there is yet a cheaper source of energy: coal. We designate coal by its chief constituents as C_xH_y, where typically $x \approx 10y$, and the (unbalanced) reaction is

$$C_xH_y + O_2 + SiO_2 \rightarrow Si + CO + H_2O. \qquad \text{rxn (2.10)}$$

Thus the first units in our silicon process are

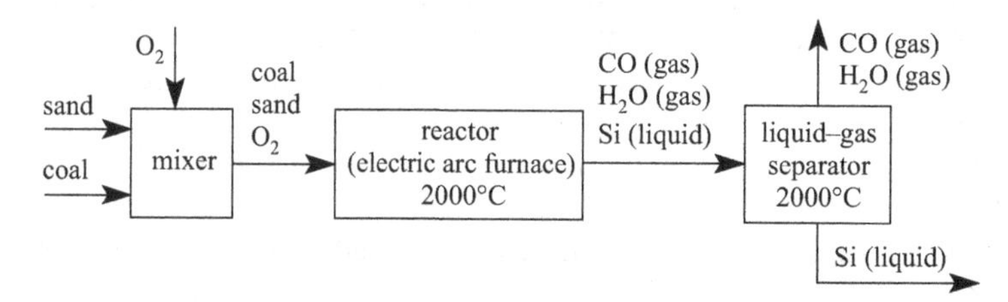

If coal is a cheaper energy source than methane, why don't we use coal in the ammonia synthesis as well? Coal contains higher concentrations of impurities, notably organic nitrogen, organic sulfur, and phosphorus, which would cause problems in the ammonia process. These impurities are not a problem in the silicon process because each reacts with oxygen and forms compounds that are gases at 2000°C and thus are easily separated from the Si liquid.

The silicon liquid from the electric arc furnace is about 98% pure. This "metallurgical-grade" silicon is used chiefly to make silicone polymers. The principal impurities – aluminum and iron – derive from the sand. How do we purify this silicon further?

One possibility is to change SiO_2 into some other chemical that can be easily purified. This is impractical because SiO_2 is a very stable material. Silica is inert to most simple chemical reactions, which is why glassware is so useful in chemistry. Another possibility is to remove Al and Fe from Si by sorting, perhaps with the assistance of magnets. But Al and Fe are not present in the Si as "lumps"; rather Fe and Al are atomically dispersed with the Si.

Distillation is often a viable alternative. As before we consult a source of chemical reference data and create a table:

Table 2.7 Boiling points for silicon processing.

Compound	Boiling point at 1 atm (°C)
CO	−191
H_2O	100
Si	2,355
Al	2,467
Fe	2,760

Clearly we may assume that CO and H_2O gases readily separate from the Si liquid. However, the boiling points for the two impurities of silicon – Al and Fe – are too close to silicon's boiling point for effective liquid–gas separation. Again, "too close" may not be obvious because the boiling points differ by at least 110°C. We only required a difference of 14°C to separate O_2 and N_2. There are two effects here. First, the ratio of the absolute boiling points – the relative volatility – (2628 K / 2740 K = 0.96) is close to 1. Also, the required purity is 1 part per billion (ppb). This is too difficult to achieve with a relative volatility of only 0.96.

How about a separation based on differences in melting points? Perhaps we can separate liquid silicon from solid Fe and solid Al. Again we look to the thermodynamic data tables and find:

Table 2.8 Boiling points and melting points for silicon processing.

Compound	Boiling point at 1 atm (°C)	Melting point (°C)
CO	−191	−205
H_2O	100	0
Si	2,355	1,410
Al	2,467	660
Fe	2,760	1,535

Unfortunately, aluminum, not silicon, has the lower melting point. Such a process would be tenuous anyway because aluminum and iron are soluble in molten silicon. Separation based on solid–liquid phase is thus impractical; how would we remove this liquid aluminum from the solid silicon? Furthermore, even if we could remove the liquid aluminum, how would we separate solid iron from liquid silicon?

The purification of Si is another example in which the problem statement needs to be examined carefully. We assumed that the problem is "remove the Al and Fe from Si." The real issue is "produce pure Si." This allows for other possibilities; consider using chemical reactivity.

In the separation of *n*-heptane and *i*-propanol we used chemical reactivity to transform the undesired component (*i*-propanol), because our only objective was to purify *n*-heptane. Selectively transforming the undesired components in Si (Al and Fe) is more difficult, if not impossible. Si, Al, and Fe are all metals, so they have similar reactivity. More importantly, the Al and Fe atoms can be reached only by removing the surrounding Si atoms. We consider here a broader application of chemistry: transform everything, isolate the Si-containing chemicals, and then reverse the reactions to obtain Si.

The chlorides of silicon, iron, and aluminum are easily prepared from the elements. Consider the following (unbalanced) reactions and boiling points (at 1 atm):

$$\begin{aligned} Si + HCl \rightarrow \quad & SiH_4 && (-112°C) \\ & + SiH_3Cl && (-30°C) \\ & + SiH_2Cl_2 && (8°C) \\ & + SiHCl_3 && (33°C) \\ & + SiCl_4 && (58°C) \end{aligned} \qquad \text{rxn (2.11)}$$

$$Al + HCl \rightarrow \quad AlCl_3 \quad (181°C) \qquad \text{rxn (2.12)}$$

$$\begin{aligned} Fe + HCl \rightarrow \quad & FeCl_2 && (1024°C) \\ & + FeCl_3 && (332°C). \end{aligned} \qquad \text{rxn (2.13)}$$

At 35°C the chlorides of aluminum and iron will condense, and only $SiCl_4$ of the silicon chlorides will condense. We add two units to our silicon process.

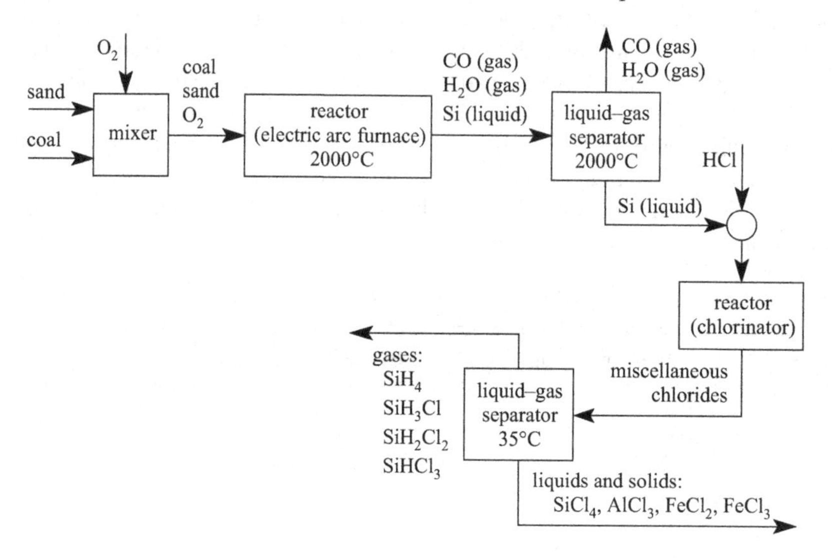

Why not separate at a higher temperature and harvest $SiCl_4$ as well? $SiCl_4$ is the only liquid in the bottom stream. The advantage of a slight increase in Si yield is outweighed by the disadvantage of conveying a stream of solids only instead of solids suspended in a liquid.

Finally, we transform a silicon chloride back into elemental silicon by the Siemens process:

$$SiHCl_3 + H_2 \rightarrow Si + 3HCl. \qquad \text{rxn (2.14)}$$

We add units to isolate $SiHCl_3$ and react with hydrogen to form silicon:

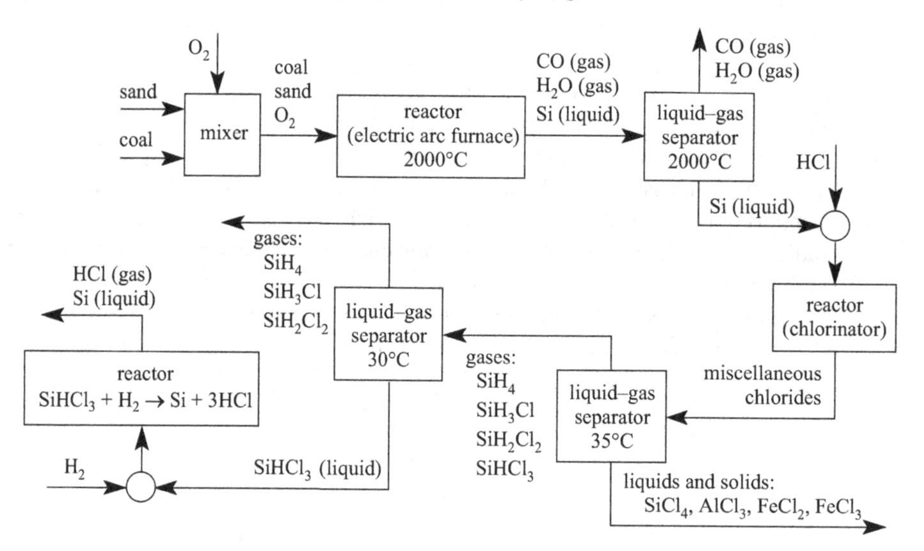

The mixture of HCl and Si is fed to an additional liquid–gas separator to yield silicon liquid with impurities less than 1 ppb. The highly pure polycrystalline silicon from the Siemens process is used to grow large Si crystals, called ingots or boules, typically 6 to

12 inches in diameter and over 3 feet long. The Si boule is sliced into wafers and integrated circuits are patterned onto the wafers to produce chips.

Chemical engineers have key roles in the semiconductor industry. Before the Siemens process, metallurgical-grade silicon was purified by solid–liquid separation, known as zone refining, invented at Bell Laboratories in 1952 by William Pfann, who earned a BS degree in chemical engineering from The Cooper Union in 1940. Andrew Grove, co-founder (1968), president (1979–1997), and CEO (1987–1998) of Intel, and *Time* magazine's 1997 Man of the Year, earned a BS degree in chemical engineering from City College of New York in 1960 and a PhD in chemical engineering from the University of California at Berkeley in 1963.

2.4 DESIGNING PROCESSES TO PRODUCE HYDROGEN

2.4.1 Hydrogen from Methane

Context: Generating electricity with fuel cells and synthesis gas (syngas).

Concepts: Using methane to "burn" water.

Defining Question: What are the advantages of a hydrogen economy?

Chemical engineering is integral to the utility industries. A facility that burns fuel to generate electricity converts chemical energy into electrical energy, usually by means of mechanical energy. The schematic below depicts the principal units in a coal-fired electricity generation plant. Chemical and mechanical engineers design reactors to efficiently burn coal in air. The burner is designed to minimize by-products such as carbon monoxide (CO) and nitrogen oxides such as NO, NO_2, and N_2O, collectively indicated as NO_x. Chemical engineers also design units to collect and convert inevitable by-products including ash and sulfur oxides, SO_2 and SO_3, represented as SO_x. The heat from the burner must be efficiently transferred to the water to produce steam. The high-pressure steam drives a turbine, which turns an electric generator.

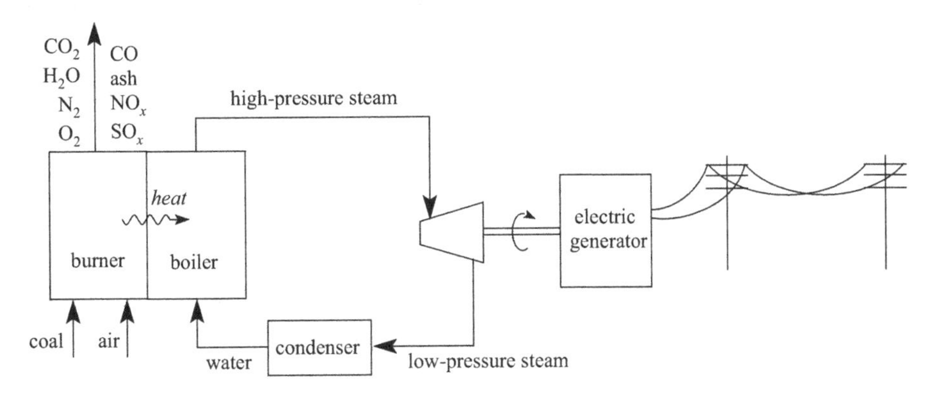

A coal-fired plant is large – a typical capacity is 1000 MW, which can service about a million residences.

As you will learn in thermodynamics, the conversion from chemical energy to thermal energy to mechanical energy and finally to electrical energy is accompanied by unavoidable energy losses at every step. Because fuel cells convert chemical energy directly to electrical energy, fuel cells are potentially more efficient. A fuel-cell electricity generation plant would be much smaller, with capacity on the order of 1 MW, and would service about 1000 homes.

Let's design a chemical process to supply fuel to a fuel cell. The fuel cell's energy source is the reaction of hydrogen and oxygen, the reverse of reaction 2.2:

$$2H_2 + O_2 \rightarrow 2H_2O(\text{gas}) + 484\,\text{kJ}. \qquad \text{rxn (2.15)}$$

When hydrogen and oxygen react, energy is released as heat. The key to producing electricity directly is to divide reaction 2.15 into three steps.

$$2H_2 \rightarrow 4H^+ + 4e^- \qquad \text{rxn (2.16)}$$

$$O_2 + 4e^- \rightarrow 2O^{2-} \qquad \text{rxn (2.17)}$$

$$4H^+ + 2O^{2-} \rightarrow 2H_2O\ (\text{liquid}). \qquad \text{rxn (2.18)}$$

The sum of reactions 2.16, 2.17, and 2.18 is reaction 2.15. The high-voltage electrons generated by reaction 2.16 are collected and sent down power lines. Utility customers use the high-voltage electrons to power various appliances, which reduces the voltage of the electrons. Similarly, high-pressure steam drives a turbine in the power plant above, which reduces the pressure of the steam. The low-voltage electrons return to the plant and react with oxygen, as given in reaction 2.17. Finally, we combine the hydrogen and oxygen ions to make water, in a kinder, gentler fashion than reaction 2.15.

Reaction 2.16 poses a challenging separation problem. Assuming we could entice a hydrogen molecule to decompose into two protons and two electrons (as in a plasma), how do we separate the protons from the electrons? Protons and electrons are both gases and neither cares to condense separately. The key is to use a third player to ferry the electrons from the hydrogen to the oxygen. One fuel cell uses carbonate ions, as follows:

$$2H_2 + 2CO_3^{2-} \rightarrow 2CO_2 + 2H_2O + 4e^- \qquad \text{rxn (2.19)}$$

$$4e^- + O_2 + 2CO_2 \rightarrow 2CO_3^{2-}. \qquad \text{rxn (2.20)}$$

As before, the sum of reactions 2.19 and 2.20 is reaction 2.15. The fuel cell in the simplified diagram below conducts reactions 2.19 and 2.20 in separate cells. The K_2CO_3 (with proprietary additives) allows carbonate ions to diffuse between the anode and cathode. The K_2CO_3 is impermeable to the non-ionic substances.

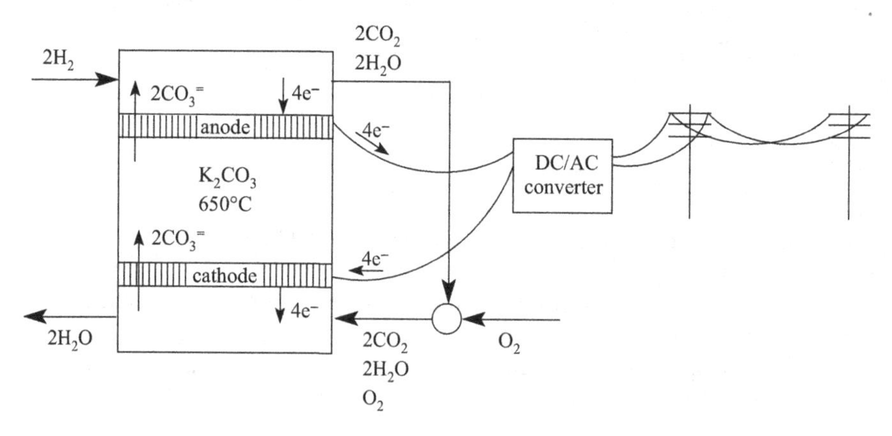

Our task as chemical engineers is:

Problem: "Provide hydrogen and oxygen to power a fuel cell."

What are our options? We could purchase hydrogen and oxygen from a chemical manufacturer and have the reactants delivered by truck, or perhaps by rail. But transporting gases is expensive, especially an explosive gas (H_2) and a flammable gas (O_2). How did we supply H_2 and N_2 for the ammonia synthesis? We produced H_2 and N_2 on site. Let's investigate producing H_2 and O_2. What is a good source? Water is inexpensive and there are no by-products when water decomposes. The decomposition reaction,

$$2H_2O \rightarrow 2H_2 + O_2, \qquad \text{rxn (2.21)}$$

requires energy. How much energy? Exactly the amount of energy generated by the fuel cell, if the fuel cell is 100% efficient. Electrolysis would consume all the electrical energy produced by the fuel cell. We need different sources for H_2 and O_2.

We can condense O_2 from the air, as we obtained N_2 (and O_2) for the ammonia process.

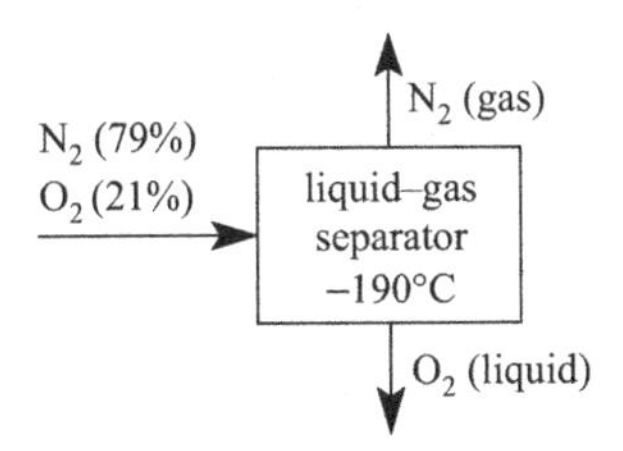

We can generate H_2 using another scheme from the ammonia process: we can "burn" water. Analogously to reaction 2.4, we have the reaction

$$CH_4 + 2H_2O \rightarrow 4H_2 + CO_2. \qquad \text{rxn (2.22)}$$

CO_2 is an acceptable by-product because CO_2 is one of the components in the fuel cell cycle, reactions 2.19 and 2.20. A process to generate H_2 is:

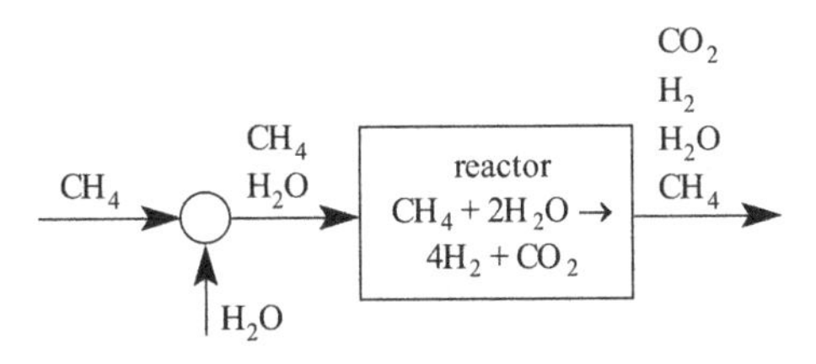

The actual process of producing H_2 from CH_4 is more complex than the reactor above and the similar reactor in the ammonia synthesis. Reaction 2.22 is the overall reaction. The actual process has an intermediate step that produces a mixture known as *synthesis gas*, or *syngas*:

$$CH_4 + H_2O \rightarrow 3H_2 + CO. \qquad \text{rxn (2.23)}$$

The CO is then reacted with H_2O:

$$CO + H_2O \rightarrow H_2 + CO_2. \qquad \text{rxn (2.24)}$$

The actual process is

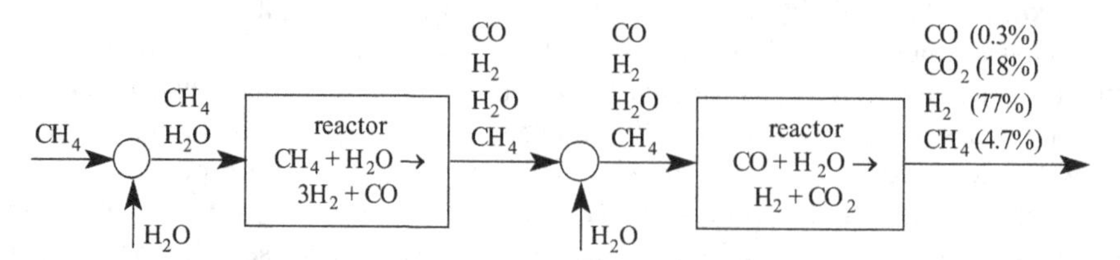

Reaction 2.23 is catalyzed by NiO particles dispersed on aluminum oxide at 500–870°C. Reaction 2.24, also known as the water–gas shift reaction, is catalyzed by iron oxide, then copper–zinc oxide on aluminum oxide, at 220–250°C.

Finally, H_2 gas is separated from the effluent of the second reactor. The CH_4 is recycled to the first reactor and the CO is recycled to the second reactor. The CO_2 may be sold as a chemical commodity if a CO_2 consumer is nearby. Otherwise the CO_2 should be sequestered, to reduce global warming. Although hydrogen from fossil fuels is not sustainable, it has an advantage over burning fossil fuels. Whereas it is impractical for individual vehicles and homes to collect and sequester CO_2 from burning fossil fuels, it is practical to collect and sequester CO_2 at a central facility and use hydrogen for energy in vehicles and homes. Reducing the greenhouse gas CO_2 is the principal advantage of the hydrogen economy.

2.4.2 Methane from Natural Gas

Context: Desulfurization.

Concepts: Absorbers and adsorbers.

Defining Question: What are the disadvantages of converting natural gas to hydrogen?

Our process to produce H_2 from natural gas is idealized: we assumed that our natural gas source is 100% methane. Typically, natural gas is only 85% to 95% methane; the balance is chiefly ethane, propane, and propene. Natural gas also contains small concentrations (<1% total) of He, CO_2, and N_2. None of these constituents will affect the process diagram above. However, natural gas also contains H_2S; up to 3 parts per million (ppm) or 0.0003% H_2S is allowed. Although 3 ppm of impurity may seem inconsequential, it is a problem. A concentration of H_2S above 0.1 ppm will deactivate the catalyst in the reactor to burn water in methane. We must remove H_2S from the natural gas to be used to synthesize hydrogen.

Let's build our fuel-cell electricity generator at the natural gas source, the wellhead. Instead of pumping natural gas to consumers, we will use it to generate electricity and "pump" electricity to consumers. But at the wellhead, the level of H_2S is even higher: 100 ppm. Our task is:

Problem: "Reduce the H_2S level in the natural gas to less than 0.1 ppm."

Again, we first consider liquid–gas separation. We prepare a table of boiling points.

Table 2.9 Boiling points for desulfurization.

Compound	Boiling point at 1 atm (°C)
CH_4	−164
H_2S	−61

Because the difference in boiling points is large, we could condense H_2S from CH_4. But we would have to cool the natural gas to very low temperatures to remove all but 0.1 ppm (0.000001%) of the H_2S. That is, H_2S remains in the gas phase even at temperatures below its boiling point. Consider water, which boils at 100°C at 1 atm. There is still water vapor in the air at temperatures below 100°C; recall a hot humid day. So H_2S could be condensed from CH_4, but it would be expensive. As a rule, cooling something 1°C costs four times as much as heating something 1°C. Engineering requires more than just a workable solution. Engineering also requires an economical solution. As stated by Arthur Mellen Wellington (1847–1895),[5]

> Engineering . . . is the art of doing that well with one dollar, which any bungler can do with two after a fashion.

We introduce here another type of separator, an *absorber*. Absorbers are prevalent in the chemical industry, second only to liquid–gas separation (distillation). An absorber extracts a substance from a gas by absorbing the substance into a liquid, such as water or oil. To explore the possibility of using a water absorber, we add a new column to our table of properties.

Table 2.10 Boiling points and solubilities for desulfurization.

Compound	Boiling point at 1 atm (°C)	Solubility in water
CH_4	−164	slightly
H_2S	−61	very

Quantitatively, 100 mL of water will absorb 440 cm^3 of H_2S gas at 1 atm and 25°C. If H_2S absorbed insufficiently in water, or if CH_4 is absorbed too strongly in water, we would seek a different liquid for our absorber.

We bubble the natural gas through water to absorb H_2S. Our absorber is shown below.

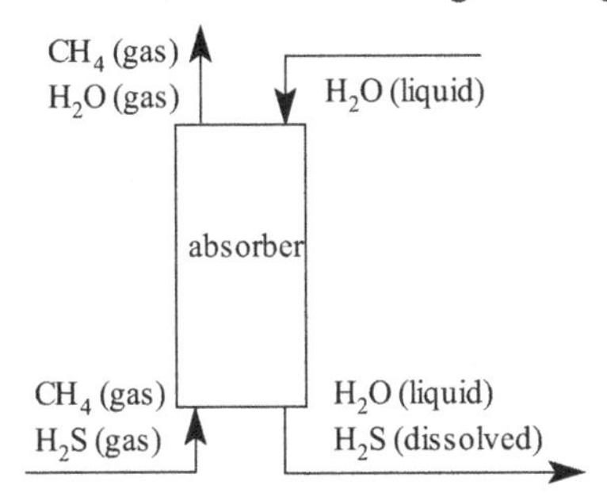

This is idealized; we have ignored the trace amount of CH_4 absorbed in the water. Note the convention in our absorber representation: the liquid stream flows down the column and the gas stream percolates up the column.

Absorbers are typically tall cylinders, also called *absorption columns*. An absorber is usually represented on a process flowsheet by a tall, narrow rectangle. We will study the inner workings of absorbers in Chapter 4.

[5] Wellington, A. M. 1910. *The Economic Theory of the Location of Railways*, 6th edn., John Wiley & Sons, New York, NY.

What shall we do with the absorber effluent? Water is cheap, so there is little economic motivation to purify and recycle the water. Can we release the effluent into a nearby stream? H_2S smells like rotten eggs (perceptible at 1 ppb), is flammable (above 4%), and poisonous (above 0.1%): one or two inhalations of H_2S causes collapse, then coma, then death (Merck & Co., 1996). We must collect the H_2S. The H_2S can be sold; it is converted to H_2SO_4 by way of SO_2. In fact, because H_2S is so noxious, it is advisable to convert it to SO_2 immediately, and then store the SO_2. Of course, we add a purge to the water recycle stream to eliminate impurities soluble in water, with boiling points comparable to water; these impurities would accumulate in the absorber–distiller loop. And the water purged must be replenished; we add a combiner for a *make-up* stream.

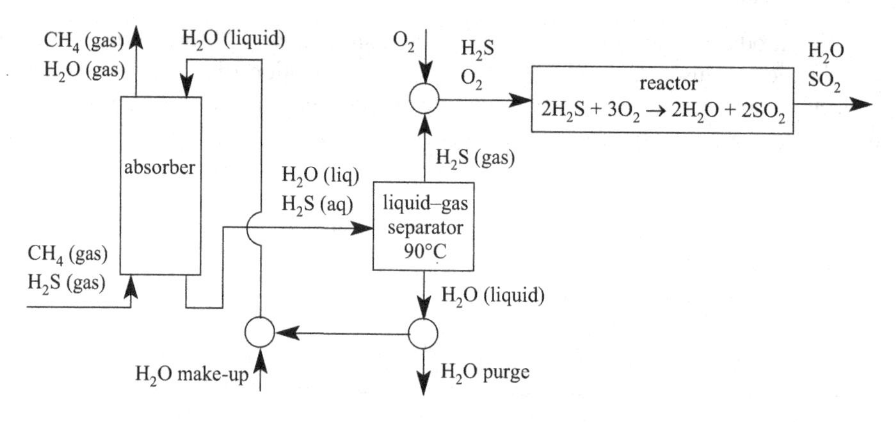

The absorber reduces the H_2S concentration to about 10 ppm (we will learn how to calculate absorber outputs in Chapter 4). In practice, absorbers use liquids with higher capacity for absorbing H_2S. A good liquid for absorbing H_2S is monoethanolamine, which can reduce the H_2S to about 1 to 3 ppm. But even with monoethanolamine, the H_2S concentration is not low enough. We need to add another unit – an adsorber.

An adsorber extracts a substance from a gas (or liquid) by adsorbing it onto a solid. The adsorbed molecule bonds to the surface of the solid. The stronger the bond, the more effectively the solid collects the substance. But a stronger bond also means it is harder to clean the surface to recycle the solid. Charcoal (activated carbon) is commonly used to adsorb impurities from drinking water. The packets of silica shipped with cameras and electronic equipment adsorb water vapor from the air. On a macroscopic level, flypaper adsorbs insects from the air.

To reduce the H_2S concentration to less than 0.1 ppm, the solid must have a strong affinity for H_2S; the surface must form a chemical bond with H_2S. An effective adsorbent for H_2S is ZnO, which adsorbs H_2S by reaction 2.25. The surface of ZnO adsorbs a H_2S molecule, a S atom exchanges with an O atom, and H_2O desorbs from the surface:

$$ZnO + H_2S \rightarrow ZnS + H_2O. \qquad \text{rxn (2.25)}$$

The ZnO surface can be rejuvenated by reaction with oxygen:

$$2ZnS + 3O_2 \rightarrow 2ZnO + 2SO_2. \qquad \text{rxn (2.26)}$$

We add units to adsorb H_2S onto ZnO, with a recycle to release SO_2 and rejuvenate ZnO.

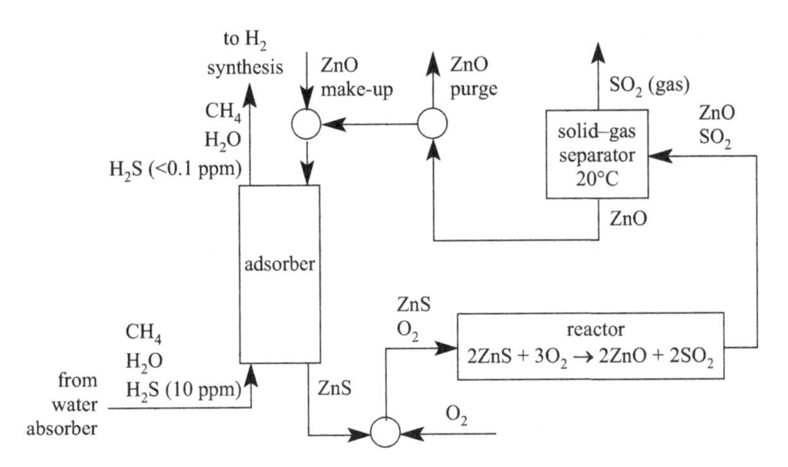

So why is the absorber and its attendant gas–liquid separator necessary? Perhaps we should just feed the CH_4 with 100 ppm H_2S directly into the adsorber. The purity of the natural gas would be the same and the design would be simpler. How else would the process change if the absorber were eliminated? The adsorber would need to remove ten times as much H_2S, and therefore it would need to be emptied and refilled ten times as often, or you would need an adsorber ten times as large. Although the design is more complex with an absorber *and* an adsorber, the design is less expensive.

2.4.3 Hydrogen from Coal

Context: Underground coal gasification.

Concepts: Using coal to "burn" water.

Defining Question: How does one extract coal from thin veins miles deep?

The US has coal resources to supply energy for 200–500 years. Coal *reserves* are available with current technology; coal *resources* are not economically accessible with traditional methods. But much of the coal is inaccessible; the coal veins are thin and deep below ground – a mile or more. Sending men down holes to harvest coal is dangerous and expensive. But harvesting deep coal is another case of defining the real problem.

The problem is not "fetch coal from thin, deep deposits." We don't need the coal, per se. We need the energy inherent in the coal. A better definition of the problem is "extract energy from coal in thin, deep deposits." The redefined problem retains the traditional solution of sending men down holes. But it also allows for a novel solution, known as underground coal gasification (UCG): bring the energy of solid coal to the surface as an energy-rich gas. React coal with steam to form syngas, represented by the following (unbalanced) chemical equation:

$$C_xH_y + H_2O\ (\text{gas}) \rightarrow H_2 + CO. \qquad \text{rxn (2.27)}$$

Coal gasification typically requires heat, so some oxygen is mixed with the steam to burn coal:

$$C_xH_y + O_2 \rightarrow H_2O + CO_2. \qquad \text{rxn (2.28)}$$

UCG eliminates the risky business of coal mining and provides syngas for energy or commodity chemicals. CO can be separated from the syngas and reacted with H_2O to produce more hydrogen. Or the syngas can be converted to methane (natural gas) by Fischer–Tropsch synthesis:

$$CO + 3H_2 \rightarrow CH_4 + H_2O. \qquad \text{rxn (2.29)}$$

Fischer–Tropsch chemistry yields higher hydrocarbons as well.

With a different catalyst, syngas can be converted to another fuel, methanol:

$$CO + 2H_2 \rightarrow CH_3OH. \qquad \text{rxn (2.30)}$$

Methanol is liquid at ambient conditions, which affords transport across oceans and transport in vehicle fuel tanks. Methanol can be converted to dimethyl ether,

$$2CH_3OH \rightarrow CH_3OCH_3 + H_2O, \qquad \text{rxn (2.31)}$$

which leads to a cavalcade of reactions that produce substitutes for the chemicals derived from crude oil.

2.4.4 Hydrogen from Thermal Energy and Water

Context: A process to synthesize hydrogen from water.
Concept: Reaction cycles.
Defining Question: How does one trade a paperclip for a house?

The processes in the two previous sections produce hydrogen from the chemical energy of natural gas and coal, fossil fuels formed from millions of years of plant growth. These processes are not sustainable; eventually the world's supply of fossil fuels will be exhausted. A sustainable process must use solar energy, geothermal energy, or nuclear energy. Each has engineering obstacles. Solar energy can be converted to electrical energy directly and indirectly. Direct conversion with a photovoltaic device has low efficiency (~12%) and is expensive. Indirect conversion has been used since the Industrial Revolution; solar heat evaporates water, which rains on elevated lands and flows to the seas. The water's potential energy can be exploited with a hydroelectric dam. But when was the last hydroelectric dam built in the USA? The public objects to filling canyons and disrupting fish migration. In recent years, a more acceptable indirect conversion of solar energy is wind turbines.

Electricity is convenient for conveying energy, but electricity is not effective for storing energy. Electricity is a perishable product created on demand; its shelf life is 1/60 s. Electricity is expensive to store, for example, as charged capacitors: alternating metal plates with positive and negative charge. Chemical compounds are effective for storing energy; the principles of quantum chemistry (which you will learn in physical chemistry) separate positive and negative charges on the atomic scale.

Let's design a process to convert thermal energy to chemical energy by driving an endothermic chemical reaction. The high-energy product(s) store the energy, easily released by reversing the reaction. If the high-energy product is a liquid or gas, it can be transported to consumers and carried in vehicles.

Let's use thermal energy to dissociate water into hydrogen and oxygen:

$$H_2O \rightarrow H_2 + \frac{1}{2}O_2. \qquad \text{rxn (2.32)}$$

Reaction 2.32 is balanced with respect to matter. For example, there are two hydrogen atoms on the left side and two hydrogen atoms on the right side. We must also balance with respect to energy. For this reaction, the molar heat of reaction, $\Delta\bar{H}_{rxn}$, is

$$\begin{aligned}\Delta\bar{H}_{rxn} &= (\text{enthalpy of products}) - (\text{enthalpy of reactants}) \\ &= \bar{H}_{H_2} + \frac{1}{2}\bar{H}_{O_2} - \bar{H}_{H_2O}.\end{aligned} \qquad (2.1)$$

From a table of enthalpy data, we calculate that $\Delta\bar{H}_{rxn}$ is +242 kJ/mol at 25°C. (This calculation is the application of the first law of thermodynamics, which we will discuss in Chapter 3.) Reaction 2.33 is balanced with respect to both mass and energy:

$$H_2O(\text{gas}) + 242\text{ kJ} \rightarrow H_2 + \frac{1}{2}O_2. \qquad \text{rxn (2.33)}$$

What temperature is needed to supply 242 kJ to each mol of H_2O that reacts? A better question is: at a given temperature, what fraction of reactants are converted to products? We appeal to the second law of thermodynamics. The extent of a reaction is determined by the change in molar Gibbs energy, $\Delta\bar{G}_{\text{rxn}}$, which can be calculated from the enthalpies and entropies of the reactants and products:

$$\Delta\bar{G}_{\text{rxn}} = \Delta\bar{H}_{\text{rxn}} - T\Delta\bar{S}_{\text{rxn}}, \tag{2.2}$$

where T is the absolute temperature and $\Delta\bar{S}_{\text{rxn}}$, the molar entropy of a reaction, is

$$\begin{aligned}\Delta\bar{S}_{\text{rxn}} &= (\text{entropy of products}) - (\text{entropy of reactants}) \\ &= \bar{S}_{H_2} + \frac{1}{2}\bar{S}_{O_2} - \bar{S}_{H_2O}.\end{aligned} \tag{2.3}$$

From a table of entropy data, we calculate that $\Delta\bar{S}_{\text{rxn}}$ is +0.045 kJ/(mol·K). If we assume the change in heat capacity from reactants to products is independent of temperature we can use $\Delta\bar{H}_{\text{rxn}}$ and $\Delta\bar{S}_{\text{rxn}}$ at 25°C and equation 2.2 to calculate $\Delta\bar{G}_{\text{rxn}}$ at any temperature.

$\Delta\bar{G}_{\text{rxn}}$ determines the relative amounts of reactants and products at chemical equilibrium:

$$\frac{\text{amount of product(s)}}{\text{amount of reactant(s)}} \propto e^{-\Delta\bar{G}_{\text{rxn}}/RT}, \tag{2.4}$$

where R is the universal gas constant, 8.314 J/(mol·K). For the dissociation of H_2O, assuming reactants and products are ideal gases,

$$\frac{P_{H_2}P_{O_2}^{1/2}}{P_{H_2O}} \approx e^{-\Delta\bar{G}_{\text{rxn}}/RT}. \tag{2.5}$$

We use equations 2.2 and 2.5 to calculate the fraction of H_2O reacted, shown in Table 2.11.

Table 2.11 Fraction of H_2O reacted at chemical equilibrium.

T (°C)	T (K)	ΔG_{rxn} (kJ/mol)	Fraction of H_2O reacted
25	298	230	10^{-27}
500	773	210	10^{-8}
1,000	1,273	190	0.00001
2,000	2,273	140	0.01
3,000	3,273	100	~0.5

The thermal source must be at least 3000°C for significant yield. Solar light can be focused with parabolic mirrors to produce temperatures above 3000°C. The solar furnace at Odeillo in the French Pyrenées-Orientales can achieve 3800°C. But what material can we use for a reactor at 3000°C? Steel is too soft above 500°C. Ceramics such as zirconia can be used at 3000°C, but ceramics are fragile and expensive. To produce a bulk chemical commodity, economics prohibits exotic reactor materials such as ceramics.

The problem is therefore "find an inexpensive, robust reactor material that retains its integrity at 3000°C." This is analogous to the design problem faced by NASA engineers at the dawn of the US space program in the 1960s: to safely return an astronaut to Earth, design a space capsule to withstand 10,000°C. This design problem was not solved until the space shuttle in 1981. The early NASA engineers redefined the problem – see the preamble to the Problem Redefinition exercises (Exercises 2.37–2.52) on pages 80–81. We, too, must redefine our problem.

Consider an analogous problem in civil engineering: "design a hydroelectric dam to accommodate the needs of migratory fish." Although fish migrating upstream can overcome small barriers (~1 ft), their success rate is near zero for even a modest dam, a 10-ft barrier. Are we restricted to dams no higher than 1 ft? No. Design a path parallel to the dam with a series of steps no more than 1 ft high.

A creative approach to a large Gibbs-energy step in an endothermic chemical reaction is to use intermediates to provide moderate Gibbs-energy steps. This strategy has led to hundreds of reaction sequences for the dissociation of water by thermal energy. General Atomics rated the "top 25 thermochemical cycles for [thermal energy from] nuclear energy," based on qualities such as the chemical separations required, the corrosiveness and toxicity of the intermediates, the ease of continuous operation (are solids involved?), and the largest Gibbs-energy step in each cycle (Brown, *et al.*, 2003; Schulz, 2003). Consider one of the top 25 reaction sequences, the sulfur–iodine cycle:

		$\Delta\bar{H}_{rxn}$ at 25°C (kJ/mol)	$\Delta\bar{S}_{rxn}$ at 25°C (kJ/mol K)	Reaction temperature	$\Delta\bar{G}_{rxn}$ at rxn temp (kJ/mol)	
	$2H_2O+SO_2+I_2 \rightarrow 2HI+H_2SO_4$	−42	−0.32	120°C	+80	rxn(2.34)
	$2HI \rightarrow H_2+I_2$	+9	−0.02	400°C	+20	rxn(2.35)
	$H_2SO_4 \rightarrow H_2O+SO_3$	+176	+0.29	850°C	−150	rxn(2.36a)
	$SO_3 \rightarrow SO_2+\frac{1}{2}O_2$	+99	+0.09	850°C	+2	rxn(2.36b)
sum	$H_2O \rightarrow H_2+\frac{1}{2}O_2$	242	+0.04			rxn(2.37)

Reaction 2.34 – the Bunsen reaction – is exothermic with a negative entropy change, so equilibrium is favored by low temperature. However, the reaction is heated to 120°C because the *rate* to equilibrium (an important consideration ignored here) is too slow at 20°C. Reactions 2.36a and 2.36b occur sequentially at 850°C. The relatively high temperature (850°C) is required by reaction 2.36a.

The four reactions constitute a cycle; all intermediates are created and consumed at equal rates. The sum of the component reactions is only reactants and products. The sum of the enthalpy of reactions 2.34, 2.35, 2.36a, and 2.36b at 25°C ($\Delta\bar{H}_{rxn}$ at 25°C) equals the enthalpy of the overall reaction. The sum of the entropy of reactions 2.34, 2.35, 2.36a, and 2.36b at 25°C ($\Delta\bar{S}_{rxn}$ at 25°C) equals the entropy of the overall reaction.

A flowsheet for a process based on gas–liquid separators for the sulfur–iodine cycle is shown below. The actual process is more sophisticated; it separates based on solubilities in water and combines reaction 2.34 (reactor 1) with the first separator, in a technique called reactive distillation.

For clarity, we have omitted purges on recycle loops. In reality, because the H_2O feed contains impurities and the reactors create by-products, every loop must be purged. We have also omitted pumps and heat exchangers.

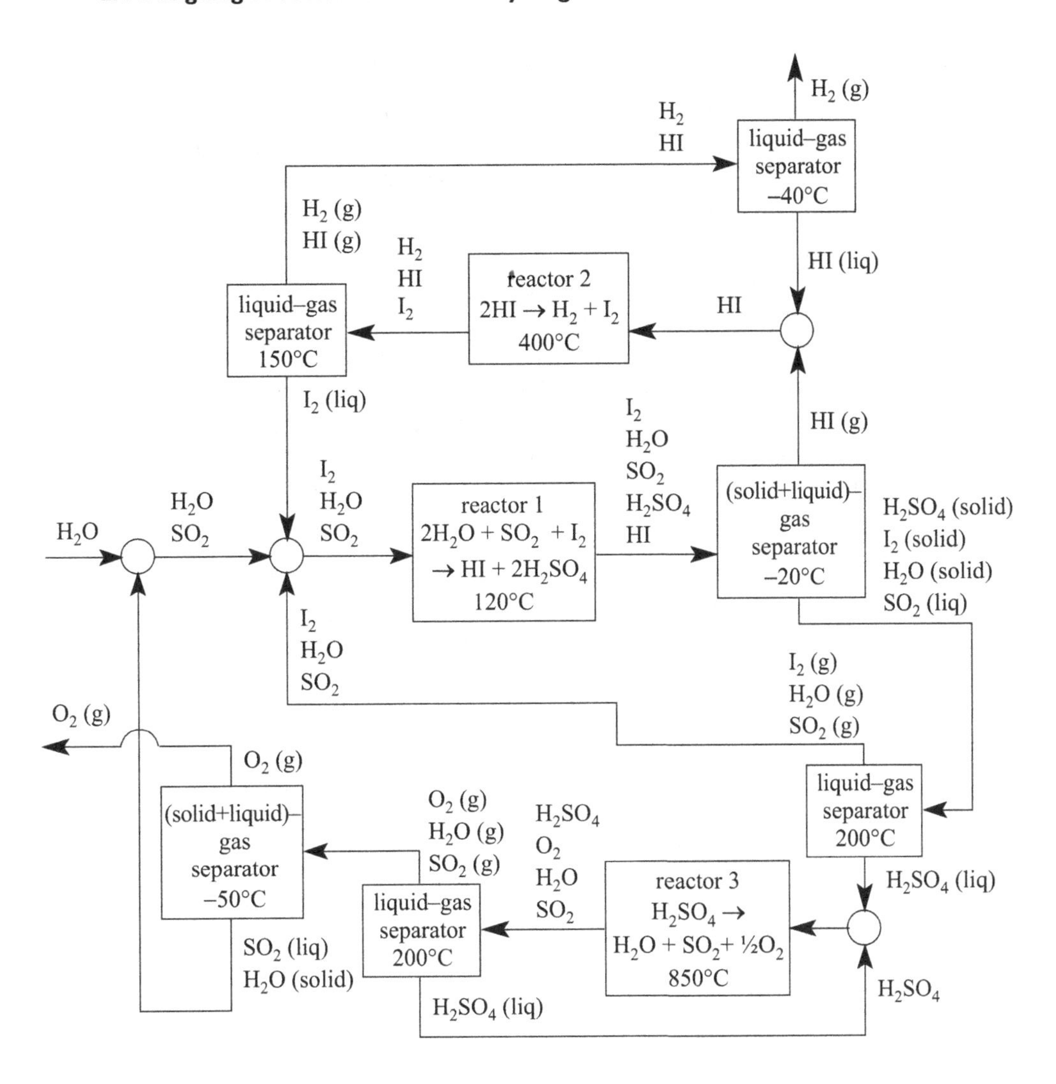

Reaction cycles are a creative solution to the problem of a large Gibbs-energy reaction and are common in the chemical industry. The large Gibbs-energy reaction may substantially increase chemical energy (large positive $\Delta\bar{H}_{rxn}$) and/or may substantially increase molecular order (large positive $\Delta\bar{S}_{rxn}$). The key to an effective design for a reaction cycle is recycling intermediates and heat.

This strategy – partitioning a large step into successive small steps – was used by Kyle MacDonald, the Canadian blogger who traded "one red paperclip" for a house. If one went door-to-door asking to trade a paperclip for a house, the odds of success are infinitesimally small. Instead, MacDonald divided the trade into smaller financial steps: paperclip to a fish-shaped pen to a hand-sculpted doorknob to a camp stove to a generator and so on. Each step moved MacDonald up (as measured in value). Because each step was small, the probability of success was perhaps 1% to 10%. Ultimately, MacDonald traded for a farmhouse in Kipling, Saskatchewan.[6]

[6] See, for example, www.npr.org/templates/story/story.php?storyId=5763262.

2.4.5 Tips for Chemical Process Design: Analyzing a Process Flowsheet

Context: Sulfur–iodine process to synthesize hydrogen from water.

Concepts: Phase lines for thermodynamic data and learning styles for assimilation (visual vs. verbal) and processing (active vs. reflective).

Defining Question: What are your natural learning styles for assimilation and processing?

Analysis of a chemical process flowsheet is an important skill in process design. How does one confront and assimilate an intricate flowsheet, such as the process for the sulfur–iodine cycle in the preceding section? Most people prefer to process information actively, by writing, highlighting, and talking. If one also prefers to assimilate information visually, one should translate from the verbal format for the substances (molecular formulas composed of Hs, Os, Cs, and Ss, each with subscripts). Let's follow the reactant, H_2O, through the process. Translate the verbal symbols "H_2O" into a visual format by highlighting every instance of H_2O with a blue pen. Now connect the blue-highlighted H_2Os to track H_2O through the process. When you encounter a reactor, use a black pen to draw an ellipse below the unit, to represent a "sink" for reactants and a "source" for products. If the reaction is incomplete, some reactant goes into the sink and some reactant passes through the reactor. Products evolve from reactor sources. Analysis of the flow is shown below. Color the circled H_2Os blue and trace the bold line in blue.

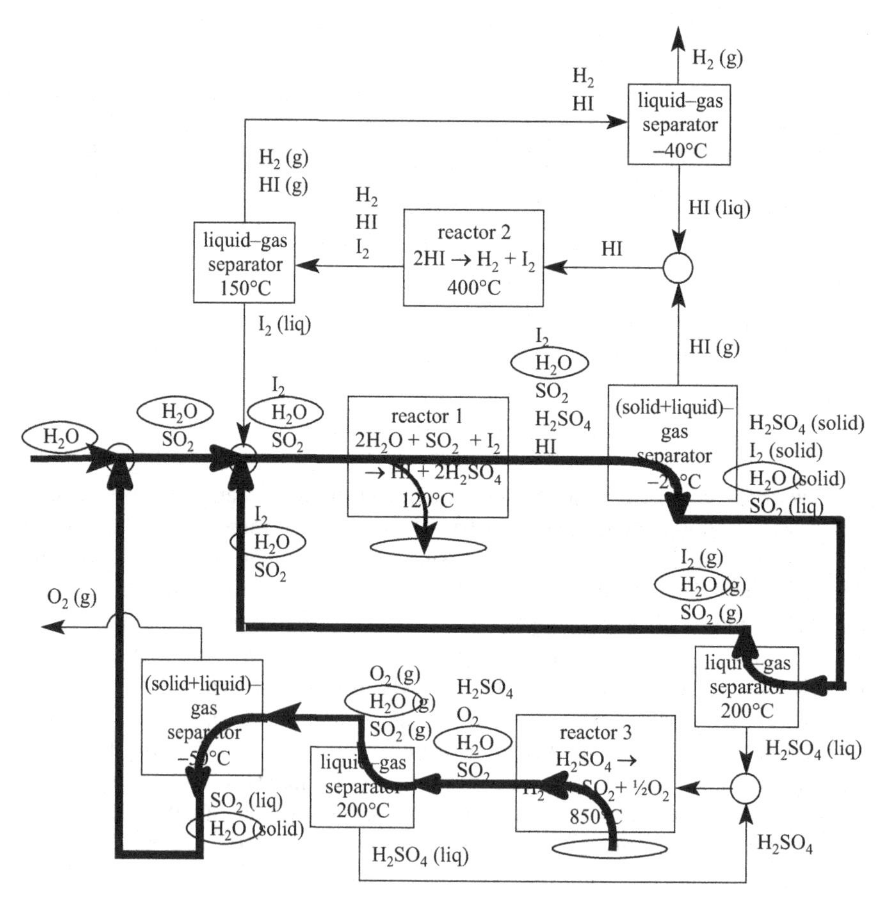

The bold line reveals that H_2O enters the process (as expected for a reactant), is partially consumed in reactor 1 and is recycled to reactor 1. H_2O is produced in reactor 3 and is combined with the reactant feed. An active processor will narrate while coloring and tracing: "The H_2O enters the process . . ." In contrast, a reflective processor will be distracted by coloring, tracing, and hearing their active-processing teammate narrate while coloring and tracing. For efficient (and enjoyable) learning, try different methods and identify your natural style. Try reflective processing to analyze the progression of an intermediate, such as SO_2. Then try active processing to analyze a different intermediate, H_2SO_4.

Are the liquid–gas separators viable? Do the substances follow the correct streams out of the separators, given the separator temperatures? To analyze a separator, one must identify the phase (gas, liquid, or solid) of every substance. If one is a visual assimilator, interpreting the verbal display of melting points and boiling points in Table 2.12 is cumbersome.

Instead, consider a visual translation of these data, shown below. We have drawn a "phase line" for each substance, with the solid, liquid, and gas regions bordered by the melting point and the boiling point. The utility of phase lines is enhanced by colors. The universal color for liquid is blue; color the liquid regions blue. Gases are at the hot end of the temperature scale and red is universal for hot; color the gas regions red. To color the solids, choose a color for aesthetics. Green is nice.

Table 2.12 Boiling points and melting points for the sulfur–iodine cycle.

Compound	Boiling point at 1 atm (°C)	Melting point (°C)
H_2	−253	−259
HI	−35	−51
H_2O	100	0
H_2SO_4	338	11
I_2	184	114
O_2	−183	−218
S	445	113
SO_2	−10	−73

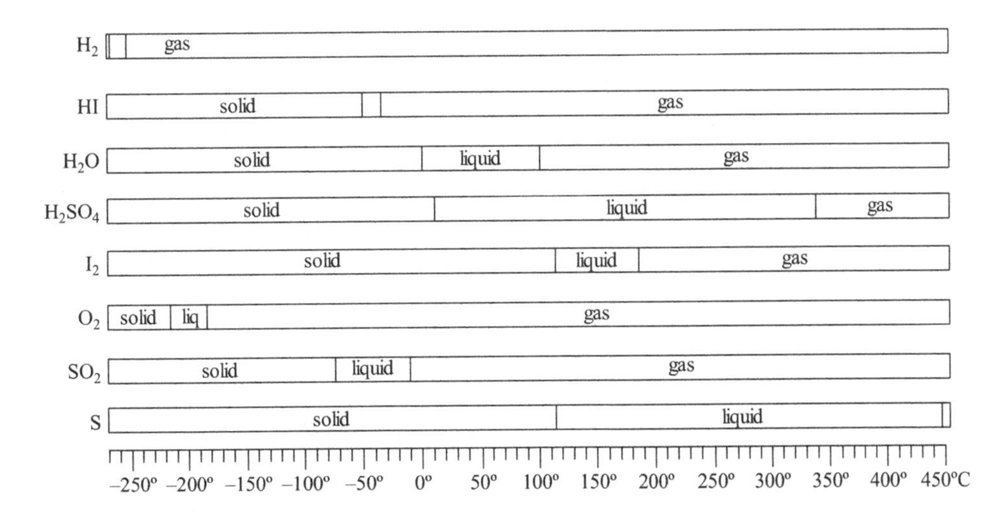

Now test your use of this graphic device. What temperature would you choose to separate HI from a mixture of HI, SO_2 H_2SO_4, H_2O, and I_2? As shown below, at −20°C, HI is a gas, SO_2 is a liquid, and H_2SO_4, H_2O, and I_2 are solids.

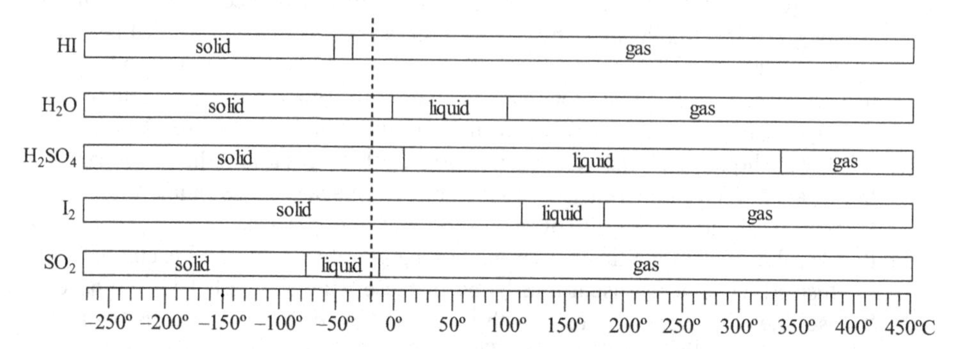

Phase lines are effective when designing a process with four or more substances. Consider drawing phase lines when you work the exercises following this chapter. Sketching phase lines consumes time, but this investment is justified (for visual assimilators) by time saved during design.

One should also verify that the chemical stoichiometry in the flowsheet agrees with the stoichiometry in the reactions. Does every reactor receive the requisite reactants and discharge the specified products? One must also check the overall process. The overall process must agree with the overall reaction. In this case, the overall process should admit only one reactant: H_2O. The process should produce only two products: H_2 and O_2. No intermediate should enter or leave the process, except in purge and make-up streams.

2.5 DESIGNING A PROCESS TO STORE HYDROGEN

2.5.1 Options for Storing Hydrogen

Context: Hydrogen storage for vehicles.

Concepts: Energy densities for fuels, designing a chemical commodity for mass consumption.

Defining Question: What are the key considerations in hydrogen for vehicles?

A few weeks after the Soviet Union won the race to send a man into space in April 1961, President Kennedy declared the US would land a man on the Moon and return him safely to Earth by the end of the decade. In July 1969, Neil Armstrong, Buzz Aldrin, and Michael Collins visited the Moon and returned safely. A few years after global warming was recognized as a threat, President George W. Bush declared in January 2003 that the US would "lead the world in developing clean, hydrogen-powered automobiles," and he committed $1.2 billion for research toward this goal. He, too, set a deadline: "Our scientists and engineers will overcome obstacles to taking these cars from laboratory to showroom, so that the first car driven by a child born today could be powered by hydrogen, and pollution-free." Extrapolating the progress to date, it is unlikely that a teenager in 2019 will drive a hydrogen-powered car. The engineering challenge of one US president was met whereas the engineering challenge of another US president has languished. Why?

2.5 Designing a Process to Store Hydrogen

A key problem is how to carry hydrogen in vehicles. The energy density of hydrogen in various forms is much lower than the energy density of current technology, gasoline.

Table 2.13 Energy densities of vehicle fuels.

Fuel	Energy density (MJ/L)
gasoline	32
H_2 gas at 20°C and 1 atm (14.7 psi)	0.01
H_2 gas at 20°C and 10,000 psi	5
H_2 liquid at −253°C	10
metal hydride, such as PdH_2	4

High-pressure hydrogen gas would require a thick-walled fuel tank, which would add considerable weight to a vehicle. Liquid hydrogen, the option adopted by BMW, would require a large fuel tank owing to the insulation, would consume approximately 30% of the fuel in liquefaction, and would lose considerable H_2 through evaporation; the BMW 7's tank empties by evaporation in 10–12 days. A metal hydride has reasonable energy density and "retains its integrity in a severe collision" – no ruptured high-pressure tank rocketing from an accident site, no flammable hydrogen cloud (Oh the humanity!), and no cryogenic freezing of a car's occupants (see *Terminator II*). Interstitial hydrides such as PdH_2 are convenient because the metal lattice retains its structure as hydrogen is withdrawn; PdH_2 has the same crystal structure as Pd. Interstitial hydrides are also convenient to recharge; the Pd lattice absorbs H_2 like a sponge. But interstitial hydrides are formed from expensive heavy metals. For a modest 10 MJ of energy from PdH_2 (a driving range of 20–30 miles or 30–50 km), the 2.5 L fuel tank would require 1100 ounces of Pd. At 2017 prices, the Pd would cost $800,000. Even fuel tanks of interstitial hydrides of industrial metals, such as titanium, vanadium, and zirconium, with fuel for 300 miles (500 km) would be prohibitively expensive and heavy.

Instead, chemical hydrides of alkali metals (LiH, NaH), alkaline metals (MgH_2, CaH_2), third-period elements (B_2H_6, AlH_3), and various mixed hydrides ($LiAlH_4$, $NaAlH_4$) have cheaper and less massive cations, with only modest compromise in energy density. A metal hydride with good overall properties is $NaBH_4$; its energy density is 2.5 MJ/L and its heat of hydrolysis (the energy lost as heat in generating hydrogen by reaction with water vapor) is low:

$$\frac{1}{4}\mathrm{NaBH_4} + \frac{1}{2}\mathrm{H_2O(g)} \rightarrow \frac{1}{4}\mathrm{NaBO_2} + \mathrm{H_2} + 80\,\mathrm{kJ}. \qquad \text{rxn (2.38)}$$

But chemical hydrides are not easily regenerated; the challenge is to synthesize $NaBH_4$ from H_2. Because $NaBH_4$ is a key reducing agent, commercial synthesis exists. The Schlesinger–Brown synthesis uses sodium metal and boron trimethoxide:

$$\mathrm{H_2} + 2\mathrm{Na} \rightarrow 2\mathrm{NaH} \qquad \text{rxn (2.39)}$$

$$4\mathrm{NaH} + \mathrm{B(OCH_3)_3} \rightarrow \mathrm{NaBH_4} + 3\mathrm{NaOCH_3}. \qquad \text{rxn (2.40)}$$

But sodium metal is formed by electrolysis, which requires substantial energy:

$$2\mathrm{NaCl} + 822\,\mathrm{kJ} \rightarrow 2\mathrm{Na} + \mathrm{Cl_2}. \qquad \text{rxn (2.41)}$$

This energy cost is tolerated for $NaBH_4$ used as a specialty chemical reagent, but is not viable for high-volume production of $NaBH_4$ to be used as an energy carrier.

2.5.2 Storing Hydrogen as Fuel for Vehicles

Context: A process to synthesize sodium boron hydride.

Concepts: Energy densities for fuels, designing a chemical commodity for mass consumption.

Defining Question: How does one conduct a continuous reaction of gas and solid reactants?

Shortly after President Bush's challenge to create a hydrogen economy, Millennium Cell, Inc. patented a low-energy scheme to recycle $NaBO_2$ to $NaBH_4$. The key is B_2H_6 as an intermediate, which avoids the reaction with elemental Na. The patented reaction cycle (Amendola, *et al.*, 2003; Ortega, *et al.*, 2003; Ortega, *et al.*, 2007) comprises three reactions that proceed at practical temperatures:

	reaction temperature	
$2NaBO_2 + 6CH_3OH + CO_2 \rightarrow 2B(OCH_3)_3 + Na_2CO_3 + 3H_2O$	150°C	rxn(2.42)
$2B(OCH_3)_3 + 6H_2 \rightarrow B_2H_6 + 6CH_3OH$	200°C	rxn(2.43)
$2B_2H_6 + 2Na_2CO_3 \rightarrow 3NaBH_4 + 2CO_2 + NaBO_2$	20°C	rxn(2.44)
sum: $NaBO_2 + 4H_2 + 300kJ \rightarrow NaBH_4 + 2H_2O$		rxn(2.45)

This new chemical cycle would allow motorists to refuel by exchanging canisters of spent fuel (solid $NaBO_2$) for canisters recharged with fuel (solid $NaBH_4$). Let's design a process to convert $NaBO_2$ to $NaBH_4$, perhaps a process that can operate at a local service station. We begin with a reactor for reaction 2.42.

$NaBO_2$ (solid)
CH_3OH (gas)
CO_2 (gas)
→ reactor 150°C →
Na_2CO_3 (solid)
$B(OCH_3)_3$ (gas)
H_2O (gas)

We must address the engineering challenge of delivering a stoichiometric mixture of solids and gases into and out of a reactor. This problem – "devise a means to transport a mixture of solids and gases" – is formidable. But the *real* problem is "devise a reactor to convert solid $NaBO_2$ to solid Na_2CO_3." This redefinition allows a better method. Rather than bringing solids to the reaction, bring the reaction to the solids. That is, start with a cylindrical reactor containing solid $NaBO_2$. The $NaBO_2$ will be granular and permeable to gas, with porous restraints at each end permeable to gas molecules but not to solid granules. Heat the reactor to 150°C and inject the gaseous reactants at one end, as shown below.

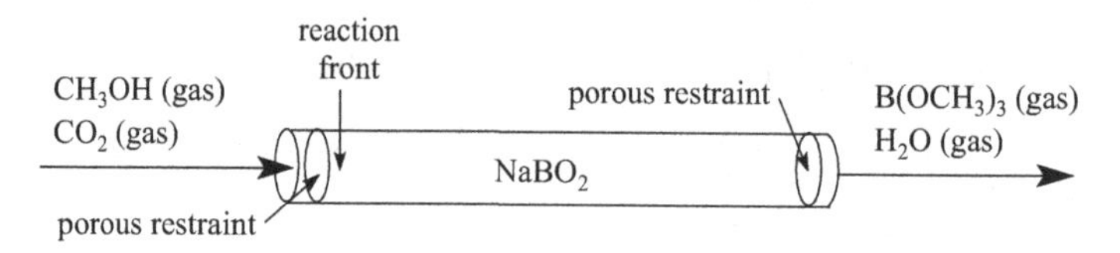

The CH_3OH and CO_2 will react with $NaBO_2$ at the inlet and produce Na_2CO_3, $B(OCH_3)_3$, and H_2O. The solid Na_2CO_3 will remain where it is created, whereas the gaseous products $B(OCH_3)_3$ and H_2O will permeate the granular $NaBO_2$ and leave the cylindrical reactor at the opposite end. Although reaction 2.42 does not go to completion, there are no unreacted reactants in the reactor effluent. The reaction is incomplete at the reaction front, but the unreacted reactants continue through the reactor where there is no product. Because reaction equilibrium is not reached, reaction 2.42 proceeds. By the end of the reactor, reactants are completely consumed. This method of shifting the local equilibrium to reach completion is the method of multistage processes, which we will adopt in Chapter 4 to purify substances through a series of liquid–vapor equilibria.

As reaction 2.42 proceeds, the canister will contain granular product Na_2CO_3 in the downstream end and reactant $NaBO_2$ in the upstream end.

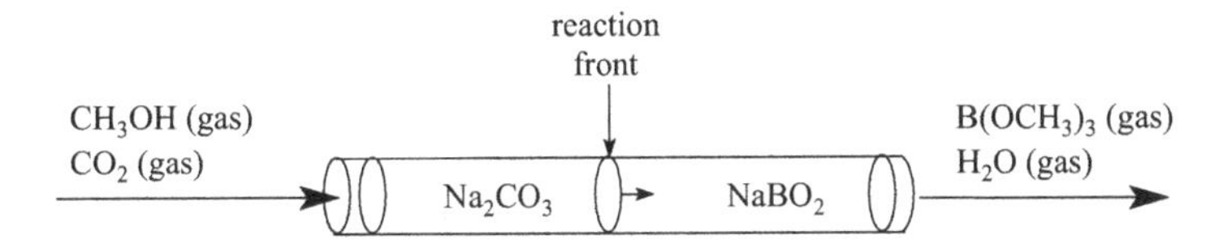

After solid reactant $NaBO_2$ is entirely converted to solid product Na_2CO_3, reactant gases CH_3OH and CO_2 will appear in the outlet stream.

CH_3OH (gas)
CO_2 (gas)
Na_2CO_3
CH_3OH (gas)
CO_2 (gas)

The reactor is then cooled to 20°C and the reactant gas is changed to B_2H_6 to conduct reaction 2.44: convert solid Na_2CO_3 to solids $NaBH_4$ and $NaBO_2$. Again, the reaction front begins at the reactor inlet. Later, the reaction front has moved along the cylindrical reactor. Reactant B_2H_6 will appear in the reactor effluent when reaction 2.44 is complete.

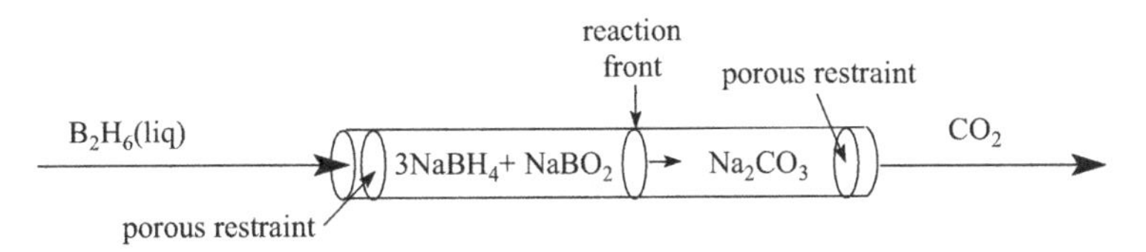

The next problem is also formidable: separate solid $NaBH_4$ from solid $NaBO_2$ to give the customer a canister with $NaBH_4$ only; no $NaBO_2$ spent fuel. Separating solids is impractical. Each particle is likely an amalgam of $NaBH_4$ and $NaBO_2$ which is inseparable. This problem cries out for redefinition, such as "conduct a reaction that yields solid $NaBH_4$ only; no $NaBO_2$." Reaction 2.44 is likely two reactions. The first converts intermediate Na_2CO_3 to product $NaBH_4$, for which H_2O is a by-product:

$$B_2H_6 + Na_2CO_3 + 2H_2 \rightarrow 2NaBH_4 + CO_2 + H_2O. \qquad \text{rxn (2.46)}$$

The H_2 in reaction 2.46 is created by the simultaneous reaction of the by-product H_2O with the product, by the same reaction that generates H_2, reaction 2.38:

$$\frac{1}{2}NaBH_4 + H_2O \rightarrow \frac{1}{2}NaBO_2 + 2H_2. \qquad \text{rxn (2.38)}$$

And the reactant possibly reacts with by-product H_2O:

$$B_2H_6 + 6H_2O \rightarrow 3H_3BO_3 + 6H_2. \qquad \text{rxn (2.47)}$$

So an alternative for reaction 2.44 must take the oxygen from Na_2CO_3 and form a substance other than H_2O – a substance that does not react with $NaBH_4$ or B_2H_6. But this problem need not render our process untenable. A fuel tank with a mixture of 3 parts $NaBH_4$ fuel and 1 part $NaBO_2$ still has an acceptable energy density.

The process below continuously converts canisters with $NaBO_2$ to canisters with 3 parts $NaBH_4$ fuel and 1 part $NaBO_2$.

Table 2.14 Boiling points and melting points for the process to recharge hydride fuel canisters.

	Boiling point at 1 atm (°C)	Melting point (°C)
H_2	−253	−259
B_2H_6	−93	−166
$B(OCH_3)_3$	67	−29
CH_3OH	−65	−198
CO_2	−79[a]	−56[b]
Na_2CO_3	decomposes	851
$NaBH_4$	–	400 (decomposes)
$NaBO_2$	1,434	966

[a] Solid sublimes to gas.
[b] At 5.6 atm.

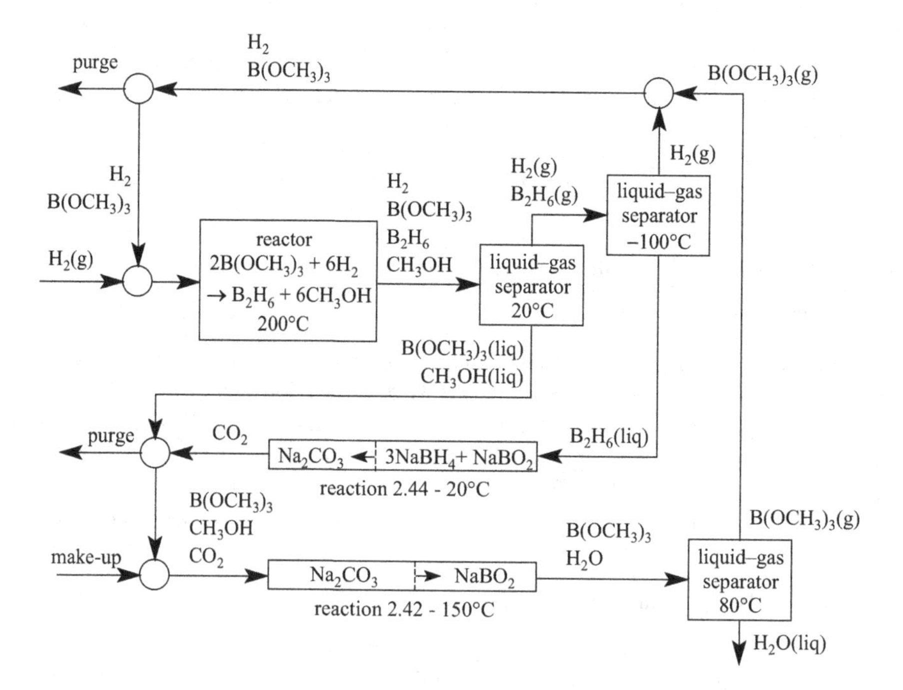

A fuel canister with $NaBO_2$ is installed in the "reaction 2.42" position. After reaction 2.42 is complete, the fuel canister with Na_2CO_3 is moved to the "reaction 2.44" position. After reaction 2.44 is complete, the canister replaces a canister with $NaBO_2$ in the customer's vehicle. With this scheme, vehicles would have several fuel canisters. When a customer visits a refueling station, $NaBH_4$-depleted canisters are replaced with recharged canisters. The fuel-depleted canisters are then recharged by the process above.

Although science and engineering for the hydrogen economy has progressed, more challenges must be overcome. Research slowed after May 2009 when President Obama canceled the Bush plan to invest \$1.2 billion, which trimmed \$100 million per year from the federal budget. Commercial enterprise has also suffered. Millennium Cell – the pioneer company in hydrogen technology that patented the breakthrough synthesis for $NaBH_4$ in 2003 – ceased in May 2008.

SUMMARY

Chemical processes can be divided into discrete components known as units. A unit performs one operation and is represented on a process flowsheet by a simple shape, such as a rectangle or circle. Descriptive shapes are used for some units, such as pumps, valves, mixers, and heat exchangers.

A complex design evolves incrementally from the key unit in the process. The key unit is often a reactor, but it can also be a specialized separation. Units are added by asking questions such as "How do I obtain the reactants needed in the reactor?" or "How do I isolate the product from the by-products and unreacted reactants?" But more important than answering the question is verifying that you have identified the *real* problem to be solved at each stage in the design. Properly identifying the true goal allows for more possibilities for solving the problem. And a longer list of options increases the chance of finding the optimal solution.

One's skill at identifying the *real* problem improves with experience. Unfortunately, formal education does not exercise problem redefinition. In fact, formal education usually does not tolerate problem redefinition. When you are assigned exercises 1, 5, and 6 at the end of a chapter, you might perceive that the *real* problem is not to complete the assignment, but to learn the material in the chapter. Your redefinition of the problem would allow you to use means other than doing the homework to attain the true goal. Most likely the other means would not be tolerated. Rather, you are encouraged to practice identifying the real problems in everyday life.

Design and analysis is efficient when one functions in one's natural style. If the design criteria and constraints are presented in a verbal format, such as a sentence, but your natural style is visual, translate the sentence to a picture. The sentence "Hydrogen and nitrogen catalyzed by iron at 500°C react incompletely to form ammonia" translates to the picture

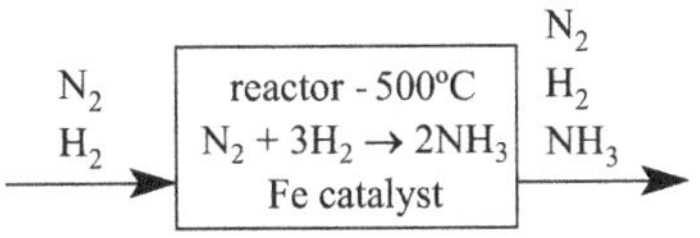

N_2, H_2 → [reactor - 500°C; $N_2 + 3H_2 \rightarrow 2NH_3$; Fe catalyst] → N_2, H_2, NH_3

Drawing is natural for active processers, to restate the problem and to sketch ideas. Active processers should list possibilities on paper. Drawing does not help reflective processers, who naturally collect and sort ideas as mental images. We introduce natural styles of other learning stages in Chapter 3.

REFERENCES

Amendola, S. C., Binder, M., Sharp-Goldman, S. L., Kelly, M. T., and Petillo, P. J. 2003, "System for Hydrogen Generation," US Patent 6,534,033, March 18, 2003.

Brown, L. C., Besenbruch, G. E., Schultz, S. K., *et al.* 2003. *High Efficiency Generation of Hydrogen Fuels Using Thermochemical Cycles and Nuclear Power*, General Atomics Project 30047, March 2002.

Fogler, H. S. and LeBlanc, S. E. 1995. *Strategies for Creative Problem Solving*, Prentice Hall, Upper Saddle River, NJ.

Luyben, W. L. and Wenzel, L. A. 1988. *Chemical Process Analysis: Mass and Energy Balances*, Prentice Hall, Upper Saddle River, NJ.

Merck & Co. 1996. *The Merck Index: An Encyclopedia of Chemicals, Drugs and Biologicals*, 12th edn., Merck & Co., Whitehouse Station, NJ.

Ortega, J. V., Kelly, M. T., Snover, J. L., Brady, J. C., and Wu, Y. 2007, "Triborohydride Salts as Hydrogen Storage Materials and Preparation Thereof," US Patent 7,214,439, May 8, 2007.

Ortega, J. V., Wu, Y., Amendola, S. C., and Kelly, M. T. 2003, "Processes for Synthesizing Alkali Metal Borohydride Compounds," US Patent 6,586,563, July 1, 2003.

Reklaitis, G. V. 1983. *Introduction to Material and Energy Balances*, John Wiley & Sons, New York, NY.

Rudd, D. R., Powers, G. J., and Siirola, J. J. 1973. *Process Synthesis*, Prentice Hall, Upper Saddle River, NJ.

Schulz, K. 2003. "Thermochemical Production of Hydrogen from Solar and Nuclear Energy," presentation to the Stanford Global Climate and Energy Project, April 2003.

Wellington, A. M. 1910. *The Economic Theory of the Location of Railways*, 6th edn., John Wiley & Sons, New York, NY.

CHEMICAL PROCESS DESIGN BIBLIOGRAPHY

Murphy, R. M. 2007. *Introduction to Chemical Processes: Principles, Analysis and Synthesis*, McGraw-Hill, New York, NY.

EXERCISES

A table of melting points, boiling points, solubilities in water, and densities appears at the end of the Chapter 2 exercises.

Process Analysis

2.1 Coal is a plentiful energy resource. The estimated US supply will last 800 years. However, the coal must be converted to a liquid to be useful for transportation fuel. A process for converting coal to diesel fuel is shown in the simplified flowsheet below.

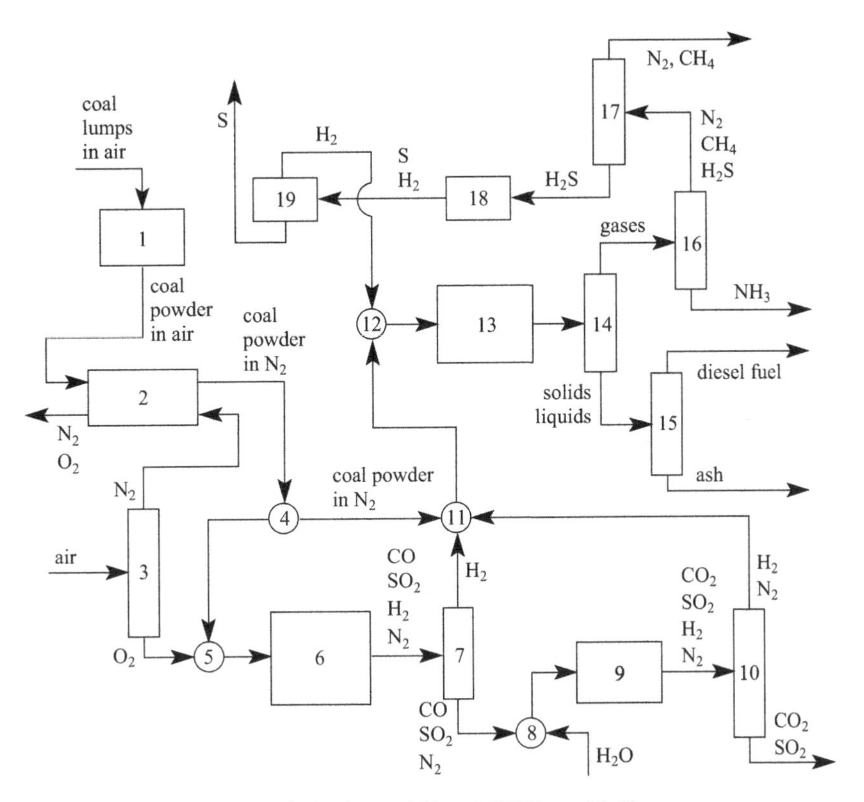

Figure adapted from a process by Luyben and Wenzel (1988), pp. 46–49.

(A) Identify the units of this process. If a unit is not one described in Chapter 2, give it a name based on its function. Coal has the generic formula $C_xH_yS_z$, where $x > y$ and $z \ll x, y$. Coal also contains some inorganic matter, called ash. Diesel fuel has the generic formula C_aH_b.

(B) Write the overall reaction for this process.

(C) Write the reactions that take place in each of the units you identify as reactors.

2.2 Ethylene oxide is a key reactant. It is used, for example, to manufacture ethylene glycol (a component of antifreeze radiator fluid) and polyethylene oxide (PEO). A simplified flowsheet of a process to synthesize ethylene oxide (C_2H_4O) from ethylene (C_2H_4) is shown below. Two reactions occur in the reactor:

$$2C_2H_4 + O_2 \rightarrow 2C_2H_4O$$

$$C_2H_4 + 3O_2 \rightarrow 2CO_2 + 2H_2O.$$

Both reactions are incomplete; neither reactant is completely consumed in the reactor. Because ethylene oxide is very soluble in water, it can be extracted by bubbling the reactor effluent gas through water. Ethylene oxide is *absorbed* into the water. Ethylene oxide is then *stripped* from the water by steam.

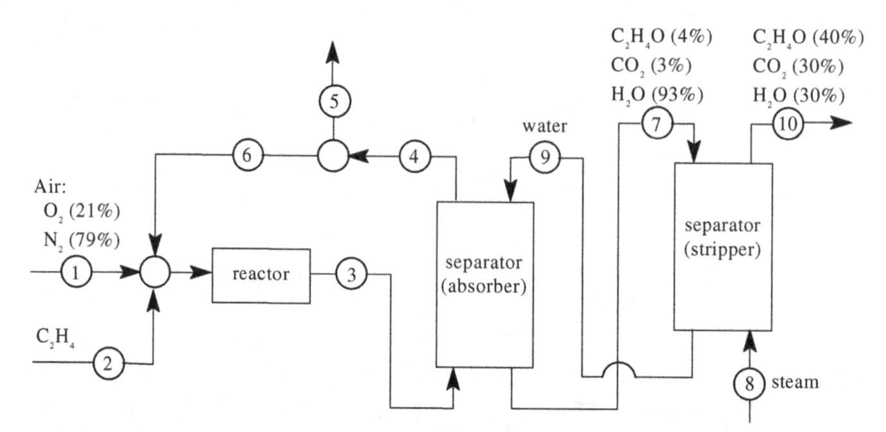

(A) What substances constitute stream 3?
(B) What substances constitute stream 4?
(C) What purpose is served by stream 5?
(D) Add one or more units to produce pure C_2H_4O from stream 10 in an energy-efficient manner. Indicate the compounds present in each stream and label each unit.

2.3 A process for producing ammonia from air, water, and methane was outlined in this chapter. Draw the flowsheet for the entire process. Label the units and list the components in each stream. Where appropriate, use the proper symbol to identify a unit operation. Finally, improve the energy efficiency of the process by adding heat exchangers before the air liquid–gas separator and the ammonia reactor.

2.4 The process shown below coats Si wafers with a thin polymer film, a crucial step in producing integrated circuits by photolithography. The process uses two reactants – monomer and Si wafers – and uses two solvents – air and acetone. Monomer molecules are linked into large polymer molecules in the reactor; an example is the reaction to form polyethylene, $n(CH_2{=}CH_2) \rightarrow (-CH_2-CH_2-)_n$, where $n > 1000$. The Si wafers and the monomer are expensive. Because the monomer is non-volatile, the air entering the dryer must be at least 200°C. The catalyst in the reactor cannot be exposed to air.

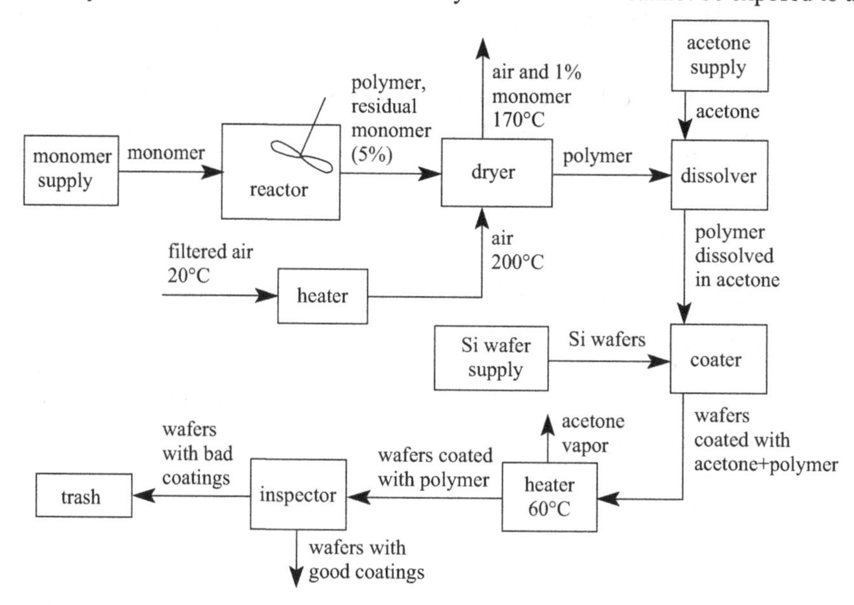

The polymer is soluble in acetone. Although air is a free raw material, the filtered air used in this process has been treated to remove particles larger than 1 μm and thus is valuable.

Improve the process with regard to environmental impact, economics, and energy efficiency. Your improvements must be in the form of unit operations, not a list of suggestions. Draw on a copy of the flowsheet above and/or redraw portions of the process. Three significant improvements will suffice. The uncorrected flowsheet may be downloaded from duncan.cbe.cornell.edu/Graphs.

2.5 Sulfur dioxide can be produced by oxidizing sulfur:

$$S + O_2 \rightarrow SO_2.$$

Because the reaction is highly exothermic, the flow to the reactor must contain cool inert gas to maintain a low temperature in the reactor. Consider the two process schemes diagrammed below. Scheme I uses N_2 as the inert gas in the reactor and Scheme II uses SO_2 as the inert gas in the reactor.

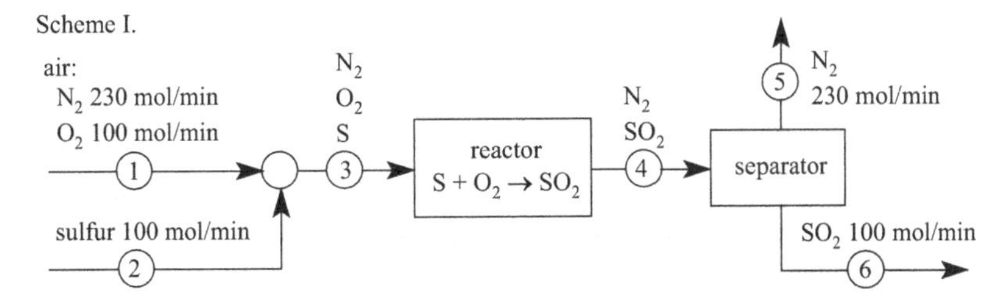

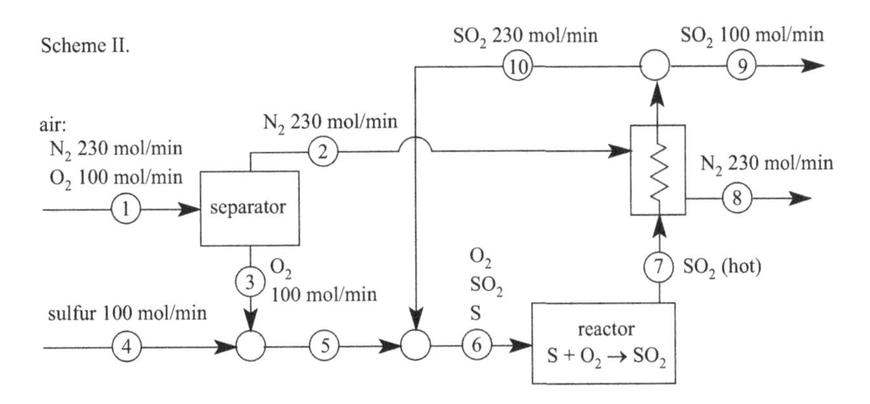

(A) In which sequence is the separation the easiest? State at least two reasons for your choice.

(B) Schemes I and II assume O_2 and S enter the reactor in exact stoichiometric ratio. Consider the consequences of a non-stoichiometric ratio of O_2 and S. In which scheme is the product purity *least* sensitive to an excess of air? Explain your choice.

Exercise adapted from Rudd, Powers, and Siirola (1973), Example 1.3–2, p. 12, and Example 3.4–1, pp. 82–84.

2.6 The steady-state process below produces cool, pure water from seawater. Find four errors in the process flow sheet. For this exercise, we assume ideal solid–liquid separation; dry solids leave the separator.

Hint: Look for errors in style as well as errors in engineering.

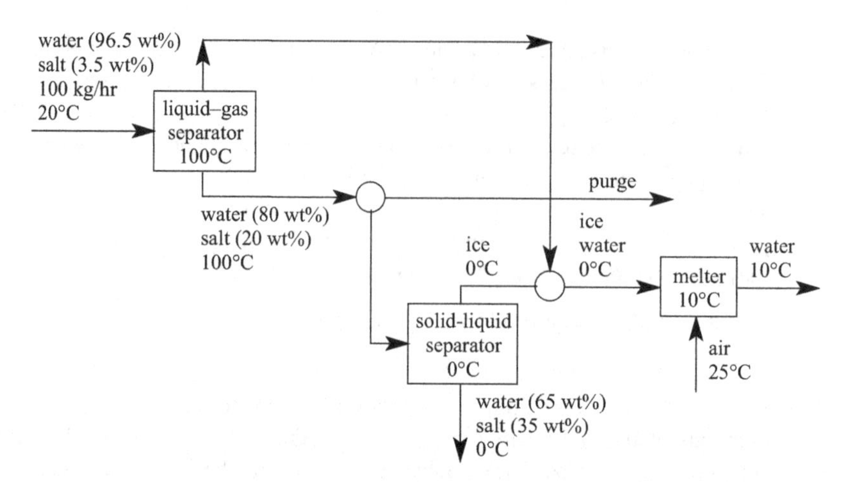

2.7 Tetrachloroethylene, $CCl_2{=}CCl_2$, is used widely as a dry-cleaning fluid and degreasing solvent. Tetrachloroethylene can be synthesized from carbon tetrachloride, CCl_4, by the following consecutive pyrolysis reactions at 800°C:

$$2CCl_4 \rightarrow CCl_3{-}CCl_3 + Cl_2$$
$$CCl_3{-}CCl_3 \leftrightarrow CCl_2{=}CCl_2 + Cl_2.$$

The first reaction is irreversible, but the reaction does not go to completion; there is CCl_4 in the reactor effluent. The second reaction is reversible and does not go to completion; there is hexachloroethane, $CCl_3{-}CCl_3$, in the reactor effluent. The Cl_2 from the first reaction impedes the forward progress of the second reaction. So rather than recycle $CCl_3{-}CCl_3$ to the reactor, the $CCl_3{-}CCl_3$ is separated and sent to a second pyrolysis reactor, but at 500°C.

Consider the process flowsheet below. Use the chemical reactions above and the table of physical properties at the end of the Chapter 2 exercises to find at least three errors in the design. Ignore errors due to omitted pumps. Assume solid–gas separations are ideal; no gas leaves with the solid. Also, find two ways to improve the process, for example, by eliminating a separator and increasing the product yield. Ignore improvements by adding heat exchangers. Finally, change the process flow diagram to fix the errors and incorporate the improvements. Note: solid $CCl_3{-}CCl_3$ is soluble in liquid $CCl_2{=}CCl_2$.

The uncorrected flowsheet may be downloaded from duncan.cbe.cornell.edu/Graphs.

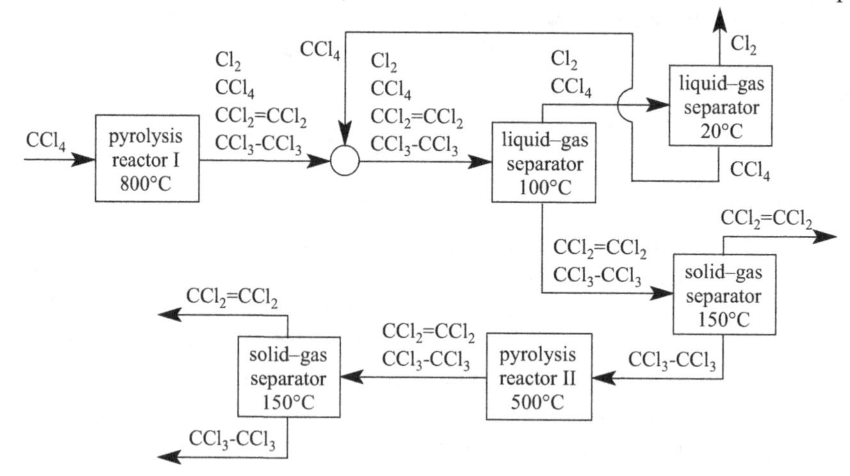

2.8 Shown below is a process for producing concentrated hydrogen cyanide, HCN, by the following overall reaction catalyzed by solid Pt:

$$2NH_3 + 3O_2 + 2CH_4 \xrightarrow{Pt} 2HCN + 6H_2O.$$

In addition, there is an undesired side reaction,

$$2O_2 + CH_4 \xrightarrow{Pt} 2CO_2 + 6H_2O.$$

The reaction is conducted with excess NH_3 and CH_4, which completely consumes the O_2. Because both reactions are highly exothermic, heat is removed from the reactor by converting water to steam. Air is used as the O_2 source. The N_2 in the air does not react. The HCN product is purified by a series of absorbers and strippers. First, an acid solution extracts unreacted NH_3 and the by-product H_2O from the product gas at 30°C. Next, the HCN is absorbed into a water solution at 20°C and the waste gas is burned as fuel. Finally, the HCN is stripped from the water solution by steam.

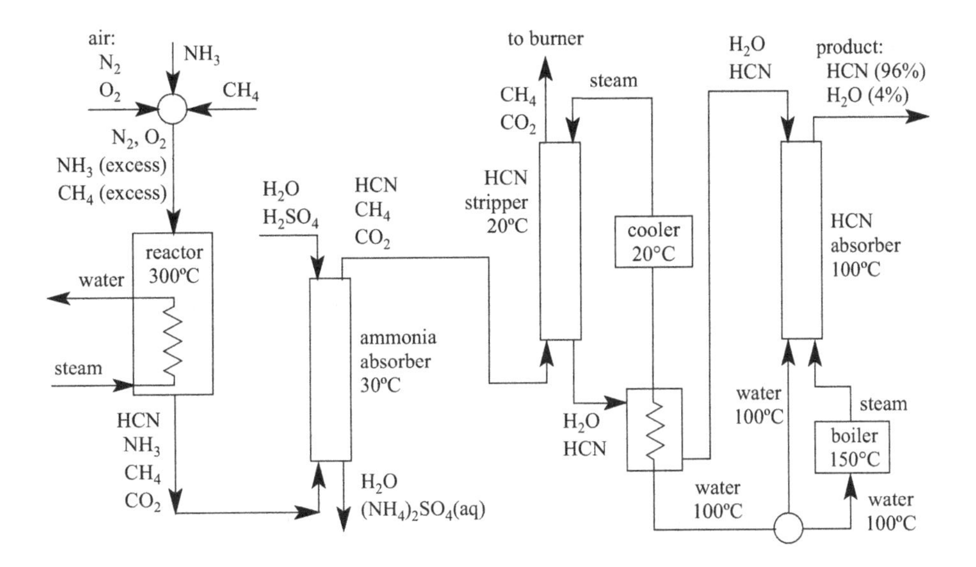

(A) Find at least seven errors in the process flowsheet. Your errors must be distinct and unrelated. For example, if two substances are missing from a stream label, that counts as *one* error.

Hint: Look for errors in style as well as deviations from the chemical reactions described above. Also, an absorber must have two distinct phases: a gas phase and a liquid phase (for example, air and water) or two immiscible liquids (for example, oil and water).

(B) As you will learn later, separations are not ideal. Small amounts of the gas leaving an absorber or stripper will be present in the liquid phase leaving the unit; the liquid stream leaving the ammonia absorber will contain a small amount of HCN. Similarly, small amounts of the liquid leaving an absorber or stripper will be present in the gas phase leaving the unit; the gas stream leaving the ammonia absorber will contain small amounts of NH_3 and H_2O vapor.

Consider only the streams leaving the HCN stripper and the HCN absorber. Describe the consequences of separator non-idealities (if any) and what actions (if any) must be taken.

(C) Find two industrial uses for our HCN product.

2.9 Carbon tetrachloride, CCl_4, was used widely in consumer products such as refrigerators, fire extinguishers, cleaners, pesticides, and fumigants. But because CCl_4 is toxic and it decomposes to substances that deplete the ozone layer, it is presently used only in industry, chiefly as a solvent. However, CCl_4 is still a high-volume chemical, with US production around 500,000 tons/year.

Carbon tetrachloride can be produced from carbon, C, and chlorine, Cl_2, via the intermediates carbon disulfide, CS_2, and sulfur monochloride (aka disulfur dichloride), S_2Cl_2. The first reaction forms carbon disulfide from elemental carbon and sulfur at 800°C in an electric arc furnace:

$$C + 2S \rightarrow CS_2.$$

This reaction goes to completion; only product leaves the reactor. Carbon disulfide and chlorine react at 130°C in the presence of a Fe catalyst to form carbon tetrachloride and sulfur monochloride:

$$CS_2 + 3Cl_2 \xrightarrow{Fe} S_2Cl_2 + CCl_4.$$

The sulfur monochloride reacts with carbon disulfide at 60°C to form more carbon tetrachloride and sulfur:

$$CS_2 + 2S_2Cl_2 \rightarrow 6S + CCl_4.$$

The second and third reactions do *not* go to completion.

(A) Show that the three reactions may be combined to the overall reaction

$$C + 2Cl_2 \rightarrow CCl_4,$$

or some multiple of this reaction, such as

$$2C + 4Cl_2 \rightarrow 2CCl_4.$$

(B) Consider the process flow diagram below. Use the reactions above and the table of physical properties at the end of the Chapter 2 exercises to find at least four errors in this process. Ignore errors due to omitted pumps. Assume solid–gas separations are ideal; no gas leaves with the solid. Change the process flow diagram to fix the errors. The uncorrected flowsheet may be downloaded from duncan.cbe.cornell.edu/Graphs.

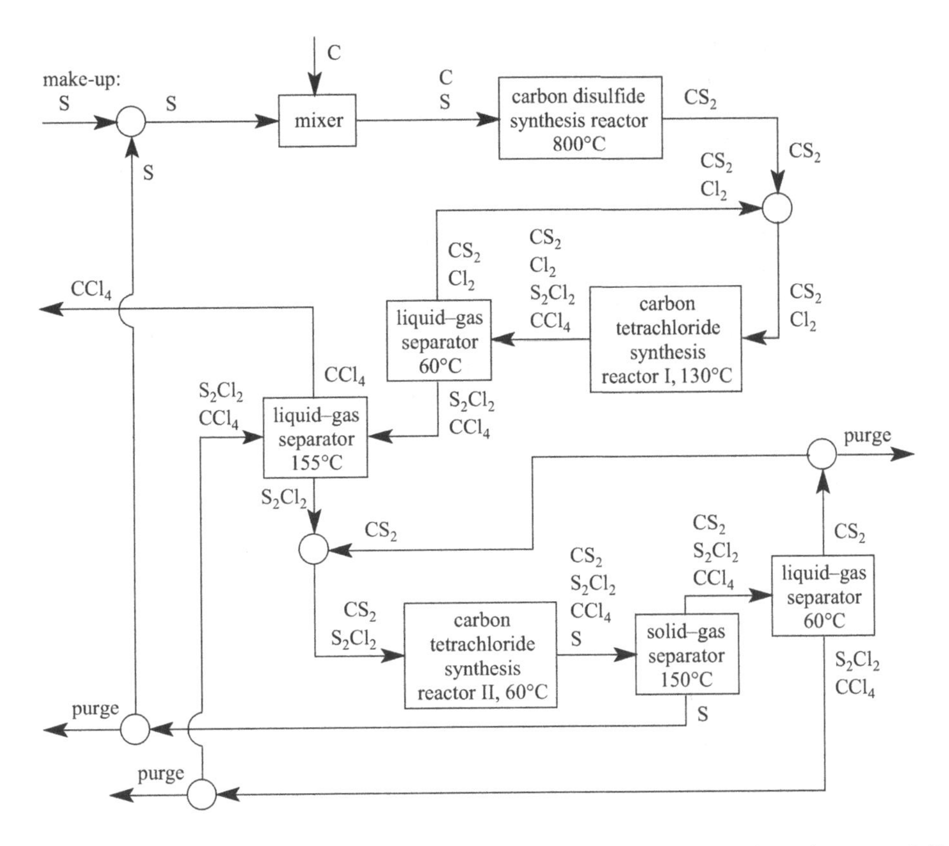

2.10 Phenol, C_6H_5OH, is a key reagent in the synthesis of polymeric resins, especially for the manufacture of plywood, home appliances, and automobiles. Because phenol is an antiseptic and an anesthetic agent, it is an active ingredient in several health products. Although concentrated phenol causes chemical burns, dilute phenol is used as a chemical exfoliant.

Phenol can be synthesized from benzene, C_6H_6, by the four-step benzenesulfonate process. The first reaction adds a sulfite group to benzene at 150°C:

$$\text{sulfonation} \qquad C_6H_6 + H_2SO_4 \rightarrow C_6H_5SO_3H + H_2O.$$

In the second reaction, the benzenesulfonic acid is neutralized to a sodium salt in an aqueous solution. The reaction is conducted at 95°C chiefly to dissolve the moderately soluble Na_2SO_3:

$$\text{neutralization} \qquad 2C_6H_5SO_3H + Na_2SO_3 \rightarrow 2C_6H_5SO_3Na + SO_2 + H_2O.$$

In the third reaction, the sulfite is replaced by reaction with sodium hydroxide at 300°C:

$$\text{fusion} \quad C_6H_5SO_3Na + 2NaOH \rightarrow C_6H_5ONa + Na_2SO_3 + H_2O\,.$$

Finally, the phenolic sodium salt is acidified in aqueous solution at 20°C to form phenol:

$$\text{acidification} \qquad 2C_6H_5ONa + SO_2 + H_2O \rightarrow 2C_6H_5OH + Na_2SO_3\,.$$

(A) Combine the four reactions to find the overall reaction. **Hint:** The overall reaction should contain only reactants, products, and by-products. It should not contain the intermediates $C_6H_5SO_3H$, $C_6H_5SO_3Na$, or C_6H_5ONa.

(B) The process flowsheet below is based on the reactions above with the following assumptions:

- Excess C_6H_6 in the sulfonation reactors completely consumes the H_2SO_4.
- Excess Na_2SO_3 in the neutralization reactor completely consumes the $C_6H_5SO_3H$.
- The fusion and acidification reactions go to completion; no reactants appear in the reactor effluents. However, H_2O is the solvent in both reactors. H_2O solvent must appear in both reactor inputs and both reactor effluents.
- Heat exchangers, heaters, coolers, and pumps have been omitted for simplicity.

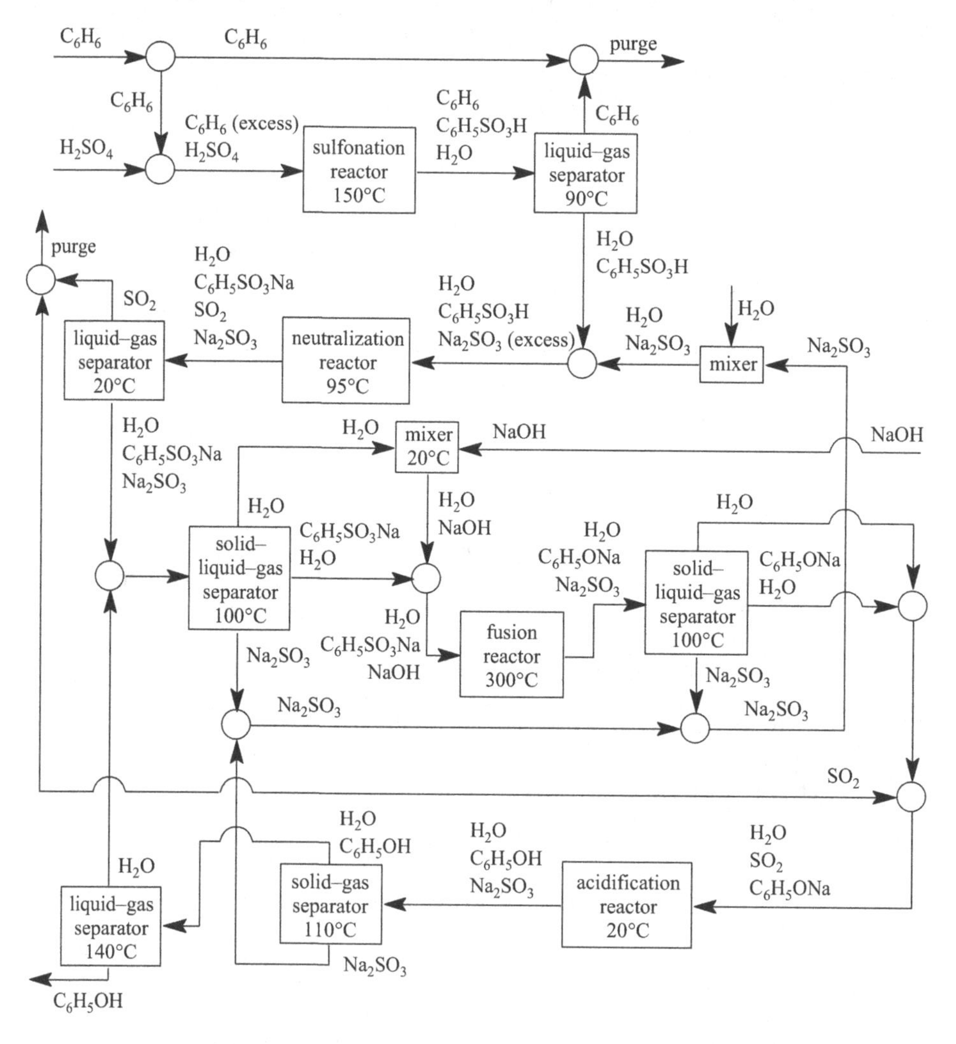

Find six errors and/or improvements. You may assume the two solid–liquid–gas separators are correct. The solution from the SO_2 separator is heated to 100°C to evaporate some of the H_2O – enough to cause the moderately soluble Na_2SO_3 to precipitate, but not so much as to precipitate the $C_6H_5SO_3Na$. Some Na_2SO_3 will remain in the H_2O–$C_6H_5SO_3Na$ solution but that is neglected. Similarly, the solution from the fusion reactor is heated to 100°C to evaporate some of the H_2O – enough to precipitate the moderately soluble Na_2SO_3, but not so much as to precipitate the very soluble C_6H_5ONa. Some Na_2SO_3 will remain in the H_2O–C_6H_5ONa solution but that is neglected. Finally, we assume ideal solid–liquid separation in both separators; dry solids leave the separators.

Finally, change the process flow diagram to fix the errors and/or improve the design. The uncorrected flowsheet may be downloaded from duncan.cbe .cornell.edu/Graphs. **Hint:** Examine individual units and streams, but also examine the process overall.

Adapted from a process by W. L. Luyben and L. A. Wenzel (1988). *Chemical Process Analysis: Mass and Energy Balances*, Prentice Hall, Upper Saddle River, NJ, pp. 46–49.

Process Design

General Guidelines for Chapter 2 Design Exercises

- Design a steady-state continuous process. Do not bother with how the process might be started. For example, the process "chickens → eggs → chickens → . . ." is continuous and at steady state. In this case we would not bother with which came first, the chicken or the egg.
- You may assume liquid–gas separations are perfect if boiling points differ by more than 3°C.
- Solid–liquid separations produce wet solids.
- Assume no reactions other than the reactions given.
- Unless instructed otherwise, you may neglect heat exchangers, heaters, coolers, and pumps.
- In commercial chemical processes, a catalyst is typically a porous and/or granular solid. The catalyst is a permanent component of the reactor; there is no catalyst in the product stream. The catalyst is not consumed in the reaction; there is no catalyst in the feed stream. Unless stated otherwise, assume no catalyst is present in a process stream.
- In most exercises, many designs are workable. Explore different options and choose your best design.

General Flowsheet Practice

- Use a discrete unit for each operation. For example, a reactor should not do double duty as a mixer and/or a separator. A reactor should have exactly one input and one output.
- Label each unit. For reactors, indicate the reaction, such as "hydrolysis reactor." For separators, indicate the physical basis of the separation, such as "liquid–gas separator." Where appropriate, indicate the temperature in the unit. You need not label combiners and splitters. You need not label mixers and heat exchangers if you use descriptive symbols (a stirring paddle or a zigzag stream, respectively).
- List the substances in every stream. Where appropriate, list the physical state of the substances – gas, solid, liquid, or aq (in aqueous solution).
- Indicate the destination of any stream that leaves the process (for example, "vent to air," "to waste treatment," or "to market").
- Use conventions to improve the readability of your flowsheet. For a liquid–gas separator the gas should leave the top and the liquid should leave the bottom. If convenient, the process should start in the upper left corner. Streams entering the process should start at the periphery and streams leaving the process should terminate at the periphery.

2.11 Sketch a flowsheet for a process to wash clothes. Your process should include units to perform the following operations:

- Sort the laundry into three types: white items, colored items, and delicate items.
- Wash laundry types separately.
- Spin dry.
- Air dry.

Label all unit operations in general terms, such as mixer, separator, heater, and cooler. Each unit on your flowsheet should perform only one operation, even though in actual practice one machine may perform multiple operations in sequence. Add the necessary streams for water, soap, bleach, and air, and list the components in all streams.

> Laundry is an extremely complex system. Those of us who do the laundry don't get enough credit for the technical decisions we make every day. Some very serious science goes into doing the laundry.
>
> K. Obendorf, Professor of Textiles and Apparel, Cornell University, in *Smithsonian Magazine*, September 1997.

2.12 Consider the recipe for Kraft Macaroni and Cheese:

> Bring 2 quarts of water to boil. Add macaroni and boil for 8 minutes. Drain and set aside. Melt two tablespoons of butter or margarine in a saucepan; add one-half cup of milk and the contents from the cheese packet. Stir until well mixed and then remove from heat. Pour sauce contents over cooked macaroni; stir well and then serve.

Draw a flowsheet for a continuous process to prepare Kraft Macaroni and Cheese for your residence hall. Label every unit and the qualitative contents of every stream. Where appropriate, indicate the temperature and/or the residence time in a unit. Use steam to provide heat.

Optional: Modify your process to produce one of the 177 tasty macaroni and cheese variations posted at www.kraftrecipes.com.

2.13 Consider a recipe for pizza dough:

- Mix four packages of baker's yeast with ¼ cup of milk and ½ cup of warm water. Gently warm and stir for 15 minutes.
- Prepare two cups of "scalded milk," that is, milk that has been heated to a boil for about 1 minute. Add 1 tablespoon of molasses and 1 tablespoon of salt to the milk after scalding.
- Make the dough by combining the yeast mixture and the scalded milk mixture with 6 cups of flour. Gently knead the dough for about 5 minutes. Let the dough rise in warm, moist air for 15 minutes. The risen dough should be about double its initial size. There should be sufficient dough to make four pizzas.
- Roll one-quarter of the dough into a flat circle, 15 inches in diameter. Lightly coat the dough with olive oil. Bake the dough at 535°F for 5 to 6 minutes.

Design a process to continuously produce prebaked pizza dough for a major grocery chain. For each unit list key quantities such as temperature and residence time of material through the unit.

2.14 Your company manufactures specialty filters, sketched below. The filter consists of an aluminum cylinder packed with threads of polymer Q. The threads are coated with an active ingredient, chemical Z. The threads are approximately 1 mm diameter and 10 cm long and are wadded into entangled balls which prevents the loss of any threads as fluid passes through the filter. The filters are produced by tamping a thread ball into an aluminum cylinder open at one end, and then crimping a cap over the open end.

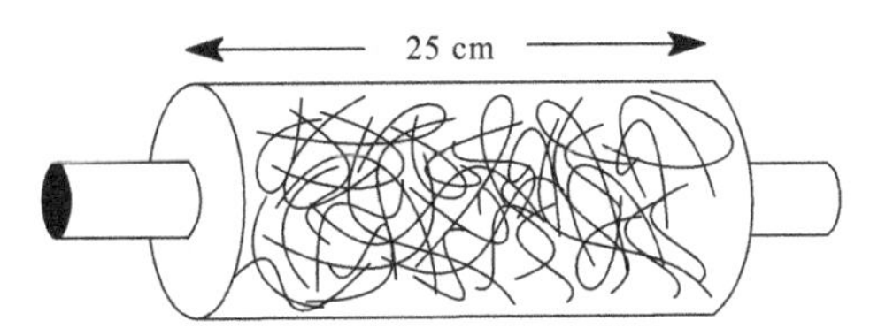

Over 10% of the filters fail to meet specifications. Filters fail for myriad reasons. Typical flaws include containers that leak, filters with too many (or too few) Q threads, and Q threads with too much (or too little) Z coating. We wish to recycle the rejected filters. *Sketch* the unit operations of a process to recover chemical Z, polymer Q, and aluminum scrap. The simplest, safest process is often the best process.

Component	Weight %	Soluble in acetone?	Soluble in water?	Melting point (°C)	Boiling point (°C)	Density (kg/m^3)
chemical Z	5	yes	yes	80	350	960
polymer Q	55	yes	no	250	decomposes at 450	870
aluminum	40	no	no	660	2,467	2,700

Acetone is volatile and highly flammable. Inhalation may cause headaches, fatigue, and bronchial irritation.

2.15 Design a process to recycle all the components in a pharmaceutical tablet. Each tablet is chiefly sugar. The tablet is an agglomeration of tiny sugar crystals, like a sugar cube. As such, the tablet is porous and may be permeated by a fluid.

Interspersed in the sugar tablets are tiny (smaller than 0.1 mm diameter), hard capsules that contain a drug. The capsules are impermeable to water, which protects the drug. The capsules melt at 95°F (35°C) to release the drug in one's stomach. The tablets are to be recycled because the dosage is incorrect; there are too many capsules per tablet.

The tablets are sealed in individual plastic pouches. An unopened pouch floats on water and dissolves in acetone. The external surfaces of the recycled pouches are dirty.

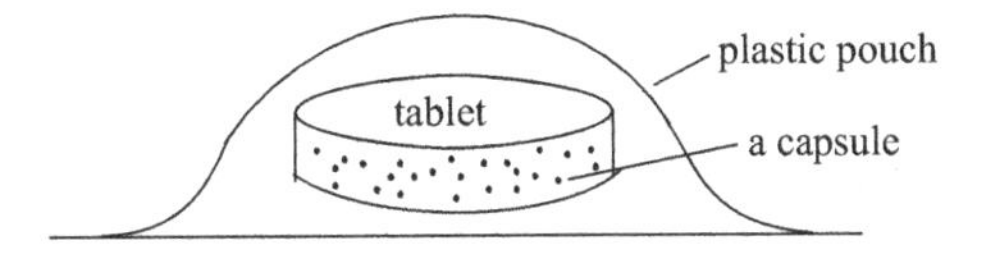

Your process should be functional, simple, energy-efficient, and environmentally benign. Acetone is volatile and highly flammable. Inhalation may cause headaches, fatigue, and bronchial irritation.

Component	Weight %	Soluble in acetone?	Soluble in water?	Melting point (°C)	Boiling point (°C)	Density (kg/m^3)
plastic pouch	42%	yes	no	185	dec at 450	860
dirt	~0.3%	yes	yes	–	–	~2,500
sugar	52%	no	yes	146	decomposes	1,560
capsule	2%	yes	no	35	85	1,140
drug	1%	yes	yes	5	decomposes	1,420

2.16 Soda ash (sodium carbonate, Na_2CO_3) is a key reactant in the production of glass and paper. Soda ash is produced from limestone and salt:

$$CaCO_3 + 2NaCl \rightarrow Na_2CO_3 + CaCl_2$$

but this reaction does not proceed as written above. Soda ash is instead produced indirectly by the ingenious route invented by Ernest Solvay (1838–1922). There are five salient reactions in this process, beginning with the decomposition of limestone into lime and carbon dioxide at 1000°C:

decomposition (1000°C):	$CaCO_3 \rightarrow CaO + CO_2$
slaking:	$CaO + H_2O \rightarrow Ca(OH)_2$
chlorination:	$Ca(OH)_2 + 2NH_4Cl \rightarrow 2NH_4OH + CaCl_2$
carbonation:	$NH_4OH + CO_2 + NaCl \rightarrow NH_4Cl + NaHCO_3$
calcination (300°C):	$2NaHCO_3 \rightarrow Na_2CO_3 + CO_2 + H_2O.$

Design a chemical process to produce soda ash by the Solvay route. Your process should minimize waste and reactants.

You may assume each reaction goes to completion; reactants are entirely converted to products. Slaking, chlorination, and carbonation occur in aqueous solutions. $CaCl_2$, NH_4OH, and NH_4Cl are highly soluble in water, whereas $NaHCO_3$ is only slightly soluble. $CaCO_3$, $Ca(OH)_2$, $NaHCO_3$, Na_2CO_3, and NH_4Cl are solids at 20°C; NH_4OH exists only in aqueous solution. Ionic solids decompose when heated: the boiling point of an ionic solid is undefined. Finally, you may assume $CaCl_2$ does not react with CO_2.

Adapted from descriptions of the Solvay process in Rudd, Powers, and Siirola (1973), *Process Synthesis*, Prentice Hall, Upper Saddle River, NJ, example 2.4–1, pp. 42–45 and by Reklaitis, G. V. 1983. *Introduction to Material and Energy Balances*, Wiley, New York, pp. 182–184, exercise 3.33.

2.17 Newly formed steel has an oxide crust that may be removed by washing in sulfuric acid:

$$\text{Fe with FeO crust} + H_2SO_4 \rightarrow Fe + FeSO_4 + H_2O.$$

The H_2SO_4 is dissolved in water and is in excess. $FeSO_4$ is soluble in water and soluble in acidic solution. We wish to recover the Fe and the H_2SO_4 by using two consecutive reactions, the first of which is

$$FeSO_4 + 2HCN \rightarrow Fe(CN)_2 + H_2SO_4.$$

The HCN is bubbled into the $FeSO_4$ solution and is in excess. Note that HCN is highly toxic. Assume the $FeSO_4$ is entirely converted to $Fe(CN)_2$. $Fe(CN)_2$ is insoluble in water and insoluble in acidic solution. In steam at 500°C, $Fe(CN)_2$ is entirely converted to FeO:

$$Fe(CN)_2 + H_2O \rightarrow HCN + FeO.$$

Solid FeO may be recycled to the smelting process, which converts iron oxides to iron metal.

(A) Design a process based on the above reactions.

(B) Assume further that the steel is contaminated with grease, which may be removed by the sulfuric acid in the generic reaction

$$\text{grease} + H_2SO_4 \rightarrow CO_2 + H_2O + SO_2.$$

CO_2 and SO_2 do not react with HCN. Modify your process to accommodate grease on the steel. Draw only the modified portion of your process; you need not redraw the entire process.

(C) Assume the steel contains no grease but is contaminated with calcium oxide. The pertinent reactions are

$$CaO + H_2SO_4 \rightarrow CaSO_4 + H_2O$$
$$CaSO_4 + 2HCN \rightarrow Ca(CN)_2 + H_2SO_4$$
$$Ca(CN)_2 + H_2O \xrightarrow{500^\circ C} CaO + 2HCN.$$

Both $CaSO_4$ and $Ca(CN)_2$ salts are soluble in water and soluble in acidic solutions. Modify your process to accommodate CaO on the steel. Draw only the modified portion of your process; you need not redraw the entire process.

Adapted from Rudd, Powers, and Siirola (1973), p. 49.

2.18 (A) Design a process to produce a nitric acid–water solution ($HNO_3 + H_2O$) from ammonia (NH_3) and air ($N_2 + O_2$), using the following reaction at 750°C:

$$NH_3 + 2O_2 \xrightarrow{Pt+Rh} HNO_3 + H_2O.$$

Assume that using air is preferable to using O_2 obtained by cryogenically separating N_2 and O_2. Cryogenic separation is expensive. N_2 in the reactor yields no by-products, although it increases the size of the reactor. The reaction conversion is less than 100%: the effluent will contain one or more reactant, as follows.

Design Goals (in decreasing importance)

- Minimize the number of process units.
- Minimize the sizes of the process units.

Design Rules

- If a stoichiometric mixture (1 mol NH_3 + 2 mol O_2) enters the reactor, the effluent will contain both NH_3 and O_2 (and N_2, HNO_3, and H_2O).
- If the reactant mixture has excess NH_3, the effluent will contain no O_2.
- If the reactant mixture has excess O_2, the effluent will contain no NH_3.

Indicate on your flowsheet which reactants (if any) are in excess.

(B) Design a process to produce a $HNO_3 + H_2O$ solution from methane, water, and air. That is, design a process to produce ammonia, and then use the ammonia to produce nitric acid. *Do not simply copy the ammonia process developed in Chapter 2.* The ammonia process can be improved when integrated with the nitric acid synthesis. Specifically, *you can avoid the costly cryogenic separation* of N_2 and O_2. **Hint:** Use excess NH_3 in the feed to the nitric acid reactor in part (B). (Excess NH_3 is not necessarily best for part (A).)

Recall, ammonia is synthesized by the following reaction at 500°C:

$$N_2 + 3H_2 \xrightarrow{\text{Fe}} 2NH_3.$$

The conversion to ammonia is less than 100%: the effluent will contain N_2 and H_2.

Generate H_2 by the method used in Chapter 2 for the fuel cell. You may assume methane can be "burned" in water in one step at 450°C:

$$CH_4 + 2H_2O \xrightarrow{\text{catalyst}\,X} CO_2 + 4H_2.$$

Furthermore, you may assume excess H_2O will consume *all* the CH_4.

Below is a summary of substances required, tolerated, and forbidden in the three reactors:

	Reactor		
Substance	Methane burner	Ammonia synthesis	Nitric acid synthesis
H_2	product	reactant	OK
N_2	OK	reactant	OK
O_2	OK	**forbidden**	reactant
CH_4	reactant	OK	**forbidden**
CO_2	by-product	**forbidden**	OK
NH_3	**forbidden**	product	reactant
HNO_3	**forbidden**	**forbidden**	product
H_2O	reactant	**forbidden**	by-product

Although O_2 is allowed in the methane burner, it will consume some of the product, H_2, by the following reaction:

$$2H_2 + O_2 \rightarrow 2H_2O,$$

and some of the reactant CH_4 by the reaction

$$CH_4 + 2O_2 \rightarrow CO_2 + 2H_2O.$$

Similarly, H_2 in the nitric acid synthesis reactor will be consumed by reaction with O_2. Thus it is wasteful to send O_2 to the methane burner or H_2 to the nitric acid synthesis reactor. If you choose to do so, verify that the waste can be justified, for example, because doing so eliminates a process unit.

2.19 (A) Air contaminated with SO_2 and NO_2 can be cleaned by bubbling the air through a suspension of $Mg(OH)_2$ in water. Two reactions occur in the suspension:

$$SO_2 + Mg(OH)_2 \rightarrow MgSO_3 + H_2O$$
$$2NO_2 + Mg(OH)_2 \rightarrow Mg(NO_2)_2 + H_2O + \tfrac{1}{2}O_2.$$

Excess $Mg(OH)_2$ removes all the SO_2 and NO_2 from the air. The magnesium solids decompose upon heating:

$$MgSO_3 \xrightarrow{200^\circ C} MgO + SO_2$$

$$Mg(NO_2)_2 + \tfrac{1}{2}O_2 \xrightarrow{250^\circ C} MgO + 2NO_2$$

$$Mg(OH)_2 \xrightarrow{350^\circ C} MgO + H_2O.$$

The second reaction above uses excess oxygen. $Mg(OH)_2$ can be regenerated by the following reaction:

$$MgO + H_2O \xrightarrow{20^\circ C} Mg(OH)_2.$$

SO_2 and NO_2 are collected separately and later converted to H_2SO_4 and HNO_3, respectively.

Design a process to produce clean air, SO_2, and NO_2 from air polluted with SO_2 and NO_2. You need not isolate pure SO_2 and NO_2; you may produce water solutions of each (to be used in parts (B) and (C) below). Use MgO, H_2O, and air as reactants. $MgSO_3$, $Mg(OH)_2$, and MgO are insoluble in water, whereas $Mg(NO_2)_2$ is highly soluble.

(B) Convert the NO_2 (or NO_2 + water solution) to a HNO_3 + water solution using the following reactions:

$$3NO_2 + H_2O \xrightarrow{20^\circ C} 2HNO_3 + NO$$

$$NO + \tfrac{1}{2}O_2 \xrightarrow[]{\substack{Pt\\250^\circ C}} NO_2.$$

The first reaction occurs only at significant concentrations of NO_2 (>10%). The reaction does not occur in the scrubber in part (A). Neither reaction goes to completion. You may use O_2 as a reactant in part (B). H_2O and O_2 are permitted in both reactors. NO is slightly soluble in water, whereas HNO_3 is soluble at all proportions.

(C) Convert the SO_2 (or SO_2 + water solution) to a H_2SO_4 + water solution:

$$SO_2 + \tfrac{1}{2}O_2 \xrightarrow{300^\circ C} SO_3$$

$$SO_3 + H_2O \xrightarrow[]{\substack{Pt\\250^\circ C}} H_2SO_4.$$

The first reaction does not go to completion. The second reaction uses excess water and completely consumes the SO_3. You may use O_2 as a reactant in part (C). H_2O and O_2 are permitted in both reactors.

2.20 H_2 can be produced from H_2O by the Ipsra Mark 4 cycle, as follows:

reaction 1:	$Cl_2 + H_2O \rightarrow 2HCl + \tfrac{1}{2}O_2$	850°C
reaction 2:	$2FeCl_2 + 2HCl + S \rightarrow 2FeCl_3 + H_2S$	120°C
reaction 3:	$2FeC13 \rightarrow Cl_2 + 2FeCl_2$	420°C
reaction 4:	$H_2S \rightarrow S + H_2$	800°C.

(A) Design a process to produce H_2 by the Ipsra Mark 4 cycle. Assume solids are transported around the process. For example, $FeCl_3$ enters the reactor for reaction 3 and $FeCl_2$ leaves the reactor for reaction 3 (with Cl_2).

Design Goals (in decreasing importance)

- Minimize the cost of producing H_2. Because the reactant cost is small, the chief cost is the energy supplied to the high-temperature reactors. To minimize cost, minimize the flow through the reactors for reactions 1 and 4.
- Minimize waste.
- Minimize the *number* of units.
- Minimize the *size* of the units.

Design Rules

- Assume reactions 2 and 3 go to completion but reactions 1 and 4 are only 90% complete.
- The input to a reactor may contain only the reactants and the products of the reaction that occurs in that reactor. For example, the input to reactor for reaction 4 may contain H_2S, S, and H_2. But this input must not contain Cl_2 or O_2.

(B) Redesign the process so solids are not transported around the process. Rather, solids remain in reactors and a reactor's input and output are reversed to change the reaction. Explain the operation of your process – how are streams redirected?

2.21 Methane, CH_4, is a plentiful energy source. Methane is easily pumped to residences and its oxidation products are chemically innocuous:

$$CH_4 + 2O_2 \rightarrow CO_2 + 2H_2O.$$

Because CO_2 is a greenhouse gas, it may be desirable to capture and sequester CO_2. However, it is ambitious and inefficient to collect CO_2 from each residence's exhaust. Rather, it may be better to convert CH_4 to H_2 at a central facility:

$$CH_4 + 2H_2O \rightarrow CO_2 + 4H_2.$$

CO_2 can be collected centrally and H_2 can be pumped to residences where it is oxidized to H_2O, which is truly innocuous.

The process to form H_2 is conducted in two steps. The first step is steam reforming at 900°C:

$$\text{steam reforming:} \qquad CH_4 + H_2O \xrightarrow{\text{Ni}} CO + 3H_2.$$

The second step is the water–gas shift reaction at 200°C:

$$\text{water-gas shift:} \qquad CO + H_2O \xrightarrow{\text{CuO–ZnO}} CO_2 + H_2.$$

Steam reforming is catalyzed by Ni, a solid that remains unchanged and is a permanent fixture in the reactor; no Ni catalyst enters or leaves the reactor. The water–gas shift is catalyzed by CuO–ZnO, also a solid, but this catalyst is deactivated by the disproportionation of CO, a side reaction:

$$\text{disproportionation:} \qquad \text{CuO–ZnO} + 2CO \rightarrow \text{C/CuO–ZnO} + CO_2.$$

Solid C deposits on the catalyst and degrades its activity. The CuO–ZnO catalyst is continuously cycled through the water–gas shift reactor and a catalyst reactivation reactor for the following reaction at 500°C:

catalyst reactivation: $C/CuO{-}ZnO + H_2O \rightarrow CuO{-}ZnO + CO + H_2$.

The CuO–ZnO catalyst is fine particles suspended in the reaction gas. A fraction of the catalyst dust is carried out with the effluent.

Design a process to produce hydrogen from methane and water, and collect the by-product CO_2.

Design Rules

- Neither steam reforming nor water–gas shift completely converts reactants to products. Each has a conversion of 80% to 90%.
- You may assume catalyst reactivation is complete if excess H_2O is present.
- To maximize conversions, do not feed reactor products into a reactor. For example, do not feed H_2 into the water–gas shift reactor.

2.22 The primary component of natural gas, methane, is a plentiful energy source and a valuable chemical feedstock. Methane resources are estimated to be as large as crude oil, energy-wise, but are far from potential customers. Because methane is a gas, transportation across oceans is costly. One solution is to condense methane by cooling to −164°C (−263°F) and transport the liquefied natural gas (LNG). Another solution is to convert methane to methanol, which is a liquid at 20°C and thus is more easily shipped:

$$2CH_4 + O_2 \rightarrow 2CH_3OH.$$

However, the *partial* oxidation of methane to methanol is difficult. When oxidized in air, methane oxidizes *completely* to carbon dioxide and water:

$$CH_4 + 2O_2 \rightarrow CO_2 + 2H_2O.$$

To oxidize methane only partially, methane is not reacted with O_2 directly. In current technology, methane reacts with H_2O over a Ni catalyst at 900°C to form a mixture of CO and H_2 known as synthesis gas, or syngas:

$$\text{steam reforming:} \quad CH_4 + H_2O \xrightarrow{Ni} CO + 3H_2.$$

The syngas is then reacted over a $ZnO{-}Cr_2O_3$ catalyst at 300°C and 300 atm to form methanol:

$$\text{methanol synthesis:} \quad CO + 2H_2 \xrightarrow{ZnO-Cr_2O_3} CH_3OH.$$

The Ni and $ZnO{-}Cr_2O_3$ catalysts are solids at their respective reaction temperatures. Each is a fixed component in the reactor.

(A) Design a process to produce methanol from methane and water.

Design Goal

- Minimize the number of units at 300 atm.

Design Rules

- Neither steam reforming nor methanol synthesis completely converts reactants to products. Each has a conversion of 80% to 90%.

- You may assume solid–gas and solid–liquid separations are ideal; the solids carry no gases or liquids.
- You may neglect heat exchangers, heaters, and coolers, but include compressors (and expanders) for the portion of the process at 300 atm.

(B) There is a side reaction in the steam-reforming reactor – CO disproportionates to C and CO_2:

$$\text{disproportionation:} \quad CO + CO \xrightarrow{Ni} C + CO_2.$$

The solid carbon deposits on the Ni catalyst and deactivates the catalyst. This problem can be avoided by using excess steam, which causes a different side reaction known as the water–gas shift reaction:

$$\text{water–gas shift:} \quad CO + H_2O \xrightarrow{Ni} CO_2 + H_2.$$

The CO_2 may be reconverted to CO in the methanol-synthesis reactor:

$$CO_2 + H2 \xrightarrow{ZnO-Cr_2O_3} CO + H_2O.$$

The CO_2 will reconvert only if there is little or no H_2O present. Also, the reconversion is not complete; some CO_2 will leave the methanol-synthesis reactor.

Augment your design from (A) to include the water–gas shift reaction. In addition to the design goal in (A), now include the following:

- Maximize product and minimize by-product(s).

2.23 The current technology to convert methane to methanol (see Exercise 2.22) requires high temperature and consumes significant energy. Researchers propose an efficient low-temperature route to methanol via methyl bisulfate, catalyzed by mercury compounds.[7] The process starts by activating a C–H bond at 180°C and 34 atm:

$$\text{activation:} \quad CH_4 + Hg(OSO_3H)_2 \rightarrow H_2SO_4 + Hg(OSO_3H)(CH_3).$$

You may assume conducting the activation reaction with excess CH_4 completely converts the $Hg(OSO_3H)_2$. The $Hg(OSO_3H)(CH_3)$ is then reacted with excess sulfuric acid at 110°C to functionalize the CH_3 group:

$$\text{functionalization:} \quad 2Hg(OSO_3H)(CH_3) + 3H_2SO_4 \rightarrow 2CH_3OSO_3H + 2H_2O + SO_2 + Hg_2(OSO_3H)_2.$$

Again you may assume the $Hg(OSO_3H)(CH_3)$ is completely converted. The methyl bisulfate is then hydrolyzed to form methanol at 90°C:

$$\text{hydrolysis:} \quad CH_3OSO_3H + H_2O \rightarrow CH_3OH + H_2SO_4.$$

If excess H_2O is present, the CH_3OSO_3H is completely hydrolyzed. To recycle the mercury catalyst to the activation reaction, the $Hg_2(OSO_3H)_2$ from the functionalization reaction is reoxidized at 80°C:

$$\text{reoxidation:} \quad Hg_2(OSO_3H)_2 + 3H_2SO_4 \rightarrow 2Hg(OSO_3H)_2 + 2H_2O + SO_2.$$

[7] Periana, R. A., Taube, D. J., Evitt, E. R., *et al.* 1993. "A mercury-catalyzed, high-yield system for the oxidation of methane to methanol," *Science*, 259, 340–343.

If excess sulfuric acid is used, the $Hg_2(OSO_3H)_2$ is completely oxidized. Finally, H_2SO_4 is reformed from SO_2 in two steps – oxidation at 450°C and hydration at 100°C:

$$\text{oxidation:} \quad 2SO_2 + O_2 \rightarrow 2SO_3$$
$$\text{hydration:} \quad SO_3 + H_2O \rightarrow H_2SO_4.$$

You may assume the oxidation of SO_2 is complete, but the hydration of SO_3 is incomplete; both reactants will appear in the product stream.

(A) Add the six reactions (or multiples of the reactions) to verify that the overall reaction is the partial oxidation of methane to methanol.
(B) Design a process based on the reaction scheme given above. All compounds are soluble in water and all compounds are soluble in H_2SO_4. All the mercury compounds are solids that decompose upon heating above 200°C. The hydrolysis of CH_3OSO_3H (third step) will proceed in the presence of H_2SO_4 and $Hg_2(OSO_3H)_2$, but SO_2 must not be present. In this process, any product of a reactor can be present in that reactor's input stream.
(C) Verify that only methane and oxygen enter your process and that only methanol leaves your process, except for minor purge streams leaving and minor make-up streams entering.
(D) Although this process may require less energy than a high-temperature process, this process has a disadvantage. Speculate why it may be difficult to win approval to produce methanol by this technology. **Hint:** Contemporary engineers design for green chemistry – chemical processes that reduce or eliminate the use or generation of hazardous substances.

2.24 Styrene (a commodity chemical) is produced in two steps. First, benzene and ethylene react at 450°C in the presence of catalyst 1 to form ethylbenzene:

$$\text{alkylation (450°C):} \quad C_6H_6 + CH_2{=}CH_2 \xrightarrow{\text{catalyst 1}} C_6H_5{-}CH_2CH_3. \qquad \text{rxn 1}$$

The ethylbenzene is then dehydrogenated at 620°C in the presence of catalyst 2 to produce styrene:

$$\text{dehydrogenation (620°C):} \; C_6H_5{-}CH_2CH_3 \xrightarrow{\text{catalyst 2}} C_6H_5{-}CH{=}CH_2 + H_2. \quad \text{rxn 2}$$

Unfortunately, side reactions occur in both reactors. In the alkylation reactor, ethylbenzene reacts with ethylene to form diethylbenzene:

$$\text{alkylation (450°C):} \quad C_6H_5{-}CH_2CH_3 + CH_2{=}CH_2 \rightarrow C_6H_4(CH_2CH_3)_2. \quad \text{rxn 1a}$$

However, the effects of side reaction 1a can be minimized (but not eliminated entirely) if an excess of benzene is present, by reaction 1b:

$$\text{alkylation(450°C):} \; C_6H_4(CH_2CH_3)_2 + nC_6H_6 \rightarrow 2C_6H_5{-}CH_2CH_3 + (n-1)C_6H_6. \quad \text{rxn 1b}$$

A nuisance in the dehydrogenation reactor is the reversal of reaction 1:

$$\text{dehydrogenation (620°C):} \quad C_6H_5{-}CH_2CH_3 \rightarrow C_6H_6 + CH_2{=}CH_2. \qquad \text{rxn 2a}$$

None of the reactions go to completion. Reaction 1 is exothermic (releases heat) whereas reaction 2 is endothermic (requires heat). Superheated steam is injected with the reactants to drive reaction 2. The steam does not affect any of the reactions and improves the catalyst performance by removing residue.

Rather than use the chemical formulas given in reactions 1 and 2, you may wish to use the more expedient (although less informative) nomenclature of benzene ≡ B, ethylene ≡ E, styrene ≡ S, ethylbenzene ≡ EB, diethylbenzene ≡ DEB, superheated steam ≡ SHS, steam ≡ S, and water ≡ W, as such:

alkylation (450°C):	$B + E \rightarrow EB$	rxn 1
dehydrogenation (620°C):	$EB \rightarrow S + H_2$	rxn 2
alkylation (450°C):	$EB + E \rightarrow DEB$	rxn 1a
alkylation (450°C):	$DEB + nB \rightarrow 2EB + (n-1)B$	rxn 1b
dehydrogenation (620°C):	$EB \rightarrow B + E$.	rxn 2a

Design a process to produce styrene.

Design Goals (in decreasing importance)

- Maximize the product and minimize by-products.
- Minimize the number of process units.
- Minimize the sizes of the process units.

You may assume all substances in this process are insoluble in water and water is insoluble in any of the substances in this process.

The idea for this exercise came from Professor C. Cohen (Cornell University, 1996).

2.25(A) Design a process to produce vinyl chloride ($CH_2{=}CHCl$) from ethylene ($CH_2{=}CH_2$) and chlorine (Cl_2). Your process should use chlorination followed by pyrolysis:

chlorination (60°C):	$CH_2{=}CH_2 + Cl_2 \rightarrow CH_2Cl{-}CH_2Cl$	rxn 1
pyrolysis (500°C):	$CH_2Cl{-}CH_2Cl \rightarrow CH_2{=}CHCl + HCl$	rxn 2a
	$CH_2Cl{-}CH_2Cl \rightarrow CH_2{=}CH_2 + Cl_2$	rxn 2b
net reaction:	$CH_2{=}CH_2 + Cl_2 \rightarrow CH_2{=}CHCl + HCl$.	

Thus only $CH_2{=}CH_2$ and Cl_2 should enter your process and only $CH_2{=}CHCl$ and HCl should leave your process, neglecting purge and make-up streams.

These compounds undergo additional reactions, which you may wish to exploit or avoid:

hydrochlorination (150°C):	$CH_2{=}CH_2 + HCl \rightarrow CH_3{-}CH_2Cl$	rxn 3
pyrolysis (500°C):	$CH_3{-}CH_2Cl \rightarrow CH_2{=}CH_2 + HCl$.	rxn 2c

Reactions 1 and 3 each require a special catalyst and do not proceed in the absence of their respective catalysts. Reactions 1 and 3 have 100% conversion of reactants into products. Reactions 2a, 2b, and 2c have about 35% conversion.

Design Goals (in decreasing importance)

- Maximize product and minimize by-products.
- Minimize the number of process units.
- Minimize the sizes of the process units.

(B) Early processes used the HCl by-product from the net reaction to hydrochlorinate acetylene ($CH{\equiv}CH$), as follows:

$$\text{hydrochlorination } (150°\text{C}): \quad CH{\equiv}CH + HCl \rightarrow CH_2{=}CHCl. \qquad \text{rxn 4}$$

Reaction 4 has about 90% conversion and requires a special catalyst – the same catalyst as reaction 3. Unfortunately, about 10% of the product $CH_2{=}CHCl$ reacts with HCl in the hydrochlorination reactor:

$$\text{hydrochlorination } (150°\text{C}): \quad CH_2{=}CHCl + HCl \rightarrow CH_2Cl{-}CH_2Cl \qquad \text{rxn 5a}$$

$$CH_2{=}CHCl + HCl \rightarrow CH_3{-}CHCl_2 \qquad \text{rxn 5b}$$

$$\text{pyrolysis } (500°\text{C}): \quad CH_3{-}CHCl_2 \rightarrow CH_2{=}CHCl + HCl \quad \text{rxn 2d}$$

$$\text{(see also rxns 2a–c)}: \quad CH_3{-}CHCl_2 \rightarrow CH_2{=}CH_2 + Cl_2 \quad \text{rxn 2e}$$

$$\text{net reaction(rxns 1 to 5)}: \quad \overline{CH_2{=}CH_2 + CH{\equiv}CH + Cl_2 \rightarrow 2CH_2{=}CHCl.}$$

The net reaction above is superior to the net reaction in (A) because there is no HCl by-product. Reactions 1 and 4 must be conducted in different reactors (each requires a different temperature and different catalyst). Reactions 2d and 2e have about 35% conversion.

Design a process to produce $CH_2{=}CHCl$ using the chemical reactions in parts (A) and (B). Use the design goals listed in part (A). You need not redraw your entire flowsheet from part (A). Integrate these new reactions into the process and redraw only the portions that change.

(C) Processes that used the chemical reactions in part (B) were made obsolete in the early 1950s by technology for the following reaction:

$$\text{oxychlorination } (275°\text{C}): \quad CH_2{=}CH_2 + 2HCl + \frac{1}{2}O_2 \rightarrow CH_2Cl{-}CH_2Cl + H_2O. \qquad \text{rxn 6}$$

Reaction 3 also occurs in the oxychlorination reactor. The chemical reactions in (A) and reaction 6 yield the *net reaction*:

$$2CH_2{=}CH_2 + Cl_2 + \frac{1}{2}O_2 \rightarrow 2CH_2{=}CHCl + H_2O.$$

Reactions 1 and 6 must be conducted in separate reactors. Reaction 6 has 100% conversion.

Design a process to produce $CH_2{=}CHCl$ using the chemical reactions in parts (A) and (C). Use the design goals listed in part (A).

(D) Technology invented in the early 1990s may render obsolete processes that use the oxychlorination reaction (rxn 6) in part (C). The new technology converts HCl into Cl_2 as follows:

$$\text{copper chlorination } (260°\text{C}): \quad 2HCl + CuO \rightarrow CuCl_2 + H_2O \qquad \text{rxn 7}$$

$$\text{copper oxidation } (350°\text{C}): \quad CuCl_2 + \frac{1}{2}O_2 \rightarrow CuO + Cl_2. \qquad \text{rxn 8}$$

Reactions 7 and 8 have 100% conversion. The chemical reactions in (A) with reactions 7 and 8 yields the *net reaction*:

$$2CH_2{=}CH_2 + Cl_2 + \frac{1}{2}O_2 \rightarrow 2CH_2{=}CHCl + H_2O.$$

Design a process to produce CH_2=CHCl using the chemical reactions in parts (A) and (D). Use the design goals listed in part (A).

2.26 Hydrazine, N_2H_4, is an important chemical precursor to agricultural chemicals such as algicides, fungicides, and insecticides, and to pharmaceuticals such as Isoniazid, a drug to treat tuberculosis. Hydrazine is also used as a rocket fuel to maintain satellite orbits; the decomposition of one liquid molecule to three gas molecules provides propulsion:

$$N_2H_4 \rightarrow N_2 + 2H_2.$$

Hydrazine may be synthesized from ammonia by the following overall reaction:

$$2NH_3 + Cl_2 + 2NaOH \rightarrow N_2H_4 + 2NaCl + 2H_2O.$$

In the Bayer process (aka the ketazine process), hydrazine synthesis consists of three steps. The first step is the synthesis of sodium hypochlorite, NaClO, in aqueous solution at 20°C:

$$\text{NaClO synthesis:} \quad 2NaOH + Cl_2 \rightarrow NaClO + NaCl + H_2O.$$

The second step is the conversion of acetone, $CH_3C(O)CH_3$, to acetazine, $(CH_3)_2C{=}N{-}N{=}C(CH_3)_2$, in aqueous solution at 35°C:

$$NaClO + 2CH_3C(O)CH_3 + 2NH_3 \rightarrow (CH_3)_2C{=}N{-}N{=}C(CH_3)_2 + NaCl + 3H_2O.$$

The third step is the hydrolysis of acetazine to hydrazine and acetone in aqueous solution at 50°C:

$$\text{hydrolysis:} \quad (CH_3)_2C{=}N{-}N{=}C(CH_3)_2 + 2H2O \rightarrow N_2H_4 + 2CH_3C(O)CH_3.$$

Hydrazine is distilled from the aqueous solution as a hydrate, $N_2H_4{\cdot}H_2O$.

Design a process to produce hydrazine hydrate from ammonia, chlorine, and sodium hydroxide.

- Use the chemical formulas C_3H_6O for acetone and $C_6H_{12}N_2$ for acetazine.
- Assume if Cl_2 is in excess in the NaClO synthesis, the NaOH is completely oxidized.
- Assume if acetone and ammonia are in excess in the acetazine synthesis, the NaClO is completely consumed.
- Assume if H_2O is in excess in the acetazine hydrolysis, the acetazine is completely consumed.
- Assume all substances in this exercise are soluble in water.

Hint: Remove the NaCl by-product before the acetazine hydrolysis reactor.

2.27 The green chemistry movement strives to replace hazardous oxidizers, such as chlorine, with innocuous oxidizers such as hydrogen peroxide. Exercise 2.26 describes the Bayer process for synthesizing hydrazine, N_2H_4. For hydrazine synthesis by green chemistry, the overall reaction is as follows:

$$2NH_3 + H_2O_2 \rightarrow N_2H_4 + 2H_2O.$$

In this modified process, only two synthesis steps are required. The first step is conversion of acetone, $CH_3C(O)CH_3$, to acetazine, $(CH_3)_2C{=}N{-}N{=}C(CH_3)_2$, in aqueous solution at 50°C:

$$H_2O_2 + 2CH_3C(O)CH_3 + 2NH_3 \rightarrow (CH_3)_2C{=}N{-}N{=}C(CH_3)_2 + 4H_2O.$$

The second step is the hydrolysis of acetazine into hydrazine and acetone in aqueous solution at 50°C:

hydrolysis: $$(CH_3)_2C{=}N{-}N{=}C(CH_3)_2 + 2H_2O \rightarrow N_2H_4 + 2CH_3C(O)CH_3.$$

The hydrazine is distilled from the aqueous solution as a hydrate, $N_2H_4{\cdot}H_2O$.

Design a process to produce hydrazine hydrate from ammonia and hydrogen peroxide. The hydrogen peroxide is available as a mixture of 70% hydrogen peroxide in 30% water.

- Use the chemical formulas C_3H_6O for acetone and $C_6H_{12}N_2$ for acetazine.
- You may assume if acetone and ammonia are in excess in the acetazine synthesis, the H_2O_2 is completely consumed.
- You may assume if H_2O is in excess in the acetazine hydrolysis, the acetazine is completely consumed.
- You may assume all substances in this exercise are soluble in water.

2.28 Hydrogen peroxide, H_2O_2, is an oxidizing agent. Because the by-product of oxidation by H_2O_2 (water) is harmless, H_2O_2 is a "green" oxidizer. There is a trend to replace hazardous oxidizers with H_2O_2, such as replacing Cl_2 in the synthesis of hydrazine (see Exercise 2.27). Hydrogen peroxide is a precursor to other "green" oxidizers, such as peroxoborate and peroxocarbonate bleaches used as detergents and used in the manufacture of paper and textiles.

Hydrogen peroxide is formed by the partial reduction of oxygen:

$$H_2 + O_2 \rightarrow H_2O_2.$$

To avoid complete reduction to H_2O, O_2 is not mixed with H_2 directly. Instead, an intermediate controls the number of electrons transferred from H_2 to O_2. An effective intermediate is 9,10-anthraquinone. The process, called auto-oxidation, consists of two reactions, both conducted in a hydrocarbon solvent. The first step is hydrogenation to 9,10-anthradiol at 20°C:

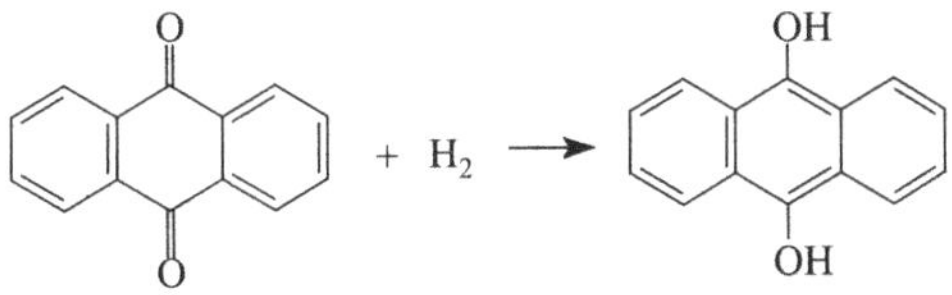

O O + H_2 → OH OH

Hydrogenation is catalyzed by solid Pd particles, which are suspended in the solvent. The Pd particles leave the hydrogenation reactor with the effluent. The second step is oxidation at 45°C to produce hydrogen peroxide and to regenerate 9,10-anthraquinone:

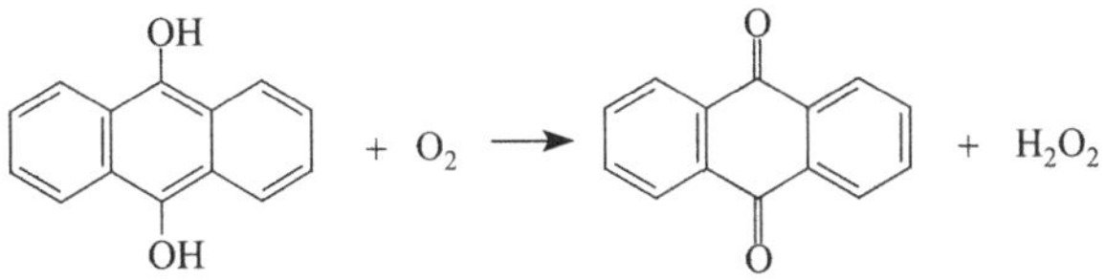

OH OH + O_2 → O O + H_2O_2

The oxidizer reactor effluent is sent to a solvent–water absorber to extract the H_2O_2; H_2O_2 is miscible in water, whereas 9,10-anthraquinone and solvent are not.

Likewise, the water does not dissolve in the solvent. The H_2O_2–water stream (25 wt% H_2O_2) is then heated to evaporate excess water and yield a product with 70 wt% H_2O_2/30 wt% H_2O.

In practice, the auto-oxidation process is optimized by fine-tuning the reactivity and solubility of the intermediate, chiefly by adding an alkyl substituent, such as $-CH_2CH_3$, to anthraquinone's aromatic rings. Because the properties of the solvent mixture are also key to optimizing the process, the composition of the solvent mixture is generally proprietary.

Design a process to produce hydrogen peroxide solution from H_2 and O_2 using auto-oxidation of anthraquinone in a generic solvent.

- Use the chemical formulas $C_{14}H_8O_2$ for 9,10-anthraquinone and $C_{14}H_8(OH)_2$ for 9,10-anthradiol.
- You may assume $C_{14}H_8O_2$ is completely hydrogenated if excess H_2 is used.
- You may assume $C_{14}H_8(OH)_2$ is completely oxidized if excess O_2 is used.

2.29 Acetanilide, shown below, is a precursor for synthesizing antibiotics such as sulfanilamide and penicillin.

O
NH-C
CH$_3$

Acetanilide (an amide) is synthesized from aniline (an amine) and acetic anhydride at 20°C:

$$C_6H_5NH_2 + (CH_3CO)_2O \rightarrow C_6H_5NHC(O)CH_3 + CH_3COOH.$$

Aniline is synthesized from benzene, chlorine, and ammonia in two sequential reactors. We first chlorinate benzene at 40°C. Unfortunately, the desired intermediate chlorobenzene reacts with chlorine to form dichlorobenzene isomers, which are useless by-products:

$$C_6H_6 + Cl_2 \rightarrow C_6H_5Cl + HCl$$
$$C_6H_5Cl + Cl_2 \rightarrow C_6H_4Cl_2 + HCl.$$

We isolate the chlorobenzene and convert to aniline in an aqueous ammonia solution at 80°C:

$$C_6H_5Cl + 2NH_3 \rightarrow C_6H_5NH_2 + NH_4Cl.$$

Acetic anhydride is synthesized by combining acetic acid and ketene (ethenone) at 700°C:

$$CH_3COOH + H_2C{=}C{=}O \rightarrow (CH_3CO)_2O.$$

Given excess acetic acid, the ketene is completely consumed. Ketene is synthesized by dehydrating acetic acid at 700°C, catalyzed by triethyl phosphate. The triethyl phosphate catalyst must be added to the reactants and separated from the reactor effluent:

$$CH_3COOH \xrightarrow{(CH_3CH_2)_3PO_4} H_2C{=}C{=}O + H_2O.$$

Unfortunately, some acetic acid decarboxylates at 700°C to methane and carbon dioxide:

$$CH_3COOH \rightarrow CH_4 + CO_2.$$

The three preceding reactions of acetic acid occur simultaneously in the same reactor.

By-product ammonium chloride is recycled by decomposition to ammonia and hydrochloric acid at 350°C. The decomposition of NH_4Cl is complete:

$$NH_4Cl \rightarrow NH_3 + HCl.$$

Because Cl_2 is a toxic gas, it is dangerous and bulky to transport. Cl_2 is generated on site from by-product HCl and reactant sodium hypochlorite in an aqueous solution (bleach) at 50°C:

$$NaClO + 2HCl \rightarrow Cl_2 + H_2O + NaCl.$$

Given excess HCl, the NaClO is completely consumed.

Design a process to produce acetanilide from benzene, sodium hypochlorite, ammonia, and acetic acid.

- You may assume gas–solid separations are perfect; no gas leaves with the solid.
- Because this process is complex, you may omit purge and make-up streams for clarity.

Hint: Because this process comprises 6 reactors and at least 11 separators, consider dividing your flowsheet into two portions: one portion synthesizes aniline from benzene, sodium hypochlorite, and ammonia, and one portion synthesizes acetanilide from acetic acid and aniline (from the first portion).

2.30 We wish to use thermal energy to generate methane from carbon dioxide and water:

$$CO_2 + 2H_2O \rightarrow CH_4 + 2O_2.$$

Because this reaction is strongly endothermic ($\Delta \bar{H}^0_{rxn} = +802\ \text{kJ/mol}$), conversion is negligible at reasonable temperatures (<2000°C). Consider the following hypothetical cycle. The hydrogen in H_2O is reduced to H_2 by oxidizing Fe^{+2} to Fe^{+3} at 1200°C:

$$3H_2O + 2FeCl_2 \rightarrow Fe_2O_3 + 4HCl + H_2.$$

The carbon in CO_2 is reduced to CH_4 by oxidizing H_2 at 200°C:

$$CO_2 + 4H_2 \rightarrow CH_4 + 2H_2O.$$

$FeCl_2$ is regenerated from Fe_2O_3 in two reactions. The first reaction converts iron oxide to iron chloride in an aqueous solution at 25°C:

$$2Fe_2O_3 + 2Cl_2 + 8HCl \rightarrow 4FeCl_3 + 4H_2O + O_2.$$

The second reaction reduces the iron at 800°C:

$$2FeCl_3 \rightarrow 2FeCl_2 + Cl_2.$$

(A) Add the four reactions (or multiples of the reactions) to verify the overall reaction of carbon dioxide and water to form methane and oxygen.

(B) Design a process to produce methane and oxygen from carbon dioxide and water.

- Assume complete conversion at the reaction temperatures stated.
- Assume solid–gas and solid–liquid separations are ideal; the solids carry no gases or liquids.

(C) Verify that only carbon dioxide and water enter your process and that only methane and oxygen leave your process, except for minor purge streams leaving and minor make-up streams entering.

2.31 Adipic acid is a key chemical commodity, used chiefly in the synthesis of nylon. The worldwide production of adipic acid is over 2 billion kg/year. Adipic acid is synthesized by partial oxidation of cyclohexane:

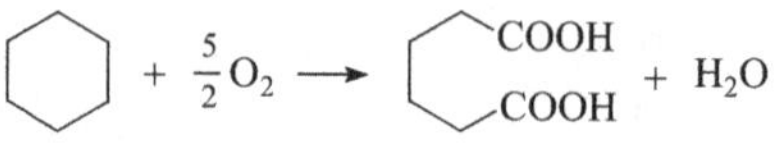

$$C_6H_{12} + \frac{5}{2}O_2 \rightarrow HOOC(CH_2)_4COOH + H_2O$$

Conventional synthesis involves organic solvents and oxidation with nitric acid, which produces toxic NO_x by-products. Consider instead a more benign synthesis that uses water as the solvent and H_2O_2 as the oxidizer (see Exercise 2.28). Like the conventional synthesis, the first step is the partial oxidation of cyclohexane to cyclohexanol in an aqueous solution with a solid Pt catalyst:

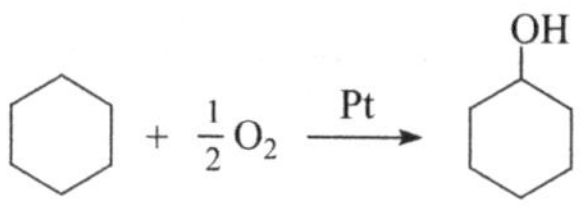

$$C_6H_{12} + \frac{1}{2}O_2 \xrightarrow{Pt} C_6H_{11}OH \qquad \text{rxn 1}$$

Cyclohexanol is then dehydrated to cyclohexene in an acidic aqueous solution:

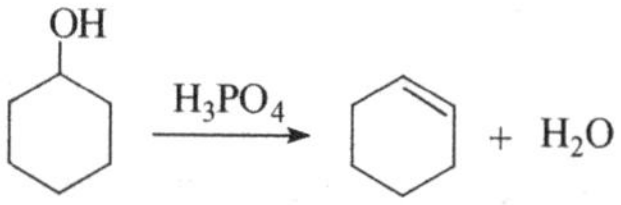

$$C_6H_{11}OH \xrightarrow{H_3PO_4} C_6H_{10} + H_2O \qquad \text{rxn 2}$$

Finally, cyclohexene reacts with H_2O_2 in an aqueous solution to form adipic acid:

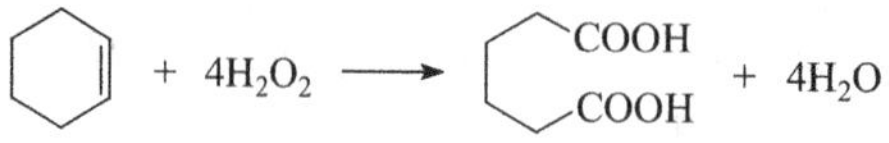

$$C_6H_{10} + 4H_2O_2 \rightarrow HOOC(CH_2)_4COOH + 4H_2O \qquad \text{rxn 3}$$

There is a complication. The partial oxidation of cyclohexane also produces cyclohexanone:

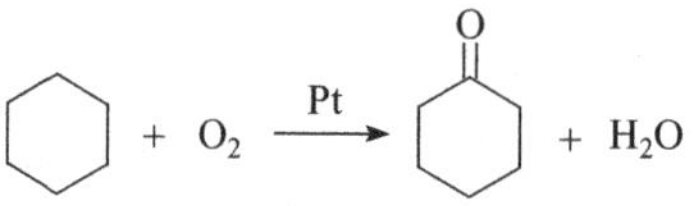

$$C_6H_{12} + O_2 \xrightarrow{Pt} C_6H_{10}O + H_2O \qquad \text{rxn 4}$$

But cyclohexanone can be reduced to cyclohexanol by the following reaction:

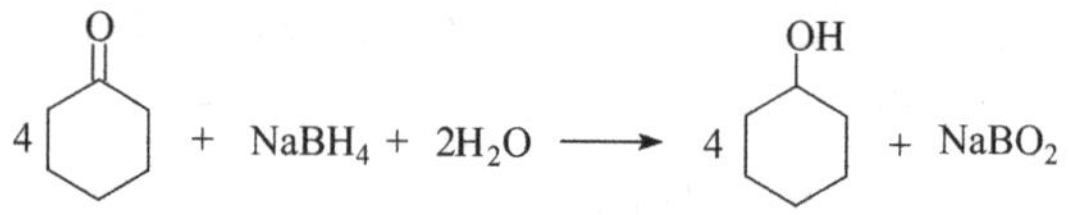

$$4\,C_6H_{10}O + NaBH_4 + 2H_2O \rightarrow 4\,C_6H_{11}OH + NaBO_2 \qquad \text{rxn 5}$$

Design a process to produce adipic acid from cyclohexane, oxygen, water, and sodium borohydride, using platinum and phosphoric acid as catalysts.

- The conversion in every reaction is less than 100%; none of the reactions go to completion.
- For reaction 5, excess cyclohexanone will consume all the $NaBH_4$.
- All reactions are in aqueous solutions at 25°C.
- Assume all substances are soluble in water.
- The solid Pt catalyst resides in the reactor; it does not enter with the feed or exit with the effluent.
- Because this process is complex, you may neglect purges and make-up streams.

2.32 Design a process to produce P by the following reaction:

$$B + H \rightarrow P.$$

B is available as a mixture with A (33 mol% A and 67 mol% B). H is produced by the following reaction:

$$A + C \rightarrow H.$$

C is available as a mixture with D (50 mol% C and 50 mol% D). Two other reactions are important:

$$B + C \rightarrow R$$
$$B + D \rightarrow Q.$$

No other reactions occur.

All reactions are spontaneous at 300°C; no catalysts are needed. Every reaction consumes all the reactants if the reactants are present in stoichiometric amounts.

Design Goals (in decreasing importance)

- Maximize the revenue from chemical products.
- Minimize the *number* of units.
- Minimize the *size* of the units.

Design Rules

- Use 3 mol/min of the *A–B* mixture: 1 mol/min A mixed with 2 mol/min B.
- Use as much *C–D* mixture as you wish.
- List the substances in every stream and state *approximate* flow rates, in mol/min. One significant figure is sufficient.

		Melting point (°C)	Boiling point at 1 atm (°C)	Market value ($/mol)
A	reactant	−67	54	1
B	reactant	−17	91	13
C	reactant	−42	37	2
D	reactant	−42	37	15
H	intermediate	16	123	2
P	product	26	162	100
Q	by-product	26	162	6
R	by-product	3	101	0

2.33 Chemicals A and X react to form a valuable chemical P:

$$X + A \rightarrow P.$$

Chemical P commands a high price because the conversion of the above reaction is low. The optimal performance of the reactor is shown below:

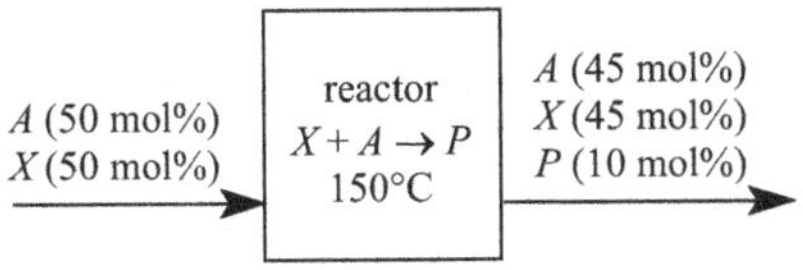

A (50 mol%)
X (50 mol%)
reactor
$X + A \rightarrow P$
150°C
A (45 mol%)
X (45 mol%)
P (10 mol%)

Chemical X is available commercially. Chemical A is available only in a mixture with chemicals B and C. Chemicals A, B, and C do not react with each other. The reactions with chemical X are

$$X + B \rightarrow \text{no reaction}$$
$$X + C \rightarrow Q.$$

Moreover, Q forms readily under optimal conditions for forming P:

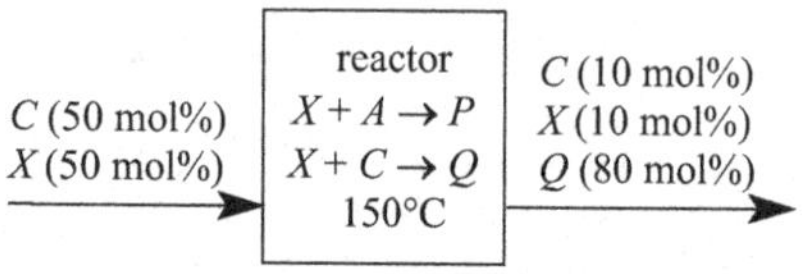

Input	reactor	Output
C (50 mol%) X (50 mol%)	$X + A \rightarrow P$ $X + C \rightarrow Q$ 150°C	C (10 mol%) X (10 mol%) Q (80 mol%)

Finally, P does not react with any of the chemicals in this process.

Design a process to produce P from two raw materials: (1) pure X; and (2) a mixture of A (30 mol%), B (5 mol%), and C (65 mol%).

Design Goals (in decreasing importance)

- Your process should be safe and environmentally benign.
- Maximize the yield of product P and any other marketable by-product.
- Minimize the number of units in your process.
- Minimize waste of material and energy in the process.

Following is a table of physical parameters at 1 atm.

	Melting point (°C)	Boiling point (°C)	Soluble in benzene?[a]	Soluble in water?	Market value ($/kg)	Comments
A	60	235	yes	no	10	non-toxic[b]
B	32	235	no	yes	30	non-toxic[b]
C	28	135	no	no	160	non-toxic[b]
P	86	120	yes	no	400	mild irritant
Q	230	800	yes	yes	20	toxic
X	120	320	yes	no	150	flammable

[a] Human toxicity of benzene is acute; it irritates mucous membranes, and death may follow respiratory failure.
[b] May be disposed of in a landfill.

2.34 Chemicals F and G react to form a valuable product P. The reaction

$$F + G \rightarrow P$$

converts 90% of the reactants, shown below.

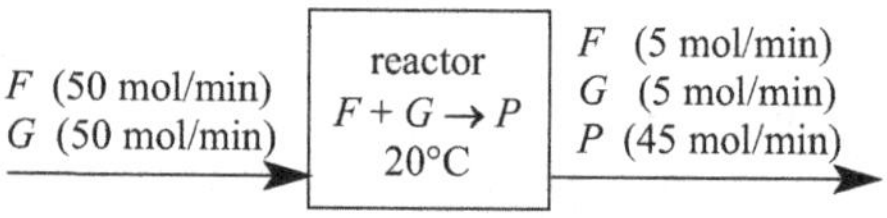

Input	reactor	Output
F (50 mol/min) G (50 mol/min)	$F + G \rightarrow P$ 20°C	F (5 mol/min) G (5 mol/min) P (45 mol/min)

F and G react to form P in a solution of F and G or in a mixture of F and G. Pure G is cheap. F is available only in a dilute aqueous solution. F is extracted from fish scales, a by-product of the halibut industry. Consequently a dilute solution of F is cheap on a basis of \$/(kg F), but it contains a reactive impurity, I. The chemical I does not react with F but reacts with G to form a less valuable by-product, B. The reaction

$$I + G \rightarrow B$$

converts 100% of the reactants, as shown below.

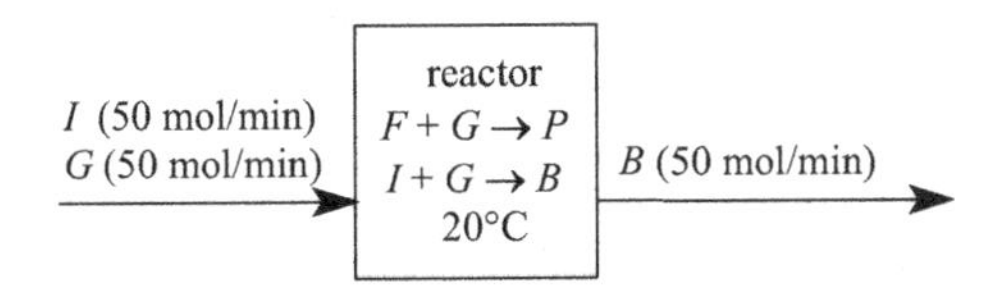

Chemicals I and G react to form B in a solution of I and G or in a mixture of I and G. Chemical G reacts preferentially but not exclusively with I over F. I reacts only with G; I does not react with F, P, B, or water. Likewise, F reacts only with G; F does not react with P, I, B, or water.

Design a process to produce P from two inputs: (1) a stream containing H_2O (1000 mol/min), F (1 mol/min), and I (2 mol/min); and (2) a stream of pure G (unlimited flow rate). In addition to listing the substances in every stream, estimate the flow rates of reactants in the stream entering the reactor. One significant figure is sufficient.

Design Goals (in decreasing importance)

- Minimize the number of units in your process.
- Maximize the yield of product P and any other marketable by-product.
- Employ easy separations.

Following is a table of physical parameters at 1 atm.

		Melting point (°C)	Boiling point (°C)	Soluble in water?	Market price (\$/kg)	Comments
F	reactant	35	125	yes	30[a]	non-toxic[b]
G	reactant	18	112	yes	5	non-toxic[b]
I	reactive impurity	62	125	yes	10	non-toxic[b]
P	product	−28	57	no	200	mild irritant
B	by-product	7	78	yes	5	non-toxic[b]
H_2O	solvent	0	100		0.002	

[a] Cost of 1 kg of F in 1000 kg of water, with 0.2 mol% I impurity.
[b] May be disposed of in landfill or wastewater.

2.35 Design a process to produce P from A (not the same A and P as in any other exercise):

$$A \leftrightarrow P.$$

The reaction is reversible; P can be converted to A. A also reacts reversibly to form Q:

$$A \leftrightarrow Q.$$

The reactions of A to form P and Q go to equilibrium, which is an equimolar mixture of A, P, and Q. The example below shows the reactor effluent when pure A enters the reactor. The numbers in parentheses are flow rates, in mol/min.

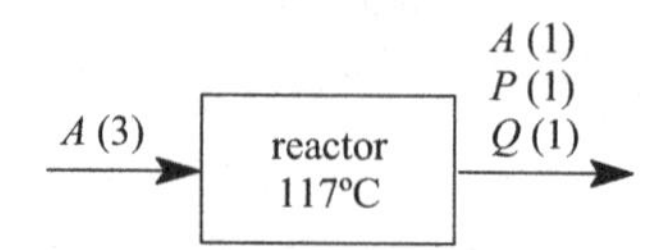

Any composition of A, P, and Q entering the reactor will leave as an equimolar mixture of A, P, and Q. For example:

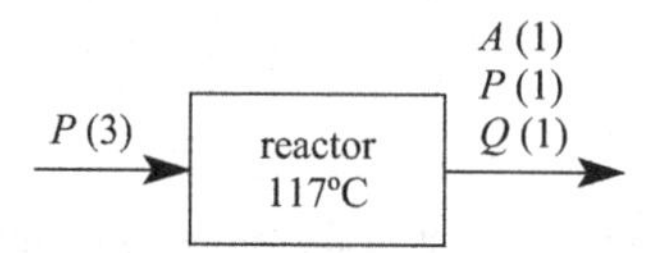

A is available only as a mixture of 90% A and 10% B. (All compositions in this exercise are in mol%.) The reactor that converts A to P (and to Q) also converts B to X:

$$B \leftrightarrow X.$$

Any composition of B and X entering the reactor will leave as an equimolar mixture of B and X.

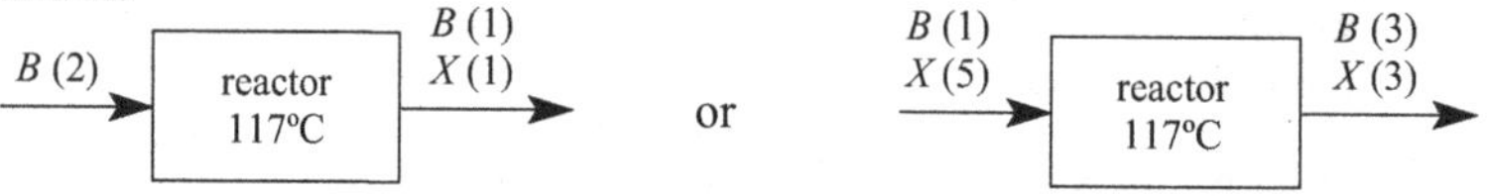

In summary, 100 mol of feed entering the reactor will produce the following output:

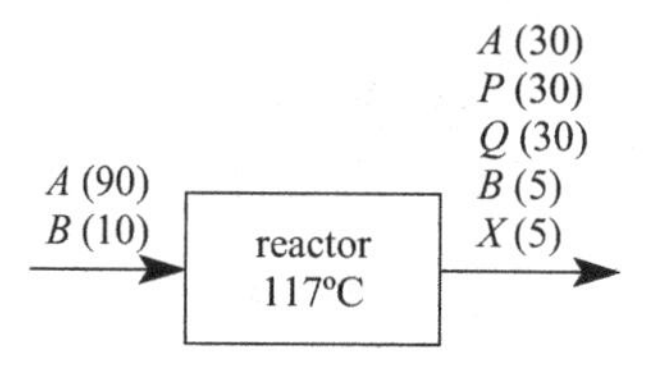

Design Goals (in decreasing importance)

- Maximize the total value of the product(s).
- Minimize the *number* of units.
- Minimize the *size* of each unit.

Design Rules

- Start with 100 mol/min of a 90:10 A:B mixture.
- List the substances and *approximate* flow rates in every stream. One significant figure is sufficient.

Following is a table of physical parameters at 1 atm.

	Molecular weight	Melting point (°C)	Boiling point (°C)	Price ($/mol)
A	120	51	147	1
B	133	51	147	20
P	120	−36	77	100
Q	120	24	123	30
X	133	−12	95	3

2.36 A good product G is produced by the following reaction:

$$M + E \rightarrow 2G. \qquad \text{rxn 1}$$

Reactants available: a liquid mixture of M (50 mol%) and X (50 mol%)
a liquid mixture of E (50 mol%) and I (50 mol%)

Compound I is inert. In a mixture of M, X, E, and I, a side reaction produces by-product B:

$$X + E \rightarrow 2B. \qquad \text{rxn 2}$$

Reactions 1 and 2 are highly exothermic. A mixture of 50 mol% M and 50 mol% E will explode. The same is true for X and E. To avoid explosion, E must not exceed 1 mol% if M and/or X are present.

All reactants and products are soluble in water. Water is inert to all reactants and products.

Reactions 1 and 2 do not go to completion at the safe reaction temperature, 20°C. That is, if M and E enter a reactor, the amounts leaving the reactor will be such that

$$\frac{(\text{mol } G)^2}{(\text{mol } M)(\text{mol } E)} = 60.$$

Two examples of reactor effluents are shown below. The first uses a stoichiometric mixture of M and E. The numbers in parentheses are flow rates, in mol/min.

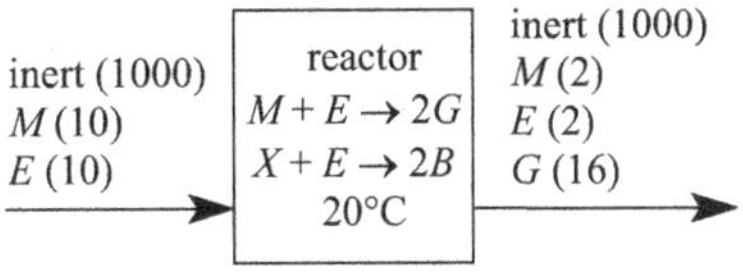

Inlet	Reactor	Outlet
inert (1000) M (10) E (10)	reactor $M + E \rightarrow 2G$ $X + E \rightarrow 2B$ 20°C	inert (1000) M (2) E (2) G (16)

The second example shows a result when M is in excess.

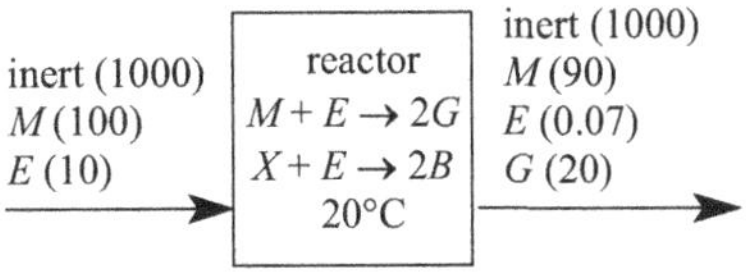

Inlet	Reactor	Outlet
inert (1000) M (100) E (10)	reactor $M + E \rightarrow 2G$ $X + E \rightarrow 2B$ 20°C	inert (1000) M (90) E (0.07) G (20)

Likewise if X and E enter a reactor, the concentrations leaving the reactor will be such that

$$\frac{(\text{mol } B)^2}{(\text{mol } X)(\text{mol } E)} = 60.$$

Reactions 1 and 2 proceed at the same rate.

Design a process to produce G.

Design Goals (in decreasing importance)

- Maximize the yield of product G and any other marketable by-product(s).
- Minimize the flow rate of the mixture of M and X.
- Minimize the number of units in your process.

Design Rules

- Use 20 mol/min of the mixture of E (50 mol%) and I (50 mol%). You may use as much of the mixture of M (50 mol%) and X (50 mol%) as you wish.
- Indicate *approximate* flow rates of the substances in each stream. One significant figure is sufficient.

Hint: In this exercise, the total number of mols is constant in both reactions. If a total of 20 mols of M and E enter a reactor, a total of 20 mols of M, E, and G will leave the reactor. (Note: this is *not* a general principle.) Also, if a total of 100 mols of M, E, X, and I enter your process, a total of 100 mols of M, E, X, I, B, and G must leave your process. Because I is inert, if 10 mol of I enter your process, 10 mol of I must leave your process.

To calculate the flow rates of M, E, and G in the reactor effluent, assume x mols of M and E react to form $2x$ mols of G. For example, if 10 mol M and 10 mol E enter the reactor,

$$\frac{(\text{mol } G)^2}{(\text{mol } M)(\text{mol } E)} = 60$$

$$\frac{(2x)^2}{(10-x)(10-x)} = 60$$

$$x \approx 8.$$

Thus the reactor effluent will contain 2 (= $10 - x$) mol M, 2 (= $10 - x$) mol E, and 16 (= $2x$) mol G.

Following is a table of physical parameters at 1 atm.

		Melting point (°C)	Boiling point (°C)	Market value ($/mol)	Comments
M	reactant	2	78	50	non-toxic
X	reactive impurity	−40	42	50	mild irritant
E	reactant	−28	47	100	non-toxic
I	inert impurity	−28	47	2	non-toxic
G	product	−54	19	300	non-toxic
B	by-product	10	92	10	non-toxic
water	solvent	0	100	0.00001	

Problem Redefinition

Exercises 2.37 through 2.52 describe scenarios in which the problem is poorly defined. Propose an improved definition of the problem for each scenario. Here are three example scenarios.

Scenario 1 Two hikers are confronted by a mountain lion.

Problem Devise a plan to prevent the lion from eating the hikers.

One hiker quickly discards her pack, removes her boots, and puts on her running shoes. "What are you doing?" asks the second hiker, "You can't outrun a mountain lion!" The first hiker responds, "I don't have to outrun the lion – I only have to outrun you."

Real Problem: Devise a way to avoid being eaten.

Adapted from Fogler, H. S. and LeBlanc, S. E. (1995). *Strategies for Creative Problem Solving*, Prentice Hall, Upper Saddle River, NJ, attributed to Professor J. Falconer, Department of Chemical Engineering, University of Colorado, Boulder.[8]

Scenario 2 An issue arose during the development of the manned space program in the early 1960s. A space capsule reentering the Earth's atmosphere is heated by air friction to thousands of degrees. This is hazardous to the capsule contents, such as its human passengers.

Problem: Find a material that can withstand the extreme heat of reentry.

This problem was eventually solved, but not until 20 years of manned space flight and the construction of the space shuttle. This problem could not be solved with 1960s technology.

Real Problem: Protect the astronauts.

One scientist considered how meteorites survive entry into the Earth's atmosphere – the meteorite surface is ablated by the heat, which protects the core. The same was done for the early space capsules – the entry side of the capsule was covered with an ablative material that absorbed the heat and vaporized.

Adapted from Fogler, H. S. and LeBlanc, S. E. (1995). *Strategies for Creative Problem Solving*, Prentice Hall, Upper Saddle River, NJ, attributed to Professor J. Falconer, Department of Chemical Engineering, University of Colorado, Boulder.

Scenario 3 The wastewater from your chemical production facility contains a infinitesimal amount of dioxin (2,3,7,8-tetrachlorodibenzo-*p*-dioxane or TCDD). Although the amount of dioxin was below the allowable level, the allowable level was decreased and your facility is now in violation.

Dioxin: teratogenic and highly toxic.

Problem: Treat the wastewater to decrease the amount of dioxin to an allowable level.

Dioxin can be extracted by treating the wastewater with charcoal. However, this process is prohibitively expensive and decreases the dioxin level to only slightly below the allowable level.

Real Problem: Decrease the amount of dioxin in the wastewater to an allowable level.

The problem was traced to a solvent that was reacting to form dioxin. The solvent was replaced with an equally effective solvent and no dioxin was produced. This solution is superior to treating the wastewater. The solution also represents a canon in the chemical process industries: design for "zero" emissions, rather than designing to comply with regulations. In this way you anticipate new regulations. Instead of working to remedy environmental hazards, don't contaminate in the first place.

[8] For this scenario the authors credit Professor J. Falconer, Department of Chemical Engineering, University of Colorado, Boulder, CO.

2.37 A shooting range adjacent to San Francisco Bay placed the targets at the water's edge so that stray bullets fell harmlessly into the bay. Normally shooting ranges recover spent bullets as metal scrap. However, recovery was impractical at the bay because the bullets sank into the muddy bottom, estimated to be 3 m of silt below 10 m of water. The lead content in the water of the bay was recently found to be abnormally high. Initially a nearby chemical plant was blamed, but close examination revealed that the plant effluent contained no lead and that no process at the plant ever involved lead.

Problem: Devise a means to remove the bullets from the bay.

2.38 Surface water in Bangladesh is contaminated with bacteria. Children died of diarrhea caused by drinking water from rivers and ponds. Beginning in the 1970s, tube wells were installed in villages; shallow aquifers (about 10 m deep) were tapped with pipes and a hand pump drew the water. In the 1990s, widespread arsenic contamination was measured in water drawn from these tube wells; 30% of the tube wells produce water with more than 50 micrograms of arsenic per liter, over five times the upper limit set by the World Health Organization. One-quarter of Bangladesh's population – 35 million people – drink water with harmful arsenic contamination.[9]

Problem: Reduce the arsenic level in tube well water to a potable level.

2.39 A medical publication addresses mercury emissions during cremation.[10] Most dental fillings are an amalgam with 50% mercury. When a body is cremated, some of the mercury is released to the air. These mercury vapors are a health hazard to the general population. The authors found that crematoria staff had 70% higher concentrations of mercury in their hair. The mercury concentration in a person's hair correlates to the mercury concentration in a person's liver and kidney.

Problem: Design a filter to remove mercury vapor from a crematorium chimney exhaust.

This report notes: "It is *estimated* [our emphasis] that one crematorium (in the UK) emits 4.453 kg of mercury per year." It is interesting that their estimate has four significant figures. This is another example of significant figure abuse by conversion (see Appendix C). The original estimate was probably 10 pounds per year, an estimate with one significant figure. The conversion 1 lb = 0.4453 kg should yield an estimate of 4 kg per year, not 4.453 kg per year.

2.40 In the past 50 years, swine agriculture in the US changed. Driven by economics and aided by technological developments and nutritional advances, large farms replaced small farms, and hog raising in full-time confinement replaced hog raising in pastures. A key innovation was a system to remove wastes efficiently. The system involves a manure lagoon, in which anaerobic microbes digest the waste.

During the past 25 years, suburbia has encroached on farmlands. Hog facility neighbors complain about the offensive odor of the pig manure, which is more pungent than cattle manure. The odor is chiefly due to ammonia and hydrogen sulfide emitted from the manure digestion. The Environmental Protection Agency (EPA) regulates concentrated animal feeding operations (CAFOs) under the

[9] See www.who.int/docstore/bulletin/pdf/2000/issue9/bu0751.pdf.

[10] Maloney, S. R., Phillips, C. A., and Mills, A. 1998. "Mercury in the hair of crematoria workers," *Lancet*, 352, 1602.

Effluent Limitations Guidelines and Standards (ELGS). Your farm's emissions of ammonia and hydrogen sulfide are too large, by a factor of ten.

Problem: Design a system to absorb these volatile malodorous digestion products.

2.41 You construct your new restaurant on a building lot in a completely developed commercial zone. After a few months of operation the restaurant is still not showing a profit. The cost of paying off the new kitchen equipment and the salaries of the kitchen employees exceeds the dining revenues. The chief reason is that not enough customers are dining at your restaurant. The customers are pleased with the food, service, and price, but complain that they have trouble finding a place to park; the restaurant parking lot is full every evening, although the dining room has empty tables. A quick study reveals that all the cars in the parking lot belong to your customers; there are no illegal parkers taking up precious spaces.

Problem: Devise a plan to accommodate more cars in the parking lot.

2.42 A hotel remodels its upper floors to convert its penthouse suites into dozens of small rooms. Although the daily rate on the smaller rooms is less, there are many more guests, which increases the hotel's revenues. However, the guests on the remodeled upper floors complain they must wait a seemingly interminable time for an elevator. The elevators are operating at peak efficiency. It's just that the number of elevators originally installed was based on a lower occupancy of the upper floors.

Problem: Devise a plan to add a new elevator to the hotel.

Ultimately the *real* problem was solved without adding a new elevator.

Adapted from Fogler, H. S. and LeBlanc, S. E. (1995). *Strategies for Creative Problem Solving*, Prentice Hall, Upper Saddle River, NJ.

2.43 An airline sought to improve service by initiating a policy of docking incoming flights at gates close to the baggage claim. This had two effects: travelers arrived in the baggage claim area much faster and travelers complained about delays in receiving baggage. Before the new policy there were seldom complaints about baggage delay or any aspects of arrival.[11]

Problem: Devise a means to deliver baggage to travelers more rapidly.

Adapted from Fogler, H. S. and LeBlanc, S. E. (1995). *Strategies for Creative Problem Solving*, Prentice Hall, Upper Saddle River, NJ.

2.44 One of the dozens of employees you supervise is best described as a "loner" but otherwise performs his tasks expertly. Lately he carries a briefcase with him *everywhere*, although no facet of his job requires items that might be kept in a briefcase. The other employees are reacting badly to the briefcase and have begun to complain.

Problem: Devise a means to eliminate the offending briefcase.

2.45 A snack food company discovered trace levels of a suspected carcinogen in one of their products. They determined that the suspected carcinogen forms during cooking; the source is an innocuous compound in the raw material, which is

[11] *The Washington Post*, A3, December 14, 1992.

present in minuscule amounts. Unfortunately, this compound cannot be removed from the raw material.

Problem: Remove the suspected carcinogen from the snack food.

This exercise was created by Samantha Neureuther, Cornell University Chemical Engineering Class of 2005.

2.46 A pneumatic system is used to transport powdered solids from a storage bin to a shipping container, as shown below.

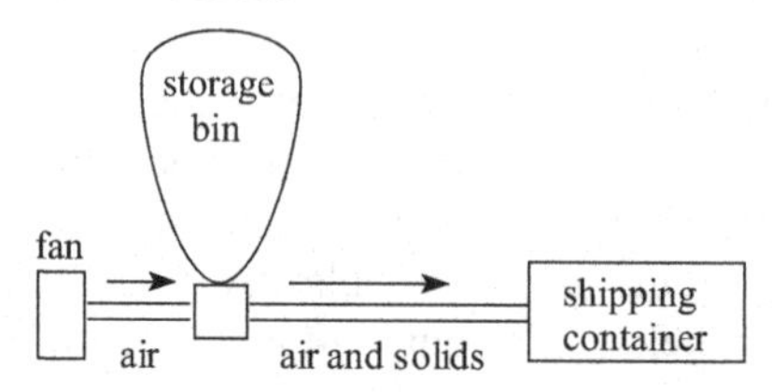

The pipe conducting the air and powdered solids plugs frequently.

Problem: Devise a means to unplug the pipe.

Adapted from Fogler, H. S. and LeBlanc, S. E. (1995). *Strategies for Creative Problem Solving*, Prentice Hall, Upper Saddle River, NJ.

2.47 Titanium dioxide is a common whitener that appears in foods, paint, paper, and health-care products such as toothpaste. A process that manufactures titanium dioxide yields an effluent of dirty water. The dirty water is treated in a clarifier, which allows the solid particles to flocculate and settle to the bottom. The clarifier was plagued by a strong odor of chlorine. An engineer assigned to the process was assigned the following problem.

Problem: Prevent chlorine from leaving the clarifier.

This exercise was created by Vivian Tso, Cornell University Chemical Engineering Class of 1995.

2.48 A stream of air contains chlorosilanes and other impurities which must be removed before the air is released. As shown in the diagram below, the stream is first passed through a liquid–gas separator, which removes the chlorosilanes. The remaining impurities are removed with a scrubber, in which air is bubbled through water. The water effluent from the scrubber was frequently plugged by a highly viscous liquid. It was determined that the viscous liquid resulted from the reaction of water and chlorosilanes.

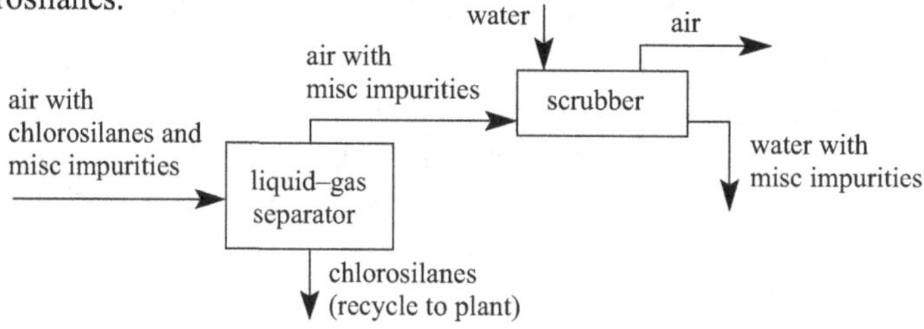

Problem: Modify the scrubber to accommodate the chlorosilanes that escape the liquid–gas separator.

This exercise was created by Elizabeth Lim, Cornell University Chemical Engineering Class of 1994.

2.49 One step in a process to manufacture integrated circuits on silicon wafers (computer chips) involves etching the surface with a plasma. The metal layer on the silicon wafer is coated with a polymer template, such that portions of the metal are exposed. A plasma above the silicon wafer selectively etches the exposed metal. When the polymer is dissolved away, a pattern of the protected metal remains. This metal pattern connects electronic devices on the silicon surface, such as transistors and diodes.

The particle concentration in the etching chamber is monitored optically because it is crucial that no particles deposit on the silicon wafer. A laser beam passed through the chamber to a detector measures the particle concentration. If a particle crosses the beam, the detector sends a signal. The signals from the detectors in several chambers are collected and analyzed by a computer, which displays the particle concentrations in each chamber. The software is updated frequently. Shortly after an update, the displays of particle levels in each chamber show no signals. The attending engineer, a Cornell University co-op student, was given the following problem.

Problem: Find the error in the software.

This exercise was created by Tom Sequist, Cornell University Chemical Engineering Class of 1995.

2.50 Chemical plants are constructed on large slabs of concrete that typically cover dozens of acres. During normal operation oil and other liquids leak onto the concrete and accumulate in various ditches. When it rains, the water washes the oil off the concrete. The runoff from the plant is therefore polluted and cannot be released to a river or lake.

Problem: Construct a holding tank to catch the water so it can later be treated and then released.

Hint: What volume of holding tank is needed to collect rain at the rate of 1 inch per hour, falling for 8 hours on a plant that covers 100 acres?

This exercise was created by Alfred Center, Cornell University Chemical Engineering Class of 1964 (BS) and 1965 (MEng). Adapted from examples 1.3-2 and 3.4-1 by Rudd, D. F., Powers, G. J. and Siirola, J. J. (1973). *Process Synthesis*, Prentice Hall, Upper Saddle River, NJ, p. 12 and pp. 82–84.

2.51 In a large chemical process, energy to run the units comes from steam. (Steam is preferable to electricity because it is less likely to spark an explosion.) Your plant has several steam-generating units in a location at the periphery. Most of the units burn natural gas (methane) but one burns a high-sulfur oil. Whereas there are negligible emissions of SO_x from the methane-burning units, the emissions of SO_x from the stack of the oil-burning unit are too high.

Problem: Install a scrubber to remove the SO_x from the offending stack.

This exercise was created by Alfred Center, Cornell University Chemical Engineering Class of 1964 (BS) and 1965 (MEng).

2.52 Your chemical plant ships product via a fleet of tankers. Your company has leased a fleet of tankers with capacity of 50,000 barrels each. These are modest tankers, neither large nor small. Your wharf has three loading docks. Each day you must ship 150,000 barrels. A tanker must be in port for one day. Much of the day in port is spent docking, connecting to the transfer lines, and disconnecting from the transfer lines. Under normal conditions, the system works well – your wharf accommodates three tankers per day, each has a capacity of 50,000 barrels, and so you ship 150,000 barrels every day.

But sometimes bad weather or shipping schedules do not allow you to have three ships in port. The shipping falls behind on those days. The deficit can never be made up because your maximum shipping rate is the shipping rate on a normal day.

Problem: Enlarge the wharf to accommodate five tankers.

This exercise was created by Alfred Center, Cornell University Chemical Engineering Class of 1964 (BS) and 1965 (MEng).

Physical Properties at 1 atm.

	Melting point (°C)	Boiling point (°C)	Solubility in water at 25°C	Density at 25°C (kg/m^3)
Al, aluminum	660	2,467	insoluble	2,700
$CaCO_3$, calcium carbonate	–	825[a]	0.01 g/L	2,710
$CaCl_2$, calcium chloride	772	>1,600	700 g/L	2,150
CaO, calcium oxide	2,614	2,850	1 g/L	3,250
$Ca(OH)_2$, calcium hydroxide	–	[a]	2 g/L	2,240
C, carbon (graphite)	–	3,652[b]	insoluble	2,250
CCl_4, carbon tetrachloride	−23	77	insoluble	1,590
CH_4, methane	−182	−164	0.002 g/L	–
CH_4O, methanol, CH_3OH	−94	65	∞	790
CH_4O_4S, methane sulfonic acid, CH_3OSO_3H	<−30	140[a]	∞	–
CO, carbon monoxide	−205	−191	40 g/L	–
CO_2, carbon dioxide	−57[c]	−79[b]	2 g/L	–
CS_2, carbon disulfide	−111	46	2 g/L	1,260
C_2Cl_4, 1,2-dichoroethylene, $CCl_2{=}CCl_2$	−19	121	insoluble	1,620
C_2Cl_6, hexachoroethane, $CCl_3{-}CCl_3$	–	186[b]	insoluble	2,090
C_2H_2, acetylene, $CH{\equiv}CH$	−81[b]	−84[d]	1 g/L	–
C_2H_3Cl, vinyl chloride, $CH_2{=}CHCl$	−154	−13	slight	–
C_2H_4, ethylene, $CH_2{=}CH_2$	−169	−104	insoluble	–
$C_2H_4Cl_2$, 1,1-dichloroethane, $CH_3{-}CHCl_2$	−97	57	5 g/L	1,180
$C_2H_4Cl_2$, 1,2-dichloroethane, $CH_2Cl{-}CH_2Cl$	−35	83	8 g/L	1,260
C_2H_4O, ethylene oxide	−111	14	∞	–
C_2H_5Cl, chloroethane, $CH_3{-}CH_2Cl$	−136	12	5 g/L	–
C_3H_6O, acetone (dimethyl ketone), $CH_3C(O)CH_3$	−95	56	∞	790
C_6H_6, benzene	5.5	80	2 g/L	880
C_6H_5OH, phenol	41	182	70 g/L	1,070
C_6H_5ONa, phenol, sodium salt	>250[a]	–	very	–
$C_6H_5SO_3H$, benzene sulfonic acid	66	–	very	–
$C_6H_5SO_3Na$, benzene sulfonic acid, sodium salt	–	450[a]	moderate	–
C_6H_{10}, cyclohexene	−104	83	insoluble	810
C_6H_{12}, cyclohexane	6	81	insoluble	780
$C_6H_{10}O$, cyclohexanone	−47	156	8.6 g/L	950

(*cont.*)

	Melting point (°C)	Boiling point (°C)	Solubility in water at 25°C	Density at 25°C (kg/m^3)
$C_6H_{11}OH$, cyclohexanol	26	162	3.6 g/L	960
$C_6H_{10}O_4$, adipic acid, HOOC–$(CH_2)_4$–COOH	152	338	24 g/L	1,360
$C_6H_{12}N_2$, acetazine, $(CH_3)_2C{=}N{-}N{=}C(CH_3)_2$	−13	113	∞	840
C_8H_8, styrene, $C_6H_5{-}CH{=}CH_2$	−31	145	0.3 g/L	910
C_8H_{10}, ethylbenzene, $C_6H_5{-}CH_2CH_3$	−95	136	0.2 g/L	870
$C_{10}H_{14}$, 1,3-diethylbenzene, C_6H_4-$(CH_2CH_3)_2$	−84	181	insoluble	860
$C_{14}H_8O_2$, 9,10-anthraquinone	286	380	insoluble	1,440
$C_{14}H_8(OH)_2$, 9,10-anthradiol	180	–	–	–
Cl_2, chlorine	−101	−35	2 g/L	–
$CuCl_2$, copper(II) chloride	620	993[a]	700 g/L	3,390
CuO, copper(II) oxide	1,326	–	insoluble	6,300
$FeCl_2$, iron(II) chloride	677	1,023	600 g/L	3,160
$FeCl_3$, iron(III) chloride	306	315[a]	700 g/L	2,900
Fe_2O_3, iron(III) oxide	1,566[a]	–	insoluble	5,240
HCl, hydrochloric acid	−115	−85	800 g/L	–
HCN, hydrogen cyanide	−14	26	∞	700
HNO_3, nitric acid	−42	83	∞	1,500
H_2, hydrogen	−259	−253	0.002 g/L	–
H_2O, ice, water, or steam	0	100	∞	1,000
H_2O_2, hydrogen peroxide	−0.4	151	∞	1,440
H_2S, hydrogen sulfide	−86	−61	7 g/L	–
H_2SO_4, sulfuric acid	10	338	∞	1,840
H_3PO_4, phosphoric acid	42	158[a]	∞	1,890
MgO, magnesium oxide	2,852	3,600	0.006 g/L	3,580
$Mg(OH)_2$, magnesium hydroxide	350[a]	–	0.009 g/L	2,360
$Mg(NO_2)_2$, magnesium nitrite	250[a]	–	high	–
$MgSO_3$, magnesium sulfite	200[a]	–	slight	1,730
$NaBH_4$, sodium borohydride	400[a]	–	reacts	1,070
$NaBO_2$, sodium metaborate	966	1,434	280 g/L	2,460
NaCl, sodium chloride	801	1,413	360 g/L	2,170
NaClO, sodium hypochlorite	(soln)	(soln)	500 g/L	–
$NaHCO_3$, sodium hydrogen carbonate		270[a]	70 g/L	2,160
NaOH, sodium hydroxide	318	1,390	1,100 g/L	2,130
Na_2CO_3, sodium carbonate	851	[a]	200 g/L	2,530
Na_2SO_3, sodium sulfite	>400[a]	–	260 g/L	2,630
NH_3, ammonia	−78	−33	900 g/L	–
NH_4Cl, ammonium chloride	−57	[a]	300 g/L	–
NH_4OH, ammonium hydroxide	−77	(soln)	moderate	–
NO, nitric oxide	−164	−152	slight	–
NO_2, nitrogen dioxide	−11	21	0.1 g/L	–
N_2, nitrogen	−210	−196	0.03 g/L	–
N_2H_4, hydrazine, $NH_2{-}NH_2$	1	114	very	1,010
$N_2H_4{\cdot}H_2O$, hydrazine hydrate	−40	119	∞	1,030
Ni, nickel	1,453	2,732	insoluble	8,900

(*cont.*)

	Melting point (°C)	Boiling point (°C)	Solubility in water at 25°C	Density at 25°C (kg/m³)
O_2, oxygen	−218	−183	0.04 g/L	–
S, sulfur	113	445	insoluble	2,070
S_2Cl_2, sulfur monochloride	−80	136	decomposes	1,680
SO_2, sulfur dioxide	−73	−10	230 g/L	–
SO_3, sulfur trioxide	17	45	decomposes	1,920
ZnO, zinc oxide	1,975	–	0.002 g/L	5,600

[a] Decomposes.
[b] Sublimation point: solid–gas transition.
[c] At 5.2 atm.
[d] At 1.2 atm.
(soln) At 25°C, substance exists in aqueous solution only.
∞ Infinitely soluble; soluble at all proportions.

3

Models Derived from Laws and Mathematical Analysis

Have you ever marveled at the beauty of an exquisite design? How did the engineers devise such a complex device or system? Surely they must have been geniuses. Although this may be true in some cases, good engineering generally does not require genius (although it certainly helps). Complex designs usually evolve from simple designs. It is important to *analyze and evaluate* the effects of each evolutionary step in a design.

How does a good design begin? Linus Pauling, who earned his BS degree in chemical engineering (Oregon State University, 1922), and received a Nobel Prize in Chemistry (1954) and a Nobel Peace Prize (1962), recommended that "The best way to get a good idea is to get a lot of ideas." We modify and extend this useful advice to "The best way to get a good design is to create many designs and choose the best." To choose the best design, one must be able to *analyze and evaluate* the designs.

How does one analyze and evaluate a design? One option is to build the system and test it. This is appropriate when the design is for a small item, such as a can opener. This approach is also appropriate for a medium-sized item, such as an automobile, if one expects to manufacture thousands of units. But clearly it is impractical to build a full-sized system when the system is very large and only a few, perhaps only one, will be built. Most chemical processes are too large and too expensive to allow one to build a full-sized model. Chemical engineers must use other means to analyze and evaluate a design. We introduce three methods in this text: mathematical modeling (this chapter), graphical analysis of empirical data (Chapter 4), and dimensional analysis/dynamic scaling (Chapter 5).

Dimensional analysis and dynamic scaling allow one to predict, for example, the performance of a full-sized system based on the performance of a small prototype of the system. One constructs a model of the system, tests the model under various operating conditions, and then applies dynamic scaling to predict how the actual system would behave, with the operating conditions scaled as well.

Rather than build a physical model of the system, one can construct a mathematical model. Rather than translate the design to a physical device, one can translate the design to equations. The equations can then be used to predict the performance of the system under various operating conditions. To translate from the description of the system (the design) to the mathematical model (the equations), one invokes various laws of nature, such as Newton's second law, $F = ma$. If our design involved a device exerting a force on a mass, we could predict the acceleration of the mass. And with the calculus, we could predict the velocity and position of the mass, given the duration of the force (an operating condition in this case). We knew qualitatively that the force would accelerate the mass. Our model would provide a quantitative description, necessary to evaluate various alternatives. As eloquently stated by Richard Alkire (Professor of Chemical Engineering,

University of Illinois), who was recognized with the Professional Progress Award of the AIChE in 1985, "A mathematical model articulates a qualitative description." A key component of chemical engineering is developing mathematical descriptions of unit operations and processes.

The third alternative, graphical analysis of empirical data, is used when the complexity of the system prohibits mathematical analysis. This alternative is usually simpler than building a prototype.

3.1 MASS BALANCES ON PROCESSES WITH NO CHEMICAL REACTION

3.1.1 Mass Balances on a Single Unit at Steady State

Context: Desalination of seawater.

Concepts: The conservation principles, closed and open systems, and steady-state systems.

Defining Question: How does one build a mathematical model? How does one start?

Desalination of seawater to water suitable for human consumption and/or agriculture is a vital operation in many parts of the world. Table 3.1 gives the typical salt content of seawater and the maximum salt content of drinking water. The chief component of sea salt is NaCl; seawater contains 1.1 wt% Na and 1.9 wt% Cl.

Table 3.1 The compositions of seawater and potable water.

	H_2O (wt%)	Salt (wt%)
seawater	96.5	3.5
potable water	>99.95	<0.05

Consider an ideal process to convert seawater to pure water, diagrammed in Figure 3.1.

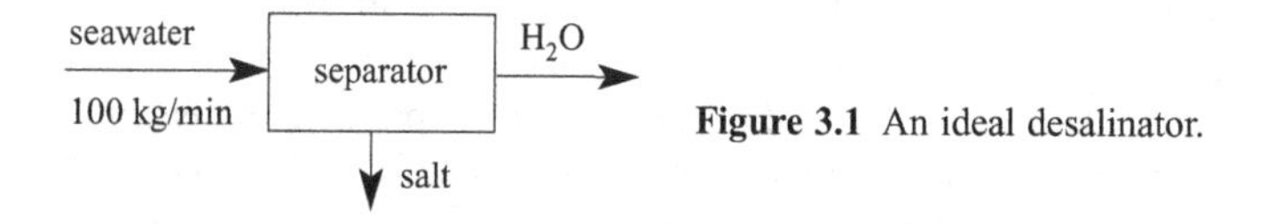

Figure 3.1 An ideal desalinator.

What are the flow rates of the water and salt streams that leave the ideal desalinator? The composition given in Table 3.1 tells us the 100 kg of seawater entering the process each minute contains 96.5 kg of water and 3.5 kg of salt. By inspection, 96.5 kg/min of water and 3.5 kg/min of salt leave the separator. How did you determine these flow rates? Probably your intuition told you that everything that enters the separator must leave the separator.

Consider a more realistic separator, shown in Figure 3.2.

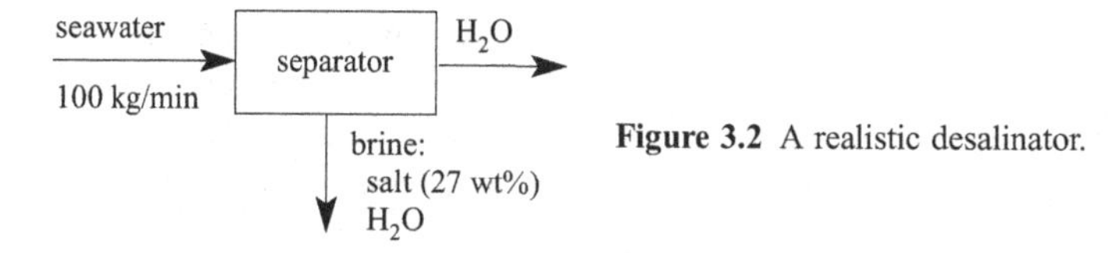

Figure 3.2 A realistic desalinator.

What are the flow rates of the water and brine[1] streams that leave the realistic desalinator? The answers are less intuitive. We need to calculate. As such, we need equations to describe the separator. To obtain these equations we apply a physical law to the separator. The first class of Great Physical Laws is the conservation principles: *certain quantities are invariant – their amount is constant.* The following quantities are conserved in any process: mass, energy, linear momentum, angular momentum, electric charge, baryon number, and strangeness.[2]

We apply the conservation of mass to model the separator. The key to applying a conservation principle is *careful accounting*. Define a system by enclosing a volume with borders.

Figure 3.3 A system, its borders, and its surroundings.

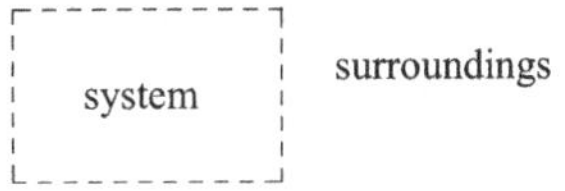

system

surroundings

If nothing crosses the borders, this is a *closed system*; conserved quantities are constant within the system. The conservation of mass translates into mathematics as equations 3.1 and 3.2. Equation 3.2 is a mathematical model of a closed system.

$$\text{mass in system at time 1} = \text{mass in system at time 2} \tag{3.1}$$

$$\text{mass}(t_1) = \text{mass}(t_2). \tag{3.2}$$

But our separator is not a closed system. Things enter and exit the system; mass crosses the borders.

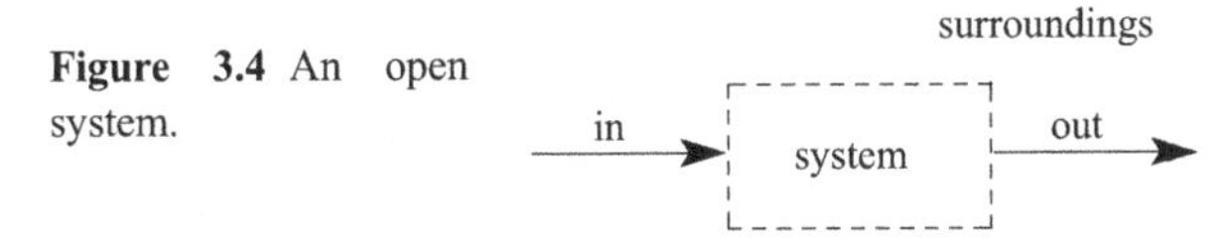

Figure 3.4 An open system.

For an open system, the conservation of mass yields

$$\begin{aligned}\text{mass}(t_2) &= \text{mass}(t_1) + (\text{mass entering from } t_1 \text{ until } t_2) \\ &\quad - (\text{mass exiting from } t_1 \text{ until } t_2).\end{aligned} \tag{3.3}$$

Assume the rate of mass flow into and out of the system is constant over the epoch t_1 to t_2. The amount of mass entering and leaving is the flow rate multiplied by the time evolved:

$$\begin{aligned}\text{mass}(t_2) &= \text{mass}(t_1) + (\text{rate of mass flow in})(t_2 - t_1) \\ &\quad - (\text{rate of mass flow out})(t_2 - t_1)\end{aligned} \tag{3.4}$$

$$\frac{\text{mass}(t_2) - \text{mass}(t_1)}{t_2 - t_1} = \text{rate of mass flow in} - \text{rate of mass flow out.} \tag{3.5}$$

We now take the limit of equation 3.5 as the time between t_1 and t_2 becomes infinitely short, that is, in the limit $t_2 - t_1 \rightarrow 0$:

[1] In this text "brine" will imply water with dissolved salt, not necessarily the same concentration as seawater.

[2] Strictly speaking, mass is not conserved; mass can be converted to energy. We will ignore nuclear processes in this text. As chemical engineers you will also use the conservation of energy, linear momentum, angular momentum, and electric charge. Most likely you will never be concerned with the conservation of baryon number or strangeness, quantities encountered in particle physics.

$$\frac{d\,(\text{mass})}{dt} = \text{rate of mass flow in} - \text{rate of mass flow out.} \tag{3.6}$$

The assumption that the flow rates are constant is no longer restrictive because all flow rates are constant over an infinitely short time period. Equation 3.6 is commonly written as

$$\text{rate of accumulation} = \text{rate in} - \text{rate out.} \tag{3.7}$$

We now assume there is no accumulation in the system. That is, we assume the system is at *steady state*; the state of the system at any spatial point, as characterized by chemical composition, temperature, and pressure, does not change. The state at a given point is steady, but not necessarily uniform; the state may differ at different spatial points. The water level of a lake is uniform. The water level of a flowing river is at steady state. Steady-state operation is effective and efficient. Steady-state processes are easier to control and do not require vessels to accumulate materials between process units. Steady-state processes also yield products with uniform quality. Steady-state operation simplifies the analysis because steady-state processes are described by algebraic equations. Systems not at steady state, called transient systems, require the calculus. We consider transient systems in Chapter 6.

For a system at steady state, the conservation of mass simplifies:

$$0 = \text{rate in} - \text{rate out} \tag{3.8}$$

$$\text{rate in} = \text{rate out.} \tag{3.9}$$

We return to the realistic desalinator in Figure 3.2 and apply the conservation of mass. We begin by defining our system. We draw system borders around the separator, a trivial step with this system. You will be tempted to omit this step. Don't. Recall that applying conservation laws requires careful accounting, which begins with drawing the system borders.

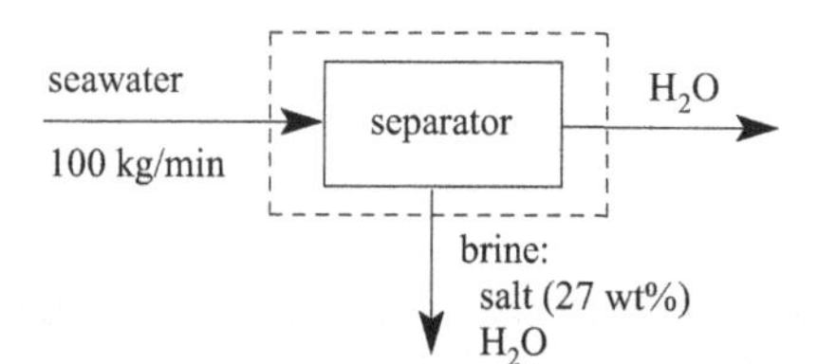

Figure 3.5 System borders for mass balances on a desalinator.

Identifying the streams with narration such as "the pure-water stream exiting the separator to the right" is impractical. It is useful to label the streams. Here we will use numbers.

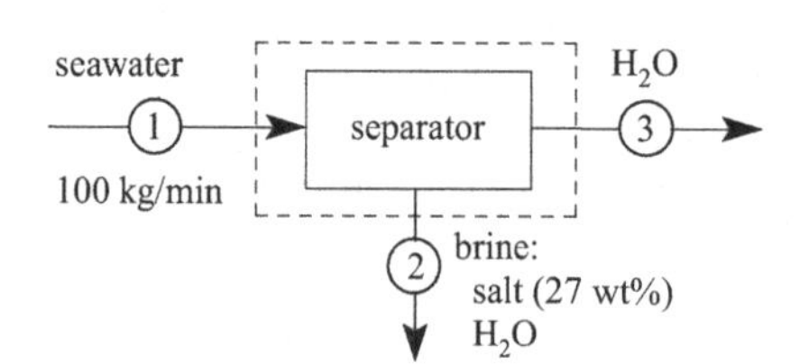

Figure 3.6 System borders and stream labels for mass balances on a desalinator.

It is useful to define variables to represent quantities such as "the flow rate of salt in stream 1." First, we assign letters for the principal components: water $\equiv W$, salt $\equiv S$, and total $\equiv T$. It is prudent to choose obvious and mnemonic letters, such as W for water, so when you see "W" in an equation you can immediately translate to "water." Labeling water as some obtuse Greek character such as ζ will complicate your analysis unnecessarily. Finally, we will use the nomenclature that $F_{i,n}$ represents the flow rate of component i in stream n. Thus, "the flow rate of salt in stream 1" translates to $F_{S,1}$.

We now translate the design into mathematics using the conservation of mass given in equation 3.9. We start with "The total flow rate of stream 1 entering equals the total flow rate of stream 2 exiting plus the total flow rate of stream 3 exiting." With our nomenclature, this translates to

$$\text{total:}\quad F_{T,1} = F_{T,2} + F_{T,3}. \tag{3.10}$$

Similarly we can write mass balances on water and salt:

$$\text{water:}\quad F_{W,1} = F_{W,2} + F_{W,3} \tag{3.11}$$

$$\text{salt:}\quad F_{S,1} = F_{S,2} + F_{S,3}. \tag{3.12}$$

Equations 3.10, 3.11, and 3.12 are related because the sum of the individual components equals the total. The total flow rate of a stream is the sum of the component flow rates. For each stream we can write an equation:

$$\text{stream 1:}\quad F_{T,1} = F_{W,1} + F_{S,1} \tag{3.13}$$

$$\text{stream 2:}\quad F_{T,2} = F_{W,2} + F_{S,2} \tag{3.14}$$

$$\text{stream 3:}\quad F_{T,3} = F_{W,3} + F_{S,3}. \tag{3.15}$$

In other words, mass is an extensive property. This is not true for the total temperature, an intensive property; $T_T \neq T_S + T_W$.

Another clue that equations 3.10 to 3.15 are not independent is that we now have six equations for six unknowns. If the equations were independent *we could calculate all flow rates with no further information.* We would not need the inlet flow rate or the composition of seawater, for example. Obviously this is not true.

Translate the process specifications given into the nomenclature for this model. The total flow rate entering via stream 1 is 100 kg/min. Thus,[3]

$$F_{T,1} = 100.\ \text{kg/min}. \tag{3.16}$$

Translate the compositions of the streams into our nomenclature. Water is 96.5 wt% of stream 1:

$$F_{W,1} = 0.965F_{T,1} = (0.965)100. = 96.5\ \text{kg/min}. \tag{3.17}$$

Similarly,

$$F_{S,1} = 0.035F_{T,1} = (0.035)100. = 3.5\ \text{kg/min}. \tag{3.18}$$

The flow rate of salt in stream 1 could have been obtained alternatively from equations 3.13, 3.16, and 3.17. Water is 100% of stream 3, so

$$F_{W,3} = F_{T,3} \quad \text{and} \quad F_{S,3} = 0. \tag{3.19}$$

And finally, salt is 27 wt% of stream 2:

$$F_{S,2} = 0.27F_{T,2}. \tag{3.20}$$

The system of equations presented here can be solved simultaneously with methods of linear algebra. However, methods of successive substitution will be sufficient for the

[3] The decimal point in "100." indicates the proper number of significant figures. If the need for the decimal point is unclear, you should study Appendix C.

applications you will encounter in this text. Indeed, successive substitution guides one's thinking. For example, we seek the flow rate of pure water. Because the total flow out must be 100 kg/min, if we knew the flow rate out via stream 2 we would know the flow rate out via stream 3. Because we know the composition of stream 2, if we knew the salt flow rate we would know the total flow rate. But salt leaves the separator only by stream 2, so we do know the flow rate of salt in stream 2. Let's follow this logic backwards from the known to the desired. Start with the salt balance, equation 3.12, and substitute equations 3.18 and 3.19. In summary, we started at the objective and traced back to something we knew:

To find $F_{W,3}$, we could use equation 3.11. We know $F_{W,1}$, but we need $F_{W,2}$.
To find $F_{W,2}$, we could use equation 3.14, but we need $F_{T,2}$.
To find $F_{T,2}$, we could use equation 3.20, but we need $F_{S,2}$.
To find $F_{S,2}$, we can use equation 3.12:

$$\text{salt:} \quad F_{S,1} = F_{S,2} + F_{S,3} \tag{3.12}$$

$$F_{S,2} = F_{S,1} - F_{S,3} = 3.5 - 0.0 = 3.5 \text{ kg/min}. \tag{3.21}$$

From the salt flow in stream 2 we can use equation 3.20 to obtain the total flow rate of stream 2. Rearrange equation 3.20 to solve for $F_{T,2}$ and substitute from equation 3.21:

$$F_{T,2} = F_{S,2}/0.27 = 3.5/0.27 = 13. \text{ kg/min}. \tag{3.22}$$

Finally, we substitute for $F_{T,2}$ in the total mass balance, equation 3.10:

$$\text{total:} \quad F_{T,1} = F_{T,2} + F_{T,3} \tag{3.10}$$

$$100. = 13. + F_{T,3} \tag{3.23}$$

$$F_{T,3} = 87. \text{ kg/min}. \tag{3.24}$$

We have our answers: 87 kg/min of pure water and 13 kg/min of brine waste. But we are not done until we check the answer. First, a qualitative check – does 87 kg/min seem reasonable? Yes – anything between 0.0 and 96.5 kg/min is reasonable. How much water leaves via stream 2? 96.5 – 87.0 = 9.5 kg/min. Is this consistent with the composition of stream 2? 9.5/13 = 0.73, which is 27% salt. Our answers check!

Compare the realistic separator to the ideal separator. As the weight fraction of salt in stream 2 decreases, or as the separator works less ideally, the output of pure water decreases. In these two examples, the salt content decreased from 100% to 27% and the output of pure water decreased from 96.5 to 87 kg/min. Often one wishes to predict the performance of a process given different operating conditions. In this case, one might wish to know the output of pure water given the salt content of stream 2, for example. Rather than repeat the above process of successive substitution for a different operating condition, it would be useful to have an equation that yields the flow rate of pure water given the salt content of stream 2.

If one defines the weight fraction of salt in stream 2 to be σ ("sigma" is the Greek symbol[4] for "s," and "s" is a good mnemonic for "salt"),

$$\sigma = \frac{F_{S,2}}{F_{T,2}}, \tag{3.25}$$

[4] Greek symbols are common in mathematical modeling, especially in chemical engineering modeling. Experience shows that Greek symbols cause anxiety for some fledgling engineers. Mathematical manipulations easily done with x and y become scary with η and θ. But avoiding Greek symbols in this text would only postpone the inevitable. Recognize this potential source of trepidation and deal with it.

one arrives at the following equation:

$$F_{W,3} = F_{T,1}\left[1 - \frac{0.035}{\sigma}\right]. \tag{3.26}$$

We recommend you verify that you can derive equation 3.26.

Again, we are not finished until we check the result. We substitute $\sigma = 0.27$ into equation 3.26 and calculate $F_{W,3} = 87$ kg/min, which agrees with the previous result.[5] It is wise to test the equation at a few other values for σ. What other values should yield obvious results? When stream 2 is entirely salt, $\sigma = 1$, all the water entering in stream 1 (96.5 kg/min) must exit via stream 3. When one substitutes $\sigma = 1$ into equation 3.26 one calculates $F_{W,3} = 96.5$ kg/min, as predicted. One final check – what is the minimum value for σ? If no pure water is produced, then stream 1 = stream 2, and $\sigma = 0.035$, and stream 3 should have no flow. Substituting $\sigma = 0.035$ again yields the expected result, $F_{W,3} = 0$. Our answer checks.

After checking an equation, explore the equation and reinforce your grasp of the equation. In this case, sketch a graph of $F_{W,3}$ vs. σ to see the tradeoff between product flow rate and waste composition.

We analyzed a simple desalinator by translating the physical description to a mathematical description. The basis for our analysis was a concept from physics – the conservation of mass. We defined a system and translated the conservation of mass into equations. We also translated process specifications – flow rates and stream compositions – into equations. We solved the equations by successive substitution.

3.1.2 Mass Balances on a Process with Several Units and a Recycle Stream

Context: Desalination by freezing.

Concepts: Strategic system borders, translating process specifications to equations.

Defining Question: How does one know if there is sufficient information to model a system?

What are our options for producing potable water from seawater? Perhaps the most obvious design is an evaporator; separate vapor from liquid, which is easy and efficient. And because water boils at a temperature much lower than salt sublimes, the vapor phase will be essentially pure H_2O. However, evaporation may require too much energy to be economically viable. Even if one uses heat exchangers and condensers to reclaim the heat from the hot brine and the steam, the process still may be too expensive – although the energy costs will be reduced, the equipment costs may be too high. Alternatively, one could use an inexpensive source of energy to evaporate the water, such as solar energy. Or, one can avoid the energy-intensive step of evaporating water. One could use membranes that pass water but exclude salt. Solar heating and membrane separation, however, are low-yield processes and are also expensive owing to the cost of the equipment. Let's consider another option – freezing. Ice formed in salt water contains negligible salt. And the energy required to freeze water (334 kJ/kg) is substantially less than the energy required to evaporate water (2260 kJ/kg).

Consider the design shown in Figure 3.7 to produce pure water from seawater by freezing.

Our design divides the process into four unit operations: cooling the seawater to form an ice + brine slush, separating the ice from the brine, washing brine from the ice with water, and melting the ice to yield water. The ice washer melts 5.0 wt% of the ice that enters. The salinity of the waste brine from the solid–liquid separator is dictated by a local ordinance, which sets

[5] Actually we calculate $F_{W,3} = 87.037037$, but we retain only two significant figures. If it is not clear why only two figures are significant, you should study *Appendix C*.

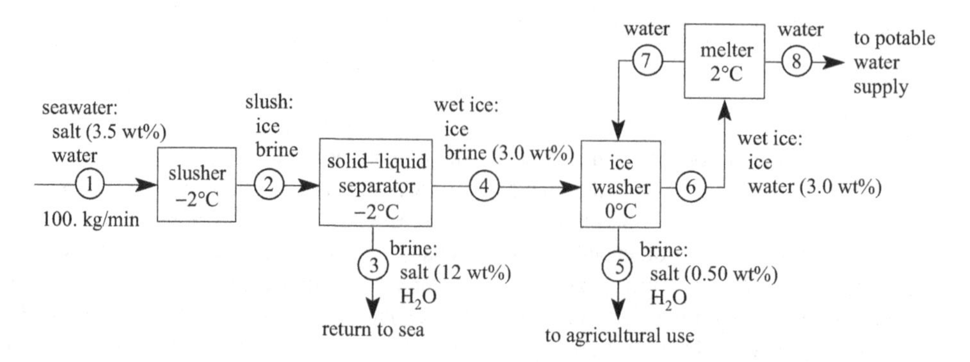

Figure 3.7 A desalination scheme based on freezing.

an upper limit of 12 wt% salt for brine discharged to the sea. The salinity of the brine in stream 5 is the upper limit for water to be used for agriculture in the local region. We wish to calculate the flow rates of four streams: the waste brine (stream 3), the brine for agriculture (stream 5), the potable water (stream 8), and the recycled water (stream 7).

Strategic system borders yield efficient analysis by avoiding unnecessary or intractable calculations. Because three of the four streams to be calculated are output streams, we can ignore the details of internal streams by drawing system borders around the entire process.

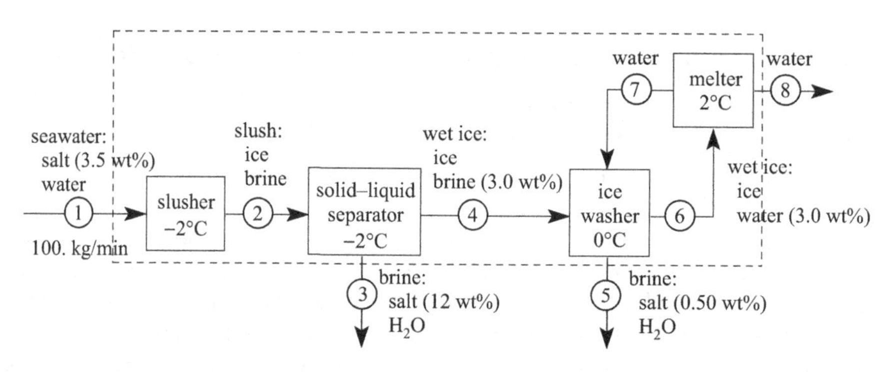

Figure 3.8 System borders for an overall mass balance.

We use the same nomenclature; water $\equiv W$, salt $\equiv S$, and total amount $\equiv T$. But now we have an additional component, ice. Do we need to define ice = I? We could, but that would require a source for the ice, such as the physical reaction

$$\text{water} \rightarrow \text{ice}. \tag{3.27}$$

It is not necessary to label ice as a pure component. Instead we will lump the ice and water into W, thus $H_2O \equiv W$ = water plus ice. Likewise we do not need to create a label for brine. Because every stream contains only salt and H_2O, we need only two component labels: W and S.

We must decide which variables to use in our model. Because the total can be calculated from components salt and water, one variable is redundant. We must select two of the three variables. We select the total flow rate $F_{T,i}$ and the salt flow rate $F_{S,i}$.

We begin our mathematical model by writing mass balances. We assume the process is at steady state, which simplifies the mass balances to "rate in = rate out." It is good practice to write this generic equation to guide one's inventory of the "in" and "out" streams:

$$\text{rate in} = \text{rate out}$$

$$\text{total:} \quad F_{T,1} = F_{T,3} + F_{T,5} + F_{T,8} \tag{3.28}$$

$$\text{salt:} \quad F_{S,1} = F_{S,3} + F_{S,5} + F_{S,8}. \tag{3.29}$$

Translate the process specifications – flow rates and compositions – to equations:

$$F_{T,1} = 100.\ \text{kg/min} \tag{3.30}$$

$$F_{S,1} = 0.035 F_{T,1} = (0.035)100. = 3.5\ \text{kg/min} \tag{3.31}$$

$$F_{S,3} = 0.12 F_{T,3} \tag{3.32}$$

$$F_{S,5} = 0.005 F_{T,5} \tag{3.33}$$

$$F_{S,8} = 0. \tag{3.34}$$

Do we have sufficient information to solve these equations? We have eight variables (four streams with two variables for each) but we have only seven equations. Unfortunately, our plan to calculate three flow rates with an overall mass balance is thwarted. (But it is usually a good strategy to start with system borders that enclose the overall process.)

If we obtain one additional specification, we can solve the overall mass balance. Let's work forward from stream 1 to calculate stream 3. Draw system borders that enclose the slusher and the solid–liquid separator, as shown in Figure 3.9.

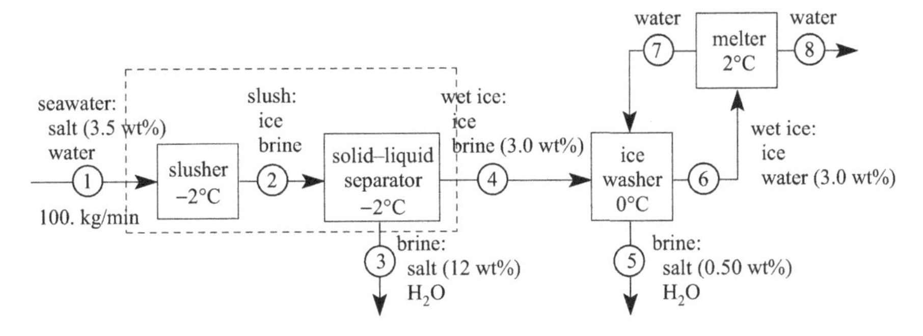

Figure 3.9 System borders for a mass balance around the slusher and solid–liquid separator.

Again we begin by writing mass balances:

$$\text{rate in} = \text{rate out}$$

$$\text{total:} \quad F_{T,1} = F_{T,3} + F_{T,4} \tag{3.35}$$

$$\text{salt:} \quad F_{S,1} = F_{S,3} + F_{S,4}. \tag{3.36}$$

We have equations 3.30, 3.31, and 3.32 for $F_{T,1}$, $F_{S,1}$, and $F_{S,3}$. Translate the composition of stream 4 to an equation – the salt is 12 wt% of the brine and the brine is 3 wt% of stream 4:

$$F_{S,4} = 0.12 \times (\text{brine in stream 4}) = 0.12 \times 0.03 F_{T,4} = 0.0036 F_{T,4}. \tag{3.37}$$

Do we have sufficient information to calculate flow rates for the system in Figure 3.9? We have six variables (three streams with two variables for each) and we have six equations. Solve by successive substitution. Substitute the salt compositions – equations 3.31, 3.32, and 3.37 – into the salt balance, equation 3.36:

$$F_{S,1} = F_{S,3} + F_{S,4} \tag{3.36}$$
$$3.5 = 0.12F_{T,3} + 0.0036F_{T,4}. \tag{3.38}$$

We need another relation between $F_{T,3}$ and $F_{T,4}$. Substitute equation 3.30 into the total mass balance equation 3.35 and solve for $F_{T,4}$:

$$100 = F_{T,3} + F_{T,4} \tag{3.35}$$
$$F_{T,4} = 100 - F_{T,3}. \tag{3.39}$$

Substitute equation 3.39 into equation 3.38 and solve for $F_{T,3}$:

$$3.5 = 0.12F_{T,3} + 0.0036(100 - F_{T,3})$$
$$F_{T,3} = 26.98 \text{ kg/min}. \tag{3.40}$$

Although the specifications have only two significant figures (see Appendix C), we do not round off until our final answer.

Equation 3.40 provides the eighth equation for the overall mass balance. By similar successive substitution of the process specifications into the total mass balance equation 3.28 and the salt balance equation 3.29, we calculate $F_{T,5} = 52.58$ kg/min and $F_{T,8} = 20.44$ kg/min.

To calculate the flow rate of the recycle stream, we need system borders that cross stream 7. To use the process specification that 5 wt% of the ice melts in the ice washer, our system borders should also cross stream 4.

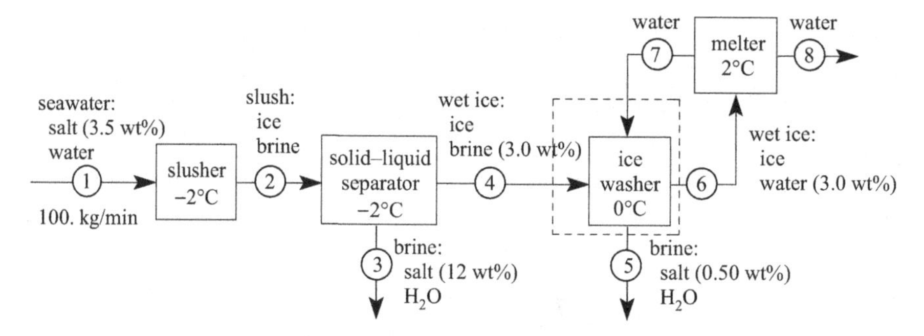

Figure 3.10 System borders for a mass balance around the ice washer.

Write mass balances around the ice washer:

$$\text{rate in} = \text{rate out}$$
$$\text{total:} \quad F_{T,4} + F_{T,7} = F_{T,5} + F_{T,6} \tag{3.41}$$
$$\text{salt:} \quad F_{S,4} + F_{S,7} = F_{S,5} + F_{S,6}. \tag{3.42}$$

Translate the process specifications to equations:

$$F_{T,4} = 73.02 \text{ kg/min} \tag{3.43}$$
$$F_{T,5} = 52.58 \text{ kg/min} \tag{3.44}$$
$$F_{S,4} = 0.0036F_{T,4} \tag{3.37}$$
$$F_{S,5} = 0.005F_{T,5} \tag{3.33}$$
$$F_{S,6} = 0. \tag{3.45}$$

We have eight variables but only seven equations. The eighth equation is subtle; it derives from the ice that flows from stream 4 to stream 6. Ice is 97 wt% of stream 4, so the flow rate of ice in stream 4 is 0.97 × 73.02 = 70.83 kg/min. Stream 6 contains 95% of the ice in stream 4, so the flow rate of ice in stream 6 is 0.95 × 70.83 = 67.29 kg/min. Ice is 97 wt% of stream 6, so the total flow rate of stream 6 is 67.29 ÷ 0.97 = 69.37 kg/min. We have our eighth equation:

$$F_{T,6} = 69.37\ \text{kg/min}\,. \tag{3.46}$$

Substitute the total flow rates in equations 3.30, 3.43, and 3.46 into the total mass balance equation 3.41 and one immediately obtains $F_{T,7}$ = 48.93 kg/min.

We present our answers by rounding each flow rate to two significant figures and drawing a box around the answers.

The waste brine (stream 3) is 27 kg/min, the brine for agriculture (stream 5) is 53 kg/min, the potable water (stream 8) is 20. kg/min, and the recycled water (stream 7) is 49 kg/min.

Check the answers. The total flow rate out is 27 + 53 + 20 = 100 kg/min, which agrees with the total flow rate in. The salt flow rate out is 0.12 × 27 + 0.005 × 53 + 0 × 20 = 3.5 kg/min, which agrees with the salt flow rate in. The flow rate of the recycle is more difficult to check. Consider the process in the melter: the brine retained by the ice in stream 4 (12 wt% salt) is diluted by the recycle to 0.5 wt% salt, a factor of 0.12/0.005 = 24. The flow rate of brine in stream 4 is 0.03 × 73 = 2.2 kg/min. Of the recycled water, 2 kg/min leaves with the washed ice and 47 kg/min dilutes the brine. The dilution is (47 + 2.2) ÷ 2.2, a factor of 22, which agrees with the factor of 24 decrease in salinity. The flow rates are reasonable.

Strategic placement of system borders is key to modeling mass flows in a process with several units. Unknowns such as recycle streams can be avoided by enclosing them inside system borders. Skill at placing system borders comes with experience. You will become adept after you work mass balance exercises at the end of this chapter.

Mass balances require careful accounting. Begin every mass balance with a template equation, such as "rate in = rate out," to guide your inventory of streams. Before attempting to solve a system of equations, verify that you have sufficient information; verify that you have at least as many equations as you have unknowns.

3.2 MASS BALANCES ON PROCESSES WITH CHEMICAL REACTIONS

3.2.1 A Chemical Reactor with a Separator

Context: The isomerization of allyl benzene to propenyl benzene.

Concepts: Mass balance equations for chemical reactions, expressions for reaction rates.

Defining Question: What is the best strategy for calculating flow rates?

Processes with chemical reactions are fundamental to chemical engineering. Mass balances on systems with chemical reactions require additional attributes, which we demonstrate in the context of a simple isomerization reaction, $A \rightarrow P$.

We wish to produce 100. kg/day of 1-phenyl propene (aka propenyl benzene, designated *P*) from reactant 3-phenyl propene (aka allyl benzene, designated *A*). The allyl benzene is dissolved in 1-hexanol (here designated *S* for solvent).

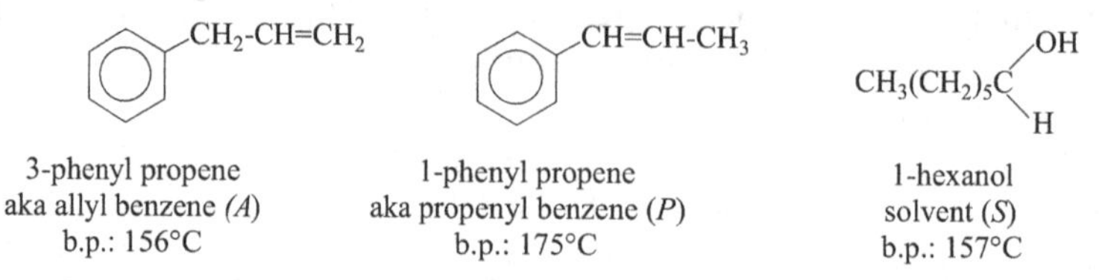

Figure 3.11 Chemical structures for reactant *A*, product *P*, and solvent *S*.

The scheme shown in Figure 3.12 comprises the minimum units: a reactor and a separator. Allyl benzene (*A*) and 1-hexanol (*S*) are provided as a mixture of 20 wt% *A* and 80 wt% *S*. The reactor effluent is equimolar in *A* and *P*. Because *A* and *S* have similar boiling points, separation is expensive. We opt to discard the unreacted allyl benzene and solvent.

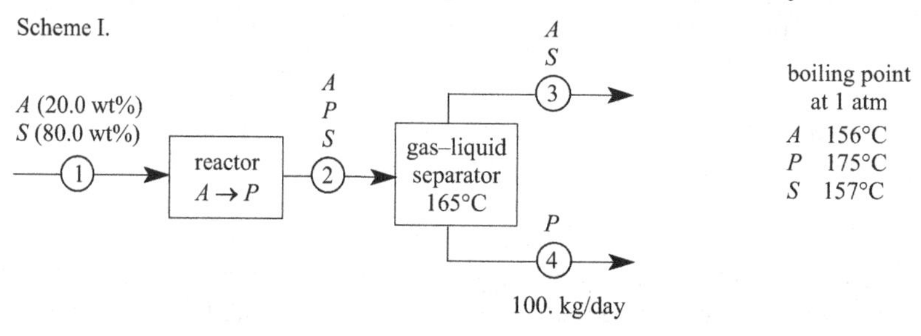

Figure 3.12 The minimal scheme to produce propenyl benzene, *P*.

To assess this design, we need flow rates of the input stream and the waste stream, to calculate reactant and disposal costs. We also need the flow rates through the reactor and separator, to calculate equipment costs.

We begin our mathematical model with a mass balance around the total process. A cursory examination reveals that our previous template equation for a mass balance at steady state, "rate in = rate out," is not valid with a chemical reaction. For example, *P* leaves the process but *P* does not enter the process. Clearly, "rate in ≠ rate out" for *P*, and for *A*.

We will present two forms of a generic mass balance equation for systems with chemical reaction at steady state. Although neither *A* nor *P* is conserved owing to the chemical reaction $A \rightarrow P$, the sum of *A* and *P* is conserved. Consider a closed reactor that initially contains *A*. Conservation of mass assures us the mass in this sealed reactor is constant even as *A* reacts to form *P*. For an open system with the reaction $A \rightarrow P$, the mass balance translates to

$$\text{rate of } A \text{ in} + \text{rate of } P \text{ in} = \text{rate of } A \text{ out} + \text{rate of } P \text{ out}. \qquad (3.47)$$

We begin by drawing system borders around the entire process as shown in Figure 3.13.

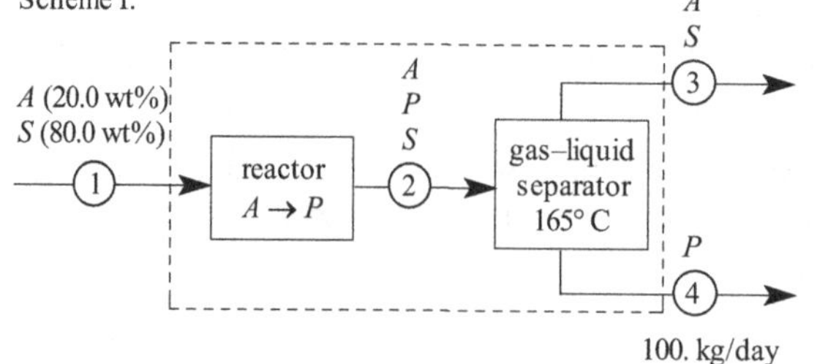

Figure 3.13 The system borders for mass balances.

We must decide which set of variables to use: the total flow rates and two of the three components, or the three components. The analysis is comparable with either set. We choose the three components A, P, and S. Write a mass balance on $A + P$, using equation 3.47 as a template:

$$\text{rate of } A \text{ in} + \text{rate of } P \text{ in} = \text{rate of } A \text{ out} + \text{rate of } P \text{ out}$$

$$F_{A,1} + F_{P,1} = F_{A,3} + F_{A,4} + F_{P,3} + F_{P,4}. \tag{3.48}$$

Write a mass balance on solvent S. Because S does not react, we use the template equation for steady state without reaction:

$$\text{rate in} = \text{rate out}$$

$$F_{S,1} = F_{S,3} + F_{S,4}. \tag{3.49}$$

Translate the process specifications to equations. Translating the composition of stream 1 to an equation that relates $F_{A,1}$ and $F_{S,1}$ is not as obvious as translating to an equation that relates $F_{A,1}$ and $F_{T,1}$. Although $F_{T,1}$ is not a principal variable, use it to derive the equation relating $F_{A,1}$ and $F_{S,1}$, as follows. Start with compositions in terms of the total flow rate, $F_{A,1} = 0.2F_{T,1}$ and $F_{S,1} = 0.8F_{T,1}$, solve each for $F_{T,1}$, and substitute into the identity $F_{T,1} = F_{T,1}$:

$$\frac{F_{S,1}}{0.8} = \frac{F_{A,1}}{0.2} \tag{3.50}$$

$$F_{S,1} = 4F_{A,1}. \tag{3.51}$$

Equation 3.51 is sensible; the flow rate of solvent is four times the flow rate of reactant A. Continue with the other, straightforward process specifications:

$$F_{P,1} = 0 \text{ kg/day} \tag{3.52}$$

$$F_{P,3} = 0 \text{ kg/day} \tag{3.53}$$

$$F_{A,4} = 0 \text{ kg/day} \tag{3.54}$$

$$F_{P,4} = 100 \text{ kg/day} \tag{3.55}$$

$$F_{S,4} = 0. \tag{3.56}$$

Do we have sufficient information? There are nine variables (three streams times three components) and we have eight equations (two mass balances and process specifications 3.51 through 3.56). We need one additional equation. We can use the process specification on the reactor: the reactor effluent is equimolar in A and P.

Chemical reactions are typically described in terms of mols. But mols are not conserved in a chemical reaction. For example, in the synthesis of ammonia, 3 mol of H_2 plus 1 mol of N_2 become 2 mol of NH_3. The molar specification must be converted to a mass basis. In this case, the conversion is trivial. Because A and P are related by the isomerization reaction $A \rightarrow P$, A and P have the same molecular weight. The reactor specification translates to

$$F_{A,2} = F_{P,2}. \tag{3.57}$$

Draw system borders around the separator and write a mass balance on P. The separator is at steady state and has no chemical reactions:

$$\text{rate in} = \text{rate out}$$

$$F_{P,2} = F_{P,3} + F_{P,4}. \tag{3.58}$$

Substitute equations 3.53 and 3.55 into equation 3.58:

$$F_{P,2} = 0 + 100 = 100 \text{ kg/day}. \tag{3.59}$$

Equations 3.57 and 3.59 give the flow rate of A in the reactor effluent:

$$F_{A,2} = F_{P,2} = 100 \text{ kg/day}. \tag{3.60}$$

Use the system borders around the separator and write a mass balance on A:

$$\text{rate in} = \text{rate out}$$

$$F_{A,2} = F_{A,3} + F_{A,4}. \tag{3.61}$$

Solve for $F_{A,3}$ and substitute equations 3.54 and 3.60:

$$F_{A,3} = F_{A,2} - F_{A,4} = 100 - 0 = 100 \text{ kg/day}. \tag{3.62}$$

Equation 3.62 is the ninth equation needed to solve the system of equations derived from the overall mass balance. Successive substitution produces the following results, with three significant figures.

> The reactant + solvent input (stream 1) is 1.00×10^3 kg/day and the waste output (stream 3) is 9.0×10^2 kg/day. The capacity of both the reactor and the separator is 1.00×10^3 kg/day.

For visual assimilators, results reported on the flowsheet are preferred, as shown in Figure 3.14 without deference to the proper number of significant figures.

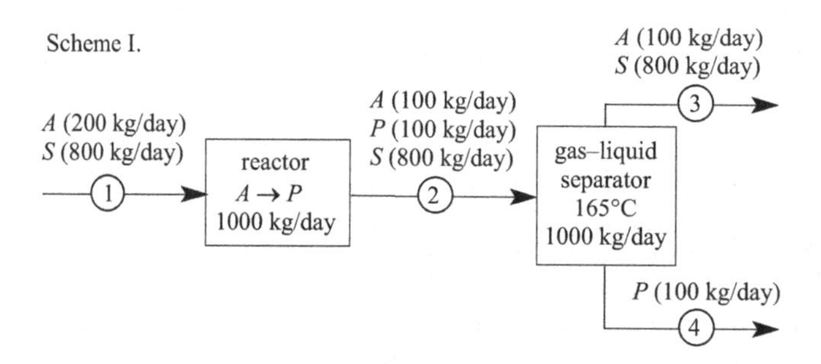

Figure 3.14 Flow rates and equipment capacities to produce 100 kg/day of product with Scheme I.

Although Scheme I is frugal with respect to equipment, it is wasteful with respect to reactant and disposal costs. Only half of reactant A is converted to product P.

The second form of a generic mass balance equation for systems with chemical reaction is in terms of the individual reacting substances. Follow the path of reactant A through the process, as shown in Figure 3.15. A enters the process, a portion of the A is consumed in the reaction "sink," and the remainder of the A leaves the process.

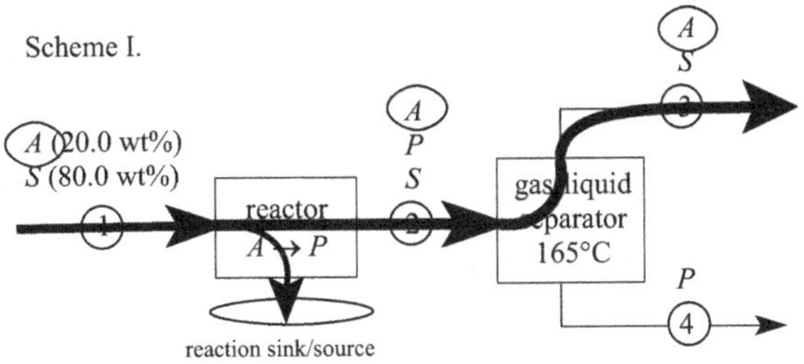

Figure 3.15 The path of A through the Scheme I design.

The path for P (not shown – draw this in a different color) begins in the reaction "source" and leaves the process. For a substance involved in a chemical reaction, the mass balance translates to:

$$\text{rate of } A \text{ in} + \text{rate } A \text{ is created} = \text{rate of } A \text{ out} + \text{rate } A \text{ is consumed.} \tag{3.63}$$

The analogous equation for P is

$$\text{rate of } P \text{ in} + \text{rate } P \text{ is created} = \text{rate of } P \text{ out} + \text{rate } P \text{ is consumed.} \tag{3.64}$$

For the reaction $A \rightarrow P$, "rate P is created" equals "rate A is consumed." Because the reaction is irreversible, both "rate A is created" and "rate P is consumed" are zero. In the general case of a reversible reaction $A \leftrightarrow P$, "rate A is created" = "rate P is consumed." In either case, adding equations 3.63 and 3.64 obtains the other form of a mass balance with chemical reaction,

$$\text{rate of } A \text{ in} + \text{rate of } P \text{ in} = \text{rate of } A \text{ out} + \text{rate of } P \text{ out.} \tag{3.47}$$

Let's analyze Scheme I with these alternative forms of the mass balance equation for chemical reactions. Use the system borders around the entire process (Figure 3.13) and write mass balances for A and P:

$$\text{rate of } A \text{ in} + \text{rate } A \text{ is created} = \text{rate of } A \text{ out} + \text{rate } A \text{ is consumed} \tag{3.63}$$

$$F_{A,1} + 0 = F_{A,3} + F_{A,4} + \text{rate } A \text{ is consumed} \tag{3.65}$$

$$\text{rate of } P \text{ in} + \text{rate } P \text{ is created} = \text{rate of } P \text{ out} + \text{rate } P \text{ is consumed} \tag{3.64}$$

$$F_{P,1} + \text{rate } P \text{ is created} = F_{P,3} + F_{P,4} + 0. \tag{3.66}$$

The mass balance written for S, equation 3.49, is still valid. We need an expression for "rate A is consumed." For every 2 mol of A that enters the reactor, 1 mol is converted to P and 1 mol passes through the reactor. Thus,

$$\text{rate } A \text{ is consumed} = \text{rate } P \text{ is created} = \frac{1}{2} F_{A,1}. \tag{3.67}$$

Substitute equation 3.67 into equation 3.65:

$$F_{A,1} + 0 = F_{A,3} + F_{A,4} + \frac{1}{2} F_{A,1} \tag{3.68}$$

$$\frac{1}{2} F_{A,1} = F_{A,3} + F_{A,4}. \tag{3.69}$$

Similarly, equations 3.67 and 3.66 yield

$$F_{P,1} + \frac{1}{2}F_{A,1} = F_{P,3} + F_{P,4}. \tag{3.70}$$

The three mass balances – equations 3.49, 3.69, and 3.70 – with the six process specifications in equations 3.51 through 3.56 provide nine equations. Successive substitution yields the same results. We did not require the equation $F_{A,2} = F_{P,2}$ derived from the reactor specification because we translated the reactor specification to derive an equation for "rate A is consumed."

Instead of deriving equation 3.67 for "rate A is consumed," we could have used equation 3.66 to calculate "rate P is created." Substitute the known flow rates into equation 3.66:

$$0 + \text{rate } P \text{ is created} = 0 + 100 + 0 \tag{3.71}$$

$$\text{rate } P \text{ is created} = 100 \text{ kg/day}. \tag{3.72}$$

Instead of equation 3.67, we have

$$\text{rate } A \text{ is consumed} = \text{rate } P \text{ is created} = 100 \text{ kg/day}. \tag{3.73}$$

In this section we used two translations of the conservation of mass to analyze a system at steady state with chemical reaction:

$$\text{total rates of reactants} + \text{products in} = \text{total rates of reactants} + \text{products out} \tag{3.74}$$

$$\begin{aligned}&\text{rate of substance 1 in} + \text{rate substance 1 is created}\\&= \text{rate of substance 1 out} + \text{rate substance 1 is consumed}.\end{aligned} \tag{3.75}$$

The choice is largely a matter of style. The second equation is useful when expressions for "rate created" or "rate consumed" are given or easily devised.

We showed two strategies for modeling the mass flow rates. These strategies were arbitrary. Other strategies are equally valid and comparably efficient. For example, one could begin with system borders around the separator. There is no "best" strategy because the advantages of any strategy are subjective. Some engineers prefer a unit-by-unit analysis and some engineers prefer to start with the overall process. It is important that you develop your style as you work the exercises at the end of this chapter.

3.2.2 A Recycle for Minimal Reactant Input and Minimal Waste Output

Context: Isomerization of allyl benzene to propenyl benzene with recycle.

Concepts: Separating and recycling unreacted reactants.

Defining Question: What are the benefits of adding a second separator?

We improve the design by separating A from the solvent and recycling A to the reactor. Because the boiling points for A and S differ by only 1°C, the separation is difficult. The second separator will be more expensive than the first separator, perhaps ten times as expensive. But the second separator also eliminates the waste stream and provides pure solvent as a by-product stream; Scheme II eliminates a disposal cost and adds a by-product revenue. The flow rate of product P is again 100 kg/day, to compare to Scheme I.

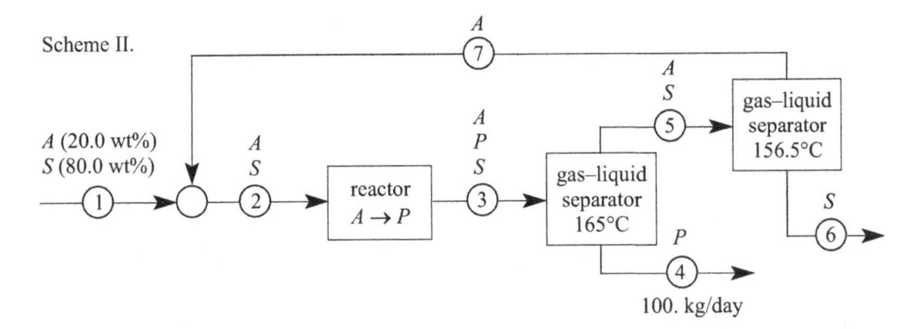

Figure 3.16 An improved scheme to produce propenyl benzene, P, by separating and recycling unreacted A.

Again, we need flow rates of the input stream and the by-product stream, and the flow rates through the reactor and the two separators.

We begin our mathematical model with a mass balance around the total process, as shown by the system borders in Figure 3.17.

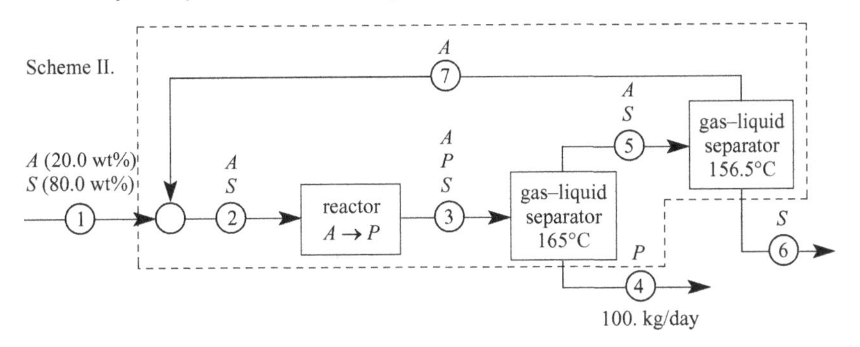

Figure 3.17 System borders for mass balances on Scheme II.

Write a mass balance on the sum of A and P:

$$\text{rate of } A \text{ in} + \text{rate of } P \text{ in} = \text{rate of } A \text{ out} + \text{rate of } P \text{ out} \tag{3.47}$$

$$F_{A,1} + F_{P,1} = F_{A,4} + F_{A,6} + F_{P,4} + F_{P,6}. \tag{3.76}$$

We will ignore S for now. Translate the process specifications to equations:

$$F_{P,1} = 0 \text{ kg/day} \tag{3.77}$$

$$F_{A,4} = 0 \text{ kg/day} \tag{3.78}$$

$$F_{P,4} = 100 \text{ kg/day} \tag{3.79}$$

$$F_{A,6} = 0 \text{kg/day} \tag{3.80}$$

$$F_{P,6} = 0 \text{kg/day}. \tag{3.81}$$

In this abridged analysis (no S, yet), we have six unknowns (three streams times two substances) and six equations. Substitute equations 3.77 through 3.81 into equation 3.76:

$$F_{A,1} + 0 = 0 + 100 + 0 + 0 \tag{3.82}$$

$$F_{A,1} = 100 \text{ kg/day}. \tag{3.83}$$

From the analysis of Scheme I, the composition of stream 1 translates to equation 3.51: $F_{S,1} = 4F_{A,1} = 4 \times 100 = 400$ kg/day. A mass balance on S around the entire process yields $F_{S,6} = F_{S,1} = 400$ kg/day.

To calculate the flow rates through the equipment, we need the flow rate of the recycle, stream 7. Let's perform an informal, but quantitative, mathematical analysis. A mass balance on P around the first separator reveals $F_{P,3} = F_{P,4} = 100$ kg/day. Because the reactor effluent contains equal flow rates of A and P, $F_{A,3} = F_{P,3} = 100$ kg/day. A mass balance on A around the first separator gives $F_{A,5} = F_{A,3} = 100$ kg/day. A mass balance on A around the second separator gives $F_{A,7} = F_{A,5} = 100$ kg/day. The flow rates are summarized in Figure 3.18, where the numbers in parenthesis are flow rates in kg/day.

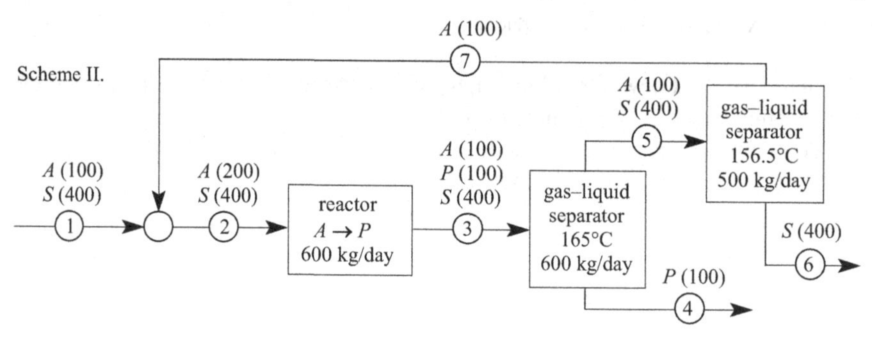

Figure 3.18 Flow rates and equipment capacities to produce 100 kg/day of product with Scheme II.

Adding a second separator reduces the reactant input by a factor of 2, eliminates the waste stream of 900 kg/day, and adds a second product: solvent at 400 kg/day. And the reactor and the first separator are 40% smaller.

3.2.3 A Purge for Moderate Reactant Input and Moderate Waste Output

Context: Isomerization of allyl benzene to propenyl benzene with a purge.
Concepts: Splitting a waste stream to recycle unreacted reactants, equations to optimize design.
Defining Question: What are the compromises of using a splitter to recycle?

Consider a design intermediate to Schemes I and II: eschew the expensive second separator, but recycle the unreacted A by splitting the waste stream, as shown in Figure 3.19. In Scheme I, the waste stream is 11.1 wt% A. In Scheme II, the solvent stream is 0 wt% A. For Scheme III, we have arbitrarily set the composition of the waste stream to be 4 wt% A.

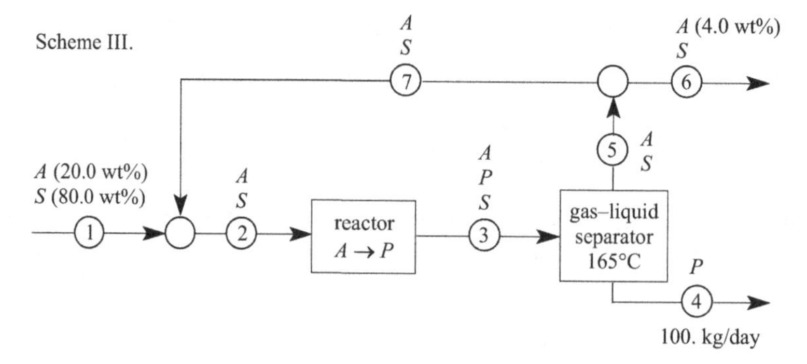

Figure 3.19 An intermediate scheme to produce propenyl benzene, P, by splitting and recycling unreacted A.

For this third scheme, we again need flow rates of the input stream and the waste stream, and the flow rates through the reactor and separator.

We begin our mathematical model with a mass balance around the total process, as shown by the system borders in Figure 3.20.

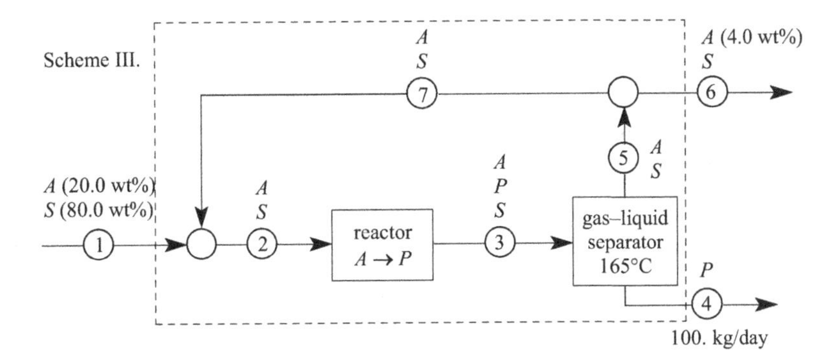

Figure 3.20 System borders for mass balances on Scheme III.

Write a mass balance on the sum of A and P:

$$\text{rate of } A \text{ in} + \text{rate of } P \text{ in} = \text{rate of } A \text{ out} + \text{rate of } P \text{ out} \tag{3.47}$$

$$F_{A,1} + F_{P,1} = F_{A,4} + F_{A,6} + F_{P,4} + F_{P,6}. \tag{3.84}$$

Write a mass balance on S:

$$\text{rate in} = \text{rate out}$$

$$F_{S,1} = F_{S,4} + F_{S,6}. \tag{3.85}$$

Translate the process specifications to equations:

$$F_{P,1} = 0 \text{ kg/day} \tag{3.86}$$

$$F_{A,4} = 0 \text{ kg/day} \tag{3.87}$$

$$F_{P,4} = 100 \text{ kg/day} \tag{3.88}$$

$$F_{S,4} = 0 \text{kg/day} \tag{3.89}$$

$$F_{P,6} = 0 \text{kg/day}. \tag{3.90}$$

As with Schemes I and II, the composition of stream 1 translates to equation 3.51:

$$F_{S,1} = 4F_{A,1}. \tag{3.51}$$

Similarly, the composition of stream 6 translates to the following equation:

$$F_{S,6} = 24F_{A,6}. \tag{3.91}$$

We have sufficient information; we have nine unknowns (three streams times three components) and we have nine equations (two mass balances and seven process specifications). Substitute the process specifications into the mass balance on $A + P$:

$$F_{A,1} + 0 = 0 + 100 + F_{A,6} + 0 \tag{3.92}$$

$$F_{A,1} = 100 + F_{A,6}. \tag{3.93}$$

Substitute the process specifications into the mass balance on S:

$$4F_{A,1} = 0 + 24F_{A,6} \tag{3.94}$$

$$F_{A,1} = 6F_{A,6}. \tag{3.95}$$

Substitute equation 3.95 into equation 3.93 and solve for $F_{A,6}$:

$$6F_{A,6} = 100 + F_{A,6} \tag{3.96}$$

$$F_{A,6} = 20 \text{ kg/day}. \tag{3.97}$$

Use the flow rate for $F_{A,6}$ to calculate the other flow rates:

$$F_{A,1} = 120 \text{ kg/day} \tag{3.98}$$

$$F_{S,1} = F_{S,6} = 480 \text{ kg/day}. \tag{3.99}$$

To calculate the equipment capacities, we need flow rates of the internal streams. The method is the same as the methods for Schemes I and II. A mass balance on P around the separator yields $F_{P,3}$ = $F_{P,4}$ = 100 kg/day. The reactor effluent specification – A and P are equimolar in stream 3 – yields $F_{A,3}$ = $F_{P,3}$ = 100 kg/day. A mass balance on A around the separator yields $F_{A,5}$ = $F_{A,3}$ = 100 kg/day. Here we use a subtle process specification: the stream into and all streams out of a splitter have the same composition. So the equation we devised for stream 6 – equation 3.91 – also applies to streams 5 and 7:

$$F_{S,5} = 24F_{A,5} \tag{3.100}$$

$$F_{S,7} = 24F_{A,7}. \tag{3.101}$$

Equation 3.100 yields $F_{S,5}$ = 2400 kg/day. A mass balance around the splitter yields $F_{A,7}$ = 80 kg/day and $F_{S,7}$ = 1920 kg/day. Finally, a mass balance around the combiner gives $F_{A,2}$ = 200 kg/day and $F_{S,2}$ = 2400 kg/day. The flow rates are summarized in Figure 3.21, where the numbers in parenthesis are flow rates in kg/day.

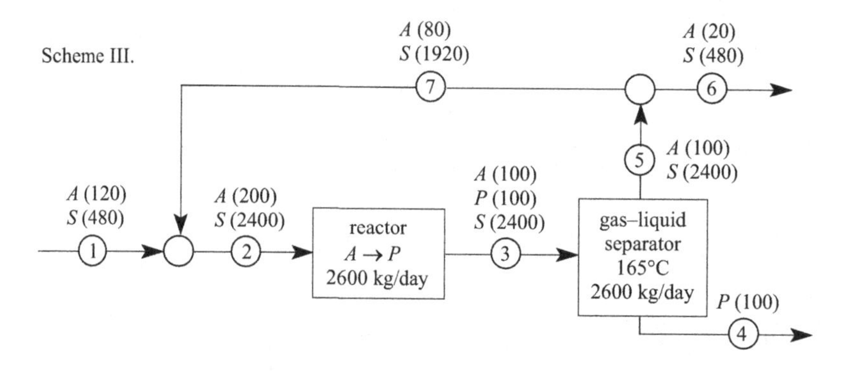

Figure 3.21 Flow rates and equipment capacities to produce 100 kg/day of product with Scheme III.

What are the compromises of using a splitter to recycle? Relative to Scheme I, Scheme III reduces the input and waste flow rates by 40%, but increases the equipment sizes by a factor of 2.6. Relative to Scheme II, Scheme III eliminates the expensive second separator, but increases the sizes of the reactor and first separator by a factor of 4.3. Which is the better scheme? We lack information (reactant costs, disposal costs, product and solvent revenue, and equipment costs) and we lack a formalism to analyze the information to identify the better scheme. We will introduce the formalism – process economics – in Section 3.7.

Suppose the economic analysis reveals that Scheme III is the better design. How do we know that Scheme III is the *best* design? That is, how do we know that 4 wt% A in the waste stream is optimal? We could repeat the analysis of Scheme III with various other concentrations of A in the waste stream. A better approach is to analyze Scheme III with an indefinite wt% A in the waste stream; define α (Greek for lowercase A) to be the weight fraction A in the waste stream. Analogous to devising equation 3.51, we begin with the identity $F_{T,6} = F_{T,6}$ and substitute $F_{A,6} = \alpha F_{T,6}$ and $F_{S,6} = (1-\alpha)F_{T,6}$:

$$\frac{F_{A,6}}{\alpha} = \frac{F_{S,6}}{1-\alpha} \tag{3.102}$$

$$F_{S,6} = \frac{1-\alpha}{\alpha} F_{A,6}. \tag{3.103}$$

We substitute equation 3.89 into equation 3.85 to obtain the obvious relation between $F_{S,1}$ and $F_{S,6}$:

$$F_{S,1} = F_{S,6}. \tag{3.104}$$

The new equations 3.103 and 3.104 with the two previous equations 3.51 and 3.93 allow us to derive equations that describe streams 1 and 6:

$$F_{A,1} = \frac{1-\alpha}{1-5\alpha} 100 \text{ kg/day} \tag{3.105}$$

$$F_{A,6} = \frac{4\alpha}{1-5\alpha} 100 \text{ kg/day} \tag{3.106}$$

$$F_{S,1} = F_{S,6} = \frac{1-\alpha}{1-5\alpha} 400 \text{ kg/day}. \tag{3.107}$$

We use these flow rates with the previous method for calculating the flow rates of internal streams to obtain the equipment capacities; the reactor capacity is the same as the separator capacity: $100 \times (1+\alpha)/\alpha$ kg/day.

Check equations 3.105 to 3.107; substitute $\alpha = 0.04$ into each equation. The flow rates are equal to the Scheme III flow rates in Figure 3.21. Substitute $\alpha = 0.111$ into each equation. The flow rates are equal to the Scheme I flow rates in Figure 3.14. The equations check.

Equations 3.105 to 3.107 are valid for $0 < \alpha < 0.2$. For $\alpha > 0.111$, the recycle flow rate (stream 7) is negative, which means the flow direction is opposite of the arrowhead; some reactant input (stream 1) bypasses the reactor and goes directly to waste.

Equations 3.105 to 3.107 are operating equations for Scheme III; given a numerical value for α, equations 3.105 to 3.107 will indicate how Scheme III operates. We can use the operating equations as *design tools* in an economic analysis to optimize Scheme III. Creating design tools is the essence of chemical engineering design.

The chemical industry consists of thousands of processes, each composed of many units. It would be impractical for you to study the operating equations for every process or every unit you might encounter during your career. It would also be unwise – how would you assimilate a new unit into your repertoire, a unit for which you have not memorized an operating equation? You don't need to memorize operating equations. Instead, you need only learn how to *create* tools to design and operate chemical engineering processes.

In the next chapter we will use graphical methods rather than mathematical expressions to create design tools. And we will use geometry, rather than arithmetic and algebra, to predict performance.

3.3 INFORMAL MASS BALANCES FOR DESIGN EVOLUTION

Context: Recycle streams for reactors and separators.

Concepts: Increase the concentration of a desired substance by recycle.

Defining Question: How does one quickly estimate flow rates and compositions?

Deriving a formal mathematical model for mass flow rates is an essential skill in characterizing a chemical process. But devising a formal mathematical model at each stage of an evolving design can be tiresome. Indeed, the meticulous accounting required for a formal mass balance is inconsistent with the creative "what if?" process. It is useful to be able to estimate flow rates for intermediate designs. In this section we present two methods. The first method removes the encumbrances of formal mass balances – defining variables, stating assumptions, drawing system borders, and counting unknowns and equations. We demonstrate informal mathematical modeling with two examples.

We have a product P mixed with a by-product Q. P and Q have identical melting points, identical solubilities in common solvents, and similar boiling points: P boils at 50°C and Q boils at 51°C. Our goal is pure P, which can be obtained by a liquid–gas separator at 50°C. Unfortunately, because the separator operates at the boiling point for P to keep Q from the product stream, half the P is lost in the liquid stream.

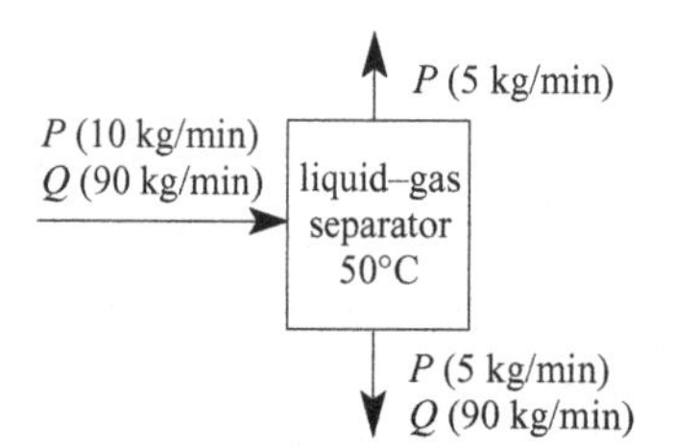

Figure 3.22 An imperfect liquid–gas separator to produce pure P from a P–Q mixture.

Let's explore the possibility of recovering some of the lost P by splitting the liquid P + Q stream and recycling to the separator. Is this a viable idea? How much P can we recover? How does the separator capacity depend on the amount of P recovered? Begin the analysis by sketching a flowsheet. We will omit stream numbers for our informal analysis. Assume we recover 9 kg/min of P. Our first informal mass balance around the entire process gives the flow rate of the waste stream, as shown in Figure 3.23.

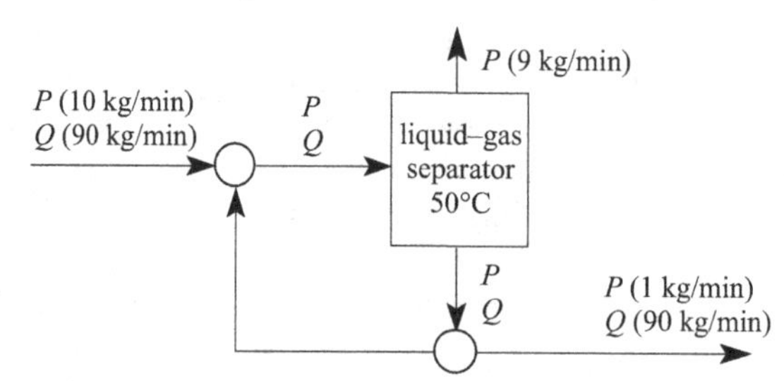

Figure 3.23 An imperfect liquid–gas separator with a recycle stream.

Because the separator loses half the P in the liquid output stream, we know the flow rate of P in the liquid output stream is 9 kg/min. Because the waste stream is split from the liquid $P + Q$ stream, these two streams have the same composition; the ratio of P:Q is 1:90. The flow rate of Q in the liquid stream is $9 \times 90 = 810$ kg/min. A mass balance around the splitter yields the flow rates in the recycle stream. A mass balance around the combiner yields the flow rates in the separator feed stream, as shown in Figure 3.24.

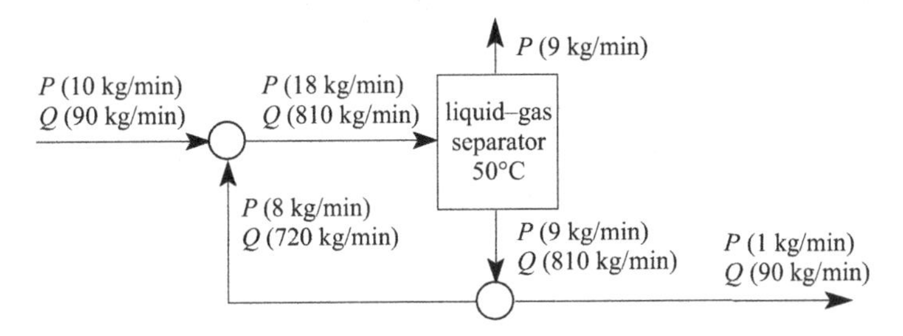

Figure 3.24 Flow rates for an imperfect liquid–gas separator with a recycle.

The recycle to recover more P increases the separator capacity by a factor of 8. What if we recover 9.9 kg/min of P? You should be able to calculate that the separator capacity must be 8839.8 kg/min (ignoring the number of significant figures).

We have another option for purifying P from the mixture of P and Q. Q can be converted to M, and M (boiling point 174°C) is easily separated from P. Although the conversion of $Q \rightarrow M$ is not complete (Q and M reach a 50:50 equilibrium), a reactor could increase the P:Q ratio before we feed to a liquid–gas separator, as shown in Figure 3.25. The $P + Q$ output has almost twice the concentration of P as the input. Can we recycle the $P + Q$ stream to further increase the P:Q ratio? What ratio can we achieve? How large will the reactor and separator be?

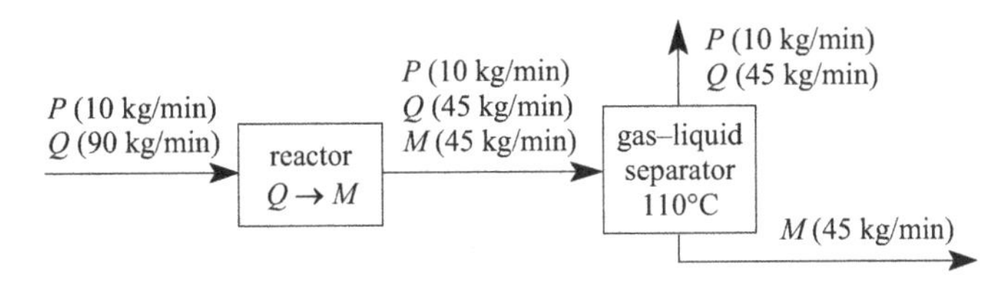

Figure 3.25 A reactor to selectively convert Q in a $P + Q$ mixture.

Again we begin by drawing a process flowsheet. Assume the P:Q ratio in our output is 10:1. A mass balance around the entire process reveals P is 10 kg/min in the output. The Q output is therefore 1 kg/min. The 90 kg/min of Q that enters must leave as Q or M; the M output is $90 - 1 = 89$ kg/min, as shown in Figure 3.26.

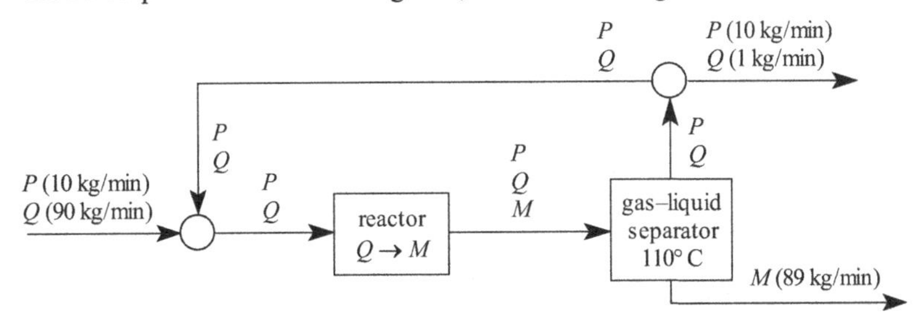

Figure 3.26 A reactor with recycle to selectively convert Q in a $P + Q$ mixture.

Our informal analysis begins with the M output stream. A mass balance around the separator gives the M flow rate in the reactor effluent: 89 kg/min. Because Q and M are the same molecular weight, and the reactor effluent is equimolar Q and M, the Q flow rate in the reactor effluent is also 89 kg/min. The Q flow rate leaving the separator is also 89 kg/min. The P:Q ratio in all splitter streams is 10:1, so the P flow rate leaving the separator is $89 \times 10 = 890$ kg/min. A mass balance around the splitter provides the flow rates of the recycle stream. A mass balance around the combiner provides the flow rates of the reactor feed stream. The flow rates are summarized in Figure 3.27.

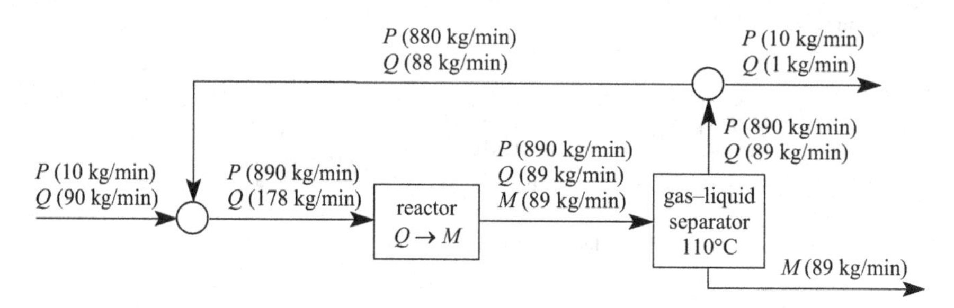

Figure 3.27 A reactor with recycle to selectively convert Q in a $P + Q$ mixture.

With no recycle, we increase the P:Q ratio by a factor of 2. With a large recycle (98.9%) we increase the P:Q ratio by a factor of 90, but we also increase the reactor and separator capacities by a factor of 10. But this feed to the process in Figure 3.24 could recover 9.9 kg/min of P with a modest separator: capacity = 119.9 kg/min. (You should verify you can quickly corroborate this flow rate.) What is the best balance of reactor and separator capacities to produce 9.9 kg/min of P? Mass balances are insufficient. We will again postpone such questions until we consider mathematical models based on process economics.

A second method for estimating flow rates in recycle systems is iteration to steady state. This method may be faster when the process specifications are not easily translated into equations. Consider another difficult separation of two isomers, X and Z. We wish to convert X to Z. A reactor quickly brings X and Z into equilibrium such that the X:Z ratio is 1:3. The liquid–gas separator produces pure Z, but 1/5 of the Z is lost in the liquid stream. This process is shown in Figure 3.28 with a nominal basis of 100 kg/min of X.

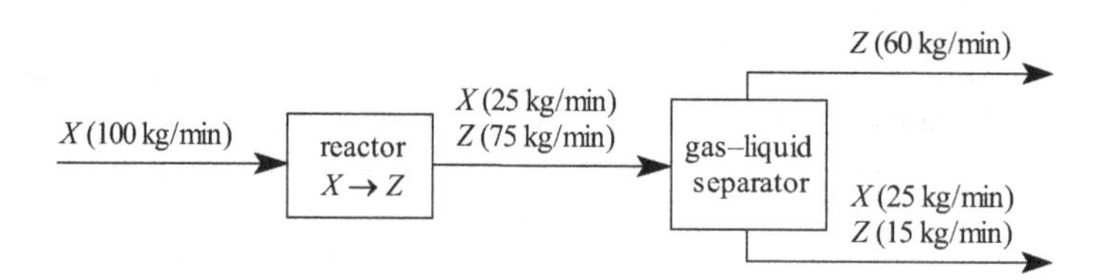

Figure 3.28 A process to isomerize X into Z and purify Z.

Can we convert all the X to Z by recycling the liquid stream? What equipment capacities will this require? Draw a recycle stream and calculate flow rates when the process starts. Initially, there is no flow in the recycle stream. The flow into the reactor is the process input only: 100 kg/min of X.

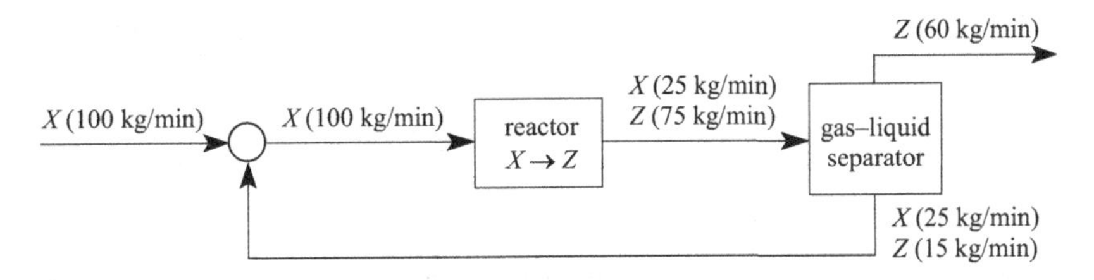

Figure 3.29 The first iteration of flow rates for a process to isomerize X into Z and purify Z.

The flow rates in Figure 3.29 are not at steady state, as revealed by a mass balance around the combiner. For the second iteration, add the recycle to the X input. Then calculate the reactor input, the reactor effluent, and the separator outputs, as shown in Figure 3.30.

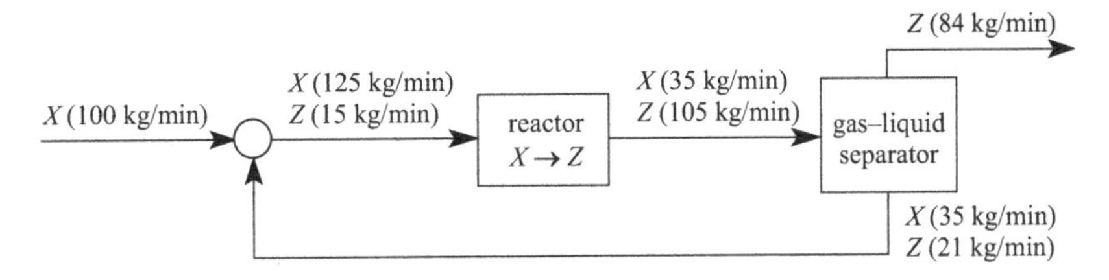

Figure 3.30 The second iteration of flow rates for a process to isomerize X into Z and purify Z.

Subsequent iterations are presented in Table 3.2. If one requires only estimated flow rates, four or five iterations are sufficient for this process. The rate of convergence depends on the process specifications. For example, if the X:Z ratio after the reactor is 1:1 and half of the Z is recycled from the separator, the product stream is only 76.3 kg/min after five iterations, and 94.4 kg/min after ten iterations. For low reactor conversions and poor separations, it is advisable to use a spreadsheet to calculate iterations.

Table 3.2 Iterated flow rates for a process to isomerize X into Z and purify Z.

	Feed	Reactor input		Reactor output		Product	Recycle	
Iteration	X	X	Z	X	Z	Z	X	Z
1	100	100	0	25	75	60	25	15
2	100	125	15	35	105	84	35	21
3	100	135	21	39	117	93.6	39	23.4
4	100	139	23.4	40.6	121.8	97.44	40.6	24.36
5	100	140.6	24.36	41.24	123.72	98.98	41.24	24.74
10	100	141.66	24.99	41.66	124.99	99.99	41.66	24.99
20	100	141.67	25	41.67	125	100	41.67	25

Designs evolve by exploring options. It is useful to be able to estimate flow rates to judge the viability and effectiveness of a process modification. An informal mass balance can provide flow rates, or reveal what additional process parameters must be specified. An iterative calculation for recycle and bypass streams can be fast, and requires only arithmetic.

3.4 MATHEMATICAL MODELING WITH MASS BALANCES: SUMMARY

Context: A procedure for mass balances.

Concepts: Tips for mass balances and mathematical modeling. Mass balances and learning styles.

Defining Question: Are mass balances really this simple?

Mathematical modeling with mass balances requires careful accounting. The following procedure summarizes the formal mass balance examples.

1. **Draw a process flowsheet and number all streams.** Write chemical components and any flow rates and compositions on each stream.
2. **Define a nomenclature** for substance names and process variables, such as a substance concentration or a splitter ratio.
3. **Identify what is to be calculated.**
4. **Decide which set of substance variables to use.** For a process with n substances, characterize each stream by (1) the total flow rate and the flow rates of $n-1$ substances, or (2) the flow rates of n substances.
5. **Define a system.** Draw borders that completely enclose one or more process units. Initially draw borders to completely enclose internal streams (especially recycle streams); initially calculate flows into and out of the overall process.
6. **Translate the process description into equations.** Write a mass balance for any substance by stating the circumstances, such as "Apply the conservation of mass, assume steady state and no chemical reactions." Use template equations such as "rate in = rate out" to guide your accounting. Translate process specifications, such as "salt is 12 wt% of stream 2," to equations.
7. **Verify the set of equations can be solved.** The number of unknowns is the number of streams that cross the system borders times the number of substances. To solve, the number of independent equations must be at least the number of unknowns. When collecting equations, watch for implied specifications, such as with splitters: the stream into and all streams out of a splitter are the same composition.
8. **Solve the system of equations.** Use successive substitution; substitute process specifications into mass balances. Linear algebra and matrix methods are useful for large systems with several substances and several streams. Spreadsheets are useful, especially for systems with recycle streams. After you gain experience with mass balance exercises, complete the spreadsheet tutorial following Exercise 3.56.
9. **Check your answers.** Substitute your answers into your initial equations to check your arithmetic and algebra. Write your answers on the process flowsheet to verify the flow rates are reasonable and mutually consistent.

For mathematical models, the method is as important as the answer. Your method must be as valid as your answer. The end does not justify the means.

3.4.1 Some Tips on System Borders

Use system borders and informal mass balances to assess design changes. For example, how could we recycle the waste liquid of heptane and propanol in the process developed in Chapter 2? Can we recycle the stream as shown in Figure 3.31? The system borders shown below have one input stream and one output stream. At steady state, the composition and flow rate of the input must equal the composition and flow rate of the output. The recycle is not viable because there is no output for propanol in the washer–melter portion.

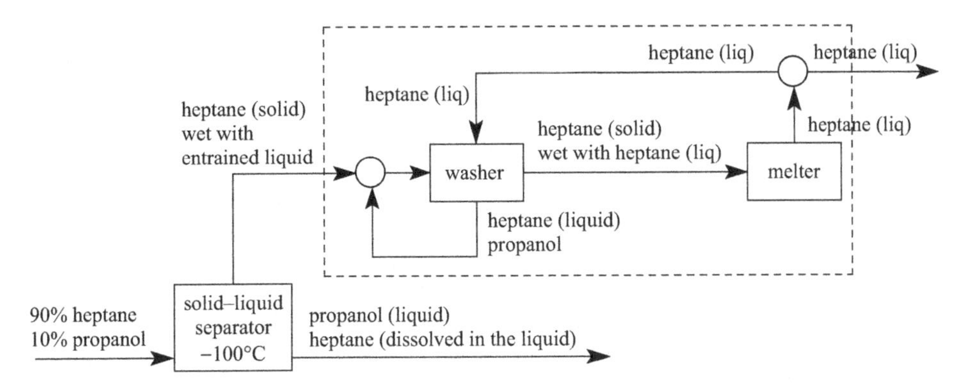

Figure 3.31 An unworkable recycle for the heptane + propanol liquid.

Instead, recycle to a combiner before the solid–liquid separator, as shown in Figure 3.32. System borders around the entire process show there are outputs for both heptane and propanol. System borders around the washer–melter (not shown) show there are outputs for both heptane and propanol. The recycle is viable.

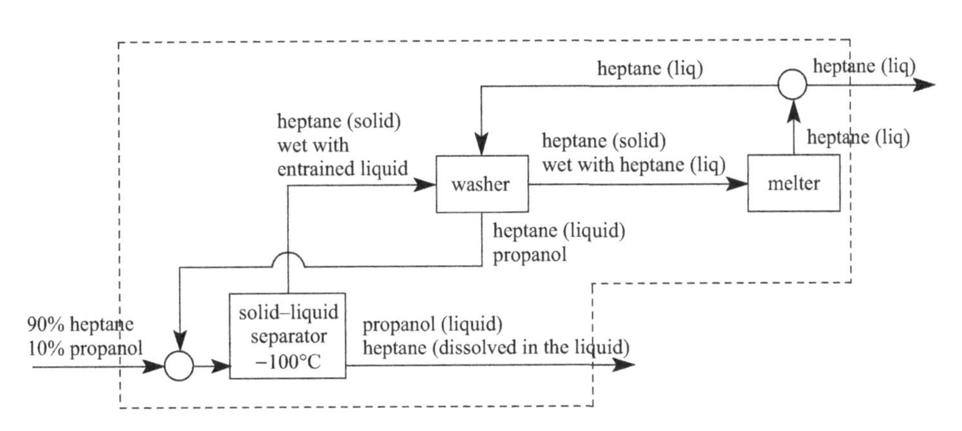

Figure 3.32 A viable recycle of the heptane + propanol liquid.

Although creativity in defining system borders is important, some creativity is not allowed. Here are two common errors.

Because only streams that cross the system borders are included in a mass balance, a system border around the entire process below allows one to ignore streams 2, 3, 5, and 7 in Figure 3.33. However, elongating the borders so a stream does not cross the border is not valid. The system borders in Figure 3.33 are poorly drawn. Although stream 6 does not cross the system borders, it should. The system borders below are invalid; the mass balance is not $F_{T,1} = F_{T,4}$.

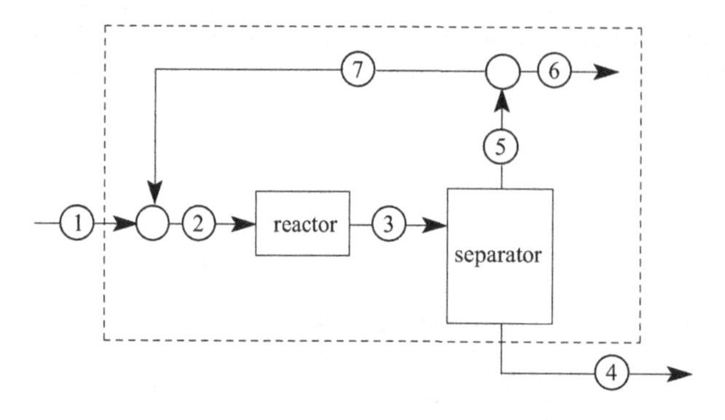

Figure 3.33 Invalid system borders to improperly exclude stream 6 from the mass balance.

Similarly, it is not valid to draw system borders through a process unit to avoid crossing a stream, such as shown in Figure 3.34. The system borders below are invalid; the mass balance is not $F_{T,1} = F_{T,6}$.

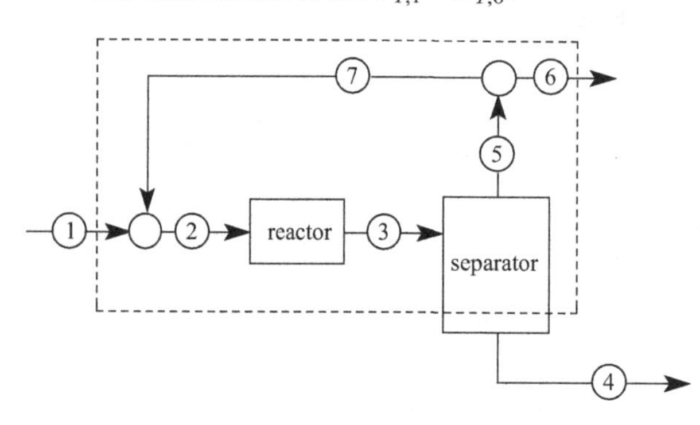

Figure 3.34 Invalid system borders to improperly exclude stream 4 from the mass balance.

3.4.2 Mass Balances and Learning Styles

As you gain experience with mathematical modeling and mass balances, you will develop your style. If you are a visual assimilator and an active processor, study the process flowsheet and plan your strategy by drawing a path for each substance. Each path should be a different color. Draw a reaction sink/source below each reactor to explicitly show the presence of a chemical reaction. Begin your model with a "tie substance" that has a simple path, such as a substance that enters by one stream only, exits in one stream only, and does not react.

Mathematical models accentuate a third dimension of learning: inductive vs. deductive organization. Deductive relations are general relations, such as the relation between the input flow rates of A and solvent in Schemes I to III: $F_{S,1} = 9F_{A,1}$. But such relations are difficult to devise for inductive organizers. The inductive approach considers definite examples to devise a general relation. For example, if the total input is 10 kg/min, what are the flow rates of A and solvent? What is the ratio of A to solvent? If the total input is 100 kg/min, what are the flow rates of A and solvent? What is the ratio of A to solvent? Another example: what is the (deductive) relation between the flow rate of A in and the flow rate of A out? Use finite quantities to probe the relationship. If 200 kg/min of A enters, 100 kg/min is converted to P, so 100 kg/min of A leaves. If 150 kg/min of A enters, 100 kg/min is converted to P, so 50 kg/min of A leaves. After a few examples (and a perhaps a plot for the visually inclined), a general relation becomes obvious: $F_{A,6} = F_{A,1} - 100$.

Inductive organizers remember deductive statements better when they translate to a specific example. Consider the deductive statement: "A solid soluble in a liquid cannot be separated by decreasing the temperature below the melting point of the solid." Devise a specific example of the statement: "Because NaCl is soluble in water, one cannot use a solid–liquid separator to separate NaCl from seawater at 20°C, even though the melting point of NaCl is far above 20°C."

Consider the following syllogism stated deductively:

No A are B.
All B are C.
Therefore, some C are not A.

Simple logic, but difficult to remember and apply for inductive organizers. Translate to a specific example.

No flying birds are penguins.
All penguins are birds.
Therefore, some birds are not flying birds.

Inductive organizers learn general statements by devising specific applications. Inductive organizers remember specific cases from which they can derive the general expression.

For inductive organizers, Schemes I to III are more difficult than the desalinator examples because Schemes I to III pose the problem deductively. The desalinator examples used "Given 100 kg/min of input, calculate the output." Scheme I to III used "What input is necessary to provide 100 kg/min of output?" In cases like Schemes I to III, convert the problem to an inductive basis. Work the problem with an arbitrary input, say 100 kg/min, and calculate the output. Because mass flow rates in process flowsheets are linear systems, the answers scale linearly – 100 kg/min of input in Scheme I produces 5 kg/min of product *P*; multiply all flow rates by 20 to produce 100 kg/min of product *P*.

Similarly, it was easier to analyze Scheme III with a definite concentration of *A* in the waste stream (1 wt% *A*), instead of an indefinite concentration (α weight fraction *A*). For inductive organizers, it is easier to first analyze a process in terms of a definite concentration to derive deductive equations like 3.105 to 3.107.

For estimating flow rates, inductive organizers will prefer the definitive numbers of iterating to steady state, instead of an informal mass balance.

With practice applying the procedure and tips above, you will find mass balances are straightforward. Of course, more complicated processes will lead to more complicated mathematical models, but the procedure really is as simple as described here.

3.5 ENERGY BALANCES ON A SINGLE UNIT WITH NO CHEMICAL REACTION

3.5.1 Energy Balances for Temperature and Phase Changes

Context: Analysis of the energy flow in the freezer–desalinator.

Concepts: Calculating energy from measurables. Calculating stream energies as differences. Decomposing a process unit into elementary energy changes. Heat of fusion.

Defining Question: How are energy balances similar to mass balances? How are energy balances different?

The chief reason we explored desalination by freezing was its potential to use less energy than evaporation. Because evaporating water requires seven times as much energy per kg as does melting ice, this seems qualitatively correct.[6] To confirm the qualitative analysis we need a quantitative analysis of the freezer–desalinator's energy requirements.

Analysis of energy flow is inherently more complicated than analysis of mass flow. Mass is a tangible quantity. Energy is elusive. In fact, energy defies definition. One can quantify mass, for example, by the force an object exerts in a gravitational field or the acceleration of an object owing to an imposed force. Energy is more difficult to quantify because energy can appear in many incarnations. To measure an object's internal energy, we must measure its thermal energy (what is its temperature?), identify its phase (is the material solid, liquid, or gas?), and determine its chemical energy (e.g. is it the object coal, or sand, or whatever?). To measure an object's potential energy, we must measure its height in a gravitational field and (if the object has electric charge) its position in an electromagnetic field. And so on for kinetic energy (its velocity), pressure energy (its pressure), etc.

Why are there so many different forms of energy? The many forms of energy evolved because of faith in the concept of the conservation of energy. When Leibnitz first proposed the principle in 1693 he included only two forms of energy: potential and kinetic. This was fine for analyzing a pendulum. But conservation of energy failed, for example, if one used a falling weight to drive a propeller that stirred and heated a fluid. Rather than discard the conservation of energy, the principle was repaired by adding an additional term to account for energy in the form of heat. Each time the conservation of energy appeared to fail, the definition of energy was expanded to include new terms and restore the principle. A more recent modification was to include the energy inherent in mass, $E = mc^2$.

The schematic of the freezer–desalinator in Figure 3.7 considers only the mass flow. We must consider the energy flow as well. There are no chemical changes. We assume energy changes associated with changes in height and pressure are negligible. For the present, we assume the chief form of energy is heat. To calculate heat, we label temperatures of the streams and units.

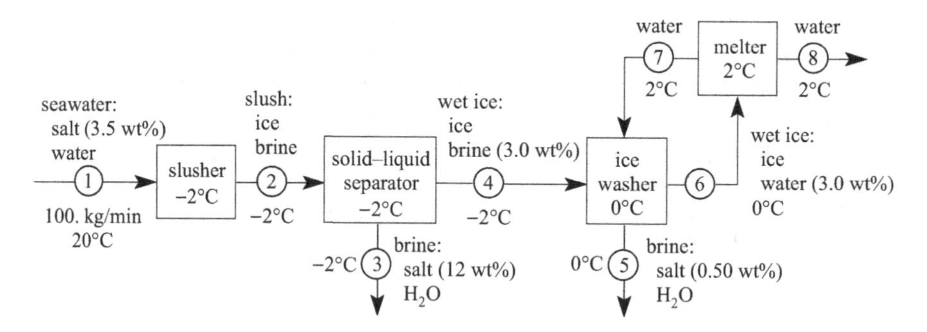

Figure 3.35 The freezer–desalinator with stream and unit temperatures.

Clearly the energy flow is not balanced in the slusher. Streams 1 and 2 have the same mass flow, so mass is conserved. But the temperature in stream 1 is higher than in stream 2, and a portion of stream 2 is converted to ice; more energy enters the slusher than leaves the

[6] Of course, we could use heat exchangers to recover some of the energy used for evaporation or freezing, but we would still lose a fraction of the energy. The energy not recovered in evaporation would be greater than the energy not recovered in freezing.

slusher. We need an outlet for the energy released by cooling and partially freezing the seawater. We indicate this with a squiggly arrow in Figure 3.36, to indicate that this stream carries only energy, no mass. Likewise the melter requires energy. Again we add a squiggly arrow. We also introduce the nomenclature q_i to represent the rate of energy flow associated with stream i. The dimensions of q are energy/time; in the mks system the units are kJ/s.

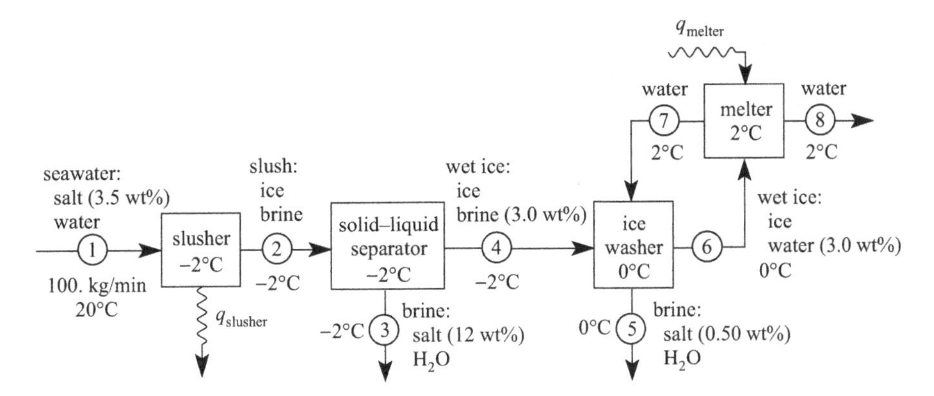

Figure 3.36 The freezer–desalinator with heat flow out of the slusher and heat flow into the melter.

Let's calculate the rate of heat flow from the slusher. We begin with the conservation of energy for a system at *steady state*,[7] which is given by equation 3.9:

$$\text{rate in} = \text{rate out.} \tag{3.9}$$

To apply the conservation of energy we must define a system. We draw borders around the slusher.

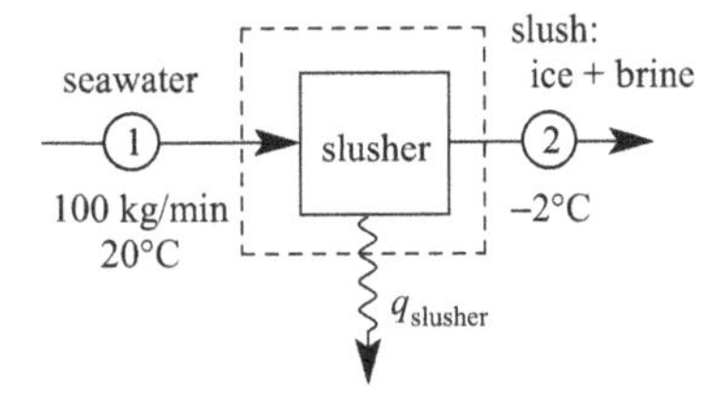

Figure 3.37 System borders for an energy balance on the slusher.

Apply the conservation of energy to this steady-state system and write an energy balance:

$$\text{rate in} = \text{rate out}$$

$$q_1 = q_2 + q_{slusher} \tag{3.108}$$

$$q_{slusher} = q_1 - q_2. \tag{3.109}$$

How do we compute the energy of streams 1 and 2? We must convert temperatures to energy. Thermodynamics provides that the change in internal energy is linearly proportional to the change in temperature. The proportionality constant is the product of two quantities: the mass and the heat capacity, which is the amount of energy needed to raise the temperature of 1 kg of a substance by 1°C. Thus from thermodynamics we use the relation

[7] Remember: energy may flow into and out of a system at steady state, but no energy accumulates in the system.

$$\text{change in internal energy} = \text{change in}\,[(\text{mass}) \times (\text{heat capacity}) \times (\text{temperature})] \tag{3.110}$$

$$\Delta U = \Delta[MC_P T]. \tag{3.111}$$

The heat capacity at constant pressure, C_P, varies from substance to substance. The heat capacity also varies with temperature, but that change is usually small. For the small temperature range in the desalinator, we assume the heat capacity is constant, which yields

$$\Delta U = MC_P(\Delta T). \tag{3.112}$$

We must convert a quantity of energy, ΔU, to a flow rate of energy, energy per time. We thus divide each side of equation 3.112 by Δt, an increment of time:

$$\frac{\Delta U}{\Delta t} = \frac{MC_P(\Delta T)}{\Delta t} = \frac{M}{\Delta t} C_P(\Delta T) \tag{3.113}$$

$$\Delta q = FC_P(\Delta T). \tag{3.114}$$

Equation 3.114 allows us to calculate the energy change associated with cooling stream 1 to stream 2 because we know F (mass flow rate) and ΔT (temperature) and we can find C_P in a table of physical constants.

Equation 3.114 illustrates another complexity of energy balances, compared to mass balances. Mass balances deal with absolute quantities; 100 kg/min enters the slusher and 100 kg/min leaves the slusher. However, it is usually not possible (nor convenient) to state an energy balance in absolute terms; for example it is not possible to state "100 kJ/min enters with the seawater and 50 kJ/min leaves with the seawater." Rather, we must consider *changes* in energy. Energy balances deal with relative quantities, the energy *difference* between the seawater entering and the seawater leaving. When we calculate equation 3.109, we will not evaluate the individual energies, q_1 and q_2, associated with streams 1 and 2. Rather we will calculate the change in energy, $q_1 - q_2$.

But there is another contribution to the slusher energy flow. We must also account for the energy change associated with forming ice from seawater. Again thermodynamics provides the necessary information: the energy change upon freezing is proportional to the amount frozen. The proportionality constant, which varies with composition, is the molar heat of fusion, $\Delta\bar{H}_{\text{fusion}}$. The flow rate of heat q associated with freezing a mass flow rate F is, analogous to equations 3.113 and 3.114,

$$\Delta q = F(\Delta\bar{H}_{\text{fusion}}). \tag{3.115}$$

Equation 3.115 again illustrates that energy balances involve relative energies and not absolute energies. The heat of fusion is the difference between the molar enthalpies of the liquid and solid phases:

$$\Delta\bar{H}_{\text{fusion}} = \bar{H}_{\text{liquid}} - \bar{H}_{\text{solid}}. \tag{3.116}$$

We will not calculate the absolute energies of salt water and ice. Rather we will calculate the energy change of freezing seawater into ice.

Preparing for the energy balance gives a different perspective on the slusher. Although the unit is trivial with regard to a mass balance, it is complicated with regard to an energy balance. That is, the slusher first cools all the seawater to −2°C from 20°C

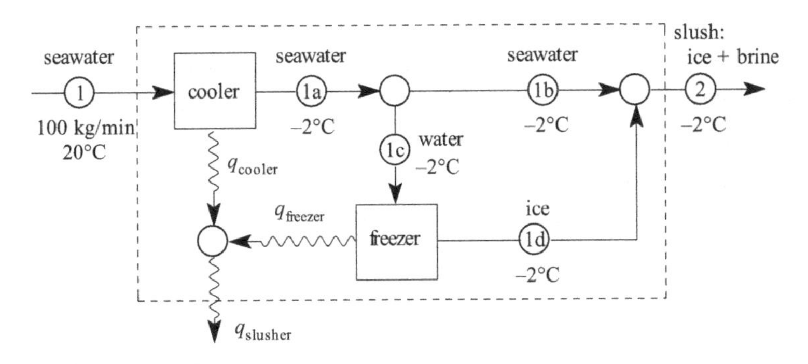

Figure 3.38 A hypothetical cooler–freezer process for the slusher.

and then the slusher freezes a portion of the seawater. The slusher seems to violate the "one-unit-one-operation" guideline of Chapter 2. A flowsheet to represent the energy changes in the slusher is shown in Figure 3.38. (We ignore the details of how the salt is excluded from the ice.)

Of course, the slusher does not operate as shown in Figure 3.38. But this is the power of thermodynamics; the outputs are independent of the path between input and output. We may devise a hypothetical process to simplify the energy calculations.

We now apply the equations from thermodynamics – equations 3.114 and 3.115 – to calculate the heat flows from the cooler and freezer of our hypothetical model for the slusher. Draw system borders around the cooler and write an energy balance:

$$\text{rate of energy in} = \text{rate of energy out}$$

$$q_1 = q_{1a} + q_{cooler} \tag{3.117}$$

$$q_{cooler} = q_1 - q_{1a}. \tag{3.118}$$

To apply equation 3.114 we use $F_{T,1} = F_{T,1a}$ and assume the heat capacity is constant between $-2°C$ and $20°C$:

$$q_{cooler} = F_{T,1}C_{P,1}(T_1 - T_{1a}). \tag{3.119}$$

We assume further that the heat capacity of seawater equals that of water, $C_P = 4.18$ kJ/(kg·°C). We may now calculate the heat flow from the cooler:

$$q_{cooler} = \left(\frac{100\ \text{kg}}{\text{min}}\right)\left(\frac{4.18\ \text{kJ}}{(\text{kg})(°\text{C})}\right)(20°\text{C} - (-2°\text{C})) = 9.2 \times 10^3\ \text{kJ/min}. \tag{3.120}$$

Let's check this answer. A numerical answer has two parts: a sign and a magnitude. Let's begin with the sign. Our answer is positive, which says heat leaves the cooler. This is correct. What about the magnitude, 9.2×10^3 kJ/min? We sympathize that quantities of "kJ/min" are probably not familiar to a nascent engineer. The answer is probably within a few orders of magnitude; 9.2×10^3 is not as suspicious as 9.2×10^{16} or 9.2×10^{-16}, for example. One approach is to translate this energy flow rate to a common phenomenon. A hot tub contains about 5000 kg of water. A mass flow rate of 100 kg/min would fill the hot tub in 50 minutes. A quick calculation shows 9.2×10^3 kJ/min would heat the water from 10°C (50°F) to 30°C (86°C) in about 45 minutes. This seems reasonable. That is, if the hot tub warmed in a few seconds, the energy flow rate would seem too high. If the hot tub warmed in a few days, the energy flow rate would seem too low.

Another way to check an energy flow rate is to convert to energy per mass, or kJ/kg. Water flowing through the cooler at 100 kg/min releases heat at a rate of 9.2×10^3 kJ/min, so the energy per mass is 92 kJ/kg. Now compare this to some rules of thumb:

2 kJ/kg will increase the temperature 1°C.
20 kJ/kg will melt a non-polar, molecular solid (butane, cyclohexane) at its melting point.
100 kJ/kg will melt a polar, molecular solid (water, ammonia) at its melting point.
400 kJ/kg will evaporate a non-polar, molecular liquid at its boiling point.
1000 kJ/kg will evaporate a polar, molecular liquid at its boiling point.
0 to 10,000 kJ/kg will be released (or absorbed) in chemical reaction.

Being rules of thumb, these numbers are rough estimates. If your answer is within a factor of 3 larger or smaller, your answer agrees with the rule of thumb. The rule of thumb for temperature change predicts 40 kJ/kg (one significant figure) for our cooler. Our answer of 92 kJ/kg is consistent.

The heat flow from the freezer can be calculated with an energy balance on the freezer:

$$\text{rate of energy in} = \text{rate of energy out}$$

$$q_{1\text{c}} = q_{1\text{d}} + q_{\text{freeezer}} \tag{3.121}$$

$$q_{\text{freezer}} = q_{1\text{c}} - q_{1\text{d}}. \tag{3.122}$$

To apply equation 3.115 we use $F_{T,1\text{c}} = F_{T,1\text{d}}$:

$$q_{\text{freezer}} = F_{T,1\text{c}}(\Delta\bar{H}_{\text{fusion}}). \tag{3.123}$$

Recall from the mass balance that $F_{T,4} = 73.02$ kg/min and stream 4 is 97 wt% ice. So the rate at which the slusher produces ice is $73.02 \times 0.97 = 70.83$ kg/min. We assume the heat of fusion of ice from seawater equals that of ice from pure water. From a table of physical data we obtain $\Delta\bar{H}_{\text{fusion,water}} = 334\ \text{kJ/kg}$:

$$q_{\text{freezer}} = F_{T,1\text{c}}\Delta\bar{H}_{\text{fusion,water}} = \left(\frac{70.83\ \text{kg}}{\text{min}}\right)\left(\frac{334\ \text{kJ}}{\text{kg}}\right) = 2.37 \times 10^4\ \text{kJ/min}. \tag{3.124}$$

The total heat flow from the slusher obtains from an energy balance on the energy combiner in our hypothetical process for the slusher:

$$\text{rate of energy in} = \text{rate of energy out}$$

$$q_{\text{slusher}} = q_{\text{cooler}} + q_{\text{freezer}}. \tag{3.125}$$

Substitute numerical values from equations 3.120 and 3.124: $q_{\text{slusher}} = 23{,}700 + 9{,}200 = 3.3 \times 10^4$ kJ/min.

Let's calculate the rate of heat flow into the melter. To translate the measurable parameters – temperature and phase – into energy we divide the melter into elementary energy changes: hypothetical units with one stream in and one stream out. Consider the hypothetical process in Figure 3.39 on the next page.

The energy changes in the hypothetical units involve one stream with one energy change; the ice melter changes the phase at constant temperature and the water warmer changes the temperature at constant phase. By analogy to the analysis of the slusher, energy balances yield the following equations:

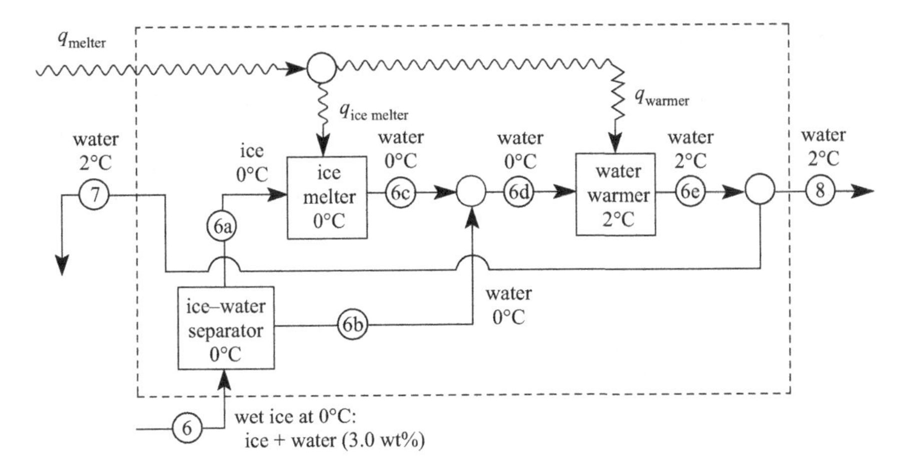

Figure 3.39 A hypothetical process for the melter.

$$q_{\text{ice melter}} = F_{T,6a}(\Delta\bar{H}_{\text{fusion,water}}) \tag{3.126}$$

$$q_{\text{warmer}} = F_{T,6d}C_{\text{P,water}}(T_{6e} - T_{6d}). \tag{3.127}$$

Recall from the mass balance that $F_{T,6}$ = 69.37 kg/min, $F_{T,7}$ = 48.93 kg/min, and $F_{T,8}$ = 20.44 kg/min. Equations 3.126 and 3.127 yield $q_{\text{ice melter}}$ = 22,500 kJ/min and q_{warmer} = 580 kJ/min. The total heat flow into the melter is q_{melter} = 23,100 kJ/min.

This section illustrates two similarities of mass balances and energy balances:

1. Both balances require a process flowsheet with numbered streams.
2. Both balances require system borders to apply a conservation principle.

This section illustrates two differences between mass balances and energy balances:

1. Energy is not measurable directly. Instead, energy is calculated indirectly from measurables such as temperature, phase, velocity, and pressure. Thermodynamics translates measurables to energy.
2. Energy is not calculated as an absolute quantity for mass flows. Rather, energy is calculated as a relative quantity: a change of energy for a mass flow. We calculate the energy change of a mass flow from changes in temperature, phase, velocity, height, and pressure. Complex energy changes may be decomposed into a hypothetical process of elementary energy changes.

3.5.2 Energy Balances for Temperature Changes with Variable Heat Capacity

Context: Analysis of the energy flow in a heat exchanger.

Concepts: Temperature dependence of heat capacity. Cocurrent vs. countercurrent devices.

Defining Question: What limits heat exchanger performance? What are the advantages of countercurrent flow?

Our freezer–desalinator needs 23,100 kJ/min for the melter and releases 32,000 kJ/min from the slusher. Chapter 2 introduced a unit to match an energy need with an energy source – the heat exchanger. Let's analyze a simple heat exchanger: a unit that cools one stream and warms another stream without phase changes in either stream. Consider one of the heat exchangers from the ammonia synthesis process in which N_2 from the N_2–O_2 liquid–gas separator cools the inlet air, as shown in Figure 3.40.

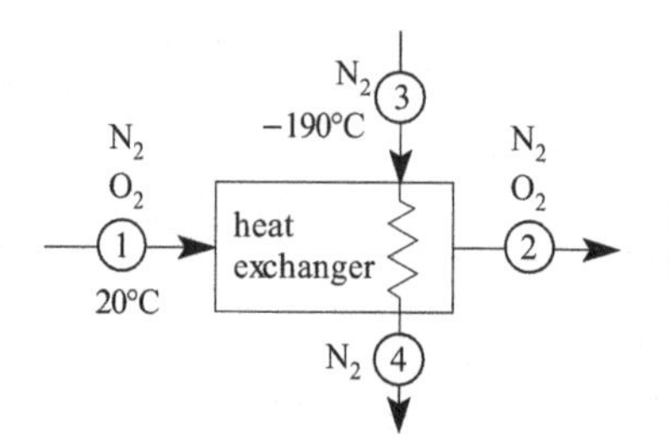

Figure 3.40 A heat exchanger from our process to synthesize ammonia in Chapter 2.

As the symbol for the heat exchanger suggests, N_2 flows inside a pipe and air flows around the outside of the pipe. The N_2 and air are isolated from each other and energy is transferred through the pipe's wall. Intuition (and later, a course on Heat and Mass Transfer) tells us the rate at which energy transfers is proportional to the surface area of the pipe. Let's assume we have a modest heat exchanger that transfers about half of the maximum; assume the temperature of the N_2 in stream 4 is −80°C. What is the temperature of the air in stream 2?

Draw a system border around the steady-state heat exchanger and write an energy balance:

$$\text{rate in} = \text{rate out}$$

$$q_1 + q_3 = q_2 + q_4. \tag{3.128}$$

To apply thermodynamics, we must group the qs into energy changes. Group the qs into differences such that each q pair has the same mass flow rate and composition:

$$-(q_4 - q_3) = q_2 - q_1 \tag{3.129}$$

$$-\Delta q_{3\to4} = \Delta q_{1\to2}. \tag{3.130}$$

Note that the differences are arranged as "stream *out* minus stream *in*." This order is key for distinguishing energy released from energy absorbed. Equation 3.130 states that the heat released by stream 3 changing to stream 4 (the minus sign indicates energy released) equals the energy absorbed by stream 1 changing to stream 2.

Translate energy differences to temperatures and heat capacities. Again assume the heat capacities are constant and use equation 3.114, $\Delta q = FC_P(\Delta T)$. Note that $F_{T,1} = F_{T,2}$ and $F_{T,3} = F_{T,4}$.

$$-F_{T,3}C_{P,\text{nitrogen}}(T_4 - T_3) = F_{T,1}C_{P,\text{air}}(T_2 - T_1). \tag{3.131}$$

We seek T_2, the temperature of the cooled air. Solve for T_2:

$$T_2 = T_1 - \frac{F_{T,3}}{F_{T,1}}\frac{C_{P,\text{nitrogen}}}{C_{P,\text{air}}}(T_4 - T_3). \tag{3.132}$$

Assume air is 79 mol% N_2. Thus the ratio $F_{T,3}/F_{T,1}$ equals 0.79. From a table of physical data, $C_{P,\text{ nitrogen}}$ = 29.11 J/(mol·K) and $C_{P,\text{ air}}$ = 29.13 J/(mol·K) for the range 100 to 300 K. Calculate T_2:

$$T_2 = 20°\text{C} - 0.79 \times \frac{29.11\ \text{J}/(\text{mol}\cdot\text{K})}{29.13\ \text{J}/(\text{mol}\cdot\text{K})}(-80°\text{C} - (-190°\text{C})) = -66.8°\text{C}. \quad (3.133)$$

Given the N_2 stream warms 110°C, the air stream cools 87°C. This assumes constant heat capacities, a reasonable approximation for small temperature changes (on the order of 10°C), as in the freezer–desalinator. But this assumption may be inaccurate for the large temperature changes in the heat exchanger (on the order of 100°C). To include the temperature dependence of C_P, we use the integral form of the energy change with temperature:

$$\int_{q_1}^{q_2} dq = F \int_{T_1}^{T_2} C_P dT. \quad (3.134)$$

The calculus in equation 3.134 is a minor complication because the temperature dependences of most heat capacities are accurately represented by polynomials such as the empirical Shomate equation,

$$C_P = a + bT + cT^2 + dT^3 + e/T^2, \quad (3.135)$$

and tables of coefficients a, b, c, d, and e are posted. Substitute equation 3.135 into equation 3.134 and integrate:

$$\int_{q_1}^{q_2} dq = F \int_{T_1}^{T_2} \left(a + bT + cT^2 + dT^3 + \frac{e}{T^2}\right) dT \quad (3.136)$$

$$q_2 - q_1 = a(T_2 - T_1) + \frac{b}{2}(T_2^2 - T_1^2) + \frac{c}{3}(T_2^3 - T_1^3) + \frac{d}{4}(T_2^4 - T_1^4) - e\left(\frac{1}{T_2} - \frac{1}{T_1}\right). \quad (3.137)$$

For N_2 in the temperature range 100 to 500 K, a = 29.0, $b = 1.84 \times 10^{-3}$, $c = -9.65 \times 10^{-6}$, $d = 1.84 \times 10^{-9}$, and e = 117 to yield C_P in units of J/(mol·K). The heat capacity of N_2 is 29.10 J/(mol·K) at −190°C and 29.01 J/(mol·K) at −80°C, a negligible 0.3% change. For air in the temperature range 100 to 500 K, a = 29.5, $b = -2.78 \times 10^{-3}$, $c = 4.53 \times 10^{-6}$, $d = 5.48 \times 10^{-9}$, and e = 1460. The heat capacity of air is 29.17 J/(mol·K) at 20°C and 29.11 J/(mol·K) at −47°C, a negligible 0.2% change.

Although the calculus for a temperature-dependent C_P is straightforward, the algebra is complicated. We cannot solve for T_2 as we did in equation 3.132. Rather we must calculate $\Delta q_{1\to 2}$ (= $q_2 - q_1$) and use an equation-solver routine to calculate T_2.

Assume a nominal flow rate of air – 100 mol/s – which produces 79 mol/s of N_2. With equation 3.137 we calculate that stream 3 absorbs heat at a rate of 345 J/s. Thus $\Delta q_{1\to 2}$ = 345 J/s. We solve equation 3.137 to find T_2 = −66.8°C, the same temperature we calculated assuming constant heat capacity.

For this heat exchanger, the temperature dependence of heat capacity had negligible effect because the two fluids are similar; air is 79 mol% nitrogen. For heat exchangers in general, temperature dependence of heat capacity has only a small effect because heat capacities increase with temperature. One substance is cooled and the other substance is warmed, so the temperature dependence somewhat cancels. However, the temperature dependence can be important for other processes, such as highly exothermic reactions.

Our modest heat exchanger cools the air and reduces the refrigeration required in the liquid–gas separator. A larger heat exchanger could cool the air more. How much cooling can we extract from the cold N_2 stream? Is the limit such that the two exit streams are the same temperature, as shown by the heat exchanger on the left of Figure 3.41? Using equation 3.132

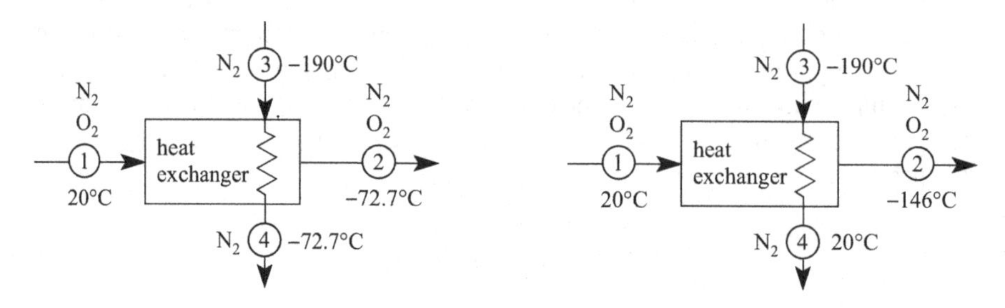

Figure 3.41 Heat exchangers to extract more cooling from the cold N_2 stream.

and setting $T_2 = T_4$, one obtains -72.7°C. Or can we exploit the cold N_2 stream more by warming the N_2 to 20°C and cooling the air stream to -146°C, as shown on the right?

The energy absorbed by the N_2 stream equals the energy extracted from the air stream for both heat exchangers. Both operations are consistent with equation 3.130, which was derived from the conservation of energy, also known as the first law of thermodynamics. But processes must also obey the second law of thermodynamics. The second law dictates heat flows from hot to cold. The modest heat exchanger clearly satisfies the second law; the N_2 stream out at -80°C is colder than the air stream out at -66.8°C, so heat flows from the air to the N_2. The larger heat exchanger on the left of Figure 3.41 seems to approach the limit of the second law; the exit N_2 and the exit air are the same temperature, so heat barely flows from the air exit stream to the N_2 exit stream. But the even larger heat exchanger on the right of Figure 3.41 is consistent with the second law. The key is an important concept in chemical engineering: *cocurrent* and *countercurrent* flows.

Let's analyze qualitatively the differences between cocurrent and countercurrent flow in a heat exchanger. Consider the simple systems shown in Figure 3.42; assume the fluids inside and outside the pipe have the same composition and the same flow rate. The only difference is the fluid temperatures.

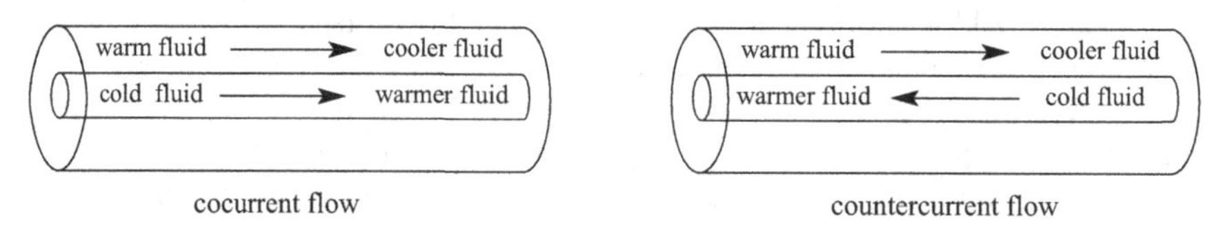

Figure 3.42 Cocurrent and countercurrent flow in a heat exchanger. In this case, the fluid being warmed moves inside the small pipe and the fluid being cooled moves outside the small pipe, but inside the outermost shell of the heat exchanger.

Which flow pattern – cocurrent or countercurrent – cools the warm fluid the most? We can estimate graphically the fluid temperatures along the pipe for each configuration. We first plot the inlet temperatures.

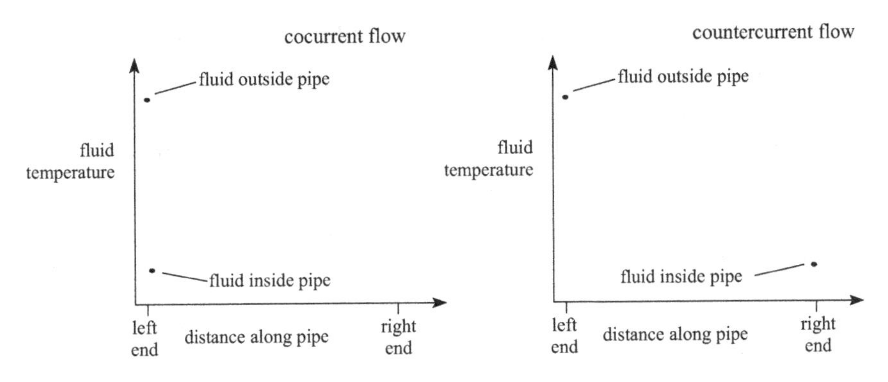

Figure 3.43 Inlet temperatures of fluids in cocurrent and countercurrent flows.

We now sketch lines from the inlet temperatures to the exit temperatures. The cold fluid warms and the hot fluid cools. The second law dictates the lines must not cross. And if the heat capacity is constant, the temperature drop in the outside fluid equals the temperature rise in the inside fluid. One arrives at the sketches shown in Figure 3.44.

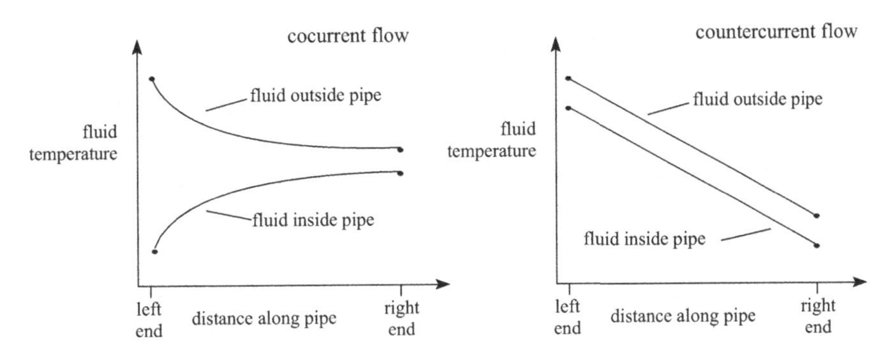

Figure 3.44 Temperature profiles of fluids in cocurrent and countercurrent flows.

Qualitatively, the countercurrent flow cools the warm fluid the most. Chemical engineers can calculate the temperature along the pipes in a heat exchanger with the material learned in Fluid Mechanics and Heat-Transfer courses. As such a chemical engineer can design a heat exchanger to meet specifications. We will not concern ourselves with those details here. For now we assume countercurrent flow in our heat exchangers. We again exploit countercurrent flow when we consider mass transfer between streams in Chapter 4.

Matching energy sources to energy needs – known as heat integration – is essential for optimizing a chemical process. Consider a process with five streams to be cooled and five streams to be heated. Which streams should be paired in heat exchangers? One optimization method maps each energy source and energy need: the amount of heat on the x axis and the temperature of the heat on the y axis. Sources and needs are matched with regard to both the first and second laws of thermodynamics. Because the limits are indicated by the closeness of "sources" lines to "needs" lines, this graphical heat integration process is known as Pinch Analysis.

3.5.3 Heat Integration: Matching Energy Needs to Energy Sources

Context: Optimizing energy use in the freezer–desalinator.

Concepts: Heat pumps, refrigeration cycles, and heat of vaporization.

Defining Question: How does an air conditioner seem to defy the second law of thermodynamics by transferring heat from a cool room to a hot outdoors?

Recall the energy analysis of the freezer–desalinator: the slusher releases 32,900 kJ/min and the melter needs 23,100 kJ/min. Can we place the slusher adjacent to the melter so energy flows from the source to the need? No, because the second law prohibits heat flow from cold to hot. The slusher is at −2°C and the melter is at 0°C. To transfer heat from cold to hot we need a heat pump.

A heat pump is similar to a fluid pump. Consider this analogy. Figure 3.45 (on the next page) shows a scheme for moving fluid from a low container to a high container. The fluid drains from the low container to a bucket at a lower level, the bucket is raised, and then the fluid is drained into the high container.

Figure 3.46 (on the next page) shows an analogous scheme for moving heat from a cool fluid to a warm fluid. The heat drains spontaneously from the cool object to a colder object, the object is warmed, and then the heat drains to the warm fluid.

Transferring another load of fluid is trivial; just lower the bucket. How does one cool the hot object to a cold object to transfer more heat? The answer will be obvious as we design a heat pump for the desalinator.

Let's design a system to remove heat from seawater at −2°C. We need a refrigerating fluid colder than −2°C. And the temperature of the refrigerant cannot rise above −2°C as it absorbs heat from the seawater. Rather than absorb heat by *warming* the refrigerant, a more effective method is to absorb heat by *vaporizing* the refrigerant. We need a refrigerant that transforms from a liquid to a vapor at a temperature less than −2°C. The refrigerant has several other requirements imposed on it as well; it must be inexpensive, have a high heat of vaporization, be non-lethal, and be innocuous (so a leak is not a threat to one's health – refrigerators once used H_2S!). Chlorofluorocarbons were ideal refrigerants until it was required that refrigerants satisfy another requirement: refrigerants should be harmless to the environment. So we consult a list of refrigerants in a handbook. Let's use a "natural" chemical as our refrigerant – ammonia.[8]

Let's arbitrarily specify that the ammonia boils at −22°C. (We could have chosen any temperature less than −2°C.) But ammonia boils at −33°C at 1 atm. How do we cause ammonia to boil instead at −22°C? We increase the pressure of the ammonia. We consult a handbook and find the saturation pressure of ammonia at −22°C is 25 psi. While in the handbook we also read that the specific heat of vaporization at −22°C and 25 psi is $\Delta\bar{H}_{\text{vap}} = 574.9$ Btu/lb. (Refrigeration data are usually in English units.) We convert to mks units:

$$\left(\frac{574.9\ \text{Btu}}{1\ \text{lb}}\right)\left(\frac{1\ \text{lb}}{0.4536\ \text{kg}}\right)\left(\frac{1.054\ \text{kJ}}{1\ \text{Btu}}\right) = 1336\ \text{kJ/kg}. \tag{3.138}$$

Note that in pure mks units the pressure should also be converted to pascals (Pa):

[8] Note that ammonia is pungent and corrosive, and its condensed vapor causes edema of the respiratory tract as well as other problems. But ammonia is natural.

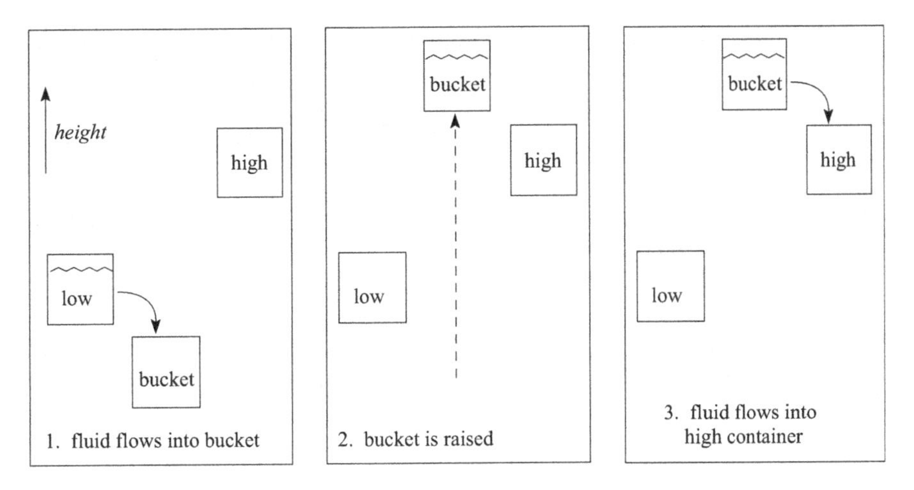

Figure 3.45 A method to transport mass from a lower container to a higher container.

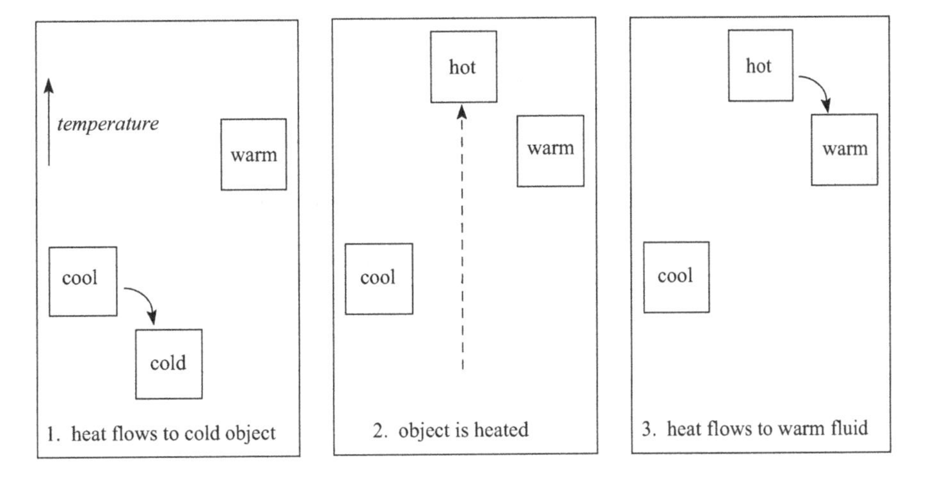

Figure 3.46 A method to transport heat from a cool fluid to a warm fluid.

$$25\text{ psi}\left(\frac{6895\text{ Pa}}{1\text{ psi}}\right) = 1.7 \times 10^5\text{ Pa}. \tag{3.139}$$

But 1.7×10^5 is an awkward number. We use the more convenient unit of atmosphere (atm) instead: 25 psi = 1.7 atm. A design for the slusher is given in Figure 3.47.

Figure 3.47 The slusher with ammonia refrigerant.

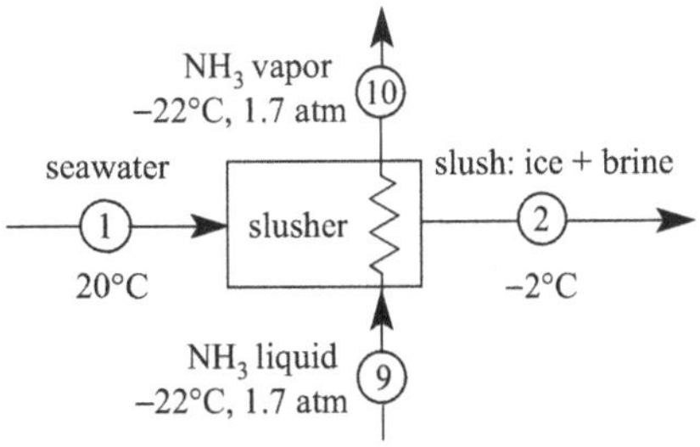

NH_3 vapor
−22°C, 1.7 atm
10
seawater
1
20°C
slusher
slush: ice + brine
2
−2°C
NH_3 liquid
−22°C, 1.7 atm
9

What flow rate of ammonia is needed to cool and partially freeze the seawater? An energy balance on the slusher yields

$$\text{rate of energy in} = \text{rate of energy out}$$

$$q_1 + q_9 = q_2 + q_{10}. \tag{3.140}$$

We rearrange the equation to "rate of energy change in seawater" on the left, "rate of energy change in ammonia" on the right:

$$-(q_2 - q_1) = q_{10} - q_9. \tag{3.141}$$

The energy released by the seawater was calculated above: 3.3×10^4 kJ/min. The right-hand side of equation 3.141 is the energy absorbed by ammonia evaporating at −22°C and 1.7 atm:

$$q_{10} - q_9 = F_{T,10}(\text{molar energy of } NH_3 \text{ vapor}) - F_{T,9}(\text{molar energy of } NH_3 \text{ liquid}) \tag{3.142}$$

$$q_{10} - q_9 = F_{T.9}\left(\Delta\bar{H}_{\text{vap, } NH_3 \text{ at 25 psi}}\right) \tag{3.143}$$

$$F_{T,9} = \frac{q_{10} - q_9}{\Delta\bar{H}_{\text{vap, } NH_3 \text{ at 25 psi}}} = \frac{3.29 \times 10^4 \text{ kJ/ min}}{1336 \text{ kJ/kg } NH_3} = 24.6 \text{ kg } NH_3/\text{min}\,. \tag{3.144}$$

Likewise we pass ammonia through the melter. We condense ammonia vapor to form liquid ammonia which releases heat. We must choose a temperature greater than 0°C for the ammonia stream. We arbitrarily choose 32°C. A handbook reveals that the saturation pressure for ammonia at 32°C is 12.2 atm and $\Delta\bar{H}_{vap} = 488.5$ Btu/lb, which converts à la equation 3.138 to 1135 kJ/kg. A design for the melter is given in Figure 3.48.

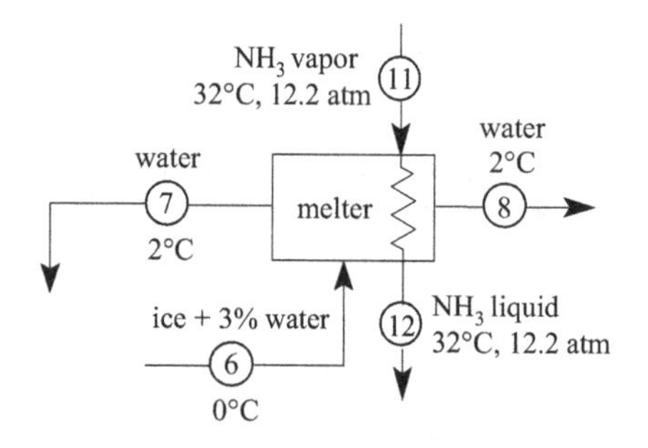

Figure 3.48 The melter with ammonia heating.

What flow rate of ammonia is needed to melt the ice? An energy balance similar to that performed on the slusher yields $F_{T,11} = 20.4$ kg/min. We recommend you verify our analysis.

We must close the ammonia loop. We connect the output from the slusher (stream 10) at 1.7 atm to the input to the melter (stream 11) at 12.2 atm by adding a compressor. We connect the output from the melter (stream 12) at 12.2 atm to the input to the slusher (stream 9) at 1.7 atm by adding a turbine, or perhaps just an expansion valve. The ammonia loop is a typical refrigeration cycle, as found in refrigerators and air conditioners. We must supply energy to drive the compressor. Some of the energy could be recovered with a turbine.

What detail of the ammonia cycle remains? The flow rate of ammonia through the slusher is 24.6 kg/min and the flow rate of ammonia through the melter is 20.4 kg/min. A mass balance on the compressor (or the expansion valve) would reveal the problem. At what rate should ammonia flow through the refrigeration cycle? If ammonia flows at 20.4 kg/min, less energy is removed from the slusher and the rate of formation of ice is insufficient. If ammonia flows at 24.6 kg/min, excess heat is supplied to the melter. What becomes of the excess heat? It warms the melted ice. We calculate that the temperature of stream 5 would be 15°C. Again, we recommend you calculate this independently.

Suppose we don't want to warm streams 7 and 8 to 15°C. The temperature of the recycle stream 7 must be close to 0°C to minimize melting ice in the ice washer. Or perhaps we wish to feed stream 8 to a drinking fountain – it would be nice to have a drink close to 0°C. What are the design options? Given an ammonia flow rate of 24.6 kg/min, we must add less energy to the melter. As you will learn in physical chemistry, the heat of vaporization decreases as the pressure increases. So you could increase the pressure in stream 11 to lower the heat of vaporization by a factor of 20.4/24.6. Alternatively, we could flow ammonia at 20.4 kg/min and increase the heat of vaporization of ammonia in the slusher by 24.6/20.4.

The process of desalination by freezing may be improved yet further. In our design, heat is pumped from the slusher to the melter by evaporating a refrigerant (ammonia) in the slusher and condensing the refrigerant in the melter. A conceptual breakthrough improved the design of the desalination process – the process fluid (H_2O) can also be used as the refrigerant. This avoids problems inherent with a foreign refrigerant. The slusher and melter are less expensive; no internal piping is needed because it is not necessary to keep the refrigerant isolated. And because the hot and cold fluids are intimately mixed, heat is transferred more efficiently.

The evolved desalinator is shown in Figure 3.49. Cooled seawater passes through a throttle valve to the slusher at low pressure. The temperature and pressure in the slusher are the triple point of seawater: the conditions at which ice, brine, and steam coexist. Heat is removed from the slusher by vaporizing water. The water vapor is compressed and heat is delivered to the melter by condensing water vapor. This is an exquisite design and yet it is a logical evolution from the design in Figure 3.7. A detailed analysis of desalination by evaporating at the triple point of seawater is presented in the textbook by Rudd, Powers, and Siirola (1973; Chapter 7, "Fresh Water by Freezing"). The optimum desalinator design depends on regional factors, including local costs of energy and the intended use for the desalinated water.

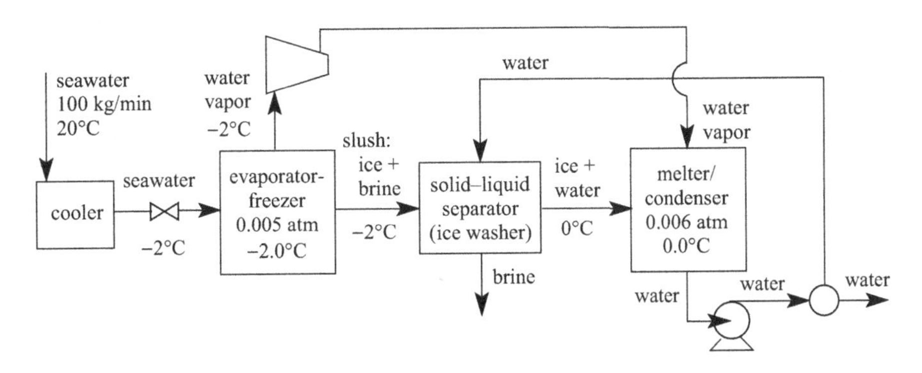

Figure 3.49 A freezer–evaporator desalinator with H_2O as the refrigerant.

Heat pumps are common: refrigerators, air conditioners, and heating units in climates with mild winters. For an outside temperature around 50°F (10°C) a heat pump to an inside temperature around 72°F (22°C) heats more efficiently than burning a fuel (or heating with electricity derived from a fuel). And every climate has access to a seasonally invariant heat source at 55°F (13°C): the geothermal heat reservoir below about 20 ft (6 m). Cornell University uses a heat pump for most of its air conditioning, using the heat sink deep in adjacent Cayuga Lake (see Exercise 3.68).

A heat pump uses mechanical energy to pump heat from a cool source to a hotter sink. The reverse of a heat pump is a heat engine, which generates mechanical energy by the flow of heat from a hot source to a cooler sink. The mechanical energy can drive a turbine to be converted to electrical energy. You will study heat pumps and heat engines in detail in your chemical engineering thermodynamics course.

3.6 ENERGY BALANCES AND CHEMICAL REACTIONS

3.6.1 Chemical Reactions with Complete Conversion

Context: The production of silicon metal from silicon dioxide.

Concepts: Heat of reaction, temperature change in an adiabatic reactor.

Defining Question: Why is an energy balance insufficient for designing a process?

The energy flow in a process with a chemical reaction is a key consideration. A chemical reaction generally releases (or absorbs) much more energy than released (or absorbed) by a change in temperature or a change in phase. We revisit the production of silicon metal from silicon dioxide from Chapter 2 to delve into the details of calculating the heat released (or absorbed) in a chemical reaction.

The reaction – balanced with respect to both mass and energy – was given in reaction 2.7:

$$SiO_2 + 911 \text{ kJ} \rightarrow Si + O_2. \qquad \text{rxn (2.7)}$$

Calculating a *heat of reaction*, $\Delta\bar{H}_{rxn}$, again requires calculating energy changes: the differences between the *heats of formation*, $\Delta\bar{H}_f$, of the products and of the reactants. A heat of formation is the difference between the enthalpy of a compound and the enthalpy of its constituent elements in their "natural" state at standard conditions, 25°C and 1 bar (=0.99 atm).

From various postings of thermodynamic data, one finds the following heats of formation: SiO_2(solid) $\Delta\bar{H}_f = -911$ kJ/mol, Si(solid) $\Delta\bar{H}_f = 0$ kJ/mol, and O_2(gas) $\Delta\bar{H}_f = 0$ kJ/mol. The heats of formation of Si and O_2 are both zero. Does this mean Si and O_2 have no enthalpy at 25°C and 1 bar? Of course not. Elements in their natural state at standard conditions are arbitrarily defined to be zero energy. The reference state could have been defined as 0 K and 0 bar. The convenience of using 25°C and 1 bar for experimental measurements is obvious. Note that some sources report heats of formation as $\bar{H}_f$. The omission of the "Δ" incorrectly suggests these are absolute energies.

The heat of reaction at standard state for an arbitrary reaction with n reactants and m products, each with stoichiometric coefficients ν_i, is

$$\Delta\bar{H}^0_{rxn} = \sum_{i=1}^{m} \nu_i \Delta\bar{H}^0_{f,\text{product } i} - \sum_{i=1}^{n} \nu_i \Delta\bar{H}^0_{f,\text{reactant } i}. \qquad (3.145)$$

For the decomposition of SiO_2, this is

$$\Delta\bar{H}^0_{rxn} = \left(1 \times \Delta\bar{H}^0_{f,Si} + 1 \times \Delta\bar{H}^0_{f,O_2}\right) - \left(1 \times \Delta\bar{H}^0_{f,SiO_2}\right) \tag{3.146}$$

$$= (0 + 0) - (-911) = +911 \text{ kJ/mol}. \tag{3.147}$$

The heat of reaction at 25°C and 1 bar is +911 kJ/mol. The reaction is strongly endothermic.

But the process developed in Chapter 2 decomposes SiO_2 at 2000°C, not 25°C, and the Si product at 2000°C is liquid, not solid. We cannot calculate $\Delta\bar{H}_{rxn}$ at 2000°C in the same manner we calculate $\Delta\bar{H}_{rxn}$ at 25°C because heats of formation are rarely posted for non-standard states. What to do?

We use thermodynamics to calculate $\Delta\bar{H}_{rxn}$ at 2000°C. We use the fact that thermodynamic states are path-independent; the change in thermodynamics properties in going from one state to another state is independent of one's path from the initial state to the final state. So we can invent a fictitious path from the initial state to the final state – a path that affords convenient calculation of the thermodynamic changes. To calculate the heat of reaction at 2000°C, we start with the reactant (SiO_2) at 2000°C, cool it to 25°C, react to Si and O_2 at 25°C, and then heat the products to 2000°C. Note that there are also phase changes: at 25°C the SiO_2 reactant is solid, which melts in the range 1600–1700°C, and at 25°C the Si product is solid, which melts at 1414°C.

We search for posted thermodynamic data and prepare a table. We will estimate $\Delta\bar{H}_{rxn}$ at 2000°C from heat capacities averaged over each temperature range. Because vitreous silica is a glass, it does not have a definite melting point or heat of fusion. We will use data for crystalline silica (quartz) in our estimate.

Table 3.3 Thermodynamic data for the decomposition of SiO_2 to Si and O_2 at 2000°C.

Substance	Melting point (°C)	Boiling point (°C)	Average heat capacity (J/(mol·K))			$\Delta\bar{H}^0_f$ (kJ/mol)	$\Delta\bar{H}_{melt}$ (kJ/mol)
			Solid	Liquid	Gas		
SiO_2 (quartz)	1,713	(decomposes)	69	81	–	−911	9.4
Si	1,414	3,265	25	27	–	0	50.2
O_2	−219	−183	–	–	29	0	–

Use the thermodynamic data with the thermodynamic relations in equations 3.119, 3.126, 3.143, and 3.145 to calculate the enthalpy change with each step in our fictitious process. You should verify you can calculate each enthalpy change in Table 3.4.

Table 3.4 Enthalpy change for the decomposition of SiO_2 to Si and O_2 at 2000°C.

Fictitious step	$\Delta\bar{H}$(kJ/mol)
Cool liquid SiO_2 from 2000°C to its melting point (1713°C): $\Delta\bar{H} = \bar{C}_P(\Delta T)$	−23
Solidify liquid SiO_2 to quartz at its melting point (1713°C): $\Delta\bar{H}_{melt}$	−9.4
Cool solid SiO_2 (quartz) from its melting point (1713°C) to standard state (25°C)	−116
Decompose solid SiO_2 to solid Si and gaseous O_2 at 25°C: $\Delta\bar{H}_{rxn}$	+911
Warm solid Si and gaseous O_2 from 25°C to the Si melting point (1414°C)	+75
Melt solid Si at its melting point (1414°C): $\Delta\bar{H}_{melt}$	+50
Warm liquid Si and gaseous O_2 from 1414°C to 2000°C: $\Delta\bar{H} = \bar{C}_P(\Delta T)$	+33
Sum	**+921**

Our estimate for the heat of reaction at 2000°C is +921 kJ/mol, 1% greater than the heat of reaction at 25°C.

Let's design a heater for the silica-decomposition reactor. The heater must supply energy to warm the reactant (SiO_2) from 25°C to 2000°C (23 + 9.4 + 116 = +148 kJ/mol) and drive the endothermic reaction (+921 kJ/mol); the heater must supply 1069 kJ/mol SiO_2.

Rather than mix coal and oxygen with the silica as used in the process in Chapter 2, let's capitalize on the abundance of natural gas from hydraulic fracturing (aka fracking). Let's design an external burner to transfer heat to the silica-decomposition reactor. Assume the natural gas is 100% methane. The reaction is

$$CH_4 + 2O_2 \rightarrow CO_2 + 2H_2O(\text{gas}) + 802\,\text{kJ}. \qquad \text{rxn (2.3)}$$

Use heats of formation and equation 3.145 to verify the heat of reaction given in Chapter 2:

$$\Delta\bar{H}^0_{\text{rxn}} = \left(1 \times \Delta\bar{H}^0_{\text{f,CO}_2} + 2 \times \Delta\bar{H}^0_{\text{f,H}_2\text{O}}\right) - \left(1 \times \Delta\bar{H}^0_{\text{f,CH}_4} + 2 \times \Delta\bar{H}^0_{\text{f,O}_2}\right) \qquad (3.148)$$

$$= (1 \times (-393.5) + 2 \times (-241.8)) - (1 \times (-74.8) + 2 \times 0) = -802\ \text{kJ/mol}. \qquad (3.149)$$

The heat of reaction at standard conditions (25°C and 1 bar) is −802 kJ/mol. The reaction is strongly exothermic; the reaction releases much heat as it progresses.

Calculating the flow rate of CH_4 and O_2 is trivial. To balance the energy, we must supply (1069 kJ/mol SiO_2)/(802 kJ/mol CH_4) = 1.3 mol CH_4/mol SiO_2. This satisfies the principle of conservation of energy, the first law of thermodynamics. But we must also consider the second law of thermodynamics. Will the heat from burning CH_4 flow to the SiO_2 decomposition reactor at 2000°C? That is, is the methane + oxygen flame at least 2000°C? We must calculate the *adiabatic* temperature of the burner.

An adiabatic process transfers no heat to its surroundings; the heat released in the reaction warms the contents of the reactor. In the actual process, a small portion of the CH_4 and O_2 reacts, which warms the unreacted CH_4 and O_2 and the small portion of products. As the reaction proceeds, the temperature increases. But calculating the temperature as a function of reaction progress is not necessary. We need only know the temperature after complete combustion of the CH_4. To get from the initial state (CH_4 and O_2 at 25°C and 1 bar) to the final state (CO_2 and H_2O at the adiabatic temperature and 1 bar), we can choose a convenient fictitious path.

We have already calculated the heat of reaction at 25°C and 1 bar. Use this energy to warm the products CO_2 and H_2O. Because this calculation needs only an estimate (is the flame temperature greater than 2000°C?), we will use heat capacities for CO_2 and H_2O averaged from Shomate equations valid over the range 0°C to 1500°C: $\bar{C}_{P,CO_2}$= 53 J/(mol·K) and $\bar{C}_{P,H_2O}$=41 J/(mol·K). The energy balance for 1 mol of CH_4 burned in O_2 is thus

$$\text{energy released in reaction of } CH_4 + O_2 = \text{energy absorbed by heating } CO_2 + H_2O \qquad (3.150)$$

$$-\Delta\bar{H}^0_{\text{rxn}} = \bar{C}_{P,CO_2}(T_{\text{adiabatic}} - 25) + 2 \times \bar{C}_{P,H_2O}(T_{\text{adiabatic}} - 25). \qquad (3.151)$$

Solve for $T_{\text{adiabatic}}$ and substitute thermodynamic data:

$$T_{\text{adiabatic}} = \frac{-\Delta\bar{H}^0_{\text{rxn}}}{\bar{C}_{P,CO_2} + 2 \times \bar{C}_{P,H_2O}} + 25 = \frac{-(-802\ \text{kJ/mol})}{53 + 2 \times 41\ \text{J/(mol·°C)}} + 25°\text{C} = 6000°\text{C}. \qquad (3.152)$$

Because our calculated $T_{adiabatic}$ above is far outside the range of C_P, our estimate is likely inaccurate. Indeed, $T_{adiabatic}$ is measured to be about 2800°C (±10°C). The point is that heating with an external $CH_4 + O_2$ burner is feasible because the burner temperature is greater than 2000°C. Heat will spontaneously flow from the external burner (at 2800°C) to the silica-decomposition reactor (at 2000°C).

But O_2 is expensive. It is cheaper to burn CH_4 in air. What is the adiabatic temperature of the reaction of CH_4 and air? The reaction is the same as reaction 2.3, but now there is an inert substance in the burner: N_2. Because air is 21 mol% O_2, every mol of O_2 is accompanied by 0.79/0.21 = 3.8 mol N_2. The energy balance for 1 mol of CH_4 burned in air is thus

$$\text{energy released in reaction of } CH_4 + O_2 = \text{energy absorbed by heating } CO_2 + H_2O + N_2 \tag{3.153}$$

$$-\Delta \bar{H}^0_{rxn} = (\bar{C}_{P,CO_2} + 2 \times \bar{C}_{P,H_2O} + 2 \times 3.8 \times \bar{C}_{P,N_2}) \times (T_{adiabatic} - 25). \tag{3.154}$$

Solve for $T_{adiabatic}$ and substitute thermodynamic data. Use the heat capacity for N_2 averaged over the range 0°C to 1500°C: $\bar{C}_{P,N_2} = 33\,\mathrm{J/(mol{\cdot}K)}$.

$$\begin{aligned} T_{adiabatic} &= \frac{-\Delta \bar{H}^0_{rxn}}{\bar{C}_{P,CO_2} + 2 \times \bar{C}_{P,H_2O} + 2 \times 3.8 \times \bar{C}_{P,N_2}} + 25 \\ &= \frac{-(-802\ \mathrm{kJ/mol})}{53 + 2 \times 41 + 2 \times 3.8 \times 33\ \mathrm{J/(mol{\cdot}°C)}} + 25°\mathrm{C} = 2100°\mathrm{C}. \end{aligned} \tag{3.155}$$

Warming the inert N_2 substantially reduces the adiabatic temperature. Again, because the calculated adiabatic temperature is outside the range of averaged heat capacities (0–1500°C), the result is suspect. Indeed, the measured $T_{adiabatic}$ for CH_4 and air is 1950°C. The point is that a CH_4 + air burner is incapable of providing the heat for the silica-decomposition reactor operating at 2000°C.

An energy balance is necessary; the energy absorbed must equal the energy released. But an energy balance is insufficient; the energy must spontaneously flow from the source to the sink. The heat source must be warmer than the heat sink.

3.6.2 Chemical Reactions with Incomplete Conversion

Context: The production of methane from synthesis gas (syngas).
Concepts: Adiabatic temperature as a function of reaction conversion.
Defining Question: How to calculate the reaction conversion for an adiabatic reactor with an upper temperature limit?

The energy of a chemical reaction is key to the design of a chemical reactor. To illustrate this concept, we consider the synthesis of methane, CH_4, from synthesis gas (aka syngas), CO and H_2. As discussed in Chapter 2, syngas can be generated by underground coal gasification. The reaction to methane creates a convenient fuel, and the CO_2 by-product can be sequestered.

The reaction is

$$2CO + 2H_2 \rightarrow CH_4 + CO_2. \qquad \text{rxn (3.1)}$$

From various postings of thermodynamic data, one finds the following heats of formation: CO $\Delta \bar{H}_f = -110.5\,\mathrm{kJ/mol}$, H_2 $\Delta \bar{H}_f = 0\,\mathrm{kJ/mol}$, CH_4 $\Delta \bar{H}_f = -74.6\,\mathrm{kJ/mol}$, and

CO_2 $\Delta\bar{H}_f = -393.5\,kJ/mol$. Use the heats of formation and equation 3.145 to calculate the heat of reaction:

$$\Delta\bar{H}^0_{rxn} = \left(1 \times \Delta\bar{H}^0_{f,CH_4} + 1 \times \Delta\bar{H}^0_{f,CO_2}\right) - \left(2 \times \Delta\bar{H}^0_{f,CO} + 2 \times \Delta\bar{H}^0_{f,H_2}\right) \tag{3.156}$$

$$= (-74.6 + (-393.5)) - (2 \times (-110.5) + 2 \times 0) = -247.1\,kJ/mol. \tag{3.157}$$

Use the heat capacities of methane and carbon dioxide averaged over the range 25°C to 2000°C to estimate the adiabatic temperature rise: $\bar{C}_{P,CH_4} = 72\,J/(mol{\cdot}K)$ and $\bar{C}_{P,CO_2} = 55\,J/(mol{\cdot}K)$:

$$\text{energy released in reaction of } CO + H_2 = \text{energy absorbed by heating } CH_4 + CO_2 \tag{3.158}$$

$$-\Delta\bar{H}^0_{rxn} = (\bar{C}_{P,CH_4} + \bar{C}_{P,CO_2}) \times (T_{adiabatic} - 25) \tag{3.159}$$

$$T_{adiabatic} = \frac{-\Delta\bar{H}^0_{rxn}}{\bar{C}_{P,CH_4} + \bar{C}_{P,CO_2}} + 25 = \frac{-(-247.1\,kJ/mol)}{(72 + 55)\,J/(mol{\cdot}°C)} + 25°C = 1970°C. \tag{3.160}$$

Given a reactor inlet temperature of 25°C, the exit temperature is 1970°C. The adiabatic temperature rise is 1945°C.

Reactor design is chiefly determined by chemical kinetics – the rate of reaction as a function of temperature. We will not calculate reaction rates here, but we will use the results of such calculations. The synthesis of methanol from syngas is unacceptably slow at 25°C. We must synthesize methane at 225°C or higher. Given an initial temperature of 225°C, the reactor will heat to 225°C + 1945°C = 2170°C.

A reactor that operates at 2170°C must be constructed of expensive materials that retain structural integrity at high temperature. Because CH_4 is a bulk chemical commodity, economics dictates that we use plain carbon steel, an inexpensive material with a maximum operating temperature of 425°C. We could design a reactor with an internal heat exchanger to remove heat. But process units that combine operations are expensive; a unit to conduct the reaction and simultaneously remove heat is more expensive than two single-operation units, a reactor and a heat exchanger. A more economical process uses a series of reactors and heat exchangers, as shown in Figure 3.50.

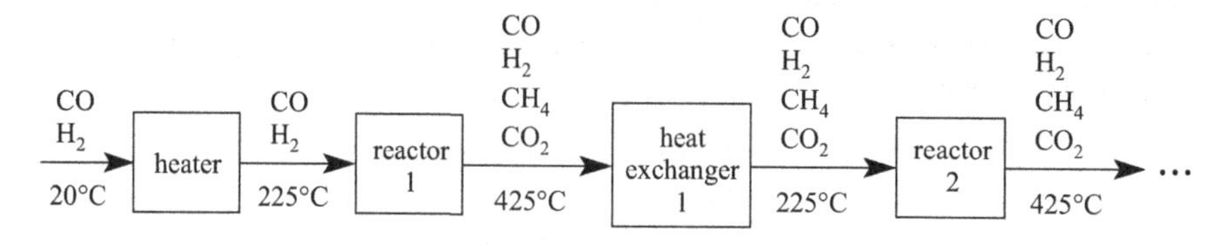

Figure 3.50 A series of adiabatic reactors and heat exchangers.

The reaction is conducted in reactor 1 until the temperature reaches 425°C. The reactor effluent is cooled to 225°C and fed to reactor 2. The reaction is conducted in reactor 2 until the temperature reaches 425°C. Additional pairs of reactors and heat exchangers are added until the desired conversion of reactants to products is achieved.

How many reactor + heat exchanger pairs are needed? We must calculate the reaction conversions in reactor 1, in reactor 2, etc.

Calculating the adiabatic temperature as a function of reaction conversion is more complicated than calculating the adiabatic temperature at complete conversion. We begin by defining the fractional conversion X:

$$\text{fractional conversion} \equiv X = \frac{\text{mols reactant consumed}}{\text{mols reactant initially}}. \tag{3.161}$$

Assume a basis for the reactor feed: 1 mol CO/min plus 2 mol H_2/min. For reactor 1 we must calculate the temperature rise when X mol/min CO reacts with $2X$ mol H_2/min to create X mol CH_3OH/min. Again we create fictitious paths of convenient thermodynamic states.

1. Split the reactor input into two streams. One stream is the portion that reacts: $2X$ mol CO/min plus $2X$ mol H_2/min. The other stream is the portion that does not react: 2 $(1-X)$ mol CO/min plus $2(1-X)$ mol H_2/min.
   1a. Cool the reacting portion from 225°C to 25°C.
   1b. React $2X$ mol CO/min with $2X$ mol H_2/min at 25°C to form X mol CH_4/min and X mol CO_2/min.
   1c. Warm X mol CH_4/min and X mol CO_2/min from 25°C to T_{final}.
2. Warm the unreacted portion $2(1-X)$ mol CO/min plus $2(1-X)$ mol H_2/min from 225°C to T_{final}.

Figure 3.51 is a flowsheet of the fictitious steps we devised to calculate the adiabatic temperature rise in reactor 1.

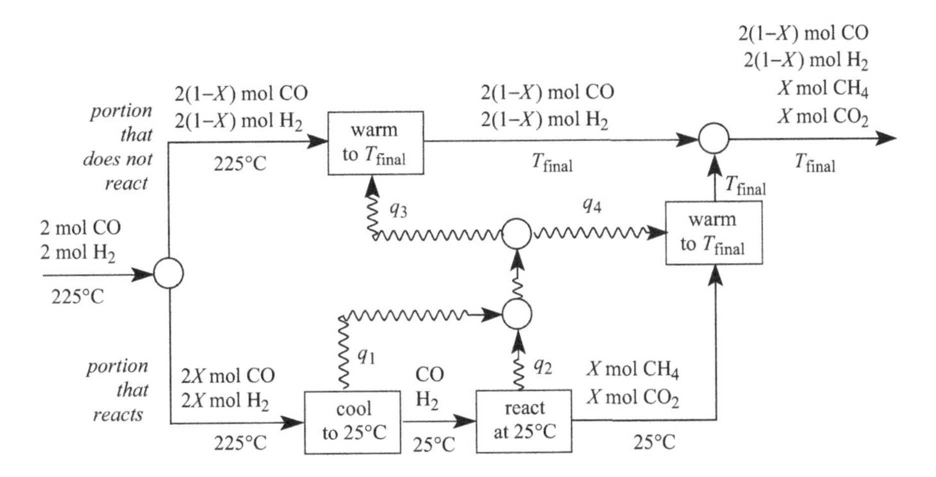

Figure 3.51 Fictitious thermodynamic steps to represent adiabatic reactor 1 with fractional conversion X.

Draw system borders around the fictitious energy combiner and the fictitious energy splitter, assume steady state, and write an energy balance:

$$\text{rate in} = \text{rate out}$$

$$q_1 + q_2 = q_3 + q_4. \tag{3.162}$$

Equation 3.162 states that the energy released by cooling a portion of the feed (q_1) plus the heat released by reacting a portion of the feed (q_2) warms the unreacted portion (q_3) and warms the product (q_4).

We begin with an accurate calculation and then approximate. We search on-line thermodynamic data and obtain polynomial representations for the heat capacities of CO, H_2, CH_4, and CO_2 valid over the range 0°C to 1500°C:

$$\bar{C}_{P,CO} = 28.95 + 4.11 \times 10^{-3}T + 3.548 \times 10^{-6}T^2 - 2.22 \times 10^{-9}T^3 \text{ J/(mol·K)} \quad (3.163)$$

$$\bar{C}_{P,H_2} = 28.84 + 7.65 \times 10^{-5}T + 3.288 \times 10^{-6}T^2 - 8.698 \times 10^{-10}T^3 \text{ J/(mol·K)} \quad (3.164)$$

$$\bar{C}_{P,CH_4} = 34.31 + 5.469 \times 10^{-2}T - 3.661 \times 10^{-6}T^2 - 1.10 \times 10^{-8}T^3 \text{ J/(mol·K)} \quad (3.165)$$

$$\bar{C}_{P,CO_2} = 36.11 + 4.233 \times 10^{-2}T - 2.887 \times 10^{-5}T^2 + 7.464 \times 10^{-9}T^3 \text{ J/(mol·K)}. \quad (3.166)$$

We then apply thermodynamics to express the fictitious energy flows in terms of measurable quantities:

$$q_1 = -\int_{225}^{25} (2X\bar{C}_{P,CO} + 2X\bar{C}_{P,H_2})dT = -2X\int_{225}^{25} (\bar{C}_{P,CO} + \bar{C}_{P,H_2})dT \quad (3.167)$$

$$q_2 = X\left(-\Delta\bar{H}^0_{rxn}\right) \quad (3.168)$$

$$q_3 = \int_{225}^{T_{final}} (2(1-X)\bar{C}_{P,CO} + 2(1-X)\bar{C}_{P,H_2})dT = 2(1-X)\int_{225}^{T_{final}} (\bar{C}_{P,CO} + \bar{C}_{P,H_2})dT \quad (3.169)$$

$$q_4 = \int_{25}^{T_{final}} (X\bar{C}_{P,CH_4} + X\bar{C}_{P,CO_2})dT = X\int_{25}^{T_{final}} (\bar{C}_{P,CH_4} + \bar{C}_{P,CO_2})dT. \quad (3.170)$$

Although we cannot obtain an expression for the final temperature in terms of the fractional conversion, we can calculate a fractional conversion as a function of the final temperature by substituting equations 3.167 through 3.170 into the energy balance, equation 3.162:

$$X = \frac{2\int_{225}^{T_{final}} (\bar{C}_{P,CO} + \bar{C}_{P,H_2})dT}{-2\int_{225}^{25} (\bar{C}_{P,CO} + \bar{C}_{P,H_2})dT - \Delta\bar{H}^0_{rxn} + 2\int_{225}^{T_{final}} (\bar{C}_{P,CO} + \bar{C}_{P,H_2})dT - \int_{25}^{T_{final}} (\bar{C}_{P,CH_4} + \bar{C}_{P,CO_2})dT}. \quad (3.171)$$

Calculating X with equation 3.171 is arduous with pencil, paper, and a calculator, but straightforward with a spreadsheet. The results are shown graphically in Figure 3.52 (on the next page).

The dashed line in Figure 3.52 reveals another feature; the result using equations 3.163 through 3.171 is well approximated by a straight line in the range 225–425°C. The least-squares best fit is obtained with an adiabatic temperature rise of 2430°C.

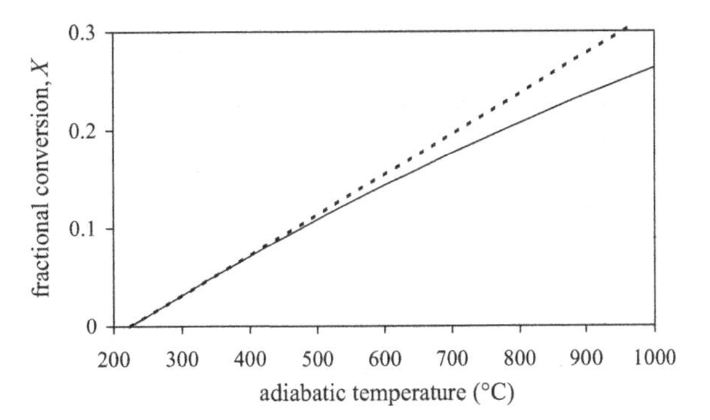

Figure 3.52 Fractional conversion of syngas to methane and carbon dioxide as a function of the final temperature in an adiabatic reactor with inlet temperature 225°C. The solid line is obtained from equation 3.171. The dashed straight line is the best fit in the range 225–425°C.

Use the straight-line approximation to design our series of reactors and heat exchangers. As shown in Figure 3.53, the first reactor reaches 425°C with a fractional conversion $X = 0.081$ (rising line with slope 1/2430°C). The output of reactor 1 is cooled to 225°C in a heat exchanger (horizontal line at $X = 0.081$), and then fed to reactor 2. Twelve adiabatic reactors are required to achieve 97% conversion.

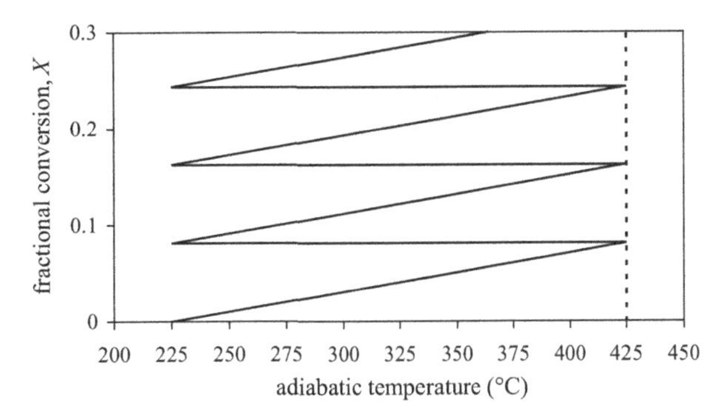

Figure 3.53 Graphical analysis of the reactor + heat exchanger series for the syngas reaction to methanol with an inlet temperature of 225°C and a maximum operating temperature of 425°C (plain carbon steel).

Of course, the design in Figure 3.50 with the temperatures given in Figure 3.53 is simplistic. The actual syngas reaction to methanol uses a more sophisticated design for heat removal, with different upper and lower temperatures as the reaction conversion increases.

To calculate the adiabatic temperature rise at a given reaction conversion, develop a mathematical model for a fictitious process similar to Figure 3.51; split the reactor input into two streams – a portion that reacts and a portion that does not. Although the temperature dependence of the component heat capacities must ultimately be considered, good estimates can be obtained with average heat capacities, and the calculation is much less complicated.

3.6.3 Chemical Reactions with Conversion Limited by Equilibrium

Context: The production of ammonia from nitrogen and hydrogen.

Concepts: Reaction conversion limited by chemical equilibrium.

Defining Question: How does one design a reactor for a reaction limited by chemical equilibrium?

To design the temperature range for a chemical reactor one must consider limits imposed by the materials used to construct the reactor (plain carbon steel limits the reactor to a maximum of 425°C) and limits imposed by chemical kinetics (the syngas reaction was unacceptably slow below 225°C). The temperature range may also be limited by chemical equilibrium. Some reactions are so strongly exothermic that the reverse reaction is negligible and one may assume the reactants are completely consumed. But all chemical reactions are reversible. For some reactions one must take into account that the reactants are not completely consumed. The fractional conversion at equilibrium, X_{eq}, for an exothermic reaction decreases with temperature; for the synthesis of methanol from syngas, X_{eq} = 0.985 at 225°C and X_{eq} = 0.862 at 425°C. The design depicted graphically in Figure 3.53 is valid for reactors 1 through 5, but the conversion in reactor 6 would not reach X = 1. In this section we consider the limit imposed by chemical equilibrium for a reaction in which the effect is more severe – the synthesis of ammonia from nitrogen and hydrogen given by reaction 2.1:

$$N_2 + 3H_2 \leftrightarrow 2NH_3. \qquad \text{rxn (2.1)}$$

We need an expression for the fractional conversion at equilibrium as a function of temperature. Recall from thermodynamics that the key is the equilibrium constant K:

$$-RT \ln K = \Delta \bar{G}^0_{\text{rxn}} = \Delta \bar{H}^0_{\text{rxn}} - T\Delta \bar{S}^0_{\text{rxn}}. \tag{3.172}$$

K is constant with respect to pressure, but varies with temperature as shown in equation 3.172, so the designation "constant" is a misnomer. The equilibrium constant is the reaction quotient Q at equilibrium, the ratio of the product activity (NH_3) divided the reactant activities (N_2 and H_2), each raised to the power of its stoichiometric coefficient:

$$K = \text{reactant quotient } Q \text{ at equilibrium} = \left(\frac{a^2_{NH_3}}{a_{N_2} a^3_{H_2}}\right)_{\text{eq}}. \tag{3.173}$$

We assume the gases in our reactor mixture are ideal, which obtains the approximation that each component's activity is the component's partial pressure divided by the standard pressure, P^0 = 1 bar = 0.987 atm:

$$K = \left(\frac{a^2_{NH_3}}{a_{N_2} a^3_{H_2}}\right)_{\text{eq}} \approx \frac{\left(\frac{P_{NH_3,\text{eq}}}{P^0}\right)^2}{\left(\frac{P_{N_2,\text{eq}}}{P^0}\right)\left(\frac{P_{H_2,\text{eq}}}{P^0}\right)^3} = \frac{P^2_{NH_3,\text{eq}}\left(P^0\right)^2}{P_{N_2,\text{eq}} P^3_{H_2,\text{eq}}}. \tag{3.174}$$

We must express the partial pressures in terms of the fractional conversion X. We assume Dalton's law,

$$P_{\text{total}} = P_{N_2} + P_{H_2} + P_{NH_3}, \tag{3.175}$$

and we assume an ideal mixture – the ratio of a component's partial pressure to the total pressure equals the ratio of the component's mol fraction. For example,

$$\frac{P_{N_2}}{P_{\text{total}}} = \frac{n_{N_2}}{n_{\text{total}}}. \tag{3.176}$$

Express each mol fraction in terms of the fractional conversion. Assume a basis: a stoichiometric mixture of 1 mol N_2 and 3 mol H_2 initially. At complete conversion, the reactor will contain 2 mol NH_3. At fractional conversion X, the reactor will contain 1 × (1–X) mol N_2, 3 × (1–X) mol H_2, and 2X mol NH_3. This is summarized in Table 3.5.

Table 3.5 Mols of reactants and products as a function of fractional conversion.

Component	Mols initially ($X = 0$)	Mols later
N_2	1	$1-X$
H_2	3	$3(1-X)$
NH_3	0	$2X$
Total mols	4	$2(2-X)$

Use equations 3.175 and 3.176 to write expressions for the partial pressures, summarized in Table 3.6.

Table 3.6 Partial pressures of reactants and products as a function of fractional conversion.

Component	Partial pressure initially ($X = 0$)	Partial pressure later
N_2	$P_{N_2} = \frac{n_{N_2}}{n_{\text{total}}} P_{\text{total}} = \frac{1}{4} P_{\text{total}}$	$\frac{1-X}{2(2-X)} P_{\text{total}}$
H_2	$P_{H_2} = \frac{n_{H_2}}{n_{\text{total}}} P_{\text{total}} = \frac{3}{4} P_{\text{total}}$	$\frac{3(1-X)}{2(2-X)} P_{\text{total}}$
NH_3	0	$\frac{2X}{2(2-X)} P_{\text{total}} = \frac{X}{2-X} P_{\text{total}}$
Total pressure	P_{total}	P_{total}

Substitute the expressions for the partial pressures in Table 3.6 into equation 3.174:

$$K \approx \frac{P^2_{NH_3,eq}(P^0)^2}{P_{N_2,eq}P^3_{H_2,eq}} = \frac{\left(\frac{X}{2-X} P_{\text{total}}\right)^2 (P^0)^2}{\left(\frac{1-X}{2(2-X)} P_{\text{total}}\right)\left(\frac{3(1-X)}{2(2-X)} P_{\text{total}}\right)^3} = \frac{16X^2(2-X)^2}{27(1-X)^4}\left(\frac{P^0}{P_{\text{total}}}\right)^2. \tag{3.177}$$

Substitute this approximation for K into equation 3.172. It would be nice to have an expression for X at equilibrium in terms of temperature. This is not possible, but we can solve for T:

$$\Delta\bar{H}^0_{\text{rxn}} - T\Delta\bar{S}^0_{\text{rxn}} = -RT \ln\left[\frac{16X^2(2-X)^2}{27(1-X)^4}\left(\frac{P^0}{P_{\text{total}}}\right)^2\right] \tag{3.178}$$

$$\frac{\Delta\bar{H}^0_{\text{rxn}}}{T} = \Delta\bar{S}^0_{\text{rxn}} - R \ln\left[\frac{16X^2(2-X)^2}{27(1-X)^4}\left(\frac{P^0}{P_{\text{total}}}\right)^2\right] \tag{3.179}$$

$$T = \frac{\Delta\bar{H}^0_{\text{rxn}}}{\Delta\bar{S}^0_{\text{rxn}} - R \ln\left[\frac{16X^2(2-X)^2}{27(1-X)^4}\left(\frac{P^0}{P_{\text{total}}}\right)^2\right]}. \tag{3.180}$$

To calculate T with equation 3.180 we need $\Delta\bar{H}^0_{\text{rxn}}$ and $\Delta\bar{S}^0_{\text{rxn}}$ as a function of temperature. From various postings of thermodynamic data, one finds the following heats of formation: N_2(gas) $\Delta\bar{H}_f = 0\,\text{kJ/mol}$, H_2(gas) $\Delta\bar{H}_f = 0\,\text{kJ/mol}$, and NH_3(gas) $\Delta\bar{H}_f = -46.1\,\text{kJ/mol}$. Use the heats of formation and equation 3.145 to calculate the heat of reaction:

$$\Delta\bar{H}^0_{\text{rxn}} = \left(2 \times \Delta\bar{H}^0_{\text{f,NH}_3}\right) - \left(1 \times \Delta\bar{H}^0_{\text{f,N}_2} + 3 \times \Delta\bar{H}^0_{\text{f,H}_2}\right) \quad (3.181)$$

$$= (2 \times (-46.1) - (1 \times 0 + 3 \times 0) = -92.2\ \text{kJ/mol}. \quad (3.182)$$

Also from postings one finds the following standard molar entropies: N_2 $\bar{S}^0 = 191.6/\text{J(mol·K)}$, H_2 $\bar{S}^0 = 130.7\,\text{J/(mol·K)}$, and NH_3(gas) $\bar{S}^0 = 192.5\,\text{J/(mol·K)}$. Analogous to the calculation of $\Delta\bar{H}^0_{\text{rxn}}$ one calculates $\Delta\bar{S}^0_{\text{rxn}} = -198.8\,\text{J/(mol·K)}$. We assume $\Delta\bar{H}^0_{\text{rxn}}$ and $\Delta\bar{S}^0_{\text{rxn}}$ are constant with respect to temperature, an approximation typically accurate to a few percent. This is equivalent to assuming the heat capacity of the reactants, $\bar{C}_{\text{P,N}_2} + 3\bar{C}_{\text{P,H}_2}$, equals the heat capacity of the products, $2\bar{C}_{\text{P,NH}_3}$, at any temperature; that is, the heat capacity of the constituent atoms is independent of the chemical arrangement of the atoms.

Like equation 3.171, calculating T with equation 3.180 is arduous with pencil, paper, and a calculator, but straightforward with a spreadsheet. Figure 3.54 shows the results.

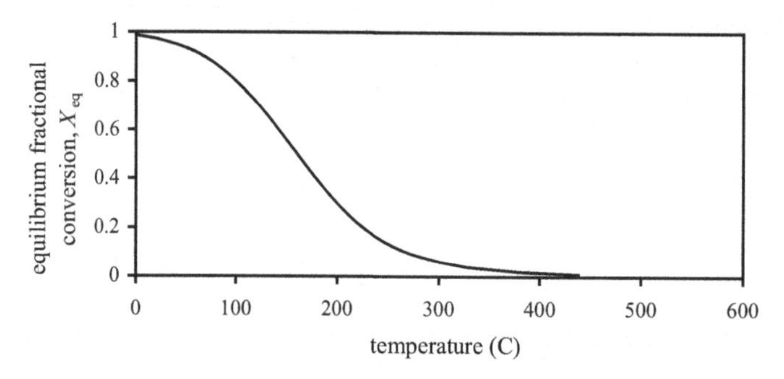

Figure 3.54 Fractional conversion of N_2 and H_2 to NH_3 at equilibrium as a function of temperature at 1 bar.

The limits imposed by chemical equilibrium are discouraging. For an appreciable reaction rate of this iron-catalyzed reaction, ammonia must be synthesized at 300°C or higher. But at 300°C, the equilibrium fractional conversion is only 0.06. The expression for the equilibrium constant K in terms of the fractional conversion X, equation 3.177, suggests a solution: K decreases with increasing pressure. We expect this from Le Chatelier's principle. The synthesis of ammonia converts 4 mol of gas to 2 mol of gas, so higher pressure shifts the equilibrium toward the product. We add equilibrium curves for 10 bar and 100 bar, shown in Figure 3.55. At 300°C and 100 bar, the equilibrium fractional conversion is an adequate 0.74.

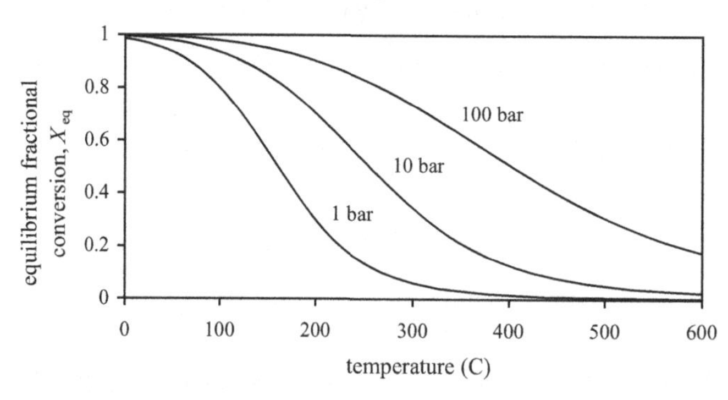

Figure 3.55 Fractional conversion of N_2 and H_2 to NH_3 at equilibrium as a function of temperature at total pressure 1 bar, 10 bar, and 100 bar.

We now design a series of pairs of adiabatic reactors at 100 bar and heat exchangers similar to the syngas process shown in Figure 3.50. But for the ammonia synthesis the minimum temperature is 300°C and the outlet temperature is dictated by chemical equilibrium. We need an expression for the adiabatic temperature as a function of fractional conversion. As we did with the syngas reactor, we will first use accurate expressions for the heat capacities as a function of temperature and then approximate with a best-fit straight line. We search on-line thermodynamic data and obtain polynomial representations for the heat capacities of N_2 and NH_3 valid over the range 0°C to 1500°C. The expression for the heat capacity of H_2 is given in equation 3.164; the expressions for N_2 and NH_2 are:

$$\bar{C}_{P,N_2} = 29.00 + 2.199 \times 10^{-3}T + 5.723 \times 10^{-6}T^2 - 2.871 \times 10^{-9}T^3 \text{ J/(mol·K)} \quad (3.183)$$

$$\bar{C}_{P,NH_3} = 35.15 + 2.954 \times 10^{-2}T + 4.421 \times 10^{-6}T^2 - 6.686 \times 10^{-9}T^3 \text{ J/(mol·K)}. \quad (3.184)$$

We then model the ammonia synthesis reactor as with the model for the methanol-synthesis reactor depicted in Figure 3.51. The expressions for the fictitious heat flows q_1 to q_4 are therefore

$$q_1 = -\int_{300}^{25} (X\bar{C}_{P,N_2} + 3X\bar{C}_{P,H_2})dT = -X\int_{300}^{25} (\bar{C}_{P,N_2} + 3\bar{C}_{P,H_2})dT \quad (3.185)$$

$$q_2 = X\left(-\Delta\bar{H}^0_{rxn}\right) \quad (3.186)$$

$$q_3 = \int_{300}^{T_{final}} ((1-X)\bar{C}_{P,N_2} + 3(1-X)\bar{C}_{P,H_2})dT = (1-X)\int_{300}^{T_{final}} (\bar{C}_{P,N_2} + 3\bar{C}_{P,H_2})dT \quad (3.187)$$

$$q_4 = \int_{25}^{T_{final}} 2X\bar{C}_{P,NH_3}dT = 2X\int_{25}^{T_{final}} \bar{C}_{P,NH_3}dT. \quad (3.188)$$

Finally, we obtain the expression for the fractional conversion as a function of the final temperature by substituting equations 3.164 and 3.183 through 3.188 into the energy balance, equation 3.162:

$$X = \frac{\int_{300}^{T_{final}} (\bar{C}_{P,N_2} + 3\bar{C}_{P,H_2})dT}{-\int_{300}^{25} (\bar{C}_{P,N_2} + 3\bar{C}_{P,H_2})dT - \Delta\bar{H}^0_{rxn} + \int_{300}^{T_{final}} (\bar{C}_{P,N_2} + 3\bar{C}_{P,H_2})dT - 2\int_{25}^{T_{final}} \bar{C}_{P,NH_3}dT}. \quad (3.189)$$

Calculating X with equation 3.189 is again arduous with pencil, paper, and a calculator, but again straightforward with a spreadsheet. The results are shown graphically in Figure 3.56 (on the next page). The temperature at complete conversion ($X = 1$) is 1207°C; the adiabatic temperature rise is 1207 – 300 = 907°C.

Similarly to the synthesis of methanol, the (arduous) result calculated with temperature-dependent heat capacities (the solid line in Figure 3.56) is well approximated by a (simple) straight line, as would be obtained with temperature-independent heat capacities. But the trick is knowing what to use for the average heat capacity. The average heat capacity for NH_3 over the range 300°C to 1200°C is 56.4 J/(mol·K), which yields an adiabatic temperature rise of 817°C, 10% lower than the accurate result. Although the

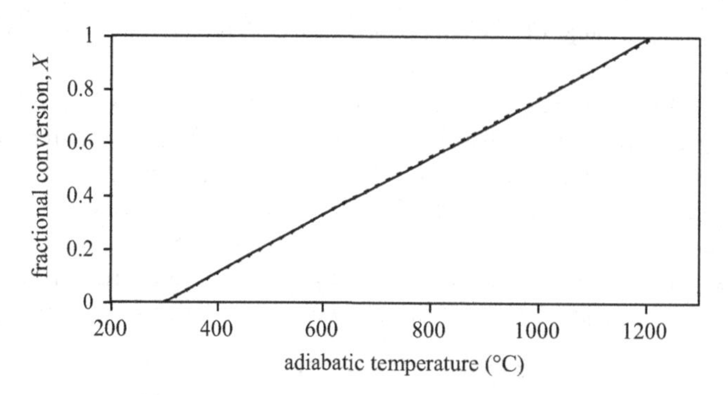

Figure 3.56 Fractional conversion of N_2 and H_2 to NH_3 as a function of the final temperature in an adiabatic reactor with inlet temperature 300°C. The solid line is obtained from equation 3.189. The dashed straight line is an approximation to the accurate result.

fractional conversion as a function of temperature is linear, temperature-dependent heat capacities must be used to obtain the adiabatic temperature rise and thus obtain the slope.

Use the straight-line approximation to design our series of reactors and heat exchangers at 100 bar. Whereas the temperature in the methanol-synthesis reactor could go to the limit imposed by the reactor construction materials (425°C), it is unwise to design a reactor to go to the limit imposed by chemical equilibrium. As the composition approaches equilibrium, the reaction rate approaches zero. So a better limit is a curve below the equilibrium curve. The dashed line in Figure 3.57 is an arbitrary $\Delta X = 0.05$ below the equilibrium curve, at which the reaction rate will be finite.

As shown in Figure 3.57, the first reactor reaches the limiting curve at 515°C with a fractional conversion $X = 0.24$ (the rising line with slope 1/907°C). The output of reactor 1 is cooled to 300°C in a heat exchanger (horizontal line at $X = 0.24$) and then fed to reactor 2. The second reactor reaches the limiting curve at 435°C with a fractional conversion $X = 0.38$. The increase in fractional conversion decreases with each subsequent reactor–heat-exchanger pair. What is the optimum number of reactor–heat-exchanger pairs? What is the optimum pressure in the reactors? Those answers derive from economic analysis, the third topic of mathematical modeling in Chapter 3.

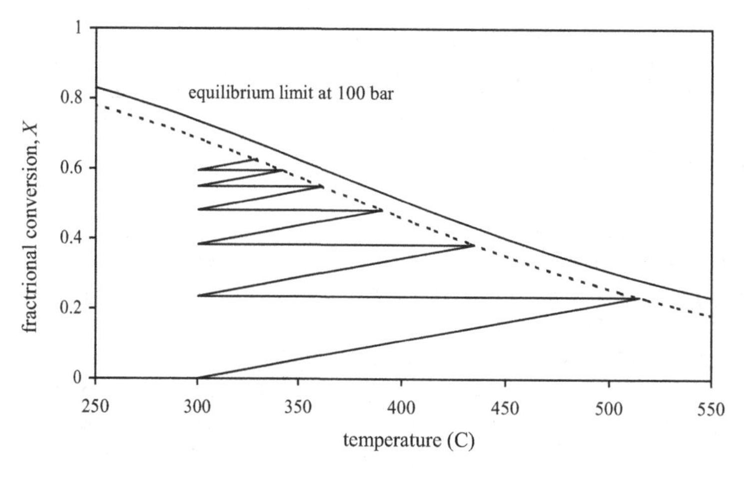

Figure 3.57 Graphical analysis of the reactor + heat exchanger series for ammonia synthesis at 100 bar. The feed to each reactor is at 300°C. The conversion in each reactor is limited to an arbitrary line 0.05 below the equilibrium limit at 100 bar.

As in the previous section, the design in Figure 3.50 with the temperatures given in Figure 3.57 is simplistic. The actual design for the synthesis of ammonia has a more sophisticated scheme to remove the heat of reaction.

To design a reactor for a reaction limited by chemical equilibrium, the key is to write the equilibrium constant K (defined in equation 3.172) in terms of the fractional conversion X. Create tables analogous to Tables 3.5 and 3.6; express the partial pressure of each component in terms of the initial partial pressure and X. Substitute the expression for K into equation 3.172 and solve for temperature. This relation yields an equilibrium line on a temperature-conversion map which defines the upper limit for a reactor trajectory line, as illustrated in Figure 3.57.

3.7 CHEMICAL PROCESS ECONOMICS

3.7.1 Economic Analysis of Operating a Chemical Process

Context: The financial aspects of operating a desalinator.

Concepts: Assets, operating costs, capital costs, revenue, depreciation, and return on investment.

Defining Question: What is the ultimate economic parameter for assessing a process?

We designed our desalinator to minimize energy requirements. Should we incorporate any modification that minimizes energy consumption? Why not? Because our true goal is to maximize economic viability. We focused on energy costs but we must also consider the cost of the equipment. We introduce here mathematical models for assessing the financial aspects of a design.

Process economics is the formalism for calculating the finances of making and selling a product. Process economics also applies to the finances of providing a service. The concept of process economics is straightforward. The accounting, however, can be complicated. But with this complication comes the opportunity for creativity. The economist John Maynard Keynes (1883–1946) remarked, "The avoidance of taxes is the only intellectual pursuit that still carries any reward."

We start by proposing the principle of conservation of assets. What is an asset? Like energy, assets can exist in many forms. An obvious form of asset is cash. But cash can be converted to other assets such as equipment. Is \$1000 cash equivalent to a piece of equipment worth \$1000? No, just as 1 kJ of thermal energy is not equivalent to 1 kJ of electrical energy. Electrical energy is a more usable form and converting thermal energy to electrical energy will cost you a fraction of the energy. Similarly, cash is a more liquid asset and converting equipment to cash may cost you a fraction of your asset.

By analogy with the principles of conservation of mass and conservation of energy, we start with the general equation

$$\text{rate of assets accumulation} = \frac{d(\text{assets})}{dt} = \text{rate of assets in} - \text{rate of assets out}. \quad (3.190)$$

As with previous conservation laws, we must define a system. The system might be your chemical company, as shown in Figure 3.58.

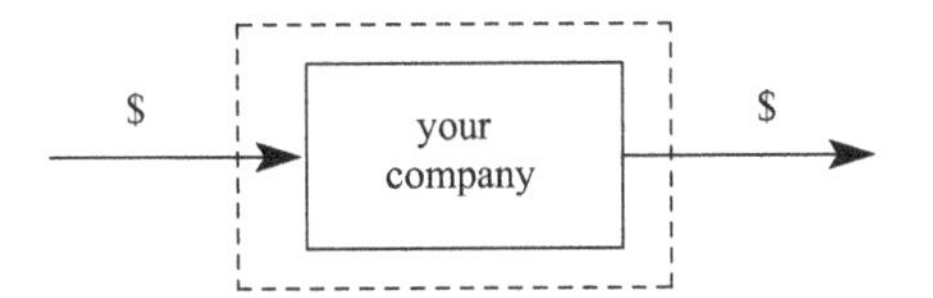

Figure 3.58 System borders drawn around a company to apply an asset balance.

The goal of every company is a positive flow of assets. Every company wants assets (such as cash) to accumulate. Writing an asset balance for a chemical process is never the simple task of counting piles of money, or subtracting the rate that money flows out from the rate that money flows in. Just as energy is measured indirectly through parameters such as temperature and pressure, assets are measured indirectly through parameters such as equipment and inventory. For energy balances we appealed to thermodynamics for equations to convert parameters such as temperature and pressure into the currency of energy. For asset balances we appeal to accounting and tax codes for relations to convert equipment and inventory into units of assets. Analogous to proportionality constants for converting temperature to kilojoules, there are proportionality constants for converting inventory to dollars. Note that the proportionality "constant" in an asset calculation is rarely constant over time.

An asset balance, equation 3.190, is usually written in the terminology of equation 3.191:

$$\text{profit (or loss)} = \text{revenue} - \text{operating costs.} \tag{3.191}$$

This is where process economics can get confusing. Operating costs include items such as paychecks to employees (assets flowing out) and the depreciation of equipment (assets consumed). The dimensions of equation 3.191 are money/time, typically expressed in dollars/year. Revenue in the chemical industry is usually derived from the sale of chemical commodities. Given a non-varying price for the chemicals and the production rate, the revenue may be calculated with equation 3.192:

$$\text{revenue} = \left(\frac{\text{product sold (kg)}}{\text{year}}\right)\left(\frac{\text{price (\$)}}{\text{kg}}\right), \tag{3.192}$$

which has units of $/year. Revenue can also be derived from services, such as remediation of a toxic waste site or licensing a patent.

Let's consider a simplified example. Assume you inherit a desalinator similar to the design in Section 3.2. You produce 100,000 gal/day of potable water that you sell at $14. per 1000 gal, plus 265,000 gal/day of brine for agriculture that you sell at $3. per 1000 gal. Your electric bill to run the desalinator is $43,000/month. The cost of labor to run the desalinator is $16.50/hr. The cost of space at the local pier is $3200 per month. In this idealized example we will neglect all other costs, such as maintenance. We assume the desalinator operates 24 hours a day, 52 weeks/year. Calculate the profit (or loss). First calculate the revenue:

$$\begin{aligned}\text{revenue} &= \left(\frac{100{,}000 \text{ gal potable water}}{\text{day}}\right)\left(\frac{14.\,\$}{1000 \text{ gal}}\right) \\ &\quad + \left(\frac{265{,}000 \text{ gal agricultural water}}{\text{day}}\right)\left(\frac{3.\,\$}{1000 \text{ gal}}\right) \\ &= \left(\frac{1400\,\$}{\text{day}} + \frac{795\,\$}{\text{day}}\right)\left(\frac{365 \text{ days}}{\text{year}}\right) = 801{,}175\,\$/\text{year}.\end{aligned} \tag{3.193}$$

Next calculate the operating costs:

$$\text{operating costs} = \text{electricity} + \text{rent} + \text{labor}$$

$$= \left(\frac{43{,}000\,\$}{\text{month}}\right)\left(\frac{12 \text{ months}}{\text{year}}\right) + \left(\frac{3{,}200\,\$}{\text{month}}\right)\left(\frac{12 \text{ months}}{\text{year}}\right) \tag{3.194}$$

$$+ \left(\frac{3 \text{ shifts}}{\text{day}}\right)\left(\frac{9 \text{ hours}}{\text{shift}}\right)\left(\frac{16.50\,\$}{\text{hour}}\right)\left(\frac{7 \text{ days}}{\text{week}}\right)\left(\frac{52 \text{ weeks}}{\text{year}}\right) \tag{3.195}$$

$$= 516{,}000\,\$/\text{year} + 38{,}400\,\$/\text{year} + 162{,}162\,\$/\text{year} \tag{3.196}$$

$$= 716{,}562\,\$/\text{year}. \tag{3.197}$$

And finally we calculate the profit from the definition in equation 3.191:

$$\text{profit (or loss)} = 801{,}175\ \$/\text{year} - 716{,}562\ \$/\text{year} = 84{,}613\ \$/\text{year}. \tag{3.198}$$

Let's make the example more realistic. Let's assume you first must purchase the desalinator, which costs \$340,000. Expenses for equipment are capital costs, not operating costs. As such, the expense of purchasing the desalinator is not included in equation 3.191 and thus does not affect the profit. But when we bought the desalinator, assets (\$340,000 cash) left the system. Why is a capital cost not included in an asset balance? This doesn't seem correct. Did we neglect something when the conservation of assets, equation 3.190, was converted to economic terminology in equation 3.191? Specifically, should not the term "rate of assets out" in equation 3.186 convert to "operating costs *plus capital costs*" in equation 3.191?

A capital expense is not a *net* flow of assets out of your company. Rather, a capital expense is a *conversion* of cash into equipment. Although cash has left the system (your company), equipment of equal value has entered the system, as shown in Figure 3.59.

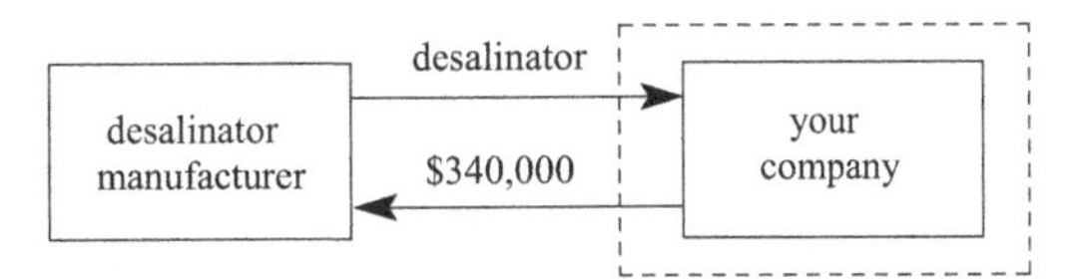

Figure 3.59 Asset flows during your desalinator purchase.

The *net* asset balance is zero. So, the asset balance, equation 3.191, is actually

$$\text{profit (or loss)} = \text{rate of assets in} - \text{rate of assets out} \tag{3.191}$$

$$= (\text{revenue} + \text{equipment purchased}) - (\text{operating costs} + \text{capital costs}), \tag{3.199}$$

and because

$$\text{equipment purchased} = \text{capital costs}, \tag{3.200}$$

equation 3.199 reduces to equation 3.191:

$$\text{profit (or loss)} = \text{revenue} - \text{operating costs}. \tag{3.191}$$

The equality in equation 3.200 is not to be taken literally. A piece of equipment worth \$1000 is not the same as \$1000 in cash. If you want to buy something, cash is a better form for one's assets. If you want to produce something, equipment is a better form for one's assets. Equation 3.200 states only that the asset value of the equipment is equal to the capital cost, in units of \$/year.

But wait, one could draw an asset flowsheet similar to Figure 3.59 for anything your company purchases. Does this mean that every purchase should be excluded from the operating cost? Consider the ammonia synthesis plant in Chapter 2, which has the overall reaction $4CH_4 + 5O_2 + 2N_2 \rightarrow 4NH_3 + 4CO_2 + 2H_2O$. The ammonia synthesis plant buys methane from a local utility, as shown in Figure 3.60.

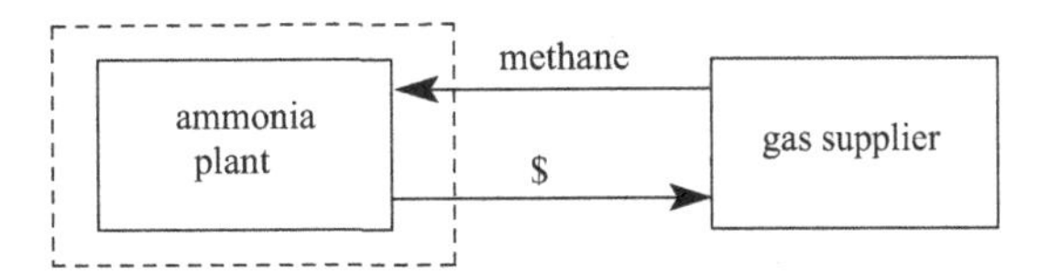

Figure 3.60 Asset flows during your methane purchase.

The similarity between Figures 3.59 and 3.60 suggests that the utility bill for methane should not be included as an operating cost, for the same reason that the payment for the desalinator was not an operating cost. But the asset balance for methane is different from the asset balance for the desalinator. Unlike the desalinator, the methane does not remain as an asset within the ammonia plant; the methane is consumed by chemical reaction. The flowsheet should show that the methane is consumed when it enters the system; the methane stream into the ammonia plant should extend to a "sink," as shown in Figure 3.61.

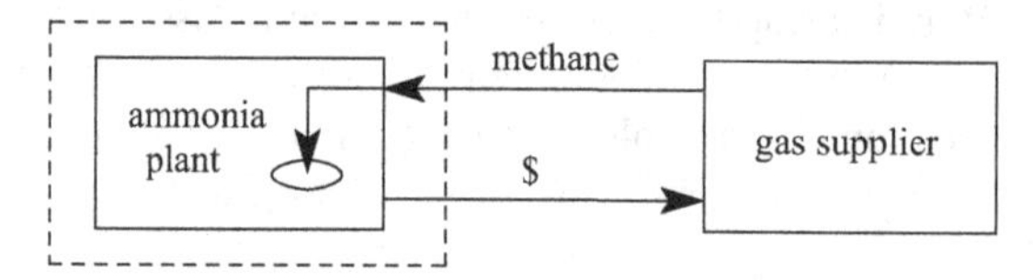

Figure 3.61 Asset flows during operation of your ammonia process.

We must modify equation 3.190 to account for assets that leave a system without crossing the system borders. We need to add another term: *rate of consumption*. Similarly, we must add a term to account for the ammonia asset that is created within the system borders. The second new term is *rate of formation*. The conservation of assets is modified as follows:

$$\frac{d(\text{assets})}{dt} = \text{rate of assets in} - \text{rate of assets out} + \text{rate of formation} - \text{rate of consumption}. \tag{3.201}$$

Equation 3.201 reverts to equation 3.191 if we redefine revenues as

$$\text{revenue} = \text{rate of assets in} + \text{rate of formation} \tag{3.202}$$

and we redefine operating costs as

$$\text{operating costs} = \text{rate of assets out} + \text{rate of consumption}. \tag{3.203}$$

For reactants, *rate of assets in* = *rate of consumption*, and there is a net decrease in assets, caused by the flow of dollars out of the system to pay for the reactants. For capital expenses, such as a piece of equipment, the asset is not consumed, at least not immediately. The rate of consumption of an asset is the distinction between items purchased as capital costs and items purchased as operating costs.

But the asset value of a capital item decreases with time. After many years of use, the desalinator will have little value other than as scrap metal. As such, even capital assets are eventually consumed. The rate of consumption of capital assets is called *depreciation*. As defined in equation 3.203, consumption owing to depreciation is an operating cost. The definition for operating costs is therefore

$$\text{operating costs} = \text{expenses} + \text{depreciation}. \tag{3.204}$$

Depreciation includes only the rate of consumption of capital assets. We do not include the consumption of a non-capital item, such as a reactant, because its rate of consumption is canceled by its rate of flow into the system.

How does one calculate the decrease in asset value of a capital item, such as a piece of equipment? Depreciation is seldom calculated by determining the actual worth of the equipment. Depreciation is usually determined by tax laws, because the rate of depreciation affects the profit and profit is taxed. Thus, it is possible to have equipment that is still useful but has been depreciated to zero value. It is also possible to have useless equipment that still has value in the eyes of the IRS.

There are several algorithms for calculating depreciation. The simplest is the straight-line method, given by

$$\text{depreciation} = \frac{\text{capital cost}}{\text{lifetime of equipment}}. \tag{3.205}$$

The lifetime of equipment is dictated by accounting conventions and tax code, not necessarily the equipment's productive life. A typical lifetime in chemical processes is 10 years.

Now that we know how to account for the capital items, let's return to the second example of calculating the profit (or loss) of buying and operating a desalinator. We will assume the same conditions as in the first example and add depreciation, assuming a straight-line 10-year rule:

$$\text{depreciation} = \frac{340{,}000\ \$}{10\ \text{years}} = 34{,}000\ \$/\text{year}. \tag{3.206}$$

Apply equations 3.191 and 3.204 to obtain a new equation for the profit:

$$\text{profit (or loss)} = \text{revenue-operating costs} \tag{3.191}$$

$$= \text{revenue} - (\text{expenses} + \text{depreciation}). \tag{3.207}$$

Substitute the asset flows from equations 3.193, 3.197, and 3.206:

$$\text{profit (or loss)} = 801{,}175\ \$/\text{year} - (716{,}562\ \$/\text{year} + 34{,}000\ \$/\text{year}) \tag{3.208}$$

$$= 50{,}613\ \$/\text{year}. \tag{3.209}$$

The profit is positive, which is necessary for economic viability. But positive profit is not sufficient. The ultimate financial parameter is the *return on investment* (ROI), defined by

$$\text{return on investment} = \frac{\text{profit}}{\text{capital cost}}. \tag{3.210}$$

Calculating the ROI for our desalinator is trivial:

$$\text{return on investment} = \frac{50{,}613\ \$/\text{year}}{340{,}000\ \$} = 15\%/\text{year}. \tag{3.211}$$

The ROI determines the economic viability of a process. Consider two designs, each with profits of $100,000 per year. Which do you build? The first design has a capital cost of $200,000 and the second design has a capital cost of $1,000,000. Thus the first design has an ROI of 50% per year and the second design has an ROI of 10% per year. Build the first design. But what if you have $1,000,000 to invest? Build five of the first design, or increase the capacity of the first design such that its capital cost is $1,000,000.

As illustrated by our analysis of the desalinator, the ROI of a process is a balance between capital costs and operating costs. Capital costs are the expenses of building. The chief capital costs, in approximate order of appearance when building a chemical process, are the costs to

- acquire a site,
- develop a site (or adapt an existing site),
- add services – plumbing and heating, ventilation, lighting and electrical, etc.,
- purchase chemical processing equipment – reactors, compressors, mixers, separators, and heat exchangers, and
- install processing equipment – piping, wiring, and controls.

Operating costs are the expenses of manufacturing. The chief operating costs, also known as plant costs or manufacturing costs, are the costs of

- materials – reactants, solvents, and fuels,
- labor,
- maintenance and repair,

- plant overhead – administration, custodial, accounting, etc.,
- distribution costs – labor, advertising, market research, etc.,
- fixed costs (fees paid even if nothing is produced) – rent, interest payments, property taxes, insurance, and
- depreciation.

3.7.2 Economic Analysis of Modifying a Chemical Process

Context: Modifying the desalinator units and the desalinator operation.

Concepts: Calculating changes in ROI caused by changes in operating costs and capital costs.

Defining Question: What is the basis of the alliterative adage "Process economics is steam vs. steel"?

Consider an alternative scheme for your desalinator company, a scheme that produces the same output but reduces the cost of labor and increases the capital cost. Assume one employee can operate three desalinators. Also assume your rented space on the pier can accommodate three desalinators. Calculate the profit and ROI of purchasing three desalinators and operating for one 9-hour shift per day, 365 days per year. Although your employee works 9 hours a day, the desalinators operate for only 8 hours a day – it takes time to start up and shut down each day.

The cost of labor in equation 3.193 decreases by a factor of 3 to 54,054 \$/year. Depreciation increases by a factor of 3 to 102,000 \$/year. The operating cost thus decreases 5.3% to 710,454 \$/year. The revenue is the same. The profit increases 79% to 90,721 \$/year. But the capital cost increases by a factor of 3, to \$1,020,000. Consequently, the ROI decreases substantially to 9% per year. Because the ROI is the ultimate financial parameter, this alternative scheme should not be adopted, even though the profit is higher.

Process economics can also assess a proposed modification to a process. Consider adding a precooler to the desalinator, as shown in Figure 3.62. A precooler would reduce the cooling cost in the slusher. But a precooler would increase the capital cost. Should we add a precooler?

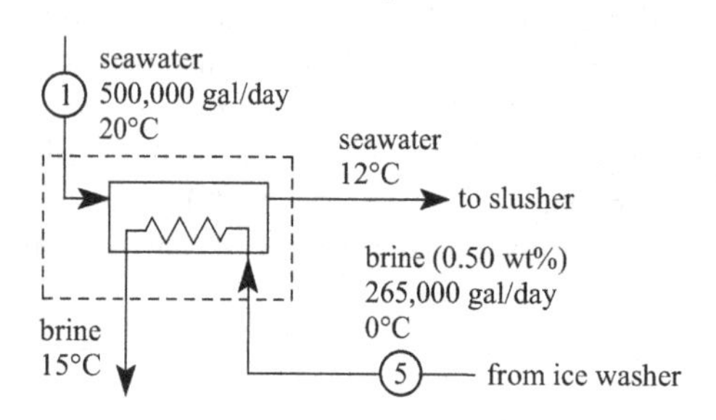

Figure 3.62 A proposed precooler for the desalinator in Figure 3.7.

Calculate the energy saved and convert to the units of the asset balance, \$/year. The energy saved is the energy we would have expended cooling the seawater to 12°C from 20°C:

$$q_{20^\circ\text{C to }12^\circ\text{C}} = F_{\text{seawater}} C_{P,\text{ seawater}} (\Delta T)_{\text{seawater}} \tag{3.212}$$

$$= \left(\frac{500{,}000 \text{ gal seawater}}{1 \text{ day}}\right)\left(\frac{1.025 \text{ kg seawater}}{1 \text{ L seawater}}\right)\left(\frac{3.79 \text{ L}}{1 \text{ gal}}\right)\left(\frac{4.18 \text{ kJ/kg}}{(\text{kg})(^\circ\text{C})}\right)$$

$$(20^\circ\text{C} - 12^\circ\text{C})\left(\frac{365 \text{ days}}{1 \text{ year}}\right) \tag{3.213}$$

$$= 2.4 \times 10^{10} \text{ kJ/year} \tag{3.214}$$

A typical heat pump (as would be used in the slusher unit) operating in the range 0°C to 20°C has a coefficient of performance of about 4; the heat pump requires 1 kJ electrical energy to pump 4 kJ of thermal energy from the seawater. Assume the electricity cost is 12¢ per kilowatt-hour (kWh), which converts to 0.033 \$/MJ. Calculate the annual cost of removing 2.4×10^{10} kJ by refrigeration:

$$\left(\frac{2.4\times10^{10}\,\text{kJ refrigeration}}{\text{year}}\right)\left(\frac{1\,\text{kJ electricity}}{4\,\text{kJ refrigeration}}\right)\left(\frac{0.033\,\$}{1000\,\text{kJ electricity}}\right)=196{,}000\,\$/\text{year}. \tag{3.215}$$

The precooler costs \$135,000 and is guaranteed to last 3 years. In addition, the precooler would reduce the capacity of the heat pump on the slusher. The smaller heat pump costs \$50,000 less than the heat pump required with no precooler.

Calculate the profit for the precooler option. There is no change in revenue – the desalinator production is the same with or without the precooler. The electricity bill decreases by 196,000 \$/year. We convince the IRS that because the precooler lasts only 3 years we should be allowed to use a lifetime of 3 years in our straight-line depreciation. (It is usually better to depreciate equipment as quickly as possible, which decreases the profits and thus decreases taxes.) Thus

$$\text{precooler depreciation} = \frac{135{,}000\ \$}{3\ \text{years}} = 45{,}000\ \$/\text{year}, \tag{3.216}$$

and the smaller heat pump on the desalinator decreases the desalinator depreciation by 50,000 \$/10 years = 5000 \$/yr. The change in profit is

$$\text{additional profit (or loss)} = \text{additional revenues} - \text{additional operating cost} \tag{3.191}$$

$$= 0-(-196{,}000\,\$/\text{year}+45{,}000\,\$/\text{year}-5{,}000\,\$/\text{year}) \tag{3.217}$$

$$= 156{,}000\,\$/\text{year}, \tag{3.218}$$

and finally, the ROI of the precooler proposal is

$$\text{return on investment} = \frac{156{,}000\ \$/\text{year}}{135{,}000\ \$} = 116\%/\text{year}. \tag{3.219}$$

The ROI strongly recommends installing the precooler. Calculate your company's overall ROI with the precooler. The profit with the precooler is $50{,}613\ \$/\text{year} + 158{,}000\ \$/\text{year} = 208{,}613\ \$/\text{year}$. The capital cost with the precooler is \$340,000 + \$135,000 – \$50,000 = \$425,000. The ROI with the precooler is 49% per year. We could have analyzed the overall ROI with the precooler to obtain the same recommendation: add the precooler.

Test your command of process economics with another example. After three years, when you are preparing to replace the precooler, a sales representative of an equipment supplier presents another option. She can deliver a precooler that performs the same as the precooler in Figure 3.62 and costs only \$105,000, but lasts only two years. The supplier for the two-year precooler warns (correctly) that the two-year precooler will decrease your profit by \$7,500/year. Should you change to the cheaper precooler? (Answer: yes – the two-year precooler's ROI is 141% per year, which is higher than the three-year precooler's ROI.)

The budget for a project is limited. One must decide how to allocate the budget between operating costs and capital costs. A major operating cost is energy, which is delivered to process units as steam. The process units are typically constructed of steel. Thus the adage "Process economics is steam vs. steel."

3.7.3 Economic Analysis for Evaluating Design Schemes

Context: Process economics of unreacted reactants: discard, separate, and recycle, or purge and recycle?
Concepts: Economy of scale and profit vs. ROI.
Defining Question: In what situations does the better process not produce the most profit?

Let's analyze the process economics of the three schemes to isomerize allyl benzene (A) to propenyl benzene (P) described in Sections 3.3, 3.4, and 3.5. Scheme I is a minimal design: a reactor and a separator to yield product, but unreacted reactants are discarded; the process with flow rates is summarized in Figure 3.14. Scheme II adds a second separator that eliminates a waste stream and yields a second product – purified solvent. Scheme II also recycles all the unreacted reactants; the process with flow rates is summarized in Figure 3.18. Scheme III is intermediate to Schemes I and II. Scheme III recycles a portion of the unreacted reactants and purges waste; the process with flow rates is summarized in Figure 3.21.

Begin with the capital cost. The capital cost of a process unit typically has at least two components. The first cost component is related to instrumentation and control. Every unit will require sensors (temperature and pressure), wiring, and control units. This cost is approximately independent of the unit's capacity, within a finite capacity range. For example, if you purchase a coffee maker, every unit has an electrical plug, an on–off switch, and perhaps a timer and temperature sensor, regardless of the coffee maker's capacity. The second cost component is the materials and labor to construct the unit. This cost is related to the size of the unit, but is not proportional to the size of the unit. A unit's capacity scales as the unit's volume, whereas the amount of material scales as the surface area. A sphere with two times the volume has only 1.6 ($= 2^{2/3}$) times as much surface area. A long cylinder with two times the volume has only 1.4 ($= 2^{1/2}$) times as much surface area. Typically, the capital cost for equipment scales as capacity raised to an exponent. A table of exponents for various types of equipment can be found in the textbook *Plant Design and Economics for Chemical Engineers* by Peters, Timmerhaus, and West (2002). Typically, the exponent ranges from 0.4 to 1.0. This is the economy of scale.

Assume the equipment prices for the process units in Schemes I, II, and III are given by the equations below, where F is the flow through the unit in kg/day:

$$\text{reactor capital cost (in \$)} = 800{,}000 + 105{,}000F^{0.6} \tag{3.219}$$

$$\text{separator I capital cost (in \$)} = 350{,}000 + 22{,}000F^{0.6} \tag{3.220}$$

$$\text{separator II capital cost (in \$)} = 750{,}000 + 430{,}000F^{0.6}\,. \tag{3.221}$$

The equipment capacities are summarized in Table 3.7.

Table 3.7 Equipment capacities for three schemes to isomerize allyl benzene (A) to propenyl benzene (P).

	Equipment capacities (kg/day)		
Process unit	Scheme I	Scheme II	Scheme III
Reactor	1,000	600	2,600
A-P separator at 165°C	1,000	600	2,600
A-S separator at 156.5°C	–	500	–

We apply equations 3.219 to 3.221 to calculate the capital costs in Table 3.8 (shown on the next page).

Table 3.8 Capital costs for the three schemes to isomerize allyl benzene (A) to propenyl benzene (P).

	Equipment capital costs ($)		
Process unit	Scheme I	Scheme II	Scheme III
Reactor	7,425,000	5,676,000	12,554,000
A-P separator at 165°C	1,738,000	1,372,000	2,813,000
A-S separator at 156.5°C	–	18,650,000	–
Total capital cost	9,163,000	25,698,000	15,366,000

The operating cost for a process unit is generally proportional to the flow rate through the unit. Assume the operating costs for the process units in Schemes I, II, and III are given by the equations below, where F is the flow through the unit in kg/day:

$$\text{reactor operating cost (in /year)} = 2{,}500F \tag{3.222}$$

$$\text{separator I operating cost (in /year)} = 780F \tag{3.223}$$

$$\text{separator II operating cost (in /year)} = 8500F. \tag{3.224}$$

We calculate the process operating costs in Table 3.9.

Table 3.9 Operating costs for the three schemes to isomerize allyl benzene (A) to propenyl benzene (P).

	Equipment operating costs ($/year)		
Process unit	Scheme I	Scheme II	Scheme III
Reactor	2,500,000	1,500,000	6,500,000
A-P separator at 165°C	780,000	468,000	2,028,000
A-S separator at 156.5°C	–	4,250,000	–
Total process operating cost	9,163,000	25,698,000	11,744,000

Assume the other major operating costs are the cost of reactants, waste disposal, labor, and depreciation. Each process operates 7 days per week for 48 weeks per year, for a total of 336 days per year. A mixture of 20 wt% A (allyl benzene) in solvent S (1-hexanol) costs $106/kg. Waste disposal costs $5/kg. Assume Schemes I and III, which have only one separator, require two operators per shift, three shifts per day. Scheme II requires three operators per shift. Each process scheme requires one supervisor. Assume the salaries with benefits are $200,000 per year for the supervisor and $90,000/year for each operator. Assume a straight-line depreciation with a 10-year lifetime. With these parameters we calculate the other operating costs in Table 3.10.

Table 3.10 Other operating costs for Schemes I, II, and III.

	Other operating costs ($/year)		
Operating cost	Scheme I	Scheme II	Scheme III
Reactants	35,616,000	17,808,000	21,340,000
Waste disposal	1,512,000	0	840,000
Labor	740,000	1,010,000	740,000
Depreciation	916,000	2,570,000	1,537,000
Equipment operating cost	9,163,000	25,698,000	11,744,000
Total operating cost	42,064,000	27,606,000	33,014,000

Product *P* (propenyl benzene) sells for \$1,300/kg and pure solvent *S* (1-hexanol) sells for \$26/kg. Each scheme produces 100 kg/day of *P*, which yields revenue of \$43,680,000/year. Scheme II also produces 400 kg/day of pure solvent, which yields additional revenue of \$3,494,000/year. The key economic parameters are listed in Table 3.11.

Table 3.11 Key economic parameters for Schemes I, II, and III.

	Revenue, profit, and ROI		
Economic parameter	Scheme I	Scheme II	Scheme III
Revenue (\$/year)	43,680,000	47,174,000	43,680,000
Profit (\$/year)	1,616,000	19,569,000	10,666,000
ROI	18%/year	76%/year	69%/year

Based on the fictitious economic parameters used here, Scheme II has the higher ROI and therefore is the better option. But recall that the weight fraction of *A* (allyl benzene) in the waste for Scheme III was arbitrarily set at 0.04, a value between 0.11 weight fraction *A* in the waste for Scheme I and zero *A* in the solvent product of Scheme II. We use equations 3.105 to 3.107 to calculate flow rates and equipment capacities for various weight fractions of *A* in the waste and use these data to calculate the economic parameters. The ROI for Scheme III is a maximum at 6.6 wt% *A* in the waste. The flow rates for 6.6 wt% *A* in the purged stream are summarized in the flowsheet in Figure 3.63 (compare to Figure 3.21).

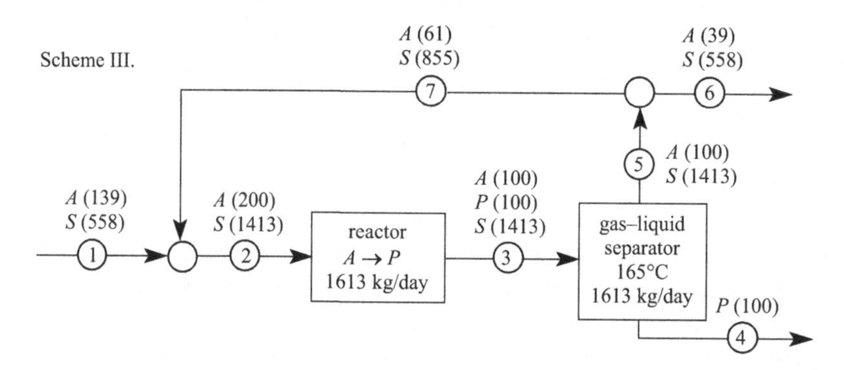

Figure 3.63 Flow rates and equipment capacities for 6.6 wt% *A* in the purged waste in Scheme III.

The total capital cost decreases to \$11,825,000 and the total operating cost is nearly unchanged at 33,057,000 \$/year; costs for reactants and waste disposal increase, but costs for operating the equipment and depreciation decrease. Thus profit is nearly constant at 10,623,000 \$ per year. But because the capital cost decreases, the ROI increases to 90.% per year. Therefore, Scheme III operating with 6.6 wt% *A* in the waste is the better option.

But the profit from Scheme II (\$19,569,000 \$/year) is much larger than the profit from the optimized Scheme III. Why is Scheme II not the better option? Why do we use ROI and not profit to select the better option? The key is that Scheme II requires a much larger investment; \$25,698,000 compared to \$11,825,000 for Scheme III. If we have \$25,698,000 to invest, we could build two Scheme III processes. Better yet, we could use the economy of scale and build a Scheme III process that produces 370 kg *P* per day, more than double a Scheme III process that costs \$11,825,000. (An investment 2.2 times larger will allow equipment with $2.2^{1/0.6} = 3.7$ times the capacity.) The ROI with this larger Scheme III would be 224% per year, substantially larger than 90% per year.

SUMMARY

Complex designs begin as simple designs. To choose between design alternatives, or to assess the impact of a design change, one must analyze and evaluate. One method of analysis is mathematical modeling.

Mathematical modeling translates a physical description to mathematical expressions. One translates by applying relations from physics, chemistry, and financial accounting. Some relations are universal laws, such as the conservation of energy. Some relations are constrained laws (constitutive equations), such as the ideal gas "law" or the capital cost of a reactor. The laws of physics and chemistry govern the behavior of a chemical process. The rules of mathematics govern the behavior of a mathematical model. As eloquently stated by Isaac Newton (1642–1727), mathematical models "subordinate the phenomena of nature to the laws of mathematics."

There are two types of laws: universal and constrained. Examples of universal laws are the conservation laws ("energy is conserved") and Newton's laws of motion (Newton's second law – "a change in motion is proportional to the impressed force"). Universal laws, as the name implies, are valid everywhere. Constrained laws are valid only in restricted domains. As such, constrained laws are not really "laws," although the terminology is common. A constrained law is more appropriately called a constitutive equation. An example of a constitutive equation is the ideal gas "law," $PV = nRT$, which is valid only at high temperature and low pressure. The adjectives "high" and "low" depend on the identity of the gas, as we shall see in Section 5.4.1.

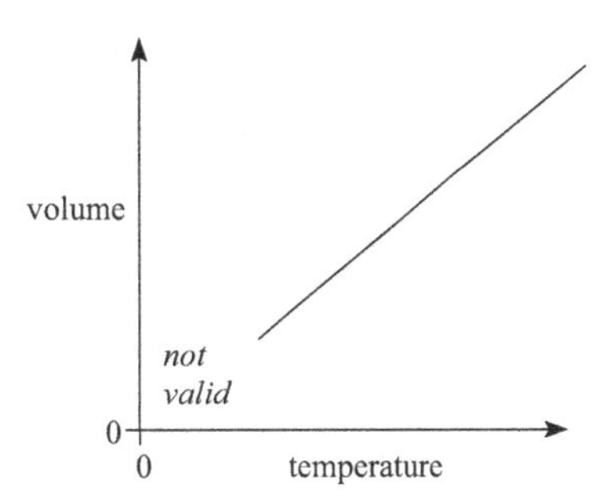

Figure 3.64 Volume of 1 mol of gas at 1 atm as a function of temperature in the region where the ideal gas law is valid. The slope of the straight line is R, the gas constant.

Another example of a constitutive equation is Stokes's "law": for a sphere of radius r moving at velocity v through a fluid of viscosity μ, the resistance force is $6\pi\mu rv$. Stokes's law is valid only at low velocity for small spheres, where "low" and "small" are relative terms that depend on the viscosity and density of the fluid. Yet another example is Henry's law: given a mixture of two liquids A and B, the partial pressure of B above the mixture is proportional to the mol fraction of B in the mixture. Henry's law is valid only for ideal solutions in which the mol fraction of B is low.

A mathematical model based on a universal law is valid universally. However, a mathematical model based on a constrained law must be constrained. A common error is to forget the constraints acquired when using a constitutive equation to translate a physical description to mathematics.

In each of the examples in this chapter, we followed three key steps. First, we studied the process to gain a qualitative understanding. What trends are expected? For example, how will the purity of the potable water depend on the amount of brine retained by the ice? What behavior is expected at extremes, such as no water in the brine stream? Second, we applied mathematical modeling to calculate a quantity or to yield operating equations. Third, we checked the result. Each step is important.

Mathematical models allow one to select the better design from various alternatives or to predict the performance of a given process as a function of operating conditions (concentrations, temperatures, etc.). Mathematical modeling can be used to derive operating equations. These can predict the performance of a process given the operating conditions. For example, equations 3.105 to 3.107 predict the flow rate in Scheme III given the weight fraction of reactant A in the purged waste stream. And equation 3.170 predicts the fractional conversion from the exit temperature in an adiabatic reactor. These equations comprise *design tools* for their respective processes.

REFERENCES

Felder, R. M. and Rousseau, R. W. 1986. *Elementary Principles of Chemical Processes*, 2nd edn., John Wiley & Sons, New York, NY.

Felder, R. M., Rousseau, R. W., and Bullard, L. G. 2016. *Elementary Principles of Chemical Processes*, 4th edn., John Wiley & Sons, New York, NY.

Harte, J. 1988. *Consider a Spherical Cow: A Course in Environmental Problem Solving*, University Science Books, Mill Valley, CA.

Himmelblau, D. M. 2003. *Basic Principles and Calculations in Chemical Engineering*, 7th edn., Prentice Hall, Upper Saddle River, NJ.

Luyben, W. L. and Wenzel, L. A. 1988. *Chemical Process Analysis: Mass and Energy Balances* Prentice Hall, Upper Saddle River, NJ.

Peters, M., Timmerhaus, K., and West, R. 2002. *Plant Design and Economics for Chemical Engineers* 5th edn., McGraw-Hill, New York, NY.

Reklaitis, G. V. 1983. *Introduction to Material and Energy Balances*, John Wiley & Sons, New York, NY.

Rudd, D. F., Powers, G. J., and Siirola, J. J. 1973. *Process Synthesis*, Prentice Hall, Upper Saddle River, NJ.

Zubizarreta, J. I. and Pinto, G. 1995. "An ancient method for cooling water explained by mass and heat transfer," *Chemical Engineering Education*, 29, 96–99.

EXERCISES

Mass Balances

3.1 Calculate the flow rate (in kg/min) and composition of the effluent from the mixer below.

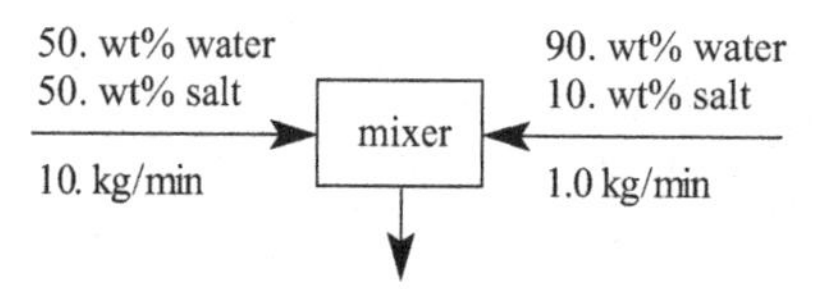

3.2 The unit below coats metal parts with plastic. A solution containing the plastic is applied to the metal. The solvent evaporates and leaves a plastic coating on the metal. Calculate the flow rates (in kg/min) of streams 1, 3, and 4.

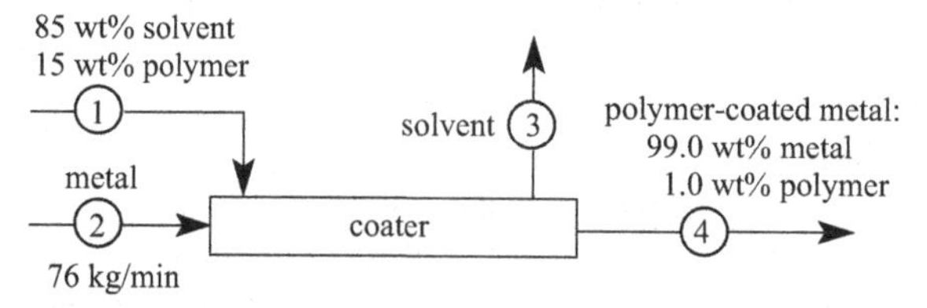

3.3 Consider two schemes for cleaning a paintbrush. In each scheme the brush contains 5.0 mL of paint and you have 0.30 L of solvent. The brush retains 10. mL of dirty solvent after cleaning. Dirty solvent is defined as solvent that contains paint.

Scheme 1: The paintbrush is washed in a container containing 0.30 L of solvent. The brush is withdrawn and the solvent on the brush evaporates.

Scheme 2: The solvent is divided into three containers in 0.10 L aliquots. The brush is washed in the first container, withdrawn, and washed in the second container, and then withdrawn and washed in the third container. The solvent retained by the brush does not evaporate before the brush is washed in the second and third containers. Finally, the brush is withdrawn from the third container and the solvent evaporates.

Calculate the amount of paint that remains on a brush with Schemes 1 and 2.

3.4 (A) The liquid–gas separator below removes some of the water from a dilute solution of dissolved solids. Calculate the flow rate (in kg/min) *and* composition of stream 3. The solids in stream 3 are completely dissolved.

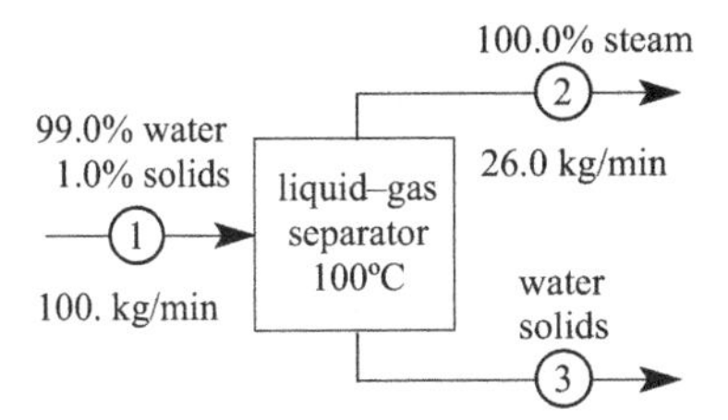

(B) Four separators are arranged in a cascade to increase the concentration of a dilute solution of dissolved solids. Calculate the flow rate (in kg/min) *and* composition of stream 9. The solids in stream 9 are completely dissolved.

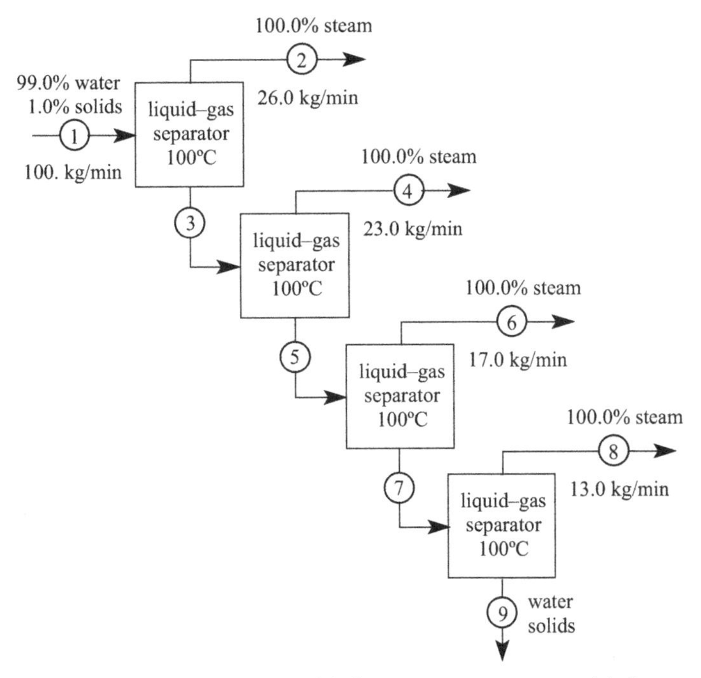

3.5 An artificial kidney, diagrammed below, removes wastes – chiefly urea, creatine, uric acid, and phosphate ions – from blood. Blood passes through tubes permeable to the wastes and a dialyzing fluid absorbs the wastes. In this exercise we consider only the removal of urea. The flow rate of blood through the artificial kidney is 0.200 kg/min. The concentration of urea is 2.15 g urea/kg blood in the input stream and 1.60 g urea/kg blood in the output stream.

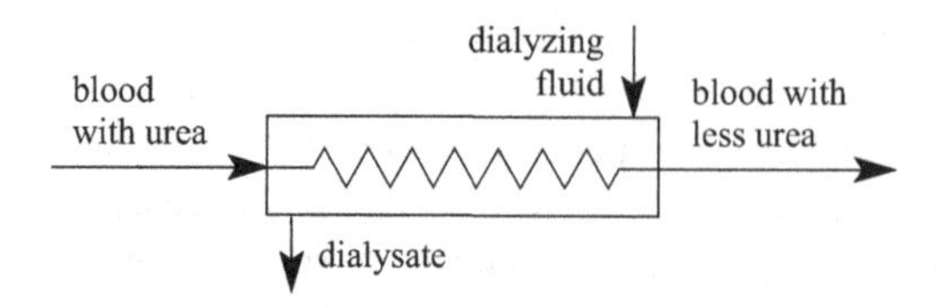

(A) Given that the flow rate of dialyzing fluid is 1.00 kg/min, calculate the concentration of urea in the dialysate (in g urea/kg dialysate).
(B) How long does it take (in minutes) to remove 5 g of urea from the blood?

Adapted by permission from Felder and Rousseau (1986), Exercise 22, p. 157.

3.6 The unit below contains a membrane permeable to both N_2 and O_2, but O_2 diffuses across the membrane faster. This separator produces O_2-enriched air, useful for remote medical applications. Because the air input is free, we only bother to extract a small fraction of the O_2, which improves the membrane performance. Calculate the flow rate of air (in mol/hr) entering the unit.

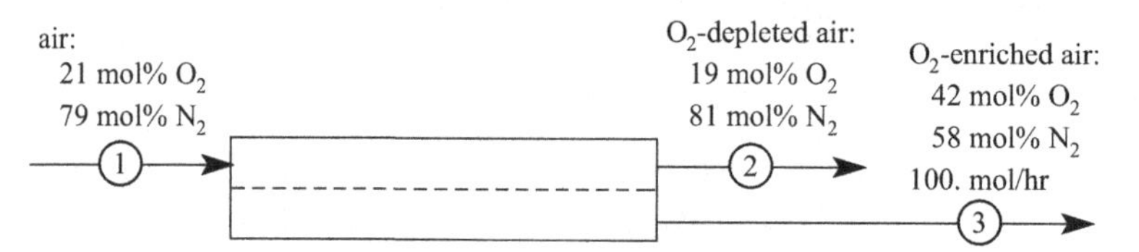

3.7 The absorber diagrammed below removes benzene from contaminated air.

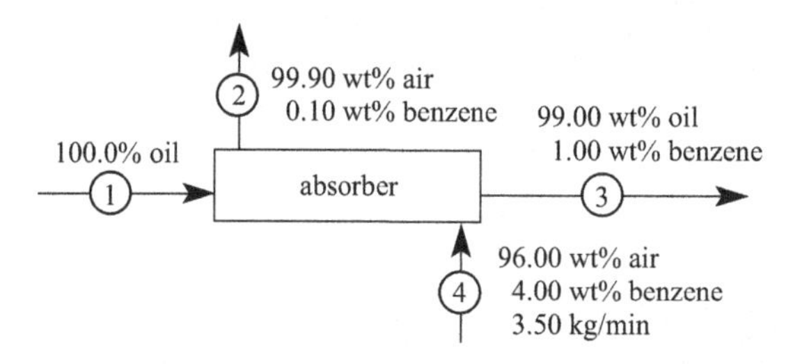

Calculate the flow rates of streams 1 and 2, in kg/min.

3.8 The separator below produces pure acetone from a mixture of acetone and toluene.

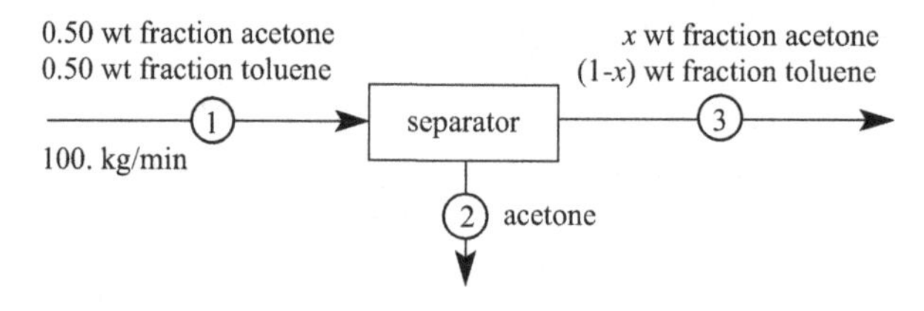

(A) Derive an equation for the flow rate of stream 2 as a function of x.
(B) Check your equation at $x = 0.50$. Does your predicted flow rate for stream 2 seem correct?
(C) Check your equation at a different value of x. Choose a value of x that yields an obvious answer for the flow rate of stream 2.

3.9 The flash drum below separates a mixture of methanol and ethanol into a vapor stream and a liquid stream.

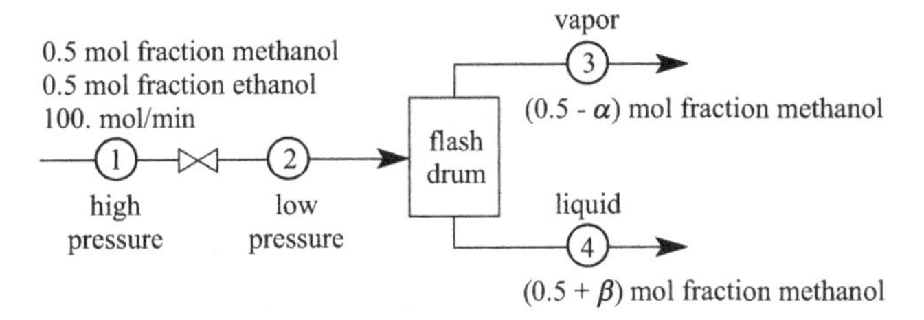

Calculate the flow rate of stream 3 (in mol/min) in terms of α and β. If you derived this correctly you proved the lever rule (see Chapter 4). You are not permitted to invoke the lever rule to calculate the flow rate of stream 3.

3.10 Orange juice concentrate can be prepared in one step as shown below.

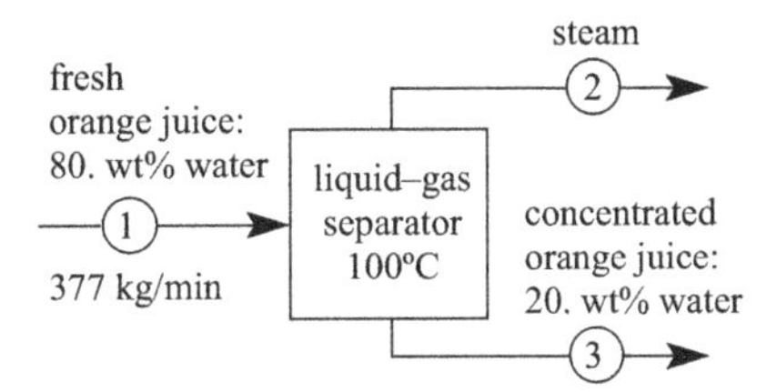

(A) Calculate the flow rate of stream 2 from the evaporator.

(B) The above process compromises the taste of fresh orange juice. The taste is improved markedly by blending a small amount of fresh orange juice with the output from the evaporator. To improve the taste, the process is modified such that the fresh orange juice stream is split; one fraction goes to the evaporator and the other fraction bypasses the evaporator and is mixed with the concentrate. The evaporator is modified to remove a larger fraction of the water entering the evaporator. Calculate the flow rate of water from the evaporator in the modified process.

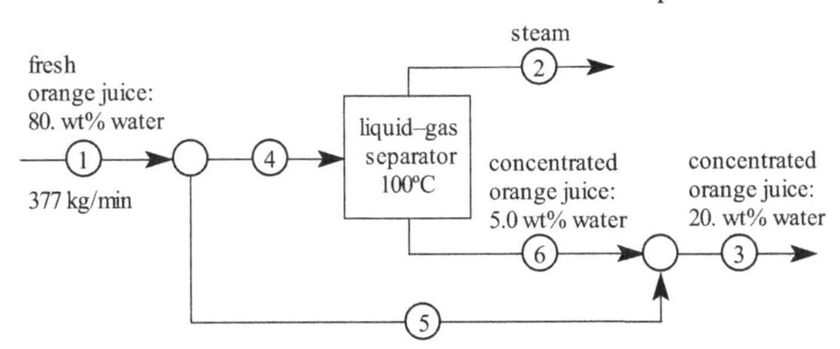

(C) Calculate the flow rates of streams 4 and 5 that will produce an orange juice concentrate (stream 3) in which 5.0 wt% of the non-water component has bypassed the evaporator.

Note: parts (B) and (C) can be simplified by judicious choices of system borders.

This exercise is based on a suggestion by G. "Chip" Bettle, Cornell University Chemical Engineering Class of 1965 (BS) and 1966 (MEng).

3.11 Your company mixes four fruit juices – orange juice, cranberry juice, apple juice, and guava juice – to produce three beverages – Red Sunshine Refresher, Mellow Morning Dee-Light, and Power-Rama – in the series of mixers shown below. Between each mixer is a splitter that removes 60 wt% of the mixture and sends the remaining 40 wt% to the next mixer to be blended further. All the juices have a density approximately equal to that of water.

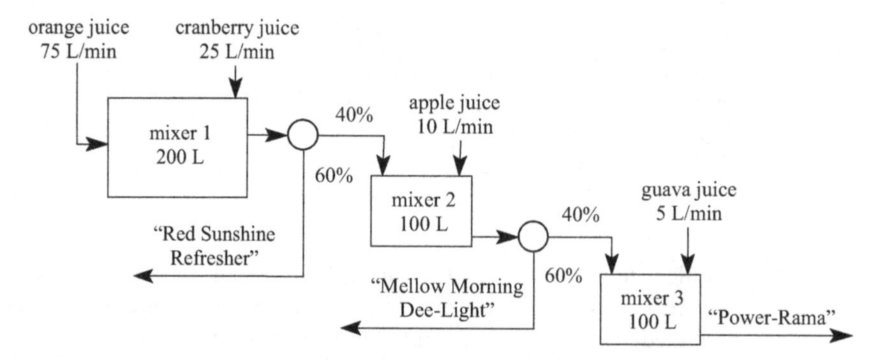

With this scheme operating at steady state, what is the flow rate and composition of each beverage? Fill in the table below and show your work separately.

Beverage	Flow rate (L/min)	Orange juice (wt%)	Cranberry juice (wt%)	Apple juice (wt%)	Guava juice (wt%)
Red Sunrise Refresher					
Mellow Morning Dee-Light					
Power-Rama					

3.12 The process below produces a detergent additive by mixing two solutions and then removing the water by air drying.

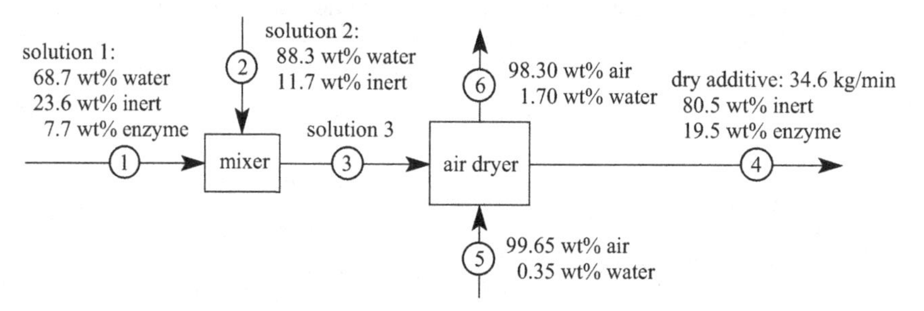

(A) Calculate the flow rates of streams 1 and 2 (in kg/min).
(B) Calculate the flow rate of stream 5.

3.13 The process below produces a detergent additive by mixing three solutions and then removing the water by air drying.

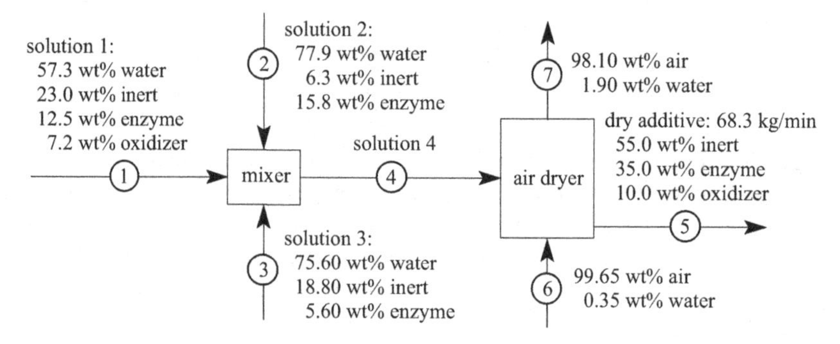

(A) Calculate the flow rate of inert substance in stream 4 (in kg/min).
(B) Calculate the flow rates of streams 1, 2, and 3, in kg/min.
(C) Calculate the flow rate of stream 6, in kg/min.

This exercise is based on a concept suggested by Jonathan Kienzle (Cornell ChemE '12).

3.14 Air at 95°F can hold a maximum of 3.52 wt% water, defined as 100% humidity at 95°F. On a hot, humid day the temperature is 95°F and the air contains 3.20 wt% water (90.9% humidity). The air is cooled to 68°F. Because air at 68°F can hold a maximum of 1.44 wt% water (100% humidity at 68°F), water condenses from the air.

Hot, humid air (95°F, 90.9% humidity) flows into a cooler at a rate 154.0 kg/min. Calculate the flow rates of the two streams leaving the cooler: (1) air at 68°F and 100% humidity, and (2) the water condensed from the air.

3.15 Air with 100% humidity is not pleasant, even at 68°F. Air with 50% humidity at 68°F can be produced by cooling some air below 68°F and then mixing with some hot, humid air at 95°F. This process is shown in the schematic below.

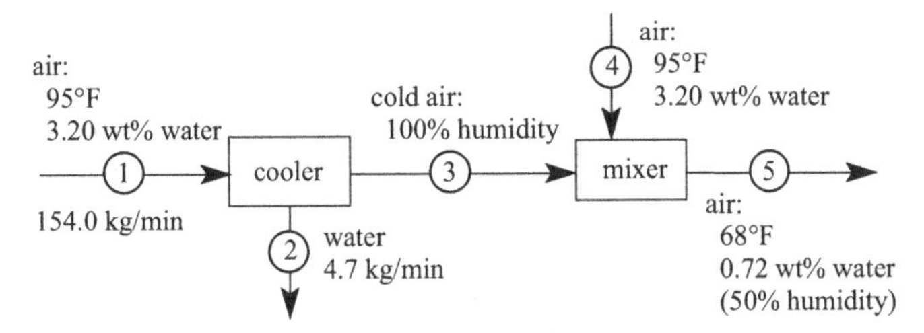

Calculate the flow rate of stream 4, the hot, humid air into the mixer.

3.16 Shown below is a portion of a process that produces styrene from benzene and ethylene (see Exercise 2.24). Benzene is recycled to an alkylation reactor and the by-product ethylbenzene is recycled to a dehydrogenation reactor. Stream 2 is 28.0% of stream 1. Also, 97.0% of the ethylbenzene in stream 3 leaves distillation column 2 via stream 4. Calculate the flow rate and composition of stream 5.

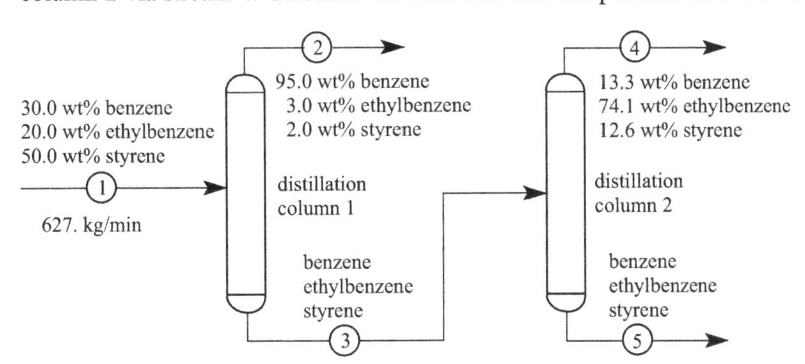

3.17 Your company manufactures an elixir that must contain exactly 1.00 wt% of active ingredient A. The elixir is prepared in 100. kg batches, which are stored. Each batch is tested for purity and wt% A, and then pumped to a process that packages the elixir into small vials. Two consecutive batches have unacceptable concentrations of A: batch 1 contains only 0.80 wt% A, whereas batch 2 contains 1.05 wt% A. Rather than discard both batches, you decide to mix batch 1 and batch 2, to produce product with exactly 1.00 wt% A.

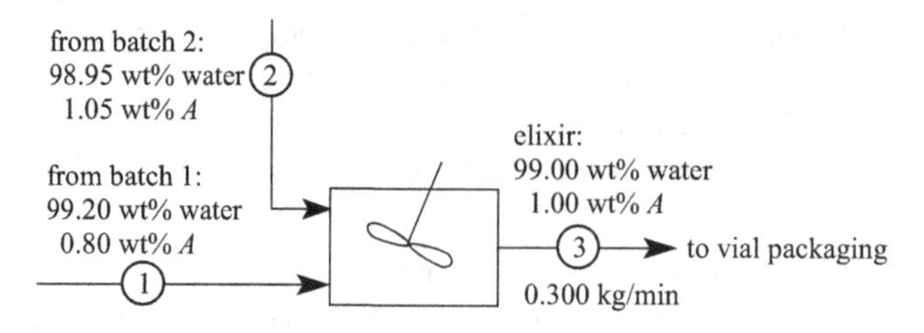

(A) Calculate the flow rates of stream 1 and stream 2 (in kg/min).

(B) Because the flow rates for streams 1 and 2 are not equal, one batch will be consumed before the other batch. Which batch is consumed first? How much of the other batch remains when one batch is exhausted? **Hint:** It may be useful to first determine the *time* at which one batch is consumed.

(C) To completely consume both batches, you decide to mix batches 1 and 2 at equal flow rates, and add a concentrated solution of A, as shown below. Calculate the flow rate of concentrated solution, stream 4.

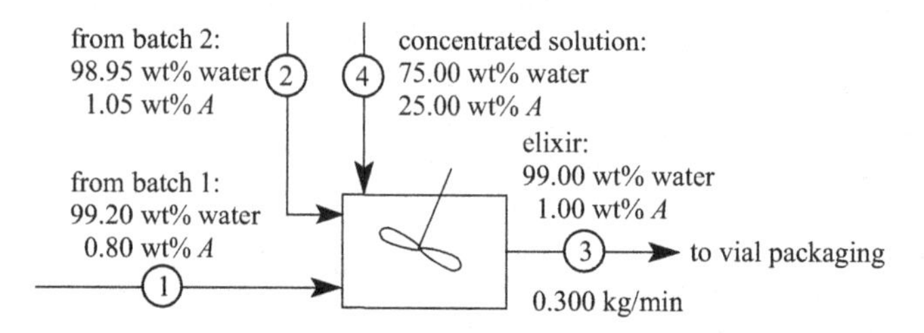

3.18 Dissolved compounds can be transferred from one liquid to another liquid by extraction, which is similar to absorption. Whereas absorption is transfer from a gas phase to a liquid phase, extraction is transfer from one liquid phase to another liquid phase. To maintain two distinct liquid phases, the liquids must be immiscible. The extractor below transfers compound A from water to oil.

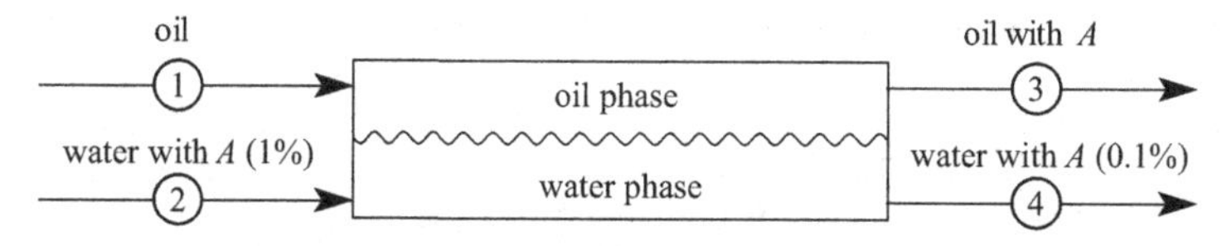

The rate at which A transfers from the water phase to the oil phase is proportional to the interfacial area between the two phases and to the amount the concentrations differ from equilibrium concentrations. For this particular system the rate of transfer is given by the following empirical equation:

$$\text{rate of } A \text{ transfer(kg}A/\text{ min)} = k\left([A]_{\text{water}} - K[A]_{\text{oil}}\right).$$

$[A]_{\text{water}}$ is the concentration of A in water, in units of (kg A)/(kg water). $[A]_{\text{oil}}$ is the concentration of A in oil, in units of (kg A)/(kg oil). The parameters k and K, determined by experiment, are $k = 46.1$ (kg water)/min and $K = 1.34$ (kg oil)/(kg water).

This equation has been used to design an extractor such that the oil and water phases leaving the extractor (streams 3 and 4) are in equilibrium. That is, the rate of A transfer is zero for the exit streams 3 and 4. Note that some A remains in the water in stream 4. The amount of A in stream 4 depends on the flow rate of oil, as you shall now calculate. (In Chapter 4 you will learn more creative ways to increase the amount of A extracted.)

Assume the flow rate of water in stream 2 is 100. kg/min, $[A]_{water,2} = 0.010$, and $[A]_{water,4} = 0.0010$; the extractor removes 90.% of the A from stream 1.

(A) Calculate the flow rate of A in the oil phase leaving the extractor (stream 3).
(B) Calculate the flow rate of oil entering the extractor (stream 1). **Hint:** Streams 3 and 4 are in equilibrium so the rate of transfer between streams 3 and 4 is zero.
(C) Now assume the extractor removes 99% of the A from stream 1, so $[A]_{water,4} =$ 0.00010 (kg A)/(kg water). All other parameters remain the same and streams 3 and 4 are again in equilibrium. Calculate the flow rate of oil entering the extractor (stream 1) needed to extract 99% of the A.

3.19 We have a precious compound (call it P) dissolved in water. We wish to isolate this compound, but distillation is impractical because P's boiling point is close to water's boiling point. Instead, we will transfer the compound to acetone (b.p. = 56°C) and evaporate the acetone to isolate P.

Because water and acetone are miscible, we cannot mix water and acetone in a single extractor (see Exercise 3.18 for a description of an extractor). Instead, we must use an intermediate liquid immiscible with both water and acetone. The oil in the flow diagram below serves as this intermediate.

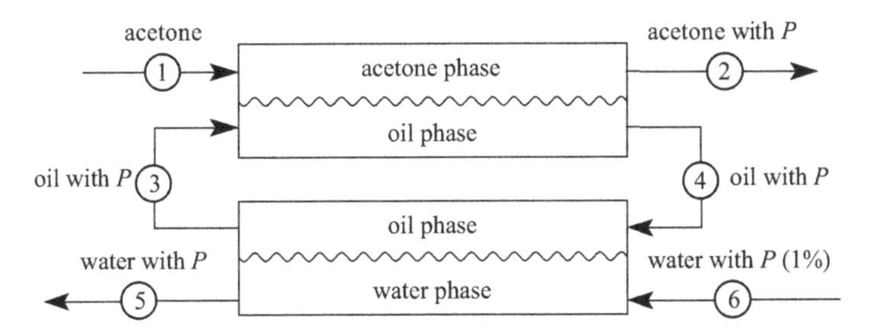

The rate at which P transfers from the water phase to the oil phase is given by the following empirical equation:

$$\text{rate of } P \text{ transfer (kg} P/\text{min)} = k_1\left([P]_{\text{water}} - K_1[P]_{\text{oil}}\right).$$

$[P]_{water}$ is the concentration of P in water, defined as (kg P)/(kg water), and $[P]_{oil}$ is the concentration of P in oil, defined as (kg P)/(kg oil). The parameters k_1 and K_1, determined by experiments on this particular extractor, are $k_1 = 85.6$ (kg water)/min and $K_1 = 1.20$ (kg oil)/(kg water).

Similarly, the rate at which P transfers from the oil phase to the acetone phase is given by a similar empirical equation:

$$\text{rate of } P \text{ transfer (kg } P/\text{min)} = k_2\left([P]_{\text{oil}} - K_2[P]_{\text{acetone}}\right).$$

$[P]_{acetone}$ is the concentration of P in acetone, defined as (kg P)/(kg acetone). The parameters k_2 and K_2, determined by experiments on this particular extractor, are $k_2 = 95.1$ (kg water)/min and $K_2 = 0.15$ (kg acetone)/(kg oil).

The extractors are sufficiently large to ensure that the two streams leaving each extractor are in equilibrium. That is, streams 2 and 4 are in equilibrium; the rate of P transfer is zero between the exit streams 2 and 4. Similarly, streams 3 and 5 are in equilibrium; the rate of P transfer is zero between the exit streams 3 and 5.

The flow rate of water in stream 6 is 100. kg/min and the flow rate of acetone in stream 1 is 800. kg/min. The oil–water extractor removes 90.% of the P from stream 6; $[A]_{water,6} = 0.010$, and $[A]_{water,5} = 0.0010$. The entire process runs at steady state.

(A) Calculate the flow rate of P in stream 2.
(B) Calculate the concentration of P in stream 3, $[P]_{oil,3}$. **Hint:** Streams 3 and 5 are in equilibrium so the rate of transfer between streams 3 and 5 is zero.
(C) Calculate the concentration of P in stream 4, $[P]_{oil,4}$. **Hint:** Streams 2 and 4 are in equilibrium so the rate of transfer between streams 2 and 4 is zero.
(D) Calculate the flow rate of oil in stream 3.
(E) How would you change the operation of this process to increase the amount of P recovered to 95%? Can this be accomplished by changing only the flow rate of the oil? Assume the flow rate and composition of stream 6 remain unchanged.

3.20 A car wash uses the process below to recycle its dirty water. Because the water for washing cars need not be pure, some dirty water is mixed with the water from the solid–liquid separator.

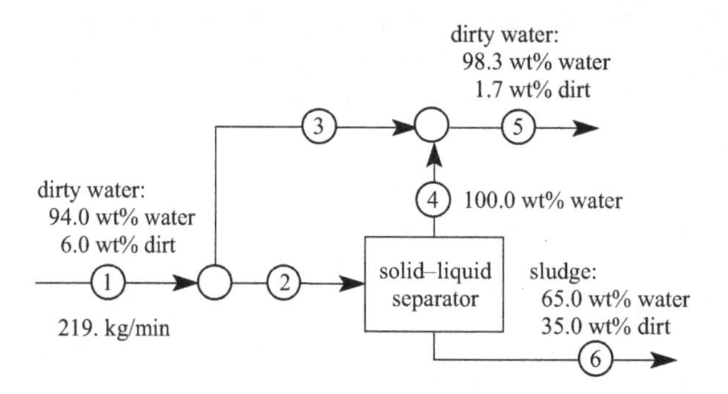

(A) Calculate the flow rate of sludge, stream 6 (in kg/min).
(B) Calculate the flow rate of less dirty water (stream 5) in kg/min.
(C) Calculate the flow rate of dirty water sent to the solid–liquid separator (stream 2) in kg/min.

3.21 A non-volatile oil absorbs hydrogen sulfide impurity from a hydrogen stream. The H_2S is distilled from the oil and the oil is recycled, as in the process below.

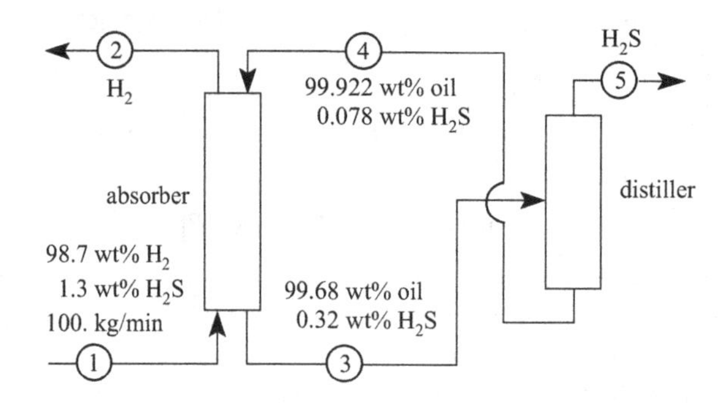

(A) Calculate the flow rate of H_2S in stream 5, in kg/min.
(B) Calculate the flow rate of oil in stream 3, in kg/min.

3.22 Mass balances are useful for analyzing large processes, such as global cycles. Consider this simplified flowsheet for the Earth's water (H_2O) cycle, where F_{ij} represents the flow rate from reservoir i to reservoir j.

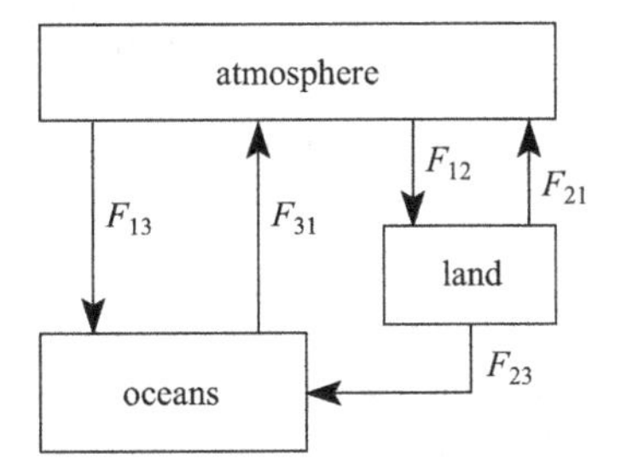

The amount of H_2O in each reservoir varies dramatically, as shown in the table below.

H_2O reservoir	Mass[a]
atmosphere	13 Eg
lithosphere (lands)	192 Eg
hydrosphere (oceans)	1,350,000 Eg

[a] 1 Eg = 1 exagram = 10^{18} g.

The Earth's water cycle is in a steady state. Precipitation from the atmosphere to the oceans is 3.83 times the precipitation from the atmosphere to land. Given F_{31} = 456 Eg/year and F_{23} = 46 Eg/year, calculate F_{12}, F_{21}, and F_{13}.

Data from J. Harte, Consider a Spherical Cow: A Course in Environmental Problem Solving (University Science Books, Mill Valley CA, 1988).

3.23 The principles of unit operations and process flowsheets can be applied to environmental systems as well. A model for the Earth's carbon cycle prior to the Industrial Revolution in 1850 is shown below.[9] The process has six units: one unit for the atmosphere, three units for the oceans (hydrosphere), and two units for the land (lithosphere). Carbon in the atmosphere is chiefly carbon dioxide, carbon in the lithosphere is chiefly cellulose, and carbon in the hydrosphere is chiefly carbonates.

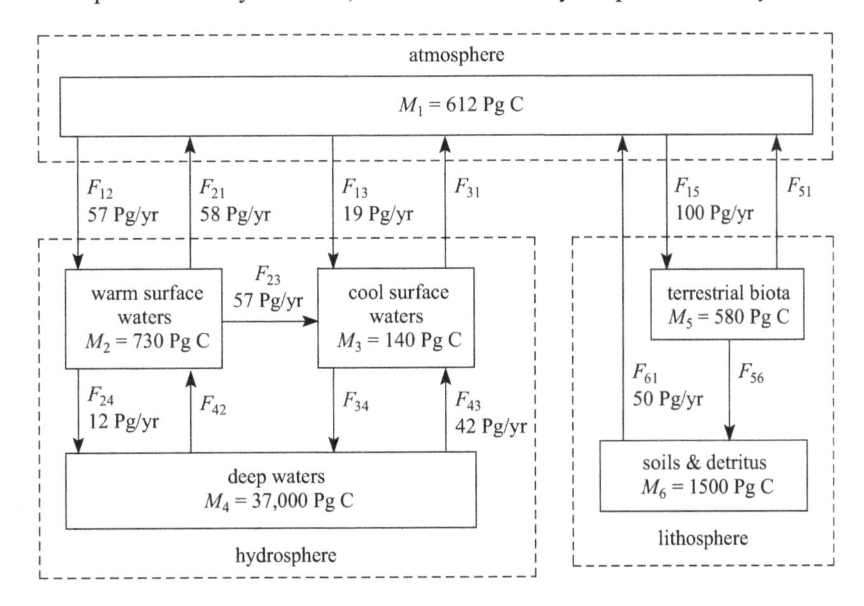

[9] The model is adapted from Schmitz, R. A. 2002. "The Earth's carbon cycle," *Chemical Engineering Education*, **36**, 296–299.

Prior to 1850, the Earth's carbon cycle was in a steady state. The model shows M_i, the mass of carbon contained in unit i in petagrams (1 Pg = 10^{15} g). The model also shows the rates carbon transfers between units as F_{ij}, the flow rate in Pg/year from reservoir i to reservoir j. The flow rates given are the steady-state values.

(A) Calculate flow rates F_{31}, F_{34}, F_{42}, F_{51}, and F_{56} in Pg/year.

(B) The Industrial Revolution introduced two activities that displaced the global carbon cycle from steady state: deforestation and the burning of fossil fuels. Deforestation is defined as forest land permanently lost, for example, to urban areas. Describe how the model should be changed to account for deforestation.

(C) Fossil fuels are a carbon reservoir estimated to contain 5300 Pg C. Describe how the model must be changed to account for burning fossil fuels.

(D) The amount of carbon in the atmosphere and its effect on the global temperature is a critical topic. The simple solution is to decrease the rate carbon enters the atmosphere due to deforestation and the burning of fossil fuels. But the problem can be redefined: balance the effects of deforestation and burning fossil fuels by increasing existing rates at which carbon leaves the atmosphere to existing reservoirs, or by creating new carbon reservoirs.

Devise a strategy to decrease CO_2 in the atmosphere, or select a proposed strategy from the current literature and show how the strategy would be modeled by the flowsheet above. For an example strategy, see Exercise 2.21. Another strategy is to harvest fast-growing grasses and bury the grasses. Or create an enormous algae pond in the Pacific Ocean in which the algae feed shellfish, and the shellfish are then sunk into deep waters. Or pump seawater to the Sahara Desert, use solar energy to desalinate it by evaporation, and then use the potable water to irrigate vast grasslands.

3.24 The steady-state process below extracts oil from soybeans.

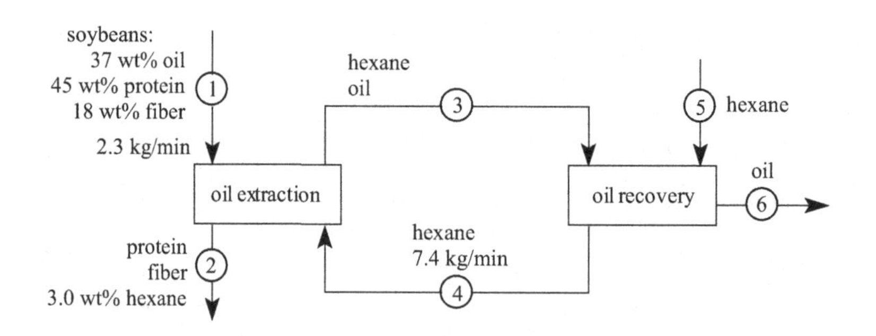

(A) Calculate the flow rate and composition of stream 2.

(B) Calculate the flow rate of stream 6.

(C) Calculate the flow rate of stream 5.

Adapted from Reklaitis, G. V. (1983). *Introduction to Material and Energy Balances*, Wiley, NY, exercise 2.29, p. 99 and Rudd, D. F., Powers, G. J., and Siirola, J. J. (1973). *Process Synthesis*, Prentice Hall, NJ, problem 7, p. 94.

3.25 Air can be dried by bubbling through sulfuric acid. The absorber shown below passes air countercurrent to a sulfuric acid solution. 99.0 wt% sulfuric acid is added to the recycled sulfuric acid solution to maintain a concentration of 94.0 wt% sulfuric acid into the absorber.

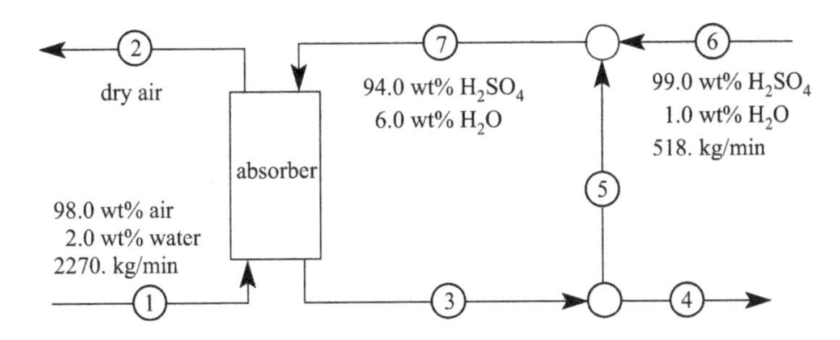

(A) Calculate the flow rate *and* composition of the effluent, stream 4.
(B) Calculate the flow rate *and* composition of the recycle, stream 5.

Adapted from Luyben, W. L., and Wenzel, L. A. (1988). *Chemical Process Analysis: Mass and Energy Balances*, Prentice Hall, Upper Saddle River, NJ, exercise 4–9, p. 107.

3.26 The process below removes water from a suspension of solid, insoluble particles of P.

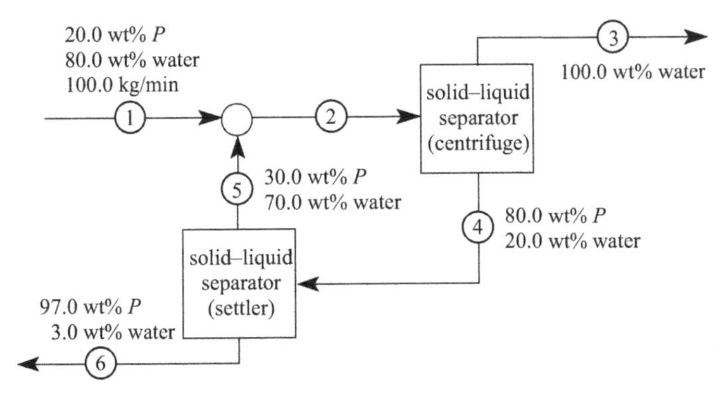

(A) Calculate the total flow rate of stream 6.
(B) Calculate the flow rate of the recycle, stream 5.

Adapted from Himmelblau, D. M. (1996). *Basic Principles and Calculations in Chemical Engineering*, 6th ed. Prentice Hall, Upper Saddle River, NJ, pp. 210–212.

3.27 The synthesis of nitric acid from ammonia produces a solution of 60 wt% HNO_3 and 40 wt% H_2O. Concentrated nitric acid (99 wt%) cannot be obtained by distillation, owing to an azeotrope at 68 wt% HNO_3. However, adding $Mg(NO_3)_2$ to dilute nitric acid allows one to distill a solution with greater than 68 wt% HNO_3, which can then be distilled to produce concentrated nitric acid, as in the process below.

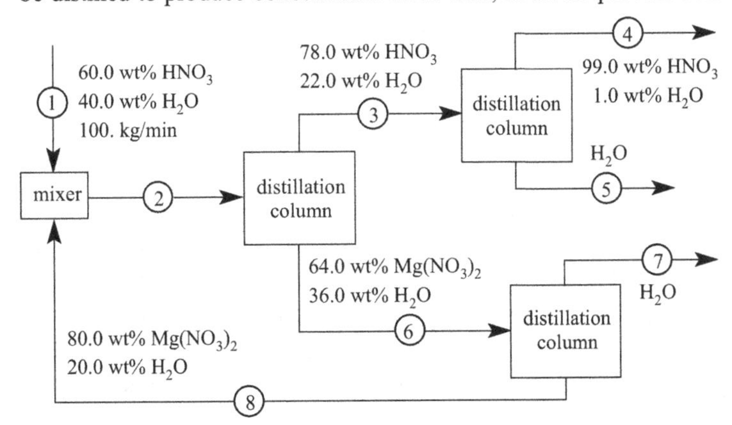

(A) Calculate the flow rate of the product, stream 4, in kg/min.
(B) Calculate the flow rate of the recycle, stream 8, in kg/min.
(C) The dilute nitric acid contains an impurity, $FeNO_3$. To prevent $FeNO_3$ from accumulating in the process, the recycle stream is purged and fresh $Mg(NO_3)_2$ is added, as in the flowsheet below. Stream 9 purges 5.0 wt% of stream 8. Calculate the composition *and* flow rate of stream 10.

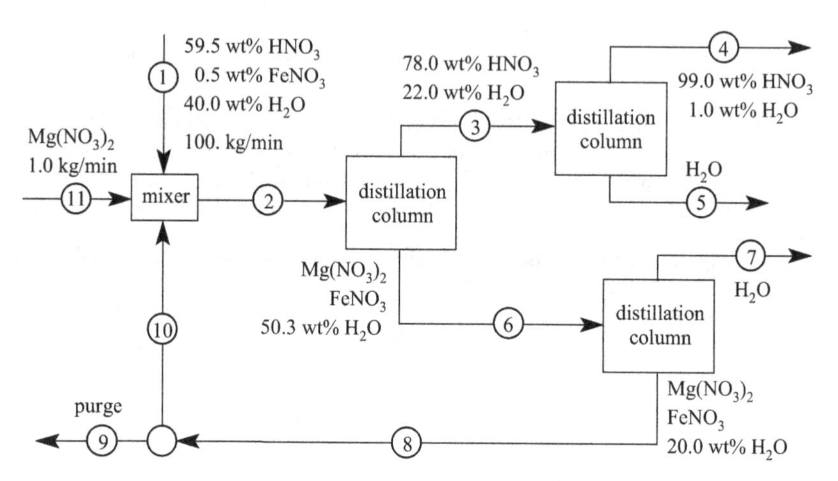

Adapted from Luyben, W. L., and Wenzel, L. A. (1988). *Chemical Process Analysis: Mass and Energy Balances*, Prentice Hall, exercise 4-28, p. 117.

3.28 Instant coffee (the soluble portion of ground roasted coffee) is produced by the process in the simplified flowsheet below. Calculate the rate of production of dried soluble coffee (stream 9). Note that the ratio of water to soluble components is the same in streams 3, 4, 5, 6, and 7.

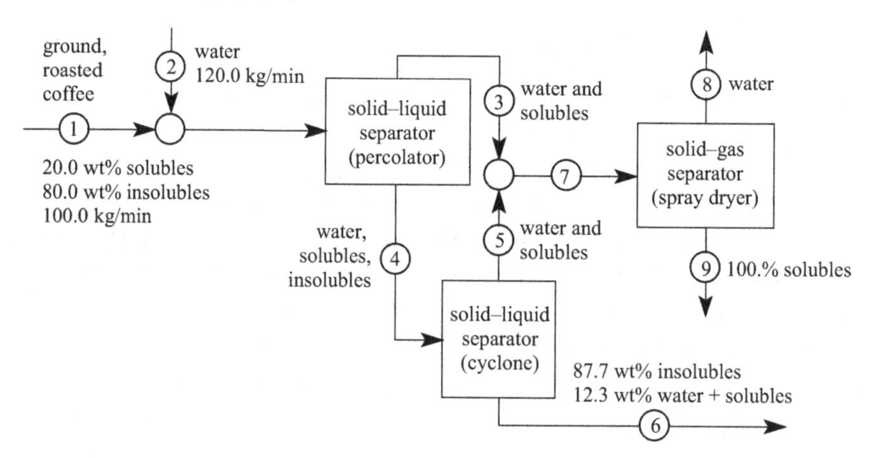

Adapted from Reklaitis, G. V. (1983). *Introduction to Material and Energy Balances*, Wiley, New York, 1983, pp. 98–99, exercise 2.27; from Felder, R. M. and Rousseau, R. W. (1986). *Elementary Principles of Chemical Processes*, 2nd ed, Wiley, New York, pp. 161–162, problem 32, and Rudd, D. F., Powers, G. J. and Siirola, J. J. (1973). *Process Synthesis*, Prentice Hall, Englewood Cliffs, NJ, pp. 69–70.

3.29 The process below extracts water-soluble solids from a mixture of soluble and insoluble solids. The solution in stream 3 is the same composition as the solution in stream 4. Likewise, the solution in stream 7 is the same composition as the solution in stream 8. A slurry is a suspension of insoluble solids in a solution. The slurry in stream 2 is not the same composition as the slurry in stream 6.

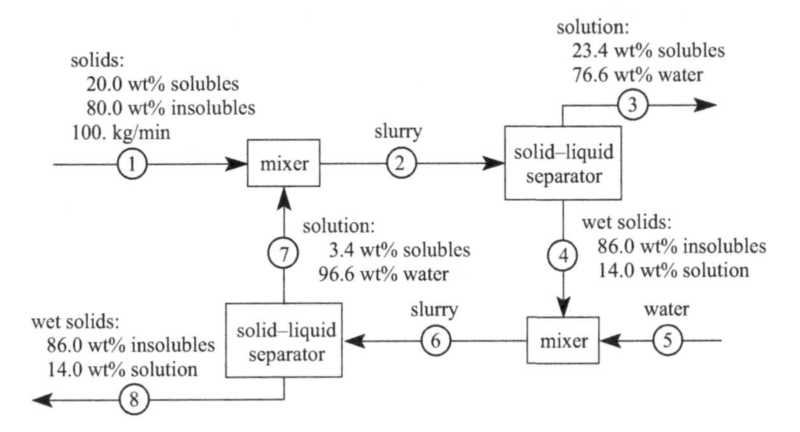

(A) Calculate the flow rate of stream 8, in kg/min.
(B) Calculate the flow rate of water, stream 5, in kg/min.
(C) Calculate the flow rate *and composition* of the slurry in stream 2, in kg/min.

3.30 Consider this process for washing cars.

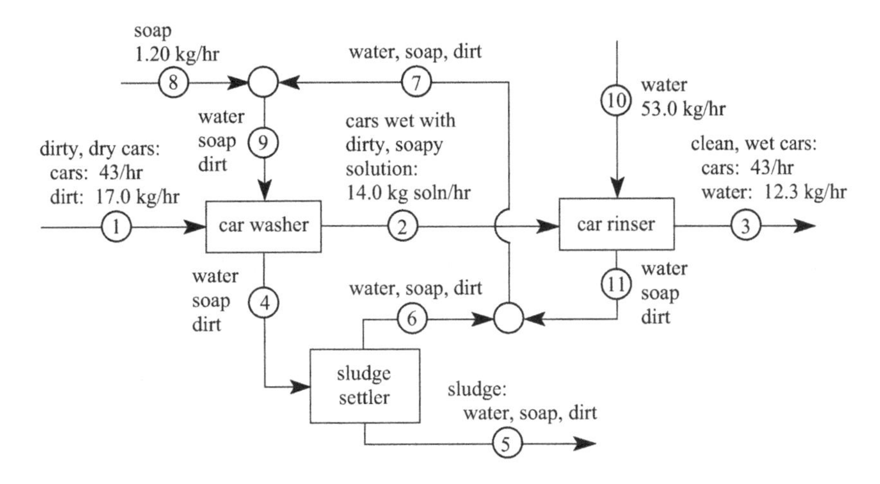

The sludge settler removes 90.0 wt% of the dirt in stream 4. The sludge settler also removes some of the soap + water solution: 20.0 wt% of the water in stream 4 is lost via stream 5 and 20.0 wt% of the soap in stream 4 is lost via stream 5.

(A) Calculate the flow rate *and* composition of stream 5.
(B) Calculate the flow rate *and* composition of stream 4.
(C) Assume the dirty, soapy solution in stream 2 is the same composition as stream 9. Calculate the flow rate of dirt in stream 2.

3.31 The process below is the first step in restoring wooden artifacts recovered from a sunken ship in salt water. The salt concentration is reduced gradually to avoid cracking the wood. To ensure that the stream entering the soaker contains 0.70 wt% salt, some salty water is recycled.

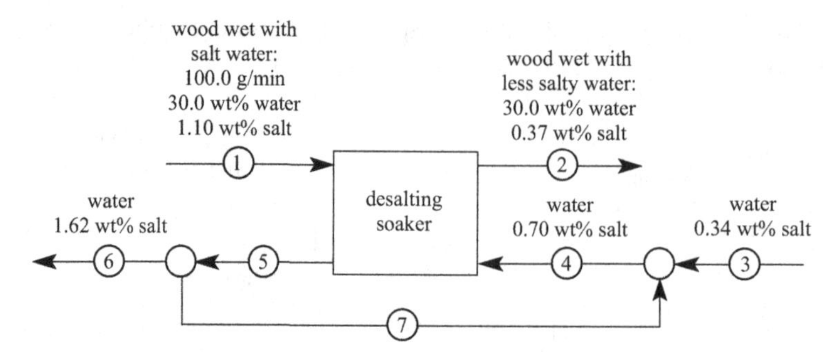

(A) Calculate the flow rate of stream 2, in g/min.
(B) Calculate the flow rate of stream 6, in g/min.
(C) Calculate the flow rate of the recycle (stream 7), in g/min.

3.32 The process below removes benzene from air by adsorption onto the surfaces of porous solids called zeolites. The adsorption unit below has a countercurrent flow of zeolites, and 1.00 kg zeolite can adsorb 0.10 kg benzene. Stream 6 purges 2% of the zeolite in stream 5. Calculate the flow rates of streams 4 and 8.

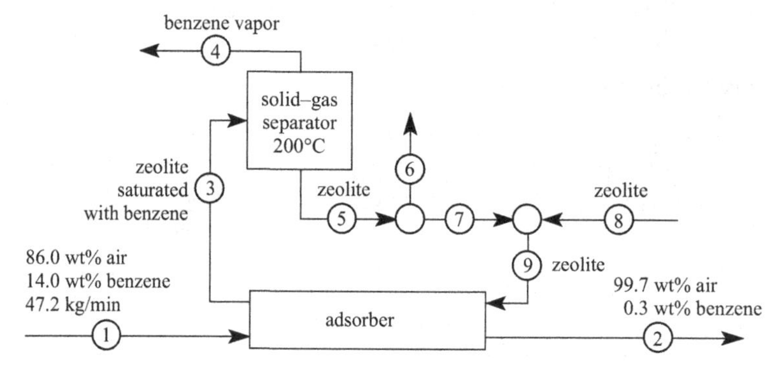

3.33 A valuable enzyme E is obtained from a biochemical process as a dilute water solution. Because E is heat-sensitive, E is purified by the three-step process below. Purge streams have been omitted.

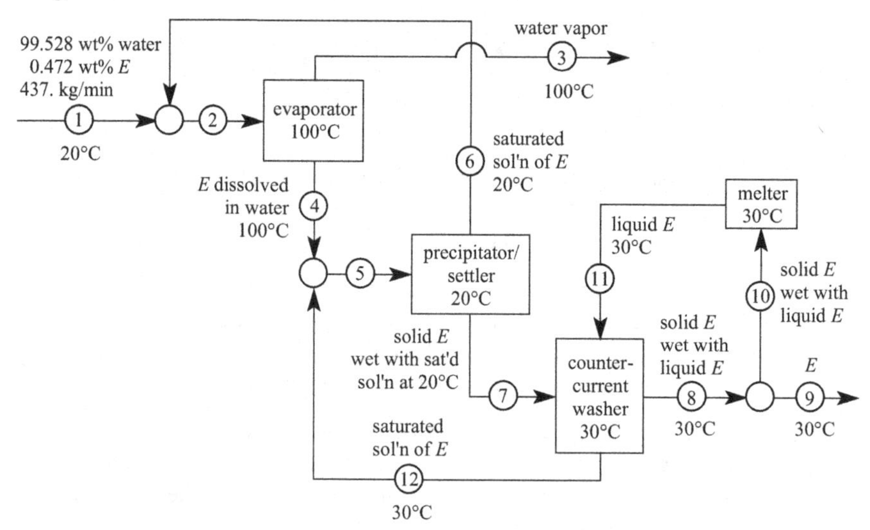

Temperature	E in a saturated solution (wt%)
20°C	5.71
30°C	7.23
100°C	31.0

(A) Calculate the flow rate of stream 3, the water vapor leaving the evaporator.
(B) The total flow rate of recycle stream 6 is 35.0 kg/min. Calculate the flow rate and composition of stream 4, the liquid leaving the evaporator.
(C) The flow rate of stream 7 is 7.06 kg/min. Calculate the flow rate and composition of recycle stream 12.
(D) Stream 7 contains solid E and dissolved E. Calculate the flow rate of solid E in stream 7.

3.34 Solid $MgSO_4$ reacts with H_2O to form three different hydrates: $MgSO_4{\cdot}H_2O$, $MgSO_4{\cdot}6H_2O$, and $MgSO_4{\cdot}7H_2O$. We want to produce a mixture of 50. wt% $MgSO_4$ and 50. wt% $MgSO_4{\cdot}7H_2O$. If one simply mixes $MgSO_4$ and H_2O, the intermediate hydrates are formed. The process below avoids producing intermediate hydrates.

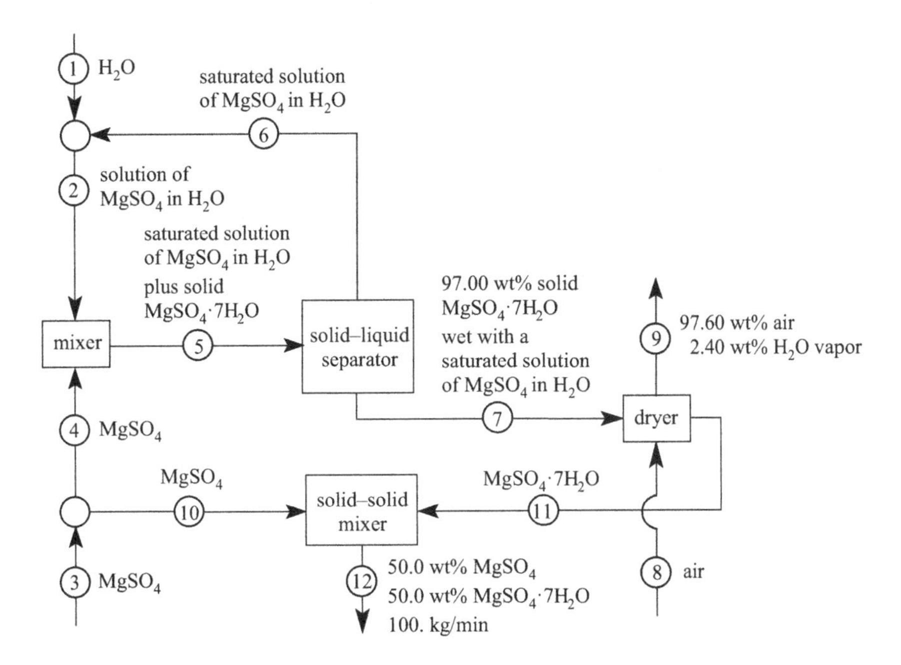

The entire process is at 20°C and is at steady state. The air in stream 8 contains no H_2O vapor. The saturated solution of $MgSO_4$ in H_2O is 27.3 wt% $MgSO_4$. The molecular weight of $MgSO_4$ is 120. amu and the molecular weight of $MgSO_4{\cdot}7H_2O$ is 246 amu.

(A) Calculate the flow rate of $MgSO_4$ entering the process (stream 3), in kg/min.
(B) Calculate the flow rate of air (stream 8), in kg/min.

3.35 The process to desalinate water is improved vastly by recycling some of the melted water to wash the ice in the skimmer, as in the flowsheet below. With an efficient skimmer, the recycled wash stream (stream 6) is only 5% of stream 5. The product stream (stream 7) contains a negligible amount of salt. Given that the brine leaving the skimmer is 7.0 wt% salt, calculate the flow rate of stream 7.

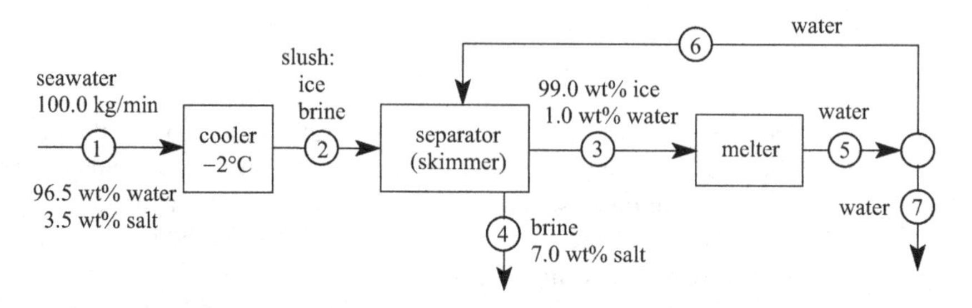

Mass Balances with Chemical Reactions

3.36 Methane (CH_4) is burned in air (21.0 mol% O_2, 79.0 mol% N_2) to yield CO_2 and water.

(A) Write a balanced chemical equation for this reaction.

(B) The inputs to the burner are 4.70 kg/min CH_4 and 92.0 kg/min air. Assume the CH_4 is completely consumed in the burner. Calculate the flow rate and composition (by weight) of the exhaust from the burner.

3.37 This exercise models the oxidation of glucose and the elimination of oxidation products. The chemical reaction is

$$C_6H_{12}O_6 + xO_2 \rightarrow yCO_2 + zH_2O.$$

(A) Balance the stoichiometry in the above chemical equation (determine x, y, and z).

(B) Consider the following flowsheet. Given that the flow rate of glucose is 0.10 kg/day, calculate the stoichiometric flow rate of oxygen, in kg/day.

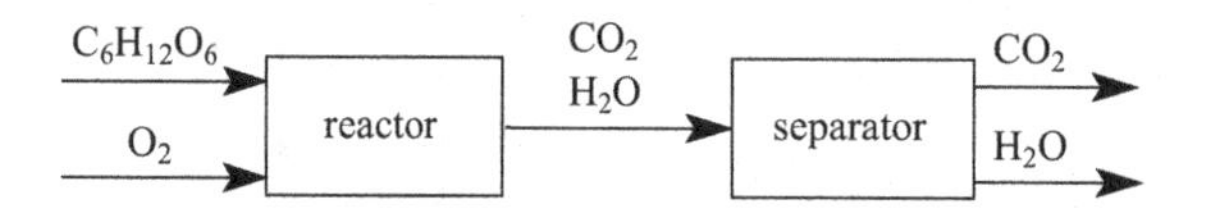

(C) Calculate the flow rate (kg/day) and composition (wt% CO_2 and wt% H_2O) in the stream leaving the reactor.

(D) Calculate the flow rates of the two streams leaving the separator.

3.38 The process below converts synthesis gas (aka syngas, a mixture of CO and H_2) to methanol, CH_3OH (see Exercise 2.22). The syngas contains an impurity of 4 mol % N_2, which leaves the process via the purge, stream 6.

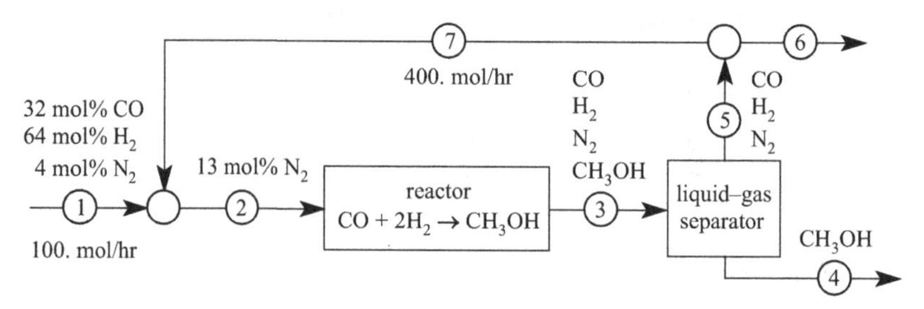

(A) Calculate the flow rate (in mol/hr) and composition (in mol%) of the reactor feed, stream 2. **Hint:** The molar ratio of CO to H_2 is 1:2 in all streams.
(B) Calculate the flow rate and composition of the purge, stream 6.
(C) Calculate the flow rate of the product, stream 4.
(D) Calculate the overall conversion; what fraction of the CO that enters the process is converted to methanol?
(E) Calculate the reactor conversion; what fraction of the CO that enters the reactor is converted to methanol?

3.39 Coal can be transformed to methanol (CH_3OH) by the process below. The partial oxidation reactor converts coal into CO_2, CO, and H_2 in molar ratio of 1:3:5:

$$\text{coal} + O_2 + 3H_2O \rightarrow CO_2 + 3CO + 5H_2.$$

CO and CO_2 are converted to methanol by the following reactions:

$$CO_2 + 3H_2 \rightarrow CH_3OH + H_2O$$

$$CO + 2H_2 \rightarrow CH_3OH.$$

To convert all the CO and CO_2 leaving the partial oxidation reactor, we need 3 mol of H_2 to react with the CO_2 and we need $3 \times 2 = 6$ mol of H_2 to react with the CO; we need a total of $3 + 6 = 9$ mol of H_2. Because only 5 mol of H_2 leaves the partial oxidation reactor, some of the CO_2 and CO are removed before the mixture is fed to the methanol reactor.

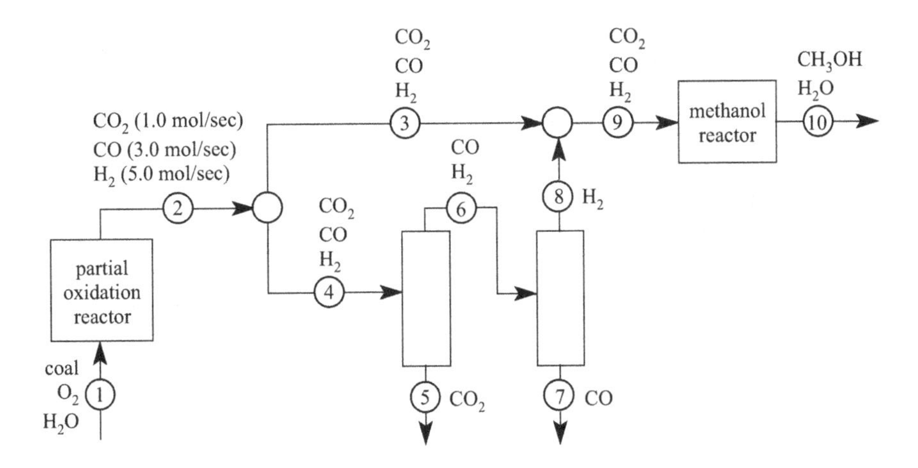

(A) What flow rate in stream 4 (mol/s) will deliver a stoichiometric mixture (CO_2:H_2 = 1:3 and CO:H_2 = 1:2) to the methanol reactor?
(B) The process to convert coal into methanol is modified slightly, as shown below. The second distillation column is changed to remove some, but not all, the CO. What composition of stream 8 will deliver a stoichiometric mixture (CO_2:H_2 = 1:3 and CO:H_2 = 1:2) to the methanol reactor?

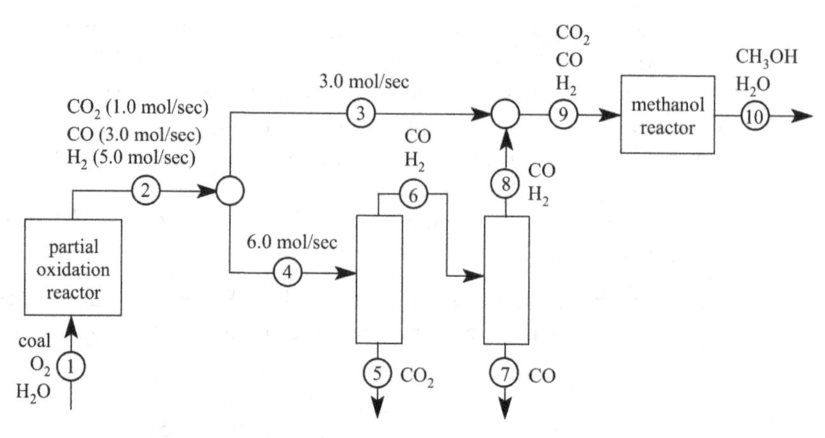

Adapted from Luyben, W. L., and Wenzel, L. A. (1988). *Chemical Process Analysis: Mass and Energy Balances*, Prentice Hall, ex. 4–32, p. 119.

3.40 1,2-butadiene (CH_2=C=CH–CH_3) is isomerized to 1,3-butadiene (CH_2=CH–CH=CH_2) in the process below. Unreacted 1,2-butadiene is separated from 1,3-butadiene and recycled to the reactor. The 1,2-butadiene contains an impurity – ethyl chloride (CH_3CH_2Cl) – which is difficult to remove by distillation. Ethyl chloride is purged via stream 7. The splitter purges 5.0 wt% of stream 5.

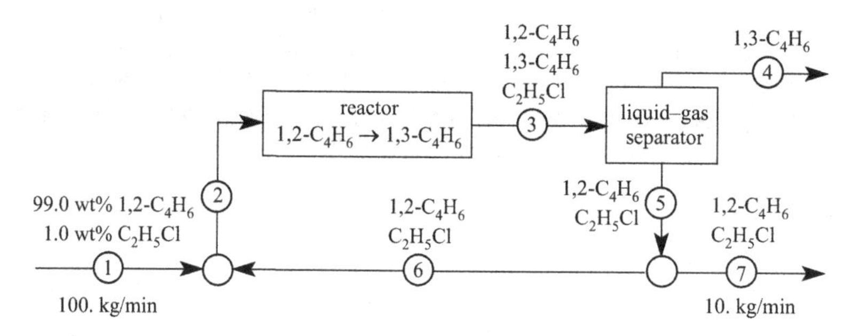

(A) Which stream has the highest flow rate of 1,2-butadiene?
(B) Calculate the flow rate of product stream 4, in kg/min.
(C) Calculate the composition of purge stream 7.
(D) Calculate the flow rate *and* composition of stream 2.

3.41 We wish to react acetylene (HC≡CH) and hydrogen chloride (HCl) to produce vinyl chloride (CH_2=CHCl), which is used to manufacture polyvinyl chloride (PVC):

$$C_2H_2 + HCl \rightarrow C_2H_3Cl \quad \text{(desired reaction)}.$$

We wish to minimize the subsequent reaction of vinyl chloride to dichloroethane:

$$C_2H_3Cl + HCl \rightarrow C_2H_4Cl_2 \quad \text{(undesired reaction)}.$$

Consider the three process schemes diagrammed below. All compositions are mol percent.

Scheme I.

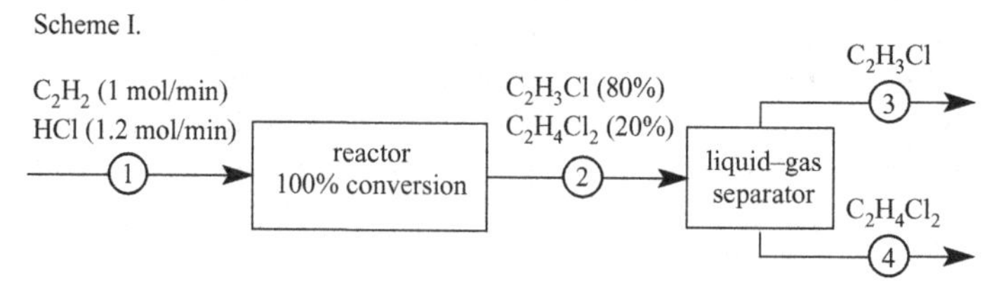

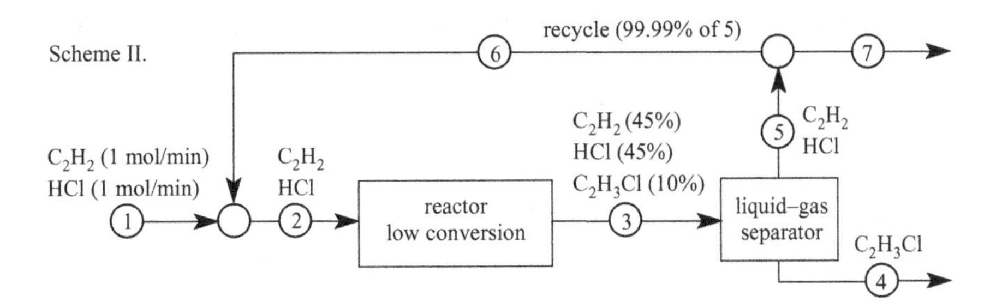

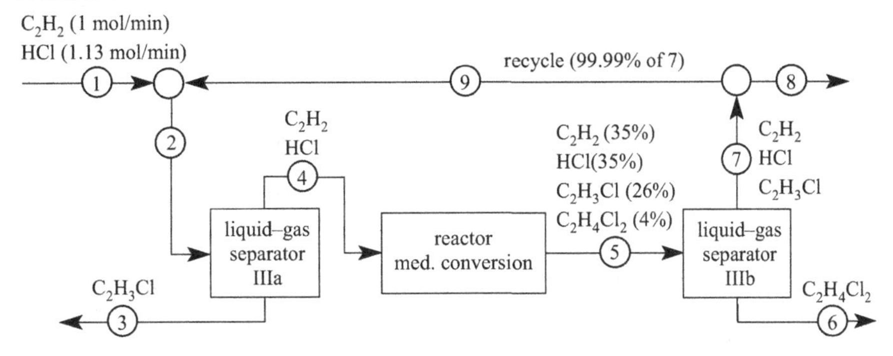

Some physical properties at 1 atm.

	C_2H_2	HCl	C_2H_3Cl	$C_2H_4Cl_2$
molecular weight	26	36.5	62.5	99
melting point (°C)	−81[a]	−115	−154	−97
boiling point (°C)	−84[b]	−85	−13	57

[a] At 1.2 atm.
[b] Sublimation of solid.

The following questions can be answered by comparing the qualitative differences in the three schemes. You need not calculate the flow rates. *Briefly* explain your qualitative analysis.

(A) Which scheme has the highest rate of production of desired product, C_2H_3Cl? *Explain.*
(B) Which scheme has the highest rate of production of undesired by-product, $C_2H_4Cl_2$? *Explain.*
(C) In which scheme is the reactor the largest, as measured by the total mol/min through the reactor? *Explain.*
(D) Which separator is the largest, as measured by the total mol/min through the separator? *Explain.*
(E) What are the approximate average temperatures in separators IIIa and IIIb? *Explain.*
(F) Improve the design of Scheme III. Your improved design must use the reactor with medium conversion. Focus on minimizing the size(s) of the separator(s). Ignore heat exchangers.

3.42 The process diagrammed below proposes to produce hydrogen bromate by the reactionx

$$3Ca(BrO_3)_2 + 2H_3PO_4 \rightarrow Ca_3(PO_4)_2 + 6HBrO_3.$$

Calculate the rate of production of $HBrO_3$. That is, calculate the flow rate of stream 7.

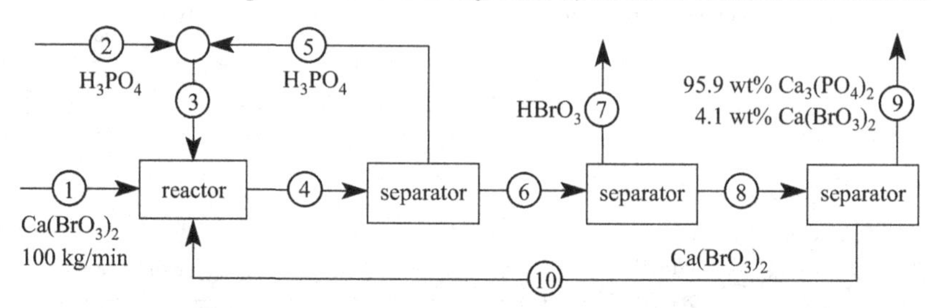

Hint: Calculate the composition of a stream in terms of cations (Ca^{2+} or H^+) and anions (BrO_3^- or PO_4^{3-}). For example, stream 1 is 13.5 wt% Ca and 86.5 wt% BrO_3.

Adapted from Reklaitis, G. V. (1983). *Introduction to Material and Energy Balances*, Wiley, New York, exercise 3.13, p. 168.

3.43 Consider two schemes for producing 100. kg/min of specialty chemical X from reactant E. E is available only as a mixture with an inert substance. The mixture of E and the inert substance is difficult to separate.

(A) The first scheme discards the unreacted E with the inert substance. Calculate the flow rates of streams 1 and 4, in kg/min.

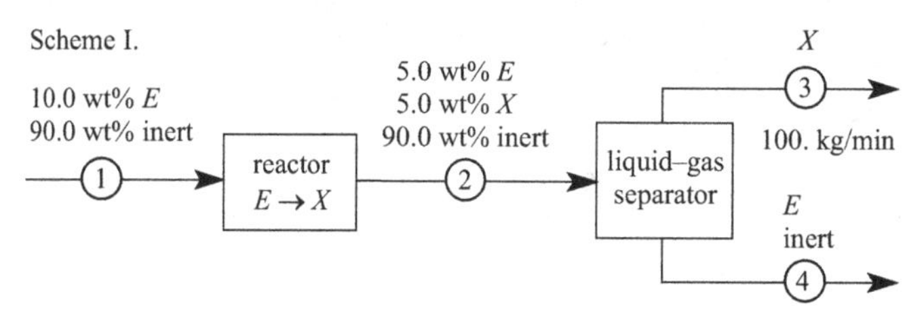

(B) The second scheme recycles the unreacted E (with some inert substance) and removes the inert substance in a purge stream (with some E). Calculate the flow rates of streams 1, 3, and 6, in kg/min.

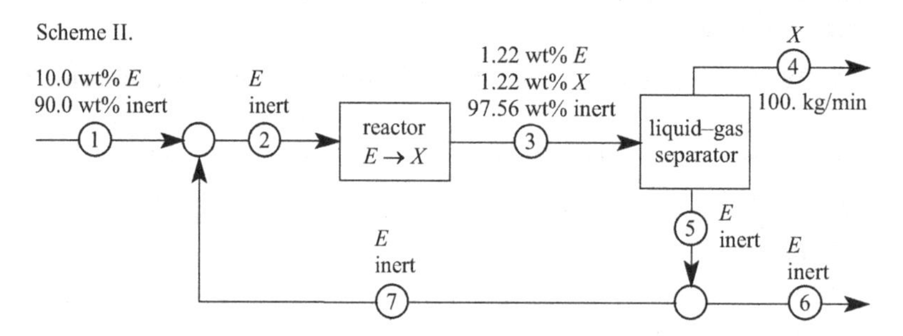

3.44 Reactant A decomposes to P (product) and B (by-product):

$$A \rightarrow P + B.$$

P and B have the same molecular weight, so 2 kg of A reacts to form 1 kg of P and 1 kg of B. The reactor converts 60% of A.

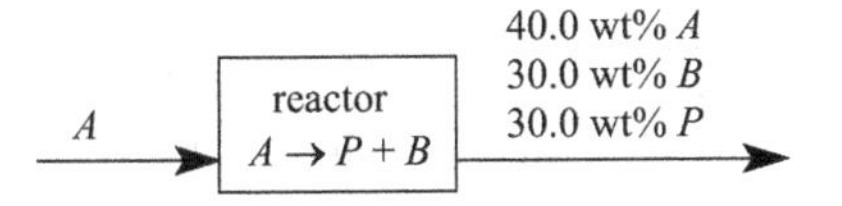

Consider three schemes for producing 100. kg/min of P. Scheme I discards the unreacted A with the by-product B.

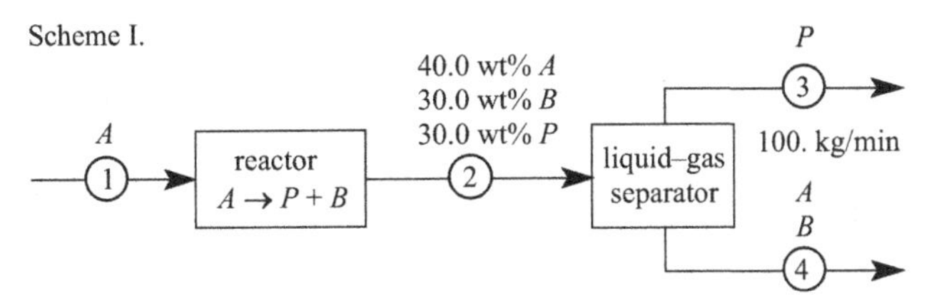

(A) Calculate the flow rate of stream 4 in Scheme I.
(B) Calculate the flow rate of stream 1, the reactant stream.

Scheme II uses a second separator to recycle the unreacted A and produce a product of pure B.

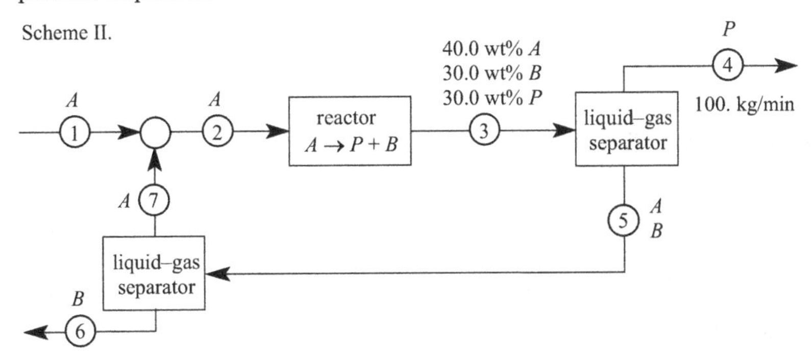

(C) Calculate the flow rate of stream 6, the by-product stream in Scheme II.
(D) Calculate the flow rate of stream 1, the reactant stream in Scheme II.

Scheme III has a recycle, but uses a purge instead of a second separator.

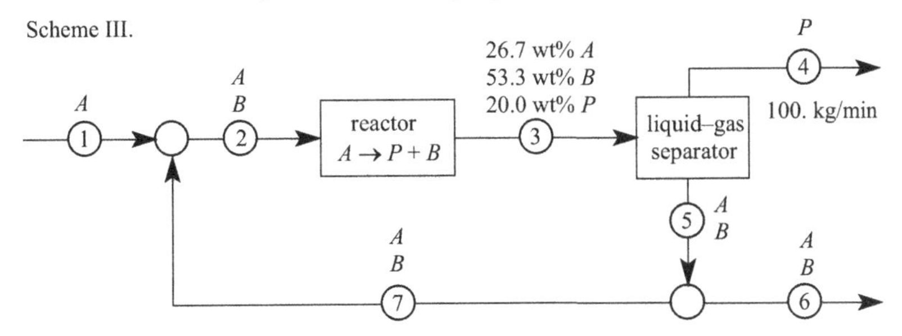

(E) Calculate the flow rate of stream 6, the purge stream in Scheme III.
(F) Calculate the flow rate of stream 1, the reactant stream in Scheme III.

3.45 Three schemes for producing 100. kg/min of 1-phenyl propene (propyl benzene, designated P) from 3-phenyl propene (allyl benzene, designated A) were discussed in Chapter 3. Recall that A is dissolved in solvent 1-heptanol (designated S) and the reactor effluent is equimolar in A and P.

Recall Scheme II recycled pure A and discarded no A, but required an expensive gas–liquid separation because the boiling points of A and S differ by only 1°C. Scheme III avoided the expensive separation and recycled a dilute stream of A, but required a large reactor.

Consider a fourth scheme, which is an intermediate between Schemes II and III. We separate A and S, but crudely. The recycled stream has 0.200 wt fraction A and the discarded stream has 0.025 wt fraction A.

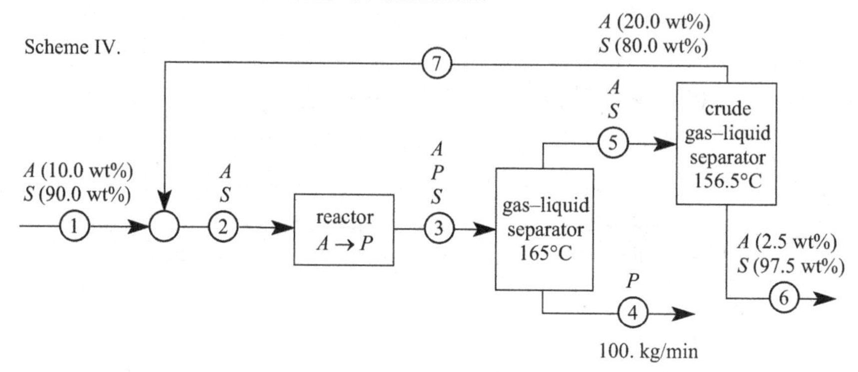

(A) Calculate the flow rate of stream 1, in kg/min.
(B) Calculate the flow rate *and* composition of stream 3.

Informal Mass Balances

To evolve the designs in the exercises at the end of this chapter a useful skill is to estimate flow rates of intermediate designs without the encumbrances of mass balance formalities. Formal mass balances are not required for Exercises 3.46 through 3.56. You may calculate the flow rates by any method. Write your answers on the process flowsheet.

3.46 Calculate the flow rates of all components in all streams in the process below.

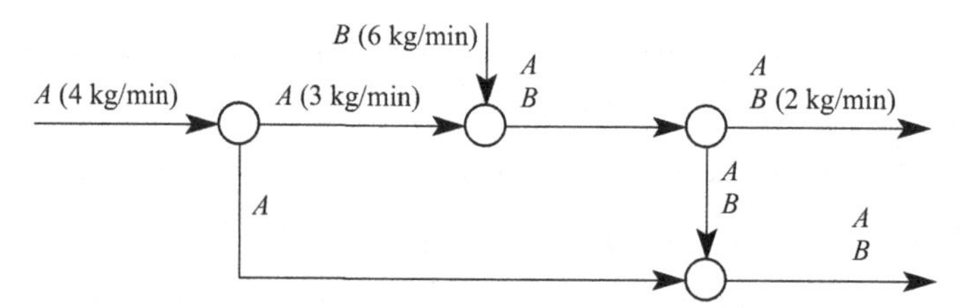

3.47 We wish to obtain pure W from a mixture of V, W, and X. Because the boiling points are similar – V (49°C), W (50°C), and X (51°C) – a separator that produces two perfect separations is prohibitively expensive. Instead, we use an imperfect separator at 50°C as in Section 3.6; all the V leaves as gas, all the X leaves as liquid, and (arbitrarily) half the W leaves as gas, half the W leaves as liquid.

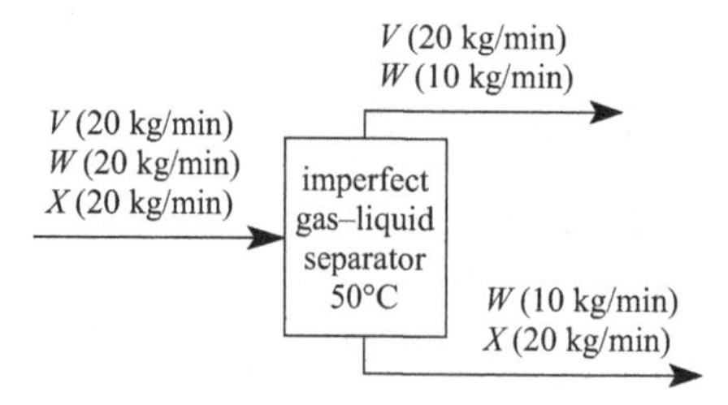

The process below uses two imperfect separators to produce pure W. Verify this process is viable by calculating the flow rates for components in all streams. Assume the flow of W is split evenly between the gas and liquid streams leaving both separators.

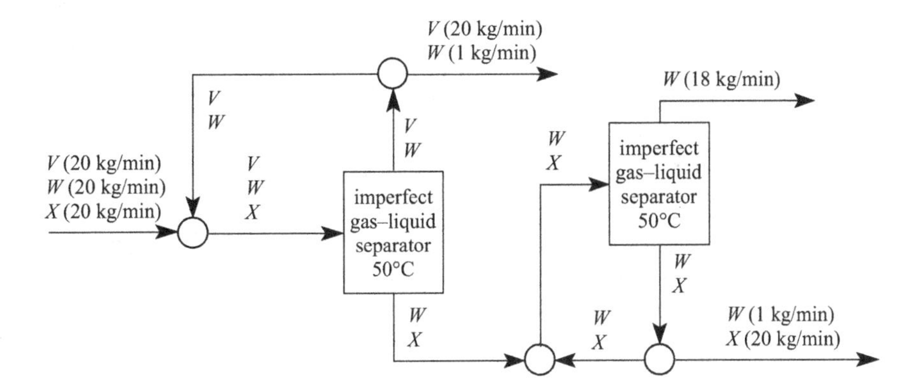

3.48 A and P cannot be separated. The process below increases the ratio of P to A by the reaction $A \rightarrow P$. The reactor converts 10% of the A that enters. Calculate the flow rates of A and P in all streams.

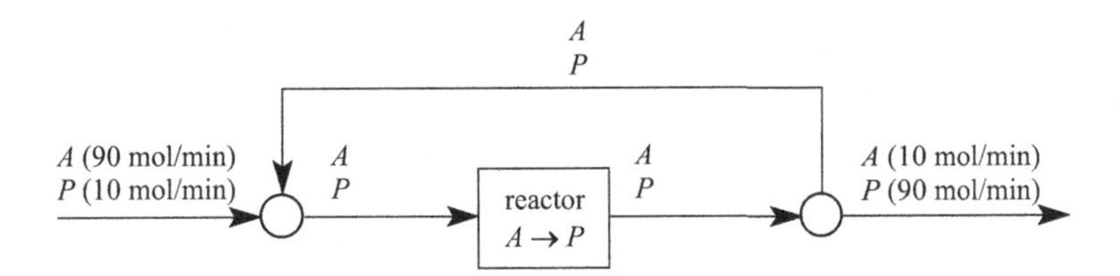

3.49 We wish to produce valuable product Z from reactant X by the reaction $X \rightarrow Y + Z$. The reactor converts 50.% of reactant X that enters. Product Z and by-product Y have the same molecular weight. Because X, Y, and Z have similar boiling points – X (79°C), Y (80°C), and Z (81°C) – a separator that produces two perfect separations is prohibitively expensive. Instead we use an imperfect separator at the boiling point of either X, Y, or Z. Assume that at the boiling point half the substance leaves as gas, half the substance leaves as liquid. The process below yields 2 kg/min of product Z from 16 kg/min of reactant X.

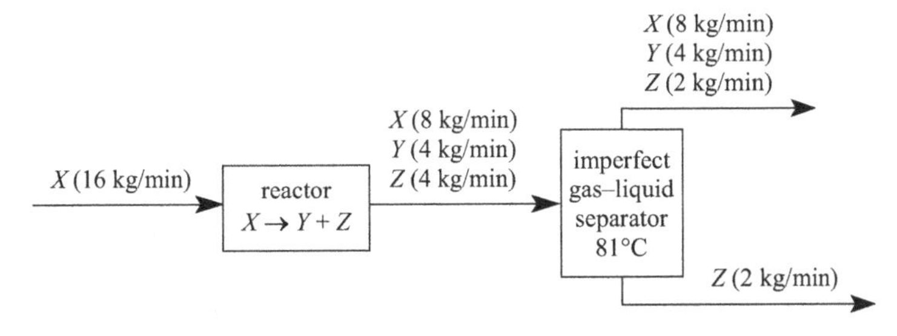

The process below is proposed to increase the yield of Z from 16 kg/min of reactant X by a factor of 3.5. Verify this process is viable by calculating the flow rate of all components in all streams.

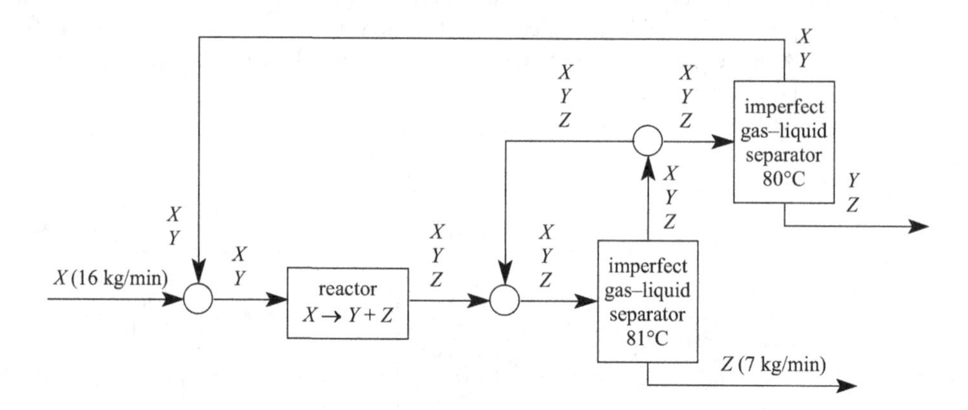

3.50 The process below produces X by the reaction $E \to X$. The reactor converts 50% of the E that enters. Product X is in reversible equilibrium with useless by-product Z. The ratio of X to Z leaving the reactor is 1:1. Calculate the flow rates of E, X, and Z in all streams.

Adapted from Felder, R. M., and Rousseau, R. W. 1986. *Elementary Principles of Chemical Processes*, 2nd ed., Wiley, New York, exercise 22, p. 157. Reproduced by permission.

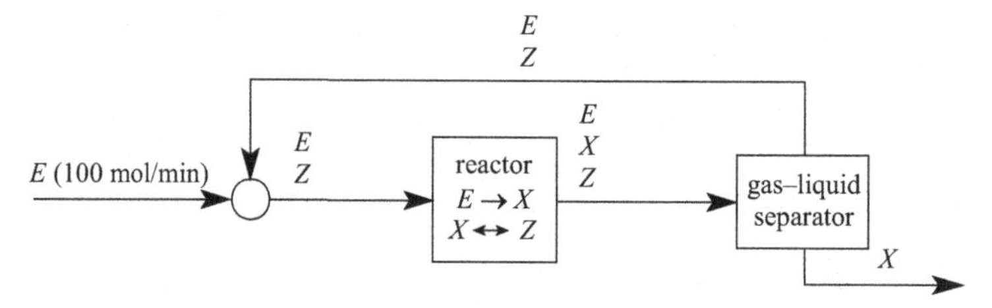

3.51 The process below produces X using the reaction $A + B \to H + X$. The reactor conversion is 40.%; if 10 mol A and 10 mol B enter the reactor, 6 mol A and 6 mol B leave the reactor.

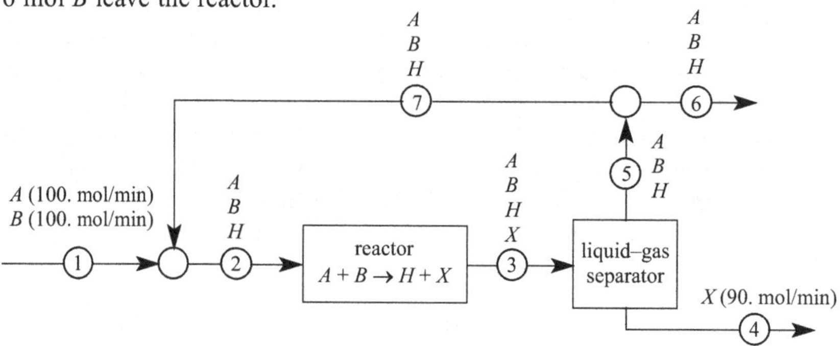

(A) Calculate the flow rate (in mol/min) of each component in stream 6.
(B) Calculate the flow rate (in mol/min) of each component in stream 3.

3.52 The process below reacts A and B to produce P. The reactor converts 50% of A and B for equal molar flow rates of A and B. Reactant A is available only as a mixture with inert substance I. Separating A from I is impractical because A and I have the same melting point and boiling point. The ratio of A to I in the feed is 10 to 1. The ratio of A to I in the purge output is 1 to 1. Calculate the flow rates of all components in all streams.

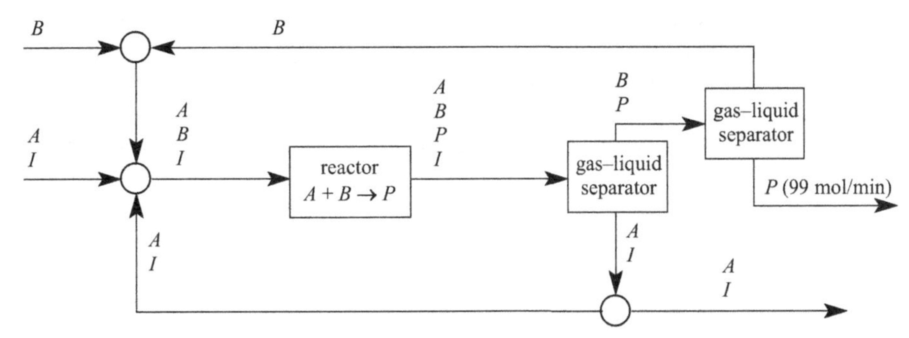

3.53 The process below reacts A and B to create P. B also reacts to form by-product Q. The output stream is treated later to isolate P. Numbers in parentheses are flow rates in mol/min. Calculate the flow rates of all components in all streams.

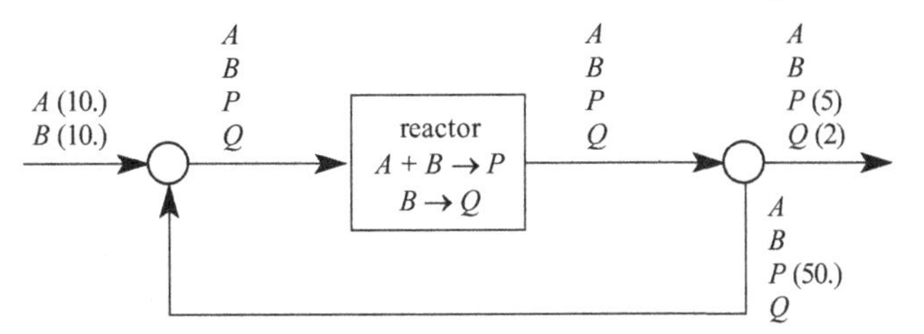

3.54 Calculate the flow rates of all components in all streams in the process below. The reactor converts 50% of the A that enters. Numbers in parentheses are flow rates in mol/min. Caution: mols are not conserved in either reaction.

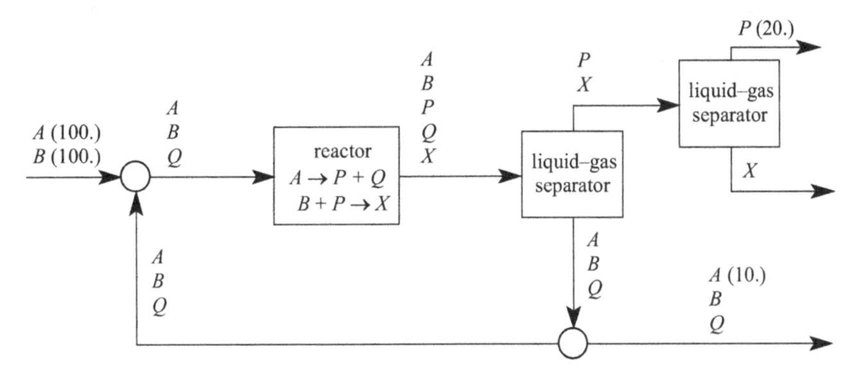

3.55 The process below creates product P from reactant A. The reactor converts 50.% of the A that enters. P equilibrates reversibly with X, such that mol P = mol X in the reactor effluent. The purge stream removes 10.% of the separator bottoms stream. Impurity I is inert. Calculate the flow rates of all components in all streams.

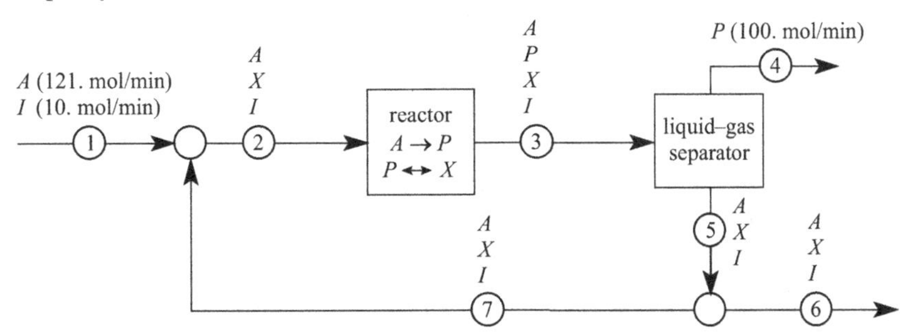

3.56 The process below reacts A and B to produce P. The reactor converts 10% of A and B for equal molar flow rates of A and B. Product P also reacts with B to form by-product X. Separating A from X is impractical because A and X have the same melting point and boiling point; X is removed from the process by a purge which also removes some A. Calculate the flow rates of all components in all streams.

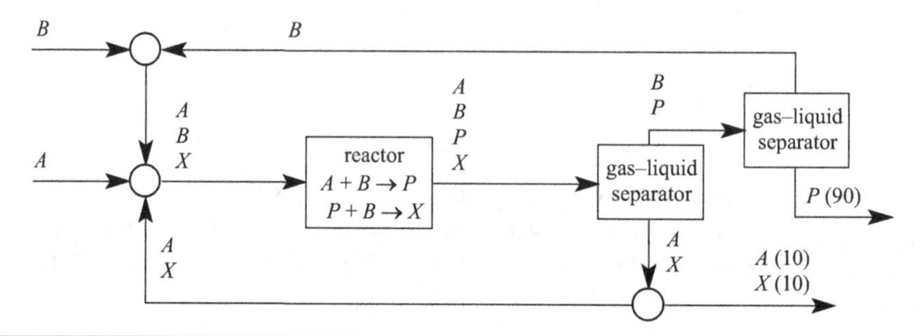

Mass Balances on Spreadsheets

Spreadsheets are adept at mass balances. The mathematical models you developed for the mass balance Exercises 3.1 through 3.56 can be solved with spreadsheets. Indeed, most commercial software to model chemical processes are embellished spreadsheets. We offer here a tutorial in two examples: a simple separation and a process with a recycle. The tutorial assumes a basic knowledge of spreadsheets, such as how to use a mouse to select a cell and how to copy and paste.

Spreadsheet Example 1

The process below separates grain extract (50 kg/min, 97 wt% water, and 3 wt% ethanol) into a product stream and a waste stream. The product stream must recover 99% of the ethanol in the grain extract. In addition, the product stream must be 50 wt% ethanol.

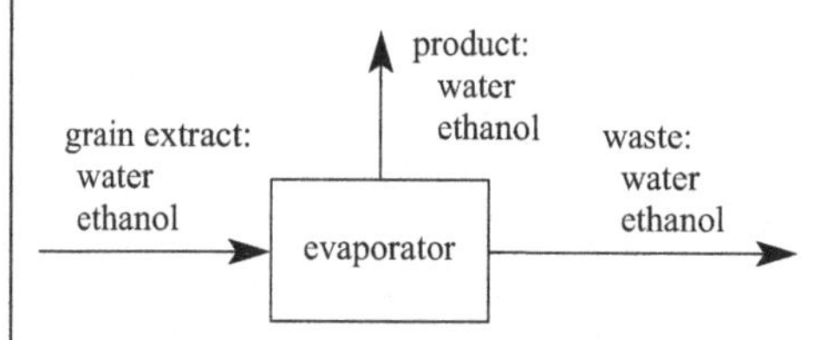

(A) Calculate the total flow rates of the product and waste streams. Also calculate the flow rates of water and ethanol in the product and waste streams.

(B) Repeat the calculation in (A), but with a grain extract composed of 98.5 wt% water, 1.5 wt% ethanol.

(C) Repeat the calculation in (A), but with the relaxed criterion that only 95% of the ethanol is recovered.

Solution to Spreadsheet Example 1

A spreadsheet solution to this exercise is shown on the following page. The steps below will lead you through the process of creating this spreadsheet. The steps will refer to locations on the spreadsheet, known as cells, by their column (an uppercase letter) and their row (a number).

1. Enter the title. Type "Example 1" into cell A1. For those new to spreadsheets, first select cell A1 with the mouse. This should highlight the border around the cell. Then type "Example 1" and hit *enter*.
2. Enter the list of process specifications. Type "Process Specifications" into cell A3. Do not be concerned that the text extends beyond cell A3. Because there is nothing in the adjacent cell,

the text appears in full. Enter the other specification descriptions in cells A4 through A7, and enter their values in cells C4 through C7 as shown in the example.

	A	B	C	D	E	F	G	H	I	J	K
1	**Example 1**										
2								**Product**	**(kg/min)**		
3	Process Specifications							water	1.485		
4	flow rate of grain extract		50					ethanol	1.485		
5	fraction of ethanol in feec		0.03					Total	2.97		
6	fraction of ethanol recove		0.99								
7	fraction of ethanol in pro(		0.5								
8				**Grain Extra**	**(kg/min)**				**Waste**	**(kg/min)**	
9				water	48.5				water	47.015	
10				ethanol	1.5				ethanol	0.015	
11				Total	50				Total	47.03	
12							**Evaporato**				
13											
14	**Example 2**										
15											
16	Process Specifications										
17	flow rate of feed		100								
18	fraction of grain in feed		0.2								
19	solution/grain ratio in slu		0.1								
20	fraction of slurry purged		0.05								
21							**Reactor**				
22				**Reactor Fee**	**(kg/min)**		**Effluent**			**Grain Extra**	**(kg/min)**
23	**Feed**	**(kg/min)**		grain	69.57		grain	52.17		water	88.46
24	grain	20.		water	84.51		water	93.21		ethanol	8.67
25	water	80.		ethanol	0.44	**yeast**	ethanol	9.14		Total	97.13
26	Total	100.		Total	154.52	**reactor**	Total	154.52	**Filter**		
27						efficiency=					
28						0.25	solution composition			**Slurry**	**(kg/min)**
29							water	0.91		grain	52.17
30							ethanol	0.09		water	4.75
31										ethanol	0.47
32										Total	57.39
33											
34											
35				**Recycle**	**(kg/min)**					**Purge**	**(kg/min)**
36				grain	49.57					grain	2.61
37				water	4.51					water	0.24
38				ethanol	0.44					ethanol	0.02
39				Total	54.52					Total	2.87

3. Draw the flowsheet for the process. Most spreadsheet software allows one to open a drawing palette or drawing toolbar to create the rectangles, circles, and arrows you will need. In Microsoft Excel, pull down the *Options* menu to list *Toolbars*. Select the *Drawing* toolbar. Practice drawing various objects, moving the objects, resizing objects, and reformatting patterns such as line widths and arrowhead sizes.

   After experimenting, create a flowsheet with the general features shown in the example spreadsheet. Place an evaporator in column G with two stream arrows on the border between rows 12 and 13 and a stream arrow extending upward in column G.
4. Label the input stream. Type "Grain Extract" into cell D8, list the components in cells D9 and D10, and type "Total" into cell D11.
5. Calculate the input flow rate and input composition. Do not copy the example; do not type "48.5," "1.5," and "50" into cells E9, E10, and E11. These cells show numbers but actually contain formulas. The formulas do not appear in the example spreadsheet printed here, only the values calculated by the formulas.

   Let's begin with what we know. The total flow rate is one of our specifications. Type the simple formula "=C4" into cell E11. The number "50" should appear in cell E11 after you hit *enter*. The flow rate of ethanol is not specified explicitly, but we can write a simple formula to calculate it. The fraction of ethanol in the feed is one of the process specifications; it appears in cell C5. Type the formula "=E11 * C5" into cell E10. What is the water flow rate? Again, it is not specified but is trivial to calculate. From a mass balance, water = total – ethanol. Translate this into a formula for cell E9; type "=E11 – E10".

It is tempting to calculate the flow rates for water and ethanol in one's head, and then type in the values. Don't. The power of spreadsheets lies in referencing to key specifications. To illustrate, change the ethanol fraction (cell C5) to 0.015. Because you typed formulas, the numbers in cells E9 and E10 change. Similarly, try changing the total flow rate of grain extract in cell C4. Now reset all specifications back to the given values.

As veterans of spreadsheets will attest, creating formulas is expedited by selecting cells rather than typing cell addresses. For example, the formula for cell E10, the ethanol flow rate, can be created by first typing "=" into cell E10. This tells the spreadsheet you are entering a formula. Then select cell E11 with the mouse. The characters "E11" will appear in cell E10. Type the symbol for multiplication, "*". Now select cell C5 with the mouse. This completes the formula. Hit *enter*.

6. Calculate the product stream. The fraction of ethanol recovered is one of the process specifications. For cell I4, type the formula "=C6 * E10". The fraction of ethanol in the product, a process specification we listed in cell C7, can be used to calculate the total flow rate. Into cell I5 type the formula "=I4 / C7". Finally, the water flow rate is obtained as before, by the formula "water = total – ethanol." Type "=I5 – I4" into cell I3.

   Because the formula for water in the product stream is the same as the formula for water in the input stream, we could have copied the formula from cell E9 into cell I3. Try this. First, delete the formula in cell I3. Select cell E9 and *copy*. Now select cell I3 and *paste*. Note that the formula pasted into cell I3 is not "=E11 – E10" as it appeared in cell E9, but is just what we wanted: "=I5 – I4". Unless one specifies otherwise, formulas use relative references, not absolute references. The formula for cell E9 is actually "subtract the number in the cell below from the number two cells below."
7. Calculate the waste stream. There are many ways to accomplish this. If you followed steps 1 through 6 you can probably do this step on your own. Try it.

Compare your formulas to the formulas we devised. The total waste flow obtains from a total mass balance on the entire process, "waste = input – product." For cell J11, we used the formula "=E11 – I5". Likewise the ethanol flow rate obtains from an ethanol mass balance on the entire process. For cell J10, we used the formula "=E10 – I4". And we entered the formula by pasting, rather than typing. We selected cell J11 and copied, and then selected cell J10 and pasted. Likewise for cell J9.

Clearly, there are many ways to calculate the waste stream. If you get the correct values, your process is correct. Indeed, there are many ways to model this process.

Your spreadsheet should now resemble the example given. Learn how to change the formats for the numbers to show only the significant figures, or perhaps one or two more than significant. Adjust the column widths to fit more of your spreadsheet onto the monitor screen.

Now complete parts (B) and (C) of Example 1. Change only the numbers in the process specifications – the numbers in column C. You need not change any formulas in your spreadsheet. These relations are correct, regardless of the process specifications. Parts (B) and (C) are intended to illustrate the utility of a spreadsheet for exploring the effects of changing process specifications.

Example 1 was a simple mass balance. You probably could have worked it with pencil and paper in less time than it took to create the spreadsheet. The advantage of spreadsheets lies in modeling more complicated processes, such as Example 2.

Spreadsheet Example 2

The grain extract in Example 1 is produced by the process below. Ethanol is synthesized by yeast in the reactor. In this idealistic process, the yeast converts 2 kg of grain into 1 kg of ethanol and 1 kg of water. A perfectly efficient yeast reactor (efficiency = 1) would convert all the grain entering the reactor. A reactor with efficiency = 0.5 would convert half the grain entering the reactor, and so on. Here are some additional specifications:

- The feed is 100. kg/min, 20 wt% grain, and 80 wt% water.
- The reactor efficiency is 0.25.
- The grain in the slurry retains 1 kg of ethanol–water solution for every 10 kg of grain.
- 5 wt% of the slurry is purged.

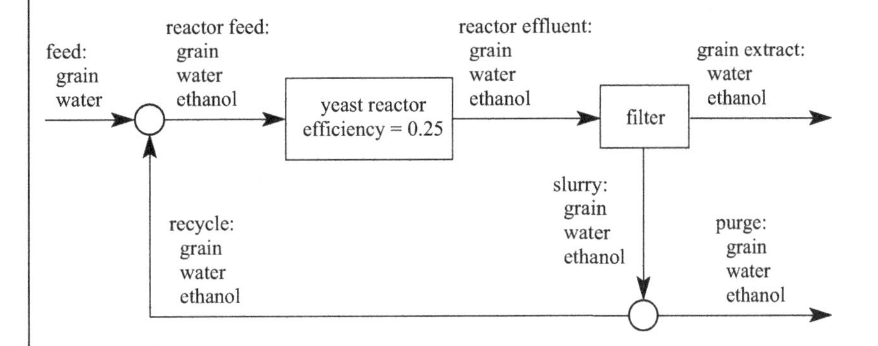

(A) Calculate the flow rates of all streams. Also calculate the flow rates of all components in each stream.
(B) Repeat the calculation in (A), but with a reactor efficiency of 0.10.
(C) Repeat the calculation in (A), but with a purge fraction of 0.01.
(D) Repeat the calculation in (A), but with a feed composed of 10 wt% grain and 90 wt% water.

Solution to Spreadsheet Example 2

We recommend you prohibit the spreadsheet from iterating on circular references until you are ready. Otherwise, you may encounter frustrating error messages as you create your spreadsheet. For Excel users, pull down the *Options* menu and open the *Calculations* window. There should be a box labeled *Iterations*. Be sure the box does not contain an "x".

1. As in Example 1, type the title and list the process specifications in column A. Type the values for the specifications in column C.
2. Draw a flowsheet for the process and label the units. This is another activity that benefits from liberal copying and pasting. We chose not to include the reactor efficiency in the list of process specifications (although it would be appropriate to do so). Rather we placed this specification in cell F28, inside the rectangle denoting the reactor.
3. Enter labels for all streams. Your experience in Example 1 showed you it is wise to maintain the same order of components in each stream. That is, the relative position of "grain," "water," and "ethanol" should be the same in all streams. Again, copying and pasting will save time and reduce drudgery.
4. Calculate the input stream in column B. Cells B24, B25, and B26 should all contain formulas in terms of cells C17 and C18. Do not simply copy the numbers from the example flowsheet. The correct numbers will appear in your spreadsheet after you have entered formulas.
5. Calculate the reactor feed flow rates in cells E23 through E26. The reactor feed is the input feed plus the recycle. Because we do not yet know the recycle flow rates, we cannot yet know the flow rates of the reactor feed. But this is not a problem. We only need to know the formulas to calculate the flow rates in the reactor feed.

   The total flow rate of the reactor feed is the total input feed plus the total recycle. The formula for cell E26 is "=B26 + E39". Because the recycle flow rate is presently zero, the total reactor feed is presently 100, and not 154.52 as shown in the completed spreadsheet.

   Enter similar formulas for the flow rates of grain, water, and ethanol in the reactor feed.

6. Calculate the reactor effluent flow rates. Again we must devise formulas for each of the component flow rates. The formulas will be in terms of the reactor feed flow rates and the reactor efficiency. The grain in the effluent is the grain not converted to ethanol and water, because the reactor efficiency is less than 1. Thus cell H23 has the formula "=E23 – E23 * F28," or "=E23 * (1 – F28)". The ethanol flow rate is half the flow rate of the grain that reacted plus the ethanol that entered the reactor (from the recycle). Recall that when 2 kg of grain is consumed, 1 kg of ethanol is synthesized. The other 1 kg is water. So the formula for ethanol in the reactor effluent, cell H25, is "=E25 + E23 * F28 / 2". Similarly, the formula for water in the reactor effluent, cell H24, is "=E24 + E23 * F28 / 2".

   The formula for the total flow rate is trivial; total in = total out. But because the component flow rates are susceptible to errors, it is better to use the total flow rate of the effluent to check the formulas for the components. Set the total equal to the sum of the components. For cell H26 enter "=H23 + H24 + H25". If the reactor effluent total does not equal the reactor feed total, one or more of your formulas in cells H23 through H25 are probably wrong.

   Again, because the recycle stream is yet to be calculated, the reactor effluent numbers in your spreadsheet will be different from the example spreadsheet given here.
7. Calculate the slurry flow rates. At this point, the next step is to calculate either the grain extract or slurry streams. The slurry is easier to calculate first because *all* the grain leaves via the slurry. The formula for the grain in the slurry, cell K29, is thus "=H23".

   Water and ethanol appear in the slurry because each kilogram of solid grain retains 0.1 kg of ethanol–water solution. We have found it is easier to first calculate the composition of the solution. This appears in cells H29 and H30. The fraction of water in the solution, cell H29, is thus "=H24 / (H24 + H25)". The fraction of ethanol, cell H30, is "= 1 – H29".

   The ratio of ethanol–water solution to grain in the slurry is the process specification in cell C19. Thus the formula for water in the slurry, cell K30, is "=K29 * C19 * H29". Similarly, the formula for ethanol in the slurry, cell K31, is "=K29 * C19 * H30".

   The total slurry flow rate is the sum of the components. The formula of cell K32 is "=K29 + K30 + K31". This formula can be pasted from cell H26.
8. Calculate the grain extract flow rates. The grain extract is the reactor effluent minus the slurry. Enter formulas for cells K23, K24, and K25.
9. Calculate the purge stream. The purge is a fraction of the slurry, as given by the specification in cell C20. Enter formulas for cells K36, K37, K38, and K39.
10. Calculate the recycle stream. The recycle is the slurry stream minus the purge. Enter formulas for cells E36, E37, E38, and E39.
11. Iterate until values in the cells converge. You have now completed the circular references, so your spreadsheet software should not complain when you ask it to iterate. For Excel users, again pull down the *Options* menus and open the *Calculations* window. Select the box labeled *Iterations*.

The spreadsheet is useful for demonstrating the dynamics of starting a process. Set the number of iterations to a small number, something in the range 1 to 10. Then manually induce iterations and watch the numbers converge.

Complete parts (B), (C), and (D) by changing the specifications in column C. As a final challenge, link the results of Example 2 to Example 1. That is, change the Example 1 specifications for "flow rate of grain extract" and "fraction of ethanol in feed" to use the values calculated by Example 2.

Use what you have learned in this tutorial to repeat some of the mass balance exercises you solved
by pencil and paper for homework assignments or in calculation sessions.

The two spreadsheet examples presented here are derived from a tutorial developed by Mary Carmen Gascó-Buisson, Cornell University Chemical Engineering Class of 1997.

Energy Balances

Unless indicated otherwise, all processes are at steady state and 1 atm. Tables of thermodynamic properties appear at the end of this section, on p. 200.

3.57 Water cools cyclohexane (C_6H_{12}) in the heat exchanger below. The water is available at 10°C and the maximum discharge temperature is 30°C. What flow rate of cyclohexane (in kg/min) can be accommodated with this unit?

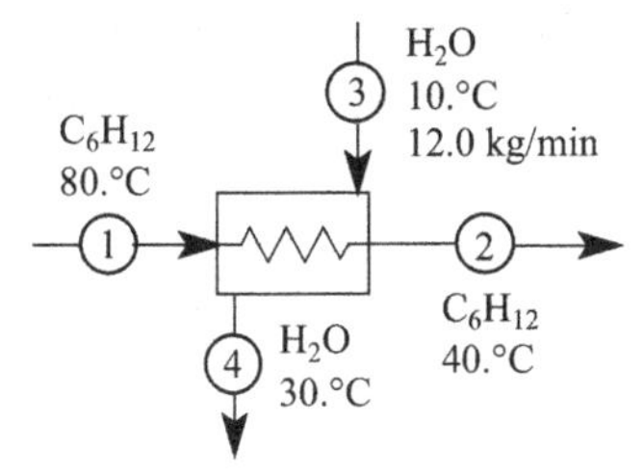

3.58 Steam heats cyclohexane (C_6H_{12}) in the heat exchanger below. What flow rate of cyclohexane (in kg/min) can be accommodated with this unit?

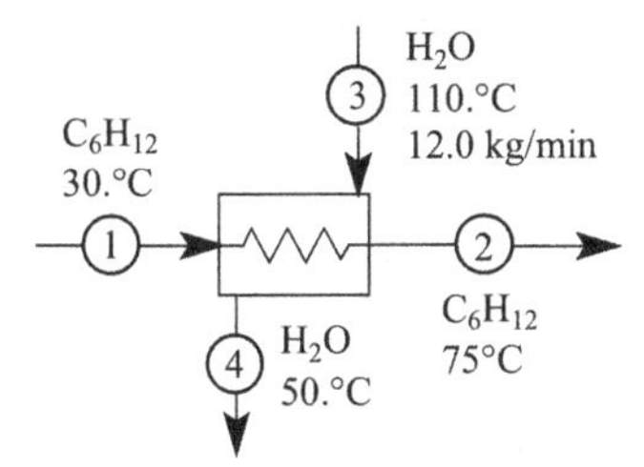

3.59 Superheated steam warms styrene ($C_6H_5CH{=}CH_2$) in the heat exchanger below. Calculate the temperature of the styrene leaving the unit.

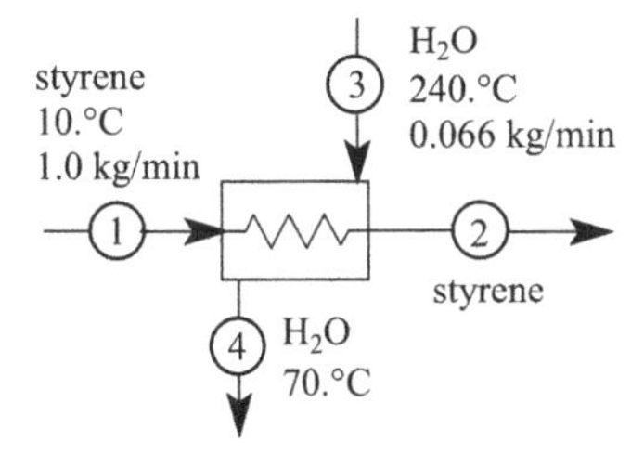

3.60 Complete the energy balance on the melter in the desalination process (shown below) to calculate the temperature of stream 5.

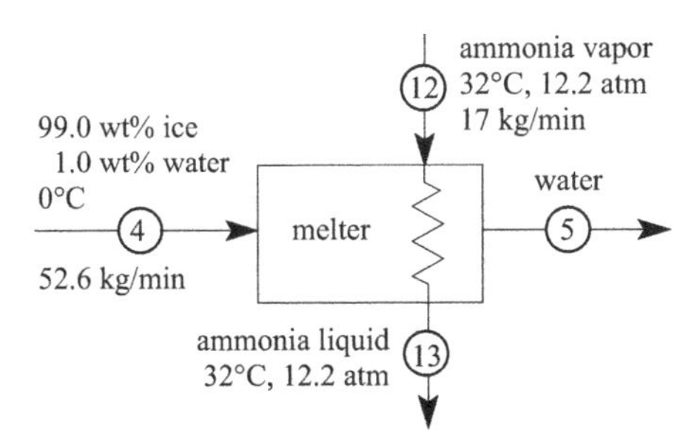

3.61 The "exquisite" desalinator in Figure 3.49 uses water as the refrigerant. Consequently the heat transfer is improved, the cost of each unit is reduced, and there is no need for a foreign refrigerant.

As shown in the flowsheet in Figure 3.49, cooled seawater is admitted to the freezer at low pressure. The freezer operates at the triple point of seawater: the temperature and pressure at which ice, brine, and steam coexist. Heat is removed from the freezer by vaporizing water. Water vapor is compressed and heat is delivered to the melter by condensing water vapor.

Recall that the sum of the flow rate of water vapor in stream 3 plus ice in stream 5 is 53 kg/min. Given an input rate of 100. kg/min of seawater, calculate the flow rate of water vapor in stream 3. You may neglect any energy flow into the freezer owing to changes in pressure.

3.62 A stream of *F*, *G*, and *H* is heated to separate *F* vapor from *G* and *H* liquid. Calculate q_{heater} in kJ/min.

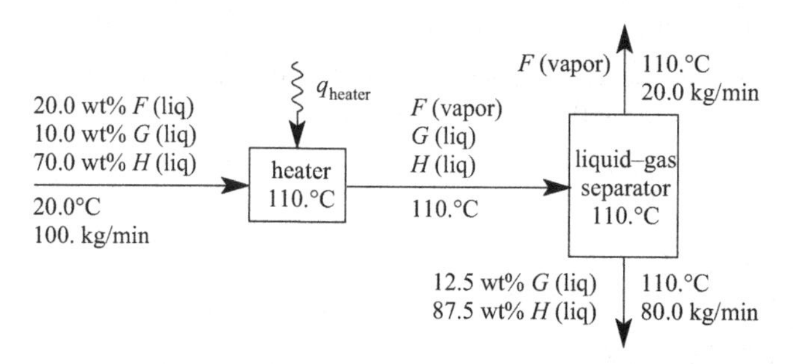

You may assume that heating a mixture of *F*, *G*, and *H* is equal to heating *F*, *G*, and *H* separately. Also, you may assume that the parameters in the table below are independent of temperature.

Some properties at 1 atm.

		$\bar{C}_P$ (kJ/(kg·°C))			
	Boiling point (°C)	Liquid	Vapor	$\Delta\bar{H}_{vap}$ (kJ/kg)	$\Delta\bar{H}_{fusion}$ (kJ/kg)
F	105.	1.60	1.20	380.	100.
G	176.	2.00	1.40	400.	120.
H	197.	2.40	1.60	420.	140.

3.63 (A) A mixture of water and styrene is separated as shown below.

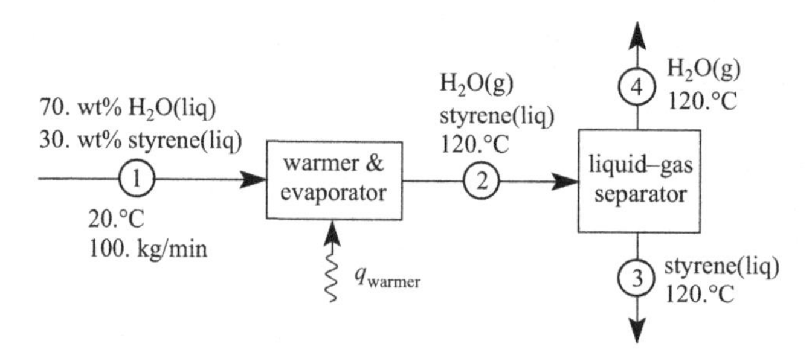

Calculate q_{warmer}, the rate that heat is delivered to the warmer + evaporator, in kJ/min. You may assume that heating a mixture of H_2O and styrene is the same as heating H_2O and styrene separately.

(B) A heat exchanger is added to reclaim energy from stream 4, as shown below.

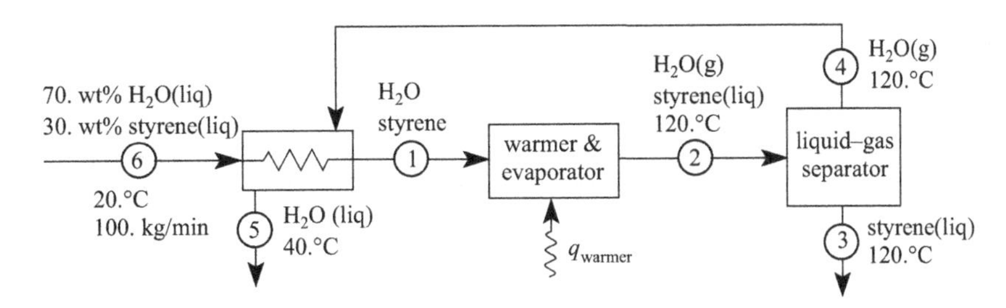

Calculate the state of stream 1. That is, calculate the temperature and the phase(s) (liquid and/or gas) of H_2O and styrene.

(C) Calculate the rate that heat is delivered to the warmer + evaporator, q_{warmer}, in kJ/min, for the process in part (B).

3.64 Water enters a boiler at 50.°C and exits as two streams: steam at 100.°C and water at 100.°C. Given that the heater supplies 4.5×10^6 J/min, calculate the flow rates (in kg/min) of streams 2 and 3.

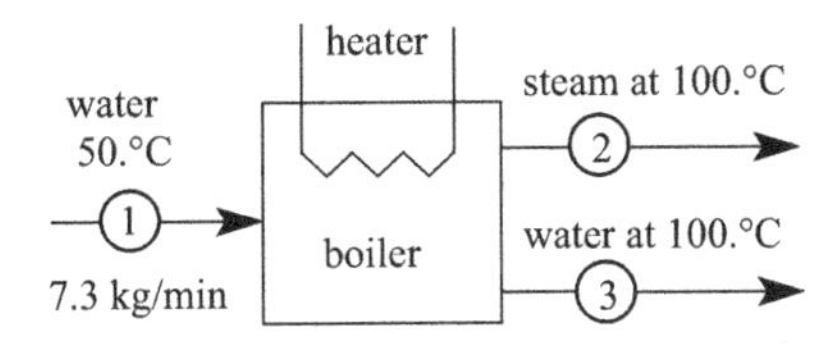

3.65 Calculate the temperature of the stream leaving the mixer below.

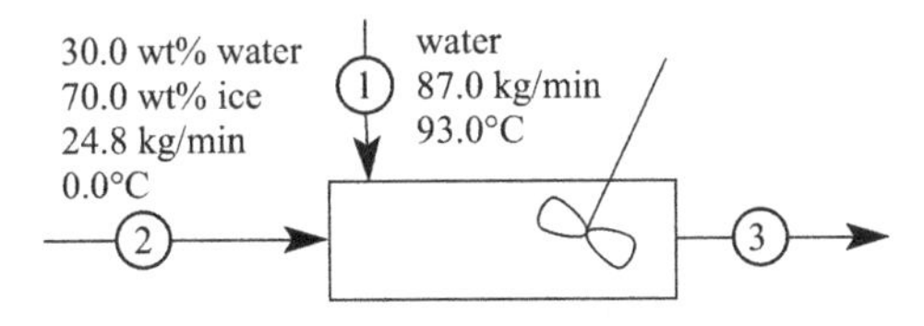

3.66 The mixer below combines ice, water, and steam. For stream 4, calculate the flow rate (in kg/min), the temperature (in °C), and the composition (wt fraction of ice, water, and steam).

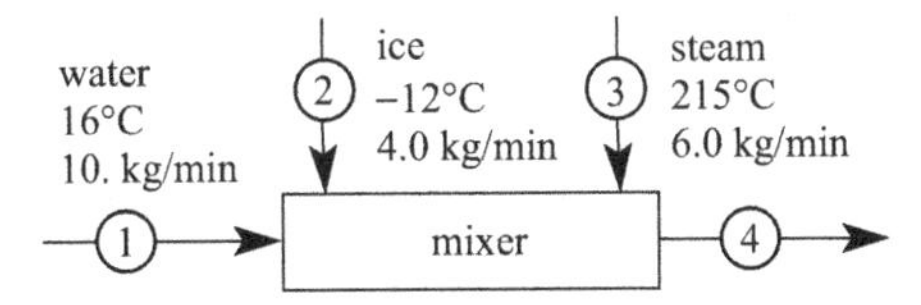

Hint: Consider the mixer as several simple units, such as melters, condensers, warmers, coolers, and combiners.

3.67 The process below produces solid ice and a mixture of ice and water (slush).

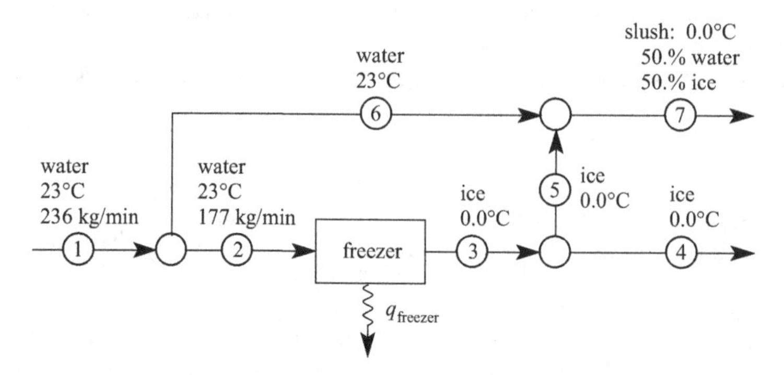

(A) Calculate $q_{freezer}$, the rate that energy is removed from the freezer, in kJ/min.
(B) Calculate the flow rate of stream 5.

3.68 Cornell University pipes chilled water to buildings for cooling and humidity control. Previously the water was chilled by electric-powered refrigerators, which used chlorofluorocarbons (CFCs) as refrigerants. As part of its sustainable energy program, Cornell replaced the refrigerators and their offending CFCs with lake-source cooling. Water is drawn from deep in Cayuga Lake where the temperature is a constant 40.°F. The cold lake water cools the chilled water to 45°F from 60.°F. The lake water is then returned to Cayuga Lake at 50.°F.

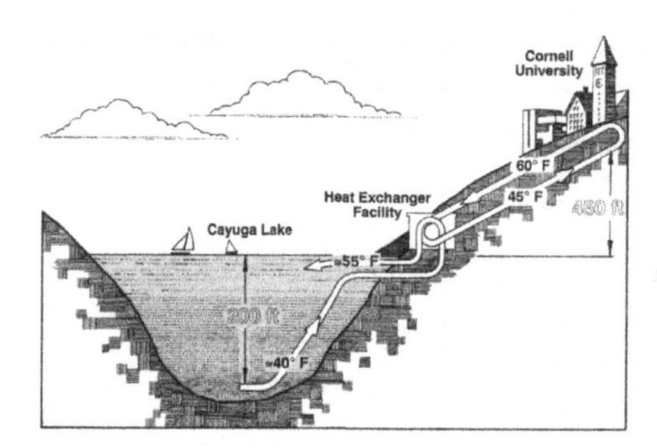

Image from the *Cornell Chronicle*, vol. 25, no. 39, 1994. www.sustainablecampus.cornell.edu/energy-lake.htm

(A) Estimate the flow rate of chilled water (in gal/hr) needed to cool 4.0×10^4 m^3/min of air to 68°F from 90.°F.
(B) Repeat the calculation in (A), but include the effect of condensing water from the air. Assume air at 90.°F and 90.% humidity (3.02 wt% water) is cooled to air at 68°F and 50.% humidity (0.72 wt% water). For this estimate, you may assume the water vapor condenses at an average temperature of 68°F and that the heat capacity of air is independent of humidity.
(C) Use the flow rate of chilled water calculated in (B) to calculate the flow rate of water from Cayuga Lake, in gal/hr.

3.69 (A) Your chemical plant proposes to discharge wastewater at 40.°C to a river. The river water is 25°C and flows at 5500. m^3/min. The river temperature must not increase more than 1.0°C. What is the maximum flow rate that the wastewater may be discharged to the river?

(B) Instead of discharging wastewater, your company proposes to use river water as the coolant in a condenser. Water will be drawn from the river, used to condense 5000. kg/min of steam at 1 atm and 100.°C, and returned to the river at 30.°C. What flow rate of water (in m^3/min) must be drawn from the river?

(C) What is the maximum flow rate water (in m^3/min) that can be drawn from the river at 25°C and returned to the river at 30.°C such that the river temperature does not increase more than 1.0°C?

3.70 The unit below is designed to cool hot, dry air by evaporating water. There is concern that the cooled air (stream 2) contains so much water that it is uncomfortable, even though it is cool.

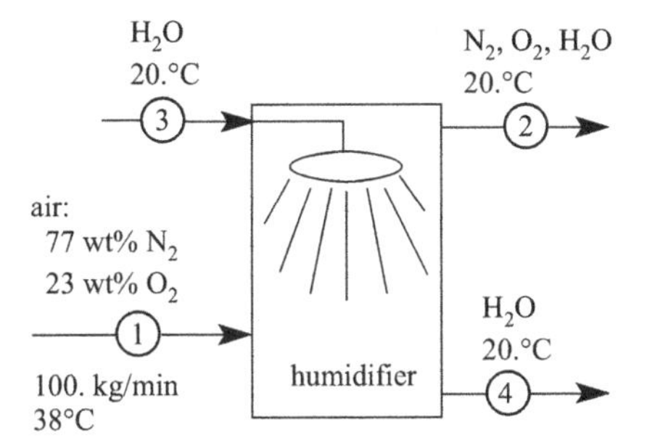

(A) Calculate the wt fraction of water in stream 2. You may assume that the temperature inside the humidifier is 20.°C. Thus the water evaporates at 20.°C. All the H_2O in stream 2 is vapor. All the H_2O in streams 3 and 4 is liquid.

(B) Air at 20°C can hold a maximum of 1.44 wt% H_2O; this is defined as 100% humidity. What is the maximum air temperature in stream 1 that can be cooled to 20.°C?

3.71 Ancient civilizations in hot, arid climates used earthenware pitchers to chill water. An earthenware pitcher was filled with water, the temperature of which was initially the same as the air. After a few hours, the water was much cooler than the air. Here is how it worked: the porous earthenware allows water to diffuse through the walls. When the water reaches the surface, it evaporates. The evaporation draws energy from the water in the pitcher, which cools the water. The lower temperature is maintained because the porous earthenware also has a low thermal conductivity.

Consider a pitcher containing 5.2 kg of water for which the rate of evaporation is 0.060 kg/hr and the rate of heat conduction into the pitcher is given by the equation

$$q_{\text{conduction}} = U(T_{\text{outside}} - T_{\text{water}}),$$

where $U = 6.0 \times 10^3$ J/(hr·K) and $T_{\text{outside}} = 39$°C. Assume the water has been in this pitcher for a long time and its temperature is constant. Calculate the temperature of the water.

You may assume:

- The water evaporates at 39°C (and the heat of vaporization is the same as at 100°C).
- The rate of evaporation is independent of the amount of water in the pitcher.

Adapted from Zubizarreta, J. I., and Pinto, G. (1995). *Chemical Engineering Education*, Spring, vol. 29, p. 96.

3.72 Chemical engineering students at the University of Michigan designed a coffee mug that keeps coffee at the ideal temperature of 72°C longer than ceramic mugs.

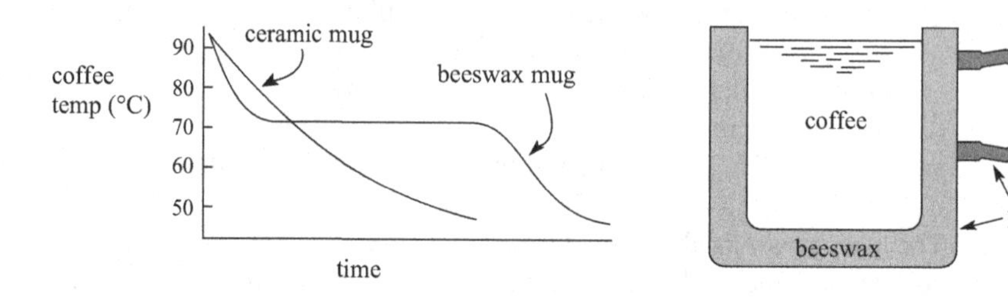

Here is how the mug works. When hot coffee is poured into the mug it melts a reservoir of beeswax sealed inside the mug. The temperature of the coffee remains at the melting point of beeswax, 72°C, until all the beeswax has solidified, after which the temperature of the coffee falls below 72°C. The coffee mug holds 250 mL of coffee and contains 80. g of beeswax. You may assume that the properties of beeswax are the same as its main component, myricyl palmitate ($CH_3(CH_2)_{14}COO(CH_2)_{30}CH_3$), and that the properties of coffee are the same as its main component, water.

Coffee at what temperature will melt all the beeswax initially at a temperature 20.°C? (You may assume that the coffee gives up heat only to the beeswax and that the heat capacity of the other components of the mug are negligible.)

3.73 Heat escapes from my house by conduction through the walls and roof. The rate of heat loss, q_{loss} in kJ/hr, is proportional to the difference between the inside temperature and the outside temperature:

$$q_{loss} = k(T_{inside} - T_{outside}),$$

where $k = 740$ kJ/(°C·hr). The volume of my house is approximately 870 m^3 and the temperature inside is uniform. The "heat capacity" of the house is 3300 kJ/°C. That is, 3300 kJ raises the temperature of my house 1°C.

(A) At what rate must I heat my house (in kJ/hr) to maintain a steady temperature of 20°C (68°F) when the outside temperature is −5°C (23°F)?

(B) The heater in my house can supply heat at a maximum rate of 5.2×10^4 kJ/hr. I wish to keep the temperature in my home at a steady 20°C (68°F). What is the lowest outside temperature (in °C) that will still allow me to maintain an inside temperature of 20°C?

(C) Air leaks into (and leaks out of) my house at a rate of 0.10 m^3/s. What heating rate (in kJ/hr) do I need to maintain a steady temperature of 20°C inside if the outside temperature is −5°C? You may assume that the volumetric flow rate of 0.10 m^3/s is referenced to 20°C.

(D) The air that leaks into my drafty house is very dry. I maintain a comfortable humidity by placing pans of water on the heaters. To maintain 50% humidity the heater must evaporate 9.0 g of water per m^3 of dry air that leaks in. Again assume air flows through my house at a rate of 0.10 m^3/s and the outside temperature is −5°C. Cold, dry air leaks in and warm, wet air leaks out. What heating rate do I need (in kJ/hr) to maintain a steady temperature of 20°C and a humidity of 50% inside my home? You may assume that the water comes out of the faucet at 20°C and evaporates at 20°C.

3.74 Water at 18°C is warmed to 45°C (stream 5) and 100°C (stream 6) by the process below.

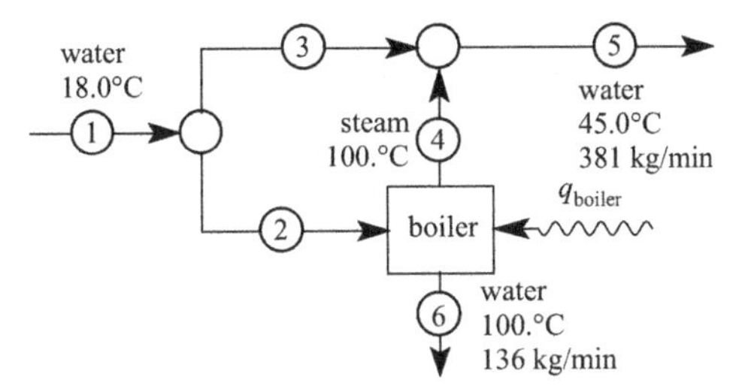

(A) Calculate q_{boiler}, the rate energy is supplied to the boiler (in kJ/min).
(B) Calculate the flow rate of stream 4, in kg/min.

3.75 The process below evaporates water from dirty water to make sludge and steam.

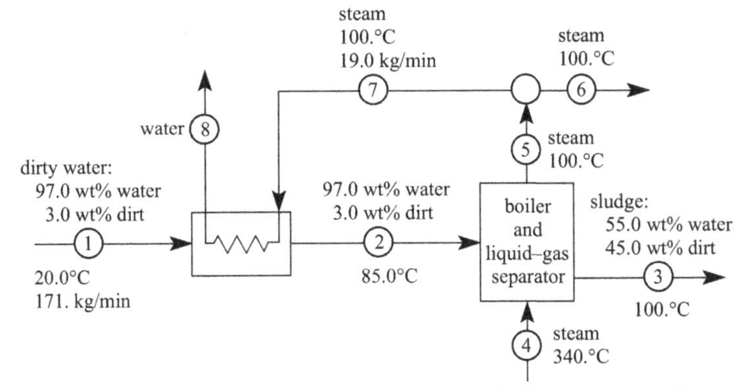

The dirt is solid insoluble particles suspended in the water. You may assume that heating a mixture of water and dirt is equal to heating water and dirt separately. In this exercise, assume the heat capacity for solid dirt is 6.95 kJ/(kg·°C).

(A) Calculate the flow rate of sludge, stream 3.
(B) Calculate the temperature of stream 8.
(C) Calculate the flow rate of stream 4.

3.76 Substance Z is cooled from 35°C to 10°C by passing it through three heat exchangers.

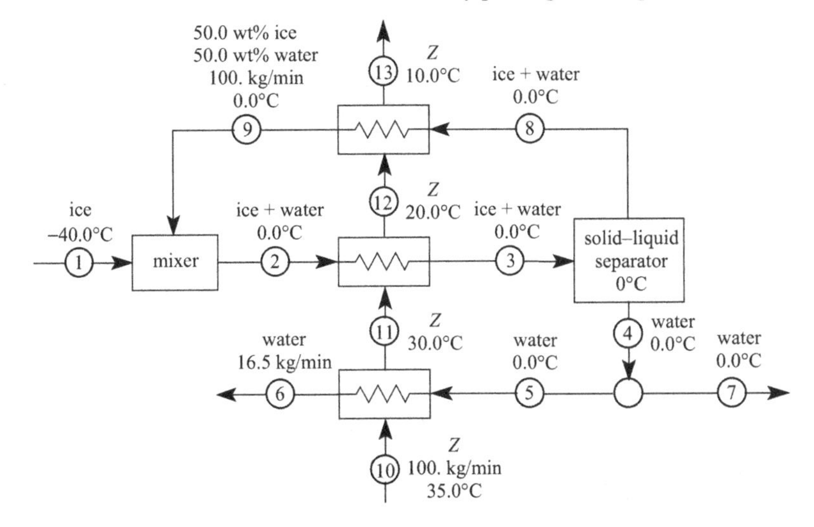

(A) Calculate the temperature of stream 6, in °C.
(B) Calculate the flow rate of stream 4, in kg/min.
(C) Calculate the flow rate and composition of stream 3.

Some properties of Z at 1 atm.

	Melting point (°C)	Boiling point (°C)	$\bar{C}_P$ (kJ/(kg·°C)) Solid	Liquid	Vapor	$\Delta\bar{H}_{vap}$ (kJ/kg)	$\Delta\bar{H}_{fusion}$ (kJ/kg)
substance Z–17		153.	1.65	3.86	1.45	1870	356

3.77 Consider this continuous process for steaming dry vegetables taken directly from a refrigerator.

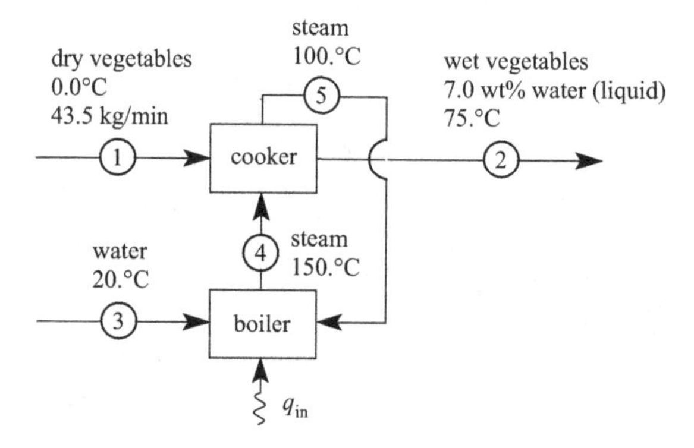

(A) Calculate the flow rate of water (stream 3), in kg/min. Assume the heat capacity of vegetables is 3.40 kJ/(kg·°C).
(B) Calculate q_{in}, the rate that energy is supplied to the boiler, in kJ/min.
(C) Calculate the flow rate of the recycle stream (stream 5).

3.78 A stream of formula X is to be gently warmed to 80°C from 30°C. Because formula X is heat-sensitive, the heating source must be exactly 100°C. The temperature of the heating source is maintained by a steam–water mixture at 1 atm.

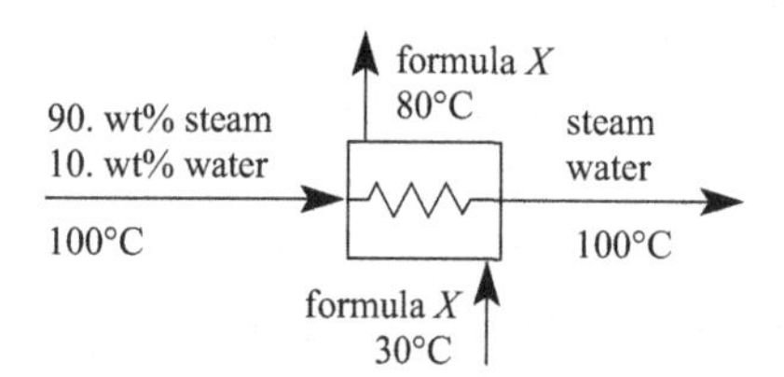

But steam is available only at 360°C and 1 atm. The process below cools the hot steam before using it to warm formula X. (Properties of formula X are given in a table following Exercise 3.88 on p. 200.)

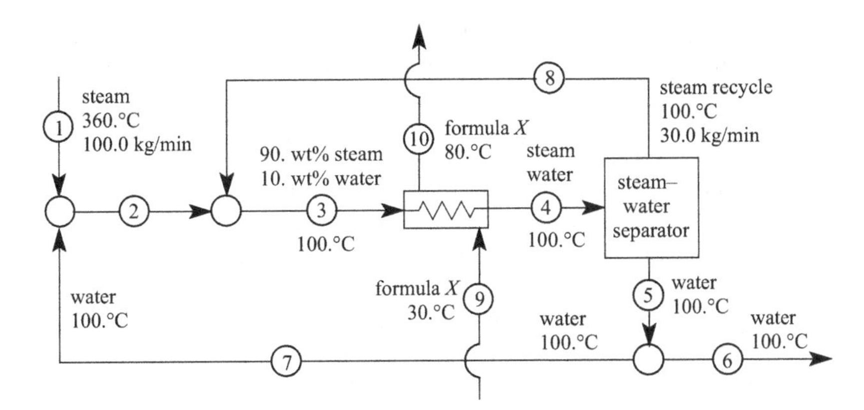

(A) Calculate the flow rate of stream 6.
(B) What flow rate of formula X (stream 9) can be accommodated by this design?
(C) What is the flow rate of the water recycle (stream 7)?

3.79 The process below creates pure water from seawater by evaporation. The waste brine (stream 4) contains sufficient liquid water to prevent solid salt.

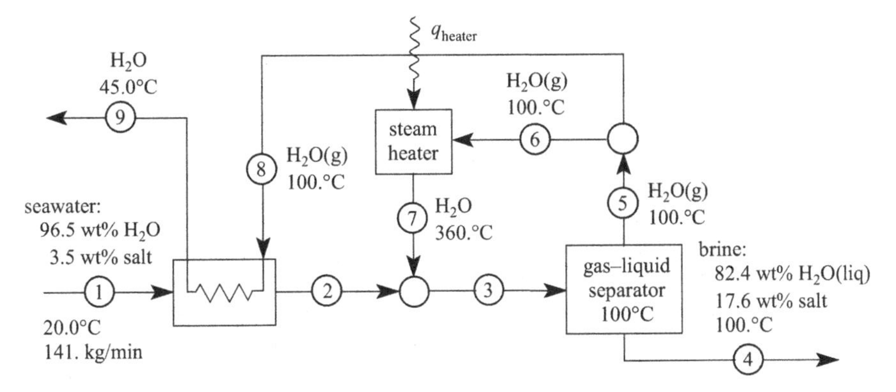

You may assume the heat capacities for seawater and brine are the same as that of pure water.

(A) Calculate the flow rate of product (stream 9), in kg/min.
(B) Calculate the temperature of the heat exchanger effluent (stream 2) and the phase(s) present in stream 2; determine the wt% liquid and the wt% vapor in stream 2.
(C) Calculate the flow rate of the recycle (stream 6), in kg/min.

Energy Balances with Chemical Reactions

3.80 Ammonia (NH_3) has potential to serve as a fuel. Ammonia can be obtained from sustainable sources, the combustion products do not harm the environment, and because ammonia is a liquid at 20°C and modest pressure, it has a high energy density (MJ/L). Assume that with catalytic converters we can burn ammonia to N_2 and H_2O by the following overall reaction:

$$4NH_3(g) + 3O_2 \rightarrow 2N_2 + 6H_2O\ (g).$$

(A) Calculate the molar heat of reaction, $\Delta\bar{H}^0_{rxn}$, at 25°C. Include a table of the heats of formation you collect for each substance. Include references – a textbook or a URL.

(B) Calculate the adiabatic temperature rise for a reactant mixture of 4 mol of NH_3 and 3 mol of O_2. Include a table of heat capacities with references. Describe your method – how you calculated an average heat capacity or how you evaluated a Shomate equation for the substances.

(C) Calculate the adiabatic temperature rise for a reactant mixture of NH_3 and air, with a stoichiometric ratio of NH_3 and O_2. Again state what heat capacities you use and your method – how you calculated an average heat capacity or how you evaluated the Shomate equation for the components.

(D) Calculate the adiabatic temperature rise for a reactant mixture NH_3 and 30. mol % excess air, necessary to ensure the complete reaction of NO intermediate products to N_2 and O_2. Again state what heat capacities you use and your method – how you calculated an average heat capacity or how you evaluated the Shomate equation for the components.

3.81 Methanol (CH_3OH) is a fuel derived from methane. Assume that with catalytic converters we can completely burn methanol to CO_2 and H_2O by the following overall reaction:

$$2CH_3OH\ (g) + 3O_2 \rightarrow 2CO_2 + 4H_2O\ (g).$$

(A) Calculate the molar heat of reaction, $\Delta\bar{H}^0_{rxn}$, at 25°C. Include a table of the heats of formation you collect for each substance. Include references – a textbook or a URL.

(B) Calculate the adiabatic temperature rise for a reactant mixture of 2 mol of CH_3OH and 3 mol of O_2. Include a table of heat capacities with references. Describe your method – how you calculated an average heat capacity or how you evaluated a Shomate equation for the substances.

(C) Calculate the adiabatic temperature rise for a reactant mixture of CH_3OH and air, with a stoichiometric ratio of CH_3OH and O_2. Again state what heat capacities you use and your method – how you calculated an average heat capacity or how you evaluated the Shomate equation for the components.

3.82 Synthesis gas (aka syngas) can be converted to methanol, which is a convenient liquid fuel and can be further converted to higher hydrocarbons via dimethyl ether. The overall reaction is

$$CO + 2H_2 \rightarrow CH_3OH\ (g).$$

(A) Calculate the molar heat of reaction, $\Delta\bar{H}^0_{rxn}$, at 25°C. Include a table of the heats of formation you collect for each substance. Include references – a textbook or a URL.

(B) Calculate the adiabatic temperature rise for a reactant mixture of 1 mol of CO and 2 mol of H_2. Include a table of heat capacities with references. Describe your method – how you calculated an average heat capacity or how you used a Shomate equation for the substances.

(C) The syngas produced from underground gasification of coal contains some CO_2. Rather than separate CO_2 from CO and H_2 (which requires cryogenic temperatures), we will synthesize methanol in the presence of CO_2 and then separate CO_2 from methanol. Calculate the adiabatic temperature rise for a reactant mixture of 1 mol of CO, 2 mol of H_2, and 1 mol of CO_2.

3.83 Hydrogen cyanide (HCN) is colorless and extremely poisonous. It is also a key reagent in the synthesis of the polymers nylon and polymethyl methacrylate (PMMA), the amino acid methionine, and the chelating agents EDTA and NTA. HCN can be synthesized by the BMA process developed by the German chemical company Degussa:

$$NH_3(g) + CH_4(g) \rightarrow HCN\ (g) + 3H_2(g).$$

The process name derives from ***Blausäure*** [hydrogen cyanide] *aus* ***Methan*** [methane] *und* ***Ammoniak*** [ammonia]. Ja?

(A) Calculate the molar heat of reaction, $\Delta \bar{H}^0_{rxn}$, and the molar entropy of reaction, $\Delta \bar{S}^0_{rxn}$, at 25°C. Include a table of the thermodynamic data you collect for each substance. Include references – a textbook or a URL.

(B) The endothermic reaction will be conducted in a series of adiabatic reactors and heaters. The inlet temperature for each reactor is 1400°C. The minimum temperature in each reactor (below which the reaction is unacceptably slow) is 900°C. How many adiabatic reactors are needed to achieve a fractional conversion of $X = 0.85$? You must justify your answer.

Assume the slope of the adiabatic temperature line on a plot of fractional conversion X vs. temperature is $-1/1420$°C in the temperature range 900°C to 1400°C. State all additional assumptions.

3.84 Urea – $(NH_2)_2CO$ – is a key nitrogen fertilizer synthesized from ammonia and carbon dioxide. The Bosch–Meiser synthesis for urea uses the following overall reaction:

$$2NH_3(g) + CO_2(g) \rightarrow (NH_2)_2CO\ (solid) + H_2O\ (g).$$

(A) Calculate the molar heat of reaction, $\Delta \bar{H}^0_{rxn}$, at 25°C. Include a table of the heats of formation for each substance. Assume the heat of formation for urea at 25°C and 1 atm is −333 kJ/mol. Include references for the data on the other substances – a textbook or a URL.

(B) The reaction is to be conducted at 150°C. Calculate the molar heat of reaction, $\Delta \bar{H}^0_{rxn}$, at 150°C. Note that the melting point of urea is 133°C. Assume the heat of melting for urea is 14 kJ/mol. Further assume the following average heat capacities in the range 25°C to 150°C: $\bar{C}_{P,NH_3} = 38\,J/(mol{\cdot}°C)$, $\bar{C}_{P,CO_2} = 40.J/(mol{\cdot}°C)$, $\bar{C}_{P,urea\ solid} = 91\,J/(mol{\cdot}°C)$, $\bar{C}_{P,urea\ liquid} = 120\,J/(mol{\cdot}°C)$, and $\bar{C}_{P,H_2O(gas)} = 34\,J/(mol{\cdot}°C)$.

(C) Calculate the adiabatic temperature rise for a reaction with reactants initially at 150°C. Use the heat capacities given in part (B).

(D) Design a process to synthesize urea in a series of adiabatic reactors and coolers. The inlet temperature for each reactor is 150°C. Because urea decomposes at about 320°C, the outlet temperature in each reactor is 280°C. How many adiabatic reactors are needed to achieve a fractional conversion of $X = 0.80$? You must justify your answer.

3.85 H_2 can reduce CO_2 to CO by the following reaction:

$$H_2 + CO_2 \rightarrow CO + H_2O\ (g).$$

The CO is separated from the H_2O, then mixed with H_2 to form synthesis gas (aka syngas).

(A) Calculate the molar heat of reaction, $\Delta\bar{H}^0_{rxn}$, and the molar entropy of reaction, $\Delta\bar{S}^0_{rxn}$, at 25°C. Include a table of the thermodynamic data you collect for each substance. Include references – a textbook or a URL.

(B) The endothermic reaction will be conducted in a series of adiabatic reactors and heaters. The inlet temperature for each reactor is 1100°C. The minimum temperature in each reactor (below which the reaction is unacceptably slow) is 900°C. What fractional conversion can be reached with two adiabatic reactors and one interstage heater? Assume each reactor is limited to the temperature range 900°C to 1100°C, or within 0.05 of the equilibrium curve. Assume the slope of the adiabatic temperature line on a plot of fractional conversion X vs. temperature is $-1/530$°C in the temperature range 900°C to 1100°C. State all additional assumptions.

3.86 Carbon disulfide is a reagent and solvent in organic synthesis, and is used as a fumigant insecticide. CS_2 can be synthesized by the following reaction:

$$2CH_4(g) + S_8(g) \rightarrow 2CS_2(g) + 4H_2S\ (g).$$

But this reaction must be conducted at temperatures higher than 445°C, and as a result the sulfur vaporizes in the form S_8. Consider a low-temperature synthesis with sulfur in the liquid state by the following reaction:

$$CH_4(g) + 4S\ (liquid) \rightarrow CS_2(g) + 2H_2S\ (g).$$

This reaction need only be conducted above the melting point of sulfur, 115°C.

(A) Calculate the molar heat of reaction, $\Delta\bar{H}^0_{rxn}$, and the molar entropy of reaction, $\Delta\bar{S}^0_{rxn}$, at 25°C for the low-temperature synthesis of CS_2. Include a table of the thermodynamic data you collect for each substance. Include references – a textbook or a URL. A good source is the data tables posted by the National Institute of Standards and Technology (NIST). Search "<substance> NIST," go to the NIST webbook, and select either "Gas Phase Thermochemistry" or "Condensed Phase Thermochemistry" for heats of formation and heat capacities.

(B) The endothermic reaction will be conducted in a series of adiabatic reactors and heaters. The inlet temperature for each reactor is 300°C. The minimum temperature in each reactor is limited by the melting point of sulfur (115°C) or a line slightly below the equilibrium fractional conversion at a given temperature. (Draw a limit line at $X = 0.05$ below the equilibrium fractional conversion line.)

How many adiabatic reactors in series are needed to achieve a fractional conversion of $X = 0.60$? You must justify your answer.

Assume the slope of the adiabatic temperature line on a plot of fractional conversion X vs. temperature is $-1/1140$°C in the temperature range 115°C to 300°C. State all additional assumptions.

Note that the activity of a condensed substance (liquid or solid) is 1. The activity of a gaseous substance may be approximated by the partial pressure of that substance relative to $P^0 = 1$ bar.

3.87 Synthesis gas (aka syngas) can be converted to methanol, which is a convenient liquid fuel and can be further converted to higher hydrocarbons via dimethyl ether. The overall reaction is

$$CO + 2H_2 \rightarrow CH_3OH\ (g).$$

(A) Calculate the molar heat of reaction, $\Delta \bar{H}^0_{rxn}$, and the molar entropy of reaction, $\Delta \bar{S}^0_{rxn}$, at 25°C for the synthesis of CH_3OH (g). Include a table of the thermodynamic data you collect for each substance. Include references – a textbook or a URL. A good source is the data tables posted by the National Institute of Standards and Technology, NIST. Search "<substance> NIST," go to the NIST webbook, and select "Gas Phase Thermochemistry" for heats of formation and heat capacities.

(B) Prepare plots of the equilibrium fractional conversion as a function of temperature at three pressures – 1 bar, 10 bar, and 100 bar – analogous to those in Figure 3.55.

(C) Choose an operating pressure in the range 1 to 100 bar and design a series of adiabatic reactors and coolers to conduct the exothermic reaction. The inlet temperature for each reactor is 200°C. The maximum temperature in each reactor is limited by a line 0.05 below the equilibrium curve.

Assume the slope of the adiabatic temperature line on a plot of fractional conversion X vs. temperature is 1/1120°C in the temperature range 200°C to 500°C. State all additional assumptions.

What is the fractional conversion at the exit of the fifth adiabatic reactor?

3.88 Which diagram of fractional conversion (X) vs. temperature could correspond to a series of adiabatic reactors and heat exchangers for the following reaction?

$$2A\ (g) + B\ (g) \rightarrow P\ (g)$$

You must justify your choice for full credit. One sentence will suffice.

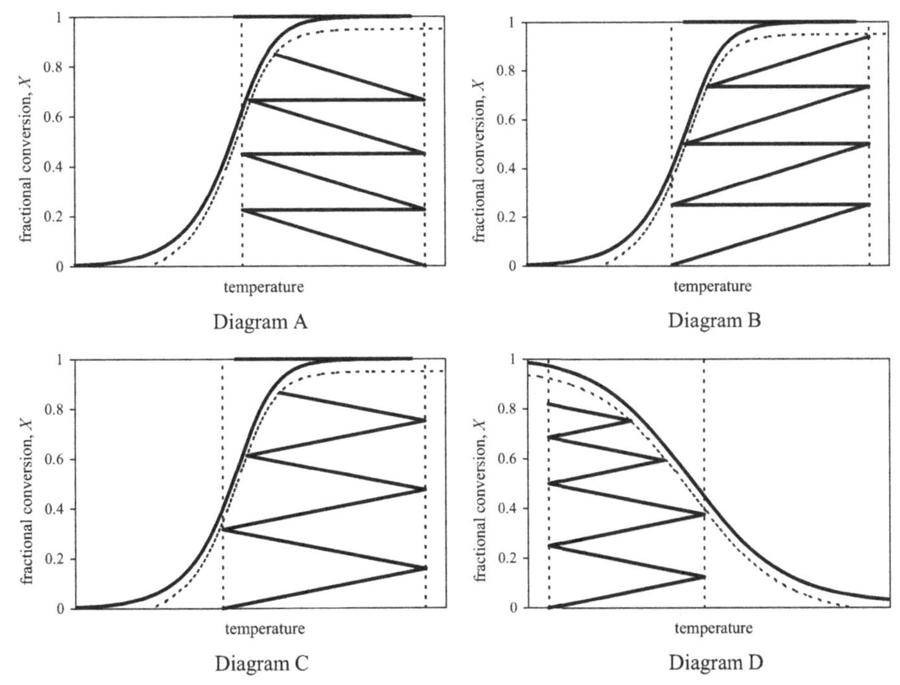

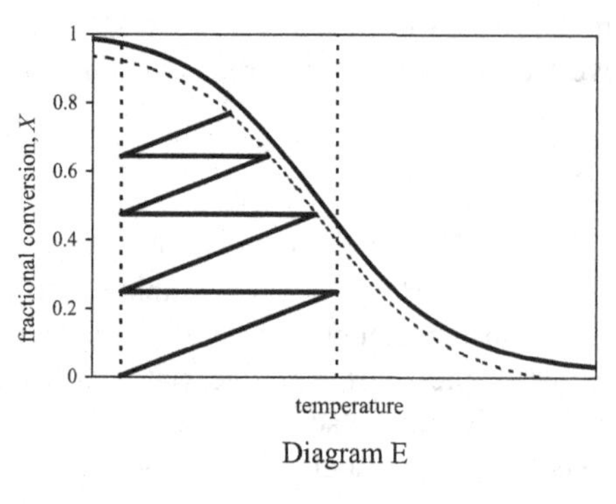

Diagram E

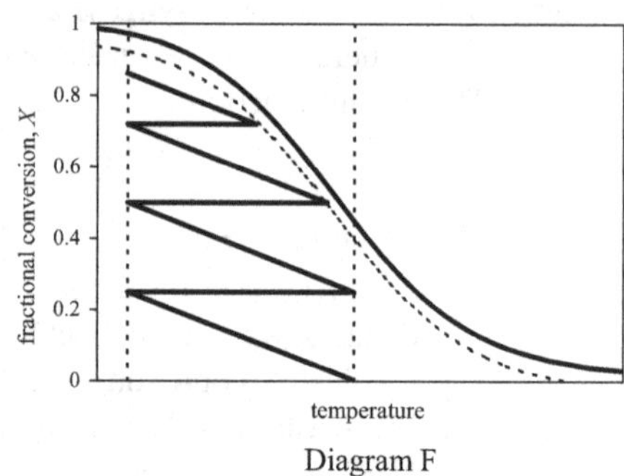

Diagram F

Thermodynamic properties for compounds at 1 atm.

	C_6H_{12}	Acetone	Myricyl palmitate	Styrene	Air	Formula X
molecular weight (g/mol)	84	58	690	104	28.8	103
C_p (J/(°C·mol))	157. (liq)	126. (liq)	1100 (sol)	183. (liq)	42. (gas)	451. (liq)
	106. (gas)	74.9 (gas)	2300 (liq)	122. (gas)		243. (gas)
boiling point (°C)	80.7	56.		145.2		103.
melting point (°C)	6.5	−95.	72	−30.6		21.
heat of vaporization (J/mol)	3.00×10^4	2.91×10^4		3.70×10^4		3.2×10^4
heat of melting (J/mol)	2.68×10^3	5.72×10^3	1.3×10^5	1.10×10^4		7.7×10^3

Thermodynamic properties for H_2O.

	$P = 1$ atm	$P = 0.005$ atm
C_p (J/(°C·mol))[a]	33. (solid)	33. (solid)
	75. (liq)	75. (liq)
	35. (vapor)	33. (vapor)
boiling point (°C)	100.0	−2.0
melting point (°C)	0.0	−2.0
heat of vaporization (J/mol)	4.1×10^4	4.5×10^4
heat of melting (J/mol)	6.0×10^3	6.8×10^3

[a] C_P varies with temperature. These values are valid to two significant figures in the range −50°C to 200°C.

Thermodynamic properties for ammonia.

	$P = 1.7$ atm	$P = 12.2$ atm
C_p (J/(°C·mol))	77 (liq)	82 (liq)
	24 (gas)	8 (gas)
boiling point (°C)	−22	32
heat of vaporization (J/mol)	2.27×10^4	1.93×10^4

Process Economics

3.89 Consider two schemes for producing B from A.

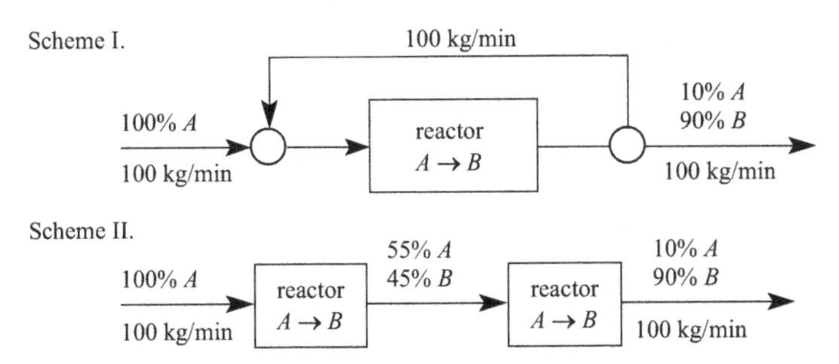

The capital cost of a reactor and the cost to operate a reactor are given by the charts below. Which scheme has the lower operating cost? Assume each process operates 24 hours per day, 7 days per week, and 50 weeks per year. You may assume straight-line depreciation over 10 years.

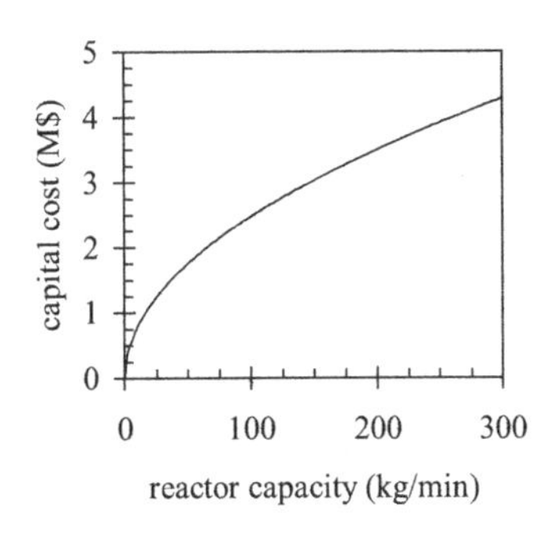

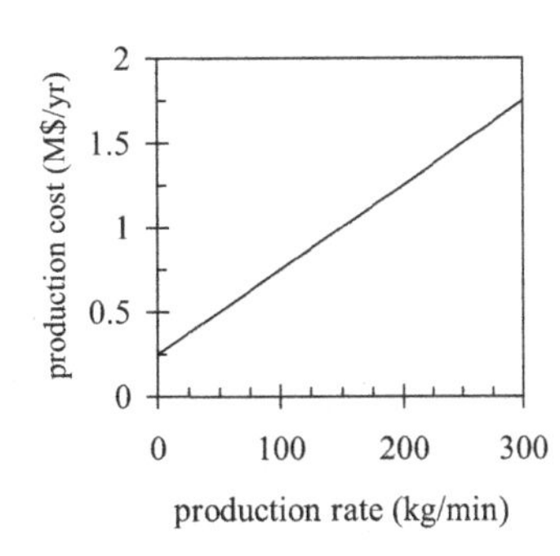

3.90 The graphs below give the capital costs and operating costs for the chemical produced by your company. Indicate the changes in these costs caused by the following three events.

(A) The property tax on your factory increases by 50%. Sketch the qualitative change in the cost curve(s) on either (or both) of the two graphs directly below.

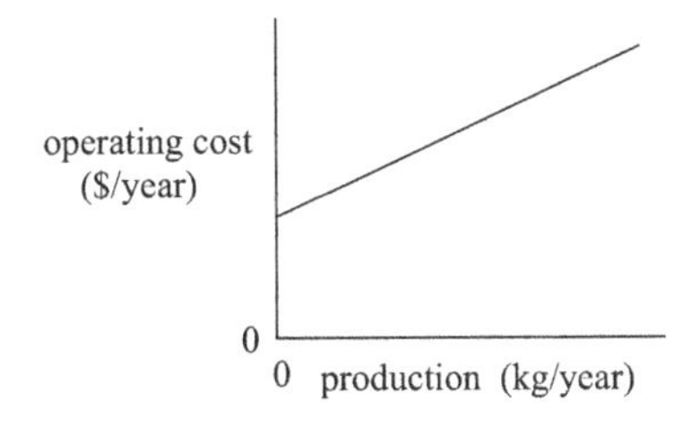

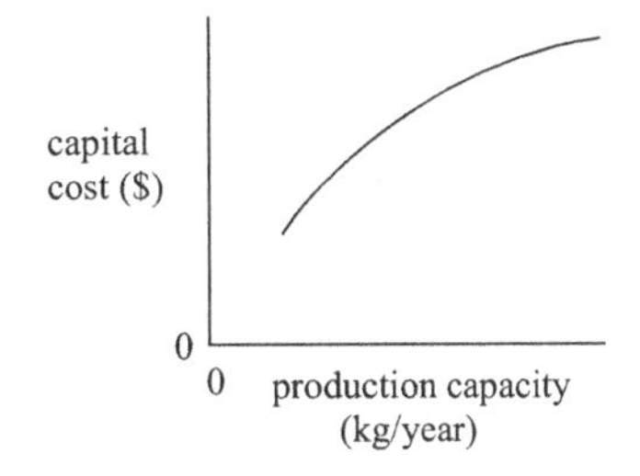

(B) The cost of chemical reactants increases by 50%. Sketch the qualitative change in the cost curve(s) on either (or both) of the two graphs directly below. **Do not** include the effect of part (A).

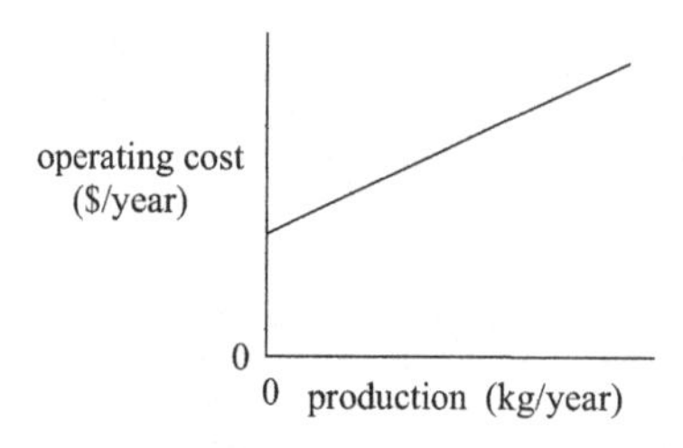

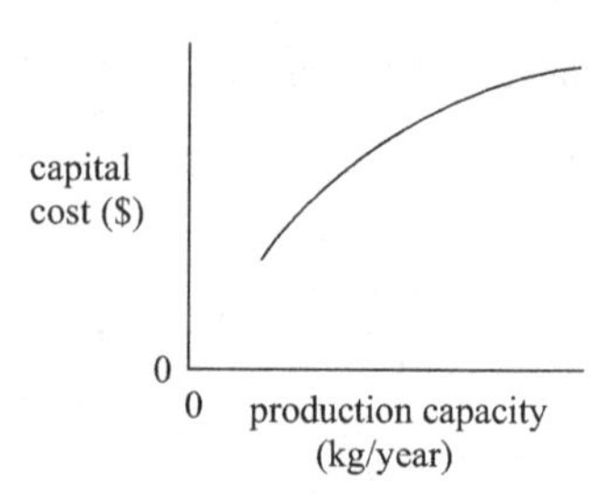

(C) The local government increases the tax on new equipment by 50%. The tax is paid when a manufacturer installs a new piece of equipment and the amount is independent of the cost of the equipment. Sketch the qualitative change in the cost curve(s) on either (or both) of the two graphs directly below. **Do not** include the effects of parts (A) and (B).

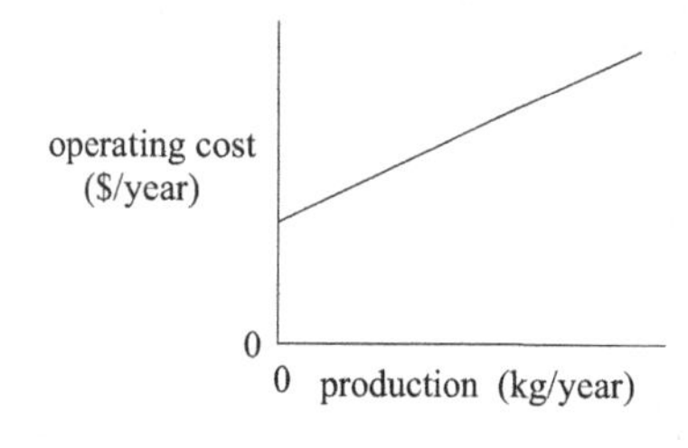

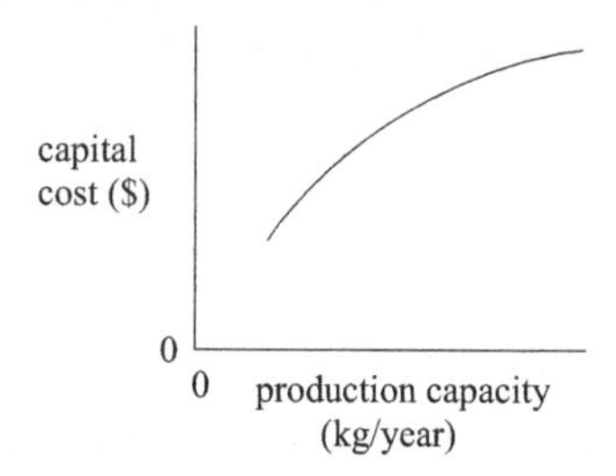

3.91 An electric company that burns coal to produce electricity must reduce the emission of sulfur compounds from its smokestacks. The company burns 17,000 tons of low-sulfur coal per year. Calculate the annual profit (in this case, annual loss) associated with the following three options:

(A) Continue to release sulfur at the present levels and pay a fine.
Fine: 2.7×10^6 \$/year .

(B) Install scrubbers to reduce sulfur emissions to the level mandated by law.
Cost of scrubbers: \$$2.3 \times 10^6$
Cost of operating scrubbers: 1.8×10^5 \$/year.

(C) Install scrubbers to reduce sulfur emissions to a level substantially below that mandated by law. This will allow the company to burn high-sulfur coal and still remain within legal limits for sulfur emissions.
Cost of scrubbers: \$$3.5 \times 10^6$
Cost of operating scrubbers: 2.1×10^5 \$/year
Low-sulfur coal: \$45/ton (2.5×10^7 Btu/ton)
High-sulfur coal: \$35/ton (2.9×10^7 Btu/ton).
You may assume a straight-line depreciation on the scrubbers, with a lifetime of 10 years.

3.92 A 1000-megawatt (MW) power plant produces an average of 20. tons of SO_2 per hour. (Wattage is a measure of power. One watt (W) equals one joule (J) per second.) 90.% of the SO_2 must be removed from the smokestack emissions. The SO_2 may be removed by reaction with limestone ($CaCO_3$) or by reaction with lime (CaO) – 1.7 tons of limestone are required to remove one ton of SO_2; 1.0 tons of lime are required to remove one ton of SO_2.

Raw materials cost \$15./ton limestone and \$60./ton lime. The capital costs for the scrubbers depend on the capacity of the power plants, rated in kW. The limestone scrubber costs \$215/kW and the lime scrubber costs \$175/kW.

Which system, the limestone scrubber or the lime scrubber, gives the lower yearly cost? That is, which system gives the least-negative profit? Assume a 10-year straight-line depreciation.

From Professor P. Harriott (Cornell University, 1993).

3.93 The steam that heats a factory is piped from a remote boiler. Insulation on the pipes will reduce the heat loss and thus reduce the heating cost. The heating cost for pipes with no insulation is \$127,000/year. The reduction in heating costs for insulation x cm thick is

$$\text{heating cost saved with } x \text{ cm of insulation} = 12{,}000\, x \ \$/\text{year}.$$

The capital cost of the insulation is

$$\text{capital cost} = \text{fixed installation cost} + \text{material cost of insulation}$$
$$\text{capital cost} = \$45{,}000 + \$2{,}100x^2.$$

Because the pipes are exposed to the weather, the insulation must be replaced every three years. What thickness of insulation (if any) will maximize the ROI?

3.94 The electric bill of your small chemical company is \$110,000/year. Your company operates only during the normal business day, when the cost of electricity is highest. Consider this scheme to reduce the electric bill: install rechargeable batteries that will be charged from midnight to 6 am (when the cost of electricity is lower) and will be discharged during the day to power the plant. The capital cost of this scheme (batteries plus installation) is \$65,000. The batteries are expected to last 5 years. The cost of maintenance and inspection of the batteries is expected to be \$4000/year. With these batteries, you predict your electric bill will drop to \$80,000/year.

Calculate the ROI for this scheme to install batteries. You may assume a straight-line depreciation on the batteries.

3.95 Assume you manage the chemical production division of a corporation. You have \$10M to invest in new capital. Which of the following projects (if any) should you fund? You may fund only one of each project. You may not, for example, use the \$10M to fund two D projects.

Project	Description	Capital cost (\$M)	Operating cost[a] (\$M/year)	Revenue (\$M/year)
A	upgrade electrical generator	9.0	11.5	13.8
B	bioconverter for waste stream	8.0	25.7	21.3
C	increase boiler efficiency	7.0	12.0	14.3
D	recover reactants from purge	5.0	3.0	4.0
E	automate shipping	3.0	2.7	3.5
F	upgrade control system	2.0	2.7	3.5
G	add heat exchanger to hot purge	1.0	1.2	1.5

[a] Includes depreciation.

3.96 You form a small company to produce the specialty chemical *Z*. You purchase equipment that will produce 100 kg *Z* per year if operated 40 hours/week. The equipment has a capital cost of $\$1.5 \times 10^5$ which includes installation. The equipment has a lifetime of 5 years. The operating costs are:

fixed costs	\$40,000 per year (rent, taxes, etc.)
payroll	you: \$120,000 per year (\$100,000 salary + \$20,000 benefits (health insurance, retirement, etc.)
	employees: \$55,000 per year per employee (\$40,000 wages + \$15,000 benefits)
variable costs	\$400 per kg of *Z* produced

(A) Three employees are needed to operate the process. Calculate the minimum selling price for the chemical *Z* (in \$/kg) that will yield a return on investment (ROI) of 20% per year.

(B) You increase production to 200 kg *Z* per year by adding a night shift of three additional employees. The salary of the night-shift employees is 25% higher than the day-shift employees. Calculate the minimum selling price for chemical *Z* that will yield a ROI of 20% per year if your company produces 200 kg *Z* per year using two work shifts.

3.97 The price of a chemical *Q* is fixed at \$235/(kg of *Q*). The cost of equipment to produce *x* kg of *Q* per year is given by the formula

$$\text{capital cost} = \$700.x - \$0.10x^2.$$

This formula is valid for pricing equipment with capacity larger than 500 kg/year and smaller than 2500 kg/year. The variable operating cost is \$30./(kg of *Q*) and the fixed operating cost is $\$1.0 \times 10^5$/year. Tax laws restrict your company to a straight-line depreciation calculated over 10 years.

(A) Calculate the profit (or loss) of producing 1000. kg of *Q* per year.

(B) Calculate the amount of *Q* that must be produced per year to yield a 20.% per year ROI.

3.98 You own a chemical plant that converts reactant *E* to product *K*. The conversion to *K* is complete, but the reaction must be conducted at high pressure. The chief expense is compressing *E* to 100 atm. To recover some of the costs, *K* is expanded through a turbine to generate electricity.

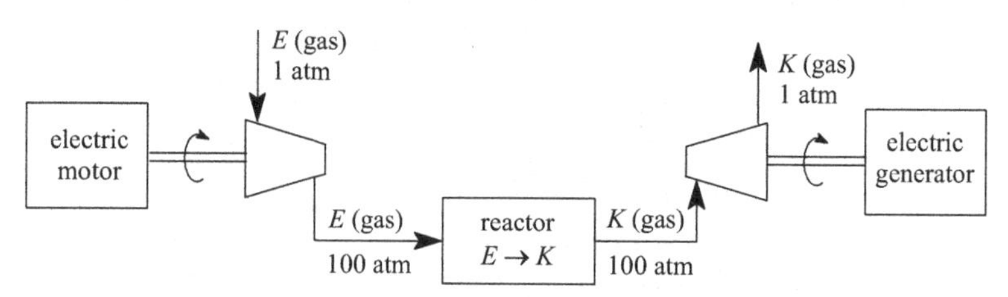

The turbine costs \$300,000. and requires \$10,000./year in operating costs (maintenance and repairs). This turbine has a lifetime of 9 years and reduces your electricity costs by \$180,000./year.

A new turbine is developed. This new turbine is more resistant to corrosion; its lifetime is 15 years and it requires only \$5000./year in operating costs. The new turbine is also more effective; it will reduce your electricity costs by \$198,000./year. This new turbine costs \$450,000.

When your present turbine reaches the end of its lifetime, should you replace it with the same model or the new model? Explain *succinctly*. It is important that you state which economic parameter(s) you (the owner) use to compare the turbines. That is, did you base your decision on operating cost, capital cost, depreciation, revenue, profit, or ROI?

3.99 Your company would like to recover the energy lost in a chemical process. The prices, costs, and savings for two different heat exchangers are shown in the table below. You may assume a straight-line depreciation. What do you recommend – Option 1, Option 2, or neither? You must justify your recommendation.

	Option 1	Option 2
heat exchanger price (\$)	10,000	16,000
maintenance costs (\$/year)	250	100
energy savings through recovered heat (\$/year)	4,250	6,000
lifetime of heat exchanger (years)	5	5

3.100 Your company will build and operate a process to supply water to a new resort hotel. The only source of water is a well. The well water is not potable, but can be treated to make potable water. Consider two schemes for supplying water. Scheme I treats all the water.

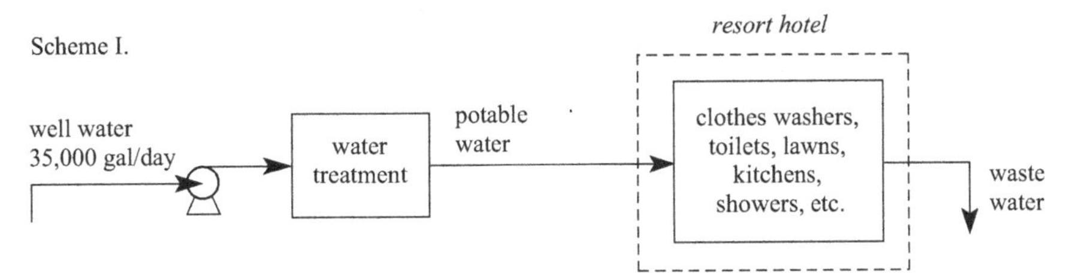

Scheme II treats only a portion of the water. The water treatment unit is smaller, but two sets of pipes are required: one for potable water and one for well water.

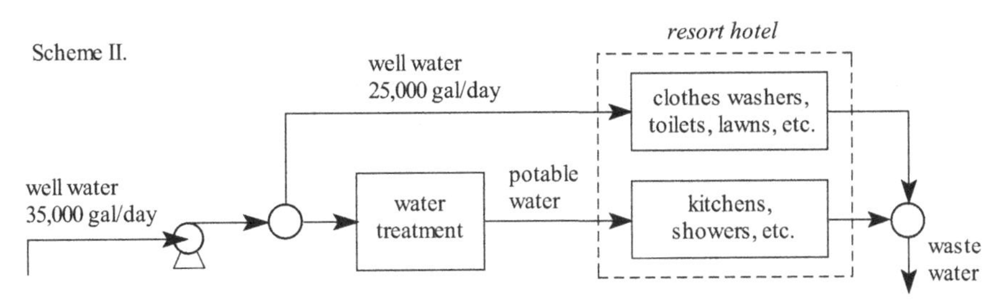

Equipment prices are given by the following formulas, where F_T is the total flow rate through the unit, in gal/day:

$$\text{pump price} = \$310. \times F_T^{0.6}$$
$$\text{water treater price} = \$900. \times F_T^{0.4}.$$

The pipes for Scheme I will cost \$320,000 and the pipes for Scheme II will cost \$576,000. The pump and water treater have 10-year lifetimes. The pipes have a 30-year lifetime. The annual operating costs for each unit are given by the following formulas:

$$\text{pump operating cost} = 0.80 \times F_T \ \$/\text{year}$$
$$\text{water treater operating cost} = 1.70 \times F_T \ \$/\text{year}.$$

The revenue for supplying water to the hotel is \$500,000/year. Which scheme (if either) should you build?

3.101 Consider this scheme for separating pure A from a mixture of A (valuable) and B (worthless).

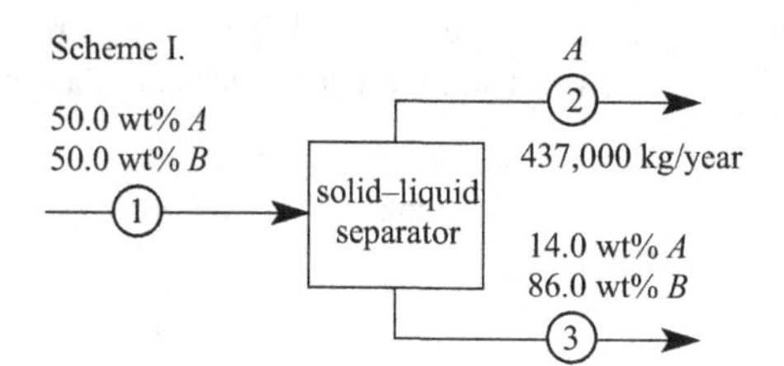

(A) Calculate the flow rates of stream 1 *and* stream 3, in kg/year.

(B) Scheme I, shown above, discards a waste stream with 14% A. Scheme II, shown below, discards less A, but requires a more expensive separator. Which scheme (if any) should our company build? Justify your choice.

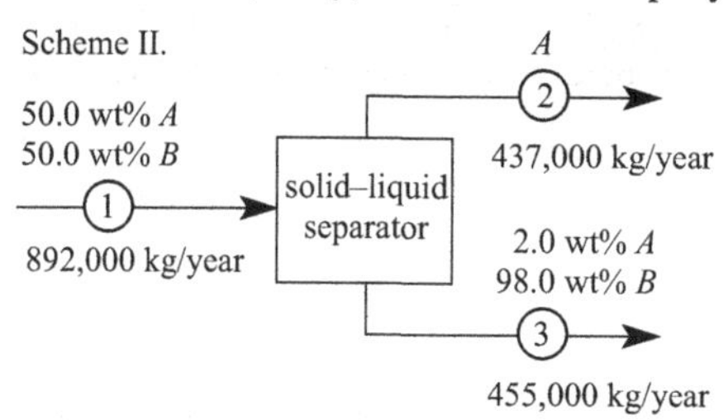

The equipment in both schemes has a lifetime of 10 years. We can sell A for \$5.50/kg. The 50–50 A–B mixture costs \$1.60/kg and disposal of the A–B waste stream costs \$0.34/kg. Scheme I has a capital cost of \$777,000, an operating cost of \$2,192,500/year, and a profit of \$211,000/year. Scheme II has a capital cost of \$1,400,000 and its expenses for labor, electricity, maintenance, and insurance are \$371,000/year.

3.102 Consider three schemes for separating two mixtures: a mixture of A and B, and a mixture of A and C. Scheme I produces A, B, and C.

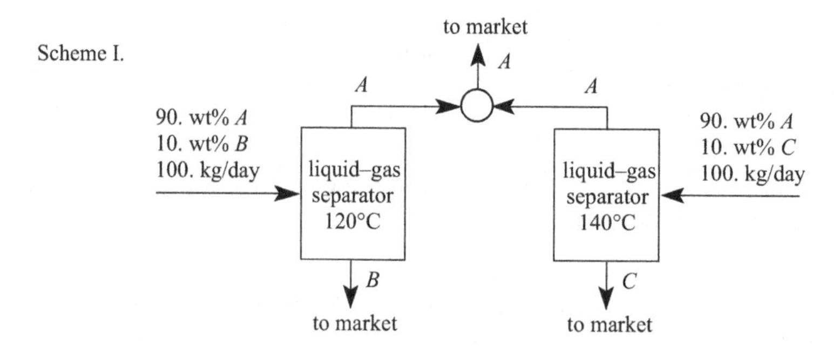

Scheme II also produces A, B, and C, but first combines the mixtures, then separates them. This reduces the size of the second separator, although the second separator operates at higher temperature.

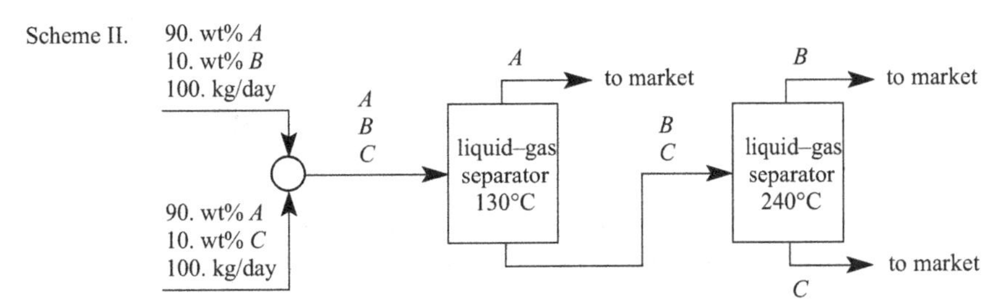

Scheme III produces only A and disposes a mixture of B and C.

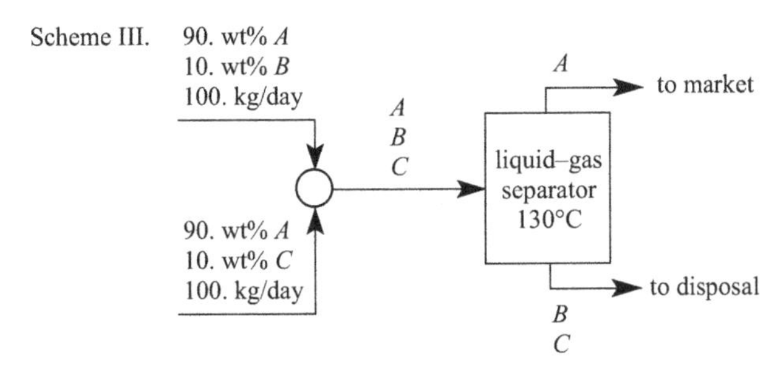

Which scheme (if any) should your company build? You must justify your choice.

	Boiling point (°C)	Market value ($/kg)
A	20	111
B	220	43
C	260	27

The 90:10 mixture of A and B costs $7.0/kg and the 90:10 mixture of A and C costs $4.0/kg. Disposal of the mixture of B and C costs $1.0/kg. The process operates 330. days per year. The equipment lifetime is 15 years.

$$\text{separator price} = \$160{,}000. \times F_{\mathrm{T}}^{0.6}$$

$$\text{separator operating cost} = \left[12.0 + 0.55\left(T_{\text{operating}} - 20\right)\right] \times F_{\mathrm{T}}\ \$/\text{day},$$

where F_T is the total flow through the separator in kg/day, and $T_{\text{operating}}$ is the operating temperature of the separator.

3.103 Consider two schemes for producing 100. kg/day of specialty chemical P from reactant A. A is available only as a mixture with an inert substance. The mixture of A and the inert substance is not easily separated. The first scheme discards the unreacted A with the inert stream.

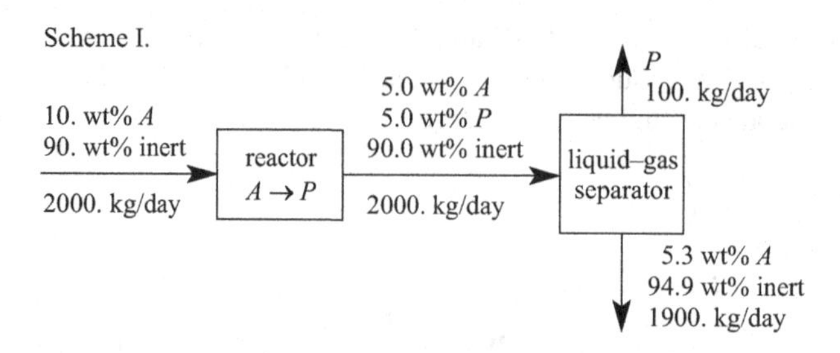

The second scheme recycles the unreacted A (with some inert substance) and removes the inert substance in a purge stream (with some unreacted A).

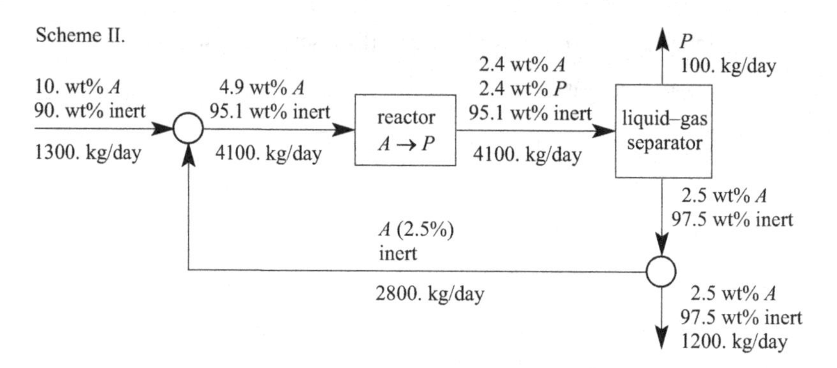

P sells for \$131/kg. The mixture of A (10%) and inert substance (90%) costs \$2.70/kg. Waste disposal costs \$0.17/kg. Assume each process operates 300. days/year and assume a straight-line depreciation over 10 years.

(A) Complete the following table:

	Scheme I	Scheme II
reactor cost	1400. k\$	2240. k\$
separator cost	750. k\$	1350. k\$
natural gas	162. k\$/year	324. k\$/year
electricity	58. k\$/year	116. k\$/year
wages	560. k\$/year	560. k\$/year
capital cost		
revenue		
operating cost		
profit		
ROI		

(B) Assume you own this process. Which scheme would you choose? Explain *succinctly*. It is important that you state which economic parameter(s) you (the owner) use to compare the schemes. That is, would you base your decision on operating cost, capital cost, depreciation, revenue, profit, or ROI?

3.104 Reactant A decomposes to P (product) and B (by-product):

$$A \rightarrow P + B.$$

The molecular weight ratio of P to B is 7:3, so 10 kg of A reacts to form 7 kg of P and 3 kg of B. The reactor converts 40% of A.

Consider three schemes for producing 100. kg/hr of P. Scheme I discards the unreacted A with the by-product B. A pump is placed before the reactor.

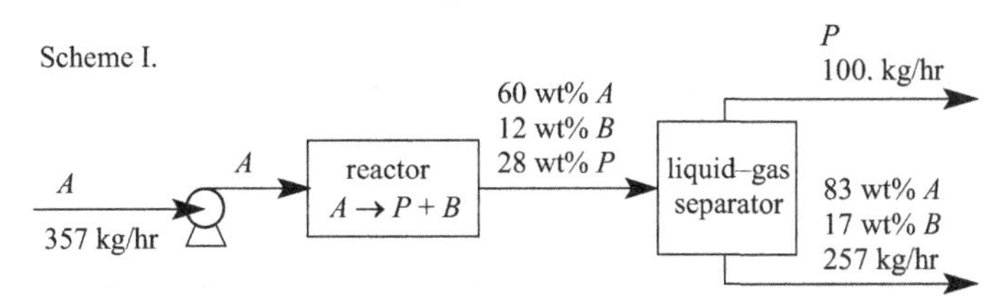

Scheme II uses a second separator to recycle the unreacted A and purify B. This adds the cost of a second separator (and a second pump), but reduces the reactant flow rate and allows one to sell the purified by-product, B. This also eliminates the cost of disposing the waste mixture of A and B.

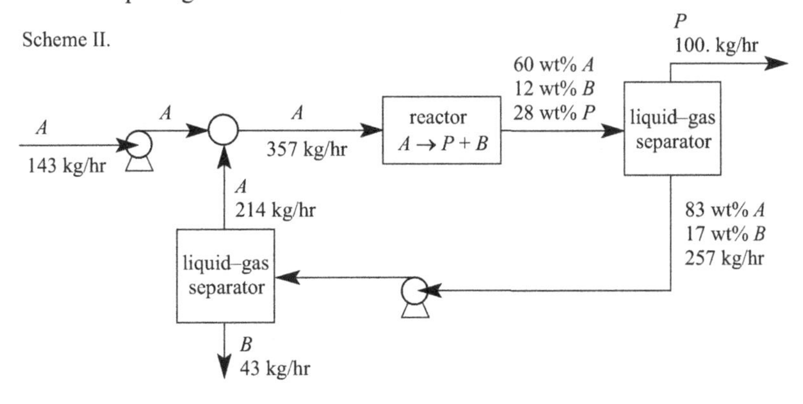

Scheme III has a recycle, but uses a purge instead of a second separator. The purge removes 20% of the stream leaving the separator.

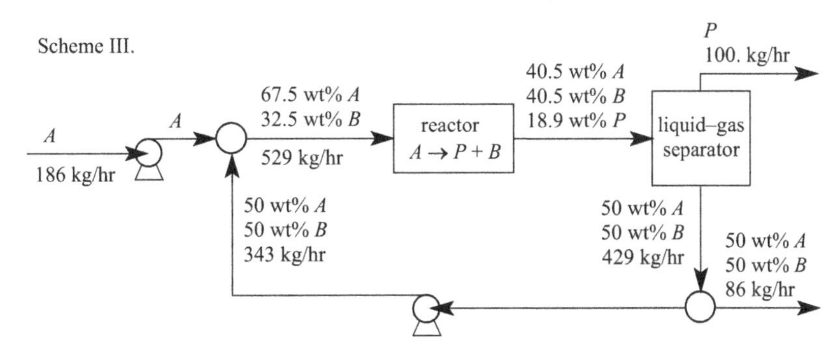

The economics are as follows. Reactant A costs \$1.47/kg. Product P sells for \$9.21/kg and by-product B sells for \$1.13/kg. Waste disposal costs \$0.11/kg. Equipment prices are given by the following formulas, where F_T is the total flow rate through the unit, in kg/hr:

$$\text{reactor price} = \$125{,}000 \times F_T^{0.6}$$
$$\text{separator price} = \$48{,}000 \times F_T^{0.6}$$
$$\text{pump price} = \$8{,}800 \times F_T^{0.33}.$$

These prices (in \$) include installation and all piping, wiring, and control instrumentation. For Scheme II, the same equation applies to both separators. The annual operating costs for each unit are given by the following formulas, where F_T is the total flow rate through the unit, in kg/hr:

$$\begin{aligned} \text{reactor operating cost} &= \$25{,}000 + \$2{,}100 \times F_T \\ \text{separator 1 operating cost} &= \$10{,}000 + \$1{,}300 \times F_T \\ \text{separator 2 operating cost} &= \$10{,}000 + \$1{,}600 \times F_T \\ \text{pump operating cost} &= \$4{,}000 + \$109 \times F_T. \end{aligned}$$

These operating costs (in \$/year) include energy costs, maintenance and replacement parts. The wages for each scheme are \$560,000/year for Scheme I, \$728,000/year for Scheme II, and \$616,000/year for Scheme III.

(A) Calculate the capital cost, operating cost, revenue, profit, and ROI for each scheme. You may assume a straight-line depreciation with a lifetime of 15 years. Assume that each process operates 330 days per year, 24 hours per day.
(B) Which is the better scheme? Briefly explain your choice.

3.105 Repeat the analysis of Exercise 3.104 with the same process schemes and flow rates, but with different costs. Reactant A costs \$1.37/kg. Product P sells for \$8.17/kg and by-product B sells for \$0.67/kg. Waste disposal costs \$0.07/kg. Relative to Exercise 3.104, the reactor is less expensive and the separator is more expensive. The purchase prices for equipment are given by the following equations, where F_T is the total flow rate through the unit, in kg/hr:

$$\begin{aligned} \text{reactor price} &= \$38{,}000 \times F_T^{0.6} \\ \text{separator price} &= \$115{,}000 \times F_T^{0.6} \\ \text{pump price} &= \$6{,}500 \times F_T^{0.33}. \end{aligned}$$

These prices (in \$) include installation and all piping, wiring, and control instrumentation. For Scheme II, the same equation applies to both separators. Relative to Exercise 3.104, the reactor costs less to operate. The major operating expenses are the separators. The annual costs of operating each unit are given by the following equations, where F_T is the total flow rate through the unit, in kg/hr:

$$\begin{aligned} \text{reactor operating cost} &= \$25{,}000 + \$600 \times F_T \\ \text{separator 1 operating cost} &= \$10{,}000 + \$2{,}500 \times F_T \\ \text{separator 2 operating cost} &= \$10{,}000 + \$3{,}200 \times F_T \\ \text{pump operating cost} &= \$2{,}000 + \$86 \times F_T. \end{aligned}$$

These operating costs (in \$/year) include energy costs, maintenance and replacement parts. The wages for each scheme are \$560,000/year for Scheme I, \$728,000/year for Scheme II, and \$616,000/year for Scheme III.

(A) Calculate the capital cost, operating cost, revenue, profit, and ROI for each scheme. You may assume a straight-line depreciation with a lifetime of 15 years. Assume that each process operates 330 days per year, 24 hours per day.
(B) Which is the better scheme? Briefly explain your choice.

3.106 Repeat the analysis of Exercise 3.104 with the same process schemes and flow rates, but with different costs. Reactant A costs \$0.47/kg. Product P sells for \$8.23/

kg and by-product B sells for \$1.53/kg. Waste disposal costs \$0.11/kg. Equipment prices are given by the following formulas, where F_T is the total flow rate through the unit, in kg/hr:

$$\text{reactor price} = \$800{,}000. + \$7{,}500. \times F_T$$
$$\text{separator price} = \$1{,}300{,}000. + \$6{,}800. \times F_T$$
$$\text{pump price} = \$36{,}000. + \$70.0 \times F_T.$$

Equipment prices (in \$) include installation and all piping, wiring, and control instrumentation. The annual operating costs for each unit are given by the following formulas, where F_T is the total flow rate through the unit, in kg/hr:

$$\text{reactor operating cost} = (25{,}000 + 2{,}100 \times F_T)\ \$/\text{year}$$
$$P - A \text{ separator operating cost} = (10{,}000 + 1{,}300 \times F_T)\ \$/\text{year}$$
$$A - B \text{ separator operating cost} = (10{,}000 + 1{,}600 \times F_T)\ \$/\text{year}$$
$$\text{pump operating cost} = (4{,}000 + 109 \times F_T)\ \$/\text{year}.$$

Operating costs (in \$/year) include energy costs, maintenance, and replacement parts. Assume a straight-line depreciation with a lifetime of 15 years. Each process operates 330 days per year, 24 hours per day. The wages are Scheme I: \$370,000/year, Scheme II: \$481,000/year, and Scheme III: \$407,000/year.

(A) Calculate the capital cost, operating cost, revenue, profit, and ROI for each scheme. You may assume a straight-line depreciation with a lifetime of 15 years. Assume that each process operates 330 days per year, 24 hours per day.
(B) Which is the better scheme? Briefly explain your choice.

3.107 Product P is obtained from reactant A. Unfortunately, some A also reacts to form by-product B. Consider two schemes to yield 100. kg/day of P. Scheme I has 100% conversion, which eliminates the need for a recycle, but has a poor ratio of product P to by-product B.

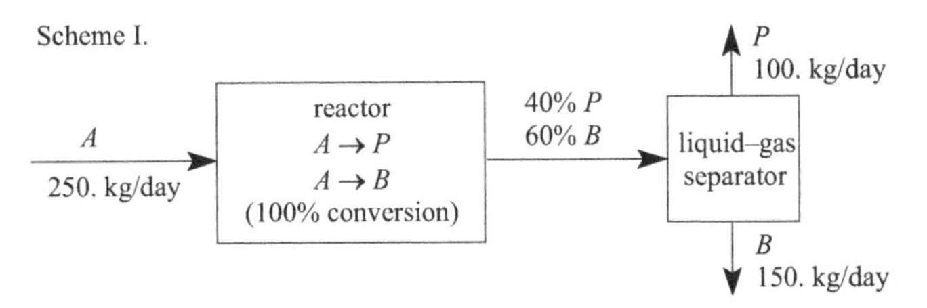

Scheme II has a better ratio of product P to by-product B, but only 25% conversion.

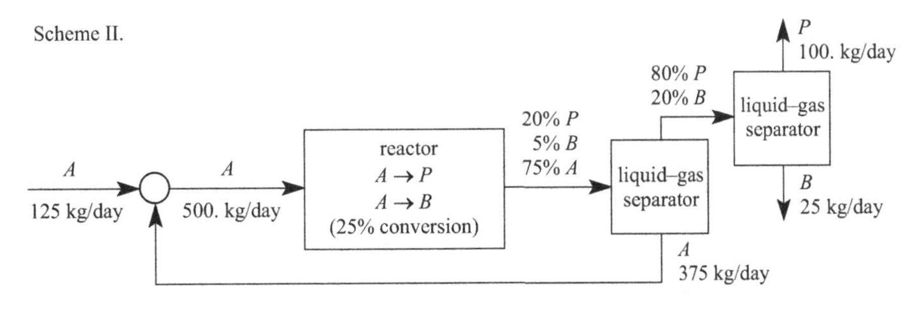

Reactant A costs \$32/kg. Product P sells for \$107/kg and by-product B sells for \$13/kg. Equipment prices (in \$) are given by the following formulas, where F_T is the total flow rate through the unit, in kg/day:

$$\text{reactor price} = \$125{,}000\sqrt{F_T}$$
$$\text{separator price} = \$48{,}000\sqrt{F_T}.$$

Annual operating costs for each unit are given by the following formulas:

$$\text{reactor operating cost} = \$15{,}000 + \$700. \times F_T$$
$$\text{separator operating cost} = \$8{,}200 + \$400. \times F_T.$$

Operating costs (in \$/year) include energy costs, maintenance and replacement parts. The wages are \$239,000/year for Scheme I and \$318,000/year for Scheme II. You may assume a straight-line depreciation with a lifetime of 15 years. Assume each process operates 365 days per year.

(A) Complete the following table.

	Scheme I	Scheme II
capital cost		
operating cost		
revenue		
profit		

(B) Which scheme (if either) do you recommend we build? Briefly explain your recommendation.

3.108 Consider the two schemes below for separating a mixture of pentane, hexane, and heptane. Their boiling points at 1 atm are 36°C, 69°C, and 98°C, respectively.

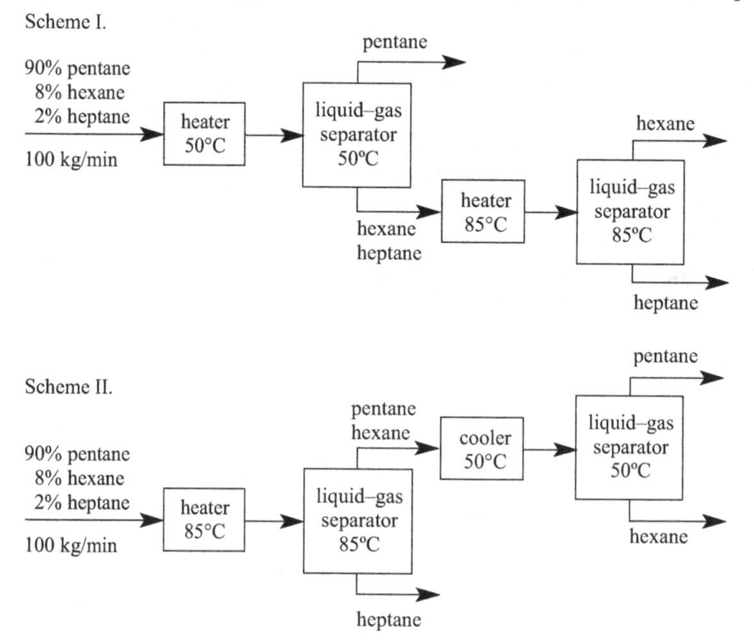

Which scheme is superior economically? Explain *succinctly.* State the basis for your choice. For example, is your choice based on the lowest depreciation? Assume the cost of a unit is directly proportional to its input capacity (in kg/min); a unit twice as large costs twice as much. Energy costs are directly proportional to the amount of energy consumed.

3.109 Your company's chemical engineers invent a reactor that converts reactant M into product Q with 100% conversion. Reactant M is available only as a mixture with inert I. Because M and I have similar boiling points, a liquid–gas separator can yield pure M, or pure I, but not both. Also, because Q and I have similar boiling points, a liquid–gas separator can yield pure Q, or pure I, but not both.

The engineers present two schemes for your consideration.

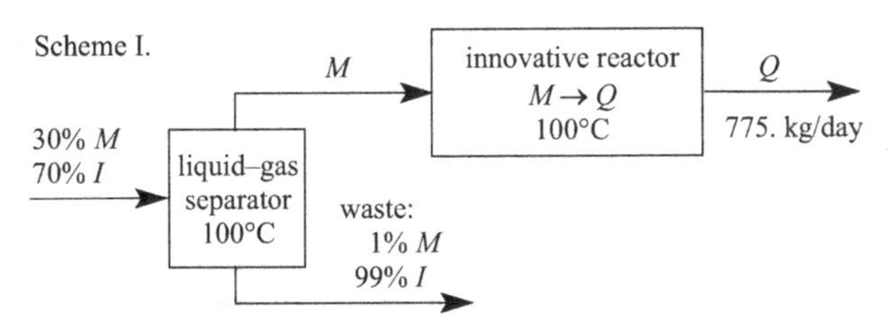

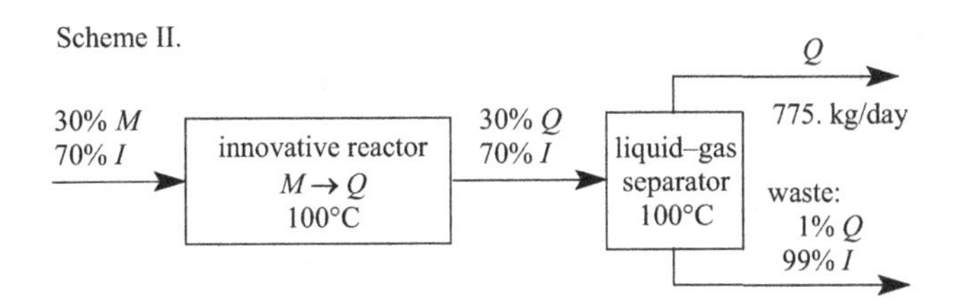

Briefly defend one of the following conclusions:

Scheme I is superior to Scheme II.
Scheme II is superior to Scheme I.
The schemes are equally viable.
There is insufficient information to identify the superior scheme.

3.110 An engineering team that you manage submits six process designs for your review. Each design produces 10^6 kg/year of the same product. Each design has a positive profit.

Process 1 has the highest profit.
Process 2 is the least expensive to build.
Process 3 has the longest equipment lifetime.
Process 4 has the lowest reactant flow rate.
Process 5 has the highest ROI.
Process 6 is the easiest to operate. This will allow for a smaller, less skilled team of operators.

(A) Which process has the lowest operating cost? If there is insufficient information, specify what additional information would be required.

(B) Which process has the least by-products and/or discarded reactants (in kg/year)? If there is insufficient information, specify what additional information would be required.
(C) Which process has the lowest depreciation? If there is insufficient information, specify what additional information would be required.
(D) Which process (if any) should you build? If there is insufficient information, specify what additional information would be required.

3.111 You buy equipment for \$523,000 that has a capacity of 340. kg/week of formula *X*. The equipment operates 50. weeks per year; there are one-week maintenance periods every 6.0 months. Formula *X* sells for \$28.7/kg. The production costs (reactants, labor, electricity, maintenance, repairs, etc.) are \$11.4/kg. Rent, insurance, and miscellaneous fees are \$62,500./year.

(A) Calculate the profit and ROI. Assume a straight-line depreciation with a lifetime of 10 years.
(B) A key part on your equipment breaks once a year, on average. When you purchased the equipment for \$523,000 you also purchased a spare part for \$24,900 (total price = \$547,900). It takes one week to replace the part, during which no formula *X* is produced. The additional costs for each repair are \$31,200 (\$6,300 for labor plus \$24,900 for a new spare part to replace the part in inventory). Calculate the profit and ROI during a year in which the part is replaced once.
(C) Because you know the key part will fail every year or so, you decide to avoid any production interruptions by replacing the part during the biannual one-week maintenance periods. Because the equipment is disassembled for maintenance, there is no additional labor cost to replace the part. The only cost of replacing the part is the price of the part, \$24,900. Calculate the profit and ROI for replacing the part twice a year during the normal maintenance periods.
(D) You decide to replace the part only after it fails, but with a different strategy. When the equipment fails, the first two days of the repair are spent disassembling the equipment and removing the broken part. Instead of having a spare part on hand, you decide to rush order the part when it fails. The rush order costs \$24,900 for the part plus \$1000 for special rush delivery. Calculate the profit and ROI for this *just-in-time* strategy during a year when the part is replaced once.
(E) A different manufacturer also sells a process to produce 340. kg/week of formula *X*. But her equipment has been proved to never fail during its lifetime. This superior piece of equipment costs \$725,000. All other parameters are as stated in part (A). Which has the better ROI – (B), (C), (D), or (E)?

3.112 Consider the imperfect separator described in Figure 3.22. Use a spreadsheet to calculate the flow rates of *P* and *Q* in all streams. Use the following process parameters:

wt fraction *P* in the input: 0.10
input flow rate: 100 kg/day
fraction of *P* recovered (pure *P* out)/(*P* in feed): 0.90
separator performance (pure *P* out)/(*P* in bottoms): 1.0 .

(A) Change the input flow rate to 200 kg/day and calculate the flow rates in all streams.
(B) Reset the input flow rate to 100 kg/day. Change the fraction P recovered to 0.99 and calculate the flow rates in all streams.
(C) Reset the fraction of P recovered to 0.90. Change the separator performance = (pure P out)/(P in bottoms) to 0.1 and calculate the flow rates in all streams. Change to 2.0 and calculate the flow rates in all streams.
(D) Reset the separator performance = (pure P out)/(P in bottoms) to 1.0. Change the wt fraction P in the input to 0.01 and calculate the flow rates in all streams. Change to 0.90 and calculate the flow rates in all streams.

Use the following economic parameters to calculate the ROI:

cost of input = 2 + 55(wt fraction P in input)^2 \$/kg
selling price for pure P = 230 \$/kg
cost of waste disposal = 0.40 \$/kg
separator capital cost = [1 + 1.2(separator performance)^2] × [100,000 + 16,000(separator capacity in kg/day)$^{0.6}$] \$
separator operating cost = 18,000 + 120(separator capacity in kg/day) \$/year
separator lifetime = 10 years

(E) Calculate the ROI using the initial processing parameters.
(F) Repeat the changes in parts (A), (B), and (C). And then reset all parameters to initial values.
(G) Change "fraction of P recovered" to obtain maximum ROI with other process parameters at initial values.
(H) Change "separator performance: (Pure P out)/(P in bottoms)" to obtain maximum ROI with other process parameters at initial values.
(I) Optimize the ROI by varying both "fraction of P recovered" and "separator performance: (Pure P out)/(P in bottoms)" with all other parameters at initial values.
(J) Optimize the ROI by varying both "fraction of P recovered" and "separator performance: (Pure P out)/(P in bottoms)" with all other parameters at initial values, but with the additional constraint that the total spent in the first year is \$600,000.

Process Design with Mathematical Modeling

General Guidelines for Chapter 3 Design Exercises

- Design a steady-state continuous process.
- Assume liquid–gas separations are perfect if boiling points differ by more than 3°C.
- Solid–liquid separations produce wet solids.
- Assume reactions occur only in reactors.
- Assume no reactions other than the reactions given.
- Unless instructed otherwise, you may neglect heat exchangers, heaters, coolers, and pumps.
- In commercial chemical processes, a catalyst is typically a porous and/or granular solid. The catalyst is a permanent component of the reactor; there is no catalyst in the product stream. The catalyst is not consumed in the reaction; there is no catalyst in the feed stream. Unless stated otherwise, assume no catalyst is present in a process stream.
- Many designs are workable. Explore different options and choose your best design.

General Flowsheet Practice

- Use a discrete unit for each operation. For example, a reactor should not do double duty as a mixer and/or a separator. A reactor should have exactly one input and one output.
- Label each unit. For reactors, indicate the reaction, such as "hydrolysis reactor." For separators, indicate the physical basis of the separation, such as "liquid–gas separator." Where appropriate, indicate the temperature in the unit. You need not label combiners and splitters. You need not label mixers and heat exchangers if you use descriptive symbols (a stirring paddle or a zigzag stream, respectively).
- List the substances in every stream and state *approximate* flow rates. One significant figure is sufficient.
- Where appropriate, list the physical state of the substances – gas, solid, liquid, or aq (in aqueous solution).
- Indicate the destination of any stream that leaves the process (for example, "vent to air," "to landfill," or "to market").
- Use conventions to improve the readability of your flowsheet. For a liquid–gas separator the gas should leave the top and the liquid should leave the bottom. If convenient, the process should start in the upper left-hand corner. It is helpful if streams entering the process start at the periphery and streams leaving the process terminate at the periphery.

3.113 The following process separates a mixture of F, G, and H and then stores each at 20°C. Improve the energy efficiency by adding two heat exchangers. That is, minimize the energy supplied by the two heaters and minimize the energy extracted by the three coolers.

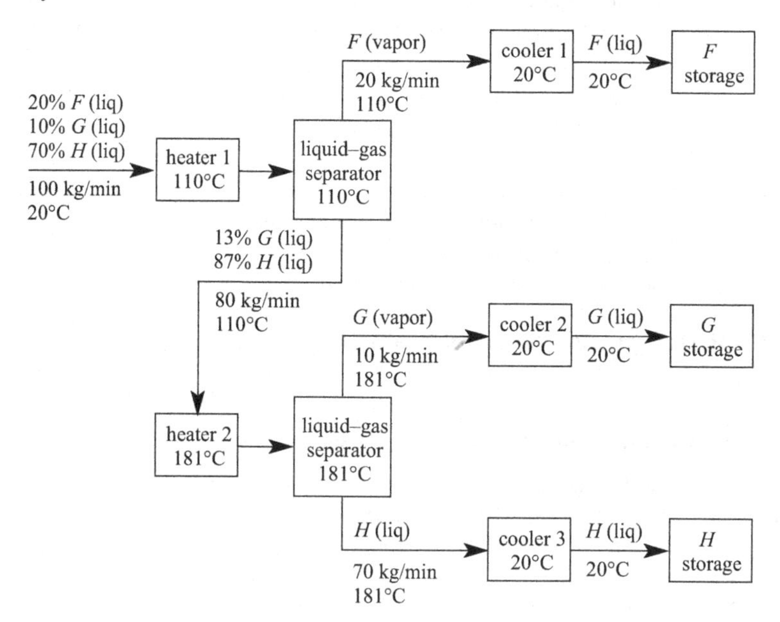

Design Rules

- Add only two heat exchangers.
- *Briefly* justify your choices for the streams passing through the heat exchangers. You may wish to consider the result of Exercise 3.50 (and similar calculations).

Following is a table of physical parameters at 1 atm.

	Boiling point (°C)	$\bar{C}_P$(kJ/(kg·°C)) Liquid	Vapor	$\Delta\bar{H}_{vap}$(kJ/kg)	$\Delta\bar{H}_{melt}$(kJ/kg)
F	105	1.6	1.2	380	100
G	176	2.0	1.4	400	120
H	197	2.4	1.6	420	140

You may assume $\bar{C}_P$, $\Delta\bar{H}_{vap}$, and $\Delta\bar{H}_{melt}$ are independent of temperature at the temperatures here.

3.114 The process below produces two chemical products, P and Q, by the following reactions. First we form the intermediate H:

$$A + B \rightarrow 2H \qquad \text{reaction is} \sim 60\% \text{complete.}$$

We then react intermediate H with C to form a product P:

$$H + C \rightarrow 2P \qquad \text{reaction is } 100\% \text{complete.}$$

We must avoid the reaction of B and C to form X:

$$B + C \rightarrow 2X \qquad \text{reaction is} \sim 90\% \text{complete.}$$

There is a complication. A and B have a parallel reaction that forms K:

$$A + B \rightarrow 2K \qquad \text{reaction is} \sim 30\% \text{complete.}$$

But we can react K with C to form a product Q:

$$K + C \rightarrow 2Q \qquad \text{reaction is } 100\% \text{complete.}$$

Because P and Q cannot be separated, we must form Q in a separate reactor.

The process below is viable and the molar flow rates (given by the numbers in parentheses) are correct. However, the process can be improved. Suggest at least **three** improvements to reduce costs or increase revenue. You might, for example, eliminate a unit, or decrease the flow rate through a unit, or decrease a reactant flow rate. You need not calculate flow rates in your improved process.

Design Rule

- Use no more than 9 mol/min of reactant A.

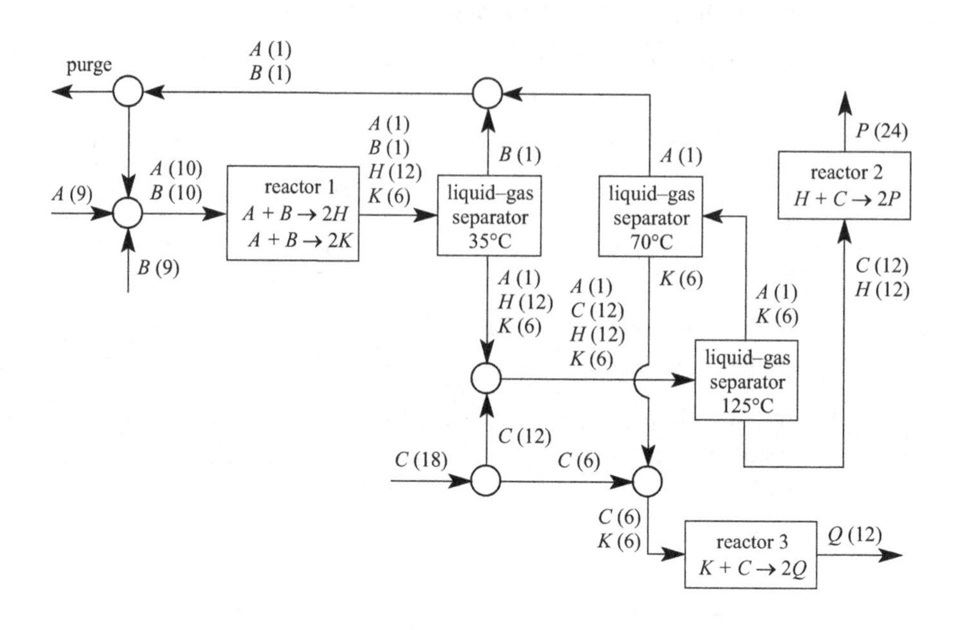

		Melting point (°C)	Boiling point at 1 atm (°C)	Market value ($/mol)
A	reactant	−85	50	3
B	reactant	−109	20	5
C	reactant	46	200	19
H	intermediate	32	160	16
K	intermediate	−33	90	41
P	product	21	115	100
Q	product	21	115	17
X	waste	−72	35	0

3.115 Design a process to produce P by the following reaction:

$$B + H \xrightarrow{\text{catalyst 1}} P.$$

The reaction of B and H occurs only in the presence of catalyst 1.

B is available only as a mixture with A (50 mol% A, 50 mol% B). H is produced by the following reaction:

$$A + C \xrightarrow{\text{catalyst 2}} H.$$

The reaction of A and C occurs only in the presence of catalyst 2, which also catalyzes the following reaction:

$$B + D \xrightarrow{\text{catalyst 2}} W.$$

C is available as a mixture with D and E (33 mol% C, 33 mol% D, 33 mol% E).

Additional reactions are possible in the presence of catalyst 3:

$$A + E \xrightarrow{\text{catalyst 3}} X$$
$$B + D \xrightarrow{\text{catalyst 3}} W.$$

Every reaction consumes all the reactants if the reactants are in stoichiometric ratios.

Crucial Fact

Catalyst 1 is deactivated by A. Catalyst 1 must not be exposed to A.

Examples of reactors with catalyst 1 are shown below. The numbers in parentheses are flow rates, in mol/min. The composition entering your reactor may be different.

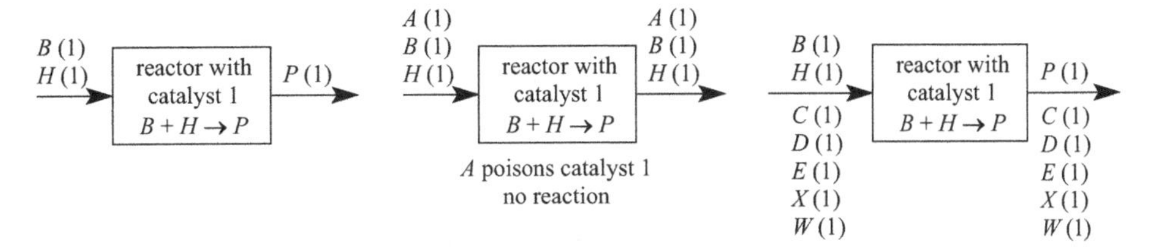

Examples of reactors with catalysts 2 and 3 are shown below. The compositions entering your reactors may be different.

A (1), B (1), C (1), D (1), E (1) → reactor with catalyst 2, $A + C \rightarrow H$, $B + D \rightarrow W$ → E (1), H (1), W (1)

A (1), B (1), C (1), D (1), E (1) → reactor with catalyst 3, $A + E \rightarrow X$, $B + D \rightarrow W$ → C (1), W (1), X (1)

Design Goals (in decreasing importance)

- Maximize the revenue from chemical products.
- Minimize the *number* of units. (Splitters and combiners do not count as units.)
- Minimize the *size* of the units.

Design Rules

- Use 4 mol/min of the A–B mixture: 2 mol A/min mixed with 2 mol B/min.
- Use as much C–D–E mixture as you wish.
- Use only one catalyst per reactor.

		Melting point (°C)	Boiling point at 1 atm (°C)	Market value ($/mol)
A	reactant	−17	91	5
B	reactant	−17	91	130
C	reactant	−42	73	20
D	reactant	−42	73	35
E	reactant	−42	73	60
H	intermediate	16	123	200
P	product	26	162	1000
W	by-product	9	147	0
X	by-product	−25	50	0

3.116 Design a process to produce P from A (not the same A and P as in any other exercise):

$$A \rightarrow P.$$

Reactant A is available only as a mixture of 90% A and 10% B. (All compositions in this exercise are in mol%.) Three types of reactors are available. Reactor type 1 converts A to P, but at moderate yield. Reactor type 2 completely consumes A, but has a side reaction,

$$A \rightarrow Q.$$

Reactor type 3 converts A to P (low yield) and completely converts B to Z:

$$B \rightarrow Z.$$

All reactions are irreversible.

Reactor Type 1

Converts ~1/2 of A to P; no Q produced. No reaction of B, P, Q, or Z. Examples of reactor type 1 are shown below. The numbers in parentheses are flow rates, in mol/hr. The composition entering your reactor may be different.

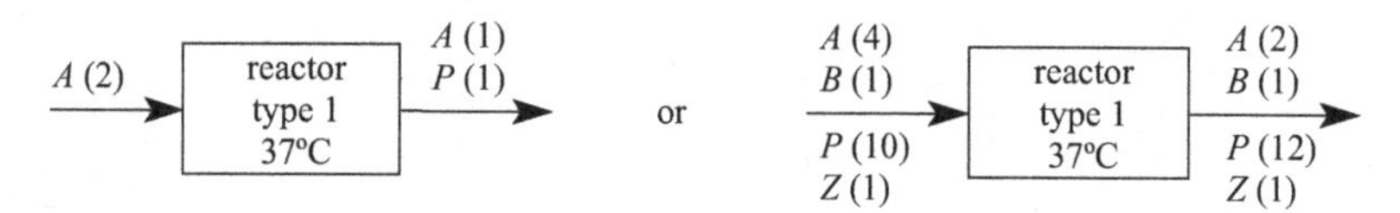

Reactor Type 2

Converts exactly 100% of A to P and Q. Ratio of P to Q is about 4:1. No reaction of B, P, Q, or Z. Examples of reactor type 2 are shown below. The composition entering your reactor may be different.

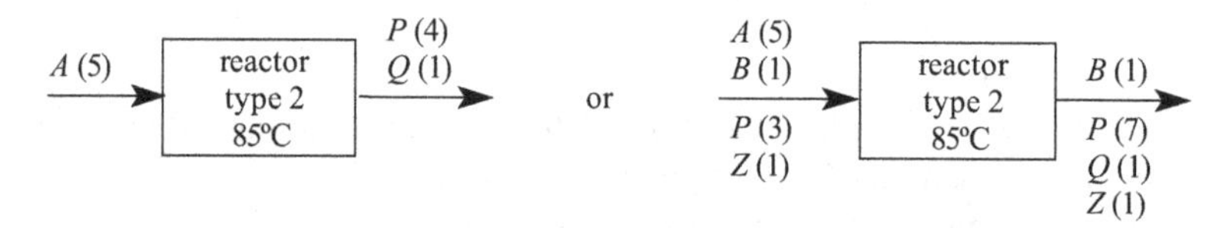

Reactor Type 3

Converts ~1/9 of A to P; no Q produced. Converts exactly 100% of B to Z. No reaction of P, Q, or Z. Examples of reactor type 3 are shown below. The composition entering your reactor may be different.

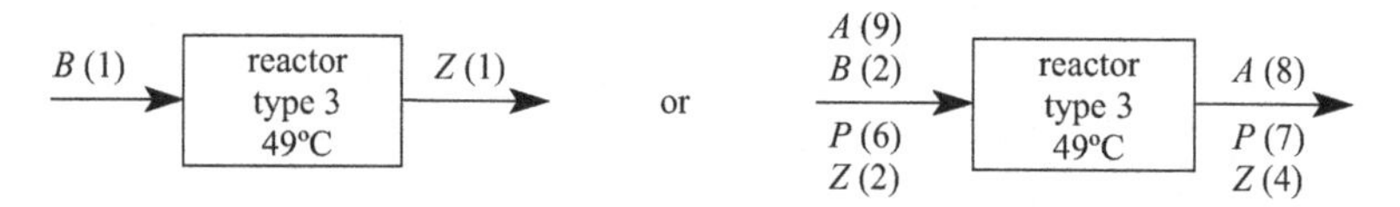

Design Goals (in decreasing importance)

- Maximize the revenue from selling pure P, as well as A, B, Q, and Z.
- Minimize the number of units.
- Minimize the size of each unit.

Design Rules

- Use 100 mol/hr of 90% A and 10% B.
- You may use one, two, or all three reactor types in your process. But use only one of each type.

Some properties at 1 atm.

	Molecular weight	Melting point (°C)	Boiling point(°C)	Price ($/mol)	Notes
A	120	−3	122	100	must not be heated above 250°C
B	133	−3	122	35	
P	120	−3	122	1000	
Q	120	18	147	20	suspected carcinogen
Z	133	−28	95	10	non-toxic

Hint: It is possible to produce P in 100% purity, with less than 1% of A converted to Q.

3.117 Design a process to produce chemical Z from reactants A and B by the following reactions:

$$B \rightarrow Q$$
$$A + Q \rightarrow 2Z.$$

B reacts concurrently to form intermediate Q and a secondary product P. The ratio of P to Q depends on the catalyst used. A reactor with catalyst 1 consumes 100% of B and the ratio of P to Q is 9 to 1. No other reactions occur. Two examples of a reactor with catalyst 1 are shown below.

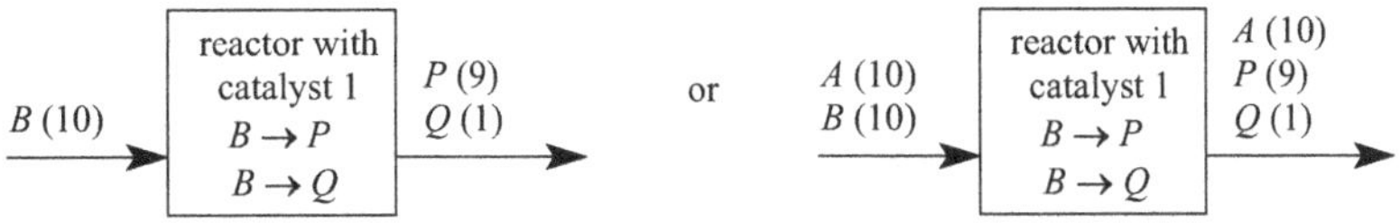

B (10) → [reactor with catalyst 1, $B \rightarrow P$, $B \rightarrow Q$] → P (9), Q (1)

or

A (10), B (10) → [reactor with catalyst 1, $B \rightarrow P$, $B \rightarrow Q$] → A (10), P (9), Q (1)

A reactor with catalyst 2 consumes 100% of B and the ratio of P to Q is 1 to 1. No other reactions occur. Two examples of a reactor with catalyst 2 are shown below.

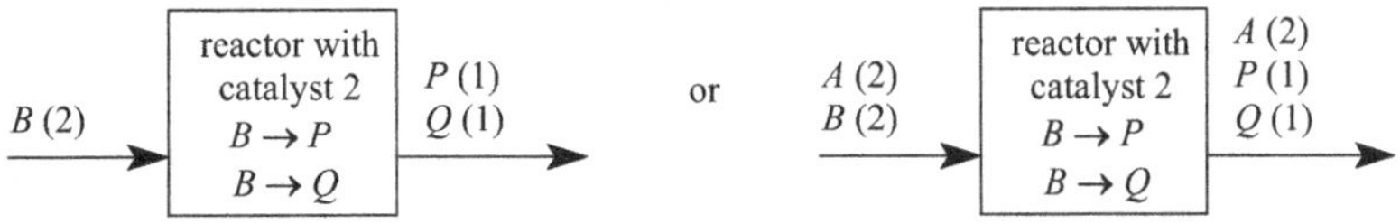

B (2) → [reactor with catalyst 2, $B \rightarrow P$, $B \rightarrow Q$] → P (1), Q (1)

or

A (2), B (2) → [reactor with catalyst 2, $B \rightarrow P$, $B \rightarrow Q$] → A (2), P (1), Q (1)

Catalyst 3 promotes the desired reaction $A + Q \rightarrow 2Z$, as well as an undesired reaction, $B + Q \rightarrow 2X$. No other reactions occur. Both reactions are incomplete. When A and Q enter a reactor with catalyst 3, the mols of reactants consumed are given by

$$\text{mol}\,A\,\text{consumed} = \text{mol}\,Q\,\text{consumed} = \frac{(\text{mol}\,A)(\text{mol}\,Q)}{\text{mol}\,A + \text{mol}\,Q} = \frac{\text{mol}\,Z\,\text{produced}}{2}.$$

So if mol A = mol Q, 50% of each reactant is consumed. If A is in large excess, Q is nearly completely consumed. Likewise when B and Q enter a reactor with catalyst 3, the mols of reactants consumed are given by

$$\text{mol } B \text{ consumed} = \text{mol } Q \text{ consumed} = \frac{(\text{mol } B)(\text{mol } Q)}{\text{mol } B + \text{mol } Q} = \frac{\text{mol } X \text{ produced}}{2}.$$

So if mol B = mol Q, 50% of each reactant is consumed. If B is in large excess, Q is nearly completely consumed. Two examples of a reactor with catalyst 3 are shown below.

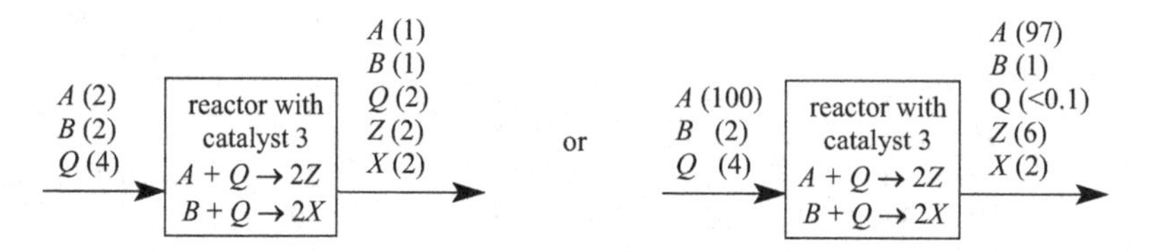

Design Goals (in decreasing importance)

- Maximize the revenue from selling product(s).
- Minimize the number of units.
- Minimize the size of the units.

Design Rules

- Use 100 mol/hr of a mixture of A (20 mol%) and B (80 mol%).
- Use only one catalyst per reactor.

	Melting point (°C)	Boiling point at 1 atm (°C)	Price ($/mol)	Comments
A	42	137	2	reactant
B	42	137	2	reactant
P	16	63	30	product
Q	42	137	20	intermediate reactant or by-product
X	−6	35	−2	must pay $2/mol to dispose
Z	29	89	100	product

3.118 Design a process to produce P from a mixture of A (50 mol%) and B (50 mol%):

$$A + B \rightarrow 2P.$$

The chemical reaction depends on the choice of catalyst, as follows.

A reactor with catalyst 1 at 111°C

$A + B \rightarrow 2P$ Catalyst 1 converts ~20% of A and B to P. Some P converts to Q.
$P \leftrightarrow Q$ The effluent ratio of P to Q will always be 4 to 1. The reaction of P to Q is reversible.

No other reactions occur.

Examples of reactors with catalyst 1 are shown below. The numbers in parentheses are flow rates, in mol/min. The composition entering your reactor may be different.

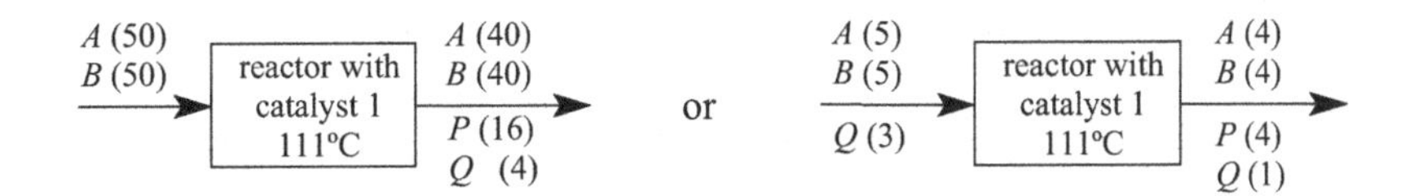

A reactor with catalyst 2 at 122°C

$A + B \rightarrow 2P$ Catalyst 2 converts 50% of A and B.
$A + B \rightarrow 2X$ The effluent ratio of P to X is about 9 to 1.
No other reactions occur.

Examples of reactors with catalyst 2 are shown below. The composition entering your reactor may be different.

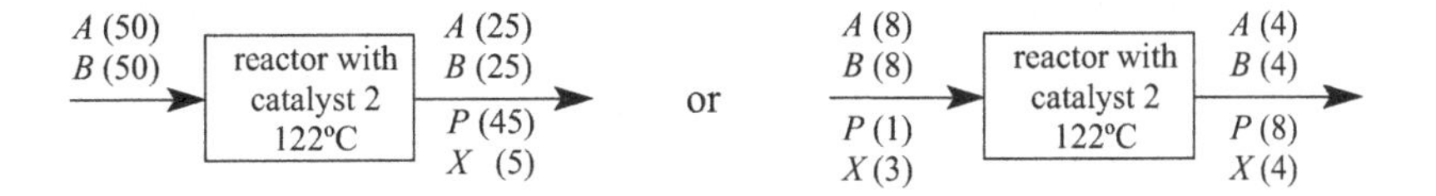

A reactor with catalyst 3 at 133°C

$2X \rightarrow A + B$ Catalyst 3 converts 100% of by-product X. Half the X converts to $A + B$, and
$X \rightarrow Q$ half the X converts to Q.
No other reactions occur.

Examples reactors with of catalyst 3 are shown below. The composition entering your reactor may be different.

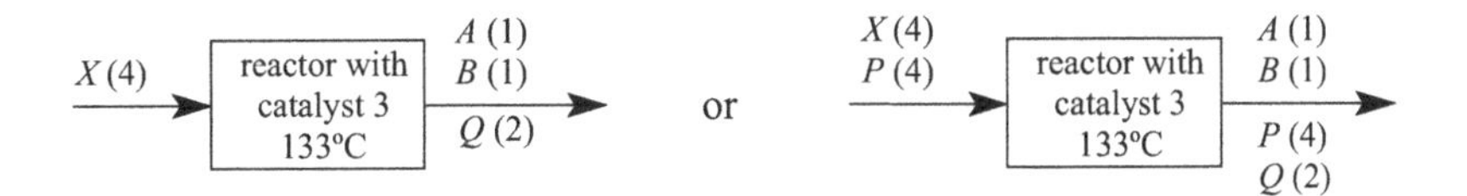

Design Goals (in decreasing importance)

- Maximize the amount of P produced.
- Minimize the *number* of units.
- Minimize the *size* of the units.

Design Rules

- Use 100 mol/min of a mixture of A (50 mol%) and B (50 mol%).
- Use only one catalyst per reactor.

		Melting point (°C)	Boiling point at 1 atm (°C)
A	reactant	−52	23
B	reactant	−42	57
P	product	16	81
Q	by-product	−55	41
X	by-product	29	102

3.119 Design a process to produce product P from reactant B, as follows:

$B \leftrightarrow P$ The reaction is reversible. $\dfrac{\text{mol } P}{\text{mol } B} \approx 0.1$ in the reactor effluent.

Any combination of B and P may enter the reactor. The ratio of B:P in the reactor effluent will 10:1. Note that *in this particular case* mols are conserved; total mols into the reactor equals total mols out of the reactor.

Complications

- Reactant B is available only as a mixture with inert substance X. The mixture composition is 100 mol B and 10 mol X. Substance X can also be sold if purified.
- The melting points of B, X, and P are identical, so liquid–solid separation is impossible. The boiling points are very close (76°C for P, 77°C for X, and 77°C for B), so gas–liquid separation is not perfect. Assume that if the boiling point of a substance equals the separator temperature, exactly half of the substance leaves via the gas stream and exactly half of the substance leaves via the liquid stream. Here are examples of separations at 76°C and 77°C; the numbers in parentheses are flow rates, in mol/min.

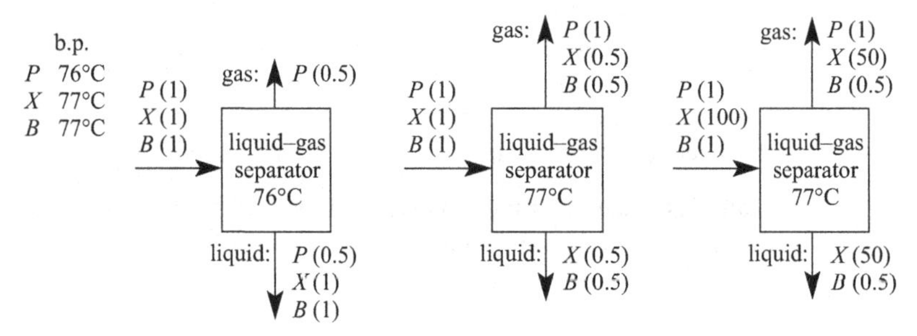

The values of chemicals P and X depend on their purity, as given in the following table.

Purity (mol%)	\$/mol of P	\$/mol of X
≥50%	1.	0.2
≥75%	2.	5.
≥98%	5.	20.
100%	10.	100.

Design Goals (in decreasing importance)

- Maximize the revenue from selling P and X. Revenue is determined by the quantity and purity of each.
- Minimize the *number* of units.
- Minimize the *size* of each unit.

Design Rules

- Use 110. mol/min of a mixture of B (100 mol/min) and X (10 mol/min).
- Use no more than one reactor, one separator at 76°C, and one separator at 77°C. Indicate the temperature in your liquid–gas separator: **use only 76°C or 77°C.**
- No process unit may have a capacity larger than 10,000 mol/min.

3.120 Design a process to produce products P and Q from reactant A, as follows:

$2A \leftrightarrow P + Q$ The reaction is reversible. $\dfrac{(\text{mol } P)(\text{mol } Q)}{(\text{mol } A)^2} \approx 1$ in the reactor effluent.

Any combination of A, P, and Q may enter the reactor. The effluent concentrations are given by the equation above. Here are two examples of reactor effluent. The numbers in parentheses are flow rates in mol/min.

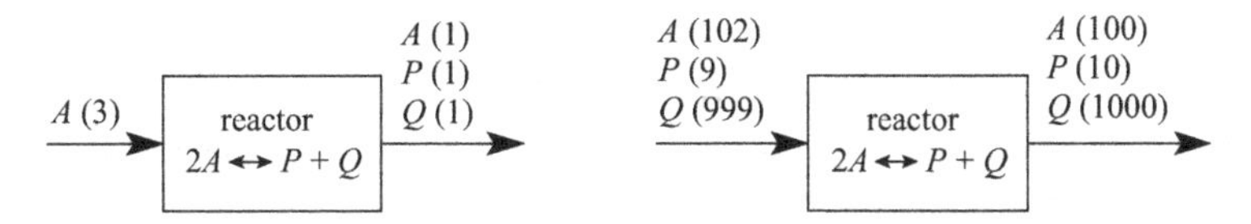

Note that *in this particular case* mols are conserved; total mols into the reactor equals total mols out of the reactor.

Complications

- Reactant A is available only as a mixture with inert substance I. The mixture is 100 mol A and 10 mol I. Substance I can also be sold if purified.
- The melting points of A, I, P, and Q are identical, so liquid–solid separation is impossible. The boiling points are very close (54°C for P, 55°C for I, 56°C for A, and 56°C for Q), so gas–liquid separation is not perfect. Assume that if the boiling point of a substance equals the separator temperature, exactly half of the substance leaves via the gas stream and exactly half of the substance leaves via liquid stream. Here are examples of separations at 54°C, 55°C, and 56°C; the numbers in parentheses are flow rates, in mol/min.

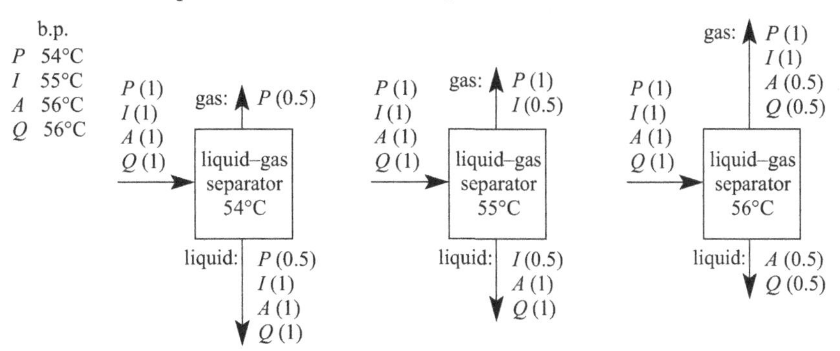

The same three separators are shown below with different inputs.

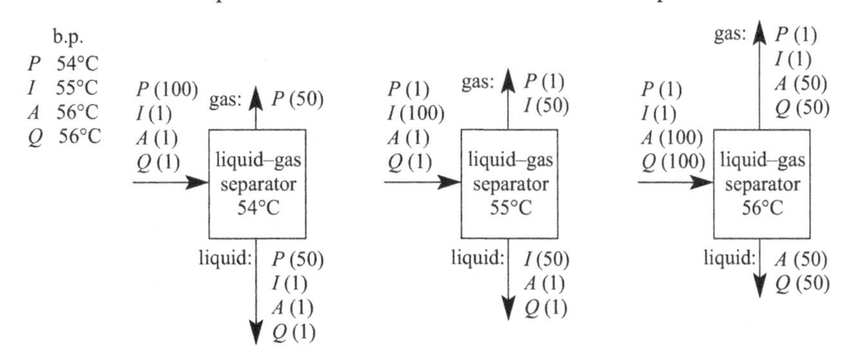

The values of chemicals P, Q, and I depend on their purity, as given in the following table.

Purity (mol%)	\$/mol of P	\$/mol of Q	\$/mol of I
≥50%	1.	0.5	0.2
≥75%	2.	2.	1.
≥98%	5.	10.	5.
100%	10.	100.	25.

Design Goals (in decreasing importance)

- Maximize the revenue from selling P, Q, and I. Revenue is determined by the quantity and purity of each.
- Minimize the *number* of units.
- Minimize the *size* of each unit.

Design Rules

- Use 110. mol/min of a mixture of A (100 mol/min) and I (10 mol/min).
- Use no more than one reactor, one separator at 54°C, one separator at 55°C, and one separator at 56°C. Indicate the temperature in your liquid–gas separator: **use only 54°C, 55°C, or 56°C.**
- No process unit may have a capacity larger than 10,000 mol/min.

Hint: The maximum revenue, about \$800/min, can be obtained with different separator and reactor configurations. If you design a process that yields the maximum revenue, you need not try other designs.

3.121 The value of chemical A depends on the purity of A, as given in the following price list.

Purity (mol% A)	\$/mol of A
≥50%	1.
≥90%	5.
≥97%	10.
≥99%	30.
100%	100.

Design a process to purify A from a mixture of A, B, and I. Because A, B, and I have the same boiling point and the same melting point, separation by phase is impossible. Consider using differences in reactivity with reactant R:

$B + R \rightarrow 2Z$ consumes 100% of B if R is in excess. The reaction is irreversible

$I + R \rightarrow$ (no reaction) I is inert.

Reactant R also reacts with A and quickly reaches equilibrium.

$A + R \rightarrow 2X$ the molar ratio of A to X is 5 to 1 at equilibrium,for excess R. The reaction is reversible.

For example, if 11 mol A and 100 mol R (excess R) enter a reactor, then 10 mol A, 2 mol X, and 99 mol R leave the reactor. You may assume reactions occur only in a reactor.

Design Goals (in decreasing importance)

- Maximize the revenue from selling A. Revenue is determined by the quantity and purity of A.
- Minimize the expense of reactant R.
- Minimize the *number* of units.
- Minimize the *size* of each unit.

Design Rule

- Use 100. mol/min of a mixture of A (55 mol%), B (44 mol%), and I (1 mol%).

Hint: It is possible to produce some A with 100% purity.

	Melting point (°C)	Boiling point at 1 atm (°C)
A	−27	37
B	−27	37
I	−27	37
R	95	287
X	24	123
Z	77	171

3.122 The value of chemicals P and Q depend on their purity, as given in the following table.

Purity (mol%)	\$/mol of P	\$/mol of Q
≥50%	10.	1.
≥90%	15.	5.
≥95%	20.	10.
≥98%	25.	30.
100%	30.	100.

Design a process to produce high-purity P and high-purity Q from a mixture of P and B. Q is obtained from B. The reversible reaction of B to Q does not go to completion:

$B \rightarrow Q$ The reaction is reversible. $\dfrac{\text{mol } Q}{\text{mol } B} \approx 0.1$ in the reactor effluent.

For example, if 11 mol B enters a reactor, there will be 10 mol B and 1 mol Q in the effluent. The reactor also reversibly disproportionates some P into X and Y, as follows:

$2P \rightarrow X + Y$ the reaction is reversible. $\dfrac{(\text{mol } X)(\text{mol } Y)}{(\text{mol } P)^2} \approx 1$ in the reactor effluent.

For example, if 3 mol P enters a reactor, there will be 1 mol P, 1 mol X, and 1 mol Y in the effluent. If 5 mol of X and 2 mol of Y enter a reactor, there will be 2 mol of P, 4 mol of X, and 1 mol of Y in the effluent. Note *in this particular case* that mols are conserved; total mols into a reactor equals total mols out of the reactor.

	Melting point (°C)	Boiling point at 1 atm (°C)
B	−47	51
P	−47	51
Q	39	116
X	39	116
Y	12	83

P and B have the same boiling point and the same melting point, so separation by phase is impossible.

Design Goals (in decreasing importance)

- Maximize the revenue from selling P and Q. Revenue is determined by the quantity and purity of each.
- Minimize the *number* of units.
- Minimize the *size* of each unit.

Design Rule

- Use 200. mol/hr of a mixture of P (50 mol%) and B (50 mol%).

Design Guidelines

- List the substances and the *approximate* flow rates (mol/hr) in every stream. One significant figure is sufficient.
- No process unit may have a capacity larger than ~2000 mol/hr.

Hint: It is possible to produce some Q with 100% purity.

3.123 Design a process to produce P by the following reaction:

$G + K \overset{\text{catalyst 1}}{\rightarrow} 2P$ The reaction is 100% complete and irreversible.

G is obtained from reactant A:

$A \overset{\text{catalyst 2}}{\longleftrightarrow} G$ The reaction is reversible. $\dfrac{\text{mol } G}{\text{mol } A} \approx 1$ in the reactor effluent.

K is obtained from reactant B:

$2B \overset{\text{catalyst 3}}{\longleftrightarrow} H + K$ The reaction is reversible. $\dfrac{(\text{mol } H)(\text{mol } K)}{(\text{mol } B)^2} \approx 1$ in the reactor effluent.

Complications

- Reactant A is available only as a 80:10 mixture with inert impurity I. I reacts with nothing.
- A and I have identical boiling points and melting points; A and I cannot be separated.
- Reactant B and intermediate K have identical boiling points and melting points; B and K cannot be separated.
- There is an additional reaction:

$G + B \overset{\text{catalyst 1}}{\rightarrow} 2X$ The reaction is 100% complete and irreversible.

Design Goals (in decreasing importance)

- Maximize the revenue from selling pure P and *any other pure substances*.
- Minimize the *number* of units.
- Minimize the *size* of each unit.

Design Rules

- Use only two inputs: (1) 90. mol/hr of a mixture of A (80 mol/hr) and I (10 mol/hr), and (2) 100. mol/hr B.
- Use only one catalyst per reactor. The catalyst is a permanent part of a reactor. You need not add catalyst and there is no catalyst in the reactor output.
- No process unit may have a capacity larger than 10,000 mol/hr.

Design Guidelines

- List the substances and *approximate* flow rates (mol/hr) in every stream. One significant figure is sufficient.

		Melting point (°C)	Boiling point at 1 atm (°C)	Market value for 100% pure ($/mol)
A	reactant	21	111	40
B	reactant	−29	78	5
I	inert impurity	21	111	20
G	intermediate	35	147	15
H	intermediate	46	153	30
K	intermediate	−29	78	10
P	product	57	168	100
X	by-product	13	131	0

Hint: Your process must produce at least 90 mol/hr of P for full credit.

3.124 Design a process to produce P from reactants A and B. The reaction sequence begins with the reaction of A to two intermediates, H and M:

$2A \rightarrow H + M$ 50% of A reacts. The reaction is irreversible.

Both intermediates H and M react with B to form product P, and a worthless by-product:

$H + B \rightarrow P + Q$ The reaction is reversible. $\dfrac{(\text{mol } P)(\text{mol } Q)}{(\text{mol } H)(\text{mol } B)} \approx 1$ in the reactor effluent

$M + B \rightarrow P + R$ The reaction is reversible. $\dfrac{(\text{mol } P)(\text{mol } R)}{(\text{mol } M)(\text{mol } B)} \approx 1$ in the reactor effluent.

Complications

- B and Q cannot be separated.
- H and R cannot be separated.

Design Goals (in decreasing importance)

- Maximize the output of pure P.
- Minimize the *number* of units.
- Minimize the *size* of each unit.

Design Rules

- Use only two inputs: (1) 2 mol/min of A, and (2) 2 mol/min of B.
- No process unit may have a capacity larger than ~100 mol/min.

Design Guidelines

- List the substances and *approximate* flow rates (mol/min) in every stream. One significant figure is sufficient.

		Melting point (°C)	Boiling point at 1 atm (°C)
M	intermediate	−23	62
B	reactant	−7	80
Q	by-product	−7	80
H	intermediate	25	95
R	by-product	25	95
A	reactant	43	134
P	product	56	176

Hints: It is possible to produce more than 1.9 mol/min of P. In this system, mols are conserved.

3.125 Design a process to produce P from reactants A and B. The reaction sequence begins with the reaction of A with intermediate X:

$A + X \rightarrow H + P$ Reaction is at 500°C

For $A : X = 1 : 1$, the reaction consumes 50% of A and 50% of X

For $A : X > 100 : 1$, the reaction consumes 100% of X

For $A : X < 1 : 100$, the reaction consumes 100% of A.

Intermediate X is obtained from intermediates H and M:

$H + M \rightarrow 2X$ Reaction is at 200°C. Reaction is 100% complete for $H : M = 1 : 1$.

Intermediate M is obtained from reactant B:

$B \rightarrow M$ Reaction is at 350°C. 25% of the B forms M.

Additional Reactions

- Unfortunately, B also reacts to form impurity I at 350°C:

$B \rightarrow I$ 25% of the B forms I. Overall, 50% of B reacts to M and I.

- For example,

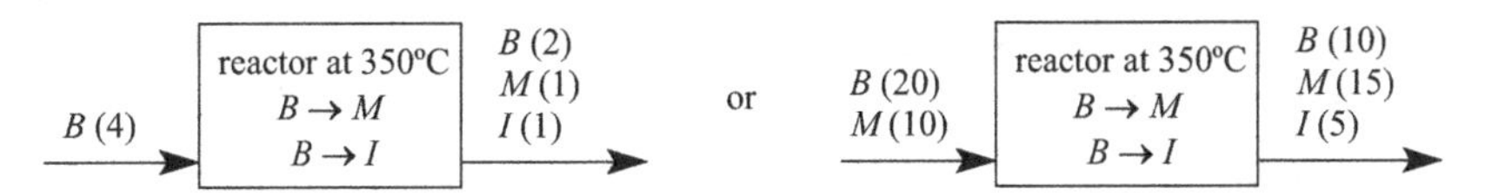

- M reacts with A at 200°C:

$$A + M \rightarrow 2Y \quad \text{Reaction is } 100\% \text{ complete for } A : M = 1 : 1.$$

- If H is also present, M reacts preferentially with A. For example,

A (1), H (1), M (1) → reactor at 200°C ($H + M \rightarrow 2X$, $A + M \rightarrow 2Y$) → H (1), Y (2)

or

A (1), H (1), M (2) → reactor at 200°C ($H + M \rightarrow 2X$, $A + M \rightarrow 2Y$) → X (2), Y (2)

- *All reactions are irreversible.*

Complications

- A and H cannot be separated.
- B and I cannot be separated.
- X and Y cannot be separated.

Design Goals (in decreasing importance)

- Maximize the output of pure P.
- Minimize the input of B.
- Minimize the *number* of units.

Design Rules

- Use only 1 mol/min of A. You may use as much B as you wish.
- No process unit may have a capacity larger than ~1000 mol/min.
- Liquid–solid separators produce wet solids.

Design Guidelines

- List the substances and *approximate* flow rates (mol/min) in every stream. One significant figure is sufficient.

Hint: In this system, mols are conserved.

		Melting point (°C)	Boiling point at 1 atm (°C)
X	intermediate	−31	67
Y	by-product	−31	67
P	product	14	92
M	intermediate	28	135
B	reactant	55	158
I	by-product	55	158
A	reactant	77	173
H	intermediate	77	173

3.126 Design a process to obtain pure A from a mixture of A, B, and I with A:B:I = 11 mols:2 mols:1 mol. Because A, B, and I have the same melting point and the same boiling point, the mixture cannot be separated by physical means. But A, B, and I have different chemical reactivities. I is inert; I reacts with nothing. A and B react to form K:

$$A + B \xleftrightarrow{\text{catalyst 1}} 2K$$ Reaction is at 200°C and requires the presence of catalyst 1

Reaction is reversible and $\dfrac{(\text{mol } K)^2}{(\text{mol } A)(\text{mol } B)} \approx 1$ in the reactor effluent.

K reacts *irreversibly* with Z to produce M and release B:

$$2K + Z \xrightarrow{\text{catalyst 2}} 2M + B$$ Reaction is at 100°C and requires the presence of catalyst 2

For $\dfrac{\text{mol } Z}{\text{mol } K} > 10$ in the reactor input, all K is consumed

For $\dfrac{\text{mol } K}{\text{mol } Z} \geq 10$ in the reactor input, all Z is consumed.

Finally, M decomposes reversibly to A and Z:

$$2M \xleftrightarrow{\text{catalyst 1}} A + Z$$ Reaction is at 200°C and requires the presence of catalyst 1

Reaction is reversible and $\dfrac{(\text{mol } M)^2}{(\text{mol } A)(\text{mol } Z)} \approx 1$ in the reactor effluent.

Design Goals (in decreasing importance)

- Maximize the output of pure A. You must produce at least 10 mol/min of pure A for full credit.
- Minimize the input of Z.
- Minimize the *number* of units.

Design Rules

- Use 14 mol/min of the $A + B + I$ mixture: 11 mol A + 2 mol B + 1 mol I. You may use as much Z as you wish.
- Use only one catalyst per reactor. The catalyst is a permanent part of a reactor. You need not add catalyst and there is no catalyst in the reactor output.

Design Guidelines

- List the substances and *approximate* flow rates (mol/min) in every stream. One significant figure is sufficient.

Hints: Mols are conserved in these reactions. Do not calculate flow rates until your final design.

		Melting point (°C)	Boiling point at 1 atm (°C)
A	reactant/product	37	118
B	reactant/impurity	37	118
I	inert impurity	37	118
K	intermediate	59	178
M	intermediate	77	194
Z	reagent	43	145

3.127 (A) Design a process to produce P from A using the reversible reaction $A \overset{\text{catalyst 1}}{\longleftrightarrow} P$. The reaction rapidly reaches equilibrium at 200°C in the presence of catalyst 1. At equilibrium, mol A = mol P.

Because A and P have the same melting point and same boiling point, A and P cannot be separated by physical means. The only means to produce pure P is to react A and P with X, which consumes all the A, but also sacrifices half the P.

$A + X \xrightarrow{\text{catalyst 2}} B$ For $\dfrac{\text{mol } X}{\text{mol } A} > 10$ in the reactor input, all A is consumed

$P + X \xrightarrow{\text{catalyst 2}} Q$ For $\dfrac{\text{mol } X}{\text{mol } P} > 10$ in the reactor input, 50% of P is consumed.

Both reactions are irreversible.

Both reactions are at 100°C and both require the presence of catalyst 2.

Design Goals (in decreasing importance)

- Maximize the output of pure P.
- Minimize the input of X.
- Minimize the *number* of units.

Design Rules

- Use 100 mol/min of A. You may use as much X as you wish.
- Use only one catalyst per reactor. The catalyst is a permanent part of a reactor. You need not add catalyst and there is no catalyst in the reactor output.
- No process unit may have a capacity larger than ~2000 mol/min.

Design Guidelines

- List the substances and *approximate* flow rates (mol/min) in every stream. One significant figure is sufficient.

		Melting point (°C)	Boiling point at 1 atm (°C)
A	reactant	33	111
P	product	33	111
X	reagent	88	177
B	by-product	55	144
Q	by-product	55	144

Be careful: mols are NOT conserved in reactions other than $A \leftrightarrow P$.

(B) Use additional reactions to increase the output of pure P.

Reactant Y reacts preferentially with A in the following reversible reaction:

$A + Y \overset{\text{catalyst 3}}{\longleftrightarrow} C$ For reactor input with mol A = mol Y, and *no* C, 50% of both A and Y react

For reactor input with *only* C, 50% of C reacts.

Reaction is at 200°C and requires the presence of catalyst 3.

The reaction of P and Y is negligible, for low concentrations of Y:

$$P + Y \xrightarrow{\text{catalyst 3}} \text{negligible reaction} \quad \text{for} \quad \frac{\text{mol } P}{\text{mol } Y} > 3 \text{ in the reactor input.}$$

Design Goals

- Same as part (A). For part (B), at least 45 mol/min of P is necessary for full credit.

Design Rules

- Same as part (A), plus you may use as much Y as you wish.

Design Guidelines

- Same as part (A).

		Melting point (°C)	Boiling point at 1 atm (°C)
Y	reagent	77	166
C	intermediate	99	222

3.128 (A) Design a process to produce P from A and B by the following reaction:

$$A + B \rightarrow P + X \quad \text{The reaction is reversible}$$

The reaction is not complete : $\dfrac{(\text{mol } P)(\text{mol } X)}{(\text{mol } A)(\text{mol } B)} \approx 1$ in the reactor effluent.

Reactants A and B are available as pure components.

Complication

- Reactant B and by-product X have identical boiling points and melting points; B and X cannot be separated.

Design Goals (in decreasing importance)

- Maximize the output of pure P.
- Minimize the *number* of units.
- Minimize the *size* of each unit.

Design Rules

- Use no more than 100 mol/min of A, and no more than 100 mol/min of B.
- No process unit may have a capacity larger than ~2000 mol/min.

Design Guidelines

- List the substances and *approximate* flow rates (mol/min) in every stream. One significant figure is sufficient.

		Melting point (°C)	Boiling point at 1 atm (°C)
A	reactant	−49	77
B	reactant	−21	155
P	product	−35	111
X	by-product	−21	155

Hint: Your process must produce at least 90 mol/min of P for full credit.

(B) Improve the yield of P by using the following additional reactions.

By-product X reacts with U in two parallel reactions:

$$X + U \rightarrow Y$$
$$X + U \rightarrow Z.$$

Both reactions are irreversible. The reactions consume all the X if U is in large excess; the reactor feed must have (mol U)/(mol X) $\geq$ 10. For excess U, (mol Y)/(mol Z) = 10 in the reactor effluent.

Substance U is available as a pure compound. You may use as much U as you wish.

Complication

- Reactant B, by-product X, and by-product Z have identical boiling points and melting points; B, X, and Z cannot be separated.

New Design Goals (in decreasing importance)

- Maximize the revenue from selling pure P.
- Minimize the consumption of reactant U.
- Minimize the *number* of units.
- Minimize the *size* of each unit.

New Design Rule

- No process unit may have a capacity larger than ~4000 mol/min.

Hint: Your process must produce at least 98 mol/min of P for full credit.
Be Careful: Mols are not conserved in the additional reactions.

		Melting point (°C)	Boiling point at 1 atm (°C)
A	reactant	−49	77
B	reactant	−21	155
P	product	−35	111
X	by-product	−21	155
U	reactant	−66	42
Y	by-product	6	222
Z	by-product	−21	155

3.129 Design a process to produce P from A and B by the following reaction:

$A + B \rightarrow P + X$ The reaction is irreversible, but slow, so the reaction is incomplete (see below).

Unfortunately, B catalyzes the isomerization of product P to worthless by-product Z:

$P \overset{B}{\leftrightarrow} Z$ The reaction is fast and reversible; mol P = mol Z in the reactor effluent.

The reactor effluent depends on the ratio of reactants A and B entering the reactor. For comparable flow rates of A and B, half the A is consumed.

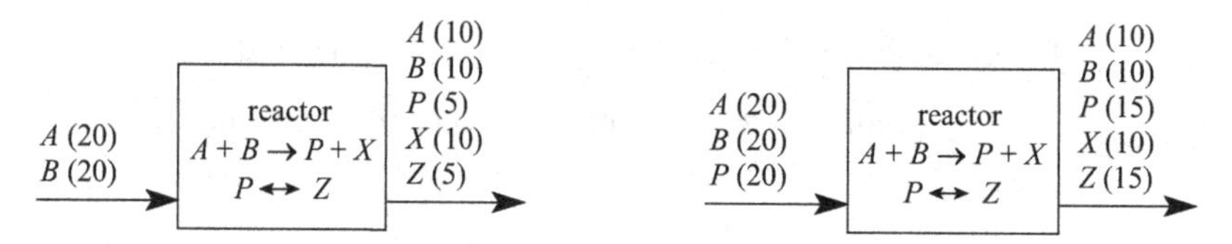

For large excess of A such that (mol A)/(mol B) $\geq$ 10, B is entirely consumed. But B still catalyzes the reaction $P \leftrightarrow Z$ before it is consumed.

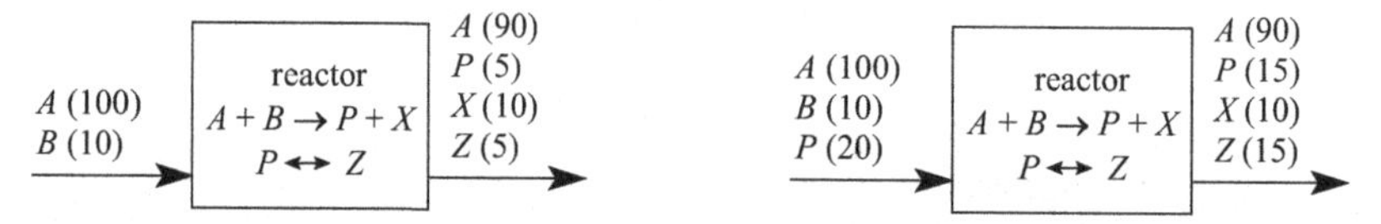

For large excess of B such that (mol B)/(mol A) $\geq$ 10, A is entirely consumed.

Complications

- Reactant A and product P have identical boiling points and melting points; A and P cannot be separated.
- Reactant B and by-product X have identical boiling points and melting points; B and X cannot be separated.

Design Goals (in decreasing importance)

- Maximize the output of pure P.
- Minimize the input of reactant B.
- Minimize the *number* of units.
- Minimize the *size* of each unit.

Design Rules

- Use no more than 10 mol/min of A. You may use as much B as you wish.
- No process unit may have a capacity larger than ~1000 mol/min.

Design Guidelines

- List the substances and *approximate* flow rates (mol/min) in every stream. One significant figure is sufficient.

		Melting point (°C)	Boiling point at 1 atm (°C)
A	reactant	−38	77
B	reactant	−67	44
P	product	−38	77
X	by-product	−67	44
Z	by-product	12	133

3.130 Design a process to produce pure P from A and B by the following reaction:

$$A + B \xrightarrow{\text{catalyst}} P + X$$

The reaction is irreversible, but slow.
The reaction is not complete (see below).

Unfortunately, B reacts with product P to form worthless by-product Z:

$P + B \xleftrightarrow{\text{catalyst}} Z$ The reaction is reversible and fast. $\dfrac{(\text{mol}\, Z)^2}{(\text{mol}\, P)(\text{mol}\, B)} \approx 1$ in the reactor effluent.

Both reactions require a catalyst. The reactor effluent depends on the ratio of reactants A and B entering the reactor. For comparable flow rates of A and B, half the A is consumed.

In: A (20), B (20) → reactor ($A + B \rightarrow P + X$; $P + B \leftrightarrow Z$) → Out: A (10), B (5), P (5), X (10), Z (5)

In: A (20), B (20), P (80) → reactor ($A + B \rightarrow P + X$; $P + B \leftrightarrow Z$) → Out: A (10), B (1), P (81), X (10), Z (9)

For large excess of B such that (mol B)/(mol A) $\geq$ 10, A is entirely consumed.

In: A (10), B (100) → reactor ($A + B \rightarrow P + X$; $P + B \leftrightarrow Z$) → Out: B (81), P (1), X (10), Z (9)

In: A (10), B (100), P (80) → reactor ($A + B \rightarrow P + X$; $P + B \leftrightarrow Z$) → Out: B (45), P (45), X (10), Z (45)

For large excess of A such that (mol A)/(mol B) $\geq$ 10, B is entirely consumed. Any Z created by the reversible reaction $P + B \leftrightarrow Z$ is eventually depleted by the excess A because excess A consumes all the B, including B created by the decomposition of Z.

In: A (100), B (10) → reactor ($A + B \rightarrow P + X$; $P + B \leftrightarrow Z$) → Out: A (90), P (10), X (10)

In: A (100), B (10), P (90) → reactor ($A + B \rightarrow P + X$; $P + B \leftrightarrow Z$) → Out: A (100), P (100), X (10)

Complications

- Reactant B and by-product X have identical boiling points and melting points; B and X cannot be separated.

- Reactant A, product P, and by-product Z have identical boiling points and melting points; A, P and Z cannot be separated.

		Melting point (°C)	Boiling point at 1 atm (°C)
A	reactant	−38	77
P	product	−38	77
Z	by-product	−38	77
B	reactant	−67	44
X	by-product	−67	44
E	reactant	2	133
Y	by-product	−45	55

Additional Reactions

- To purify P, Z can be removed by reaction with E to produce Y. Both E and Y are easily separated from P.
- $Z + E \rightarrow Y$ The reaction is irreversible. No catalyst is required.
- Z is entirely consumed for excess E such that (mol E)/(mol Z) ≥ 10.
- E reacts with no other substances.

Design Goals (in decreasing importance)

- Maximize the output of pure P.
- Minimize the input of reactants B and E.
- Minimize the *number* of units.
- Minimize the *size* of each unit.

Design Rules

- Use 10 mol/min of A. You may use as much B and E as you wish.
- No process unit may have a capacity larger than ~1000 mol/min.
- Liquid–solid separators produce wet solids.

Design Guidelines

- List the substances and *approximate* flow rates (mol/min) in every stream. One significant figure is sufficient.

Hint: For full credit it is necessary to produce more than 1 mol/min of P.

3.131(A) Design a process to produce pure P from A and B by the following reaction:

$A + B \leftrightarrow P + Q$ The reaction is fast and reversible. $\dfrac{(\text{mol } P)(\text{mol } Q)}{(\text{mol } A)(\text{mol } B)} \approx 1$ in the reactor effluent.

Complication

- Unfortunately, reactant B and worthless by-product Q have identical boiling points and melting points; B and Q cannot be separated by physical means.

Additional Reactions

- $B + D \rightarrow X$ For large excess of D such that (mol D)/(mol B) ≥ 10, B is entirely consumed.

- $Q + E \rightarrow Z$ For large excess of E such that (mol E)/(mol Q) ≥ 10, Q is entirely consumed.

Design Goals (in decreasing importance)

- Maximize the output of pure P.
- Minimize the input of reactants B, D, and E.
- Minimize the *number* of units.
- Minimize the *size* of each unit.

Design Rules

- Use no more than 10 mol/min of A. You may use as much B, D, and E as you wish.
- No process unit may have a capacity larger than ~1000 mol/min.
- Liquid–solid separators produce wet solids.

		Melting point (°C)	Boiling point at 1 atm (°C)
A	reactant	−34	77
B	reactant	−57	55
D	reactant	2	166
E	reactant	−9	122
P	product	−83	33
Q	product	−57	55
X	product	−24	111
Z	by-product	−17	99

Design Guidelines

- List the substances and *approximate* flow rates (mol/min) in every stream. One significant figure is sufficient.

(B) Assume the same reactions as part (A), but now by-product Q is valuable, almost as valuable as P. The design goals are the same as (A), except the primary goal is now to maximize the production of pure P *and* pure Q.

This exercise is based on a suggestion by Jonathan Kienzle, Cornell University Chemical Engineering Class of 2012.

3.132 Design a process to produce pure X from D and E by the following reaction:

$D + E \rightarrow X$ Reaction is irreversible but slow; reaction consumes half of whichever reactant is less.

X reversibly isomerizes to Z:

$X \leftrightarrow Z$ In the reactor, X and Z immediately equilibrate such that mol X = mol Z.

The reactions are conducted in an oil solvent at 70°C, such that either D or E is less than 10 mol%. Furthermore, if D and E are together in a stream, either D or E must be less than 10 mol%. Two reactor scenarios are shown below. Numbers in parentheses are flow rates, in mol/min.

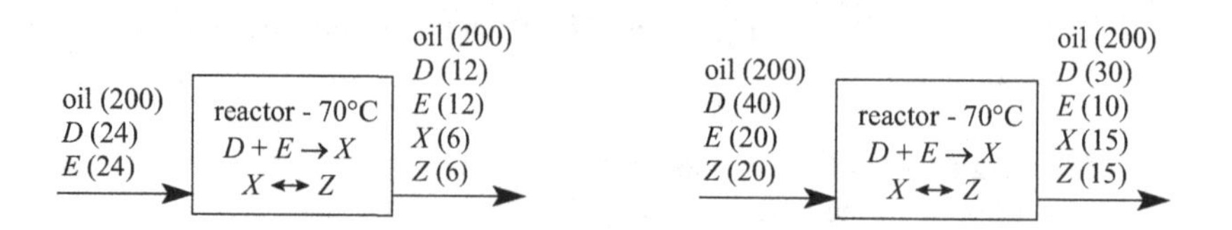

Complications

- Product X and reactant E have identical boiling points and melting points. Both X and E are soluble in oil. E is insoluble in water, whereas X is soluble in water.
- By-product Z and reactant D have identical boiling points and melting points. Both Z and D are soluble in oil. Z is insoluble in water, whereas D is soluble in water.

D and X will apportion equally between oil and water phases. Two oil–water absorber scenarios are shown below.

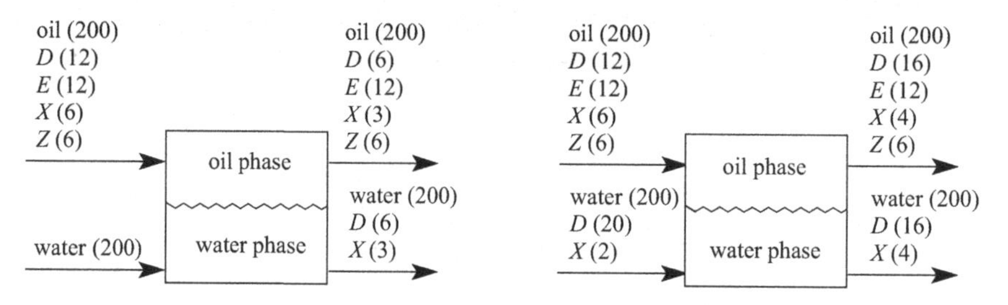

Design Goals (in decreasing importance)

- Maximize the production of pure X.
- Minimize the consumption of E, oil, and water.
- Minimize the *number* of units.
- Minimize the *size* of units.

Design Rules

- Use 24 mol/min of D. You may use as much E, oil, and water as you wish.
- For the oil–water absorber, you must use equal molar flow rates of oil and water.
- Water is not permitted in the reactor.

Design Guidelines

- List the substances and *approximate* flow rates (mol/min) in every stream.

		Melting point (°C)	Boiling point at 1 atm (°C)	Soluble in oil?	Soluble in water?
D	reactant	16	140	yes	yes
Z	by-product	16	140	yes	**no**
E	reactant	−5	80	yes	**no**
X	product	−5	80	yes	yes
oil	solvent	6	220	–	**no**
water	solvent	0	100	**no**	–

Engineering Calculations

3.133 If one travels at 55 miles per hour in a car with a fuel tank capacity of 30.4 liters and a fuel efficiency of 23 miles per gallon, how many minutes can one travel before the fuel tank is empty?

3.134 On January 4, 1994, a storm deposited 18 inches of snow on the 200 scenic acres of the Cornell University campus. An article in the *Cornell Chronicle*, January 13, 1994, claimed the storm "dumped what seemed like tons of the stuff" on Cornell. Use the rule of thumb that 12 inches of snow melts to 1 inch of water to assess the *Chronicle*'s claim. Did the *Chronicle* exaggerate or understate the snowfall on the Cornell campus?

3.135 Optical fibers are fine threads of silica drawn from a rod of quartz, called a boule. One end of the boule is heated to its melting point and a thread of silica 0.125 mm in diameter is drawn from the melt. By careful control of the boule temperature and the drawing tension, long fibers of uniform diameter can be drawn.

(A) Calculate the length of fiber (in km) that can be drawn from one boule 1.0 m in length and 2.5 cm diameter.

(B) The fiber is drawn at a rate of 10 m/s. Calculate the time required to draw the boule into fiber.

(C) The drawn fiber is wound onto a cylindrical spool of diameter 20 cm and height 10 cm. Estimate the diameter of the spool after all the fiber is wound onto the spool.

3.136 Optical fibers are silica threads with a protective polymer coating. Given that the silica fiber diameter is 0.125 mm and the coating is 0.050 mm thick, how many kilograms of polymer are needed to coat 26 km of fiber? (Density of polymer = 1740. kg/m^3.)

3.137 In the preparation of integrated circuits (chips), silicon wafers are coated with a photolithographic polymer. The wafers are 8 inches in diameter and the coating is 1000 Å thick. Each day of a five-day workweek, 4000 wafers are coated. How many liters of polymer are needed to coat wafers for a year?

3.138 Calculate the annual global CO_2 flux from human metabolism. You may assume an average person consumes 2500 Cal/day, Earth's population is 6×10^9 people, and 1 g glucose ($C_6H_{12}O_6$) yields 1.6×10^4 J. Note that 1 Cal = 1000 cal = 4184 J.

3.139 The *New York Times* (April 7, 2005) reported that discarded plastic bags have become a problem in Nairobi, Kenya. In addition to being blight on the landscape, the bags are a health hazard because they "provide a breeding place for malarial mosquitoes, helping spread one of the continent's major killers." Kenya's National Environmental Management Authority and the Kenya Institute for Public Policy Research and Analysis estimated "more than 100 million light polythene bags, many of them thinner than 30 microns, are handed out each year in Kenyan supermarkets, which is more than

4000 tons of the bags every month." Does 100 million bags per year equal 4000 tons per month? Assume the typical bag material is 50 microns thick ($\approx$ 2 mil) and the density of polyethylene (aka polyethene and polythene) is 930 kg/m^3. Also assume "ton" means a metric ton in this context; 1 metric ton = 1000 kg.

4

Models Derived from Graphical Analysis

Complex designs evolve from simple designs. Each step in the evolution is guided by analysis and evaluation. One method of analysis – mathematical modeling – entails translating the physical or chemical description into equations. For example, in Chapter 3 we applied the conservation of mass (a universal law) to translate steady-state systems into the equation "rate in = rate out." For other systems one might apply the ideal gas law (a constrained law) and translate into the expression "$PV = nRT$."

Mathematical modeling is based on a symbolic representation of quantities. For example, one might describe a separator input as $F_{\text{acetic acid}} = 32$ mol/min, $F_{\text{ethanol}} = 14$ mol/min, and $T = 34°C$. One can also represent a system graphically. For example, the separator input could be represented by a coordinate on a plot with two axes: mol fraction and temperature. Whereas we used algebra to develop the mathematical model of a system at steady state, we will use geometry to develop a graphical model.

Mathematical modeling is generally better for analyzing systems in which the underlying principles are known. Graphical modeling is generally better for systems described by empirical data. For example, what if we wanted to model a system that operates at conditions outside the ideal gas law's range of validity? How might we model, for example, a gas at high pressure and low temperature? One could search for a constitutive equation that applies in these limits; the basic course in chemical engineering thermodynamics offers several constitutive equations for specific conditions or types of mixtures. But to apply these equations one needs to know parameters specific to each gas or each mixture. Suppose these parameters have never been reported. What is one to do? One approach is to measure data at the conditions of interest. For example, one might measure the volume of 1 mol of the gas at temperatures and pressures relevant to your process. You could then extend the plot from the ideal region (Figure 3.64) to the plot shown in Figure 4.1.

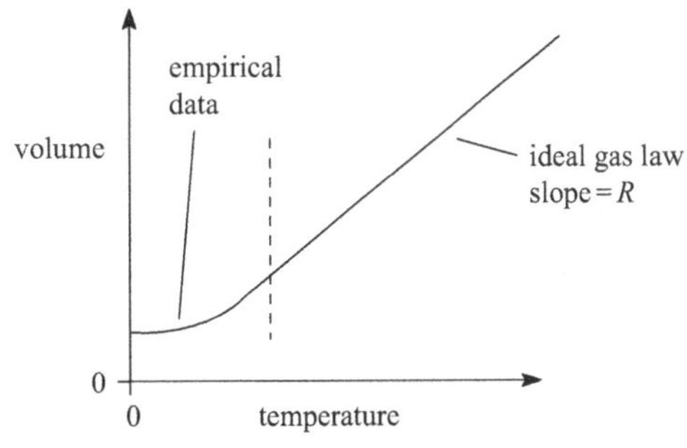

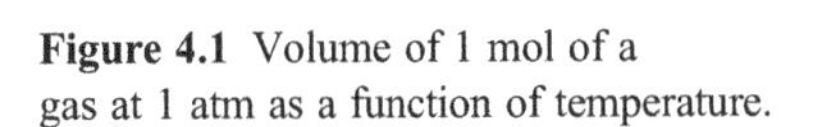
Figure 4.1 Volume of 1 mol of a gas at 1 atm as a function of temperature.

Although we may not understand the fundamental phenomena at low temperatures and/or high pressures, we can still use the data to model a process – for example, to pressurize the gas – and for other systems in regions outside the validity of constrained laws.

After introducing some basic skills – tie lines and the lever rule – the context for graphical analysis will be separators. Separating mixtures is often 50% to 90% of the capital cost of a chemical process. The separation units contribute a significant portion of the operating costs as well. Indeed, the cost of producing a chemical commodity is often determined by the difficulty of purifying the commodity. The value of pharmaceuticals, precious elements, and rare isotopes is generally independent of the costs of reactants and reactors, for example. The selling price of a chemical commodity is inversely proportional to the concentration of the commodity, as shown in Figure 4.2, which is known as the Sherwood plot. Remarkably, the correlation spans ten orders of magnitude!

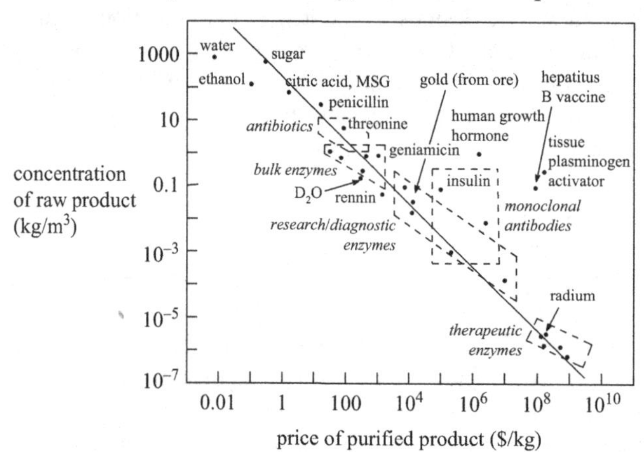

Figure 4.2 The Sherwood plot. Adapted from a similar figure (Nystrom 1984) and discussed in an article by Lightfoot (1988). Additional data taken from texts by King (1971), figures 1–16; Sherwood, Pigford, and Wilke (1975), figure 1.2; and Blanch and Clark (1995), figure 6.2.

4.1 TIE LINES, MIXING LINES, AND THE LEVER RULE

4.1.1 Graphical Mass Balances

Context: Mixing and separating two substances.

Concepts: The mass number line, mixing line segments, and the lever rule.

Defining Question: Where does one place the endpoints and fulcrum on a mixing line?

In Chapter 3 we developed a mathematical model for a mass balance. Here we develop a graphical model for a mass balance. Consider a mixer at steady state. Two streams with two components (A and B) enter a mixer.

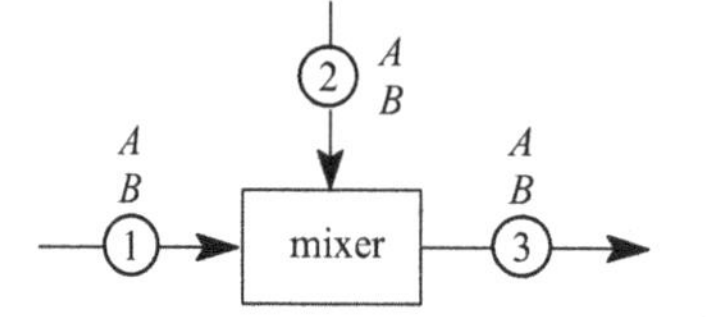

Figure 4.3 A mixer for streams containing A and B.

To develop a graphical mass balance, we begin with a mathematical mass balance. As in Chapter 3, $F_{T,i}$ is the mass flow rate of stream i and $F_{A,i}$ is the mass flow rate of A in stream i. Define $f_{A,i}$ as the mass fraction of A in stream i, such that

$$F_{A,i} = f_{A,i} F_{T,i}. \tag{4.1}$$

Write a mass balance on A and then substitute equation 4.1 for $i = 1, 2,$ and 3, to obtain an expression in terms of total flow rates and stream compositions:

$$\text{rate of } A \text{ in} = \text{rate of } A \text{ out} \tag{4.2}$$

$$F_{A,1} + F_{A,2} = F_{A,3} \tag{4.3}$$

$$f_{A,1}F_{T,1} + f_{A,2}F_{T,2} = f_{A,3}F_{T,3}. \tag{4.4}$$

From a total mass balance we obtain $F_{T,3} = F_{T,1} + F_{T,2}$. Substitute into equation 4.4 and rearrange:

$$f_{A,1}F_{T,1} + f_{A,2}F_{T,2} = f_{A,3}(F_{T,1} + F_{T,2}) \tag{4.5}$$

$$(f_{A,1} - f_{A,3})F_{T,1} = (f_{A,3} - f_{A,2})F_{T,2}. \tag{4.6}$$

Equation 4.6 represents a balanced lever. This is the apparatus you mastered in your elementary school playground, also known as a seesaw, or a teeter-totter, or a teeterboard, or a dandle board. The seesaw is balanced with mass $F_{T,1}$ at one end and mass $F_{T,2}$ at the other end. The total length of the seesaw is $f_{A,1} - f_{A,2}$. The fulcrum is a distance $f_{A,1} - f_{A,3}$ from mass $F_{T,1}$ and a distance $f_{A,3} - f_{A,2}$ from mass $F_{T,2}$.

To see the analogy to a seesaw, consider a graphical representation for the mass balance in equation 4.6. Draw a number line for the mass fraction of A, which extends from $f_A = 0$ (pure B) to $f_A = 1$ (pure A), as shown in Figure 4.4.

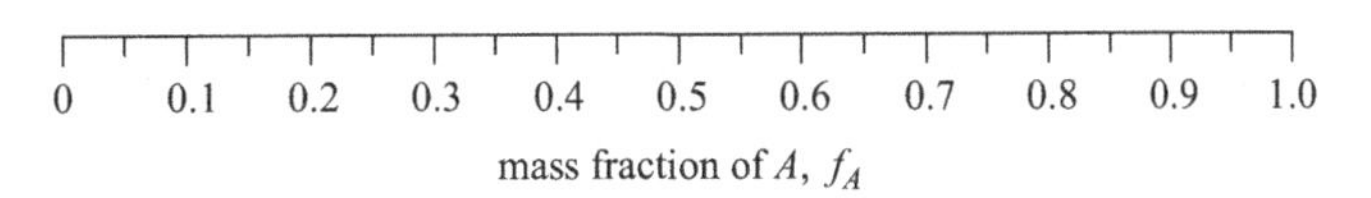

Figure 4.4 A number line for mass balances on A.

Let's work a specific example. Set $f_{A,1} = 0.7$, $f_{A,2} = 0.1$, and $f_{A,3} = 0.3$. If $F_{T,2} = 10$ kg/min, what are the flow rates of streams 1 and 3? Draw a seesaw above the number line with one end at $f_{A,1}$, the other end at $f_{A,2}$, and the fulcrum at $f_{A,3}$. This is drawn in Figure 4.5.

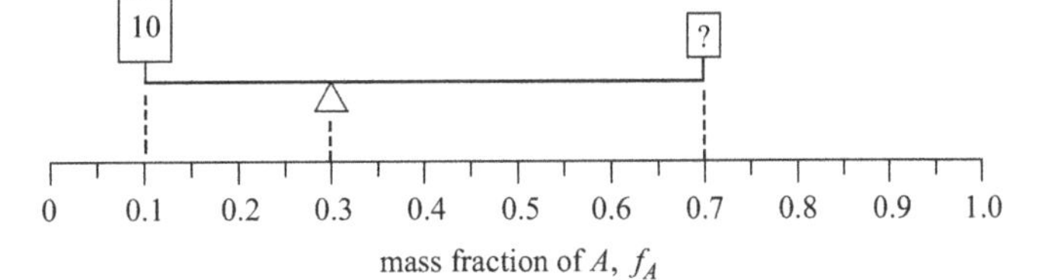

Figure 4.5 The mass balance seesaw for $f_{A,1} = 0.7$, $f_{A,2} = 0.1$, and $f_{A,3} = 0.3$.

What mass must be placed at $f_{A,1} = 0.7$ to balance the seesaw? Qualitatively, the mass at $f_{A,1}$ must be smaller than the mass at $f_{A,2}$. Quantitatively, the ratio of $F_{T,1}$ to $F_{T,2}$ is inversely proportional to the lengths of the lever arms. To see this, start with equation 4.6 and rearrange to a ratio:

$$\frac{F_{T,1}}{F_{T,2}} = \frac{f_{A,3} - f_{A,2}}{f_{A,1} - f_{A,3}} = \frac{\text{length of left lever arm}}{\text{length of right lever arm}} \tag{4.7}$$

$$F_{T,1} = \frac{f_{A,3} - f_{A,2}}{f_{A,1} - f_{A,3}}F_{T,2} = \frac{0.3 - 0.1}{0.7 - 0.3}10 = \frac{1}{2}10 = 5 \text{ kg/min}. \tag{4.8}$$

The right lever arm is twice as long as the left lever arm, so the right mass is half the left mass. The flow rate of stream 1 is ½(10) = 5 kg/min. The flow rate of stream 3 (the fulcrum) is the total mass on the seesaw: 15 kg/min.

Let's work another example. Calculate the composition of stream 2 that must be mixed with stream 1 to yield stream 3.

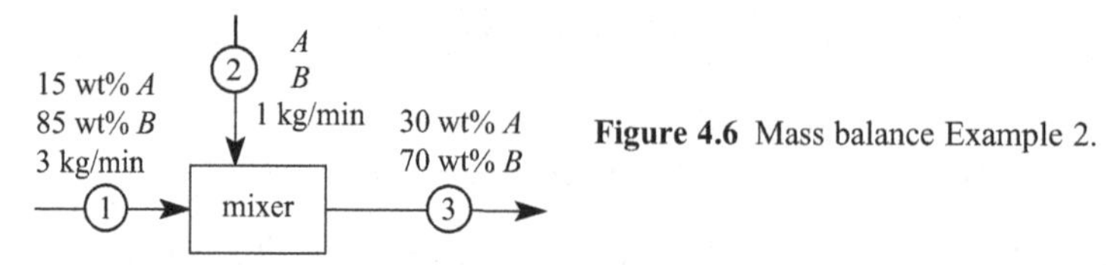

Figure 4.6 Mass balance Example 2.

To model Example 2 graphically, draw a seesaw with 3 kg on the left end and 1 kg on the right end. The left end is at $f_{A,1} = 0.15$ and the fulcrum is placed at the composition of stream 3, $f_{A,3} = 0.3$. What length of the right lever arm balances the seesaw?

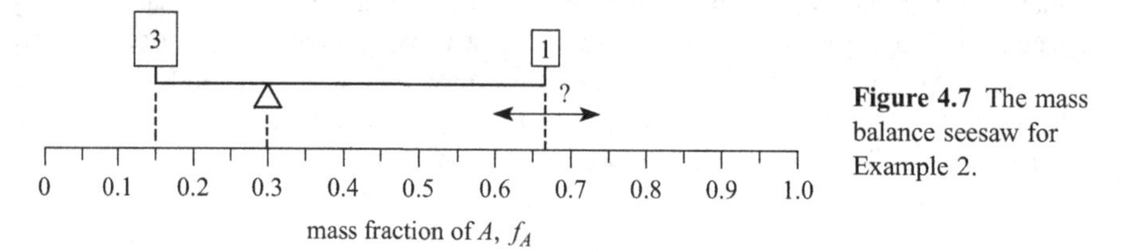

Figure 4.7 The mass balance seesaw for Example 2.

The left mass is three times the right mass, so the length of the right lever arm must be three times the length of the left lever arm, $3 \times (0.3 - 0.15) = 0.45$. The left mass is at $0.3 + 0.45 = 0.75$. Stream 2 must contain 75 wt% *A*.

This graphical method is known as the *lever rule*. The seesaw board is called the *mixing line*. We will use the lever rule to analyze several different systems in this chapter. For example, the lever rule can be used to analyze separators. Consider the separator in Figure 4.8. What are the flow rates of streams 2 and 3?

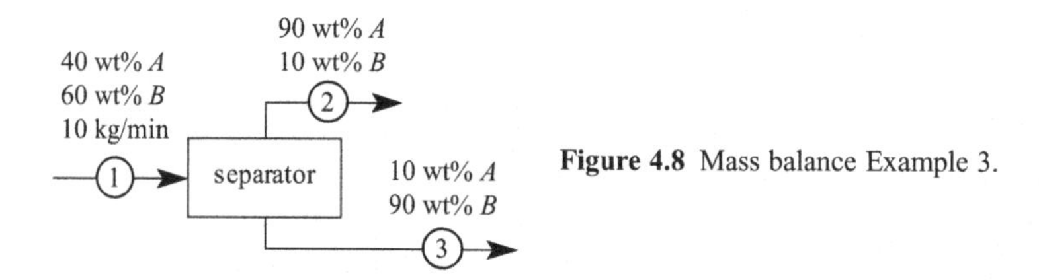

Figure 4.8 Mass balance Example 3.

To model Example 3 graphically, draw a mixing line (seesaw) with ends at 0.1 and 0.9, and the fulcrum at 0.4. How do we distribute 10 kg to balance the seesaw?

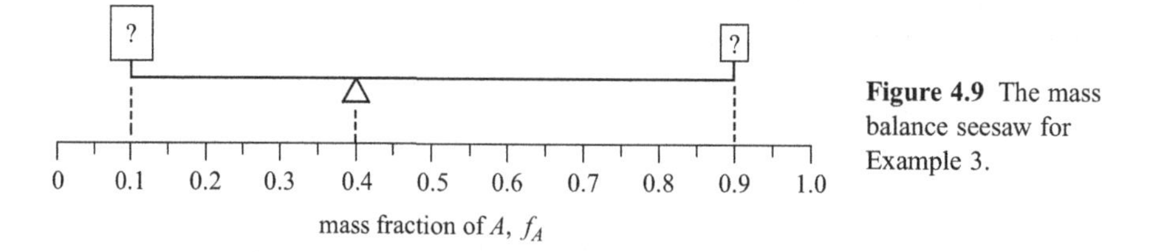

Figure 4.9 The mass balance seesaw for Example 3.

The ratio of the lever arm lengths is $(0.4 - 0.1) / (0.9 - 0.4) = 0.3/0.5 = 3/5$. Thus the ratio of the masses must be 3:5 and the total mass must be 10 kg/min. The mass on the left end of the fulcrum (stream 3) is 10 × 5/(3 + 5) = 6.25 kg/min and the mass on the right end (stream 2) is 10 × 3/(3 + 5) = 3.75 kg/min.

A mass balance on a two-component mixer can be represented graphically by a lever on a mass fraction line segment, with its endpoints as the two input streams and its fulcrum at the composition of the output mixture. Similarly, a mass balance on a two-component separator can be represented graphically by a lever on a mass fraction line segment, with its endpoints as the two output streams and its fulcrum at the composition of the input mixture. The masses that balance the lever are proportional to the flow rates of the two streams entering the mixer or the two streams leaving the separator.

4.1.2 Graphical Energy Balances

Context: Mixing ice, water, and steam.

Concepts: The energy number line and mixing line segments. Two-phase segments and tie lines.

Defining Question: What differentiates a mixing line from a tie line?

Admittedly, graphical mass balances are not easier than mathematical mass balances. On the other hand, we shall see that graphical *energy* balances can be considerably easier, because the graphical method automatically converts temperature and phase to energy.

A linear scale can also be used for graphical energy balances. For example, consider a linear scale for the specific energy of liquid H_2O. Assume the heat capacity is 4.17 kJ/(kg·°C) and is constant over the range 0°C to 100°C. Set the reference to be water at 0°C; arbitrarily assign $\bar{H}(0°C) = 0$ kJ/kg for water at 0°C. The specific enthalpy of water at temperatures above 0°C is calculated relative to this reference. For example, the specific enthalpy of water at 60°C is $\bar{H}(60°C) = \bar{H}(0°C) + 4.17$ kJ/(kg·°C) $\times (60 - 0°C) = 250$ kJ/kg. Figure 4.10 shows a linear scale for the specific energy of water with a temperature scale.

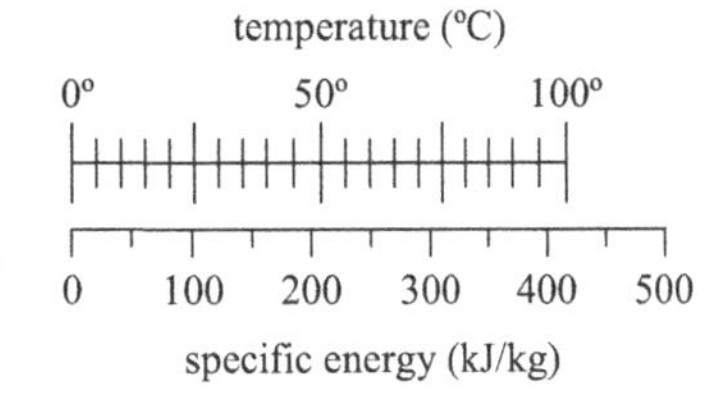

Figure 4.10 A number line for energy balances on water, with a parallel scale for temperature.

This linear scale for specific energy can be used to calculate energy balances graphically. For example, one can apply the lever rule to determine the temperature of the mixer output below.

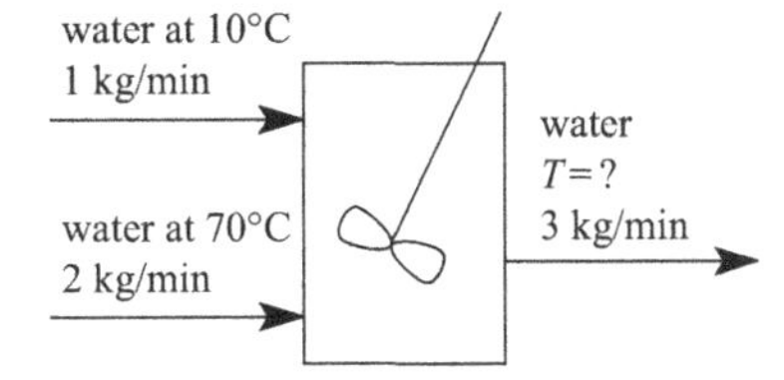

Figure 4.11 Energy balance Example 1.

To perform an energy balance graphically, draw a lever between the two input streams at 10°C (42 kJ/kg) and 70°C (292 kJ/kg), and place masses at the each end of the lever to represent the flow rates, as shown in Figure 4.12.

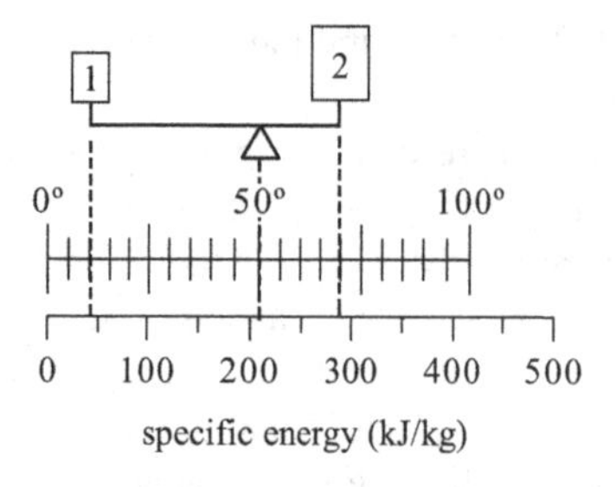

Figure 4.12 The lever rule solution to energy balance Example 1.

To determine the output temperature, we apply the lever rule; we measure lengths on the specific-energy scale, in units of kJ/kg:

length of mixing line = 292 − 42 = 250 kJ/kg
length of 70°C arm of mixing line = 250 × 1/(1 + 2) = 83 kJ/kg
location of fulcrum = 292 − 83 = 209 kJ/kg.

The fulcrum is at 209 kJ/kg, which corresponds to 50°C. The temperature of the output stream is 50°C.

Of course, it is equally valid to measure the mixing line with a ruler and use units of cm rather than kJ/kg. Any linear scale – kJ/kg, cal/g, Btu/lb, or cm – is valid, and a ruler is more convenient (and more accurate).

Graphical energy balances are convenient for systems with phase changes, such as liquid to vapor. We extend the specific-energy scale to H_2O vapor at 100°C by adding $\Delta\bar{H}_{\text{vap}}$ to H_2O liquid at 100°C: $\bar{H}(100°\text{C}, \text{vapor}) = \bar{H}(100°\text{C}, \text{liquid}) + \Delta\bar{H}_{\text{vap}} = 417 + 2280 = 2697\ \text{kJ/kg}$. We then continue to higher energy by using the heat capacity of H_2O vapor, which we will assume is 1.94 kJ/(kg·°C) and is constant over the range 100°C to 300°C.

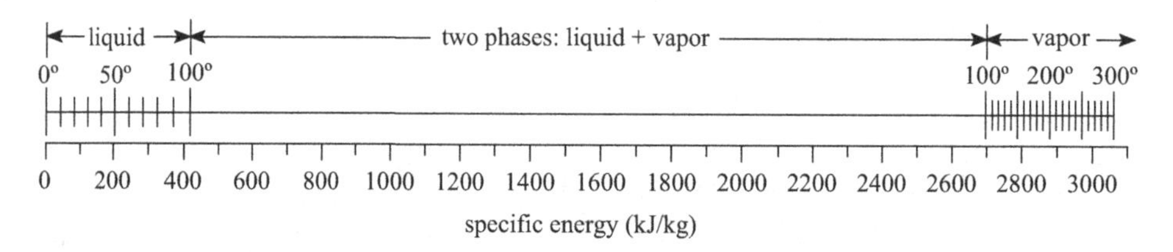

Figure 4.13 A number line for energy balances involving water and steam, with scales for temperature and phase.

Use the energy line in Figure 4.13 to analyze the more challenging energy balance in Example 2.

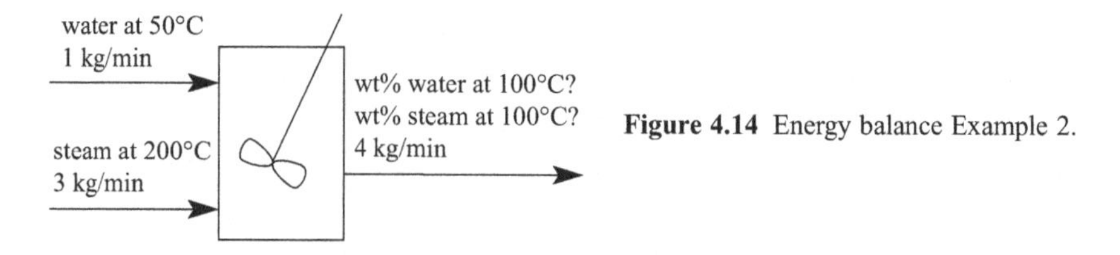

Figure 4.14 Energy balance Example 2.

Again draw a lever between the two input streams: water at 50°C (209 kJ/kg) and steam at 200°C (2891 kJ/kg), and place masses at the each end of the lever to represent the flow rates. Apply the lever rule:

length of mixing line = 2891 − 209 = 2682 kJ/kg
length of steam arm of mixing line = 2682 × 1/(1 + 3) = 671 kJ/kg
location of fulcrum = 2891 − 671 = 2220 kJ/kg.

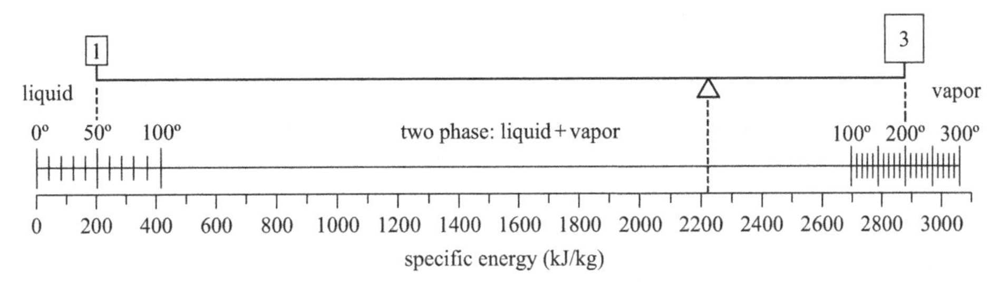

Figure 4.15 A mixing line (seesaw) drawn with ends at the input streams and fulcrum at the output.

The output stream lies in the liquid + vapor portion of the energy line. A point in a two-phase segment indicates a mixture of two phases. To determine the relative amounts of the two phases, we draw a line segment to the endpoints of the two-phase line segment, as shown in Figure 4.16. This lever that represents two phases in equilibrium is called a *tie line*.

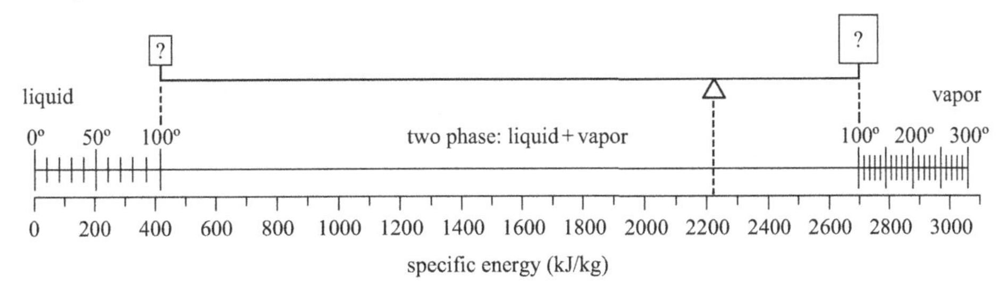

Figure 4.16 A tie line between the two phases in equilibrium: water at 100°C and steam at 100°C.

Apply the lever rule to distribute the total flow (4 kg/min) and balance the lever with the fulcrum at 2220 kJ/kg:

length of tie line = 2280 kJ/kg (= $\Delta\bar{H}_{vap}$)
length of steam arm of tie line = 2697 − 2220 = 477 kJ/kg
amount of water at 100°C = (477/2280) × 4 kg/min = 0.21 × 4 kg/min = 0.84 kg/min
amount of steam at 100°C = 4 − 0.84 = 3.16 kg/min.

The mixer output is 4 kg/min, 21% water and 79% steam at 100°C.

To complete our specific-energy line for H_2O, we extend the scale to lower temperatures, to solid H_2O: $\bar{H}(0°\text{C}, \text{solid}) = \bar{H}(0°\text{C}, \text{liquid}) - \Delta\bar{H}_{\text{melt}} = 0 - 333$ kJ/kg. We then continue to lower energy by using the heat capacity of H_2O solid, which we will assume is 1.83 kJ/(kg·°C) and is constant over the range 0°C to −200°C (as shown in Figure 4.17 on the next page).

Consider the mixer shown in Figure 4.18 (on the next page). Calculate the temperature and phase(s) of stream 3. If two phases are present, calculate the weight fraction of each phase.

Draw a lever between the two input streams at −50°C (−425 kJ/kg) and 150°C (2794 kJ/kg), and place equal masses at the each end of the lever. Because the flow rates of streams 1 and 2 are equal, the lever balances with the fulcrum at the midpoint of the lever, at (2794 + (−425))/2 = 1185 kJ/kg (as shown in Figure 4.19 on the next page).

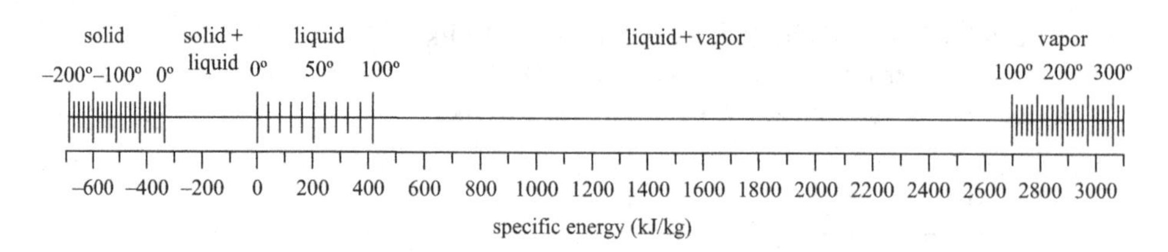

Figure 4.17 A number line for energy balances involving ice, water, and steam, with scales for temperature and phase. Larger energy line templates may be downloaded from duncan.cbe.cornell.edu/Graphs.

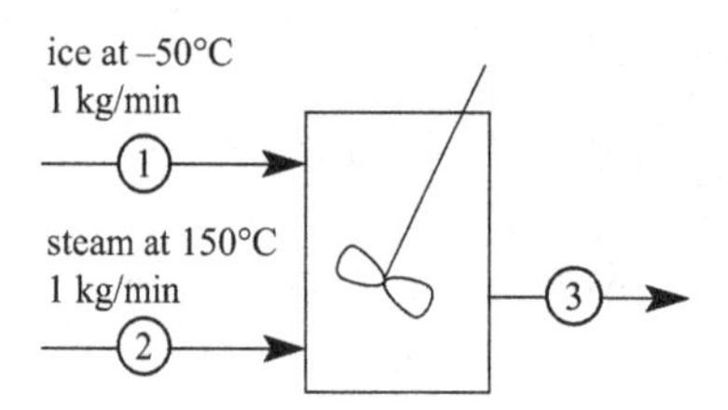

Figure 4.18 Energy balance Example 3.

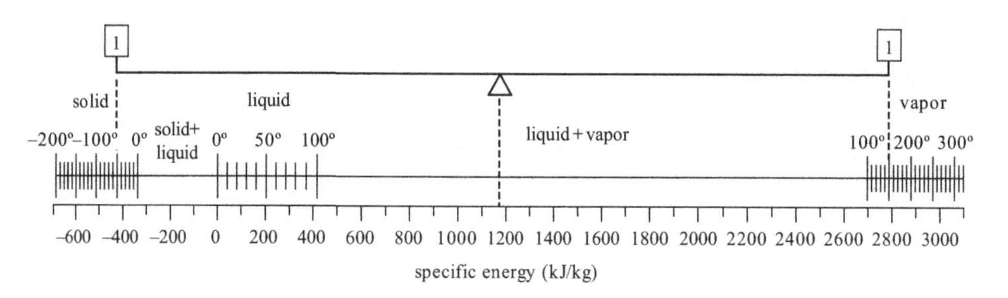

Figure 4.19 A mixing line representation of the energy balance in Example 3.

Because the fulcrum (stream 3) lies in the two-phase liquid + vapor region, we draw a tie line from the endpoints of the two-phase segment (water at 100°C and steam at 100°C).

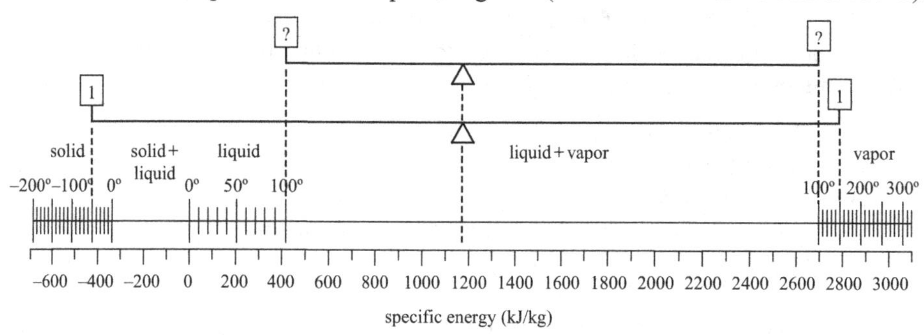

Figure 4.20 A tie line to determine the water and steam flow rates at 100°C.

Apply the lever rule to determine how to distribute the total flow (2 kg/min) to balance the lever at 1185 kJ/kg:

length of tie line = 2280 kJ/kg (= $\Delta \bar{H}_{vap}$)
length of steam arm of tie line = 2697 − 1185 = 1512 kJ/kg
amount of water at 100°C = (1512/2280) × 2 kg/min = 0.66 × 2 kg/min = 1.33 kg/min
amount of steam at 100°C = 2 − 1.33 = 0.67 kg/min.

The temperature of stream 3 is 100°C. Stream 3 is 66 wt% water and 34 wt% steam.

Consider the mixer shown in Figure 4.21. Calculate the flow rate of ice in stream 1 required to produce water at 50°C.

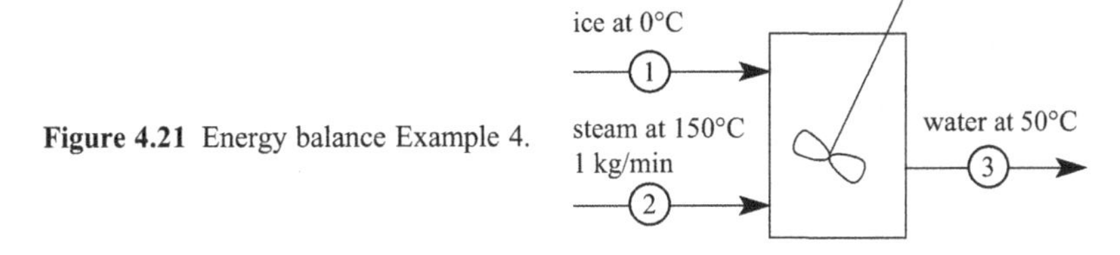

Figure 4.21 Energy balance Example 4.

The mixer in Example 4 is represented by a lever with ends at ice (0°C) and steam (150°C) and the fulcrum at water (50°C). A mass of 1 kg is placed on the steam end.

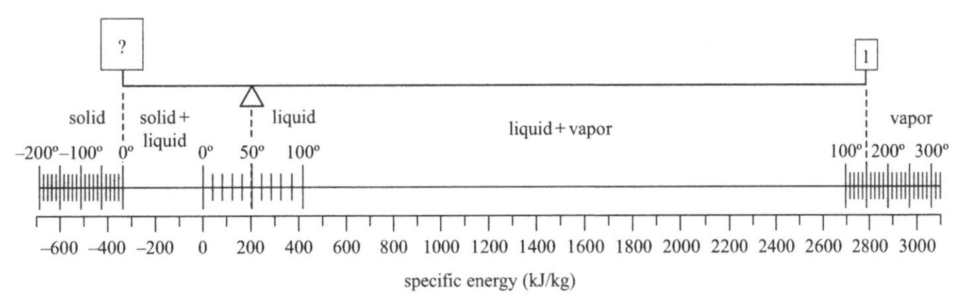

Figure 4.22 A mixing line representation of the energy balance in Example 4.

Calculate the mass that must be placed at the ice end to balance the lever:

length of mixing line = 2794 − (−333) = 3127 kJ/kg
length of steam arm of mixing line = 2794 − 209 = 2585 kJ/kg
length of ice arm of mixing line = 209 − (−333) = 542 kJ/kg
amount of ice at 0°C = (2585/542) × 1 kg/min = 4.77 kg/min.

The flow rate of ice in stream 1 is 4.77 kg/min.

The lever rule can also be applied to analyze a mixer with three or more input streams, such as the example in Figure 4.23 (on the next page).

The mixer in Example 5 is represented by a lever with three masses: ends at ice (−30°C) and steam (200°C), and a mid-seesaw mass at water (50°C).

The mixture (stream 4) corresponds to the fulcrum location that balances the seesaw. The fulcrum is also known as the center of mass or first moment. Define ε_i as the specific energy of stream i. For a system with n masses such that mass M_i is a distance ε_i from a reference point,

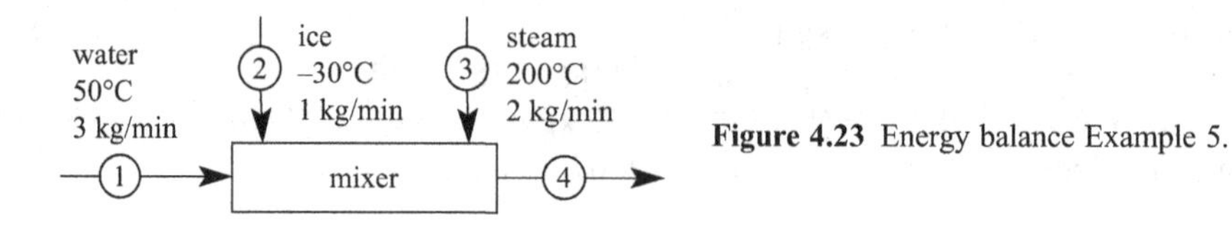

Figure 4.23 Energy balance Example 5.

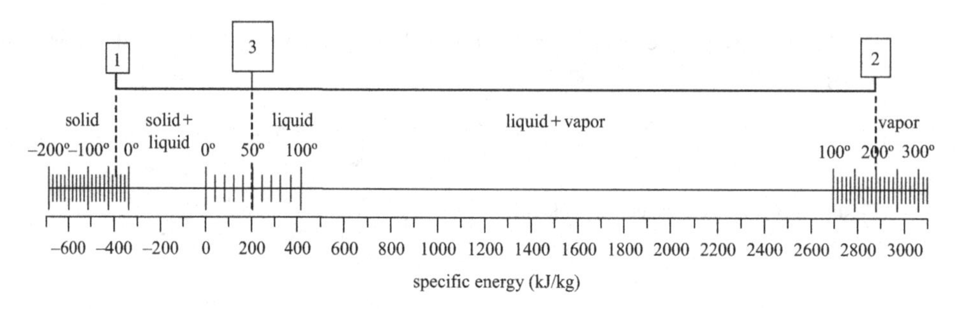

Figure 4.24 A mixing line representation of the energy balance in Example 5.

$$\text{center of mass} = \frac{\sum_i^n \varepsilon_i M_i}{\sum_i^n M_i}. \tag{4.9}$$

Substitute the numerical values for Example 5. There are three masses, so $n = 3$. Stream 1 is 3 kg/min of water at 50°C, so $\varepsilon_1 = 209$ kJ/kg and $M_1 = 3$ kg/min. Stream 2 is 1 kg/min of ice at -30°C, so $\varepsilon_2 = -390$ kJ/kg and $M_2 = 1$ kg/min. Stream 3 is 2 kg/min of steam at 200°C, so $\varepsilon_3 = 2890$ kJ/kg and $M_3 = 2$ kg/min.

$$\text{center of mass} = \frac{\varepsilon_1 M_1 + \varepsilon_2 M_2 + \varepsilon_3 M_3}{M_1 + M_2 + M_3} \tag{4.10}$$

$$= \frac{\left(\frac{209\ \text{kJ}}{\text{kg}}\right)\left(\frac{3\ \text{kg}}{\text{min}}\right) + \left(\frac{-390\ \text{kJ}}{\text{kg}}\right)\left(\frac{1\ \text{kg}}{\text{min}}\right) + \left(\frac{2890\ \text{kJ}}{\text{kg}}\right)\left(\frac{2\ \text{kg}}{\text{min}}\right)}{\frac{3\ \text{kg}}{\text{min}} + \frac{1\ \text{kg}}{\text{min}} + \frac{2\ \text{kg}}{\text{min}}}$$

$$= 1000\ \text{kJ/kg}. \tag{4.11}$$

The center of mass (fulcrum) lies in the two-phase liquid + vapor region; stream 4 is a mixture of water at 100°C and steam at 100°C. Use the center-of-mass formula to calculate the amounts of water (M_{water}) and steam (M_{steam}) in stream 4. For water at 100°C, $\varepsilon_{\text{water}} =$ 417 kJ/kg. For steam at 100°C, $\varepsilon_{\text{steam}} = 2697$ kJ/kg. From a mass balance on the mixer, $M_{\text{water}} + M_{\text{steam}} = 6$ kg/min, or $M_{\text{steam}} = 6 - M_{\text{water}}$. Substitute into the equation for the center of mass with two masses:

$$\text{center of mass} = \frac{\varepsilon_{\text{water}} M_{\text{water}} + \varepsilon_{\text{steam}} M_{\text{steam}}}{M_{\text{water}} + M_{\text{steam}}} \tag{4.12}$$

$$1000\ \text{kJ/kg} = \frac{\dfrac{417\ \text{kJ}}{\text{kg}} M_{\text{water}} + \dfrac{2697\ \text{kJ}}{\text{kg}} (6 - M_{\text{water}})}{\dfrac{6\ \text{kg}}{\text{min}}} \tag{4.13}$$

$$6000 = 417 M_{\text{water}} + 16180 - 2697 M_{\text{water}} \tag{4.14}$$

$$M_{\text{water}} = 4.5\ \text{kg/min}\,. \tag{4.15}$$

Stream 4 is a mixture of water and steam at 100°C. The flow rate of water is 4.5 kg/min and the flow rate of steam is 1.5 kg/min.

The energy balance number line is a convenient graphical tool. The line automatically converts observable parameters, such as temperature (units of °C) and phase, into specific energy (units of kJ/kg). For example, one can use the specific-energy line to read that steam at 200°C has energy 2890 kJ/kg, relative to 0 kJ/kg for water at 0°C.

The number line allows one to estimate flow rates, or to check a calculation. For example, if in Example 5 we had incorrectly calculated M_{water} = 1.5 kg/min and M_{steam} = 4.5 kg/min, it is obvious this is wrong. A lever with a fulcrum at 1000 kJ/kg is not balanced with 1.5 kg/min water and 4.5 kg/min steam.

The relative lengths of the line segments on the number line show the relative magnitudes of energy changes owing to phase changes and temperature changes. For example, the length of the liquid + vapor two-phase region is almost seven times the length of the solid + liquid two-phase region, and about five times the length of the liquid line segment. When steam condenses, a large amount of energy is released – much more than when water cools or when water freezes. Imagine using steam to heat a reactant at 20°C. What is the added benefit of using steam at 200°C compared to steam at 100°C? The energy line shows that the benefit is marginal: steam at 200°C provides about 10% more energy. Would you have realized that before seeing the energy line for H_2O?

Let's summarize graphical mass balances and graphical energy balances to this point. A mixer or a separator at steady state can be represented by a balanced lever. To balance a lever with two masses, the ratio of the two input flows is inversely proportional to the ratio of the lever arm lengths. To balance a lever with three or more masses, use the formula for the center of mass (aka the first moment).

When the specific energy of a system (in kJ/kg) lies in a two-phase line segment (solid + liquid or liquid + vapor) the system is a mixture of two phases in equilibrium. To calculate the amounts of the two phases, place the fulcrum at the system's specific energy and draw a tie line that spans the two-phase line segment. For any two-phase mixture of water at 100°C and steam at 100°C, for example, the tie line is always the same length but the location of the fulcrum changes, depending on the relative amounts of the two phases.

The lever rule can be used for any conserved quantity. We have used it here for mass and energy. More precisely, we have used the lever rule for *specific* mass ((mass of *A*)/(total mass of *A* and *B*)) and *specific* energy (energy/mass). If volume is conserved in a particular system (such as an ideal mixture), a linear scale of specific volume (= volume/mass = 1/density) could be used to determine the specific volume of a mixture of two components. Specific quantities may also be expressed on a molar basis, such as mol *A*/(mol *A* + mol *B*), or energy/mol, or volume/mol.

4.1.3 Graphical Mass Balances for Single-Stage Liquid–Vapor Separations

Context: A flash drum for separating benzene + toluene mixtures.

Concepts: Pressure–composition (P–(x,y)) and temperature–composition (T–(x,y)) phase diagrams. The dew point and the bubble point.

Defining Question: What differentiates a graph from a map?

We now apply graphical mass balances to separating a mixture in which the components have comparable volatility – for example, a mixture of benzene (which boils at 80°C at 1 atm) and toluene (which boils at 111°C at 1 atm).

A flash drum, shown in Figure 4.25, is a simple unit to effect a crude separation. A liquid mixture is throttled through a valve into a vessel with lower pressure. The mixture "flashes" into two phases – vapor and liquid. One expects the vapor is enriched with the more volatile component (e.g. benzene) and the liquid is enriched in the less volatile component (e.g. toluene). Flash drums can also operate by increasing the temperature or by both decreasing the pressure and increasing the temperature.

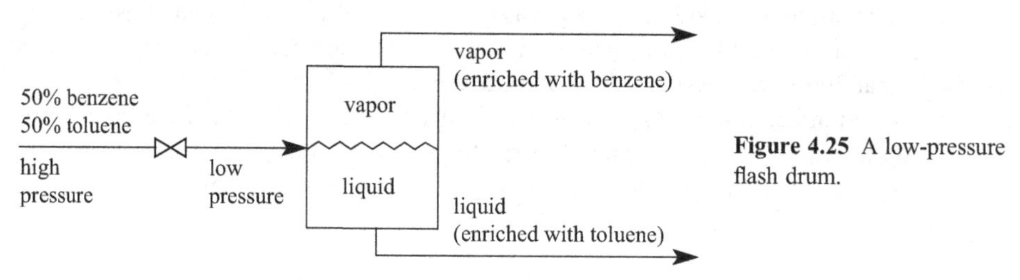

Figure 4.25 A low-pressure flash drum.

Given an operating pressure in the flash drum, what are the compositions of the vapor and liquid streams? Or, at what pressure should we operate the flash drum to achieve a desired separation? What is the best separation we can expect from this simple unit?

Because benzene and toluene are common solvents, one can find in the literature the thermodynamic data needed to model the flash drum. However, it is instructive to consider how one might measure the phase(s) present (vapor, liquid, or liquid + vapor) for benzene + toluene mixtures at various compositions, temperatures, and pressures at equilibrium. As always, it is wise to start with an extreme – pure benzene. So we seal some benzene in a cylinder fitted with a thermometer, a pressure gauge, and a viewing port. The top of the cylinder is a piston, as shown in Figure 4.26.

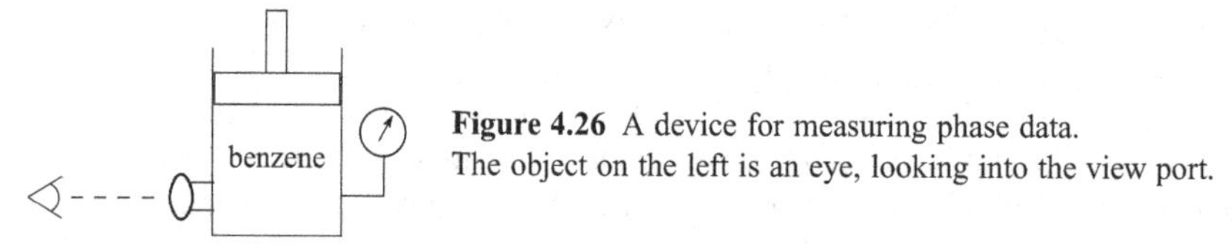

Figure 4.26 A device for measuring phase data. The object on the left is an eye, looking into the view port.

We arbitrarily decide to systematically increase the pressure while maintaining the temperature at 80°C. We record the phase(s) of the benzene. Table 4.1 reports the results of our experiment.

Table 4.1 Phase data for 1 mol benzene at 80°C.

Pressure (atm)	View port shows:	Phase(s) present
0.27	transparent gas	vapor
0.77	transparent gas	vapor
0.99	transparent gas	vapor
1.00	mist	vapor + liquid
1.00	liquid below gas	vapor + liquid
1.00	liquid below gas	vapor + liquid
1.00	liquid below gas	vapor + liquid

Table 4.1 (*cont.*)

Pressure (atm)	View port shows:	Phase(s) present
1.00	liquid below gas	vapor + liquid
1.00	liquid only	liquid
1.38	liquid only	liquid

We perform the same experiment with the other extreme, pure toluene. Some of the data we measure are given in Table 4.2.

Table 4.2 Phase data for 1 mol toluene at 80°C.

Pressure (atm)	Phase(s)
0.20	vapor
0.30	vapor
0.39	vapor
0.40	vapor + liquid
0.40	vapor + liquid
0.40	vapor + liquid
0.40	vapor + liquid
0.40	vapor + liquid
0.40	liquid
0.42	liquid
0.55	liquid

How do we plot these data so they are useful for designing a flash drum? We have *two* independent variables – composition (mol fraction benzene) and pressure – and one dependent variable – phase(s). The usual situation of one independent variable and one dependent variable can be accommodated on a standard x–y plot with a line representing the correlation. How do we show phase as a function of composition *and* temperature? Perhaps a three-dimensional plot? Three-dimensional plots are difficult to prepare and difficult to use. A better option is to prepare a two-dimensional phase diagram, which is in the general category of a "map."

A *map* is a convenient means of representing the state of a system as a function of two independent variables. The independent variables on a conventional map are "distance east" and "distance north" from a reference point, as shown in Figure 4.27.

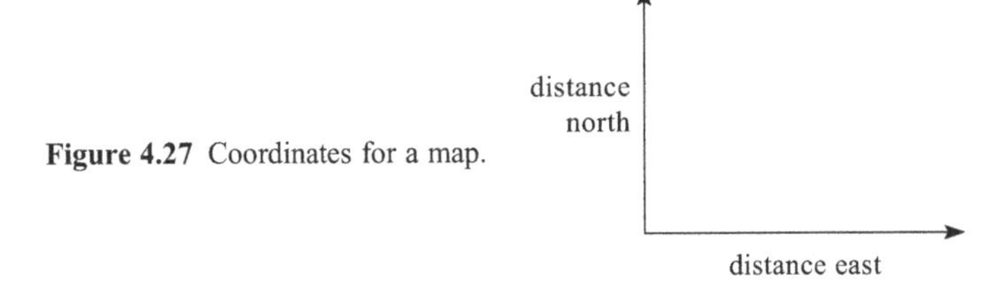

Figure 4.27 Coordinates for a map.

Figure 4.27 is not a graph. A graph, such as shown in Figure 4.28, has an independent variable, typically x, a dependent variable, typically y, and a line to represent $y = f(x)$. Given a value of x, one can read a value of y from the line on the graph. Clearly, no function "$y = f(x)$" exists for the map in Figure 4.27. If you tell us your distance east from a reference point, there is no function that tells us your distance north. Rather, you must tell us your distance east from a reference point *and* your distance north from a reference point. Then we could use a physical map to determine if you are in a lake, on a hill, or in a forest. If the map is a topographical map, such as shown in Figure 4.29, your east–north coordinates will tell us your elevation above sea level.

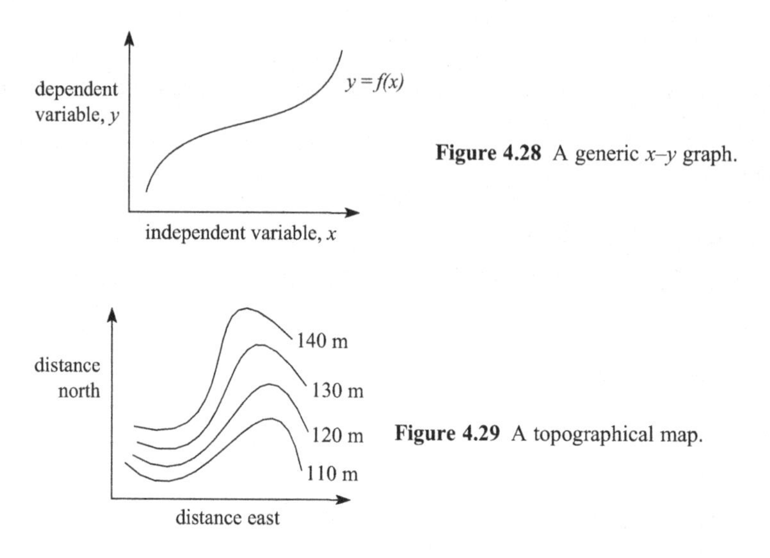

Figure 4.28 A generic x–y graph.

Figure 4.29 A topographical map.

For our measurements of benzene + toluene mixtures, the coordinates of the map should be the composition of the mixture and the pressure of the system. By convention, the composition is plotted on the x axis, as shown in Figure 4.30. Also by convention, the composition is expressed in terms of the more volatile component. In our mixture, benzene is the more volatile. As with all new maps it is important to get one's bearings. Pure benzene is the right axis of the map; pure toluene is the left axis. The data for pure benzene in Table 4.1 and the data for pure-toluene data in Table 4.2 are plotted on the benzene + toluene phase map in Figure 4.30.

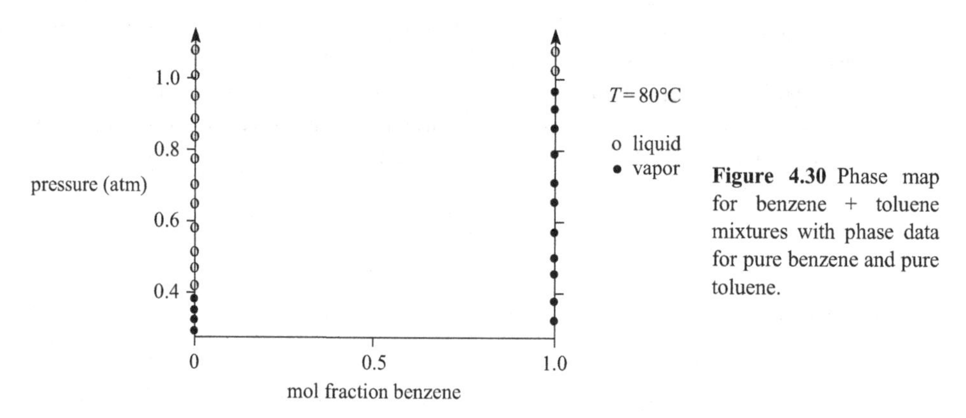

Figure 4.30 Phase map for benzene + toluene mixtures with phase data for pure benzene and pure toluene.

We proceed to a mixture of 50 mol% benzene and 50 mol% toluene. But now we have additional data to measure – the compositions of the vapor and liquid phases. So we add an instrument to the device in Figure 4.26 to measure compositions. We also record the amount of each phase given in Table 4.3 on the next page.

Table 4.3 Phase data for a mixture of 0.5 mol benzene + 0.5 mol toluene at 80°C.

		Vapor phase		Liquid phase	
Pressure (atm)	Phase(s)	Amount (mol)	mol% benzene	Amount (mol)	mol% benzene
0.31	vapor	1.0	0.5		
0.37	vapor	1.0	0.5		
0.45	vapor	1.0	0.5		
0.47	vapor + liquid	0.999	0.50	0.001	0.12
0.51	vapor + liquid	0.79	0.59	0.21	0.18
0.55	vapor + liquid	0.59	0.67	0.41	0.26
0.61	vapor + liquid	0.35	0.76	0.65	0.36
0.69	vapor + liquid	0.001	0.83	0.999	0.50
0.70	liquid			1.0	0.5
0.78	liquid			1.0	0.5
0.96	liquid			1.0	0.5

We plot these data on our phase map for benzene + toluene mixtures, in Figure 4.31. At 80°C for 50:50 benzene:toluene the first drop of liquid forms at 0.47 atm. At 80°C for 50:50 benzene:toluene the last bubble condenses at 0.69 atm. In contrast to pure substances, the start and end of the liquid–vapor coexistence are at different pressures, at a given temperature.[1]

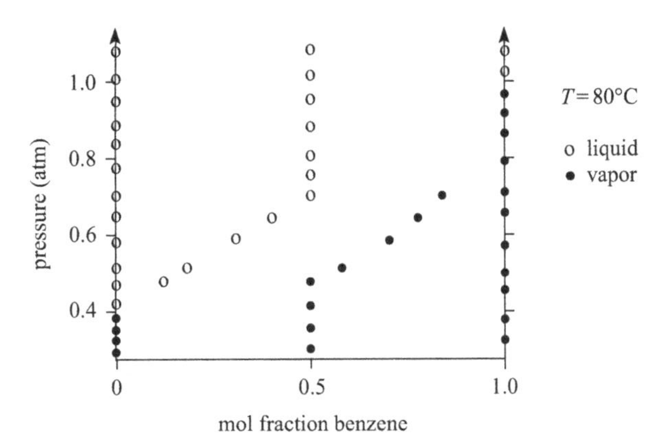

Figure 4.31 Phase map for benzene + toluene mixtures with phase data for pure benzene, pure toluene, and 50:50 benzene:toluene.

We also conduct experiments with 25:75 benzene:toluene and 75:25 benzene:toluene. These data are added to our phase map in Figure 4.32 (shown on the next page).

The data begin to define the boundaries on our map. After experiments with more compositions, the map in Figure 4.33 on the next page emerges.

The map in Figure 4.33 has a two-phase region, which corresponds to mixtures of liquid and vapor. The two-phase *area* on our map is similar to the two-phase *line segment* on the number line for energy balances in Figures 4.15 and 4.16. If a mixture's composition and pressure lie in the two-phase region, then liquid and vapor coexist. The compositions of the liquid and vapor phases are found by drawing a horizontal tie line from the liquid border to the vapor border, as illustrated in Figure 4.34. The tie line must be horizontal because the liquid and vapor phases are at equilibrium, and thus are at the same pressure.

[1] Some mixtures form an azeotrope at certain compositions for which the start and end of the liquid–vapor coexistence are at the same pressure at a given temperature. Ethanol and water form an azeotrope at 89.4 mol% ethanol.

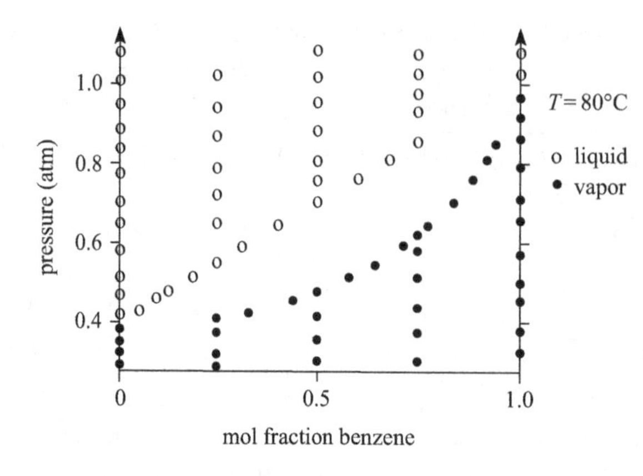

Figure 4.32 Phase map for benzene + toluene mixtures with phase data for pure benzene, pure toluene, 25:75 benzene:toluene, 50:50 benzene:toluene, and 75:25 benzene:toluene.

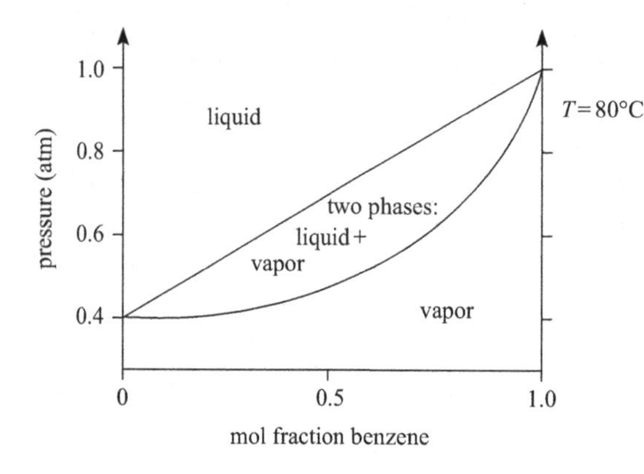

Figure 4.33 The pressure–composition phase map for benzene + toluene mixtures at 80°C.

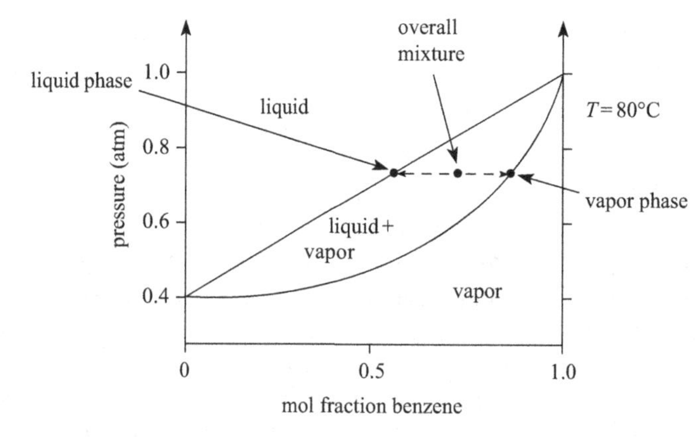

Figure 4.34
Finding the liquid and vapor compositions for a coordinate in the two-phase liquid + vapor region.

Figure 4.34 is useful for defining two terms: the dew point and the bubble point. If we increase the pressure of a vapor mixture, we follow a vertical path upward until the first iota of liquid forms. This is the dew point for this particular composition. If we continue to increase the pressure, the system is represented by tie lines across the liquid + vapor region until the last iota of vapor remains. This is the bubble point for this particular composition. Note that the dew point and bubble point are not unique points, as their names might suggest. Rather the dew point can be anywhere on the border between vapor and liquid + vapor. This border can also be called the dew point line. Likewise, the bubble point can be anywhere on the border between liquid and liquid + vapor. This border can also be called the bubble point line. For pure substances, the temperature of the dew point is the same as the temperature of the bubble point, and both equal the boiling point.

The mol fraction of component i in the liquid phase is defined as x_i. The mol fraction of component i in the vapor phase is defined as y_i. It follows that

$$x_{\text{benzene}} \equiv \frac{\text{mols of benzene in the liquid phase}}{\text{total mols in the liquid phase}} \tag{4.16}$$

$$y_{\text{benzene}} \equiv \frac{\text{mols of benzene in the vapor phase}}{\text{total mols in the vapor phase}}. \tag{4.17}$$

To find the relative amounts of the liquid and vapor we apply the lever rule to the tie line. Consider a magnified view of Figure 4.34's tie line, shown in Figure 4.35, and apply the lever rule.

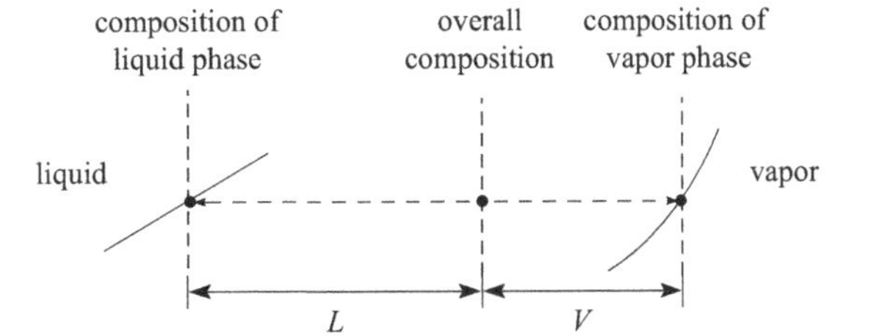

Figure 4.35 Determining the liquid and vapor compositions for a coordinate in the liquid + vapor region.

The length of the tie line segment to the liquid border is defined as L. The length of the tie line segment to the vapor border is defined as V. The total length of the tie line is T, so $T = L + V$. The amounts of the liquid and vapor phases are

$$\text{mol fraction liquid} = \frac{\text{mols in the liquid phase}}{\text{total mols in the system}} = \frac{V}{T} \tag{4.18}$$

$$\text{mol fraction vapor} = \frac{\text{mols in the vapor phase}}{\text{total mols in the system}} = \frac{L}{T}. \tag{4.19}$$

Let's apply equations 4.18 and 4.19 to the data measured for the 50:50 benzene:toluene mixture. We started with all vapor and increased the pressure. At 0.47 atm a tiny drop of liquid condensed. A tie line at this pressure has $L \gg V$. From equations 4.18 and 4.19 we calculate that the system is mostly vapor. The tie line and the experimental data agree. As we increased the pressure, more liquid formed and the tie line shifted to the right. Eventually $L \approx V$ and there are equal amounts (in mols) of liquid and vapor. Finally, at 0.69 atm, only a tiny amount of vapor remains. And the tie line is such that $L \ll V$. From equations 4.18 and 4.19 we calculate that the system is mostly liquid. Again the tie line and the experimental data agree.

A constant-temperature flash drum can be analyzed with the pressure–composition phase map for benzene + toluene mixtures and the lever rule. For example, we can determine the pressure needed to produce a vapor with 75 mol% benzene, the output flow rates, and the composition of the liquid stream.

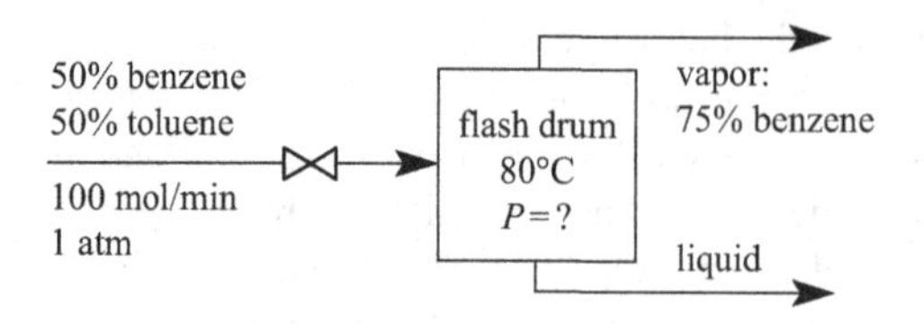

Figure 4.36 A low-pressure flash drum.

The graphical solution is shown in Figure 4.37. Because the vapor is 75 mol% benzene, we know the tie line (the solid horizontal line across the liquid + vapor region) must intersect the vapor border at $y_{\text{benzene}} = 0.75$. The endpoint of the tie line on the liquid border gives the composition of the liquid phase: $x_{\text{benzene}} = 0.38$. Extending to the ordinate axis reveals the pressure: 0.62 atm. With a ruler we measure $T = 23$ mm, $L = 8$ mm, and $V = 15$ mm. From equations 4.18 and 4.19 we calculate flow rates of $100 \times 8/23 = 35$ mol/min in the vapor and $100 \times 15/23 = 65$ mol/min in the liquid.

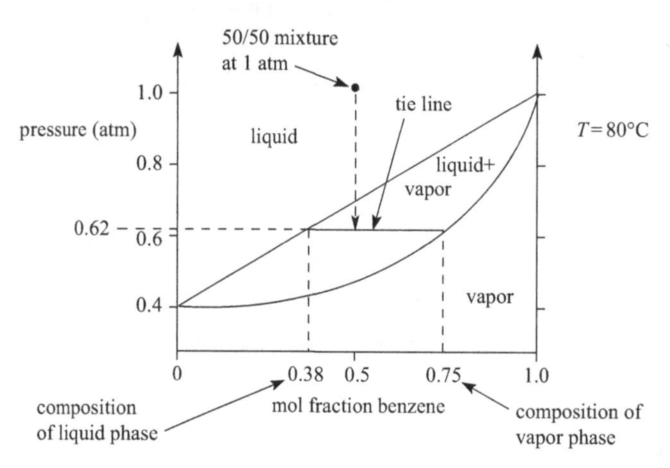

Figure 4.37 Graphical analysis of the low-pressure flash drum at 80°C.

Let's check our graphical analysis with a mol balance on benzene. (Note that a mol balance obtains from a mass balance with the conversion factor, for example, 1 mol benzene/78 g benzene.) We have

$$\text{molar flow rate of benzene in} \stackrel{?}{=} \text{molar flow rate of benzene out} \tag{4.20}$$

$$(100\,\text{mol/min})(0.50) \stackrel{?}{=} (65\,\text{mol/min})(0.38) + (35\,\text{mol/min})(0.75) \tag{4.21}$$

$$50\,\text{mol/min} \approx 51\,\text{mol/min}. \tag{4.22}$$

Our graphical analysis agrees with a mathematical analysis.

What if we want a higher mol fraction of benzene in the vapor stream? Our map tells us that the highest benzene content we can achieve with this design is about 83 mol%, and at the maximum concentration, the lever rule tells us that the flow rate of the vapor distillate is nearly zero.

A liquid stream may be flashed to a liquid + vapor mixture by increasing the temperature at constant pressure, as shown in Figure 4.38.

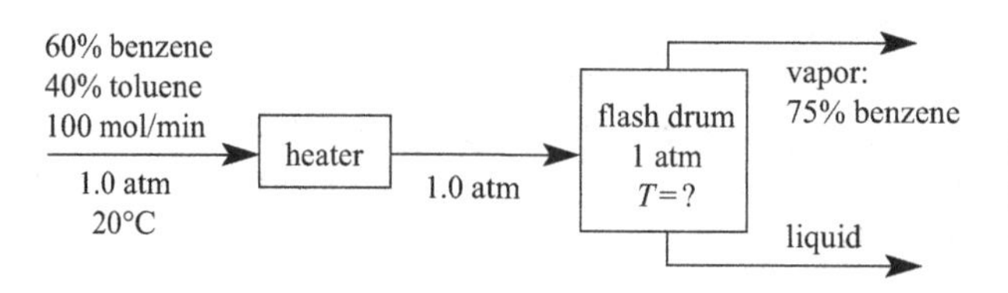

Figure 4.38 A constant-pressure flash drum.

To analyze a constant-pressure flash drum, we need a different phase diagram. We need a phase diagram with composition and temperature as the independent variables, and pressure constant at 1 atm. We conduct new experiments with the device in Figure 4.26 and systematically increase the temperature while maintaining the pressure at 1 atm. We plot the data and obtain the temperature–composition phase diagram shown in Figure 4.39.

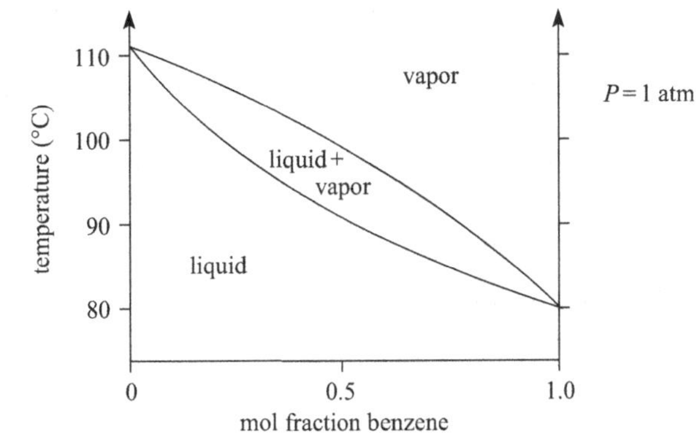

Figure 4.39 The temperature–composition phase diagram for benzene + toluene mixtures at 1 atm.

It is good practice to examine a new map before navigating with the map. Like the pressure–composition phase diagram, the temperature–composition phase diagram has three regions: liquid, vapor, and two-phase liquid + vapor. For pure benzene (right axis), the transition from liquid to vapor (the boiling point) is at 80°C, as expected. For pure toluene (left axis), the transition from liquid to vapor is at 111°C, also as expected. Contrary to the pressure–composition map, the vapor region is above the two-phase region and the liquid region is below the two-phase region. For a benzene + toluene system corresponding to a point in the two-phase liquid + vapor region, the two phases are found by drawing a horizontal tie line, and the relative amounts are calculated by applying the lever rule. The tie line is horizontal because the liquid and vapor phases are at equilibrium, and thus are the same temperature.

To find the operating temperature for the flash drum in Figure 4.38, we locate 75 mol% benzene on the vapor border and draw a horizontal tie line, as shown in Figure 4.40. Draw a horizontal tie line to the liquid border; this is the liquid phase. The composition of the liquid phase is 50 mol% benzene. Extend the tie line to the left axis to read the temperature: 91°C.

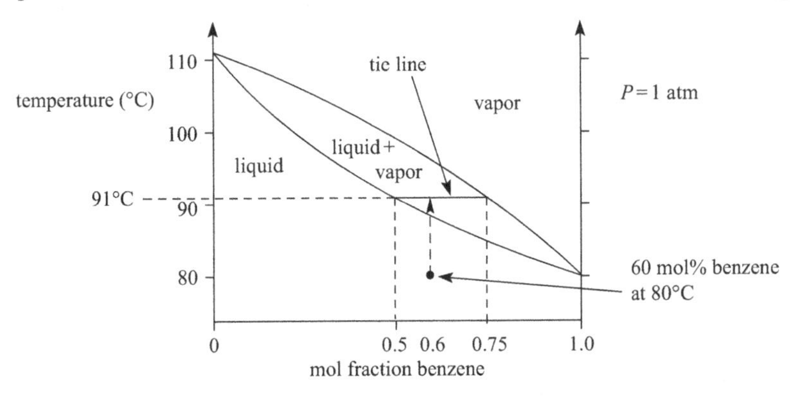

Figure 4.40 Analysis of the constant-pressure flash drum in Figure 4.38.

Calculate the flow rates. Place the fulcrum at the overall composition (60 mol% benzene) and apply the lever rule to balance the tie line. The liquid stream is 60 mol/min and the vapor stream is 40 mol/min. Check these flow rates. Does it make sense that the flow rate of the liquid stream is larger than the flow rate of the vapor stream?

In summary, tie lines and the lever rule we developed for specific mass number lines in Section 4.1.1 may be applied to analyze the compositions and flow rates for single-stage liquid–vapor separators, called flash drums. The tie lines must be parallel to the specific mass line (the x axis), which is assured because the liquid and vapor phases are in equilibrium, and thus at the same temperature and pressure. We encourage you to repeat the graphical analyses in Figures 4.37 and 4.40. Download quantitative pressure–composition and temperature–composition phase diagrams for benzene + toluene mixtures from duncan.cbe.cornell.edu/Graphs, carefully draw the tie lines, and apply the lever rule.

4.1.4 Combined Mass and Energy Balances on Two-Component Mixtures

Context: Benzene + toluene mixtures.

Concepts: Enthalpy–composition (H–(x,y)) phase maps. Lines of constant temperature (isotherms).

Defining Question: Why are tie lines on H–(x,y) phase maps not horizontal or even parallel?

The temperature–composition phase diagram in Figure 4.39 can be converted to a map for combined mass balances and energy balances. To accommodate mass balances our new map will have the same scale on the x axis: mol fraction benzene. To accommodate energy balances we convert the y axis from a temperature scale to a specific-enthalpy scale, in units of kJ/mol. The blank map is shown in Figure 4.41.

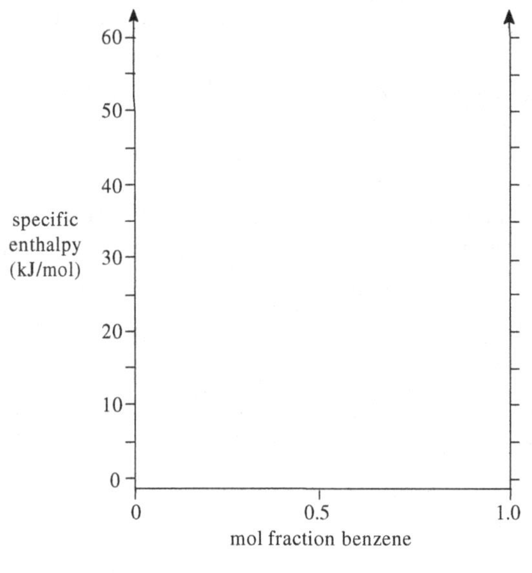

Figure 4.41 The coordinate system required for combined mass balances and energy balances.

We need to calibrate the specific-enthalpy axes with temperatures and phases, as we did for the energy lines in Section 4.2. The vertical axis on the right side of Figure 4.41 is an energy line for pure benzene. For benzene, C_P = 138 J/(mol·°C) for liquid in the range 20°C to 80°C, C_P = 121 J/(mol·°C) for vapor in the range 80°C to 250°C, and $\Delta\bar{H}_{vap} = 30.8\,\text{kJ/mol}$ at 80°C. We use the method that produced the temperature and phase scales for H_2O (see Figure 4.13); we arbitrarily assign $\bar{H} = 0$ to benzene at 20°C, which places the upper and lower limits of benzene's two-phase liquid + vapor region at 8.3 kJ/mol and 39.1 kJ/mol. A temperature and phase scale for benzene is added to the right side of the map in Figure 4.42. For toluene C_P = 170 J/(mol·°C) for liquid in the range 20°C to 111°C, C_P = 147 J/(mol·°C) for vapor in the range 111°C to 200°C, and $\Delta\bar{H}_{vap} = 33.5\,\text{kJ/mol}$ at 111°C. We arbitrarily assign $\bar{H} = 0$ to toluene at 20°C, so we calculate that the upper and lower limits of toluene's two-phase liquid + vapor region are at 15.5 kJ/mol and 48.9 kJ/mol. A temperature and phase scale for toluene is added to the right side of the map in Figure 4.42.

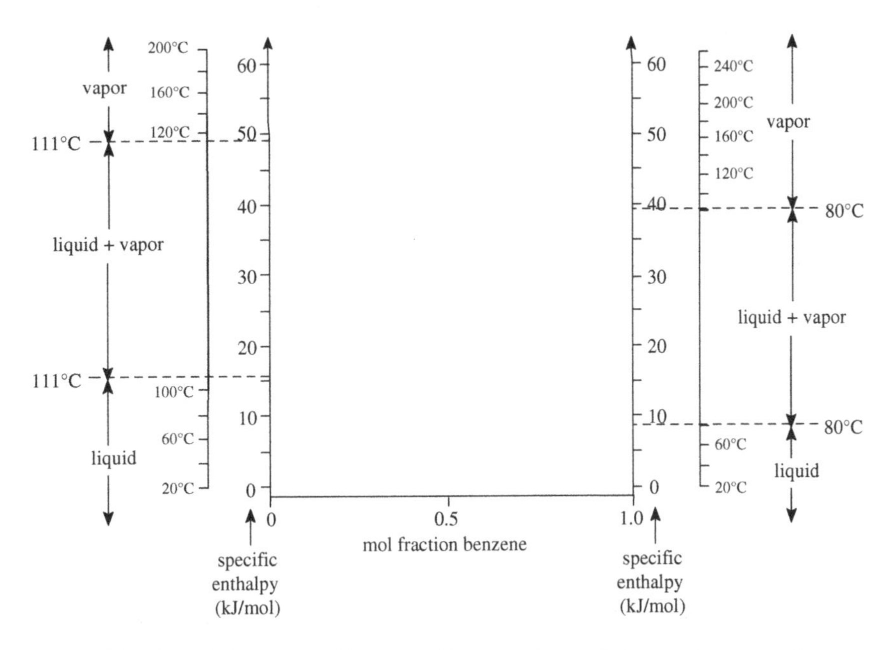

Figure 4.42 An enthalpy–composition map with temperature and phase scales for pure benzene and pure toluene.

We can now create temperature and phase scales for mixtures of benzene and toluene. If we assume benzene and toluene form an ideal mixture (a reasonable assumption in this case), the phase borders and the lines of constant temperature (isotherms) are straight lines. The border between the liquid region and the two-phase liquid + vapor region is a line from the left axis at 15.5 kJ/mol (toluene liquid at its boiling point, 111°C) to the right axis at 8.3 kJ/mol (benzene liquid at its boiling point, 80°C), shown in Figure 4.43 on the next page. The border between the vapor region and the two-phase liquid + vapor region is a line from the left axis at 48.9 kJ/mol (toluene vapor at its condensation point, 111°C) to the right axis at 39.1 kJ/mol (benzene vapor at its condensation point, 80°C). Draw an isotherm at 20°C; connect toluene liquid at 20°C ($\bar{H} = 0$) to benzene liquid at 20°C ($\bar{H} = 0$). This is the straight, horizontal dashed line at $\bar{H} = 0$. Draw an isotherm at 60°C; connect toluene liquid at 60°C ($\bar{H} = 6.8\,\text{kJ/mol}$) to benzene liquid at 60°C ($\bar{H} = 5.5\,\text{kJ/mol}$). This straight line has a slight negative slope because the heat capacity for liquid toluene is larger than the heat capacity for liquid benzene. Similarly, we can connect any temperature from the toluene scale to the same temperature on the benzene scale by a straight line, but we must connect liquid to liquid and vapor to vapor. In Figure 4.43 we add isotherms in the vapor region at 120°C, 160°C, and 200°C.

Let's check that the specific-enthalpy–composition map in Figure 4.43 makes sense. Let's calculate the energy needed to evaporate a 50:50 mixture at its boiling point. This is the distance from the liquid border to the vapor border along a vertical line at $x_B = 0.5$. We measure the distance with a ruler: 37 mm. We calibrate the energy scale: 60 kJ/kg = 68 mm. Thus 37 mm × (60 kJ/kg)/68 mm = 33 kJ/mol. We use a mathematical energy balance and

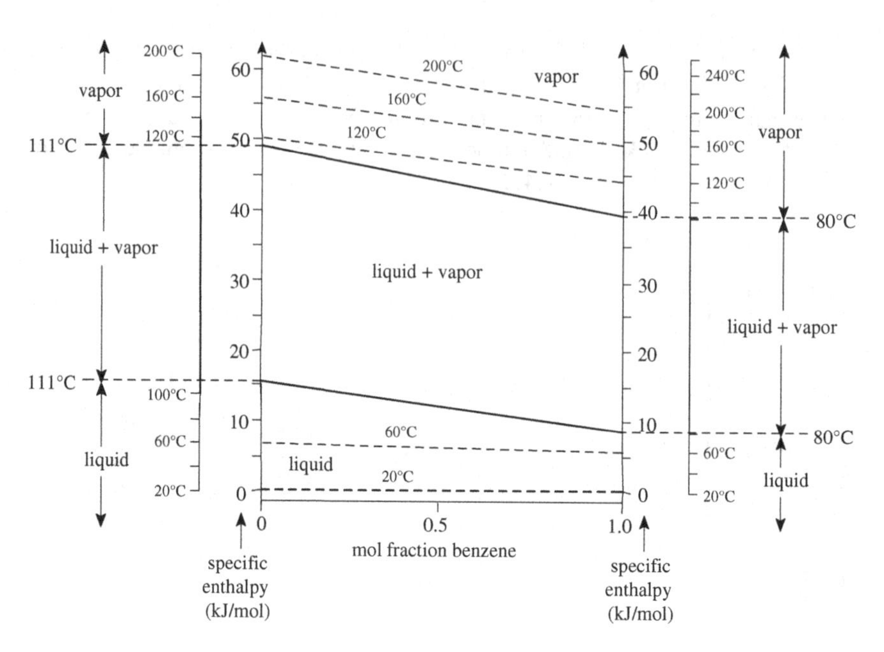

Figure 4.43 An enthalpy–composition map with temperature and phase scales for pure benzene and pure toluene.

calculate the energy needed to evaporate 0.5 mol benzene at 80°C plus the energy needed to evaporate 0.5 mol toluene at 111°C. We calculate 32 kJ. Our map makes sense. Note: this check is valid because we assume benzene and toluene form an ideal mixture.

If we start with a 50:50 mixture at 20°C and add energy, what is the temperature of the mixture when it reaches the two-phase liquid + vapor border; what is the bubble point of a 50:50 mixture? And what is the temperature of the 50:50 mixture when we complete the journey across the two-phase liquid + vapor region and reach the vapor border; what is the dew point of a 50:50 mixture? These questions are answered easily with the temperature–composition map in Figure 4.39. We need to transfer information from Figure 4.39 to our enthalpy–composition map.

We align the temperature–composition map with our enthalpy–composition map (they are conveniently the same width), as shown in Figure 4.44 on the next page. Draw the tie line at 100°C on the temperature–composition phase diagram. Transfer this tie line to the enthalpy–composition diagram by drawing vertical lines downward; connect the bubble point line to the bubble point line and connect the dew point line to the dew point line. Draw a tie line between these points on the enthalpy–composition diagram; this is a tie line at 100°C. Unlike previous tie lines, this tie line is not horizontal. Liquid and vapor in equilibrium have the same temperature, but not the same enthalpy. Next, we extend the 100°C isotherm across the vapor region by connecting the intersection on the dew point line with the right axis at 100°C. Finally, we extend the 100°C isotherm across the liquid region by connecting the intersection on the bubble point line with the left axis at 100°C.

We use the same graphical method to draw tie lines across the two-phase region at 85°C, 90°C, 95°C, and 105°C as shown in Figure 4.45 two pages hence. There are two other implicit tie lines across the two-phase liquid + vapor region. The tie line at the boiling point of benzene (80°C) lies on the border on the right side. The tie line at the boiling point of toluene (111°C) lies on the border on the left side (the *y* axis). Note that the tie lines are not parallel. Starting on the left side, the tie line at 111°C is exactly

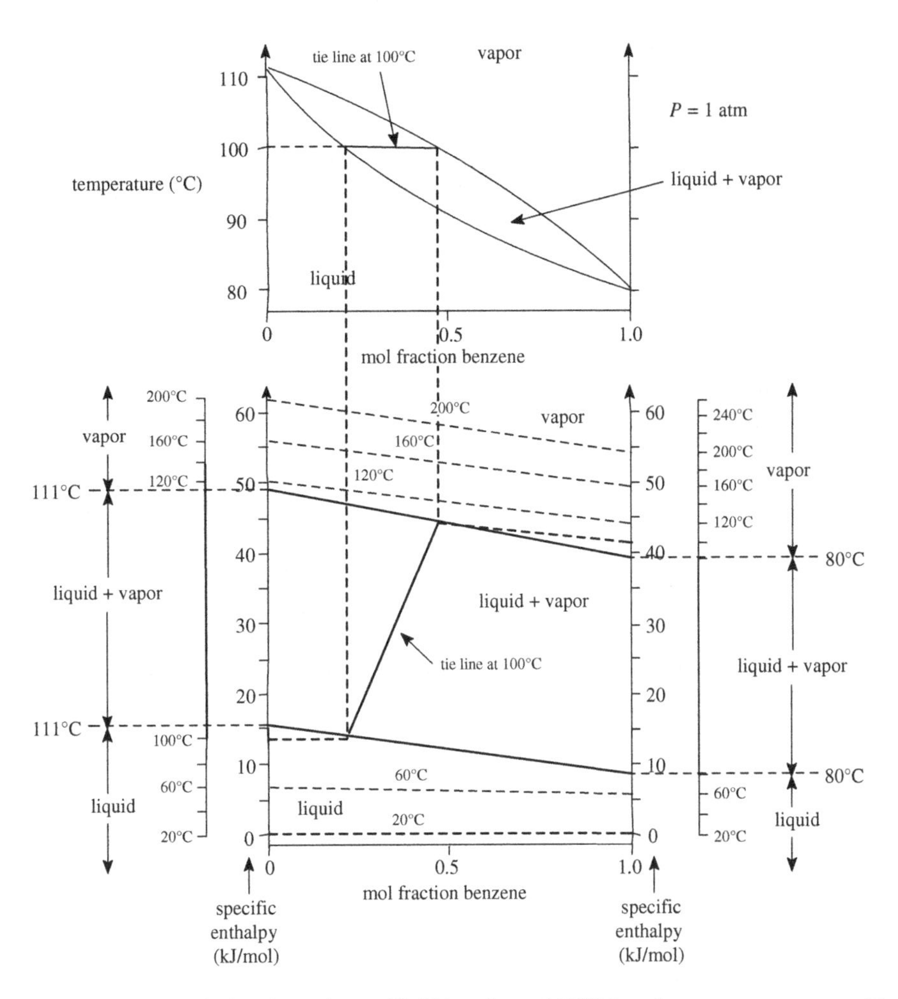

Figure 4.44 Transferring dew points and bubble points at 100°C from the temperature–composition map to the enthalpy–composition map.

vertical (pure toluene). The tie lines slant away as temperature decreases, finally returning to exactly vertical at 80°C (pure benzene).

Let's use the enthalpy–composition diagram to determine the energy flow rate into the heater in Figure 4.38. Our starting point is the intersection of a vertical line at 60 mol% benzene and the isotherm at 20°C, as shown in Figure 4.46 on the next page. Our destination is a tie line across the liquid + vapor region with an endpoint on the vapor border at 75 mol% benzene. The tie line is about 1/5 the distance between the tie lines at 90°C and 95°C. The temperature of the mixture is 91°C, consistent with the graphical analysis in Figure 4.40. Measure the vertical distance from the starting point to the fulcrum of the tie line: 21 mm. Convert to kJ/mol: 21 mm × (60 kJ/kg)/53 mm = 24 kJ/kg. Because the flow rate is 100 mol/min, the energy demand in the heater is 2400 kJ/min.

One can also apply the lever rule to the tie line in Figure 4.46. The fulcrum is at 60 mol% and the endpoints are at 50 mol% and 75 mol%. One can measure the tie line lengths by projecting the line onto the x axis or the y axis. Or one can lay a ruler on the tie line and measure the lengths. We obtain 60 mol/min liquid and 40 mol/min vapor. Check the result by verifying that the lever would appear balanced with 60 mol on the liquid end and 40 mol on the vapor end. It helps to rotate your textbook so the tie line appears horizontal.

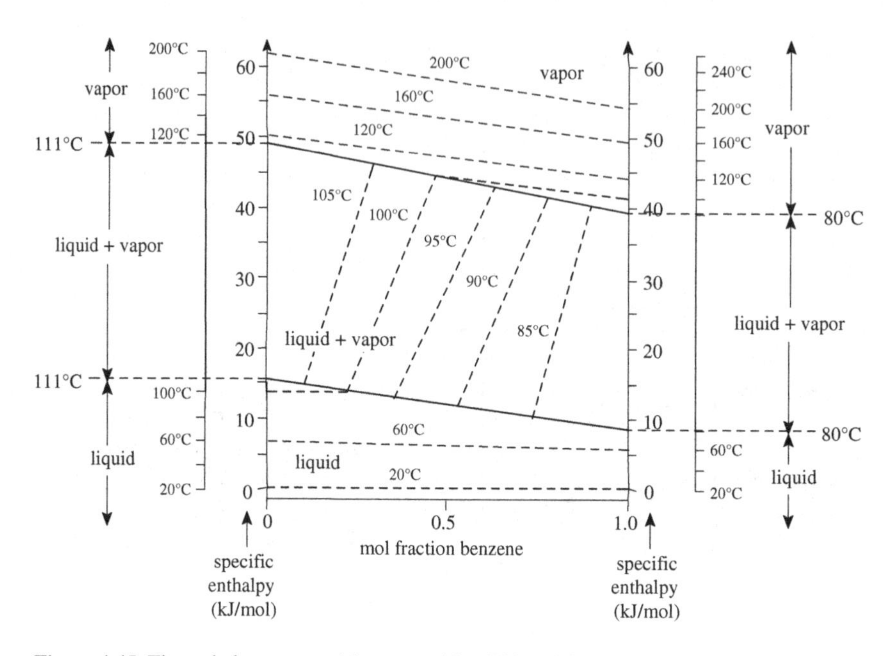

Figure 4.45 The enthalpy–composition map with additional tie lines across the two-phase liquid + vapor region.

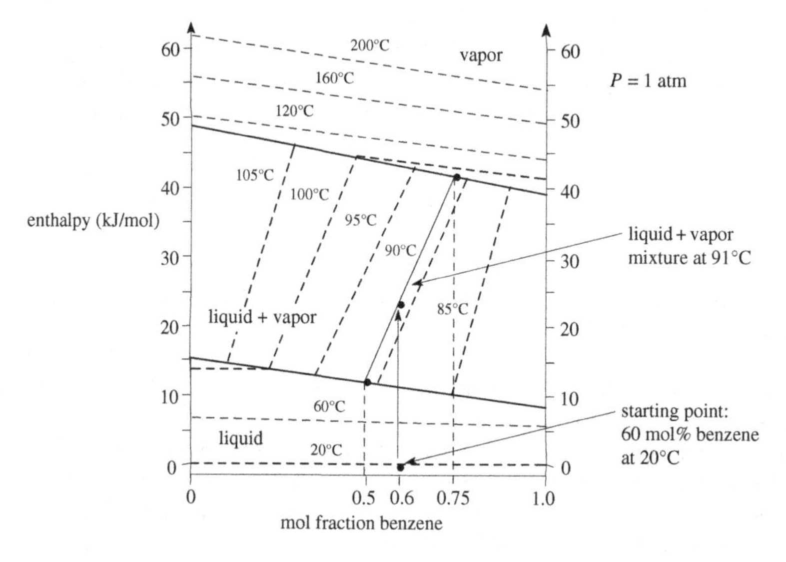

Figure 4.46 An graphical construction to obtain the energy flow rate into the heater in Figure 4.38.

In summary, combined mass and energy balances can be obtained graphically on a map with conserved quantities on two perpendicular axes. In this case, we developed enthalpy–composition maps with specific mass on the x axis and specific enthalpy on the y axis. Because both axes are conserved quantities, tie lines and mixing lines may be drawn at any angle. Again the lever rule applies.

Enthalpy–composition diagrams are especially useful for combined mass balances and energy balances on non-ideal mixtures, as evidenced by curved borders for the liquid + vapor region, such as ammonia–water mixtures (see Exercise 4.34) and sulfuric acid–water mixtures (see Exercise 4.36).

We created the enthalpy–composition phase diagram from thermodynamics parameters and by assuming an ideal mixture. Of course, we could have conducted experiments to prepare the phase diagram. We could have loaded 1 mol of a benzene + toluene mixture into a cell at 20°C, added 1 kJ, and then recorded the temperature and phase(s). If we repeat this 60 times, we would have data for the scale 0 to 60 kJ/mol. We would then load the cell with a different composition, start at 20°C, and repeat.

4.2 OPERATING LINES FOR TWO-PHASE SYSTEMS

4.2.1 Single-Stage Absorbers

Context: Absorbing benzene from air into oil.

Concepts: Maps for two-phase systems. Calculating a path to equilibrium.

Defining Question: What determines the slope of a path to equilibrium?

Another important separation method is absorption – a gaseous substance is absorbed from a non-condensible gas into a non-volatile liquid. An example is the absorption of benzene vapor from air into an oil. During absorption, the substance moves from the gas phase to the liquid phase. This movement is an example of *mass transfer*. To move between phases, the pollutants must cross the liquid–gas interface. Increasing the interfacial area increases the rate of mass transfer. An easy way to increase the interfacial area is to bubble the gas through the liquid. This concept is effective in absorbing oxygen from air into water in a fish tank, and it is effective in absorbing a pollutant such as benzene from air into oil.

Contaminated air is bubbled through oil and the oil absorbs benzene, as shown in Figure 4.47. For now we will ignore the rate of mass transfer from the gas phase to the liquid phase. We will assume the bubble size and the time a bubble spends in the oil are such that mass transfer is complete when a bubble reaches the top of the absorber. In other words, we will assume the air + oil + benzene system reaches equilibrium in our absorber. But note: when mass transfer is complete there will still be benzene in the air. How much benzene? That is what we will calculate in this section.

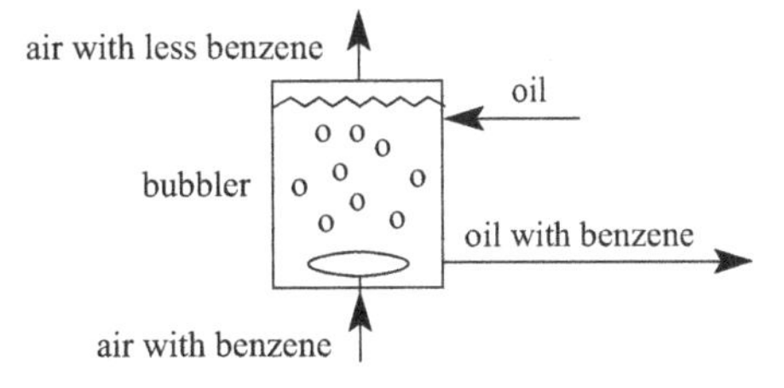

Figure 4.47 An absorber for removing benzene from air.

But first, we complete our absorber flowsheet. Of course, the oil should be recycled. Because the oil is non-volatile, it is a simple matter to distill benzene vapor from the oil as shown in Figure 4.48 on the next page.

And, of course, one would want to consider adding a heat exchanger so the hot oil leaving the distiller could heat the oil + benzene mixture entering the distiller as shown in Figure 4.49 on the next page.

We assume the bubbler is designed such that streams 2 and 4 are in thermodynamic equilibrium. Given a flow rate of polluted air, say 1 kg/min, what flow rate of oil is needed to reduce the benzene content of the air to an acceptable level? It is illustrative to begin with a qualitative analysis. We start with extremes – assume the air leaving the bubbler contains *no* benzene. Thus all the benzene that enters with the polluted air leaves with the oil. Assume the flow rate of benzene is 0.01 kg/min in both streams 1 and 4 and the equilibrium concentration of benzene in the oil is 1 kg benzene/100 kg oil. The flow rate of oil is therefore 1 kg/min. If a different oil can hold twice as much benzene, the flow rate of that oil would be half as large, and so on. We need to know the oil's capacity for absorbing benzene.

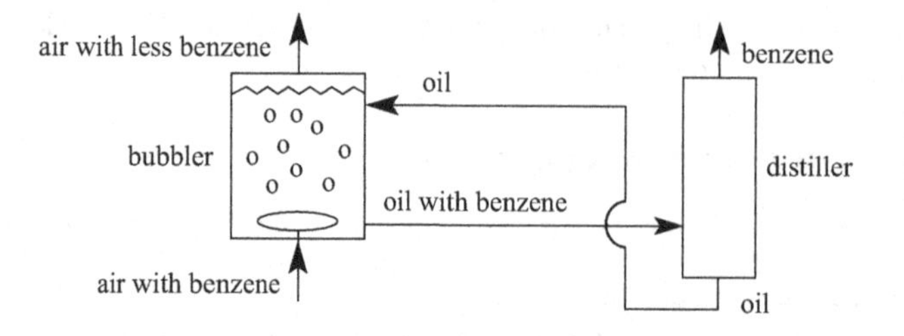

Figure 4.48 An absorber for removing benzene from air, with oil recycle.

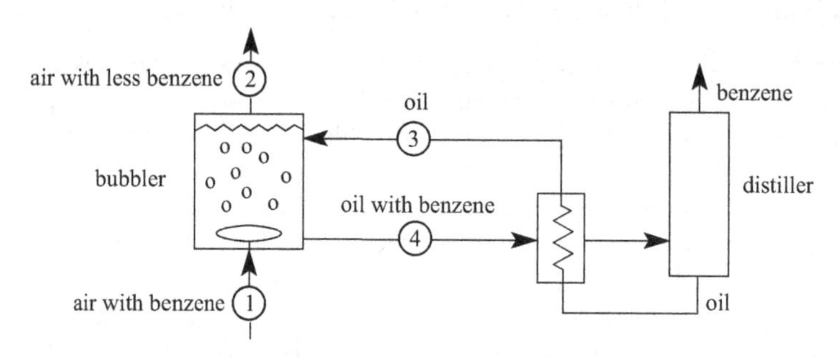

Figure 4.49 An absorber for removing benzene from air, with oil recycle and heat recovery.

Assume no data are available, so we perform some simple experiments. We fill a container with air and oil. We then inject a known quantity of benzene into the container and wait for the benzene to equilibrate between the two phases. We then measure the concentration of benzene in the air (or in the oil) and calculate the other concentration using a mass balance. (You might verify you could calculate this.)

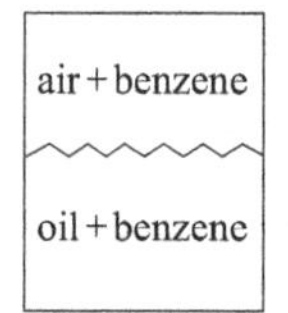

Figure 4.50 An apparatus for measuring the absorption of benzene in an oil.

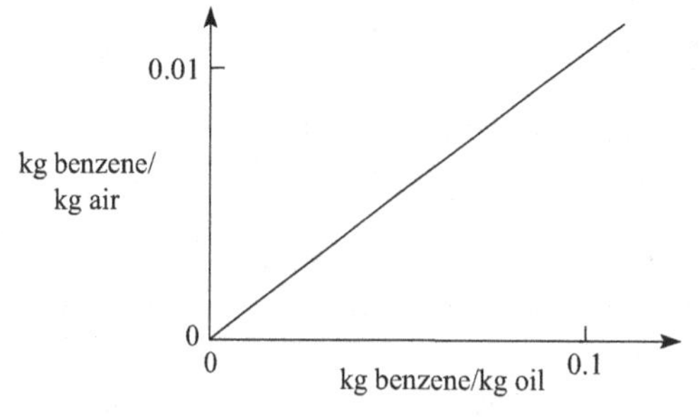

Figure 4.51 A map of the vapor–liquid two-phase system of (air + benzene)–(oil + benzene).

We inject more benzene into the container and repeat the process with the temperature constant. We plot the data as shown above in Figure 4.51, which looks very much like a standard x–y graph. That is, given a weight ratio of benzene in oil, the line indicates the weight ratio of benzene in air, for a system at equilibrium.

But Figure 4.51 is actually a map, with three regions. The first region is the line – this represents systems at equilibrium. What is the region below the line? Obviously, the region corresponds to systems not at equilibrium. We have two choices for non-equilibrium: excess benzene in the oil or excess benzene in the air. Mark an arbitrary point in the region below the equilibrium line. How will a system at this point move toward equilibrium? It will move upward and to the left. How do these directions translate on this map? "Upward" means the system will increase the amount of benzene in the air. "To the left" means the system will decrease the amount of benzene in the oil. Thus the region below the line represents "excess benzene in the oil." Similarly, the region above the equilibrium line represents "excess benzene in the air." These regions are labeled on the map in Figure 4.52.

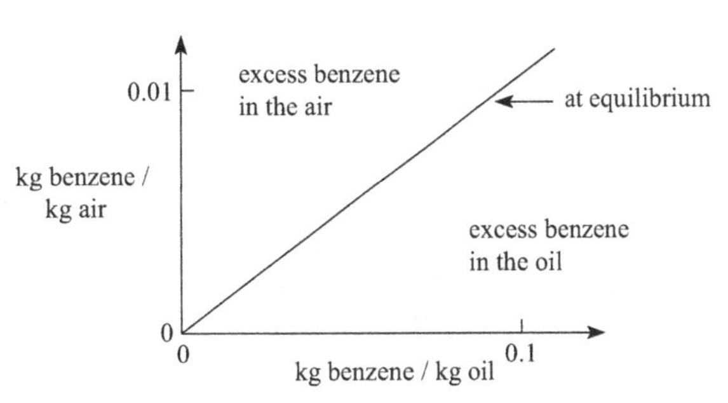

Figure 4.52 Labeled regions on a map of the vapor–liquid two-phase system of (air + benzene)–(oil + benzene).

Compare the map in Figure 4.52 to the pressure–composition map (Figure 4.33), or the temperature–composition map (Figure 4.39). Every point on the map in Figure 4.52 corresponds to a two-phase system, but only points on the line correspond to equilibrium. Every point on the pressure–composition map and the temperature–composition map corresponds to a system at equilibrium, but only points in the liquid + vapor region correspond to two-phase systems. (Test your understanding of this concept – which type is the enthalpy–composition map in Figure 4.45?) Because the map in Figure 4.52 is inherently different, we will use a different graphical method. Rather than tie lines, we will use operating lines.

Before we analyze the bubbler, let's start with a simpler system: a closed system of air, oil, and benzene. Assume there is initially no benzene in the oil. We begin by locating our starting point on the map in Figure 4.53. We choose an arbitrary point on the ordinate. Check this – is our starting point in the correct region of the map? Yes – it is in the region "excess benzene in the air." What will be the concentrations of benzene in the air and in the oil at equilibrium? The system must move to the equilibrium line. But where on the equilibrium line? What else do we need to know?

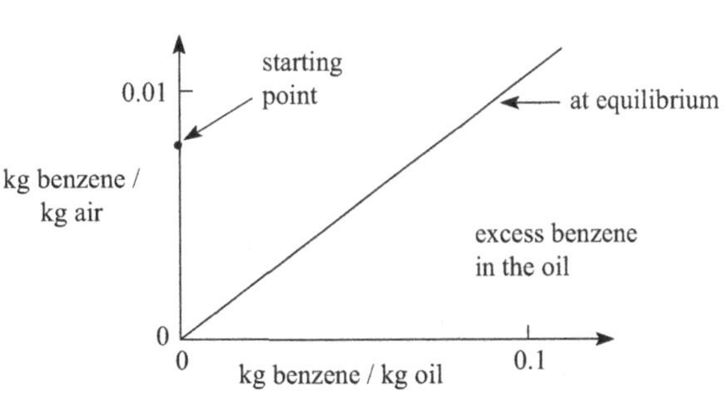

Figure 4.53 Starting point on the map for a closed system with benzene in the air only; no benzene in the oil.

Again, let's consider extremes. Consider a system with ample air and only a tiny drop of oil. The tiny drop of oil absorbs an even tinier amount of benzene. When this system comes to equilibrium only a infinitesimal amount of benzene has left the air. The concentration of benzene in the air remains essentially constant. The path on the map is shown in Figure 4.54.

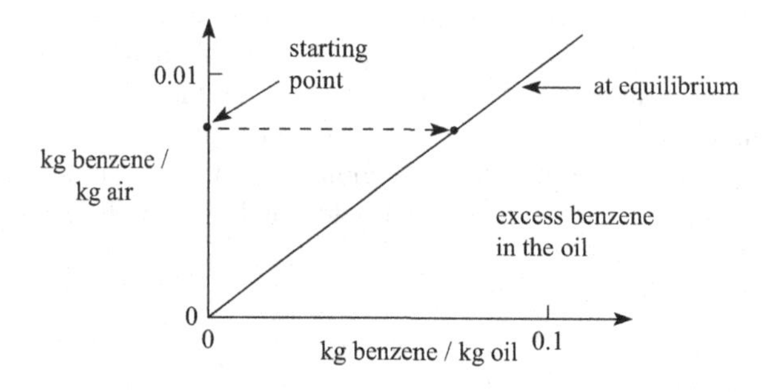

Figure 4.54 The path to equilibrium for copious air and a minuscule amount of oil.

Now consider the converse – a tiny amount of air and ample oil. The oil has a large capacity for benzene. The benzene content of the air will fall nearly to zero whereas the benzene content of the oil will increase barely because there is so much oil to dilute the benzene. This path, shown in Figure 4.55, follows the ordinate downward.

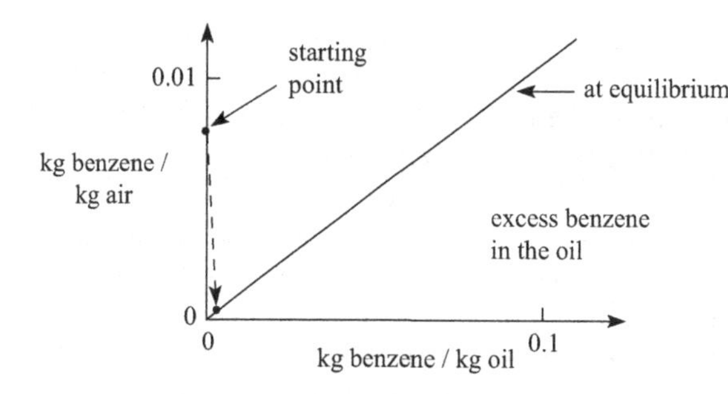

Figure 4.55 The path to equilibrium for a minuscule amount of air and copious oil.

Note what we have done to familiarize ourselves with the new map. We found our starting point and we explored some paths corresponding to extreme circumstances. This is a good practice to adopt.

Our quantitative analysis suggests the relative size of the air and oil phases will determine the path to the equilibrium line. Copious air corresponds to a horizontal path to the right. Copious oil corresponds to a vertical path downward. An intermediate ratio of air to oil would yield a path between these two extremes.

We now apply the conservation of mass to this closed system to calculate the slope of the path from the initial point to the equilibrium line. We need some nomenclature to represent quantities such as "kg benzene per kg air." Define M_X as the mass of component X. Conservation of mass requires that the amount of benzene in the system initially equals the amount of benzene in the system at any time later:

$$\left((M_{\text{benz}})_{\text{vapor}} + (M_{\text{benz}})_{\text{liquid}}\right)_{\text{initially}} = \left((M_{\text{benz}})_{\text{vapor}} + (M_{\text{benz}})_{\text{liquid}}\right)_{\text{at equilbrium}}. \quad (4.23)$$

We now multiply each term by strategically chosen ratios equal to 1 (such as $M_{\text{air}}/M_{\text{air}}$) and substitute $w_{\text{benz/air}}$ for the weight ratio of benzene to air and $w_{\text{benz/oil}}$ for the weight ratio of benzene to oil. This gives

$$\left((M_{\rm B})_{\rm vapor}\right)_{\rm init}\left(\frac{M_{\rm air}}{M_{\rm air}}\right)+\left((M_{\rm B})_{\rm liq}\right)_{\rm init}\left(\frac{M_{\rm oil}}{M_{\rm oil}}\right)$$
$$=\left((M_{\rm B})_{\rm vapor}\right)_{\rm eq}\left(\frac{M_{\rm air}}{M_{\rm air}}\right)+\left((M_{\rm B})_{\rm liq}\right)_{\rm eq}\left(\frac{M_{\rm oil}}{M_{\rm oil}}\right) \tag{4.24}$$

$$\left(w_{\rm benz/air}\right)_{\rm init}M_{\rm air}+\left(w_{\rm benz/oil}\right)_{\rm init}M_{\rm oil}=\left(w_{\rm benz/air}\right)_{\rm eq}M_{\rm air}+\left(w_{\rm benz/oil}\right)_{\rm eq}M_{\rm oil} \tag{4.25}$$

$$-\left(\left(w_{\rm benz/air}\right)_{\rm eq}-\left(w_{\rm benz/air}\right)_{\rm init}\right)M_{\rm air}=\left(\left(w_{\rm benz/oil}\right)_{\rm eq}-\left(w_{\rm benz/oil}\right)_{\rm init}\right)M_{\rm oil} \tag{4.26}$$

$$\frac{\left(w_{\rm benz/air}\right)_{\rm eq}-\left(w_{\rm benz/air}\right)_{\rm init}}{\left(w_{\rm benz/oil}\right)_{\rm eq}-\left(w_{\rm benz/oil}\right)_{\rm init}}=-\frac{M_{\rm oil}}{M_{\rm air}}. \tag{4.27}$$

We have assumed implicitly that no air dissolves in the oil and no oil evaporates into the air.

Let's plot the starting point and equilibrium point. A line from the initial point to the equilibrium point is the hypotenuse of a right triangle whose sides are the terms in the ratio of equation 4.27.

The numerator in equation 4.27 is the rise of our path to the equilibrium line. The denominator in equation 4.27 is the run of the path. The ratio, rise/run, is the slope of the line from the initial conditions to the equilibrium conditions. The slope of the line is $-M_{\rm oil}/M_{\rm air}$, the ratio of oil to air. Let's check if this equation agrees with our qualitative analysis. In the limit of almost no oil, $M_{\rm oil} \approx 0$, so $-M_{\rm oil}/M_{\rm air} \approx 0$, and a slope of 0 is a horizontal line. Check. In the limit of infinite oil, $M_{\rm oil} \approx \infty$, so $-M_{\rm oil}/M_{\rm air} \approx -\infty$, and a slope of $-\infty$ is a vertical line directed downward. Check.

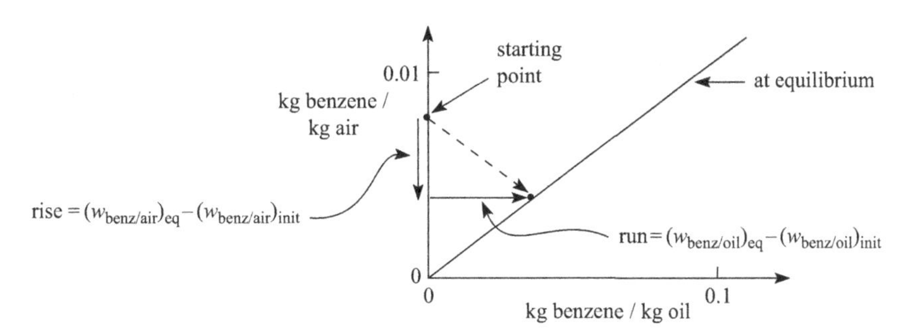

Figure 4.56 The path to equilibrium for $M_{\rm air}$ kg of air and $M_{\rm oil}$ kg of oil.

So we can predict graphically the equilibrium state of any non-equilibrium closed system of air, oil, and benzene. Note that equation 4.27 also applies to systems below the equilibrium line: systems with excess benzene in the oil.

In this section we developed a second type of thermodynamic map. T–(x,y), P–(x,y), and H–(x,y) maps have three regions: liquid, vapor, and liquid + vapor. Absorber maps represent liquid + vapor only. T–(x,y), P–(x,y), and H–(x,y) maps represent systems at equilibrium; if a system is not equilibrated these maps may not be used. Absorber maps apply to equilibrated systems as well as to non-equilibrated systems. Equilibrated systems lie on a line and non-equilibrated systems are in regions above and below this line. A point on a T–(x,y), P–(x,y), or H–(x,y) map represents a liquid or vapor stream. A point on an absorber map represents two streams: one liquid and one vapor.

Consequently graphical analysis on these two types of thermodynamic maps differs. T–(x,y), P–(x,y), and H–(x,y) maps use tie lines that connect two phases (liquid and vapor) in equilibrium and mixing lines that represent two phases mixed. Mass balances obtain from the lever rule. Absorber maps use trajectory lines drawn from a point for two streams initially (two streams entering a single-stage separator) to a point representing two streams at equilibrium (two streams leaving a single-stage separator). Mass balances obtain from the slope of the trajectory line.

4.2.2 Multistage Absorbers

Context: Absorbing benzene from air into oil, continued.

Concepts: Operating lines for absorbers and strippers.

Defining Question: Why is an operating line a convenient graphical tool for analyzing multistage cascade absorption?

Now, how about open systems at steady state, such as the bubbler? The derivation above can be repeated with mass flow rate F substituted for mass M, and the result is

$$\frac{\left(w_{\text{benz/oil}}\right)_{\text{out}} - \left(w_{\text{benz/oil}}\right)_{\text{in}}}{\left(w_{\text{benz/air}}\right)_{\text{out}} - \left(w_{\text{benz/air}}\right)_{\text{in}}} = -\frac{F_{\text{oil}}}{F_{\text{air}}}. \tag{4.28}$$

This equation assumes the two streams leaving the bubbler are in equilibrium.

Let's apply the design tool in equation 4.28 to a more complicated absorber: the two-stage absorber shown in Figure 4.57. Assume the oil-to-air flow rate ratio is the same in each bubbler. Use the graphical techniques to determine the concentration of benzene in the air leaving the second bubbler.

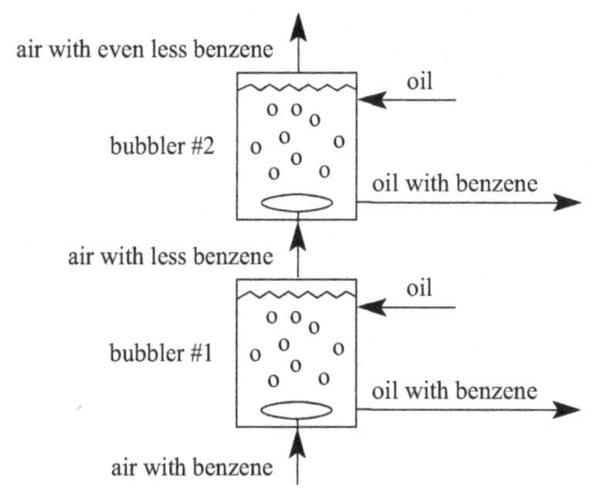

Figure 4.57 A two-stage absorber for absorbing benzene from air.

The first bubbler is trivial to analyze. It is the same problem we solved for the closed system. The solution is shown in Figure 4.58.

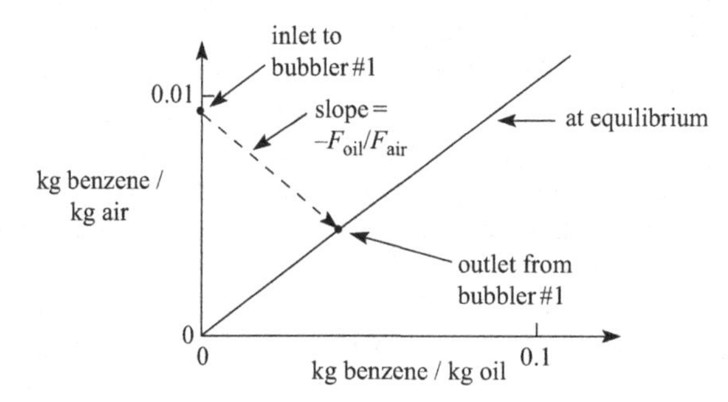

Figure 4.58 Graphical analysis of the first bubbler in a two-stage absorber.

But what is the initial point for the second bubbler? We need two coordinates: the mass ratio of benzene to the air entering bubbler #2 (which is given by the outlet from bubbler #1) and the mass ratio of benzene to oil entering bubbler #2 (which is zero). The inlet condition for bubbler #2 is obtained by moving horizontally from the equilibrium line to the y axis. From here we move toward the equilibrium line at a slope $-F_{oil}/F_{air}$, as shown in Figure 4.59.

How would one analyze a four-stage absorber designed like the process in Figure 4.57? Just add two more steps to the map in Figure 4.59. One can appreciate that this method is very efficient for exploring different designs or different flow rates. For example, given that one has a limited flow of fresh oil to share between the bubblers, how would you adjust the relative flow rates to minimize the benzene in the outlet air? Increase the flow to bubbler #1 and decrease the flow to bubbler #2? Or vice versa?

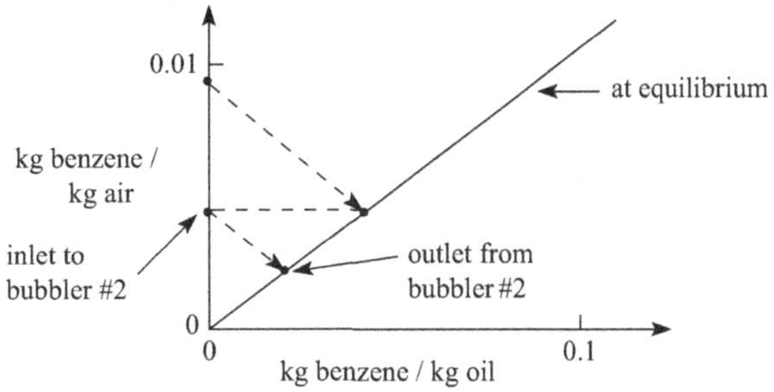

Figure 4.59 Graphical analysis of the second bubbler in a two-stage absorber.

As illustrated in Exercise 4.38, a multistage absorber removes more benzene from the air than does a single-stage absorber, given the same total flow of oil through both systems. However, an even better design exists. Can you guess what it is? Think of the counter-current mechanism of the heat exchanger we analyzed in Chapter 3. Each stream, the hot and the cold, entered the heat exchanger at only one port and exited at only one port. Analogously, a better multistage absorber does not inject fresh oil at each stage. Rather, the oil leaving the top stage is fed into the stage below, and so on to the bottom stage. Likewise, the air bubbles up through each stage in succession, as shown in Figure 4.60.

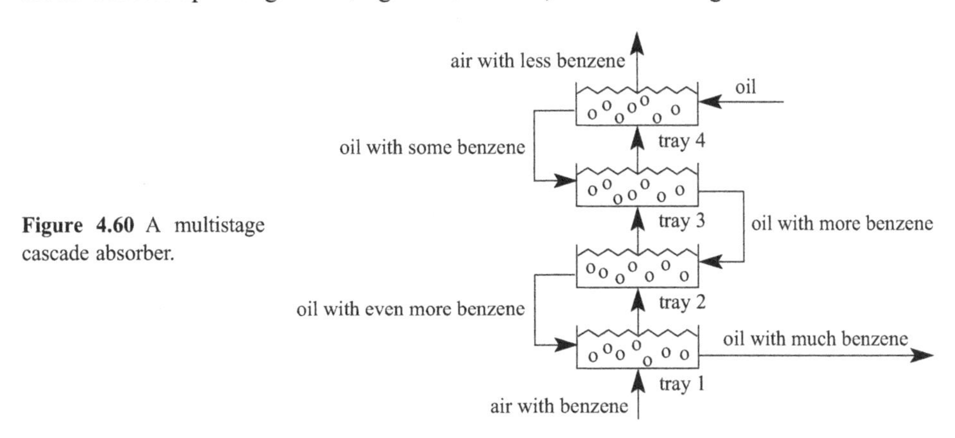

Figure 4.60 A multistage cascade absorber.

How do we analyze the process in Figure 4.60? We could use the graphical method developed for multistage absorbers, as shown in Figure 4.59. However, this method is awkward for analyzing *cascade* absorbers. To start at the bottom tray we need to know the concentration of benzene in the air entering the bottom tray (no problem) and the concentration of benzene in the oil entering the bottom tray (problem). We don't know the composition of the oil stream entering the bottom tray or the composition of any intermediate stream. We only know the compositions of the streams entering and leaving the overall process.

If we were to apply a mass balance to every stage in the cascade and then plot the results on our map, the pattern in Figure 4.61 would emerge. The inlet coordinates to the bottom tray give the upper point on the dashed line. The equilibrium condition leaving the bottom tray (tray 1) is the point on the equilibrium line labeled "1". And so on for trays 2, 3, and 4. The difficulty is that subsequent trays don't lie on the y axis and thus cannot be plotted without a complete analysis of the cascade unit. However, as suggested by the dashed line, these points lie on a line, called the operating line. This will always be the case. Before you can ask "why?" let's derive an equation for an operating line.

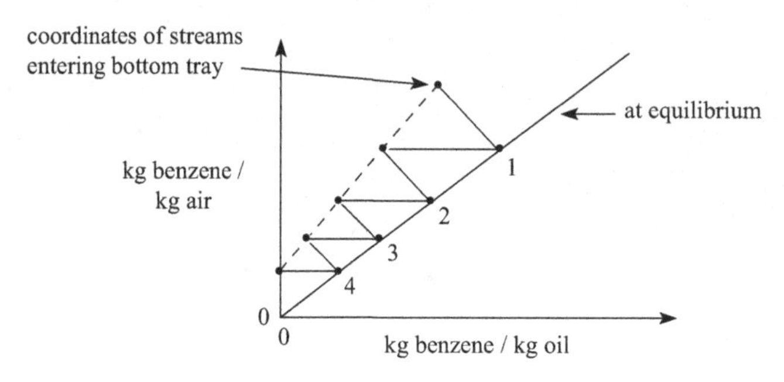

Figure 4.61 Analysis of a multistage cascade absorber.

To derive the equation for any line, we can use the coordinates of two points on the line. But we do not want to use the points in Figure 4.61, because we do not know those points. Instead, we want to plot points derived from what we know. Two known points are the (x,y) pair given by the outlet oil and the inlet air (bottom stage) and the pair given by the inlet oil and the outlet air (top stage). Let's derive an equation that contains these points. We start with a mass balance around the entire multistage absorber:

$$\text{rate of benzene in} = \text{rate of benzene out} \tag{4.29}$$

$$\text{benzene in polluted air} = \text{benzene in oil effluent} + \text{benzene in cleaned air} \tag{4.30}$$

$$\left((F_{\text{benz}})_{\text{vapor}}\right)_{\text{in}} = \left((F_{\text{benz}})_{\text{liquid}}\right)_{\text{out}} + \left((F_{\text{benz}})_{\text{vapor}}\right)_{\text{out}} \tag{4.31}$$

$$\left(w_{\text{benz/air}}\right)_{\text{in}}(F_{\text{air}})_{\text{in}} = \left(w_{\text{benz/oil}}\right)_{\text{out}}(F_{\text{oil}})_{\text{out}} + \left(w_{\text{benz/air}}\right)_{\text{out}}(F_{\text{air}})_{\text{out}}. \tag{4.32}$$

As before, we assume that air does not dissolve in oil and vice versa, so $F_{\text{in}} = F_{\text{out}}$ for air and oil. Equation 4.32 thus simplifies to

$$\left(w_{\text{benz/air}}\right)_{\text{in}} = \left(\frac{F_{\text{oil}}}{F_{\text{air}}}\right)\left(w_{\text{benz/oil}}\right)_{\text{out}} + \left(w_{\text{benz/air}}\right)_{\text{out}}. \tag{4.33}$$

If one repeats the mass balance on the top tray, one obtains a similar equation:

$$\left(w_{\text{benz/air}}\right)_{\text{into top tray}} = \left(\frac{F_{\text{oil}}}{F_{\text{air}}}\right)\left(w_{\text{benz/oil}}\right)_{\text{out of top tray}} + \left(w_{\text{benz/air}}\right)_{\text{out}}. \tag{4.34}$$

Indeed, if one performs a mass balance around a system that includes the top tray and the subsequent trays below, down to tray n, one finds a similar relationship:

$$\left(w_{\text{benz/air}}\right)_{\text{into nth tray}} = \left(\frac{F_{\text{oil}}}{F_{\text{air}}}\right)\left(w_{\text{benz/oil}}\right)_{\text{out of nth tray}} + \left(w_{\text{benz/air}}\right)_{\text{out}}. \tag{4.35}$$

Equations 4.33, 4.34, and 4.35 all have the form of a straight line, $y = mx + b$, where m is the slope and b is the y-intercept. The coordinates formed by the (x,y) pair ($w_{\text{benz/oil}}$ leaving a tray,

$w_{benz/air}$ entering a tray) lie on a straight line. This line has slope F_{oil}/F_{air} and y-intercept given by the weight ratio of benzene to air leaving the cascade absorber, $(w_{benz/air})_{out}$. This line is the operating line, although not the same operating line as shown in Figure 4.61.

Let's use the operating line to analyze a multistage cascade absorber. We begin in Figure 4.62 by drawing the operating line based on its slope, F_{oil}/F_{air}, and its intercept, $(w_{benz/air})_{out}$.

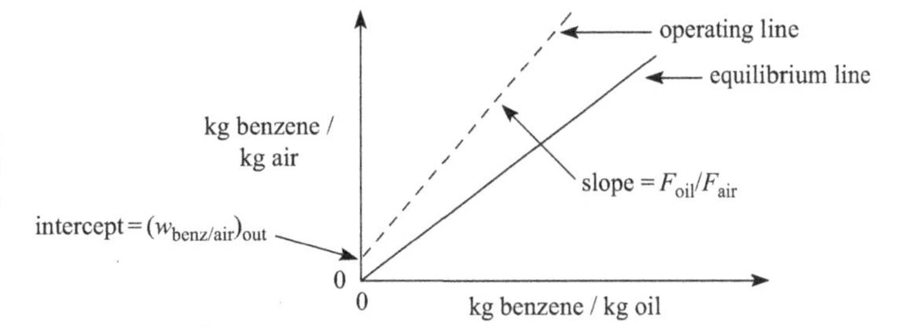

Figure 4.62 Analysis of a multistage cascade absorber: drawing the operating line.

We now determine the compositions of the liquid and vapor phases in equilibrium in the top stage. This coordinate must lie on the equilibrium line. And the $w_{benz/air}$ coordinate is the same as the $w_{benz/air}$ coordinate of the stream leaving the top stage, the point where the operating line intercepts the y axis. Thus the equilibrium in the top stage is directly to the right of the y-axis intercept, as shown in Figure 4.63.

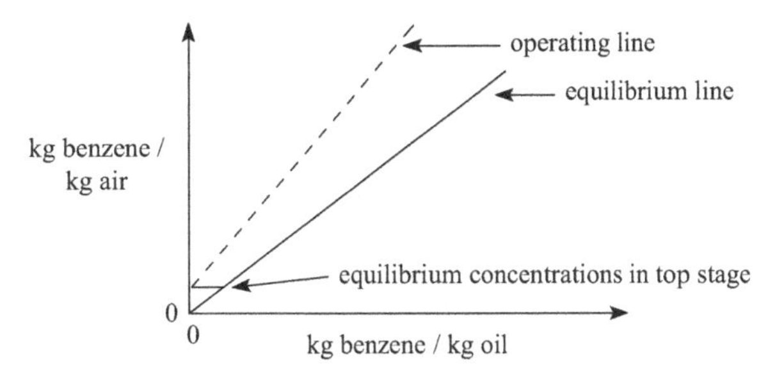

Figure 4.63 Analysis of a multistage cascade absorber: stepping off the first stage.

Next we find the next point on the operating line. The x coordinate of this point is the concentration of benzene in the oil leaving the top stage, which is the x coordinate of the equilibrium point we just plotted. So the point we seek lies on the operating line directly above the point we just plotted on the equilibrium line. From here we draw a horizontal line to find the equilibrium condition in the second stage down, as shown in Figure 4.64.

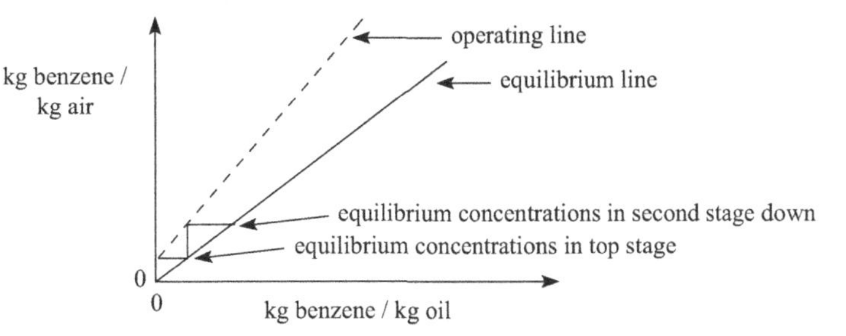

Figure 4.64 Analysis of a multistage cascade absorber: stepping off the second stage.

We continue to step off the stages in our four-stage unit, which yields the analysis shown in Figure 4.65. The coordinates corresponding to the equilibrium stages are labeled 1 through 4. This method is much more convenient than the method shown in Figure 4.60 because one steps vertically and horizontally, there are no slopes to calculate after the operating line is drawn.

This analysis predicts we can obtain the desired purity of air with an input composition labeled "calculated input" in Figure 4.65.

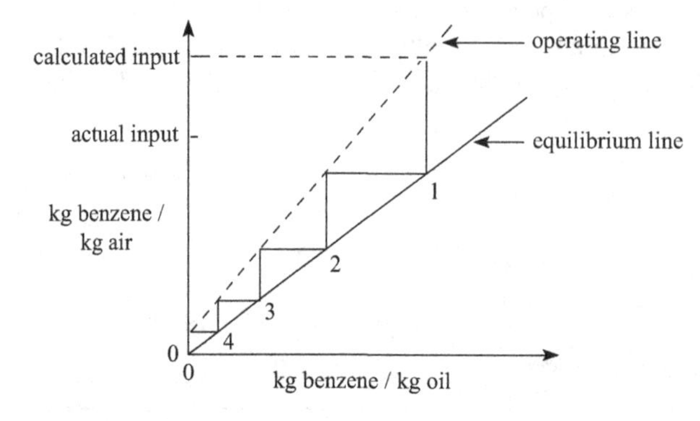

Figure 4.65 Analysis of a multistage cascade absorber, with four stages stepped off.

But what if the actual concentration of benzene in the incoming air (labeled as "actual input" in Figure 4.65) is less polluted than the "calculated input"? Could we make more efficient use of our absorber? Yes! We could change the slope of the operating line, by changing the flow rates of air and oil, until the stages we stepped off matched the actual concentration, as shown in Figure 4.66.

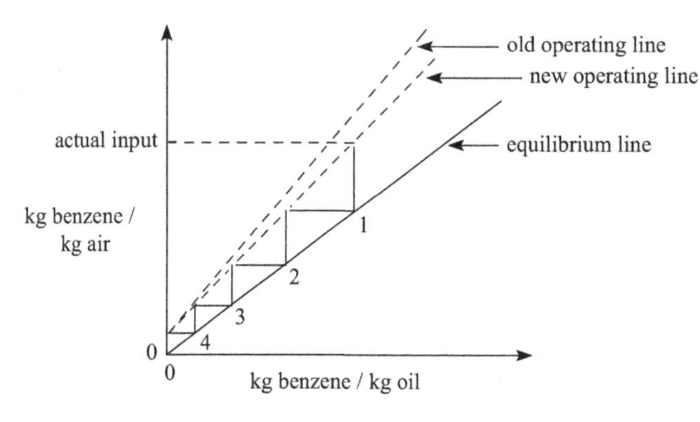

Figure 4.66 Analysis of a multistage cascade absorber.

To match the actual inlet concentration, we decreased the slope of the operating line. So how should the operation of the absorber be changed? F_{oil}/F_{air} should be decreased. Because the flow rate of polluted air is presumably constant, this means we can use a lower flow rate of oil. Thus we will reduce our operating cost. What is the minimum flow rate of oil? We decrease the slope of the operating line until it intersects the equilibrium line at the composition of the inlet polluted air, as shown in Figure 4.67. The intersection of an operating line and an equilibrium line is called a *pinch point*.

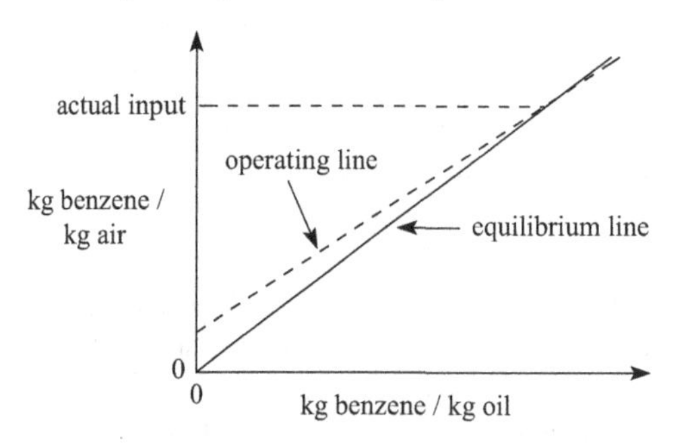

Figure 4.67 Operating line with minimum slope for a multistage cascade absorber.

How many stages are needed to operate with this minimum flow rate of oil? If one steps off stages, one finds that an infinite number are needed to reach the intersection of the operating line and the equilibrium line. Thus we reduce operating cost, but we increase capital cost (to infinity).

A multistage countercurrent absorber is conveniently analyzed with an operating line on an absorber thermodynamic map. The operating line may be drawn from two points – one point for the adjacent liquid–vapor streams at the top of the unit and one point for the adjacent liquid–vapor streams at the bottom of the unit – or from either point plus the slope, given by the ratio of the liquid flow rate to the vapor flow rate. The stages in the unit are determined by stepping horizontally and vertically from the operating line and the equilibrium line until the steps span the length of the operating line segment.

The capital costs and operating costs are generally inversely related, as sketched in Figure 4.68. At either extreme, small decreases in one cost require large increases in the other cost. The curved line shown in Figure 4.68 is for well-designed absorbers. Although one cannot go below the line, a poor design could lie substantially above the curve. If the air does not bubble effectively through the oil, for example, one could design a column with high operating cost *and* high capital cost.

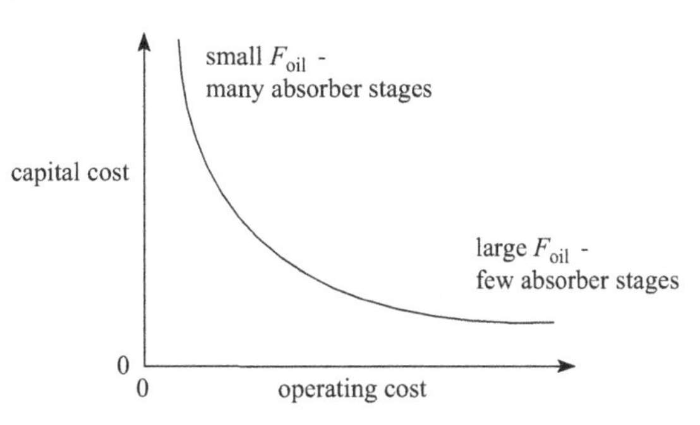

Figure 4.68 The tradeoff between operating cost and capital cost for a multistage cascade absorber.

Figure 4.68 is yet another type of design map. The curve represents efficiently designed absorbers for the particular operating conditions. A different design – perhaps a different type of perforated tray – would have a different curve. A cost map with many curves corresponding to many different designs would be useful in choosing the optimum.

4.2.3 Multistage Liquid–Vapor Separations

Context: Cascading flash drums.

Concepts: Analysis of recycles for cascading flash drums.

Defining Question: How does one translate a T–(x,y) phase map to a map for operating lines?

In Section 4.1.3, just after Figure 4.37, we found that the maximum purity of the vapor stream in Figure 4.36 was 83 mol% benzene. How can we obtain a vapor stream with more than 83 mol% benzene? Recall how we improved the performance of the absorber: we added stages. So we add another flash drum to the vapor stream, as shown in Figure 4.69. The second stage takes in vapor and gives out vapor and liquid streams. We must increase the pressure to condense some of the vapor, so we add a compressor before the second stage. But what is the pressure in the second stage?

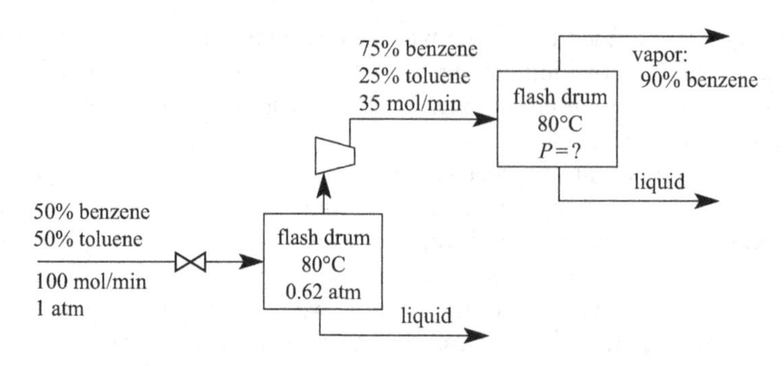

Figure 4.69 A two-stage flash system.

Figure 4.70 shows the analysis of this two-stage flash system. We estimate that the second flash drum will operate at 0.78 atm, and the flow rate of the vapor at 90 mol% benzene is 15 mol/min. Study the graphical analysis in Figure 4.70 and verify this analysis.

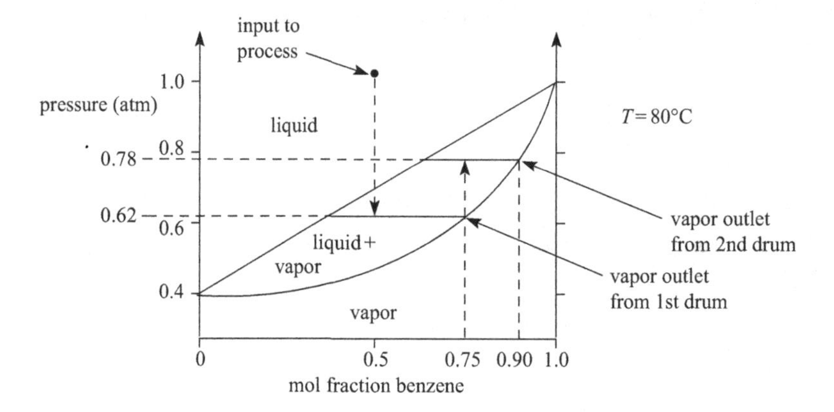

Figure 4.70 Graphical analysis of the two-stage flash system in Figure 4.69.

What is the composition of the liquid stream leaving the second flash drum? What do you propose we do with this liquid stream? Recycle it to the first drum? Good idea. In fact, assuming we also wanted to purify the toluene stream, we would add a third flash drum to the liquid output of the first flash drum. And what would we do with the vapor effluent of this third flash drum? That's right – recycle it to the first flash drum. If we continue in this manner we soon have a complicated process of flash drums that accepts 50:50 benzene:toluene and produces very pure benzene and very pure toluene, with no other waste streams. Such a process is shown in Figure 4.71 on the next page. (Compressors and expansion valves have been omitted.)

It is impractical to analyze the multistage flash drum process with the graphical method we devised for a single flash drum. The problem is the recycle streams – it is awkward to determine the net composition of the flow into a particular drum. For example, how would you account for the liquid effluent from the second unit in Figure 4.69 being recycled to the first unit? Yet multistage distillation should be no more difficult to analyze than multistage absorption. What is the problem?

The problem is our map. We have the wrong coordinates for analyzing multistage countercurrent flow. The map that worked well for the absorber had "benzene in the liquid

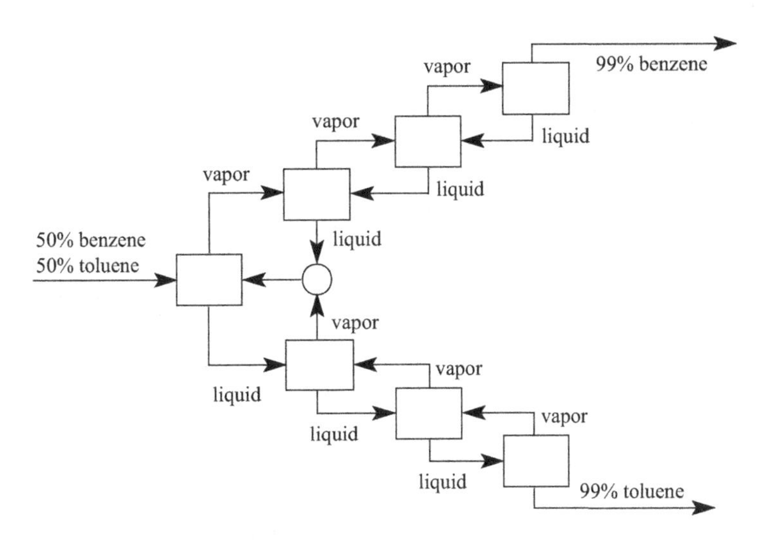

Figure 4.71 A multistage flash drum process.

(oil)" on the abscissa and "benzene in the vapor (air)" on the ordinate. Every point on the absorber map corresponded to a two-phase liquid + vapor system. We need to transcribe our map in Figure 4.70 to a map of "mol fraction benzene in the liquid (x_{benzene})" on the abscissa and "mol fraction benzene in the vapor (y_{benzene})" on the ordinate. Specifically, we need to map the two-phase liquid + vapor region to a map that corresponds to liquid + vapor at all points.

Figure 4.72 shows three points transcribed from the plot on the left to the plot on the right. Point 1 corresponds to pure benzene. Point 2 corresponds to pure toluene. Point 3 is an arbitrary point that corresponds to $x_{\text{benzene}} = 0.50$ and $y_{\text{benzene}} = 0.83$. In this manner many pairs of points are transferred to the new plot.

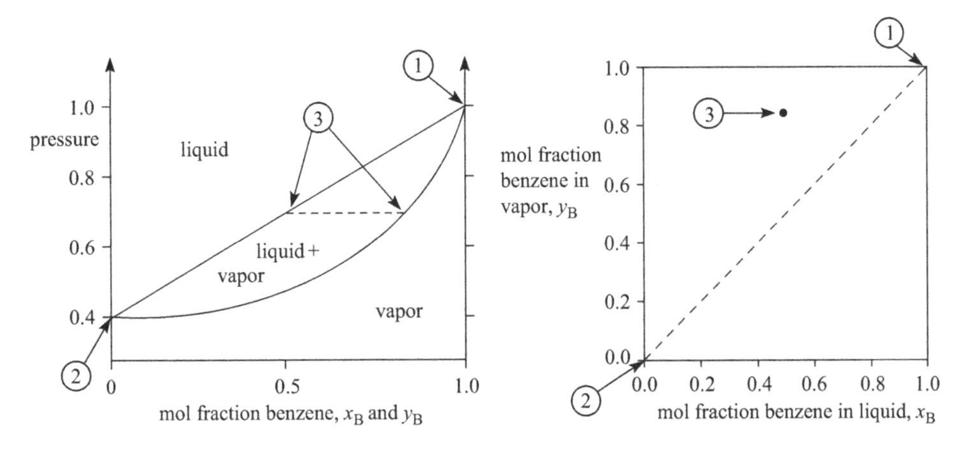

Figure 4.72 Transcribing thermodynamic data from an "equilibrium everywhere" map to a "liquid + vapor phase everywhere" map.

Eventually a line can be drawn through the data, as shown in Figure 4.73 on the next page. The dashed diagonal line is to guide one's eye – it does not represent anything physical . . . yet.

What is a good practice when one encounters a new map? Label the landmarks. Attempt to label some landmarks and regions on the map in Figure 4.73. Then inspect the

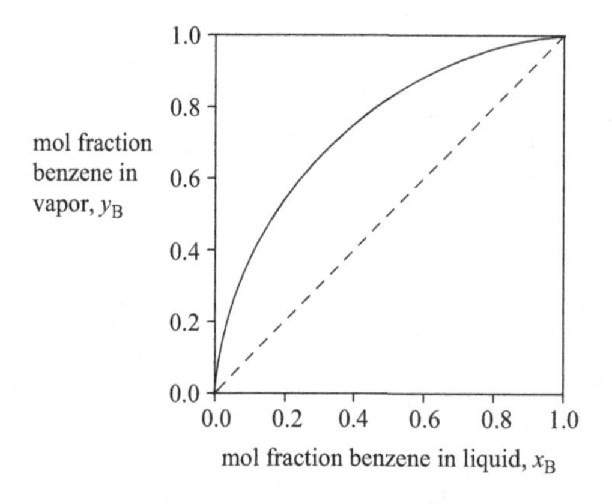

Figure 4.73 Our new map for benzene + toluene mixtures, an x–y diagram.

labeled map in Figure 4.74. Generally, phase equilibria data are plotted so the equilibrium line lies above the diagonal. That is, the more volatile component (the component with the lower boiling point) is chosen as the basis for x and y.

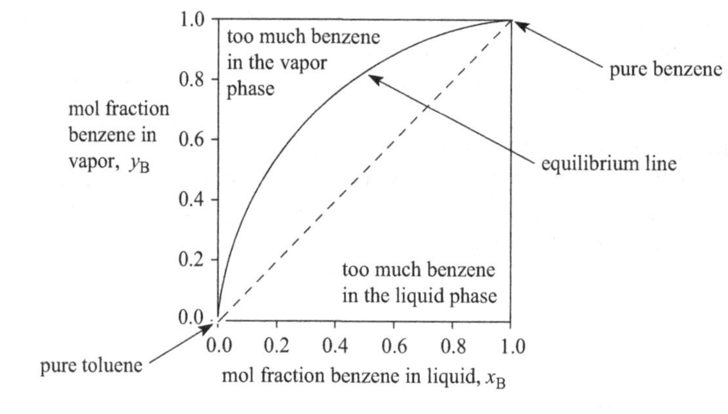

Figure 4.74 Our new map for benzene + toluene mixtures, with labels.

In Figure 4.75 (on the next page) are shown x–y maps for benzene mixed with other chemicals (not toluene). What is the volatility of the second component, compared to benzene?[2] Compared to toluene?[3] Hint: Use the maps in Figures 4.74 and 4.75 to determine the composition of the vapor above a 50:50 liquid mixture. Which benzene mixture will be easier to separate by distillation?

If a mixture is not ideal, the phase-equilibrium maps are not symmetric, as the maps in Figures 4.74 and 4.75 are. In fact, the equilibrium line may cross the diagonal. Maps of vapor–liquid equilibrium data for non-ideal mixtures are shown in Exercises 4.56, 4.106, 4.107, 4.109, 4.110, and 4.113.

Let's use the map in Figure 4.74 to analyze benzene + toluene systems. As we did with the absorber, let's begin with a closed system. Consider a two-phase system of benzene and toluene at equilibrium in a closed vessel, as shown in Figure 4.76. A known

[2] In both diagrams, the equilibrium line is above the $x = y$ line, so $y > x$ everywhere on the equilibrium line. Benzene is the more volatile in both mixtures.

[3] In the diagram on the left, the equilibrium line is closer to the $x = y$ line than in Figure 4.75. The second component in the left graph is more volatile than toluene. Similarly, the second component in the right graph is less volatile than toluene.

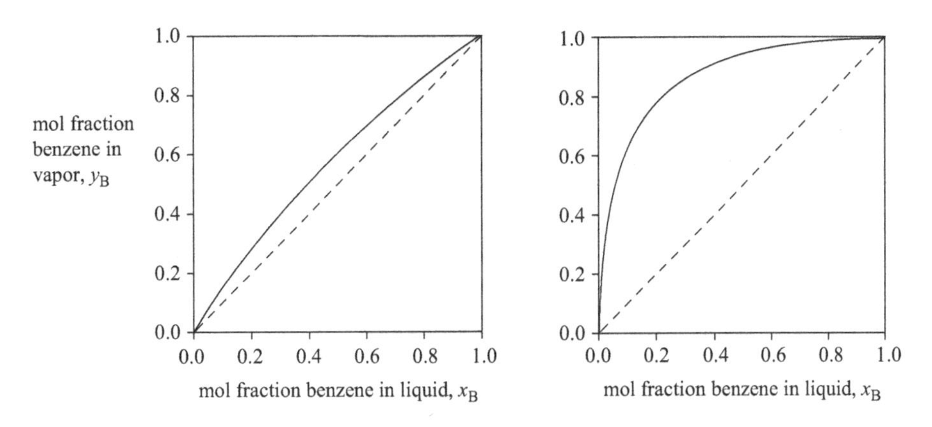

Figure 4.75 Vapor–liquid equilibrium diagrams for benzene mixed with other compounds.

quantity of liquid benzene is added to the vessel. Indicate on the benzene + toluene phase-equilibrium map this system before and after the liquid benzene is added.

The system begins at a point on the equilibrium line and then moves horizontally to the right, as shown in Figure 4.76. Why horizontally? Because only the liquid composition changes initially; the vapor composition is unchanged. Why to the *right*? Because the mol fraction of benzene in the liquid *increases*.

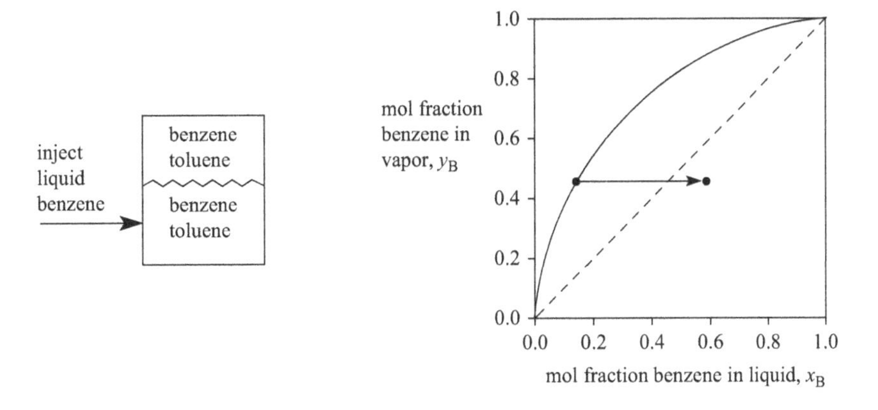

Figure 4.76 Analysis of a closed benzene + toluene system.

How will the system evolve from this point? It will return to equilibrium by decreasing the mol fraction of benzene in the liquid and thus increasing the mol fraction of benzene in the vapor. How does this translate into a direction on our map? The system proceeds to the left and upward. At what angle? Recall how we analyzed the two-phase system of air + benzene over oil + benzene. Consider the extreme case of an infinite amount of liquid and a tiny amount of vapor: x_{benzene} would decrease only infinitesimally and y_{benzene} would increase a lot, and the path would be almost straight up. What about an infinite amount of vapor and a tiny amount of liquid? Yes – the path would be almost horizontal to the left. And what about a system with V mols of vapor and L mols of liquid? Take a guess. That's correct – the system would move to the equilibrium line at a slope of $-L/V$. You can verify this by applying a mol balance to the closed system, similar to the mass balance we applied to the absorber.

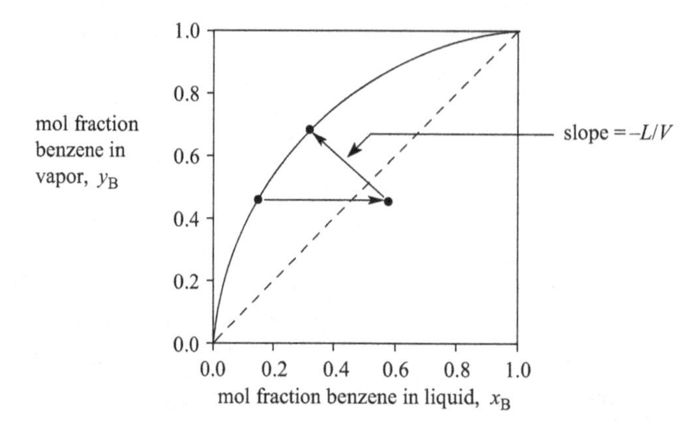

Figure 4.77 Re-equilibration of the closed system after liquid benzene is injected.

Our motivation for creating the new map of the vapor–liquid equilibrium data was to analyze the multistage countercurrent distillation process diagrammed in Figure 4.71. Let's redraw this process so it is similar to the multistage countercurrent absorber in Figure 4.60.

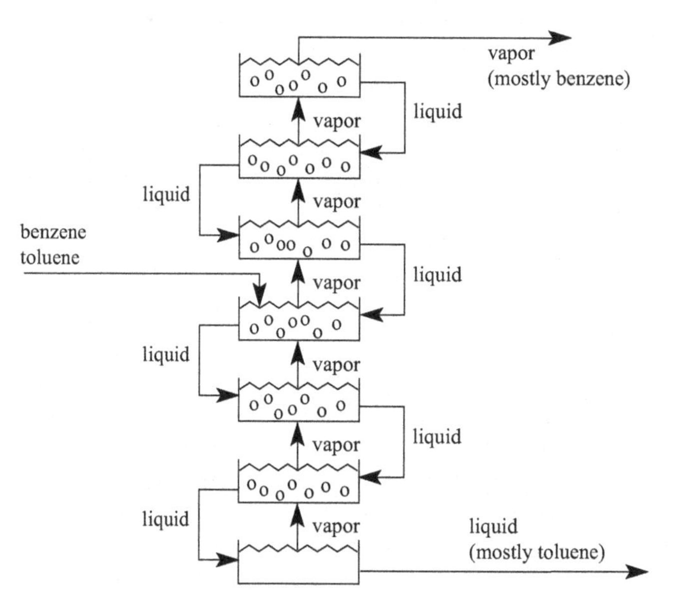

Figure 4.78 A multistage distillation process.

Multistage distillation is similar to multistage absorption, except that there is only one input, which enters somewhere between the top and bottom stage (not necessarily the middle stage). An energy balance on the distillation process in Figure 4.78 would reveal problems with the top and bottom stages. Consider the bottom stage – a liquid stream enters and divides into a liquid stream and a vapor stream. We need to add energy to the bottom stage to vaporize some of the liquid. Consequently, below the bottom stage is an evaporator, or boiler. An overall energy balance of the system would reveal that we need to remove the energy added by the evaporator. We withdraw energy at the top to condense some of the vapor that enters. Above the top stage is a condenser. The multistage distillation process is shown in Figure 4.79.

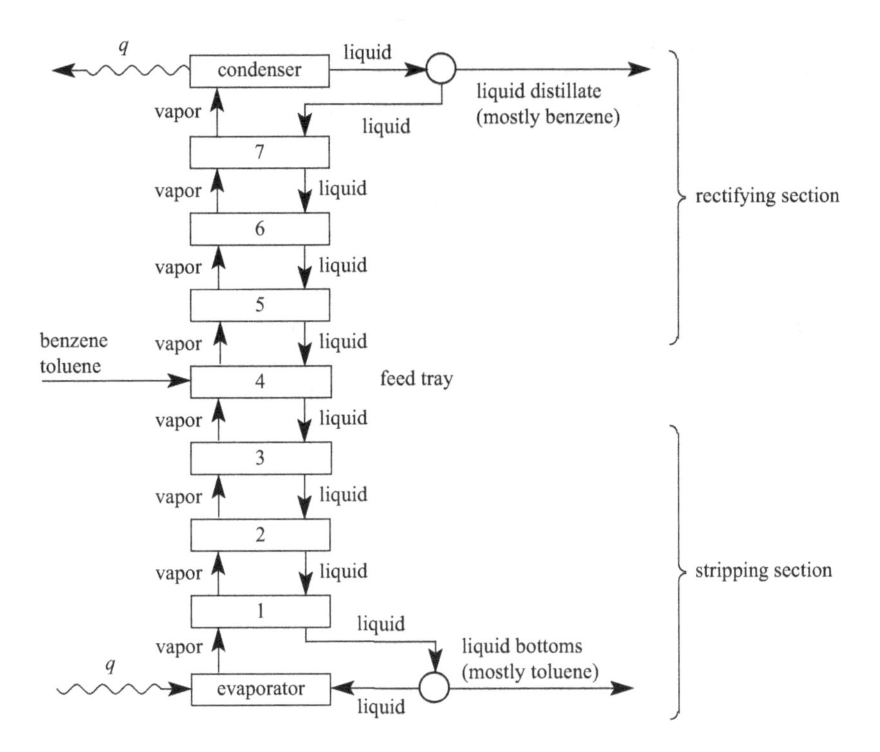

Figure 4.79 A multistage distillation process.

4.2.4 Multistage Cascading Flash Drums

Context: A distillation column.

Concepts: The McCabe–Thiele method.

Defining Question: How does the reflux ratio relate to the process economics balance between steam and steel?

An actual multistage distillation process consists of a series of *trays* (aka stages) with holes in the bottom of each to allow the vapor to bubble up through the liquid. The liquid flows over the edge of a tray onto the tray below. The trays are round and are stacked in a tube, forming a column. For this reason distillation units are commonly called distillation *columns*. Some additional nomenclature is introduced in Figure 4.79. The tray to which the input is injected is called the *feed tray*. The trays below the feed tray (including the evaporator) constitute the *stripping section*. The trays above the feed tray (including the condenser) constitute the *rectifying section*.[4] Finally, the output from the top of the column is the *distillate* and the output from the bottom of the column are the *bottoms*.

Distillation is an integral component in the chemical process industry. About 90% to 95% of all separations are done by distillation. It is estimated there are 40,000 distillation columns in operation in the United States alone, representing a capital investment of $8 billion (Humphrey, 1995).

[4] The complement to the stripping section should logically be called the absorbing section. But the established terminology is "rectifying section."

The graphical method for analyzing distillation columns is similar to the graphical method for analyzing multistage countercurrent absorbers: find a reference point, draw an operating line, and step off the stages. The difference is that we must perform this method twice – once for each section of the column.

But first we must derive the graphical model for a distillation operating line. The derivation is similar to the operating line for an absorber (equations 4.29 through 4.35), but compositions for distillation are in terms of mol fraction of the volatile component, rather than mass ratio of the absorbed substance. Begin by drawing system borders around one or more liquid + vapor equilibrium stages in Figure 4.79 (not including the feed stage), as shown in Figure 4.80. Label the vapor stream leaving the top stage V1 and label the liquid stream entering the top stage L1. Label the vapor stream entering the bottom stage V2 and label the liquid stream leaving the bottom stage L2, as shown in Figure 4.80.

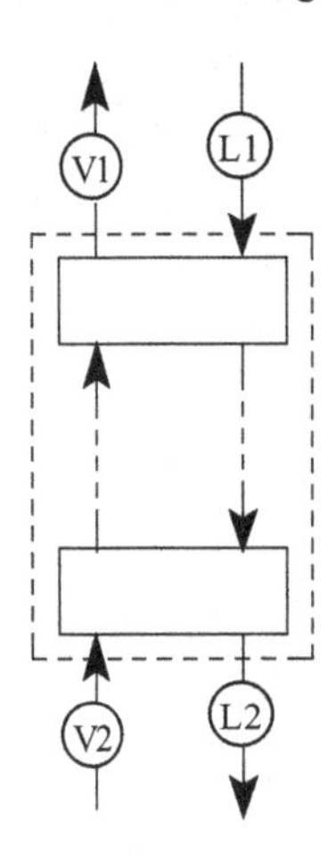

Figure 4.80 System borders around two or more stages in a distillation column. The borders must not include the feed stage.

Assume steady state and no chemical reaction. Thus a mass balance on benzene (B) translates to a mol balance on B. Define F as flow rate in mol/min.

$$\text{molar flow rate of B in} = \text{molar flow rate of B out} \tag{4.36}$$

$$F_{\text{B,V2}} + F_{\text{B,L1}} = F_{\text{B,V1}} + F_{\text{B,L2}} \tag{4.37}$$

$$F_{\text{B,V2}} - F_{\text{B,V1}} = F_{\text{B,L2}} - F_{\text{B,L1}}. \tag{4.38}$$

Define V as the total molar flow rate of a vapor stream, in mol/min. Define L as the total molar flow rate of a liquid stream, in mol/min. A combined overall balance on energy and mass requires $V_{\text{in}} = V_{\text{out}}$ and $L_{\text{in}} = L_{\text{out}}$. Set the total molar flow rates of both streams V1 and V2 equal to V. Set the total molar flow rates of both streams L1 and L2 equal to L:

$$F_{\text{total,V1}} = F_{\text{total,V2}} = V \tag{4.39}$$

$$F_{\text{total,L1}} = F_{\text{total,L2}} = L. \tag{4.40}$$

Recall that y_{B} is the mol fraction of B in the vapor and x_{B} is the mol fraction of B in the liquid. Express the molar flow rate of B in each stream in terms of the total molar flow rate and the mol fraction of B:

$$F_{\text{B,V1}} = y_{\text{B,V1}}V,\ F_{\text{B,V2}} = y_{\text{B,V2}}V,\ F_{\text{B,L1}} = x_{\text{B,L1}}L,\ F_{\text{B,L2}} = x_{\text{B,L2}}L. \tag{4.41}$$

Substitute the four relations above into the mass balance, equation 4.38:

$$y_{B,V2}V - y_{B,V1}V = x_{B,L2}L - x_{B,L1}L \tag{4.42}$$

$$(y_{B,V2} - y_{B,V1})V = (x_{B,L2} - x_{B,L1})L \tag{4.43}$$

$$\frac{L}{V} = \frac{y_{B,V2} - y_{B,V1}}{x_{B,L2} - x_{B,L1}} = \frac{\text{rise}}{\text{run}}. \tag{4.44}$$

Equation 4.44 obtains for any collection of consecutive equilibrium stages in the stripping section or the rectifying section. Therefore all adjacent liquid + vapor pairs between stages lie on a line of slope L/V on a map of y_B vs. x_B.

Let's use the distillation operating line to analyze the distillation column. We start with the stripping section. We will use a point on the operating line and the slope of the operating line to draw the operating line. The coordinates of x_B leaving a stage and y_B entering that stage lie on the operating line, the graphical representation of the mass balance derived in equations 4.36 through 4.44. A convenient point is the x_B, y_B pair corresponding to the streams leaving and entering the bottom stage, respectively. Note that the bottoms composition is the same as the composition of the liquid entering the evaporator in Figure 4.79. And because the vapor stream is the only stream that leaves the evaporator, its composition must be the same as the entering liquid. The liquid leaving the bottom stage is the same composition as the vapor entering the bottom stage. Our reference point for the operating line is thus *on the diagonal* of the vapor–liquid equilibrium map.

We need the slope of the operating line for the stripping section. Is the ratio L/V less than 1 or greater than 1? That is, is the liquid flow rate out of a stage smaller or larger than the vapor flow rate into the stage? Again consider the liquid stream leaving the bottom stage and the vapor stream entering the bottom stage. The liquid stream is split: some leaves the distillation column as the bottoms product and some is recycled to the bottom stage via the evaporator. Therefore the liquid flow rate exiting the bottom stage must be greater than the vapor flow rate entering the bottom stage. Therefore $L > V$, and L/V is greater than 1 throughout the stripping section.

We assume an arbitrary slope greater than 1 and draw an operating line for the stripping section in Figure 4.81. Note that the stripping operating line starts near the lower left of the map, near the coordinates of pure toluene.

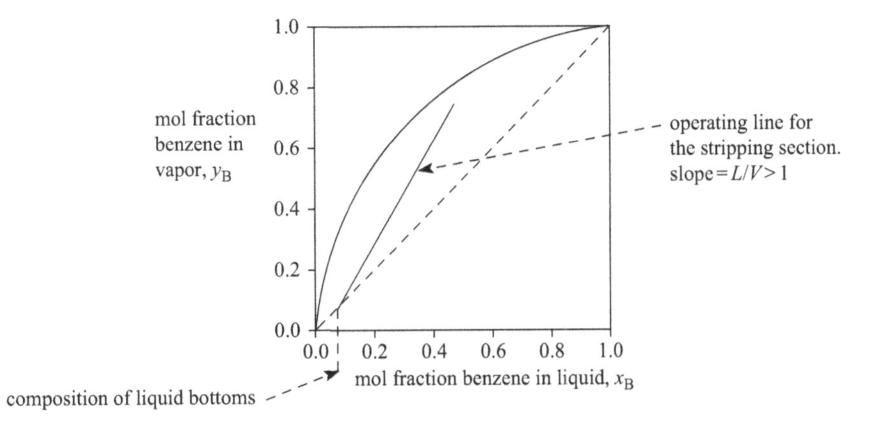

Figure 4.81 The operating line for the stripping section.

We use the stripping-section operating line to step off the stages as shown in Figure 4.82. The stages are numbered on the equilibrium line. Where do we stop stepping? The answer comes from the analysis of the rectifying section.

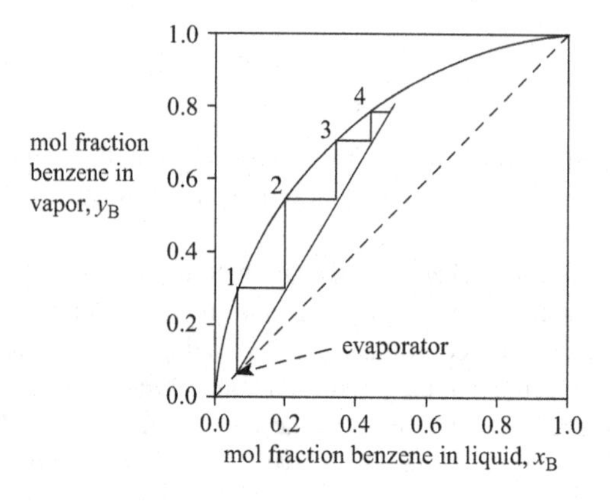

Figure 4.82 Analysis of the stripping section of a multistage distillation process.

The rectifying section is analyzed in a manner similar to the stripping section. Note that the reference point for the rectifying operating line is on the diagonal and near the coordinates of pure benzene. An analysis of the top stage reveals that $L < V$. Thus L/V is less than 1 throughout the rectifying section. We assume an arbitrary slope less than 1 and step off stages in the rectifying section, as shown in Figure 4.83.

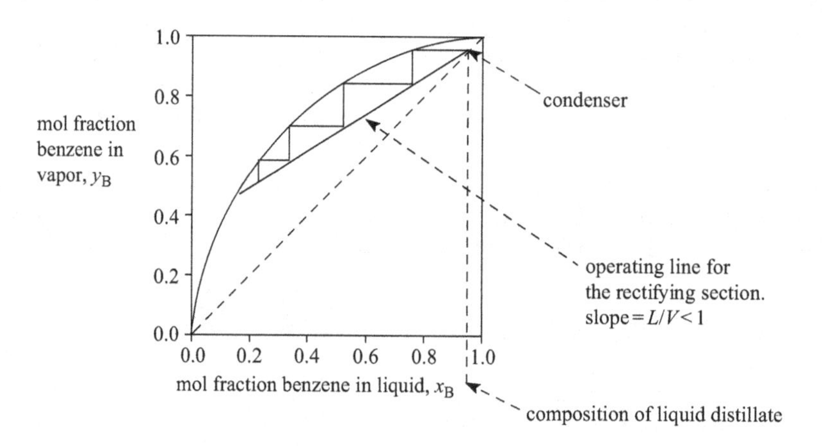

Figure 4.83 Analysis of the rectifying section of a multistage distillation process.

Figures 4.82 and 4.83 are combined in Figure 4.84. When does one switch from the stripping-section operating line to the rectifying-section operating line? The transition is the location of the feed tray. In the analysis shown in Figure 4.84, the feed tray is the third tray in the column. We will assume in this text that the feed is a vapor–liquid equilibrated stream. In this case we have a feed with $x_{\text{benzene}} = 0.35$ and $y_{\text{benzene}} = 0.70$.

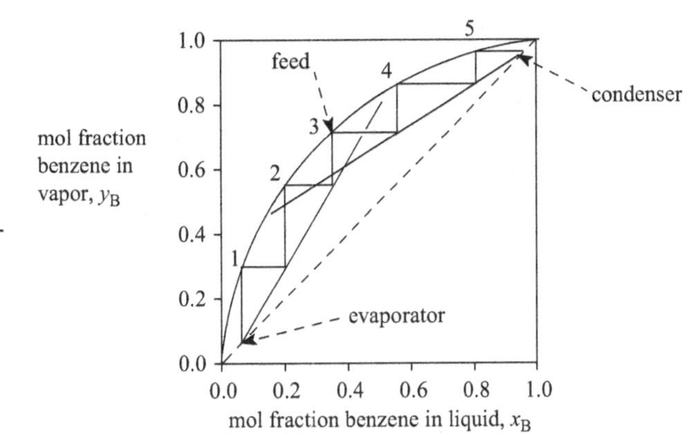

Figure 4.84 Analysis of a multi-stage distillation process.

What if the feed has a higher concentration of benzene, say $x_{benzene} = 0.5$ and $y_{benzene} = 0.8$? The analysis is shown in Figure 4.85. The feed tray is now tray 5. Note that we now have more trays in the stripping section (four) and fewer trays in the rectifying section (one). That makes sense: because the feed has a higher concentration of benzene, it is easier to make the benzene-rich product and more difficult to make the toluene-rich product.

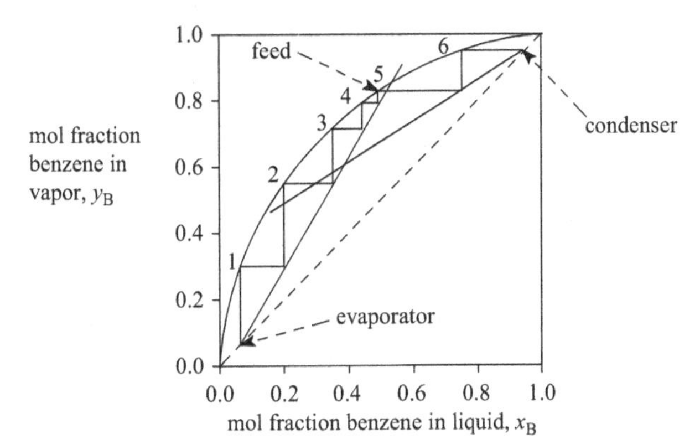

Figure 4.85 Analysis of a multi-stage distillation process.

As stated above, we consider here only feeds in which the vapor and liquid are in equilibrium. Thus the feed point will lie on the equilibrium line. The more general cases are treated in the chemical engineering course on separation processes. In the examples shown here, we started counting trays at the bottom (or top) and the feed composition conveniently matches one of the intersections with the equilibrium line. In other cases (such as in the exercises at the end of this chapter), the feed composition is less conveniently matched to the diagram. In these cases, you might consider starting at the feed composition and stepping off trays upward toward the condenser and downward toward the evaporator.

Note that we would need *many* more trays to distill feed mixtures with benzene concentrations greater than shown in Figure 4.85. The stripping-section operating line and the equilibrium line intersect at a *pinch point*. It would take an infinite number of trays to reach the pinch point. Moreover, it is impossible to distill a feed with a composition above the pinch point. Does this mean one cannot produce 99% toluene from a 90:10 benzene:toluene feed? How might one redesign a column to accommodate a 90:10

benzene:toluene mixture? We need to move the pinch point. We cannot change the equilibrium line.[5] So we change the operating line by changing the ratio L/V. In fact, we can decrease this ratio until $L/V \approx 1$ for both the stripping and rectifying sections. The analysis in Figure 4.86 is repeated for $L/V \approx 1$ in both sections.

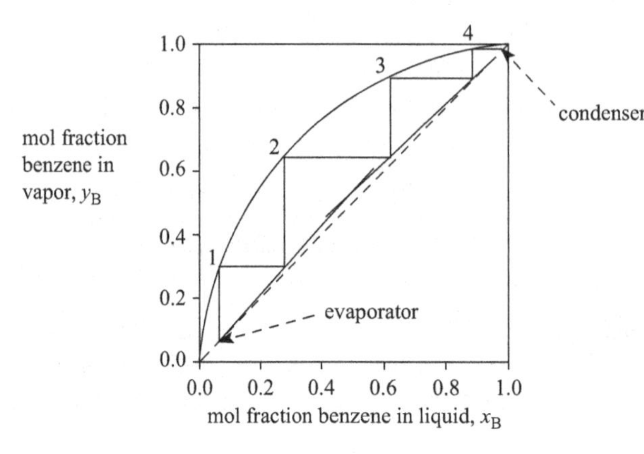

Figure 4.86 A multistage distillation process with nearly minimum L/V for the stripping and rectifying sections.

Figure 4.86 indicates that we could accommodate feed of any composition with a four-tray distillation column. That seems better than the columns analyzed in Figures 4.84 and 4.85. Why bother with L/V ratios other than 1? Well, consider this: if $L/V = 1.001$ in the stripping section, for example, and we are drawing 1 mol/min of bottoms, what is the flow rate of liquid down the column in the stripping section? Roughly, V is about 1000 mol/min. So although you only need four trays, the column must have an enormous diameter to handle 1000 mol/min of vapor flow. Instead, for $L/V = 2$, and again drawing 1 mol/min of bottoms, the liquid flow rate in the column is only 1 mol/min. Changing L/V from 1.001 to 2 doubles the number of trays, but it reduces the column's cross-sectional area by a factor of 1000! Because the cost of a distillation column is roughly proportional to the amount of steel it contains, changing L/V from 1.001 to 2 will reduce the capital cost by a factor of $\sqrt{1000} \approx 32$. The operating cost is also reduced – much more steam is required to evaporate 1000 mol/min than to evaporate 1 mol/min.

This graphical method simplifies design of a column (to specify the number of trays) and analysis of the effect of changing the operating parameters in an existing column – feed composition, bottoms composition, distillate composition, or L/V ratios in the stripping and rectifying sections. Can you imagine analyzing distillation columns mathematically? Neither could Warren McCabe and Ernest Thiele when they were graduate students at MIT in 1925. They developed the graphical procedure shown in Figures 4.81 through 4.86, known as the McCabe–Thiele method (McCabe and Thiele, 1925). Nowadays, distillation columns are analyzed with commercial software, which is efficient but also quite opaque. (Chemical engineering students using such software in the senior design course routinely propose distillation columns with two trays, 3 m in height, and 100 m in diameter.) The performance of a column, such as the change in composition from tray to tray, or the existence of pinch points, is easily visualized with a McCabe–Thiele diagram. If you use commercial software, prepare a McCabe–Thiele plot of your results. (Your manager will ask to see it anyway.

[5] Actually, we *can* change the equilibrium line, by changing the pressure. This is the basis of separations called pressure-swing processes (see Exercise 4.107). But *how* will the equilibrium line change? That's a subject of the course in chemical engineering thermodynamics. For now, you would have to return to the lab and measure more data.

And when you become a manager, always ask to see the McCabe–Thiele diagram yourself.) In Chapter 5 we will examine the internal workings of distillation columns to design the tray diameters and tray spacings.

4.3 TRAJECTORIES ON PURE-COMPONENT PHASE DIAGRAMS

4.3.1 Mapping Solid–Liquid–Gas Phases of a Pure Component

Context: Benzene phases as a function of temperature and pressure.
Concepts: P–T and P–V phase diagrams of pure substances.
Defining Question: How does a three-dimensional P–V–T surface of phases project onto two-dimensional P–T and P–V phase maps?

In Section 4.2.1 we used an absorber to remove benzene from air. You may have wondered why we did not use a flash drum, in which the benzene-contaminated air is cooled to condense benzene as a liquid or solid, as shown in Figure 4.87.

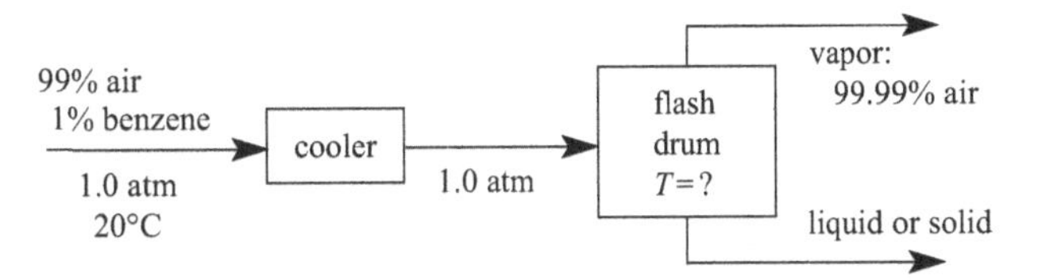

Figure 4.87 Condensing benzene from air by decreasing the temperature at constant pressure.

What temperature is needed to remove 99 mol% of the benzene from the air? To answer this question, we need a temperature–composition phase diagram for benzene + air mixtures. But we would discover that maps similar to Figures 4.39 and 4.73 for benzene + toluene mixtures are ill-suited for analyzing benzene condensation from air. Consider the temperature–composition and liquid–vapor x–y diagrams for benzene + toluene mixtures shown in Figure 4.88. Benzene and toluene are comparably volatile: the ratio of the boiling points for toluene (111°C = 384 K) and benzene (80°C = 353 K) is 384/353 = 1.09.

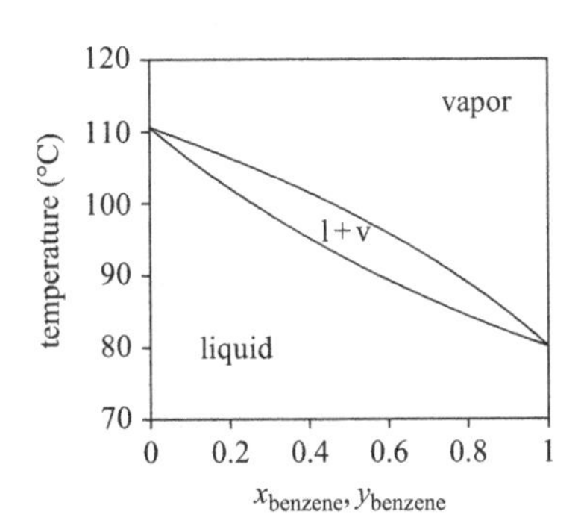

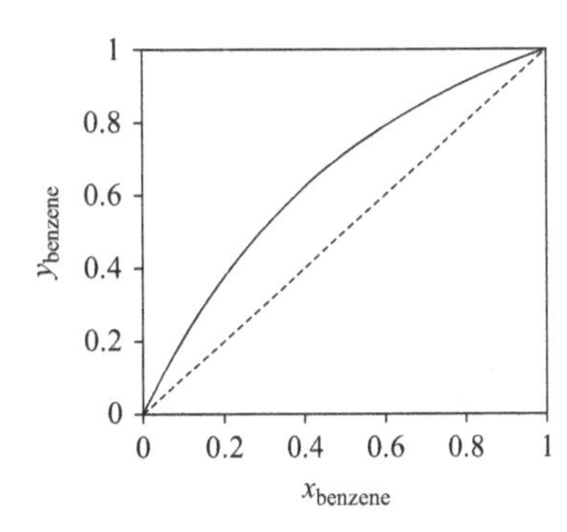

Figure 4.88 Temperature–composition and x–y diagrams for benzene + toluene mixtures at 1 atm.

Consider the temperature–composition and x–y diagrams for furfural + chloroform mixtures shown in Figure 4.89. The ratio of the boiling points for chloroform (162°C = 435 K) and furfural (61°C = 334 K) is 435/334 = 1.3. Because the relative volatilities are more different than for benzene and toluene, the liquid + vapor region is larger and the equilibrium line is farther from the diagonal.

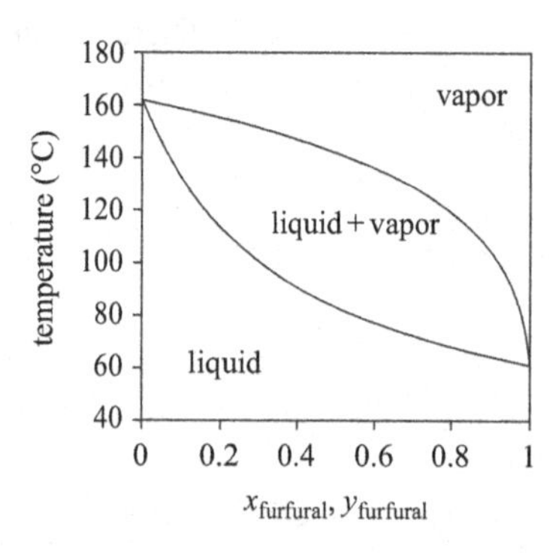

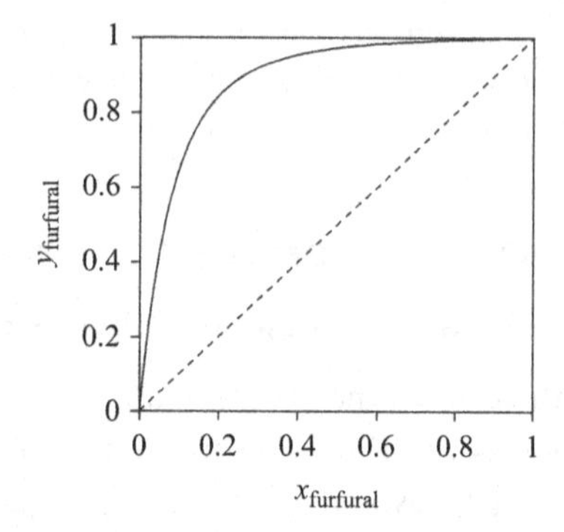

Figure 4.89 Temperature–composition and x–y diagrams for furfural + chloroform mixtures at 1 atm.

Consider the temperature–composition and liquid–vapor x–y diagrams for nitrogen + benzene mixtures shown in Figure 4.90. The ratio of the boiling points for benzene (80°C = 353 K) and nitrogen (−196°C = 77 K) is 353/77 = 4.6.

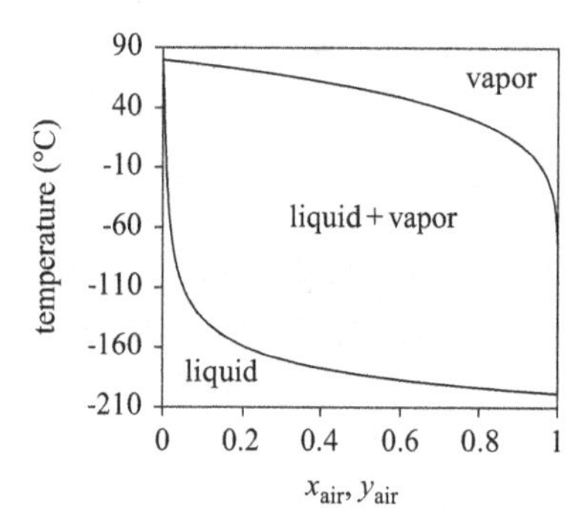

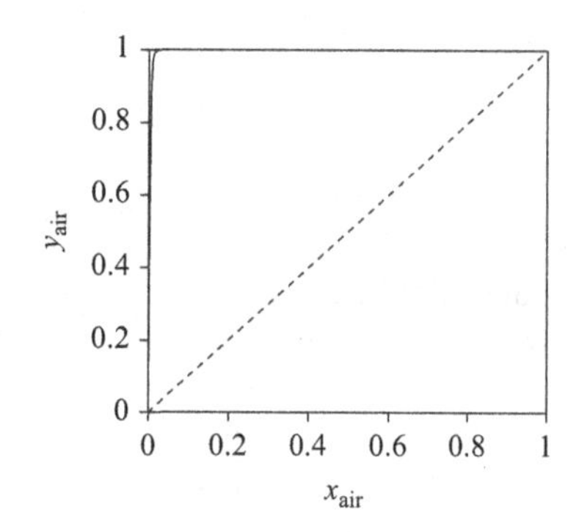

Figure 4.90 Temperature–composition and x–y diagrams for nitrogen + benzene mixtures at 1 atm.

The data for benzene + air mixtures are difficult to analyze graphically when plotted in the forms of Figure 4.90. Consider an analysis of the process in Figure 4.87 on the temperature–composition phase diagram in Figure 4.90. The overall mixture is represented by a tie line with its fulcrum at 99 mol% air. At what temperature is the endpoint on the vapor border at 99.99 mol% air? Because the vapor border is so close to the y axis, it is difficult to locate 0.9999 on the phase diagram. We need to plot the data differently.

For mixtures with very different boiling points, such as nitrogen and benzene, we will assume that none of the more volatile component (nitrogen) condenses for temperatures far above its boiling point (−196°C). Graphically, we assume the liquid border in the left side of Figure 4.90 lies on the y axis at 0 mol% air. For air – assumed to be nitrogen and oxygen only – this is a reasonable assumption for temperatures as low as −60°C. With this assumption we can use phase data for pure benzene to model the condenser in Figure 4.87.

We will prepare a phase diagram for pure benzene. We repeat the experiments with the device in Figure 4.26, but we measure an additional quantity: the specific volume of the benzene, in units of m^3/mol. Again we systematically decrease the volume while holding the temperature constant. Table 4.4 presents the results of one such experiment.

Table 4.4 Phase data for 1 mol benzene at 80°C.

Molar volume (m^3/mol)	Pressure (atm)	View port shows:	Phase(s)
10.14	0.27	transparent gas	vapor
3.67	0.77	transparent gas	vapor
2.92	0.99	transparent gas	vapor
2.897	1.00	mist	vapor + liquid
1.04	1.00	liquid below gas	vapor + liquid
0.49	1.00	liquid below gas	vapor + liquid
0.062	1.00	liquid below gas	vapor + liquid
0.0013	1.00	liquid below gas	vapor + liquid
0.0000887	1.00	liquid only	liquid
0.0000887	1.38	liquid only	liquid
0.0000885	96.57	liquid only	liquid

For safety, we dare not exceed 100 atm in our device. At this limit, the benzene is still entirely liquid. The experiment is repeated at different temperatures. At lower temperatures, the first evidence of liquid is detected at higher volumes; at 50°C, a minuscule liquid film forms at 8.134 m^3/mol and 0.357 atm. At even lower temperatures, solid benzene is detected in the liquid at high pressures. And at even lower temperatures, the first phase to condense is solid benzene.

How do we plot these data so they are useful for designing a condenser? We have *two* independent variables – molar volume and temperature – and two dependent variables – pressure and the phase.

Let's map the phase data we collected for benzene at 80°C. Shown in Figure 4.91 are the data from one series of measurements at a given temperature, such as are presented in Table 4.4. Our qualitative map uses logarithmic scales (see Appendix D) for both pressure and molar volume.

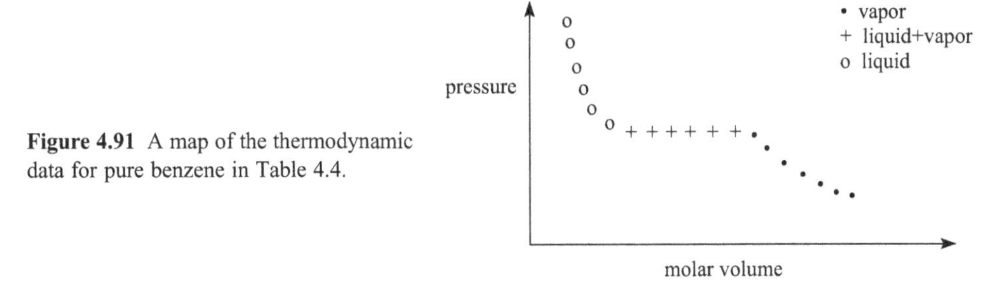

Figure 4.91 A map of the thermodynamic data for pure benzene in Table 4.4.

After we plot data from experiments at other temperatures, as shown in Figure 4.92, regions on the map become evident.

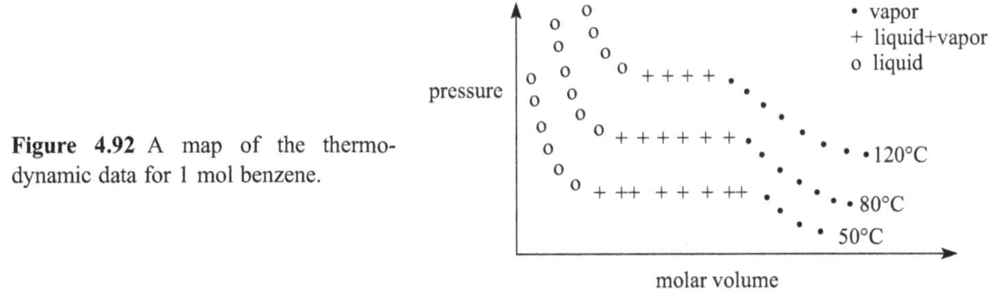

Figure 4.92 A map of the thermodynamic data for 1 mol benzene.

At high molar volume and low pressure, benzene is entirely vapor. At low molar volume and high pressure, benzene is entirely liquid. In between is the region of "liquid + vapor."

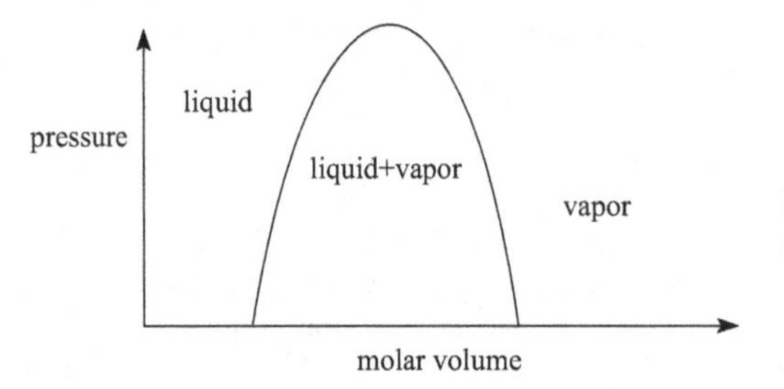

Figure 4.93 A pressure–volume phase diagram for 1 mol benzene.

Lines of constant temperature can be added to the map, as shown in Figure 4.94. These lines are analogous to the constant elevation lines on a topographic map. In the vapor region, a constant-temperature line shows that the pressure increases as the volume decreases. If the vapor is ideal, the relation is $P = (nRT)/V$. In the liquid region, a line of constant temperature is very steep. Why? Because a very large pressure increase is needed to decrease the volume of a liquid at constant temperature.

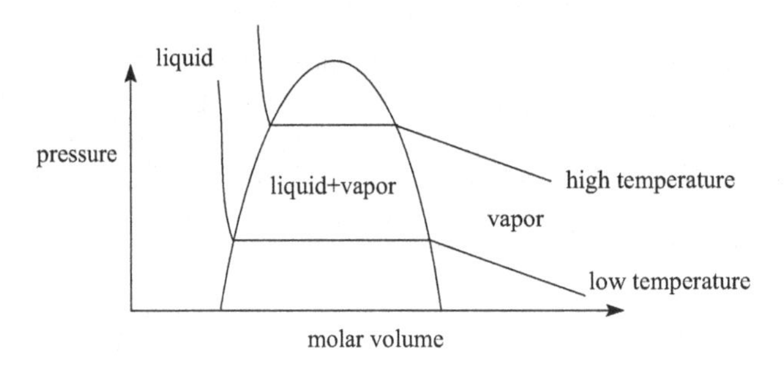

Figure 4.94 A pressure–volume phase diagram for benzene, with isothermal contours.

Note that the dew point and the bubble point for a pure substance are at different points on a pressure–volume map. If we follow an isotherm from the vapor region to smaller molar volumes, the dew point is the condition where the first iota of liquid forms; the point where we reach the border to the liquid + vapor region. If we continue to reduce the volume, the bubble point is the condition at which the last iota of vapor remains; the point where we reach the liquid border. As with mixtures, the dew point and bubble point for a pure substance are not unique points, as their names might suggest. Rather, the dew point can be anywhere on the border between vapor and liquid + vapor. Likewise, the bubble point can be anywhere on the border between liquid and liquid + vapor.

The maps we prepared in Figures 4.93 and 4.94 are useful in modeling processes whose operating parameters are pressure and molar volume. But we want to vary the pressure and *temperature* to operate our condenser. So we also need a map with axes of pressure and temperature. Let's get our bearings on this new map. Where do you expect the regions corresponding to vapor, liquid, and solid? Vapor should be at high temperature and low pressure. Solid should be at low temperature and high pressure. Liquid should lie in between. Qualitatively, our diagram looks like Figure 4.95 on the next page. (We continue to use a logarithmic scale for pressure but a standard scale for temperature.)

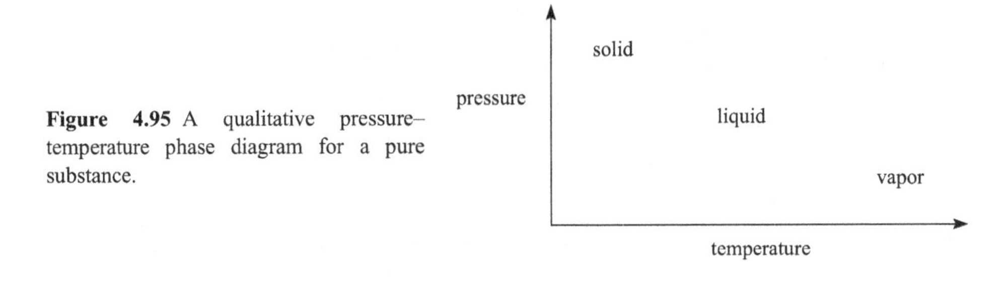

Figure 4.95 A qualitative pressure–temperature phase diagram for a pure substance.

When we plot our experimental data, we arrive at the phase diagram shown in Figure 4.96. In contrast to the phase diagram in Figure 4.91, two-phase systems such as liquid + vapor mixtures are represented here by a line, not a region; and likewise for the other two-phase systems, solid + vapor and solid + liquid. The three two-phase lines meet at the triple point, the unique condition at which solid, liquid, and vapor coexist. The triple point is usually at a temperature slightly colder than the melting point at 1 atm, but at a much lower pressure, typically less than 0.01 atm.

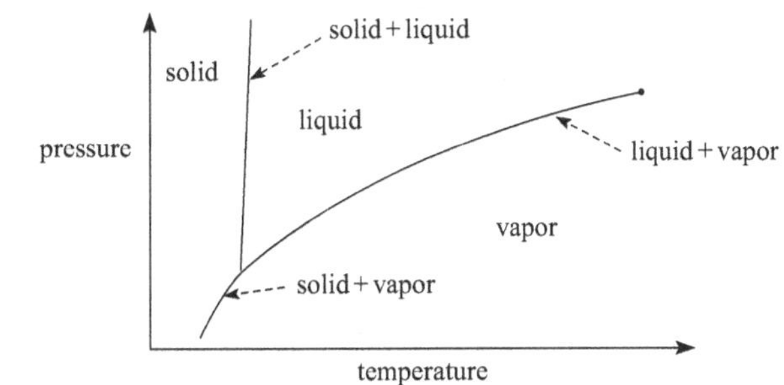

Figure 4.96 A pressure–temperature phase diagram for a pure substance.

The solid + liquid line extends upward indefinitely. However, the liquid + vapor line terminates at a condition called the critical point. Liquid and vapor are indistinguishable above the critical point. Vapor transforms into liquid without condensation; the vapor gets denser and denser until it is a liquid. Why? That's thermodynamics – we warned you in the first chapter that we would occasionally have to postpone "why?" until later in the curriculum. The critical point is at very high temperature and pressure for substances that are liquid at 25°C and 1 atm. For H_2O, the critical temperature is 374°C and the critical pressure is 218 atm. For benzene, the critical point is 289°C and 46.8 atm.[6]

Most separators based on phase use liquid and vapor, as opposed to solid and vapor. Liquid, vapor, and liquid + vapor are three distinct areas on pressure–volume phase maps of pure components, which provides a clear visual representation of the state of the system. However, pressure and *temperature* are more useful operating variables. But liquid + vapor systems on a pressure–temperature phase map are represented by a line, because the two phases at equilibrium have the same pressure and temperature. This complicates analysis of condensers, as we shall learn in the next section.

[6] The critical point for carbon dioxide – a gas at 20°C and 1 atm – is accessible at modest conditions: 31°C and 73 atm. Supercritical carbon dioxide is a key industrial solvent and is used to decaffeinate coffee beans. See Exercises 4.29 and 4.98.

4.3.2 Condensation from a Non-condensible Gas

Context: Separation of benzene from air by condensation.

Concepts: *P–T* and *P–V* phase diagrams of pure substances.

Defining Question: There is no benzene vapor in a system of pure benzene below its boiling point, so why does a mixture of benzene and air always have some benzene vapor, even below the benzene boiling point?

We now use the maps in Figures 4.93 and 4.96 to determine how to condense benzene from the vapor phase to the liquid phase. Let's first do this for pure benzene and then we will add the (minor) complication of a mixture with air.

First, we find the point on the map that corresponds to benzene vapor and label this on the map, as shown in Figure 4.97.

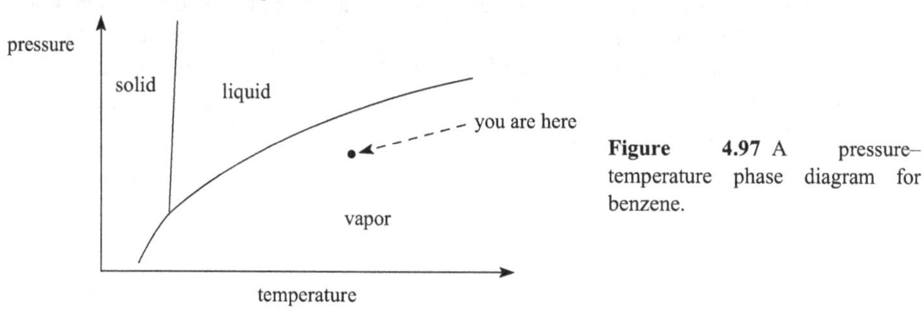

Figure 4.97 A pressure–temperature phase diagram for benzene.

What is our destination? To condense benzene, we must travel to the liquid + vapor line or to the solid + vapor line. Our map tells us we must lower the temperature and/or raise the pressure. Let's first lower the temperature at constant pressure. The process is diagrammed in Figure 4.98.

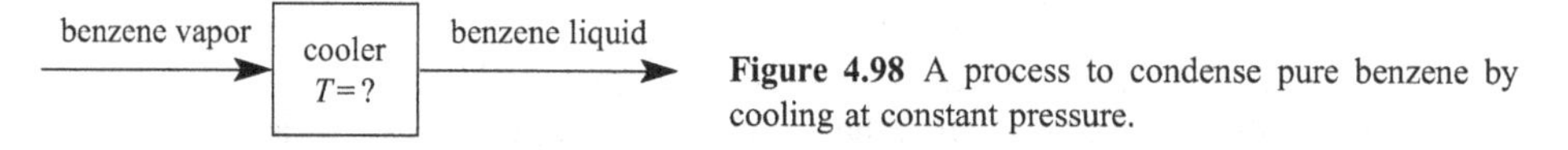

Figure 4.98 A process to condense pure benzene by cooling at constant pressure.

The path on the phase map is shown in Figure 4.99. Liquid benzene begins to condense when we reach the liquid + vapor line. Below $T_{condense}$ the pure benzene is all liquid. The map indicates the minimum temperature for our condenser. Any temperature below $T_{condense}$ will condense all the liquid (at the particular unspecified pressure).

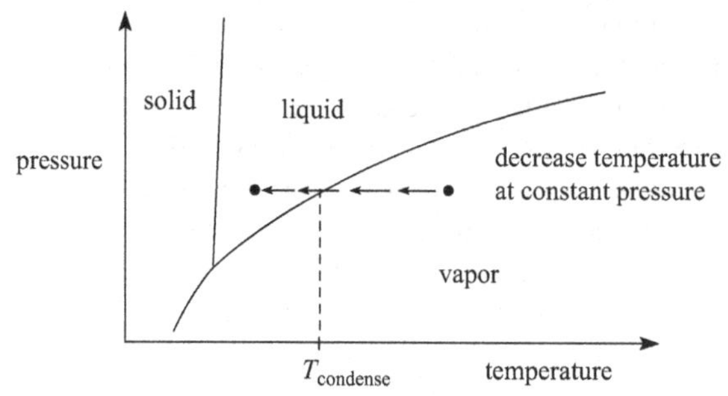

Figure 4.99 A path corresponding to decreasing the temperature at constant pressure.

Likewise we could condense the benzene by increasing the pressure at constant temperature, as in the process diagrammed in Figure 4.100 and as shown by the route in Figure 4.101 (both on the next page). Again, liquid benzene begins to condense when we reach the liquid + vapor line. When the pressure is increased above $P_{condense}$, the pure benzene is entirely liquid.

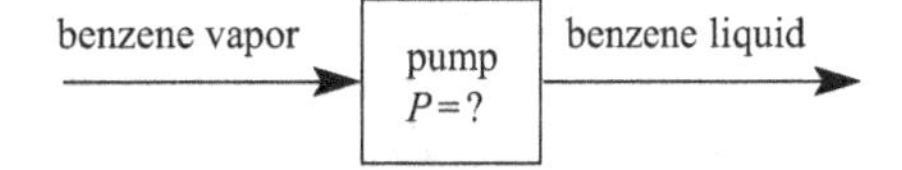

Figure 4.100 A process to condense pure benzene by increasing the pressure at constant temperature.

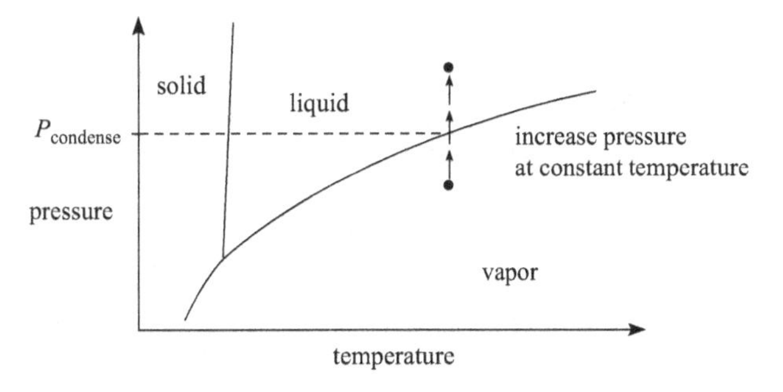

Figure 4.101 A path corresponding to increasing the pressure at constant temperature.

We now consider benzene vapor in air. Air and benzene are very different. At 1 atm, pure benzene condenses at 80°C, whereas air condenses at about −193°C. We will assume benzene and air form an "ideal solution," so our data for the vapor pressure of benzene can be used for the partial pressure of benzene in air. We further assume air is not affected by the presence of a small fraction of benzene.

Let's quantify our design. Assume we have 2 mol% benzene in air at 1 atm and 20°C. Because benzene and air are assumed to behave ideally, the partial pressure of benzene is proportional to the mol fraction of benzene:

$$\frac{\text{partial pressure of benzene}}{\text{total pressure}} = \frac{\text{mols of benzene vapor}}{\text{total mols vapor}} \tag{4.45}$$

$$\frac{P_{\text{benzene}}}{P_{\text{total}}} = y_{\text{benzene}} \tag{4.46}$$

$$P_{\text{benzene}} = (0.02)(1.0\text{ atm}) = 0.02\text{ atm}. \tag{4.47}$$

We locate our starting point on the map, as shown in Figure 4.102. Note that we changed the y-axis label to remind us that benzene is present as a mixture with air. The pressure on this map is the partial pressure of benzene, not the total pressure of the mixture.

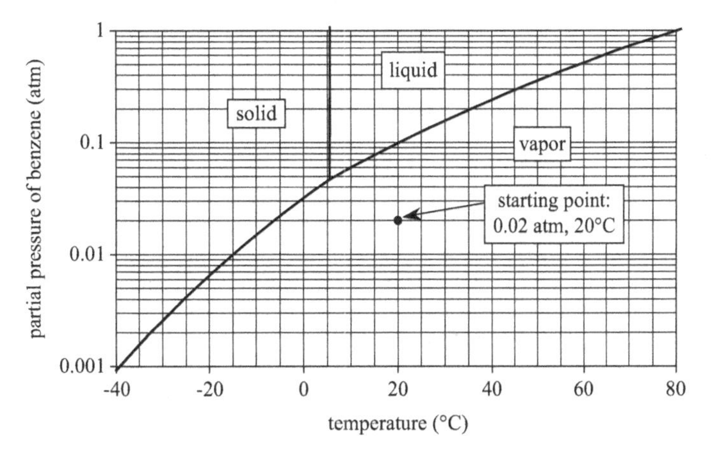

Figure 4.102 The starting point for 2 mol% benzene in air.

Whereas it was possible to condense pure benzene entirely to liquid or solid, it is not possible to condense all the benzene from air by cooling. Even if we cool below the dew point for benzene at 0.02 atm, some benzene remains in the vapor phase. Consider water vapor in air. A temperature decrease may cause some water vapor to condense (as rain or snow), but not all the water vapor is condensed; the relative humidity does not decrease to zero below the dew point. Because some benzene vapor will remain in the air, our destination on the phase map will be on a two-phase line: on the liquid + vapor line or on the solid + vapor line. This is important to remember.

We first design a process to condense benzene by decreasing the temperature. Such a process is diagrammed in Figure 4.103. Let's determine the temperature decrease needed to condense 90% of the benzene vapor.

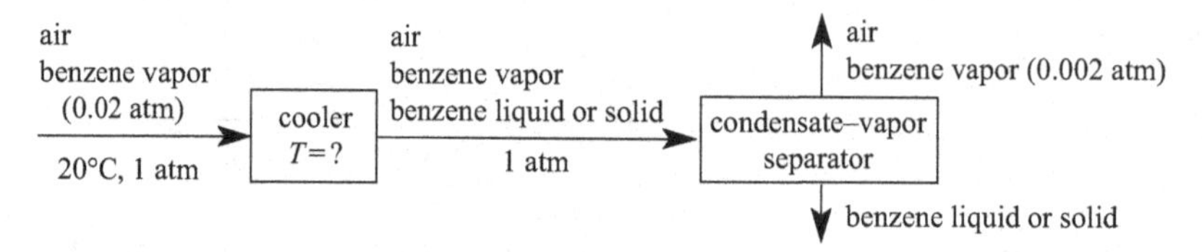

Figure 4.103 A process to condense benzene from air by cooling at constant total pressure.

We now use the pressure–temperature map in Figure 4.102 to determine the temperature of our destination. Our condenser maintains 1 atm total pressure, so we plot a path of constant pressure. We move horizontally on the map until we reach the two-phase boundary. The path is horizontal because the partial pressure of benzene does not decrease until some benzene condenses and decreases the mol fraction of benzene in the vapor. The first contact with the two-phase border is the dew point. Note that we reach the border of the vapor region at the solid + vapor line; the benzene condenses as a solid. For our system, the dew point at 1 atm total pressure is −6°C.

How does the path continue after reaching the solid + vapor line? The temperature continues to decrease, so the path is to the left. But recall that the system is a mixture of benzene solid and benzene vapor. So we are constrained to move along the solid + vapor line. As we move along the line, the partial pressure of benzene decreases. Benzene began to condense at a partial pressure of 0.02 atm. Therefore 90% will be condensed when we reach 0.002 atm. We continue cooling the system until we reach 0.002 atm, which is at −33°C, as shown in Figure 4.104. Our graphical analysis predicts that the condenser should operate at −33°C. We then remove the solid benzene from the system and warm the air + benzene vapor to 20°C.

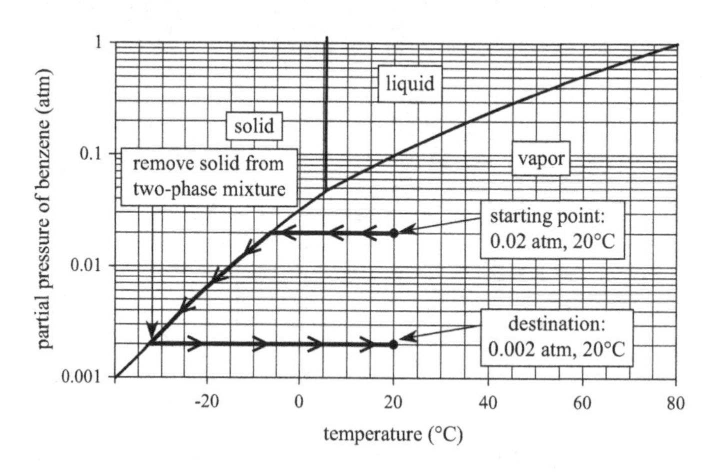

Figure 4.104 Condensing benzene from air by cooling at constant total pressure.

However, our process in Figure 4.103 is not a satisfying design. Benzene condenses as a solid. How will we scrape the solid benzene from the cold condenser surfaces? How will we separate vapor and solid such that there is no vapor in the solid stream? How will we transport solid benzene from the condenser?

Let's consider a different design; let's use the graphical design tools to determine the pressure needed to condense the benzene from the air at constant temperature. The process in Figure 4.105 compresses the benzene + air mixture to condense 90 mol% of the benzene. The liquid is separated from the vapor (at high pressure) and the air is then expanded to 1 atm.

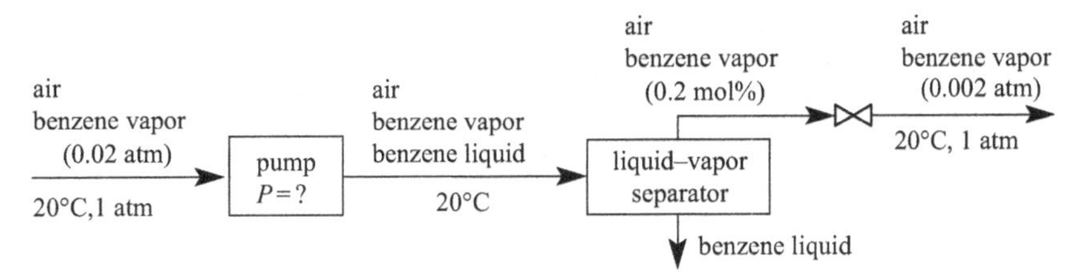

Figure 4.105 A process to condense benzene from air by increasing the pressure at constant temperature.

What is the operating pressure of the process? Again we start at 0.02 atm and 20°C. But now we proceed vertically (because temperature is constant) until we reach the liquid + vapor line. The upward path in Figure 4.106 shows that the dew point for benzene is 0.1 atm at 20°C. Thus to condense any benzene we must increase its partial pressure to 0.1 atm, a factor of 5, which means we must increase the *total* pressure by a factor of 5 from 1 atm to 5 atm to condense the first iota of benzene. This is a key concept.

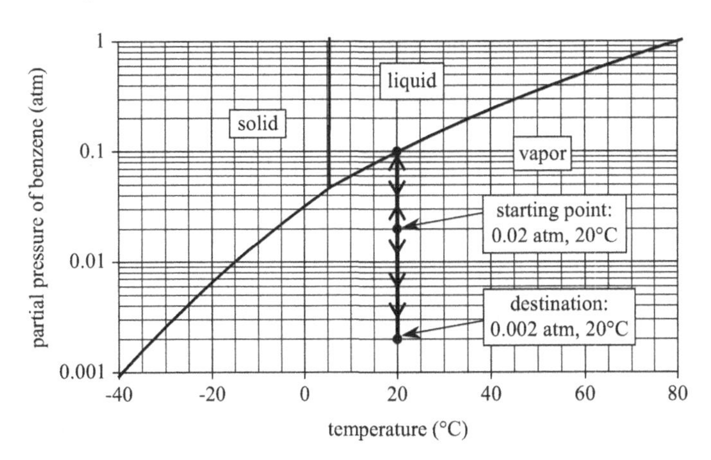

Figure 4.106 Condensing benzene from air by increasing the pressure at constant temperature.

To condense more benzene, we must increase the pressure further. Recall that pure benzene condensed entirely to liquid when we increased the pressure above the dew point; we crossed into the liquid region on the phase map. With an air + benzene mixture we must remain on the liquid + vapor line. And with the further constraint of constant temperature, we are obliged to remain at the point where we first contacted the liquid + vapor line.

How much higher must we raise the pressure to achieve the design specification of 0.2 mol% benzene in the air? Or, phrased differently: what is the total pressure if the partial pressure of benzene is 0.1 atm and benzene is 0.2 mol%? We use equation 4.48:

$$\frac{P_{\text{benzene}}}{P_{\text{total}}} = y_{\text{benzene}} \tag{4.48}$$

$$P_{\text{total}} = \frac{P_{\text{benzene}}}{y_{\text{benzene}}} = \frac{0.1 \text{ atm}}{0.002} = 50 \text{ atm}. \tag{4.49}$$

We increase the total pressure to 5 atm (upward path in Figure 4.106 to 0.1 atm partial pressure) and then remain at the point on the liquid + vapor line while the total pressure is increased to 50 atm (0.1 atm partial pressure), drain the condensed benzene liquid from the system at 50 atm (remain at the point on the liquid + vapor line), and then decrease the total pressure to 1 atm (downward path to the destination).

Although it is less obvious, the process in Figure 4.105 is not a satisfying design. Although the benzene condenses as a liquid, which overcomes the unpleasantness of condensing benzene as a solid in Figure 4.103, the cost of compressors and high-pressure separators would be onerous.

Let's compare the visual effectiveness of our graphical analyses in Figures 4.104 and 4.106. The path in Figure 4.104 clearly illustrates the extent of condensation. The farther we move downward along the solid + vapor line, the more benzene condenses. In contrast, the graphical analysis in Figure 4.106 is obtuse. The system remains at the same point on the map regardless of the amount of benzene condensed. We need a map that better shows the progress of condensation at constant temperature.

The condensation in Figure 4.106 remains at one point because the two map coordinates – temperature and partial pressure – are constant as the benzene condenses. To move on a map, at least one coordinate must change. What is changing? Well, if we used a device such as shown in Figure 4.26 to increase the total pressure at constant temperature, we would depress the piston to decrease the volume. Volume is changing. Let's use a map with pressure–volume coordinates, as shown in Figure 4.94.

We begin in the vapor region of the map and follow a constant-temperature path to the liquid + vapor line, as shown in Figure 4.107. As the volume decreases at constant temperature, the pressure increases, as one would expect from the ideal gas law, $PV = nRT$.

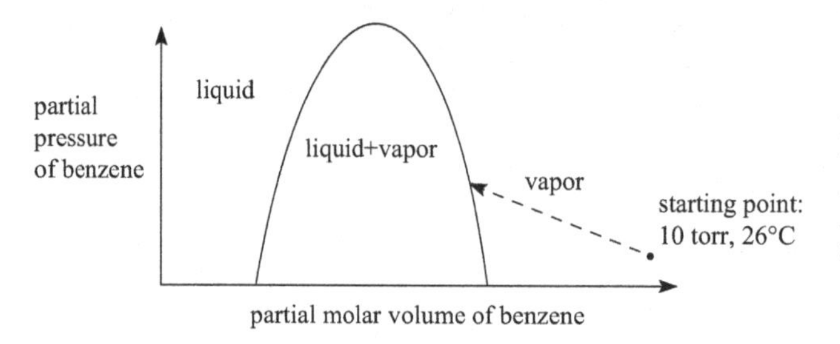

Figure 4.107 Increasing the total pressure (by decreasing the volume) at constant temperature until the first drop of benzene condenses.

When the path reaches the liquid + vapor border, the first drop of benzene condenses. This is the dew point. Again the total pressure increased by a factor of

5 to 5 atm and the partial pressure of benzene increased by a factor of 5, from 0.02 atm to 0.1 atm.

What happens when we increase the total pressure by an additional factor of 2? Does the partial pressure of benzene increase by a factor of 2, to 0.2 atm? No. The phase map in Figure 4.106 tells us the partial pressure of benzene cannot exceed 0.1 atm at 20°C. Visualize the pressure increase in terms of the system volume. How did we increase the total pressure by a factor of 2? We decreased the total volume by a factor of 2 (assuming the ideal gas law applies and neglecting the volume of the benzene that condensed). What path on the map in Figure 4.107 corresponds to increasing the total pressure by a factor of 2? The partial pressure of benzene is constant at 0.1 atm (the path is horizontal) and the volume decreases by a factor of 2 (the path is to the left). We have moved into the two-phase region of vapor + liquid mixtures, as shown in Figure 4.108. Recall that the volume scale is logarithmic; a factor of two does not move one-half the distance to the pressure axis.

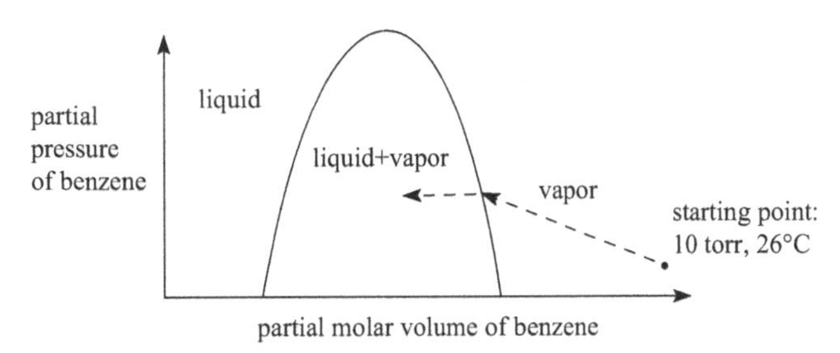

Figure 4.108 Increasing the total pressure at constant temperature to condense about half the benzene.

So the partial volume of the benzene vapor decreased by a factor of 2. But the partial pressure of benzene remained constant at 0.1 atm and the temperature remained constant at 20°C. How can this be if $PV = nRT$? If V changes, something else must change. In this case, n (the mols of benzene in the vapor phase) decreased by a factor of 2. How did this happen? Half of the benzene in the vapor phase condensed to the liquid phase. The distance traveled across the liquid + vapor region illustrates the extent of condensation at constant temperature, in contrast to the pressure–volume map, for which the all condensation occurred at a single point.

Systems are generally described by several parameters. You must choose the map style appropriate to your needs. Consider a road map of Ithaca, New York, or Berkeley, California. Each map shows road locations in terms of east–west/north–south coordinates. Most road maps don't show elevations. Likewise, most temperature–pressure maps don't show volumes. Elevation changes can be very important when bicycling. Volume changes can be important when designing process equipment. Just as one soon learns that traveling west on Buffalo Avenue in Ithaca or west on Hearst Avenue in Berkeley entails a decrease in elevation, one should be aware that increasing the pressure at constant temperature entails a decrease in volume. Crossing the liquid + vapor border on a pressure–temperature map decreases the volume by a factor of about 1000 – like driving off a cliff on a road map.

An alternative to two-dimensional maps such as Figures 4.92 and 4.96 is a three-dimensional map. Figure 4.109, on the next page, shows a map with three axes: temperature, pressure, and volume. For a pure substance, the valid combinations of T, P, and V form a three-dimensional surface.

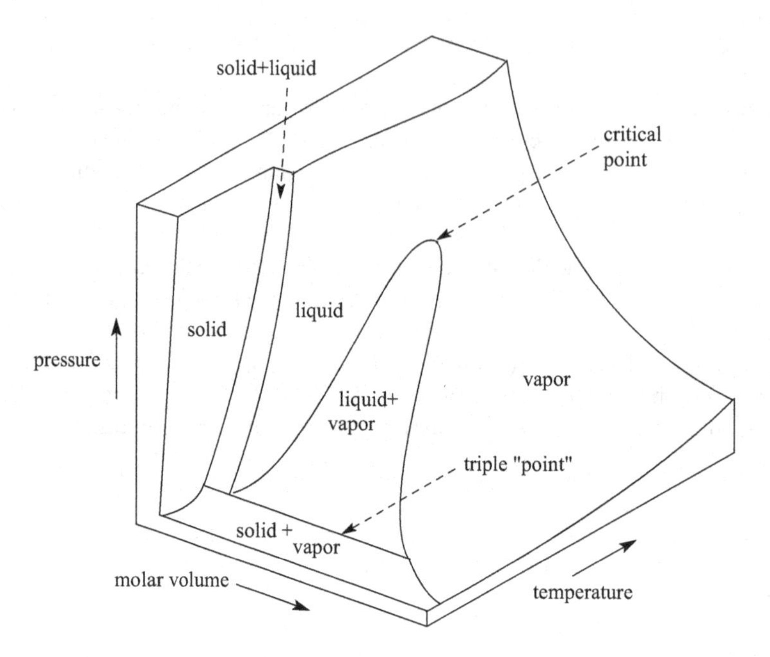

Figure 4.109 A three-dimensional phase map of a pure substance. Adapted from Sandler (1989), p. 219, figure 5.6.

If you project the solid object in Figure 4.109 onto the *P–T* plane, a two-dimensional map like Figure 4.96 results. Similarly, if you project the solid object onto the *P–V* plane, Figure 4.92 results. Actually, Figure 4.92 is only the upper portion of the map. A larger *P–V* phase map with lower pressures and lower volumes is shown in Figure 4.110. Two-phase mixtures, such as solid + vapor, appear as regions on a *P–V* phase map. As with any new map, it is helpful to identify landmarks. You are advised to locate the critical point and the triple "point" on the phase map in Figure 4.110.

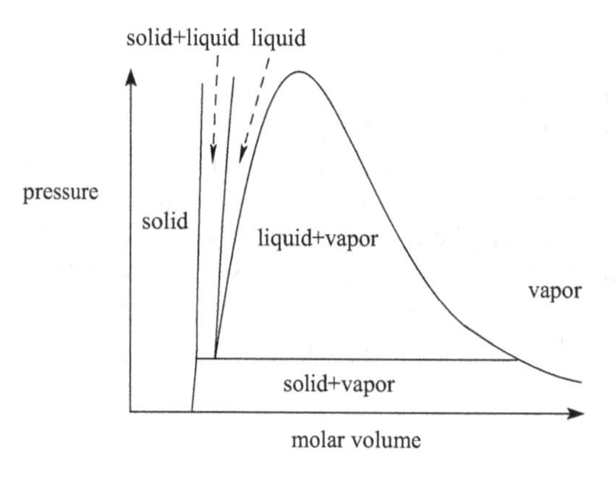

Figure 4.110 A pressure–volume phase diagram for a pure substance.

We can use the phase maps in Figures 4.96, 4.109, or 4.110 to evaluate any combination of pressure and temperature to operate our condenser. It is generally preferable to condense organic liquids by reducing the temperature. Moreover, condensation is often used as a pretreatment for adsorption: lower temperatures favor adsorption of organics on adsorbents. Furthermore, lower temperatures help decrease the humidity of the air before it reaches the adsorber; water often competes with the organic for adsorption sites on the adsorbent. Condensation is effective for high concentrations of organic pollutants, where the partial pressure of the pollutant can be substantially decreased by only modest changes in temperature. We arrive at the more general scheme for condensing organic compounds shown in Figure 4.111.

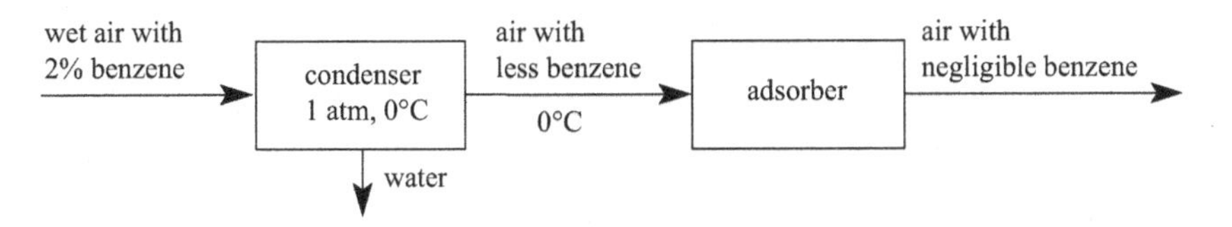

Figure 4.111 A general scheme for removing benzene from air.

The water removed by condensation will have traces of benzene and must be treated (with a charcoal adsorber, for example) before being released.

SUMMARY

Graphical methods are useful tools for process analysis based on experimental data. Experimental data are effectively displayed on two-parameter maps and interpolation is straightforward. The maps are useful for translating parameters such as temperature and composition into other parameters, such as phase (solid, liquid, or vapor) and energy.

Maps are useful devices for representing data. You need not be familiar with every map of technical data, just as you need not know every street map in the world. When you see a new street map, you can immediately find your way because you have learned how to read street maps. If you develop some basic skills in reading engineering maps, you will seldom be lost. The procedures for reading street maps apply to engineering maps. When you encounter a new map, first get your bearings; label extremes on the map – left upper corner, lower right corner, etc. Determine the map's type – for example is the map "equilibrium everywhere" or "two-phase liquid + vapor everywhere"? Then find your present location and your destination on the map. Finally, explore alternative routes to get from here to there.

While applying graphical analysis to analyze the separation processes, we developed several graphical devices, such as operating lines, tie lines, and the McCabe–Thiele method. These devices are *design tools* for analyzing separation processes, analogous to the operating equations obtained by mathematical analysis.

REFERENCES

Blanch, H. W. and Clark, D. S. 1995. *Biochemical Engineering*, Marcel Dekker, New York, NY.

Brown, G. G., Foust, A. S., Katz, D. L., *et al.* 1950. *Unit Operations*, John Wiley & Sons, New York, NY.

Felder, R. M. and Rousseau, R. W. 1986. *Elementary Principles of Chemical Processes*, 2nd edn., John Wiley & Sons, New York, NY.

Gmehling, J. and Onken, U. 1977. *Vapor–Liquid Equilibrium Data Collection*, Dechema, Frankfurt.

Gmehling, J., Onken, U., and Arlt, W. 1982. *Vapor–Liquid Equilibrium Data Collection: Organic Hydroxy Compounds – Alcohols*, Dechema, Frankfurt.

Gmehling, J., Onken, U., and Rarey-Nies, J. R. 1988. *Vapor–Liquid Equilibrium Data Collection: Aqueous Systems – Supplement 2*, Dechema, Frankfurt.

Hougen, O. A. and Watson, K. M. 1946. *Chemical Process Principles Charts*, John Wiley & Sons, New York, NY.

Hougen, O. A., Watson, K. M., and Ragatz, R. A. 1954. *Chemical Process Principles. Part I*, 2nd edn., John Wiley & Sons, New York, NY.
Humphrey, J. L. 1995. "Separation processes: Playing a critical role," *Chemical Engineering Progress*, 91, 31–35.
King, C. J. 1971. *Separation Processes*, McGraw-Hill, New York, NY.
Lightfoot, E. N. 1988. "Recovery of potentially valuable biologicals from dilute solution," in *Chemical Engineering Education in a Changing Environment*, ed. S. I. Sandler and B. A. Finlayson, AIChE Publications, New York, NY.
McCabe, W. L. and Thiele, E. W. 1925. "Graphical design of fractionating columns," *Industrial & Engineering Chemistry,* 17, 605–611.
Nystrom, J. M. 1984. *Product Purification and Downstream Processing*, 5th Biennial Executive Forum, A. D. Little, Boston, MA.
Perry, R. H. and Green, D. W. 1984. *Perry's Chemical Engineers' Handbook*, 6th edn., McGraw-Hill, New York, NY.
Sandler, S. I. 1989. *Chemical and Engineering Thermodynamics*, John Wiley & Sons, New York, NY.
Sherwood, T. K., Pigford, R. L., and Wilke, C. R. 1975. *Mass Transfer*, McGraw-Hill, New York, NY.

EXERCISES

Most solutions to the following exercises require you to draw on a diagram of thermodynamic data. You may wish to draw your solutions on copies of the diagrams, which may be downloaded from duncan.cbe.cornell.edu/Graphs. The downloaded copies are larger and have finer scales.

All exercises apply concepts and skills introduced in this chapter. Some exercises use maps and graphs similar to those in the text, and some exercises use maps and graphs not covered in the text. The goal is to master graphical methods, not to memorize maps. If one learns how to read a road map by studying one's home city, one should be able to apply these techniques to other road maps. Likewise, if one learns how to use tie lines for liquid + vapor two-phase regions on temperature–composition maps, one should be able to apply these techniques to other two-phase regions (for example, liquid + solid) and on maps with other coordinates (for example, enthalpy–composition maps).

Linear and Logarithmic Scales

4.1 Plot the data given in (A), (B), and (C) to determine the functional relation between the two variables. That is, obtain an equation to express the second variable as a function of the first variable. Each set of data will yield a straight line when plotted on one of three types of graph paper – normal, semilog, or log–log (see Appendix D).

(A) Fluid mechanics. The frictional force on a flat plate (F) as a function of the velocity (v) of the fluid flowing past the plate.

v (cm/s)	F (dynes)
4.0	1.35
5.0	1.8
10.	5.3
20.	15.
45.	50.
70.	98.

(B) The Leibnitz experiment to measure the conversion of potential energy to heat. The temperature change of a water bath (ΔT) is measured as a function of the mass (M) that falls a given distance.

M (kg)	ΔT (mK)
10.	12.
25.	30.
40.	49.
55.	64.5
62.5	75.
75.	90.

(C) Chemical kinetics. The concentration of reactant A in a batch reactor as a function of time, t.

t (min)	[A] (mol/m^3)
0.	90.
10.	50.
30.	15.
40.	8.5
55.	3.5
70.	1.5

Graphical Energy Balances on Pure Substances

4.2 Shown below is a temperature scale and a linear specific-energy scale for pure substance Q. The temperature tick marks are at intervals of 20°C.

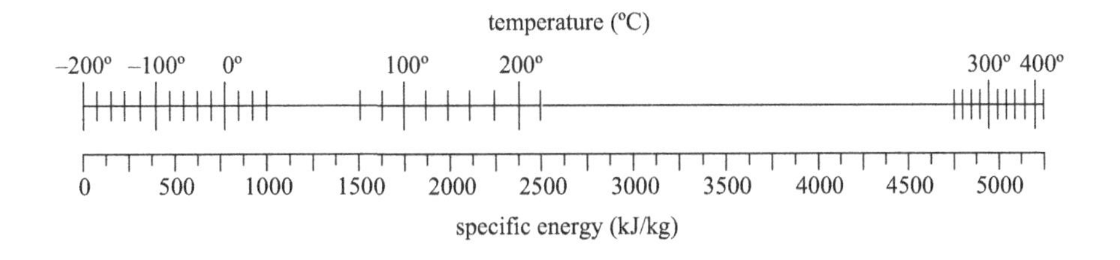

(A) At what temperature does Q melt, in °C?
(B) What is the heat of vaporization, $\Delta \bar{H}_{vap}$, of Q at 1 atm, in kJ/kg?
(C) What is the heat capacity of liquid Q, $\bar{C}_{P,liquid}$, in kJ/(kg·°C)? You may assume the heat capacity of liquid Q is independent of temperature.

4.3 (A) Draw a specific-energy–phase line for phenol (C_6H_5OH) at 1 atm, similar to the specific-energy–phase lines we used for graphical energy balances with H_2O at 1 atm. On the linear scale below, label the regions from "solid" to "vapor." Set specific energy = 0 kJ/kg for liquid phenol at its freezing point. *You need not draw a temperature scale.* If a region is outside the range on the number line below, state the location of the border(s) to this region, in kJ/kg.

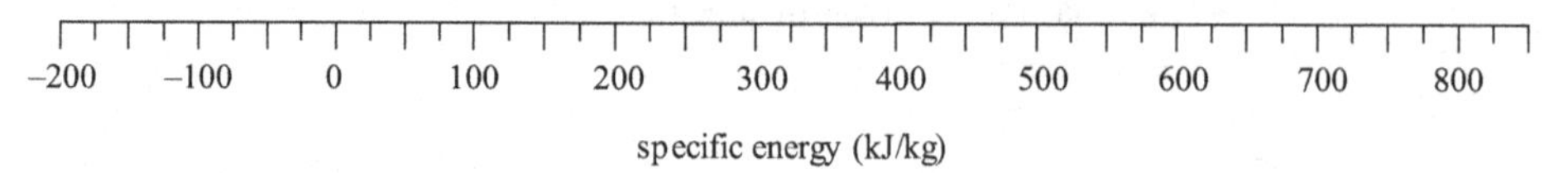

(B) Indicate on your phase line the locations that correspond to the following three temperatures: 0°C, 111°C, and 181°C.

Physical data for phenol:

melting point = 40.9°C at 1 atm
boiling point = 181.9°C at 1 atm
critical temperature = 421.1°C
critical pressure = 60.5 atm

$$\Delta\bar{H}_{\text{melt}} = 122.4 \text{ kJ/kg}$$

$$\Delta\bar{H}_{\text{vap}} = 432.6 \text{ kJ/kg}$$

$$\bar{C}_{\text{P,solid}} = 1.432 \text{ kJ/(kg·°C)}$$

$$\bar{C}_{\text{P,liquid}} = 1.354 \text{ kJ/(kg·°C)}$$

$$\bar{C}_{\text{P,vapor}} = 1.100 \text{ kJ/(kg·°C)}$$

molecular weight = 94.1 amu
density = 1.058 g/cm^3 at 20°C.

4.4 Construct a graphical solution to Exercise 3.64.

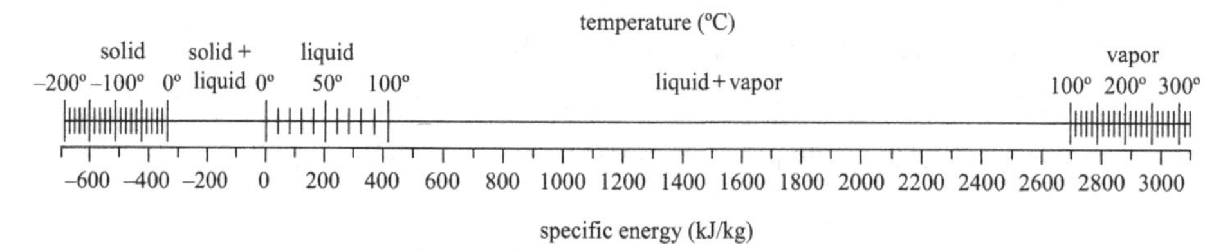

4.5 Construct a graphical solution to Exercise 3.65.

4.6 Construct a graphical solution to Exercise 3.66.

4.7 Use graphical methods to calculate the flow rate of stream 2.

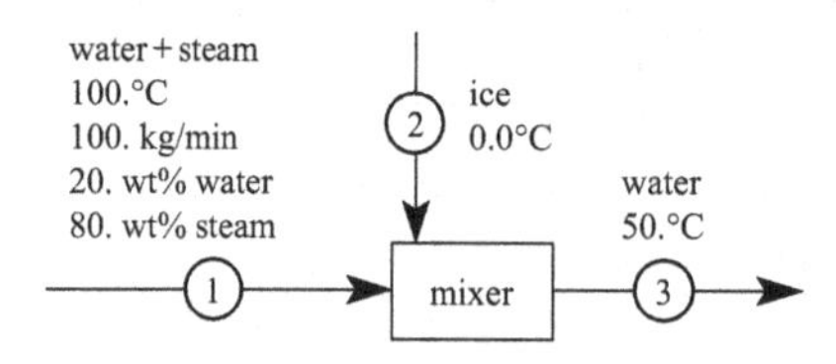

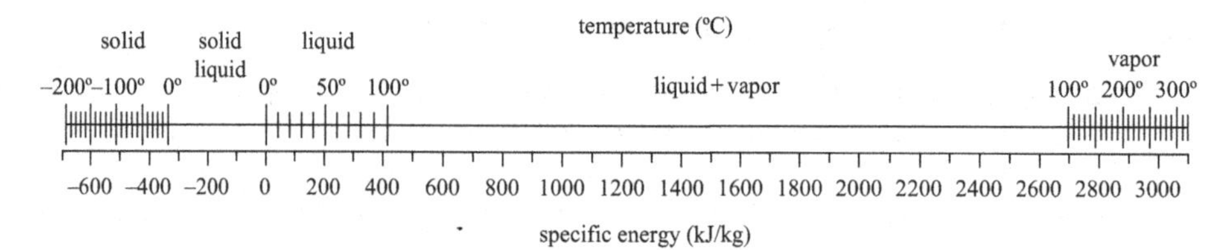

4.8 Use graphical methods to calculate the composition of stream 2; calculate the wt fractions of water and steam in stream 2.

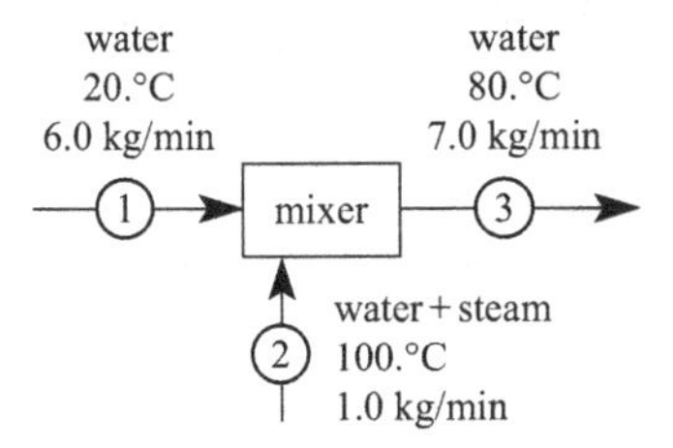

4.9 Use graphical methods to calculate the flow rate of stream 1 needed to produce an output stream with no vapor.

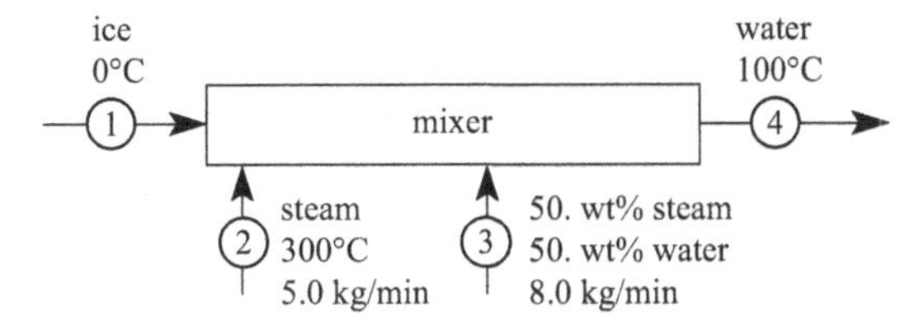

4.10 Use graphical methods to calculate the temperature of stream 5. If stream 5 is a mixture of two phases (ice + water or water + steam), calculate the relative fractions of each phase.

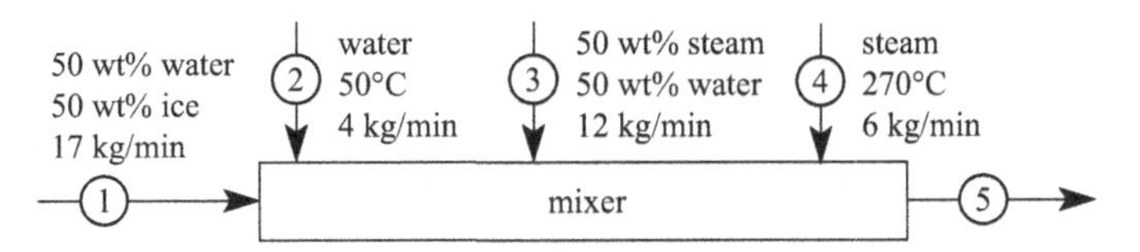

4.11 Use graphical methods to calculate the flow rate of cooling water (stream 3) needed to condense and cool stream 1. **Hint:** Use the H_2O energy line to calculate the specific-energy change (kJ/kg) for stream 1 to stream 2, and for stream 3 to stream 4.

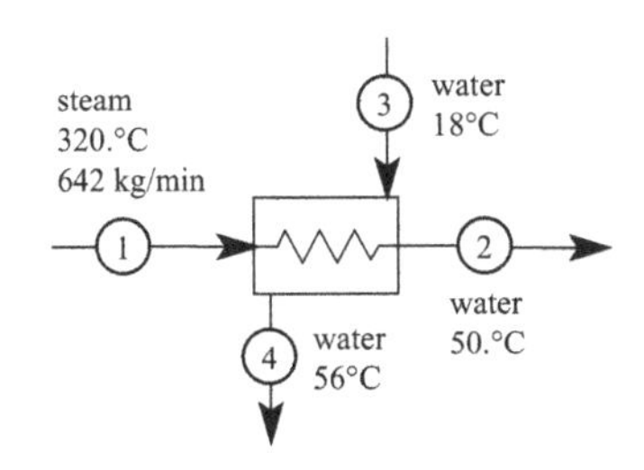

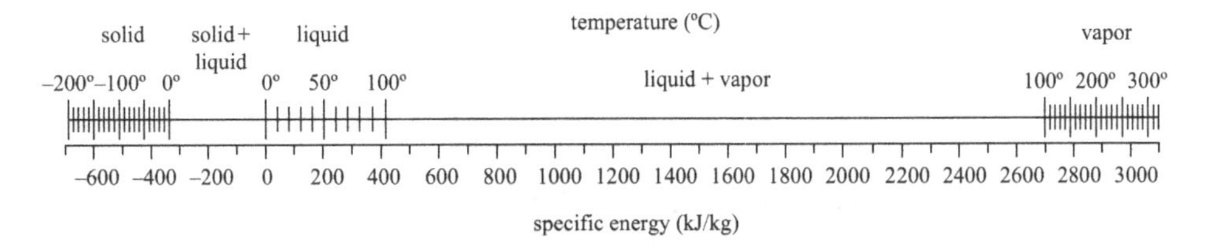

4.12 Construct a graphical solution to Exercise 3.67.

Graphical Mass Balances on Temperature–Composition and Pressure–Composition Phase Diagrams

4.13 The unit below treats a mixture of tetrachloromethane (CCl_4) and chlorobenzene (C_6H_5Cl) by flash distillation.

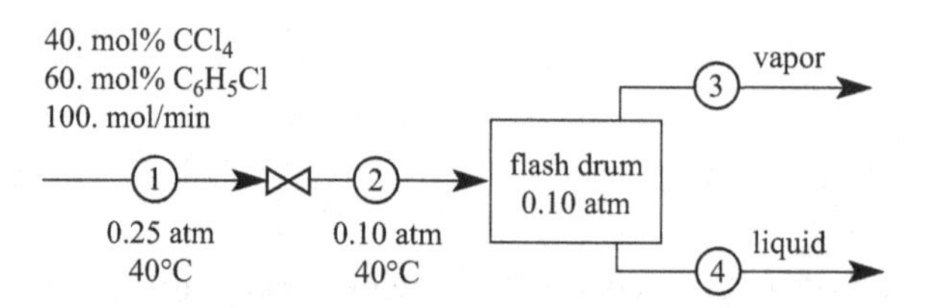

Use the CCl_4-C_6H_5Cl phase diagram below to determine the compositions (mol fractions CCl_4 and C_6H_5Cl) and flow rates (mol/min) of streams 3 and 4. Also, label on the CCl_4–C_6H_5Cl phase diagram the phases corresponding to each region and indicate the coordinates of (1) the stream before the expansion valve, (2) the vapor stream, and (3) the liquid stream.

Pressure–composition phase diagram for CCl_4 + C_6H_5Cl mixtures at 40°C.

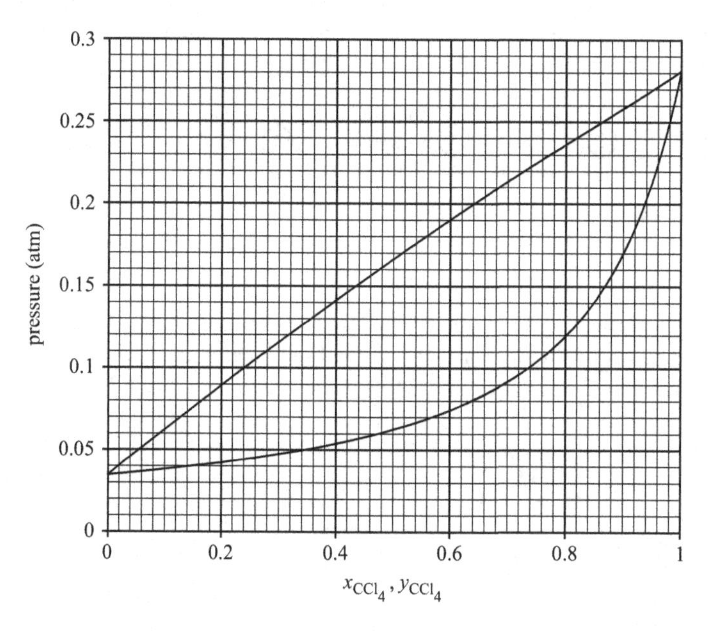

4.14 Use the phase diagram for CCl_4 + C_6H_5Cl mixtures shown in Exercise 4.13 to determine the flow rate and composition of stream 4 leaving the flash drum shown below. Also determine the pressure in the flash drum.

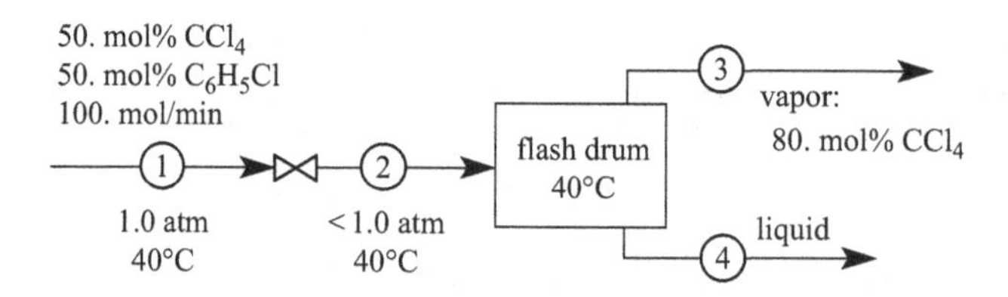

4.15 The flash drum shown below separates a 50:50 mixture of acetonitrile (ACN) and 2-methylpyridine (MP) at 98°C and 1 atm.

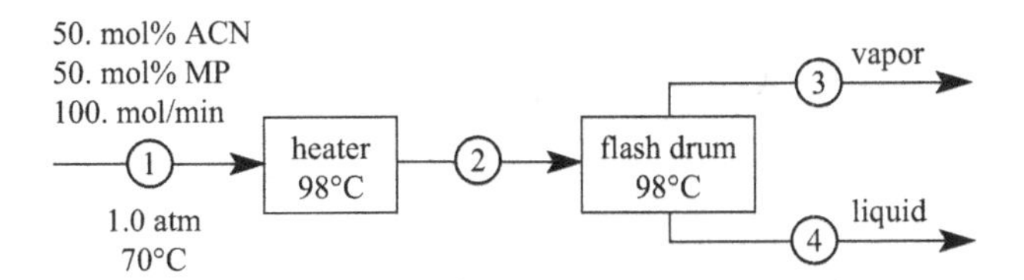

(A) Determine the composition of stream 3 and the composition of stream 4.
(B) Use the lever rule to determine the flow rate of stream 3 and the flow rate of stream 4.

Temperature–composition phase diagram for acetonitrile + 2-methylpyridine mixtures at 1 atm.

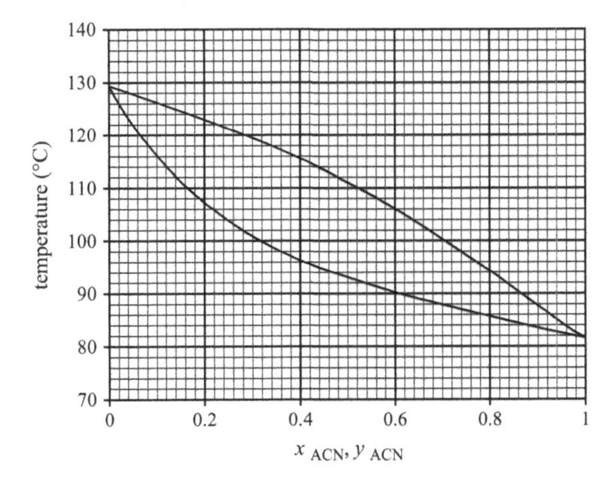

(C) A second flash drum is added. What is the temperature in the second flash drum?

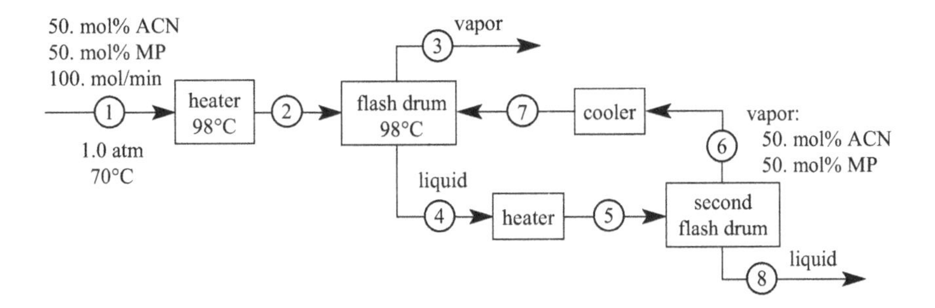

(D) What is the composition of stream 8 (in mol% ACN)?
(E) Calculate the flow rate of stream 8 *and* the flow rate of stream 7. You may use the lever rule and/or mass balances. Note that the flow rate of stream 4 may be changed by the addition of the second flash drum.

4.16 A mixture of M and K is treated with the flash drum shown below.

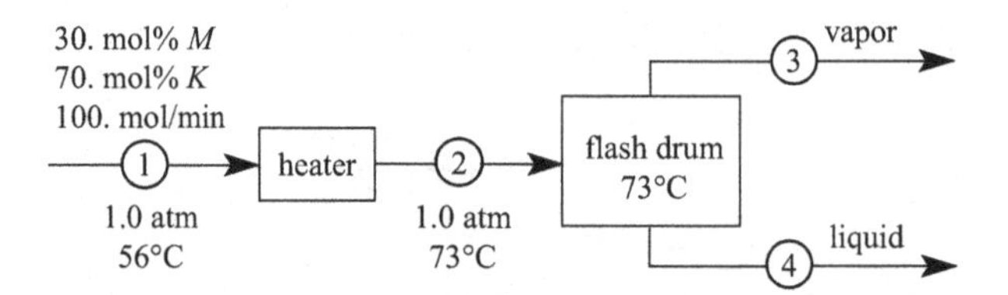

Use the phase diagram shown below for $M + K$ mixtures to determine the compositions (mol fractions M and K) and flow rates (mol/min) of streams 3 and 4. Also, label on the M–K phase diagram the phases corresponding to each region and indicate the coordinates of streams 1, 3, and 4.

Temperature–composition phase diagram for $M + K$ mixtures at 1 atm.

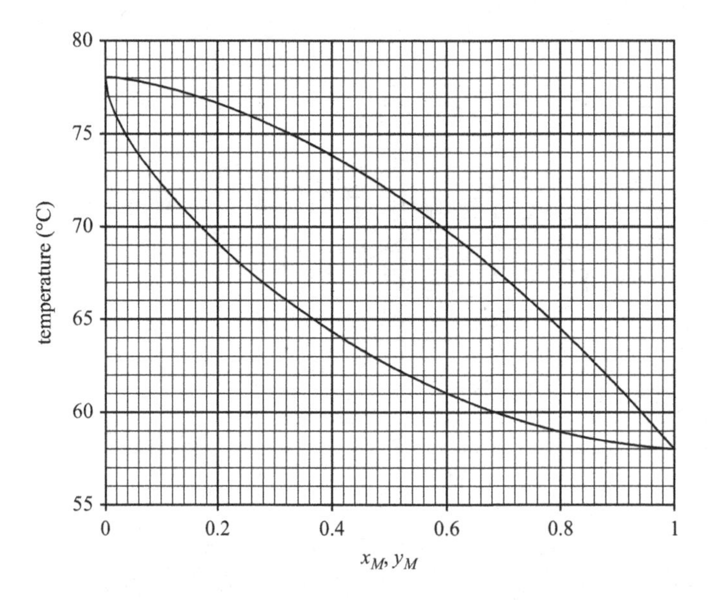

4.17 The flash drum diagrammed below produces an equimolar vapor of M and K.

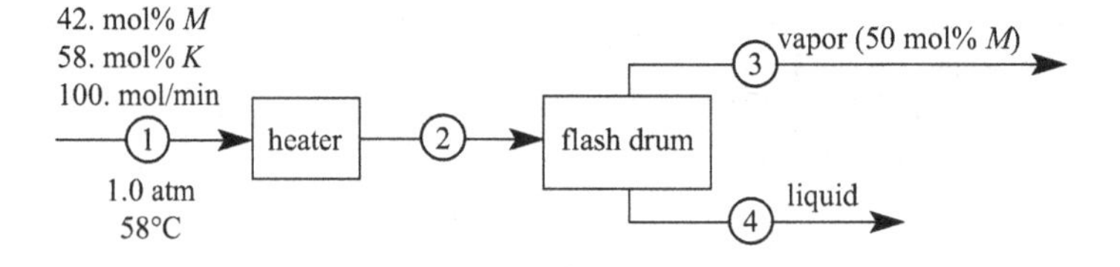

(A) What is the composition of stream 4 (in mol% M)?
(B) What is the flow rate of stream 4 (in mol/min)?
(C) What is the temperature in the flash drum?
(D) The temperature is changed to maximize the concentration of M in stream 3. What is the maximum possible concentration of M in stream 3?

4.18 The process diagrammed below produces an equimolar mixture of M and K at 100. mol/min from either stream 3 (vapor) or stream (4), depending on the composition of the input (stream 1), which varies over time.

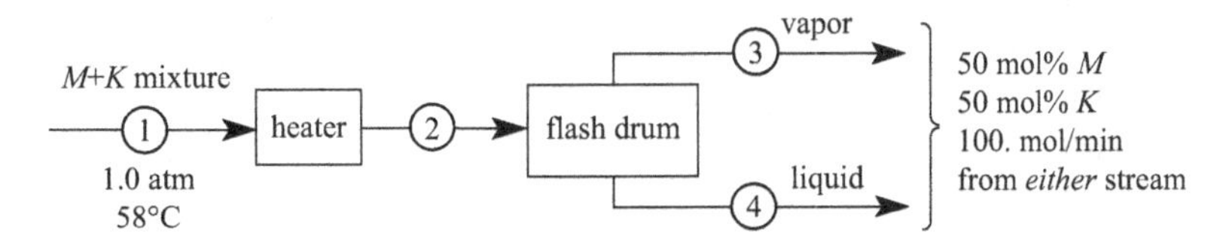

Assume the equimolar product is the vapor stream (stream 3).

(A) What is the range of compositions permitted in stream 1?
(B) What is the composition of stream 4 (in mol% M)?
(C) What is the temperature in the flash drum?

Instead assume the equimolar product is the liquid stream (stream 4).

(D) What is the range of compositions permitted in stream 1?
(E) What is the composition of stream 3 (in mol% M)?
(F) What is the temperature in the flash drum?

Assume further that the vapor flow rate (stream 3) is 190. mol/min.

(G) What is the flow rate and composition of stream 1?

4.19 Use the experimental data given below to determine the flow rate (mol/min) and composition of stream 7 in the process diagrammed below. Both flash drums operate at 1.0 atm.

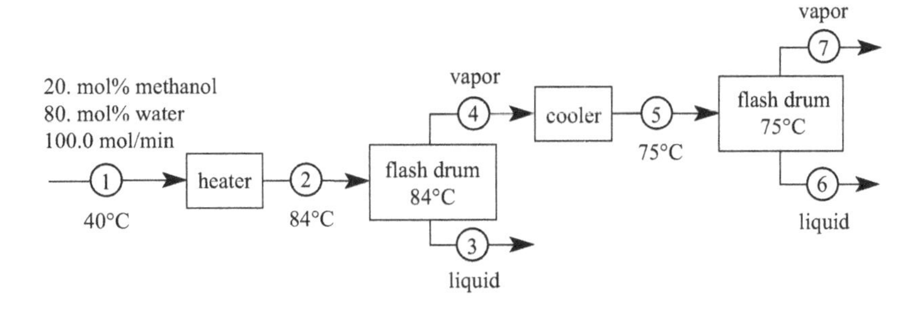

The data below were measured with methanol + water mixtures at 1.0 atm. These data may be downloaded from duncan.cbe.cornell.edu/Graphs.

T (°C)	x_{methanol}	y_{methanol}
100.0	0.0	0.0
98.4	0.012	0.068
96.9	0.020	0.121
95.8	0.026	0.159
95.1	0.033	0.188
94.1	0.036	0.215

(*cont.*)

T (°C)	$x_{methanol}$	$y_{methanol}$
92.2	0.053	0.275
90.0	0.074	0.356
88.6	0.087	0.395
86.9	0.108	0.440
85.4	0.129	0.488
83.4	0.164	0.537
82.0	0.191	0.572
79.1	0.268	0.648
78.1	0.294	0.666
76.5	0.352	0.704
75.3	0.402	0.734
74.2	0.454	0.760
73.2	0.502	0.785
72.0	0.563	0.812
70.9	0.624	0.835
69.2	0.717	0.877
68.1	0.790	0.910
67.2	0.843	0.930
66.9	0.857	0.939
65.7	0.938	0.971
65.0	1.0	1.0

Experimental data from the compilation by Gmehling, J. and Onken, U. (1977). *Vapor-Liquid Equilibrium Data Collection*, Dechema, Frankfurt, Germany, vol. 1, p. 60.

4.20 A mixture of P and Q is treated in the process below.

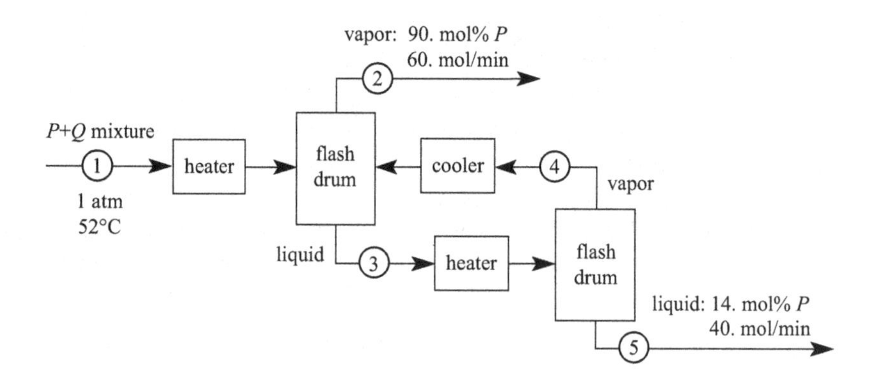

(A) What is the composition of stream 3?
(B) What is the flow rate of stream 4?
(C) What is the composition of stream 1?
(D) Given this arrangement of flash drums, what is the maximum mol% P in stream 2?
(E) If stream 2 is maximum mol% P, what is the minimum mol% P that can be achieved in stream 5?

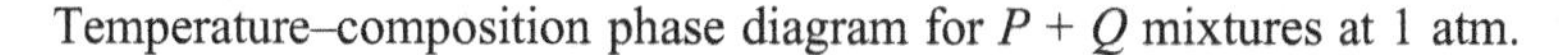
Temperature–composition phase diagram for $P + Q$ mixtures at 1 atm.

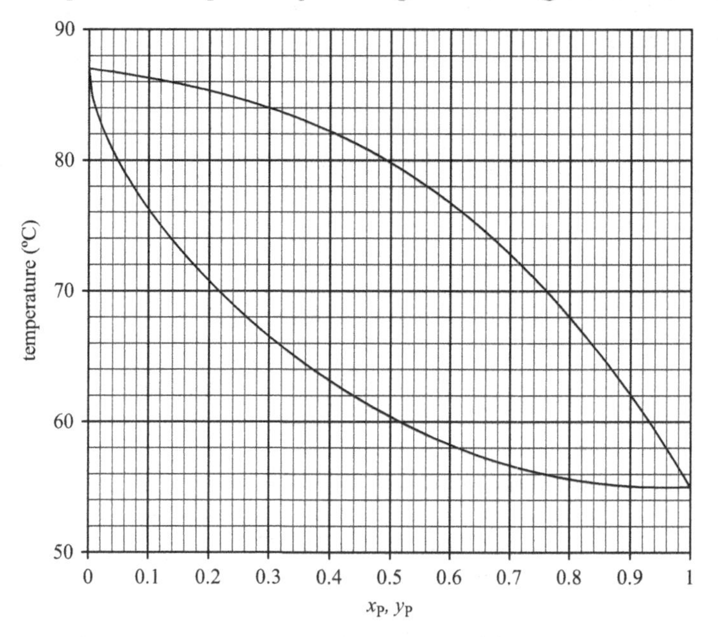

4.21 A mixture of A and B is treated in the flash drum shown below.

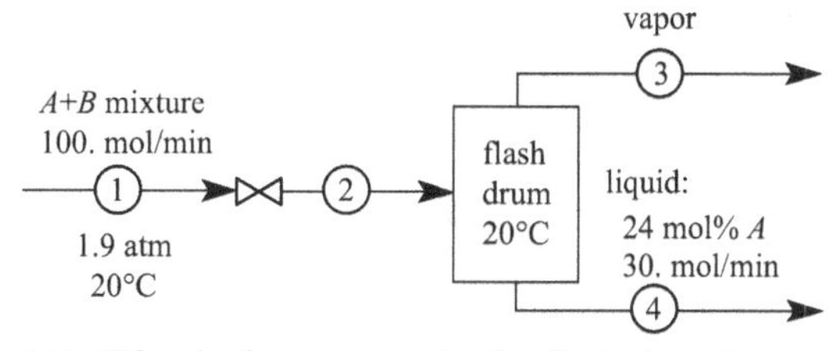

(A) What is the pressure in the flash drum?

(B) What is the composition of the vapor in stream 3?

(C) Use the lever rule to calculate the composition of the feed in stream 1. You may use a mass balance to check your answer, but you must demonstrate the lever rule for full credit.

Pressure–composition phase diagram for $A + B$ mixtures at 20°C.

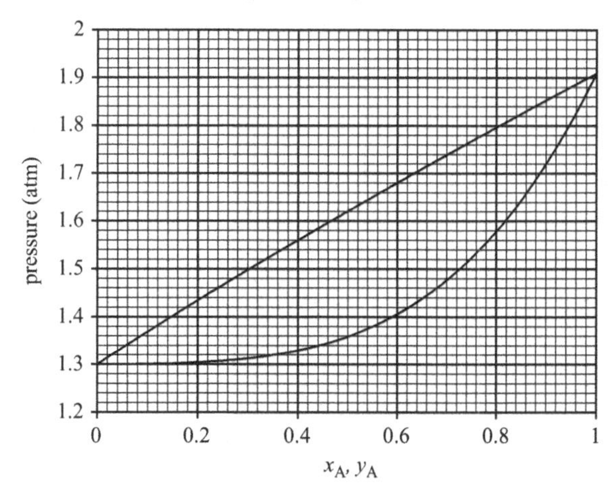

4.22 Consider the following process to separate a mixture of dimethylamine (DMA) and 1-propanol (P). The pressure in each flash drum is 1 atm. Heaters, coolers, and pumps are not shown.

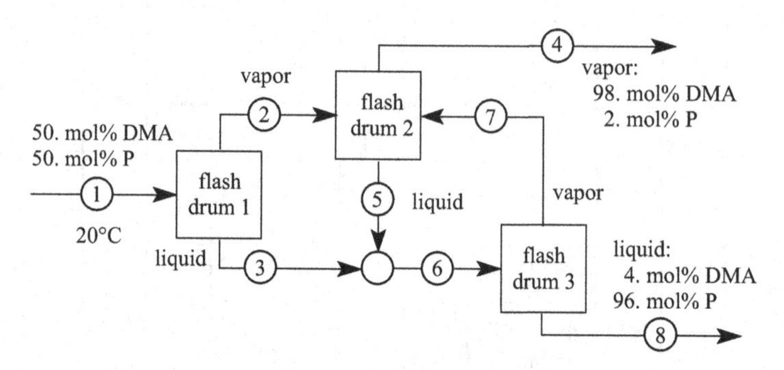

(A) What is the temperature in flash drum 2?
(B) What is the temperature in flash drum 3?
(C) Is this process viable? That is, can this process separate the input stream into two outputs with the compositions stated? Draw tie lines on the temperature–composition phase diagram to justify your conclusion.
(D) If this process is not viable, fix it. You may add flash drums and/or redirect streams as needed to fix the process. If the process is viable, improve it. You may eliminate flash drums and/or redirect streams as needed to improve the process. Your new process must start with 50 mol% DMA and produce two outputs: one with ≥98 mol% DMA and one with ≤4 mol% DMA.

Temperature–composition phase diagram for dimethylamine (DMA) + 1-propanol (P) mixtures at 1 atm.

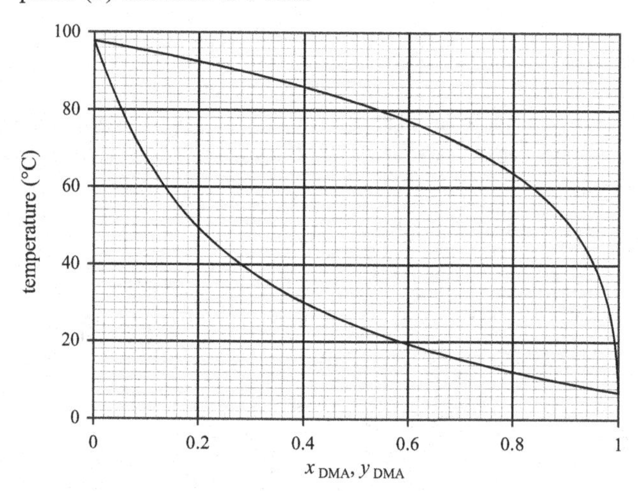

4.23 Consider the following process to treat a mixture of dimethylamine (DMA) and 1-propanol (P) to produce 99 mol% DMA. Is this separation scheme viable? Validate viability with a graphical model or explain why the process is not viable.

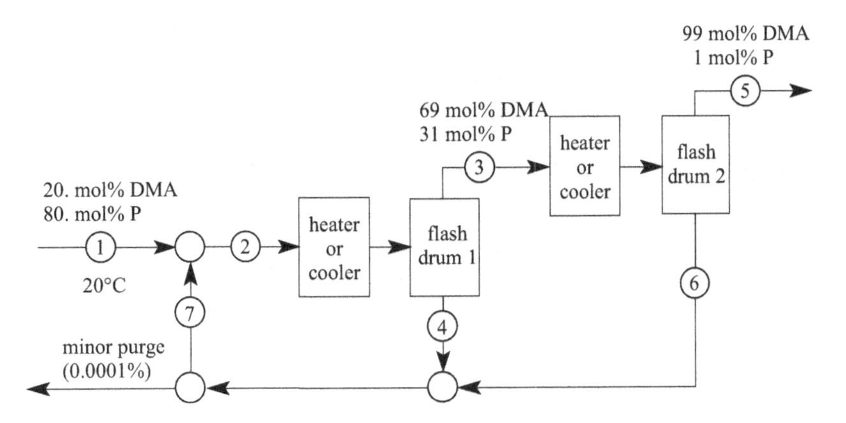

4.24 Consider the following process of three flash drums to separate a mixture of hexane (H) and chlorobenzene (CB). The pressure in each flash drum is 1 atm. Heaters, coolers, and pumps are not shown.

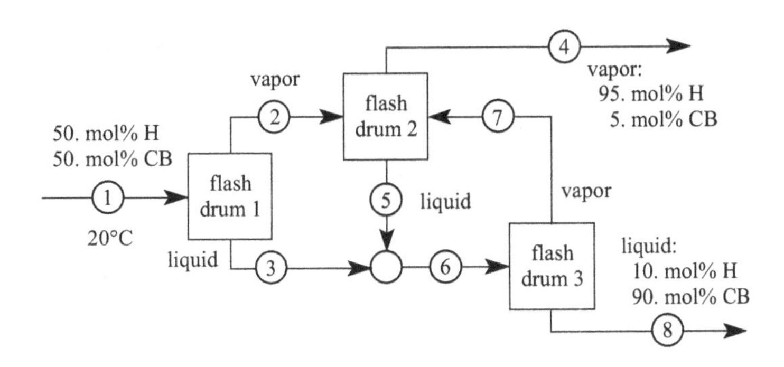

(A) What is the temperature in flash drum 2?
(B) What is the temperature in flash drum 3?
(C) Is this process viable? That is, can this process separate the input stream into two outputs with the compositions stated? Draw tie lines on the temperature–composition phase diagram to justify your conclusion.
(D) If this process is not viable, fix it. You may add flash drums and/or redirect streams as needed to fix the process. If the process is viable, improve it. You may eliminate flash drums and/or redirect streams to improve the process. Your new process must start with 50 mol% H and produce two outputs: one with $\geq$95 mol% H and one with $\leq$10 mol% H.

Temperature–composition phase diagram for hexane (H) + chlorobenzene (CB) mixtures at 1 atm.

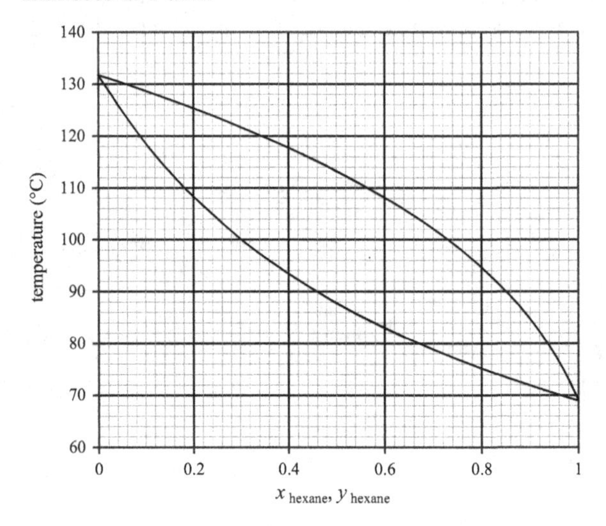

4.25 Apple juice with 10. wt% sugars is concentrated to 40. wt% sugars. Distillation is unacceptable because boiling degrades the taste. Rather, water is removed by freezing, as in the process below.

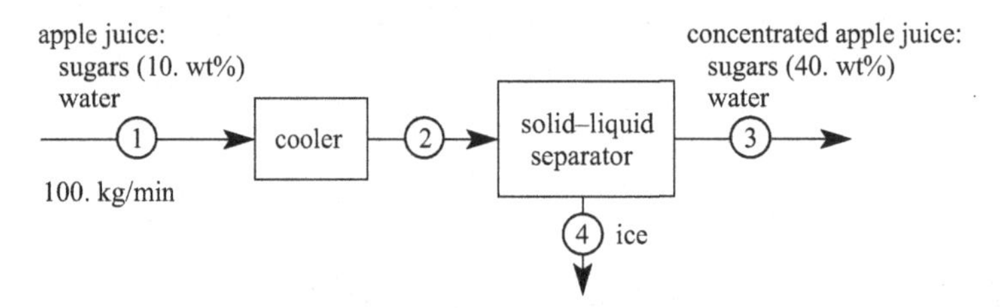

(A) What is the temperature in the cooler (and the solid–liquid separator)?

(B) What is the flow rate of ice in stream 4? You may assume ideal solid–liquid separation; the ice retains no liquid.

(C) Because the solid–liquid separation is expensive, the process is modified so some apple juice bypasses the separator, as below. The cooler's temperature in the process below may be different from the cooler's temperature in the process above. What is the flow rate of ice in stream 4?

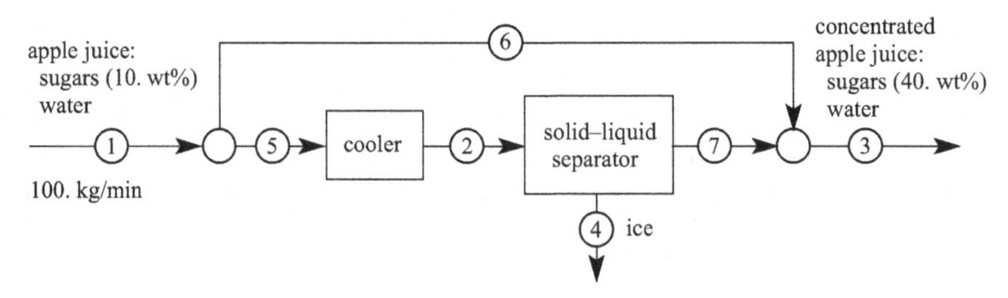

(D) Calculate the *maximum* flow rate in the bypass, stream 6. **Hint:** What temperature in the cooler will maximize the flow rate of stream 6?

Temperature–composition phase diagram for water + sugars mixtures.

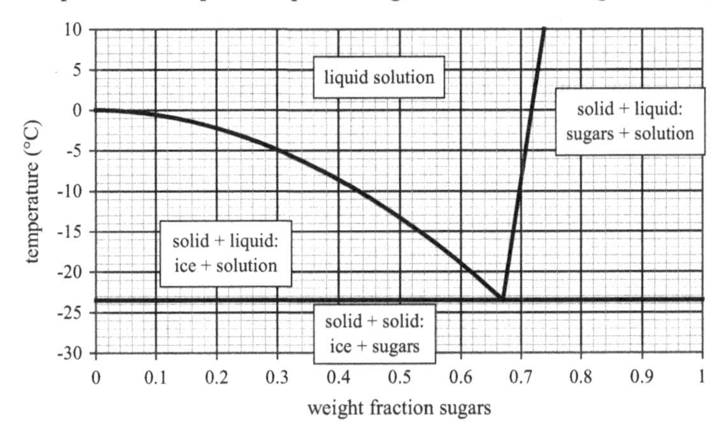

4.26 Shown below is the phase diagram for water and 1-butanol. A liquid mixture with very little butanol is homogeneous. Likewise, a liquid mixture with very little water is homogeneous. For an intermediate composition below 70°C, butanol and water are immiscible and the mixture separates into two liquid phases – one liquid is mostly water and one liquid is mostly butanol. The other two-phase regions are liquid + vapor mixtures. The compositions of the two phases at equilibrium are given by a tie line and the quantities of the two phases are given by the lever rule.

Temperature–composition phase diagram for 1-butanol + water mixtures at 1 atm.

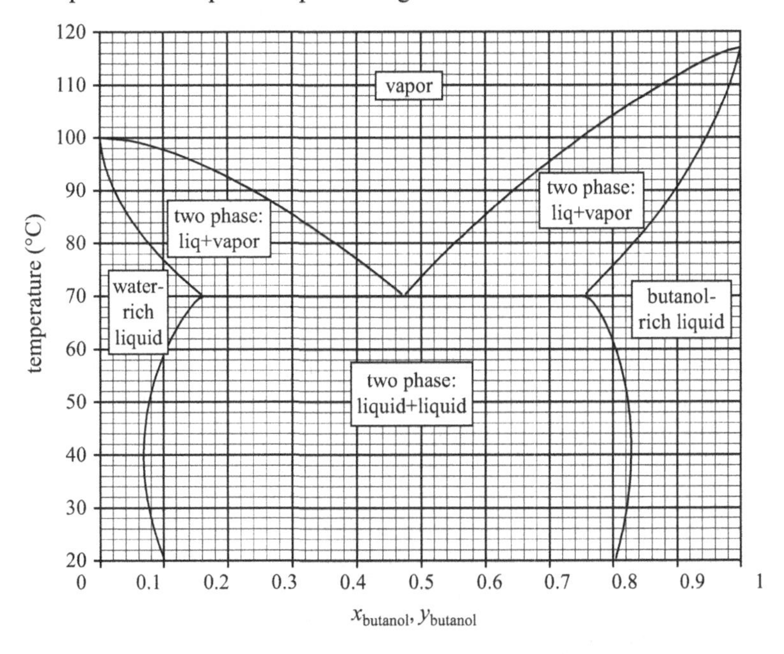

(A) Label on the diagram the boiling point of pure water.
(B) Label on the diagram the boiling point of pure butanol.
(C) Complete the following table for a mixture of 0.90 mol butanol and 0.10 mol water. If only one phase is present, leave the last three columns blank.

		First phase			Second phase		
T(°C)	1 or 2 phases?	Liquid or vapor?	Mol fraction butanol	Total mols	Liquid or vapor?	Mol fraction butanol	Total mols
60							
80							
105							
120							

(D) Complete the following table for a mixture of 0.65 mol butanol and 0.35 mol water. If only one phase is present, leave the last three columns blank.

		First phase			Second phase		
T(°C)	1 or 2 phases?	Liquid or vapor?	Mol fraction butanol	Total mols	Liquid or vapor?	Mol fraction butanol	Total mols
60							
80							
105							
120							

4.27 As discussed in Section 4.1, mass balances can be done graphically, with a number line. Although graphing has little advantage for mixtures with two components, this exercise demonstrates the utility of graphical mass balances for mixtures with three components.

Whereas compositions of two-component mixtures can be represented by a line, compositions of three-component mixtures require an additional dimension. A common format is an equilateral triangle, shown below for water + methanol + ethanol mixtures. A coordinate on this ternary diagram corresponds to a unique composition. Because water, methanol, and ethanol are mutually miscible, every coordinate on this ternary diagram corresponds to a homogeneous liquid.

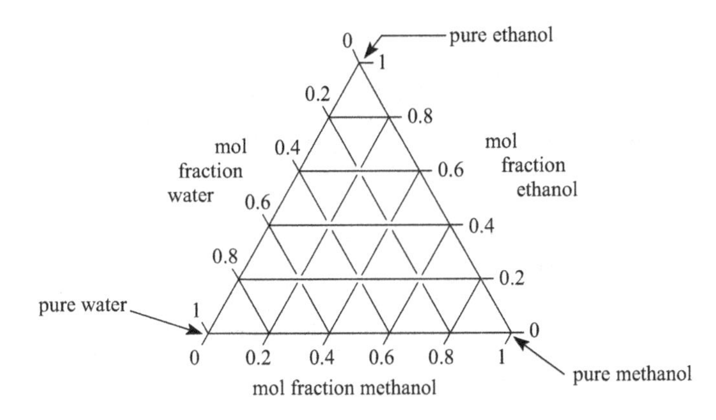

Horizontal lines give the mol fraction of ethanol.

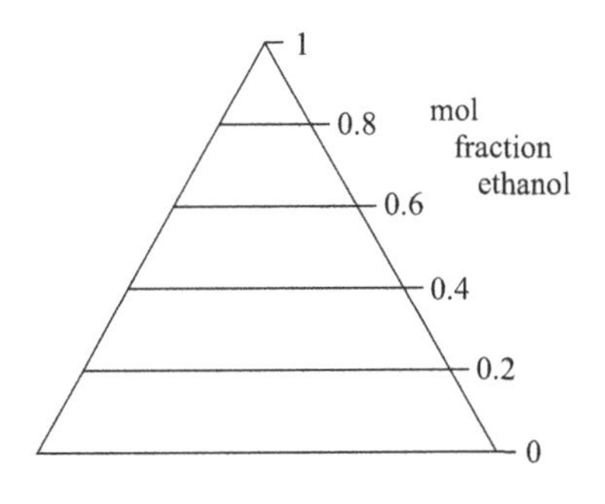

Lines parallel to the left side give the mol fraction of methanol.

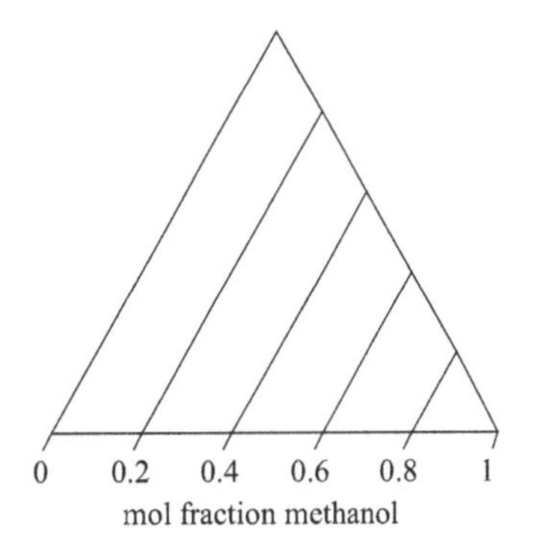

Lines parallel to the right side give the mol fraction of water.

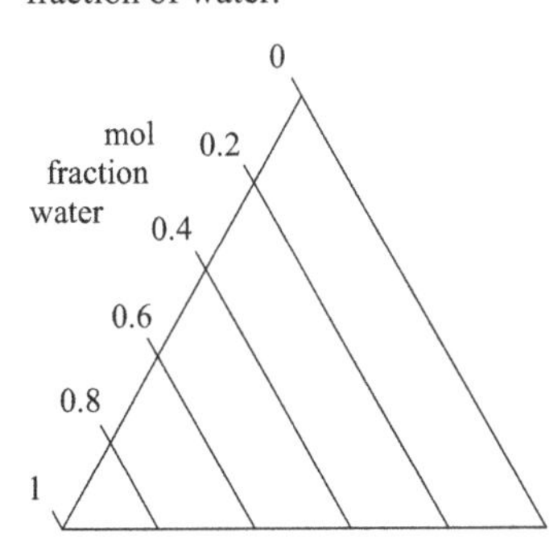

The point below corresponds to a mixture of 10% water, 30% ethanol, and 60% methanol.

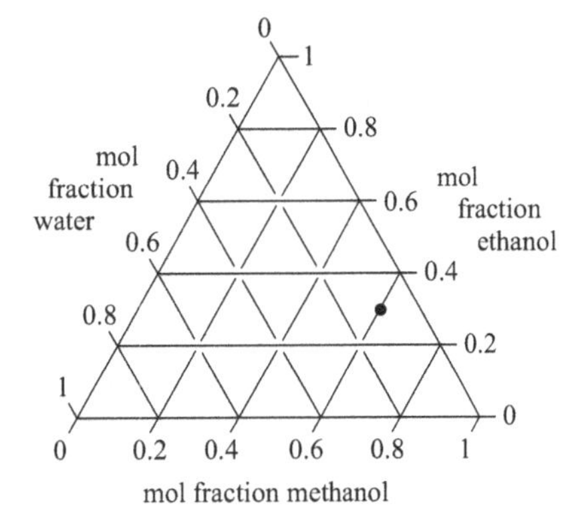

(A) Use a mixing line to calculate the composition of stream 3 in the process diagram below, as follows. Locate the coordinates of streams 1 and 2 on the ternary diagram below. Connect the points with a mixing line. The point corresponding to stream 3 is the intersection of the mixing line and the ethanol composition line at 0.58.

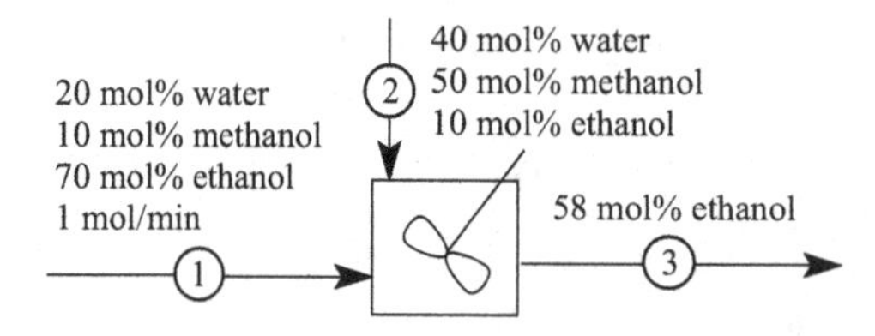

(B) Apply the lever rule to the mixing line drawn in part (A) to determine the flow rate of stream 2 (in mol/min). You may wish to rotate the ternary diagram so the mixing line is horizontal. You may measure the mixing line arms with a ruler or project the mixing line onto an axis.

Ternary diagram for water + methanol + ethanol mixtures at 20°C and 1 atm.

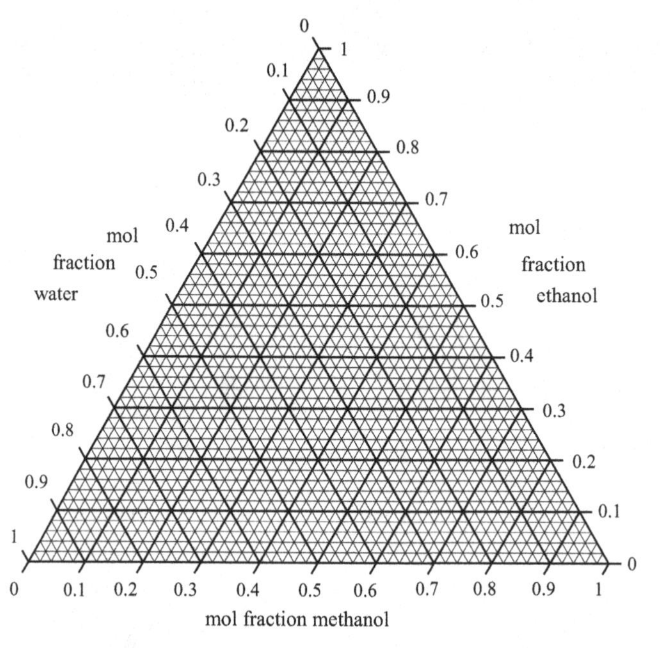

4.28 Mixtures of partially immiscible liquids separate into two liquid phases for some compositions. For example, hexane and methylcyclopentane (MCP) are miscible for all compositions. Also, aniline and MCP are miscible for all compositions. However, aniline and hexane are immiscible for some compositions. The ternary phase diagram for MCP + hexane + aniline at 45°C and 1 atm is sketched below.

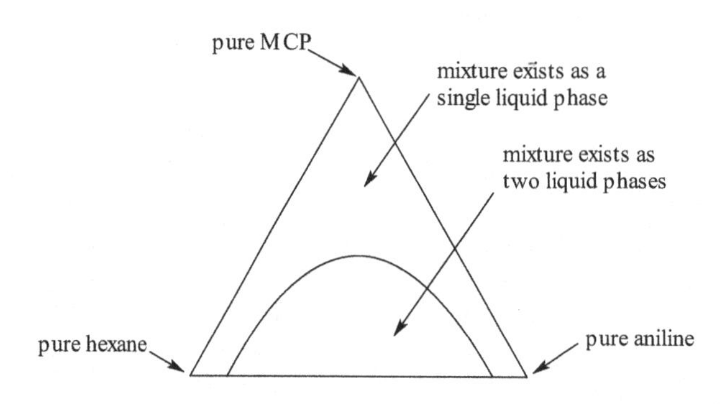

The compositions of the two liquid phases in equilibrium are obtained by a horizontal tie line, as shown below. In this example, a mixture of 15% MCP, 60% hexane, and 25% aniline separates into two liquid phases; one phase is 15% MCP, 75% hexane, and 10% aniline, and the other phase is 15% MCP, 10% hexane, and 75% aniline.

Ternary diagram for MCP + hexane + aniline mixtures at 20°C and 1 atm.

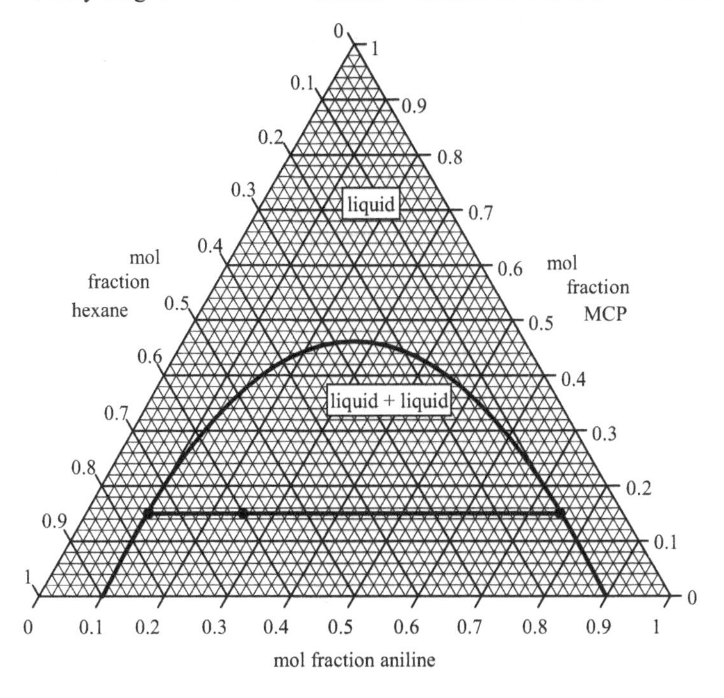

Calculate the compositions and flow rates of streams 4 and 5 in the process below.

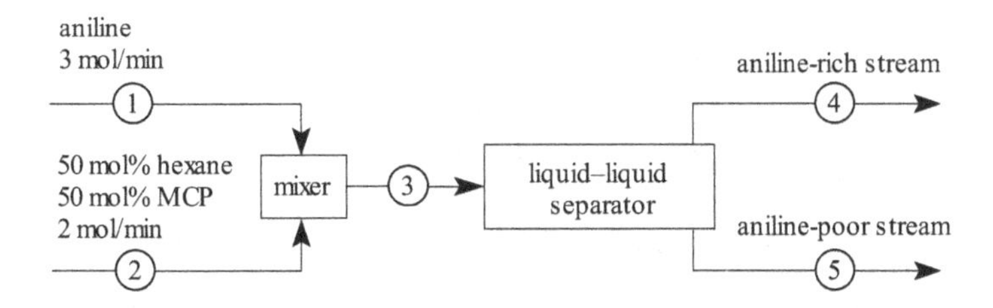

4.29 CO_2 above its critical pressure (72.9 atm) and critical temperature (31°C) – supercritical CO_2 – is useful for selective extraction; aromatic compounds are generally soluble in supercritical CO_2 whereas polar compounds are generally insoluble in supercritical CO_2.

(A) Consider a ternary mixture of an aromatic compound A, a polar compound P, and supercritical CO_2. As shown in the ternary phase diagram below, there are two regions: a single liquid region and a two-phase liquid + liquid region. Some representative tie lines are shown. Label on the ternary diagram the points (or lines) that correspond to (1) pure A, (2) pure P, and (3) pure CO_2.

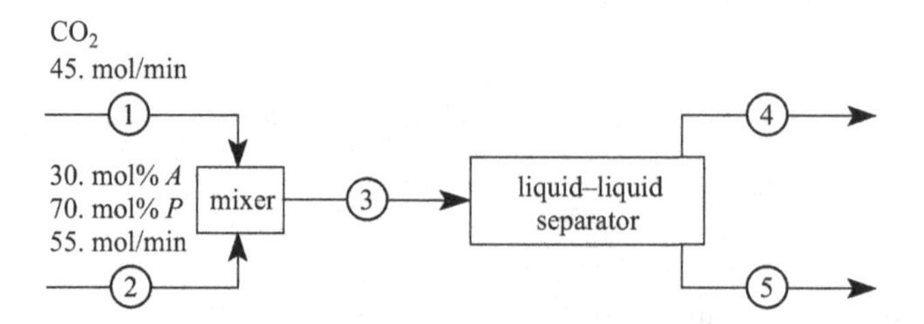

(B) A vessel contains 26 mol supercritical CO_2, 30. mol A, and 44 mol P (100. mol total). Calculate the amount and the composition of each liquid phase.

(C) Calculate the flow rates and compositions of streams 4 and 5 in the process below.

Ternary diagram for supercritical CO_2 + A (aromatic) + P (polar) mixtures.

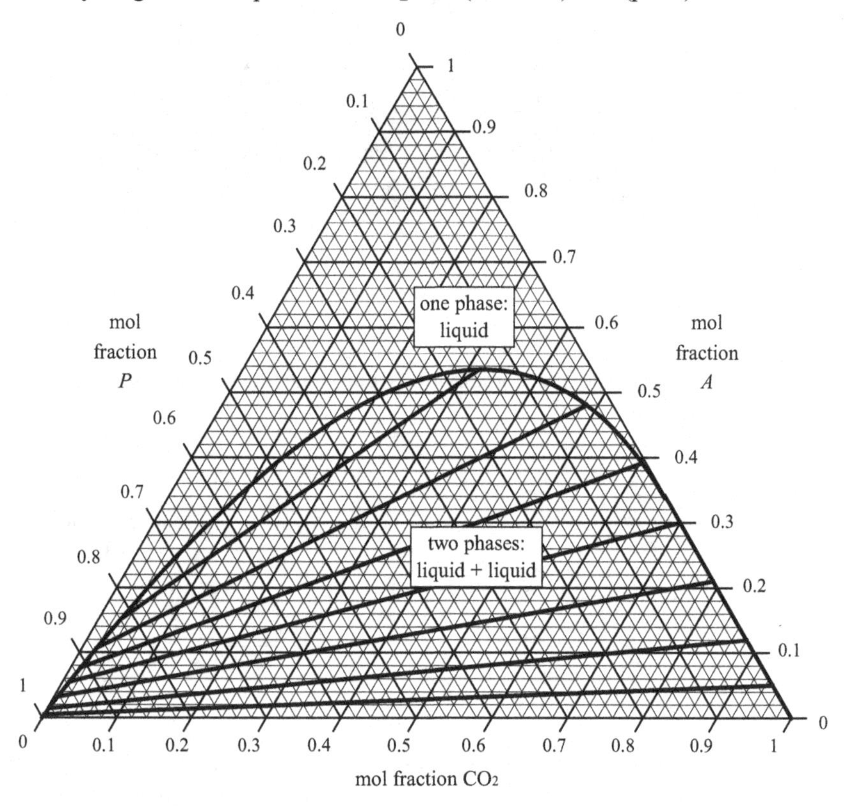

4.30 The three ternary diagrams below show mixtures of liquid A and liquid B with different extracting liquids; X, Y, or Z.

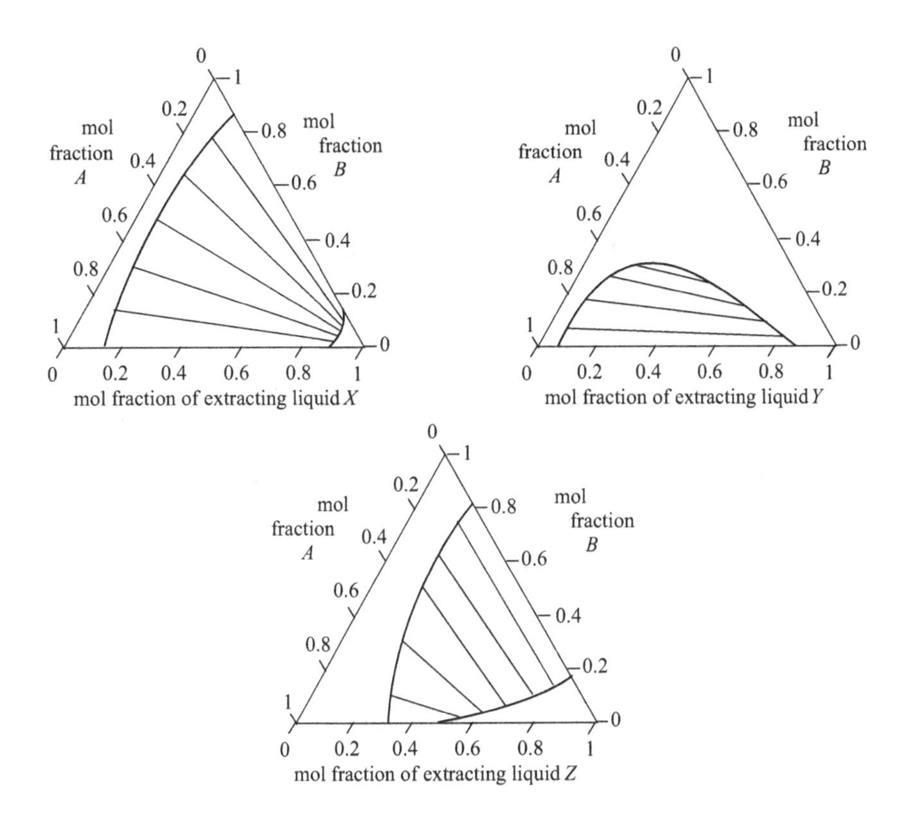

Each ternary diagram has two types of regions: two-phase immiscible liquid (with representative tie lines) and single-phase liquid (no tie lines).

(A) Use the ternary diagrams to decide whether the following mixtures will be one liquid phase or two immiscible liquid phases. If two phases are present, state the approximate composition of each phase. All compositions are on a molar basis.

			If two liquid phases, complete these columns	
	Label point on a ternary diagram as:	One liquid phase or two liquid phases?	Approx. composition of first phase?	Approx. composition of second phase?
50% *A*, 50% *B*	1			
50% *A*, 50% *X*	2			
50% *B*, 50% *Z*	3			
40% *A*, 20% *B*, 40% *X*	4			
5% *A*, 5% *B*, 90% *Z*	5			

We need to separate a mixture of 50 mol% *A* and 50 mol% *B*. Distillation is impractical because *A* and *B* have the same boiling point. Instead, we propose an extraction process that begins as below.

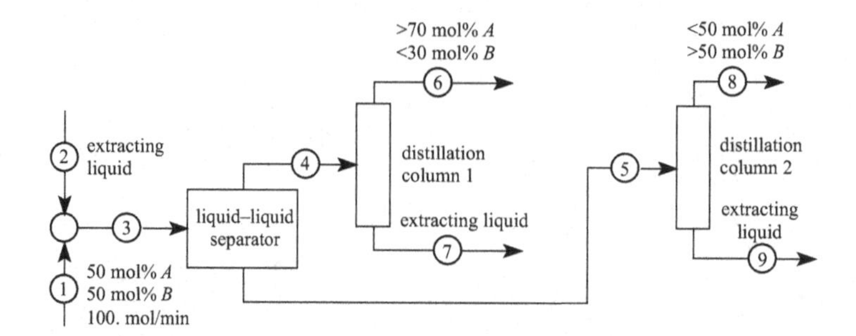

(B) Use the ternary diagrams above to select an extracting liquid (X, Y, or Z) that will maximize the mol% A in stream 6 and will maximize the mol% B in stream 8. *Justify your choice.*

(C) Specify a flow rate for stream 2 that yields approximately equal flow rates in streams 4 and 5.

(D) Given the flow rate you specified in (C), what are the compositions of streams 4 and 5?

(E) What are the compositions of streams 6 and 8?

4.31 The four preceding exercises used ternary diagrams plotted on equilateral triangles to analyze three-component mixtures. Below is a different representation of three-component mixtures – n-heptane (H), methylcyclohexane (MCH), and aniline (A). The wt fraction of H is the x axis; the wt fraction of MCH is the y axis. The wt fraction of A is implied; a mixture with 40% H and 50% MCH implicitly has 10% A.

There are three regions on the map: two regions of single-phase liquids, separated by a two-phase liquid + liquid region. The dashed lines in the two-phase region are tie lines. Additional tie lines may be drawn by interpolating between the dashed lines.

Ternary diagram for heptane + methylcyclohexane + aniline mixtures at 20°C and 1 atm.

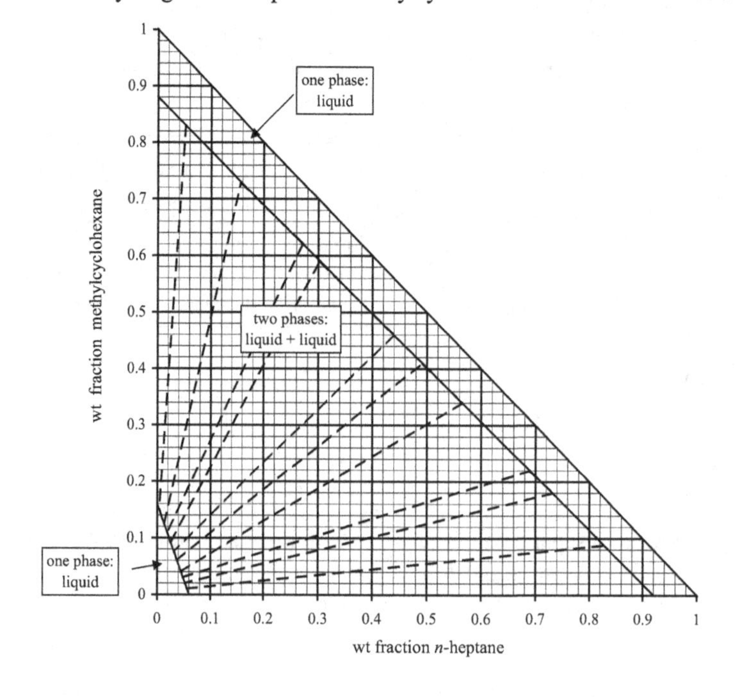

(A) Label on the map the points (or regions) that correspond to: (1) pure *n*-heptane (H), (2) pure methylcyclohexane (MCH), (3) pure aniline (A), (4) aniline-rich liquid, and (5) aniline-poor liquid.
(B) Sketch the tie line for two-phase mixtures that contain no *n*-heptane.
(C) A mixture with 10% H, 40% MCH, and 50% A will separate into two liquid phases. Determine the compositions of the two liquid phases.

4.32 The phase diagrams for $L + M$ mixtures and $L + N$ mixtures are shown below.

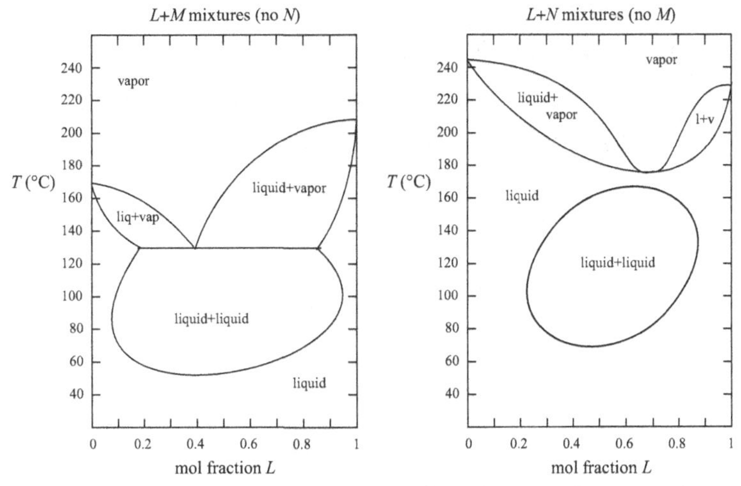

(A) Shown below is a ternary diagram for mixtures of L, M, and N at a particular temperature. What temperature does this ternary diagram represent: 40°C, 60°C, 80°C, 100°C, 120°C, 140°C, 160°C, 180°C, 200°C, 220°C, or 240°C?

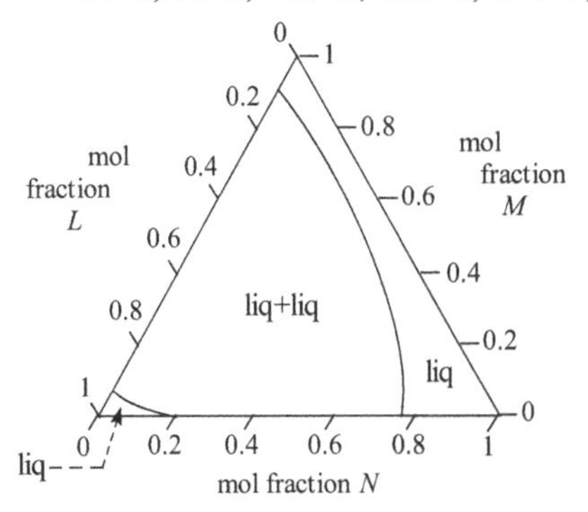

(B) Use the ternary diagram below and sketch the approximate phase boundaries (if any) for 40°C lower than the temperature in (A).

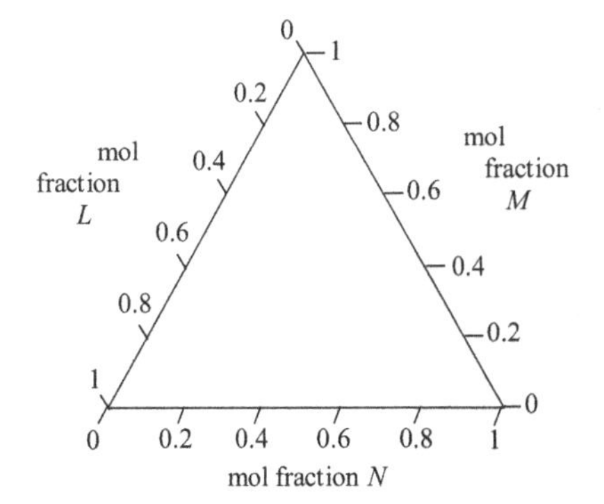

Combined Mass and Energy Balances on Two-Component Mixtures: Enthalpy–Composition Phase Diagrams

4.33(A) Use the temperature–composition and enthalpy–composition phase diagrams for mixtures of G and H below to determine the enthalpy (in kJ/mol) of liquid and vapor phases in equilibrium at 87.5°C. **Hint:** Use a tie line on the temperature–composition phase diagram at 87.5°C to determine the liquid and vapor compositions. Use these compositions to draw a tie line on the enthalpy–composition phase diagram.

Temperature–phase diagram for $G + H$ mixtures at 1 atm.

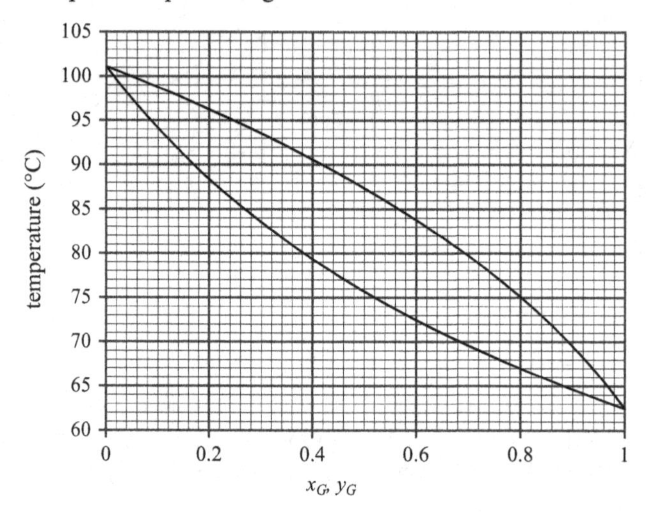

Enthalpy–composition phase diagram for $G + H$ mixtures at 1 atm.

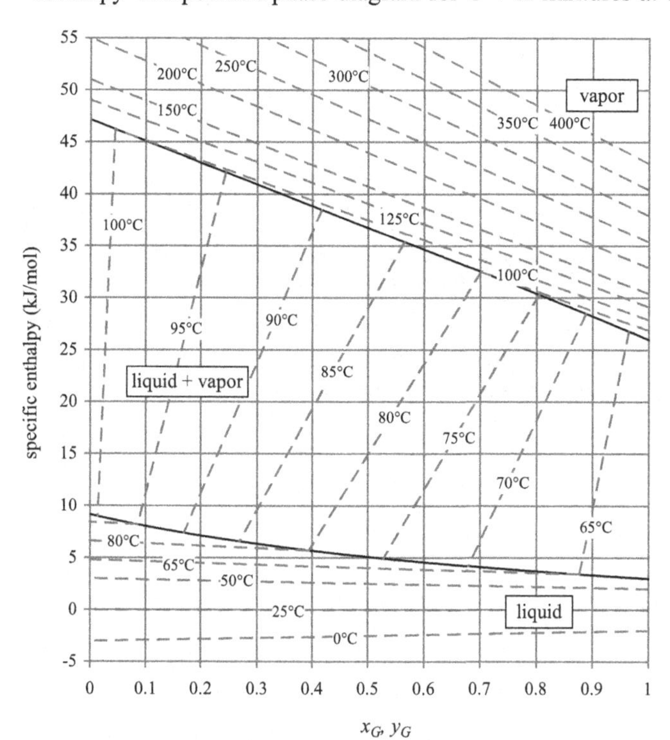

(B) Calculate the energy needed to heat 10 mol of liquid with 20 mol% G at 0°C to vapor at 125°C.

(C) One mol of 50 mol% G at 125°C is mixed with 1 mol of 90 mol% G at 400°C. Calculate the temperature and composition of the resulting mixture.

(D) One mol of H liquid at 0°C is mixed with 1 mol of G vapor at 200°C. Calculate the temperature and composition of the resulting mixture.

4.34(A) Use the enthalpy–composition phase diagram below for ammonia + water mixtures at 300 psia to calculate the composition, temperature, and phase(s) of stream 3 in the process below. The entire process is at 300 psia (20.4 atm).

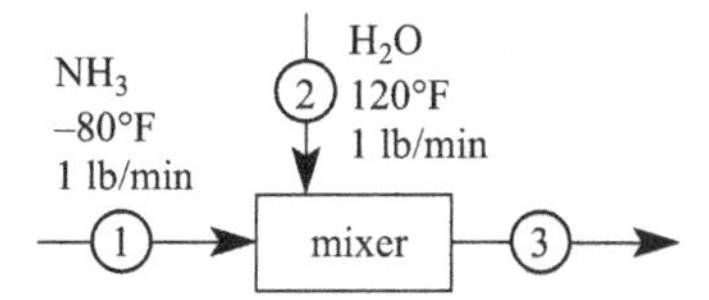

Enthalpy composition phase diagram for $NH_3 + H_2O$ mixtures at 300 psia (20.4 atm).

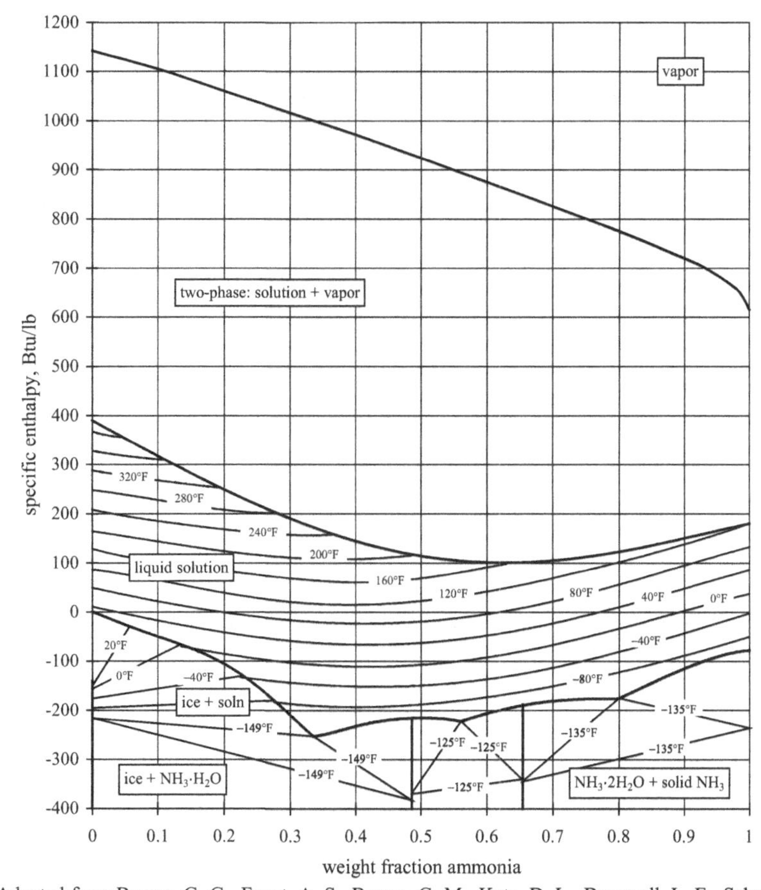

Adapted from Brown, G. G., Foust, A. S., Brown, G. M., Katz, D. L., Brownell, L. E., Schneidewind, R., Martin, J. J., White, R. R., Williams, G. B., Wood, W. P., Banchero, J. T., and York, J. L. (1950). *Unit Operations*, Wiley, New York, page 592.

(B) Use graphical methods to calculate the composition, temperature, and phase(s) of stream 6 in the process below. The entire process is at 300 psia (20.4 atm).

Hint: You will need to use the temperature–composition phase diagram to construct a liquid–vapor tie line at the temperature of stream 6.

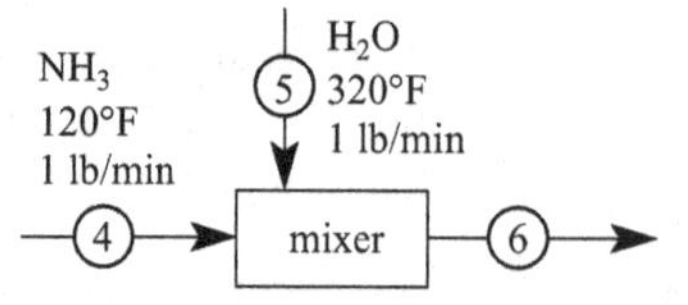

Temperature–composition phase diagram for $NH_3 + H_2O$ mixtures at 300 psia (20.4 atm).

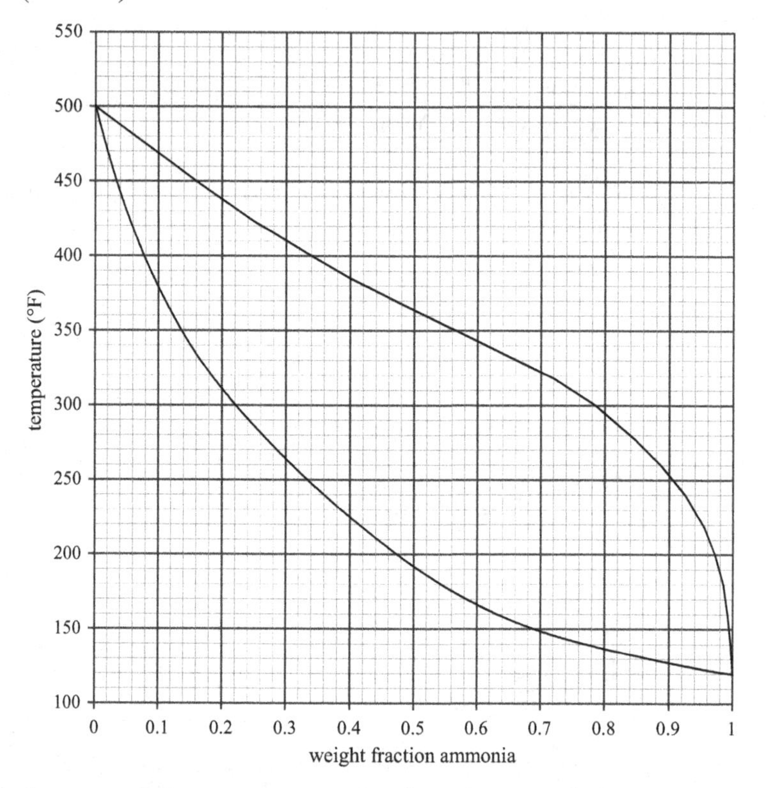

4.35 In the process below, every stream contains only ammonia (NH_3) and/or H_2O. The entire process is at 300 psia (20.4 atm).

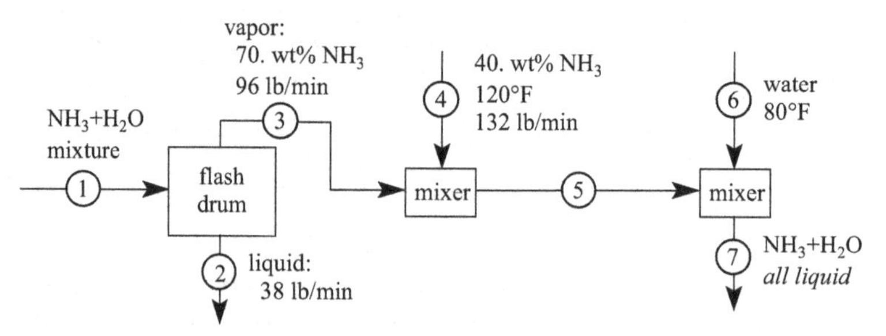

(A) What is the temperature in the flash drum? Use the phase diagrams in Exercise 4.34 or download from duncan.cbe.cornell.edu/Graphs.

(B) What is the composition of stream 2?

(C) What is the composition of stream 1?

(D) Use the temperature–composition phase diagram in Exercise 4.34(B) to draw tie lines across the solution + vapor two-phase region of the enthalpy–composition phase diagram at four temperatures: 350°F, 300°F, 250°F, and 200°F.

(E) What is the approximate temperature (±10°F) of stream 5? You must use the enthalpy–composition diagram in Exercise 4.34(A) for graphical energy balances. Do not use the temperature–composition diagram.

(F) What is the minimum flow rate of stream 6 (water at 80°F) to ensure that stream 7 will be entirely liquid?

4.36 Graphical energy balances are particularly useful for substances that release heat upon mixing, such as water and sulfuric acid. The liquid and liquid + vapor regions on the enthalpy–composition phase diagram are separated by a dark arc; "liquid" is below the border and "liquid + vapor" is above the border. The arcs in the liquid region are isotherms. The straight lines in the liquid–vapor region are tie lines. The tie line endpoints at the vapor border are off the map.

Enthapy–composition phase diagram for water + sulfuric acid mixtures at 1 atm.

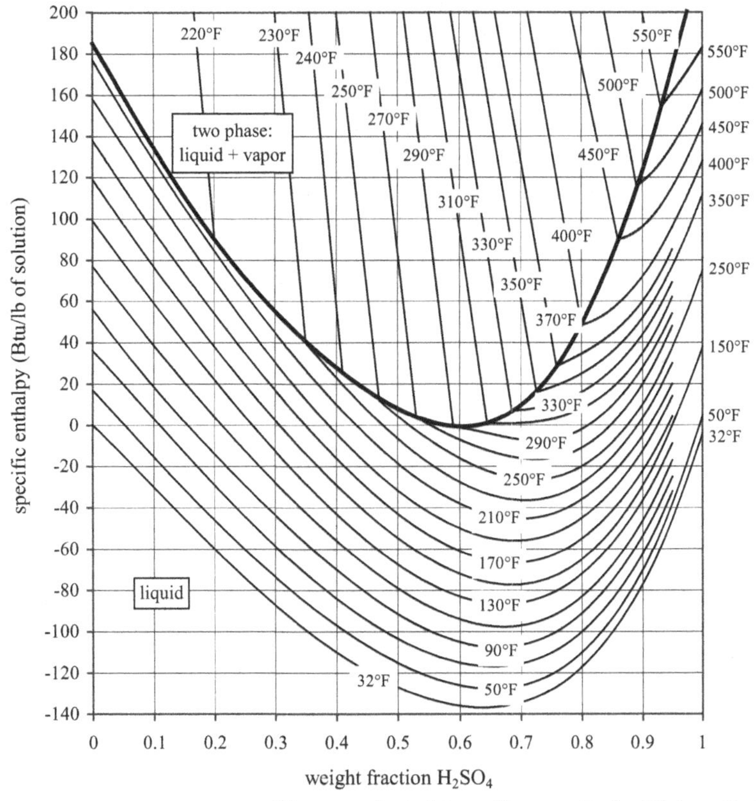

Diagram adapted from Hougen, O. A., Watson, K. M. (1946). *Chemical Process Principles Charts*, Wiley, New York.

(A) A liquid with 20. wt% H_2SO_4 at 32°F is warmed to 130°F. Calculate the change in specific enthalpy, in Btu/lb.

(B) 100. lbs of 20. wt% H_2SO_4 at 32°F is mixed with 100. lbs of 90. wt% H_2SO_4 at 250°F. Calculate the temperature and composition of the resulting mixture. Mixing sulfuric acid and water releases heat. The mixture's temperature may be higher than the average of the initial temperatures.

(C) 30. lbs of water at 150°F is mixed with 70. lbs of H_2SO_4 at 150°F. Determine the temperature and composition of the resulting mixture.

4.37 (A) Use the enthalpy–composition phase diagram for acetonitrile + nitromethane mixtures to sketch the borders of the two-phase liquid + vapor region on a temperature–composition diagram. The boiling points of acetonitrile and nitromethane at 1 atm are 81°C and 101°C, respectively. A blank plot appears after the enthalpy–composition phase diagram.

(B) Consider the mixer below. Plot on the enthalpy–composition phase diagram the points that correspond to streams 1, 2, 3, and 4.

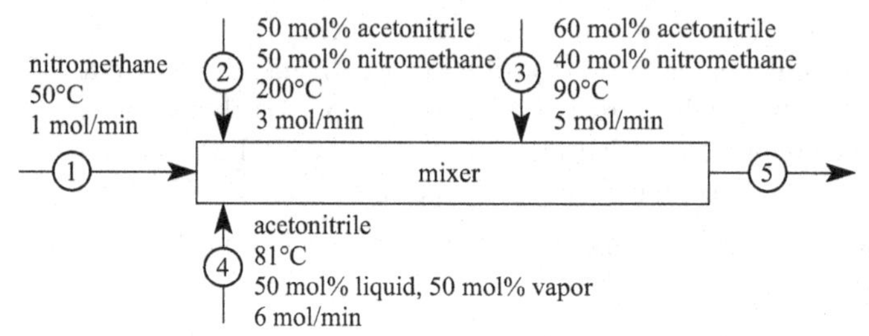

(C) Use a graphical energy balance to find the temperature and phase(s) of stream 5. If two phases are present, find the composition and amount of each phase. You have two options for your graphical solution. You may consider the streams pairwise; mix streams 1 and 2, then mix the result with stream 3, then mix the result with stream 4. With this option you will need to draw four tie lines. Or you may find the center of mass of the weighted trapezoid with corners at streams 1, 2, 3, and 4. Find the energy center of mass and find the composition center of mass. The intersection is stream 5.

Enthapy–composition phase diagram for acetonitrile + nitromethane mixtures at 1 atm.

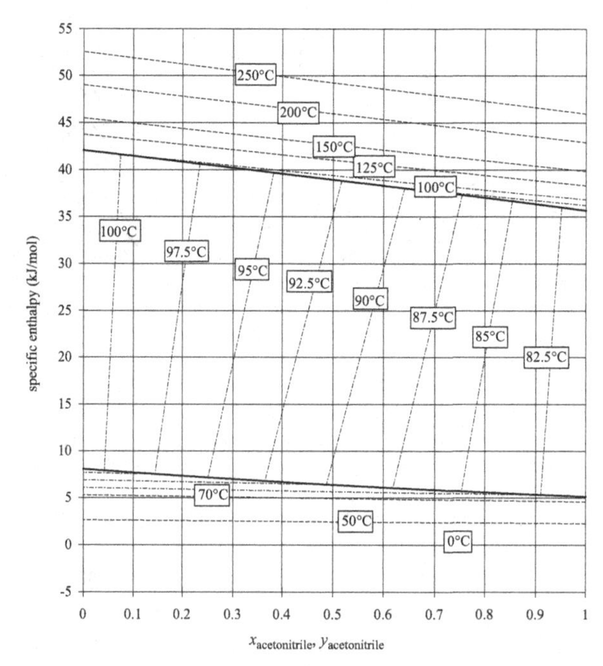

Temperature–composition phase diagram for acetonitrile + nitromethane mixtures at 1 atm.

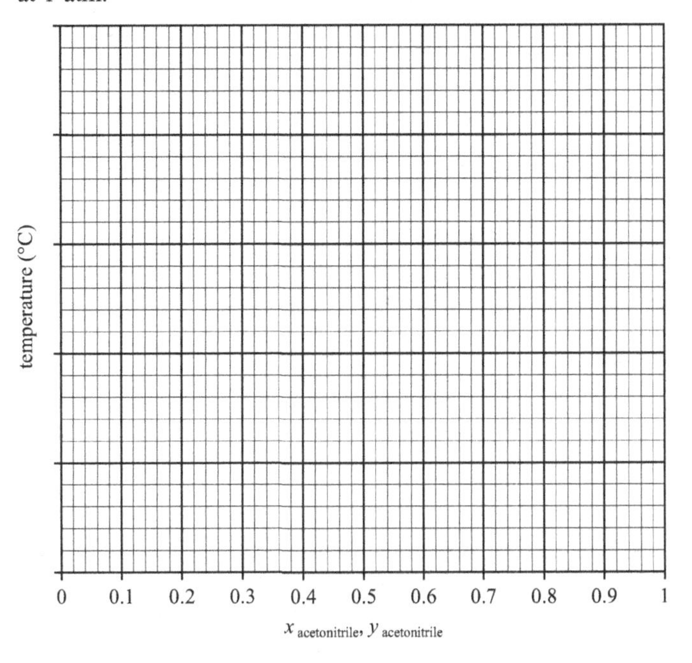

Operating Lines for Multistage Countercurrent Separators: Absorbers and Strippers

4.38 Consider three different configurations of absorbers for removing an organic pollutant P from air. In each configuration, the amount of P in the air is reduced from 4.0 wt% P to 1.0 wt% P. For this particular system, the equilibrium line is straight:

$$\left(\frac{\text{kg of } P}{\text{kg of air}}\right)_{\text{eq}} = \frac{1}{10}\left(\frac{\text{kg of } P}{\text{kg of oil}}\right)_{\text{eq}}$$

$$w_{P/\text{air}} = \frac{w_{P/\text{oil}}}{10}.$$

(A) Consider the single-stage absorber shown below. The oil and air are well mixed and the streams that exit the unit are in equilibrium with each other. Use a graphical method to compute the flow rate of oil.

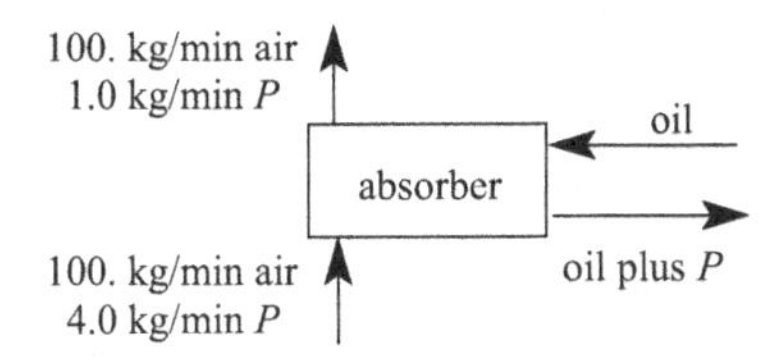

(B) Consider the three-stage absorber on the left below. The streams that exit each absorber are in equilibrium with each other. The flow rate of oil into each stage is not necessarily the same. Use a graphical method to compute the total flow rate of oil.

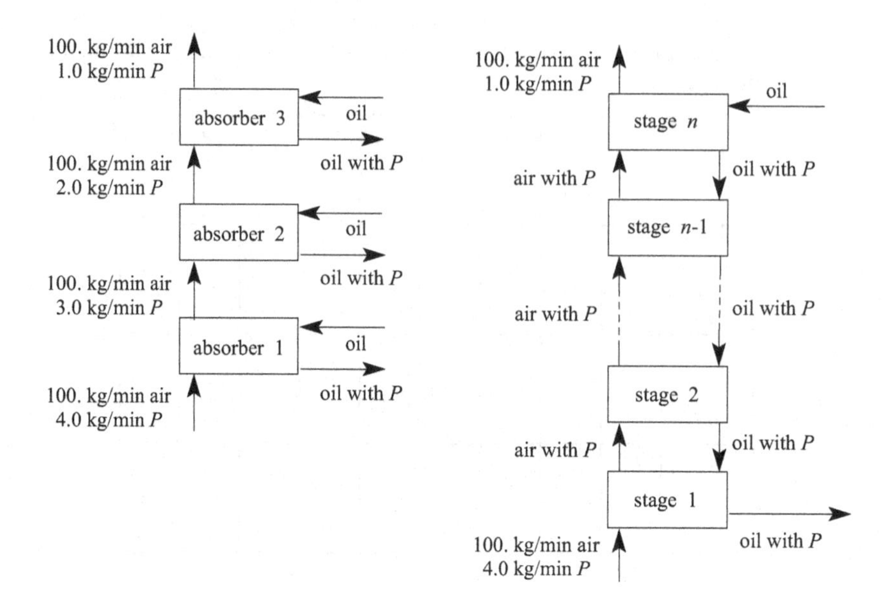

(C) Consider a countercurrent multistage absorber such as the one shown on the right. The flow rate of oil is 10 kg/min. Again, the streams that exit each absorber are in equilibrium with each other. Use a graphical method to calculate the number of stages required.

(D) Which design – (A), (B), or (C) – uses the least oil? Hypothesize as to why.

4.39 Your lungs can be modeled as an absorber that transfers O_2 from the air to your blood.

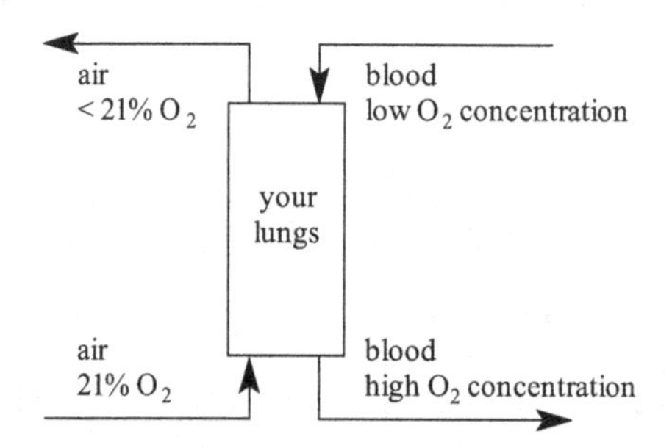

In this simple model we will assume:

- the lungs are a single-stage absorber
- the flows through the lungs are continuous and at steady state
- the O_2 + air and O_2 + blood streams leaving the lungs are in equilibrium.

The map below is a graphical analysis of lungs for a normal breathing rate and a normal pulse rate. Because you will use this map to graph your answers to parts (A)–(D) of this exercise, you are advised to download copies of the map from duncan.cbe.cornell.edu/Graphs.

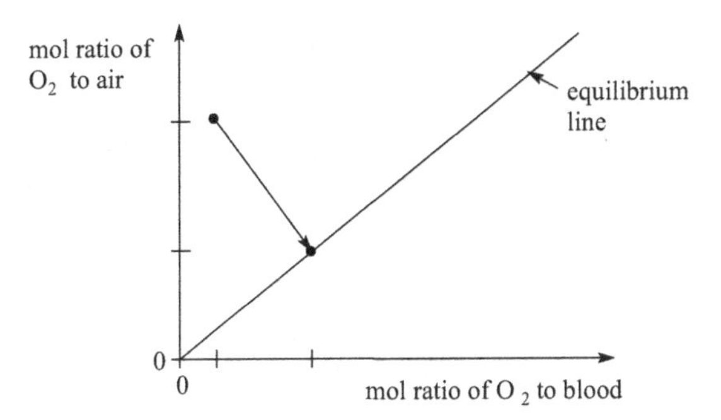

(A) Indicate the following on the map:
 1. mol ratio of O_2 to air the inlet air
 2. mol ratio of O_2 to air in the outlet air
 3. mol ratio of O_2 to blood in the inlet blood
 4. mol ratio of O_2 to blood in the outlet blood.

(B) Sketch on the map a graphical analysis of hyperventilation, in which the breathing rate doubles but the pulse rate is unchanged. Assume the mol ratio of O_2 to blood in the inlet blood is unchanged.

(C) You are transported to a planet whose atmosphere has three-fourths the O_2 concentration of Earth's. Draw a graphical analysis on the map to determine what you must do to achieve the same mol ratio of O_2 to blood you had on Earth.

(D) In an emergency, a saline solution (water with salt) is administered intravenously to someone who has suffered a severe loss of blood. Consider a person who has lost half his or her blood. This person's blood has half the capacity to absorb oxygen. Draw a graphical analysis on the map to determine what this person must do to achieve the same mol ratio of O_2 in diluted blood.

(E) Your lungs also serve as a stripper that transfers CO_2 from your blood to the air. Sketch a graphical analysis of this stripper on the map below. The concentration of CO_2 in the air is essentially zero (0.03 mol%). Assume that the blood leaving the lungs contains CO_2.

 Indicate on your sketch:
 1. mol ratio of CO_2 to air in the outlet air.
 2. mol ratio of CO_2 to blood in the inlet blood.
 3. mol ratio of CO_2 to blood in the outlet blood.

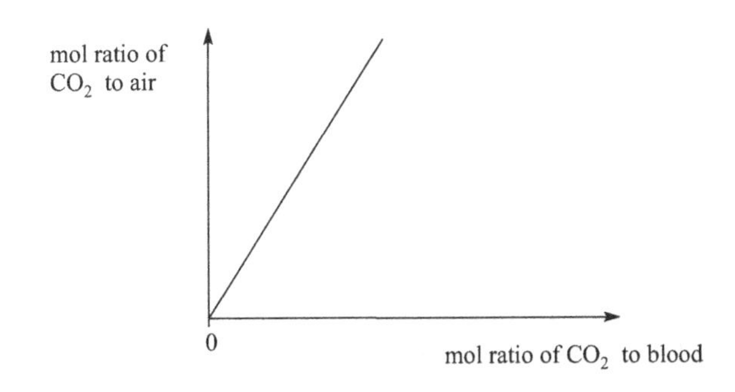

We are indebted to Professor Richard Seagrave for suggesting the concept for this exercise.[7]

4.40 As described in the previous exercise, your lungs can be modeled as an absorber that transfers O_2 from the air to your blood. The map below is a graphical analysis of lungs for a normal breathing rate and a normal pulse rate.

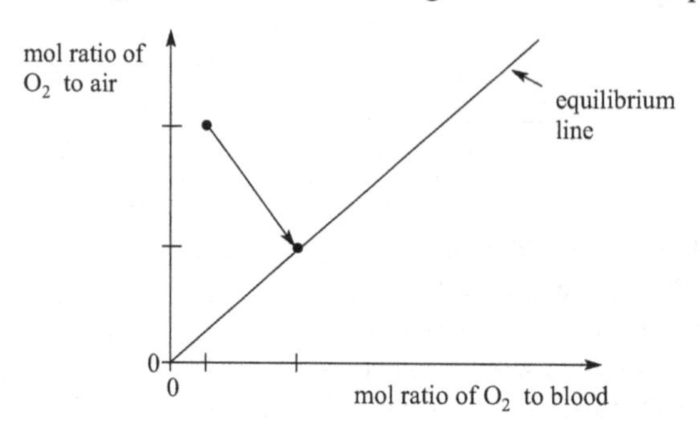

Some professional athletes – notably cyclists and long-distance runners – enhance their blood's aerobic capacity by adding oxygen carriers. Weeks before a competition, they withdraw blood and then extract and freeze their red blood cells. In the interim weeks, their body restores the red blood cell level in their body to normal. Immediately before the competition, they thaw and reinfuse their extracted red blood cells. For this exercise, assume these cheaters enhance the oxygen capacity of their blood by 50%.

On the map above, indicate the mol ratio of O_2 to blood in the blood leaving the lungs of a cheater with normal pulse rate and *double* the normal breathing rate.

This exercise was suggested by Matt Ferguson, Cornell University Chemical Engineering Class of 2016.

4.41 A breathalyzer is an instrument that estimates the amount of ethanol in a person's blood by the amount of ethanol in that person's exhaled air. In this exercise we will model a person's lungs as a stripper that transfers ethanol from the blood to the air.

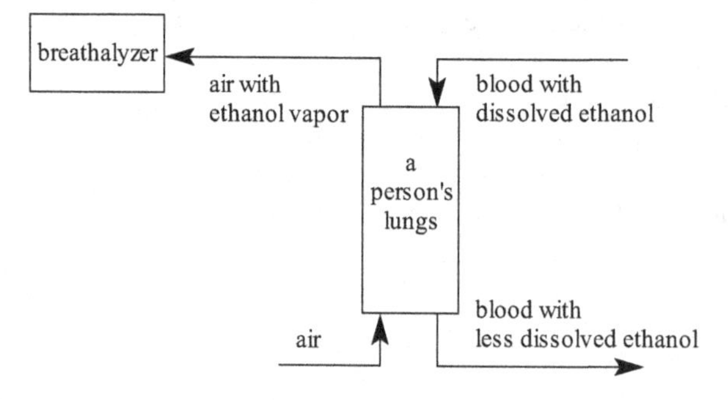

In this simple model we will assume that the lungs are a single-stage stripper, the blood flow and air flow through the lungs are continuous and at steady state, and the "air with ethanol vapor" and the "blood with less dissolved ethanol" streams leave the lungs in equilibrium.

[7] Similar interesting applications to biological systems can be found in Seagrave's text: Seagrave, R. 1971. *Biological Applications of Heat and Mass Transfer*, Iowa State University Press, Ames, IA.

(A) Sketch on the map below a qualitative analysis for a normal breathing rate and a normal pulse rate. Indicate the following on the map:

1. mol ratio of ethanol to air in the inlet air
2. mol ratio of ethanol to air in the outlet air
3. mol ratio of ethanol to blood in the inlet blood
4. mol ratio of ethanol to blood in the outlet blood.

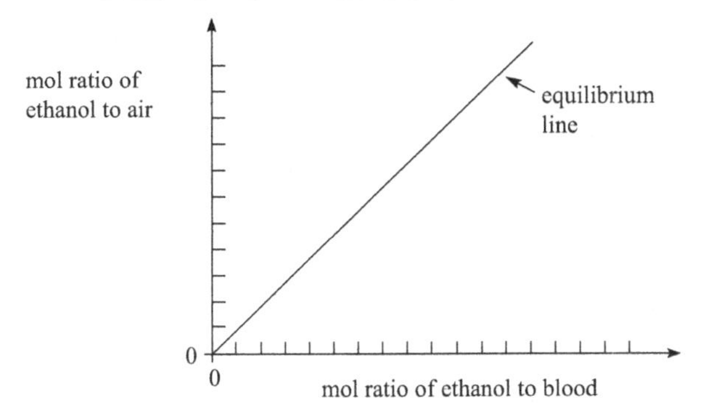

(B) Assume you have been hired as an expert witness to analyze breathalyzer results. Assume the person in part (A) is breathing twice as fast as normal, but with a normal heart rate. Start with the inlet blood ethanol content you assumed in part (A) and predict the ethanol content of the exhaled air. Also indicate the blood ethanol content predicted if the breathalyzer assumes a normal breathing rate.

(C) Assume the person in part (A) has a heart rate double the normal rate (the person is nervous), but has a normal breathing rate. Start with the inlet blood ethanol content you assumed in part (A) and predict the ethanol content of the exhaled air. Also indicate the blood ethanol content predicted if the breathalyzer assumes a normal heart rate.

Disclaimer: This simple analysis is not intended to advise one how to defeat a breathalyzer. There is no way to defeat a breathalyzer. If a person registers a non-negligible blood ethanol content, that person will be required to submit to a blood test, even if the breathalyzer result is below the legal limit. This simple analysis is only meant to demonstrate the effect of assumptions inherent in any instrumental measurement. There is only one accepted behavior: don't drink and drive.

4.42 A chemical process produces a valuable chemical X dissolved in cyclohexane. Because X and cyclohexane have similar boiling points, distillation is impractical. Instead, the X is extracted from cyclohexane by water in a liquid–liquid extractor, as shown below. The extractor has four equilibrium stages. Separators with gas and liquid phases are absorbers or strippers; separators with two immiscible liquid phases are extractors. (Purge streams have been neglected.)

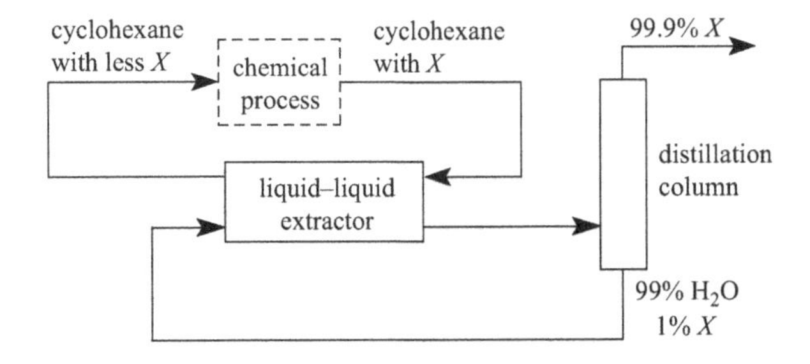

Shown below are equilibrium lines for three different temperatures, such that $T_A < T_B < T_c < 100°C$. You may assume water and cyclohexane are mutually insoluble. At which temperature should we operate the liquid–liquid absorber? You must justify your choice with a written explanation and/or a graphical analysis.

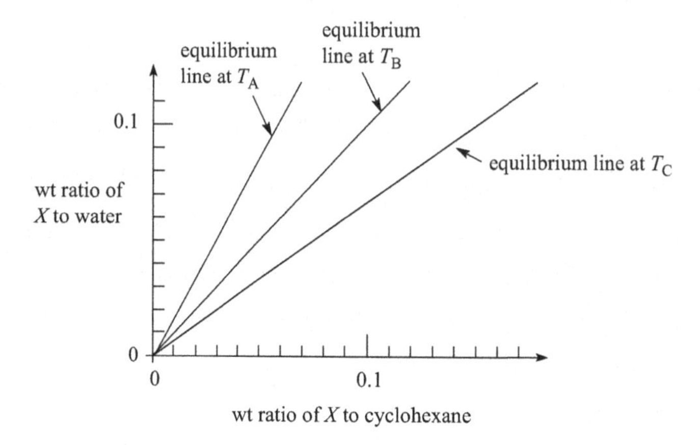

4.43 A portion of an analysis of a four-stage countercurrent oil–air absorber that absorbs acetone from air is shown below. You may need to draw one or more additional lines to answer the following questions.

Acetone/oil–acetone/air solubility diagram.

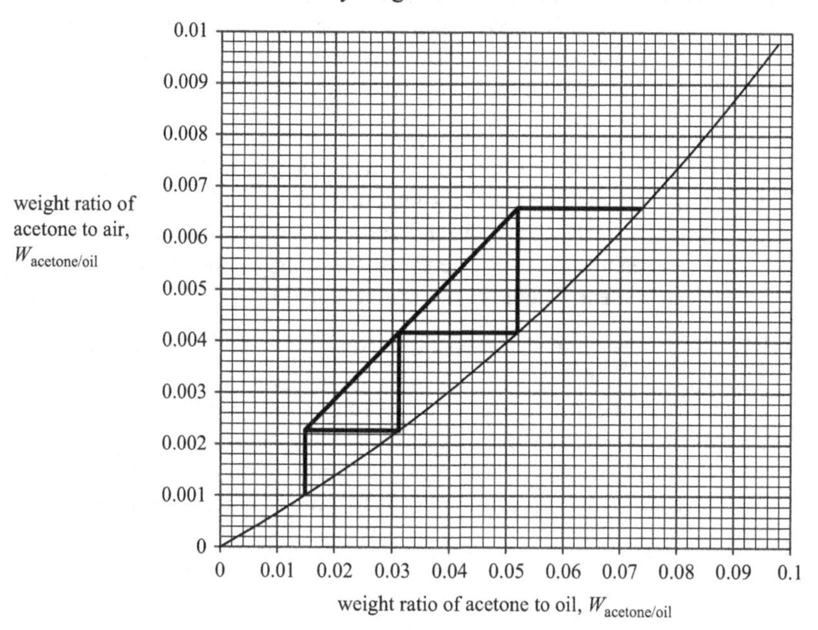

(A) What is the weight ratio of acetone to oil in the oil *leaving* the absorber?
(B) What is the weight ratio of acetone to air in the air *entering* the absorber?

(C) What is the weight ratio of acetone to oil in the oil *entering* the absorber?
(D) What is the weight ratio of acetone to air in the air *leaving* the absorber?
(E) The flow rate of air is 57 kg/min. What is the flow rate of oil?

4.44 Consider an absorber that extracts Q from air into oil.

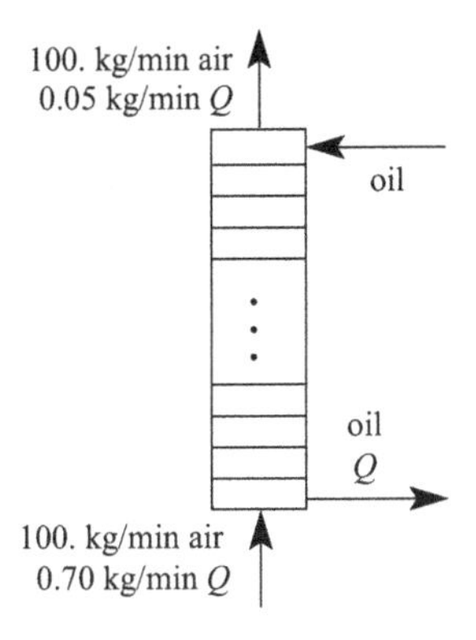

(A) What flow rate of oil is needed if the absorber has only one equilibrium stage?
(B) What is the minimum flow rate of oil?
(C) Calculate the number of stages needed if the flow rate of oil is 25% higher than the minimum determined in part (B).
(D) What flow rate of oil is needed if the absorber has four equilibrium stages?

Q/oil–Q/air solubility diagram.

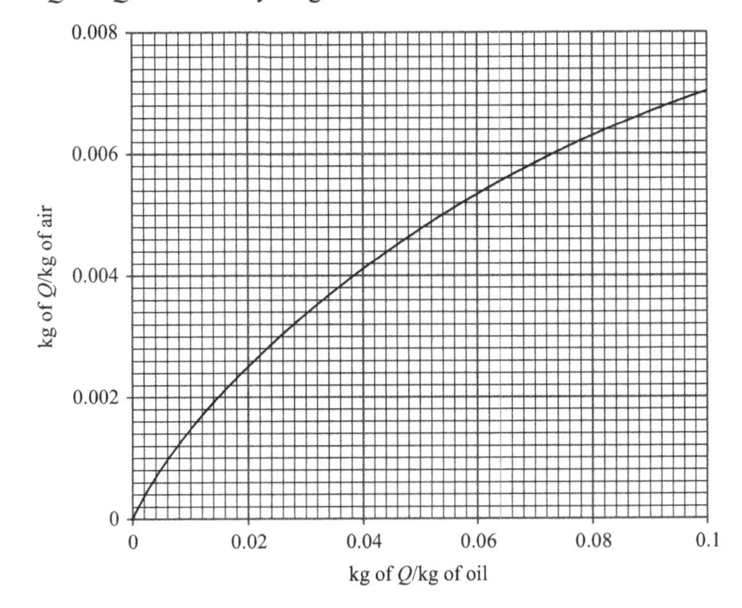

4.45 Consider the absorber with the flow rates and compositions given in the diagram below. This is the same Q and oil as in Exercise 4.44.

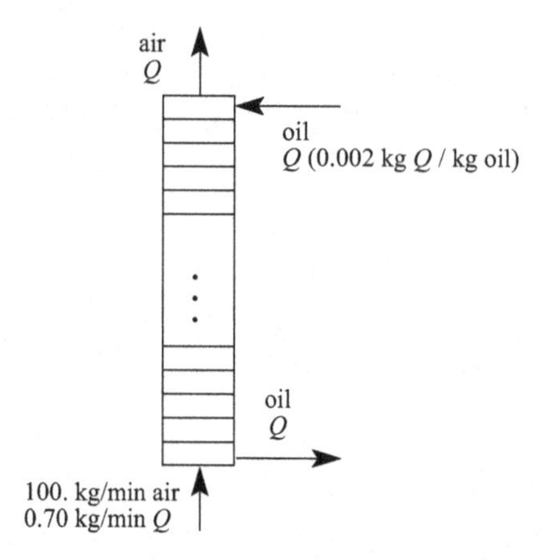

(A) Given that the composition of the oil stream leaving the bottom is 0.033 kg Q/kg oil and the air stream leaving the top is 0.0008 kg Q/kg air, calculate two quantities: (1) the number of equilibrium stages in the absorber and (2) the flow rate of oil entering the column.

(B) Given that the composition of the oil stream leaving the bottom is 0.033 kg Q/kg oil, what is the lowest possible concentration of Q in air (in units of kg Q/kg air) leaving the top? You may use any flow rate of oil entering the top and you may use any number of stages.

(C) Given that the composition of the oil stream leaving the bottom is 0.07 kg Q/kg oil, what is the lowest possible concentration of Q in air (in units of kg Q/kg air) leaving the top? Again you may use any flow rate of oil entering the top and you may use any number of stages.

4.46(A) The countercurrent multistage absorber shown below removes K from air. How many stages are required *and* what is the mass fraction of K in the oil effluent?

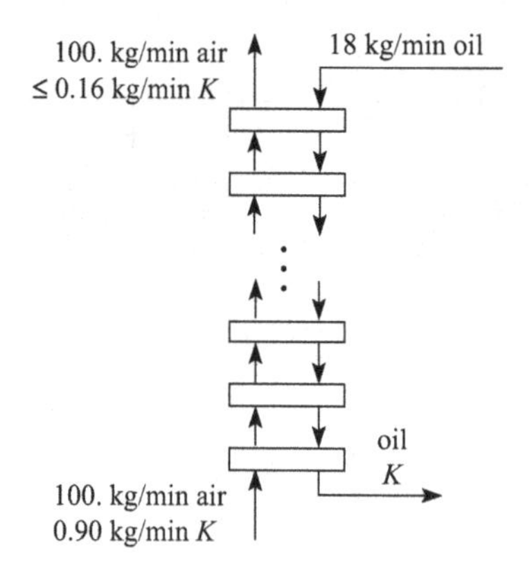

(B) A different countercurrent multistage absorber, shown below, removes *K* from air. In this case, the oil entering the absorber is not pure; the oil contains some *K*. How many stages are required?

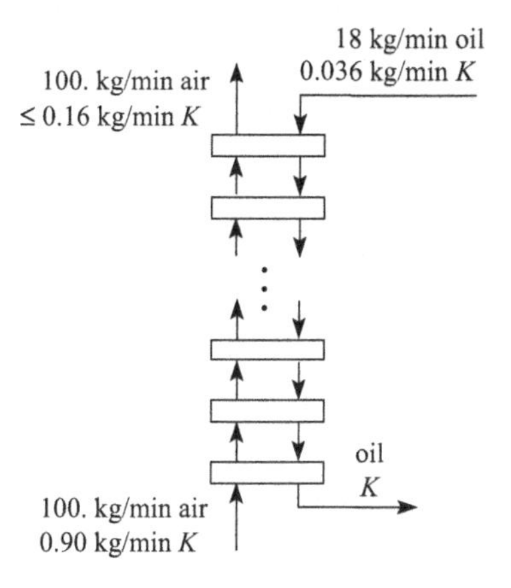

(C) A yet-different countercurrent multistage absorber, shown below, removes *K* from air. Pure oil enters the column at the top. Oil leaves the column at the bottom *and* after an intermediate stage. Given the flow rates and compositions shown here, how many stages are required?

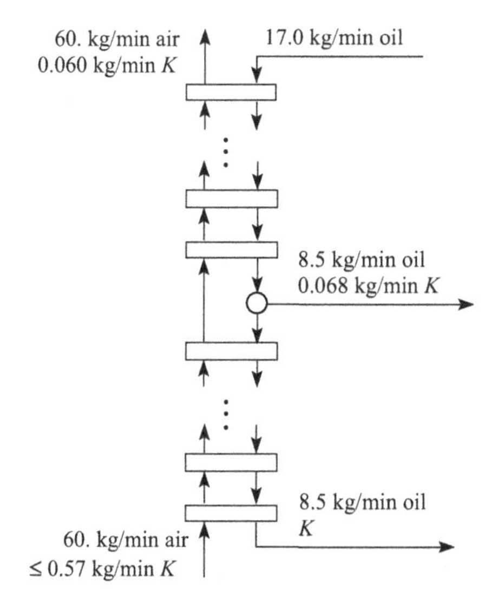

K/oil–K/air solubility diagram.

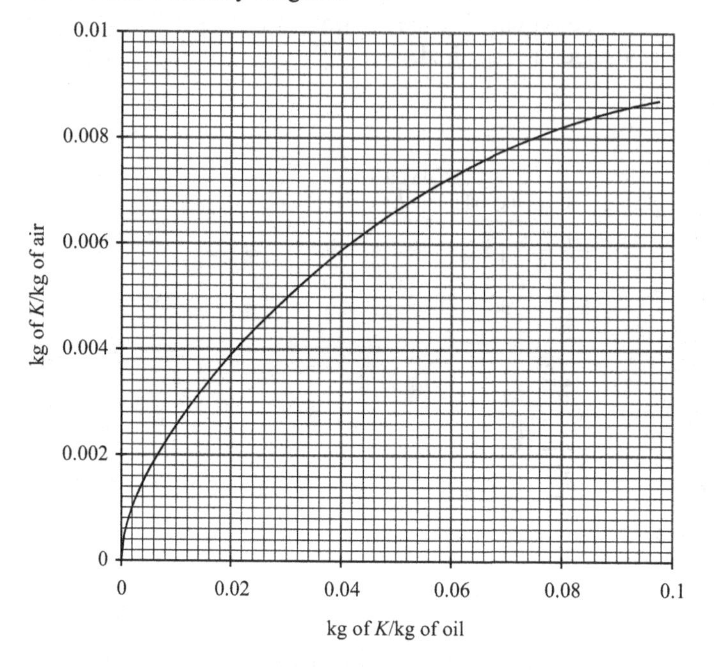

4.47 This absorber cleans air polluted with acetone.

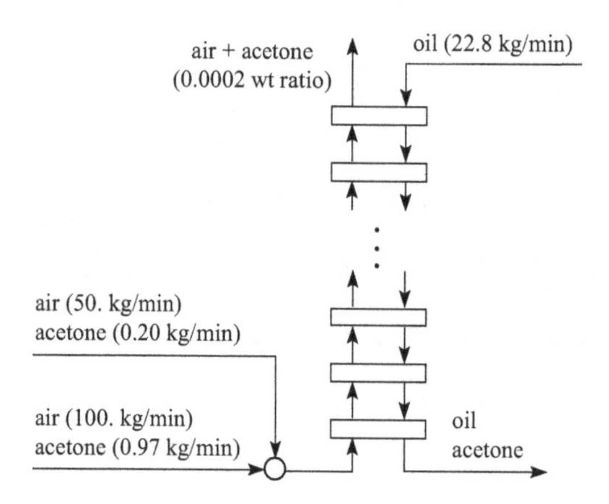

(A) Calculate the number of equilibrium stages in the absorber.
(B) Calculate the weight ratio of acetone to oil in the oil leaving the absorber. You must use a graphical method.
(C) Consider a different absorber to clean the two air streams, shown below. The air with less acetone enters the absorber column between two equilibrium stages. Calculate the number of stages in the absorber.

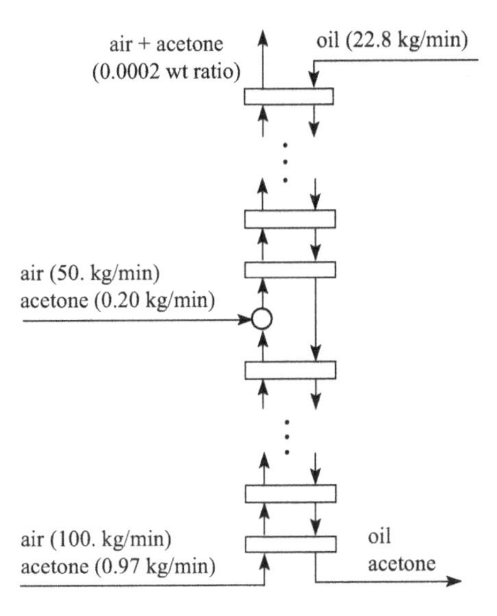

(D) If the top stage is stage 1, between which two stages should the less polluted air enter?

Acetone/oil–acetone/air solubility diagram.

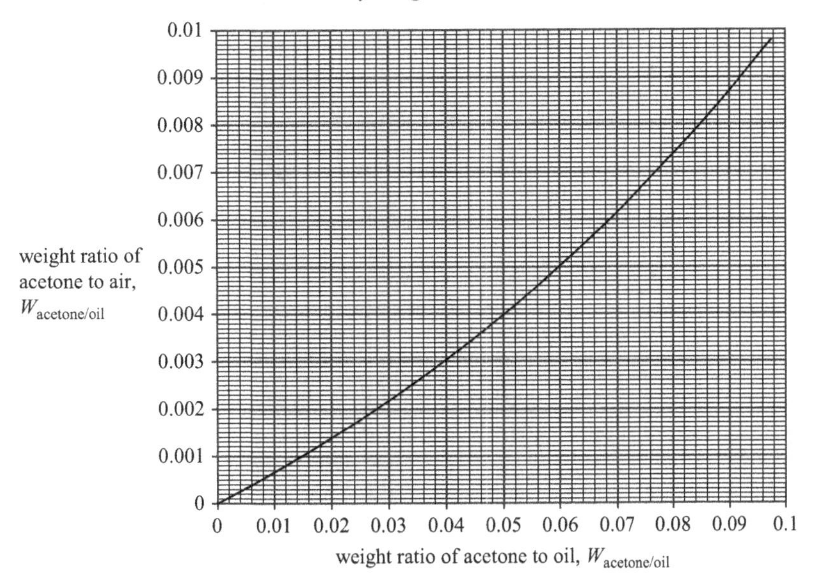

4.48(A) The absorber below cleans air polluted with chloroform. How many equilibrium stages are required? What is the wt ratio of chloroform in the oil stream leaving the absorber?

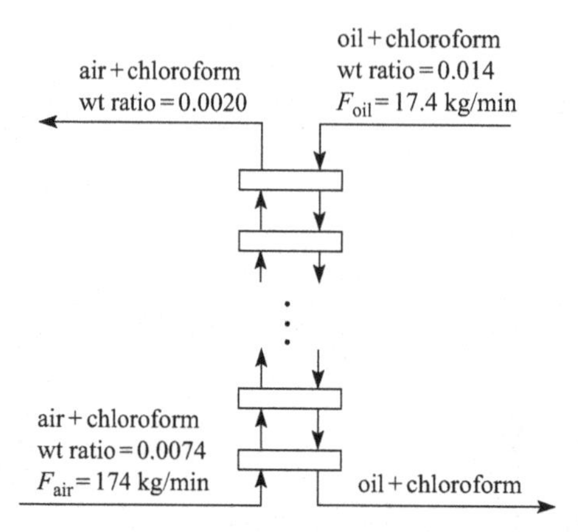

(B) Assume the flow rate of oil + chloroform (wt ratio = 0.014) is unlimited. What is the minimum weight ratio of chloroform in the cleaned air? How many equilibrium stages are required to attain the minimum weight ratio of chloroform in the cleaned air?

Chloroform/oil–chloroform/air solubility diagram.

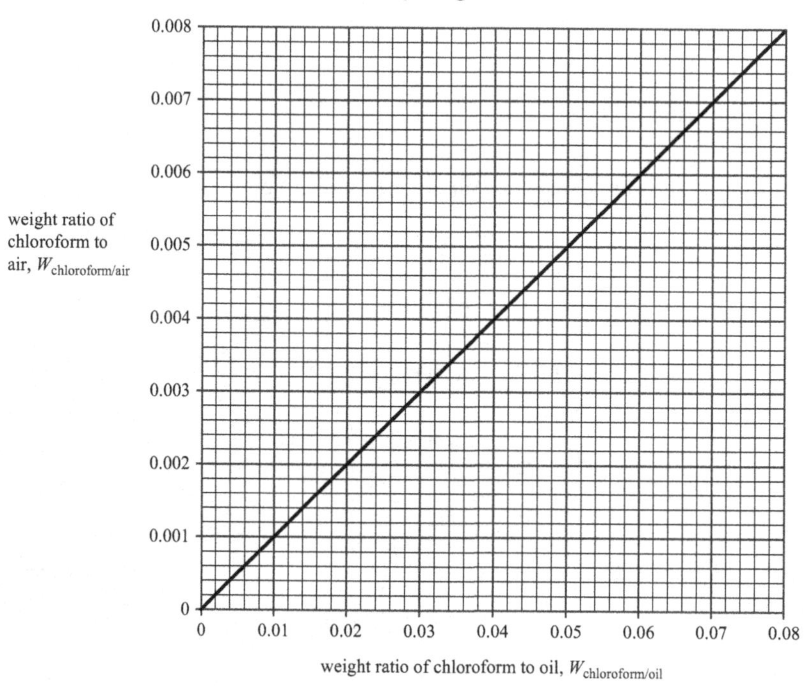

(C) Design an absorber to clean the air in part (A) to 0.0006 wt ratio of chloroform to air. To produce cleaner air, you may use the oil stream in part (A) plus an oil stream with less chloroform: 0.002 wt ratio of chloroform to oil. You may mix the two oil streams, or use either oil stream or both oil streams individually.

Summary

- Input air: 0.0074 wt ratio chloroform to air, F_{air} = 174 kg/min
- Output air: 0.0006 wt ratio chloroform to air

- Input oil stream 1: 0.014 wt ratio chloroform to oil, F_{oil} = 17.4 kg/min
- Input oil stream 2: 0.002 wt ratio chloroform to oil, F_{oil} = 17.4 kg/min.

Design Goals (in decreasing importance)

- Clean the air to 0.0006 wt ratio chloroform to air.
- Minimize the number of equilibrium stages.

4.49 Match the absorbers and strippers below with the graphical models A to F that follow them. You may assume that the lower left-hand corner is the origin (0, 0) and that the thermodynamic maps follow the convention of liquid on the x axis (abscissa) and vapor on the y axis (ordinate).

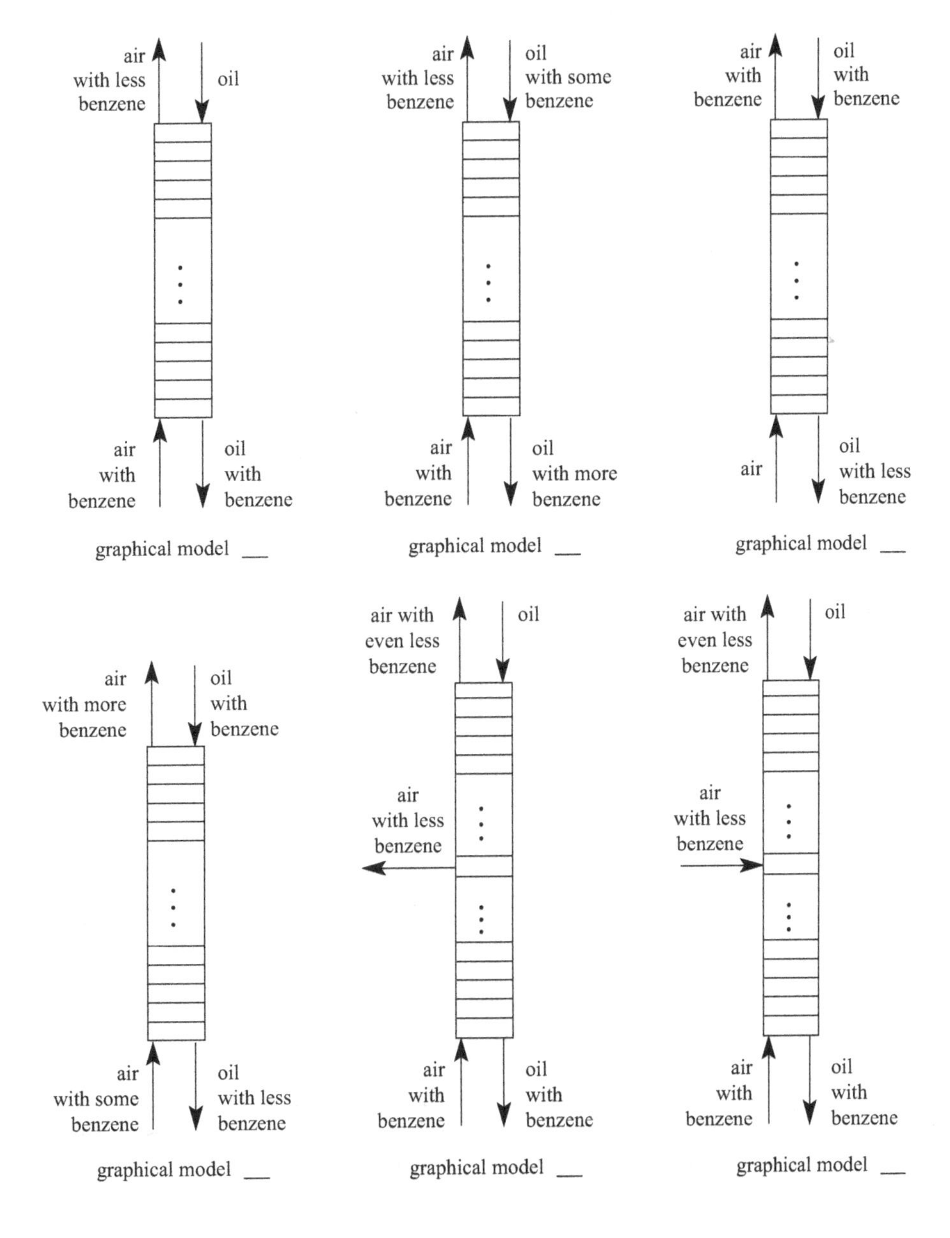

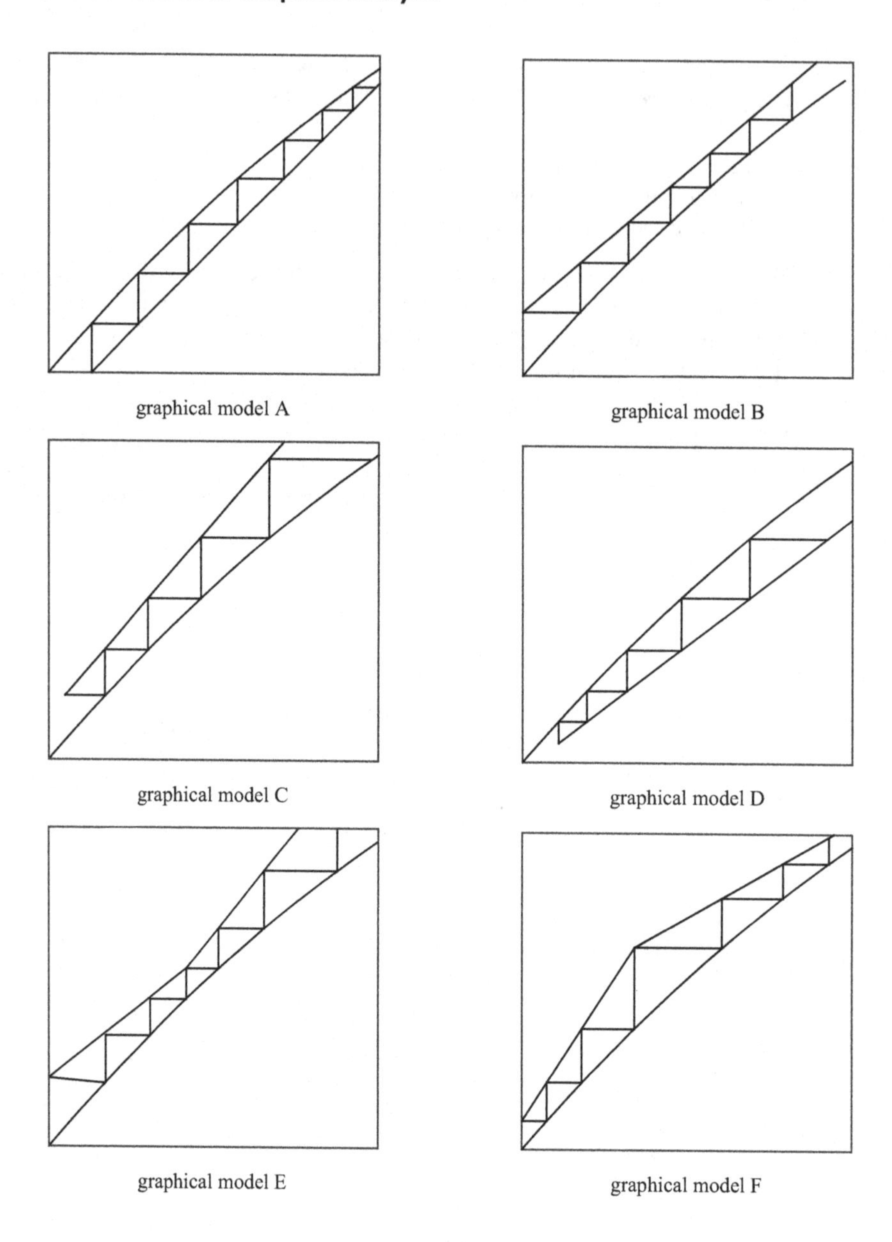

4.50 Groundwater is contaminated occasionally by anthropogenic activities. A chemical may leak from a storage tank into the soil and be carried to an aquifer by rainfall. An example is the leakage of gasoline into groundwater from storage tanks underneath filling stations. Public health, safety, and environmental concerns have spawned designs for removing these trace chemicals.

Two designs are considered here. Both involve the adsorption of benzene onto activated carbon, a substance similar to charcoal and relatively inexpensive to produce and regenerate. In the first scheme, the benzene adsorbs directly from the contaminated water onto the activated carbon.

First Scheme

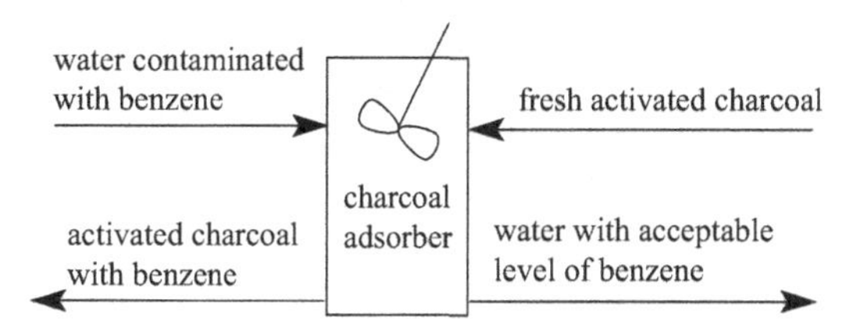

This adsorber is a well-mixed slurry of water and activated carbon. The concentration of benzene in the water and the concentration of benzene adsorbed on the activated carbon are uniform throughout the adsorber.

Second Scheme

Air strips the benzene from the groundwater. The contaminated air then passes through a bed of activated carbon, which adsorbs the benzene from the air.

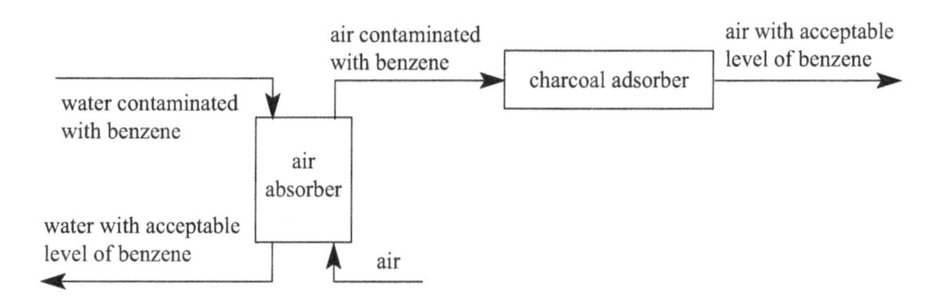

The adsorber in the second scheme is a packed bed of activated carbon. The air is not mixed in the adsorber and therefore the concentration of benzene in the air varies with position in the adsorber, as well as with time, because the adsorber is charged initially with fresh activated carbon. Until the activated carbon saturates with benzene, the activated carbon absorber reduces the benzene/air concentration by a factor of 1000. When the outlet concentration of benzene reaches a critical level, breakthrough has occurred and the activated carbon is replaced.

Use the graphical data below to calculate the capacity and cost of each scheme. Which scheme has the lower capital cost? Which scheme has the lower operating cost? Calculate the profit (or loss) associated with each scheme. Which scheme is best overall?

Process specifications:

- The process must treat 100 gallons of water per minute.
- The water is contaminated with 20 ppm (by weight) benzene.
- The water leaving each process has a residual concentration of 20 ppb (by weight) benzene.
- The air leaving the second scheme contains essentially zero benzene.
- The equipment is depreciated by a 10-year, straight-line method.

Gas-phase adsorption isotherms for benzene over activated carbon.

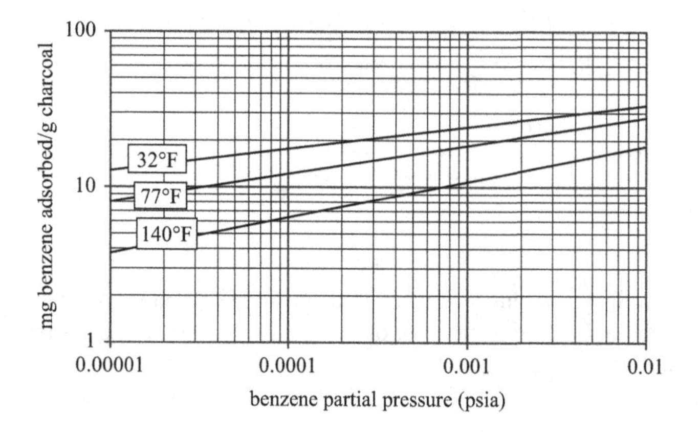

Figure adapted from tabulated data from Calgon Corporation, Pittsburgh, PA (a subsidiary of Merck & Co).

Liquid-phase adsorption isotherm for benzene over activated carbon.

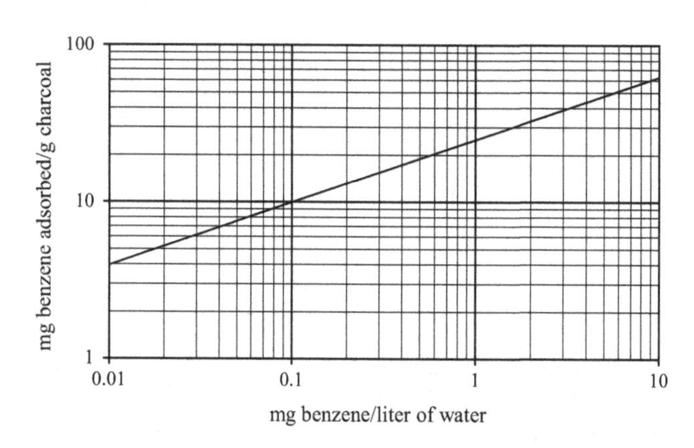

Figure adapted from tabulated data from Westates Carbon, Inc., Los Angeles, CA.

Here are some facts obtained from colleagues at Levine–Fricke, Engineers, Hydrogeologists, and Applied Scientists, an environmental engineering firm in Emeryville, California:

- A water–activated-carbon adsorber with a capacity of 50 gallons of water per minute holds 1800 pounds of activated carbon and costs $9800, which includes the cost of pipes, pumps, valves, controls, etc.
- Activated carbon costs $2.50 per pound. (Most of the cost is for transportation.)
- An air absorber with a capacity of 25 gallons of water per minute costs $20,000, which includes the cost of pipes, pumps, valves, controls, etc.
- The air absorbers use 350 cubic feet of air to strip the benzene from 25 gallons of water.

- An air–activated-carbon adsorber with a capacity of 350 cubic feet of air per minute holds 2200 pounds of activated carbon and costs $6800, which includes the cost of pipes, pumps, valves, controls, etc.

Finally, here are some conversion factors:

$$1\,\text{gal} = 3.8\,\text{L}, \quad 1\,\text{kg} = 2.2\,\text{lb}, \quad 1\,\text{ft}^3 = 28.3\,\text{L}, \quad 1\,\text{mol of benzene} = 78\,\text{g}.$$

Standard pressure is 1 atm.

4.51 Examine the process in the second scheme of Exercise 4.50. *Sketch* on a plot similar to the one below a graphical analysis of the multistage air–water column.

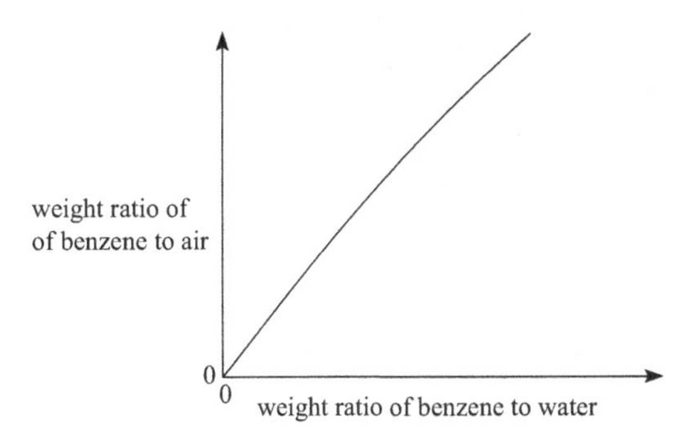

Your sketch need only be *qualitative*. Ignore all *quantitative* compositions and flow rates. Just sketch a generic analysis of a unit in which clean air strips benzene from contaminated water. Label the following on your sketch:

- the equilibrium line
- the operating line
- the inlet *and* outlet compositions of the air
- the inlet *and* outlet compositions of the water.

Finally, state how many stages your analysis predicts.

4.52 Water can be extracted from a dilute solution by osmotic separation, as follows.[8] Consider a tank divided into two sections by a porous membrane. The membrane allows only water to pass. One side of the tank is filled with a dilute sugar solution and the other side is filled with a concentrated salt solution. Water diffuses through the membrane until equilibrium is reached.

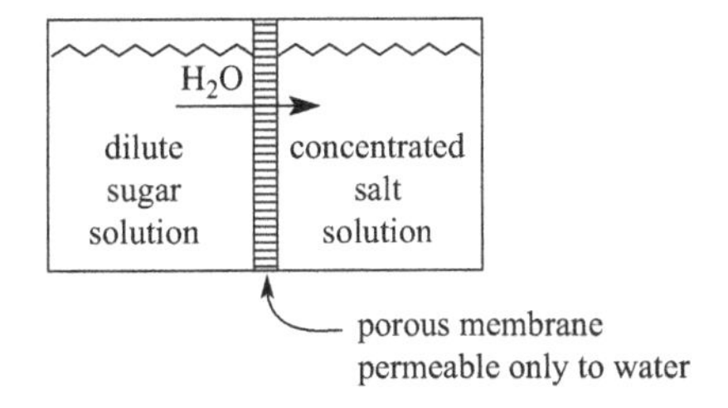

[8] *Chemical Engineering Progress*, July 1998, p. 49.

Because osmotic separation is performed at room temperature, it can be more energy-efficient than distillation. Osmotic separation is also better than distillation if the solution (such as orange juice – see Exercise 3.10) is harmed by heat.

Here we use osmotic separation to remove water from a sugar solution. We use the tank shown above to observe equilibria at several different concentrations. We then plot the ratio of water/sugar vs. the ratio of water/salt, and draw a smooth line through the data, as shown below.

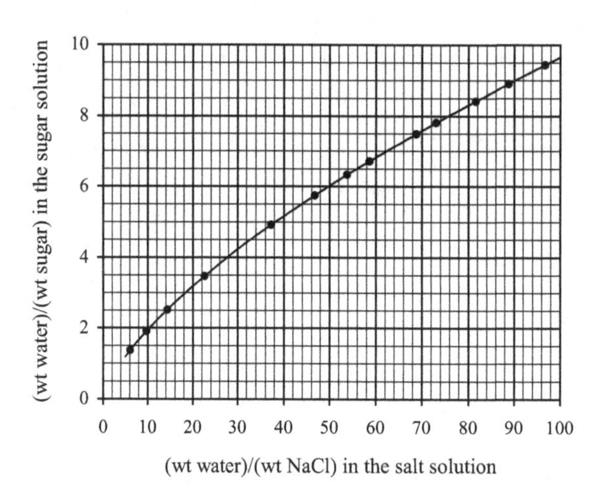

(A) Label the points, lines, or regions that correspond to the following conditions:
 1. sugar solution + salt solution at equilibrium
 2. excess water in the salt solution
 3. excess water in the sugar solution
 4. all NaCl (no water) in the salt "solution."

(B) An osmotic separator is similar to an absorber. The map above can be used to analyze osmotic separators, such as the one shown below. In osmotic separation, a non-equilibrium system moves to equilibrium along a path with a slope of $-F_{salt}/F_{sugar}$ on a map of (wt water/wt sugar) vs. (wt water/wt salt). F_{salt} is the flow rate of salt through the osmotic separator (in kg/min), which is unchanged from inlet to outlet. F_{sugar} is the flow rate of sugar through the osmotic separator (in kg/min), which is unchanged from inlet to outlet. F_{salt} and F_{sugar} are *not* the solution flow rates.

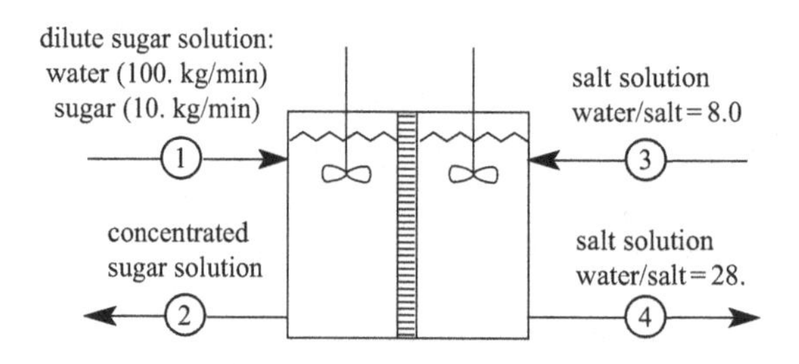

Use the map above to calculate the water/sugar ratio in stream 2. You may assume that the sugar solution and the salt solution reach equilibrium in the separator.

(C) Calculate the flow rate of the salt solution (salt plus water) in stream 3, in kg/min.

(D) Use a mass balance to calculate the flow rate of water across the membrane, in kg/min.

(E) Consider the three-stage countercurrent osmotic separator shown below. For osmotic separation, the slope of an operating line is F_{salt}/F_{sugar}. Draw an operating line and step off stages to calculate the ratio of water/sugar in the dilute sugar solution.

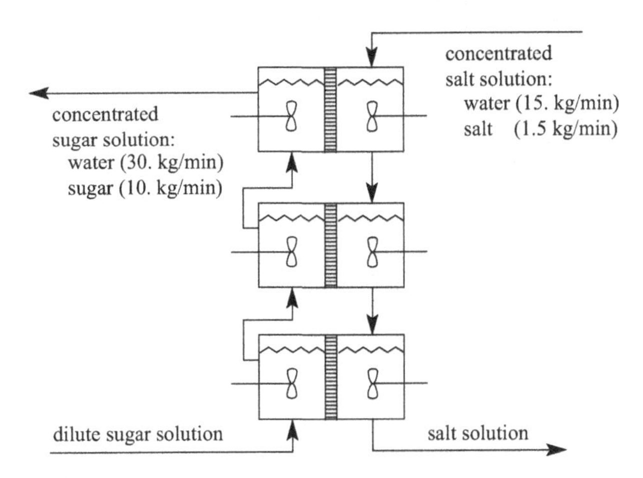

(F) Calculate the ratio of water/salt in the outlet salt solution.

4.53 A company produces syrup and glue. The syrup needs to be cooled and the glue needs to be warmed. A conventional shell and tube heat exchanger (see Figure 5.22) is impractical because both liquids are too viscous to be pumped through the tubes. Instead you decide to use a large tank with a divider in the middle, as shown below.

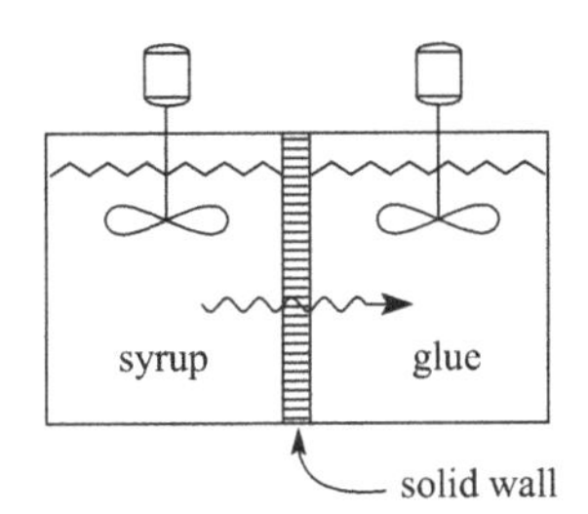

This unit is similar to an absorber. Glue absorbs heat from syrup like oil absorbs benzene from air.

(A) We will use the map below to analyze this unit graphically. Label the regions on the map that correspond to "thermal equilibrium," "too much heat in the syrup," and "too much heat in the glue."

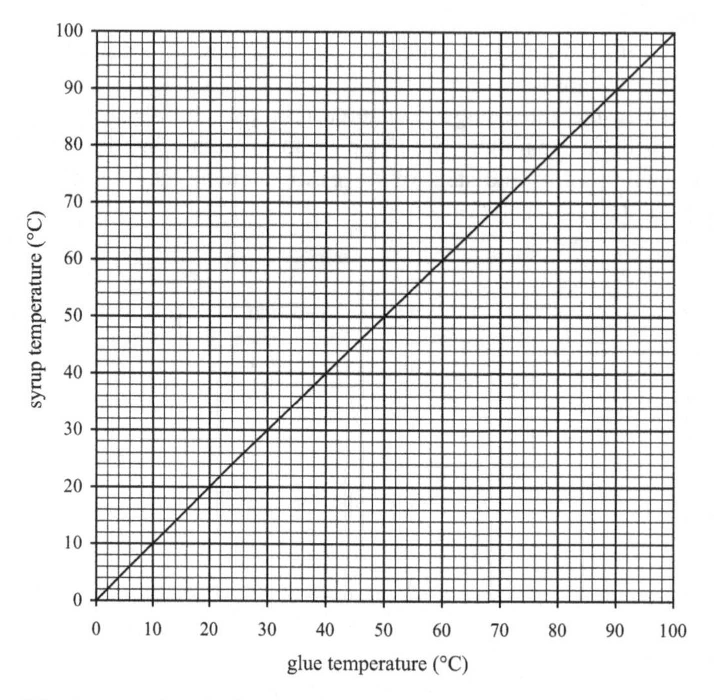

(B) Assume that the heat exchanger initially contains syrup at 85°C in one side and glue at 25°C in the other side. Plot the point that corresponds to this initial condition.

(C) Further assume that the heat exchanger contains a large amount of 85°C syrup and a tiny amount of 25°C glue. Plot the point this system will reach at thermal equilibrium.

(D) Instead assume that the heat exchanger contains a tiny amount of 85°C syrup and a large amount of 25°C glue. Plot the point this system will reach at thermal equilibrium.

(E) The heat exchanger is modified so that syrup and glue continuously flow in and out. The syrup and glue reach thermal equilibrium; they leave the heat exchanger at equal temperatures.

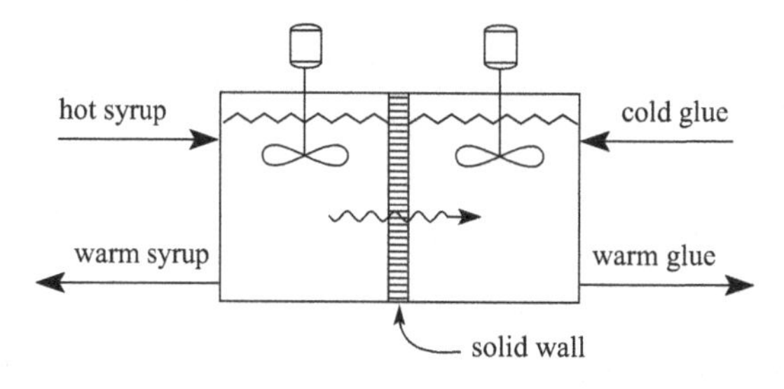

An energy balance on this single-stage heat exchanger yields the following equation:

$$\frac{T_{\text{syrup,out}} - T_{\text{syrup,in}}}{T_{\text{glue,out}} - T_{\text{glue,in}}} = -\frac{F_{\text{glue}} C_{\text{P,glue}}}{F_{\text{syrup}} C_{\text{P,syrup}}}.$$

Draw a graphical solution to calculate F_{glue}, given the following conditions: $T_{\text{syrup, in}} = 88°\text{C}$, $T_{\text{glue,in}} = 26°\text{C}$, $T_{\text{syrup, out}} = T_{\text{glue, out}} = 42°\text{C}$, $F_{\text{syrup}} =$ 33.7 kg/min, $C_{\text{P, glue}}/C_{\text{P, syrup}} = 0.67$.

(F) The inlet temperatures change to $T_{\text{syrup, in}} = 94°\text{C}$, $T_{\text{glue, in}} = 44°\text{C}$. The flow rates do not change. Use a graphical solution to calculate the outlet temperature. You must construct graphical solutions to (E) and (F) to receive full credit.

(G) We propose to replace the single-stage heat exchanger with a multistage countercurrent heat exchanger. Each stage comes to thermal equilibrium; the two streams leaving a stage are the same temperature.

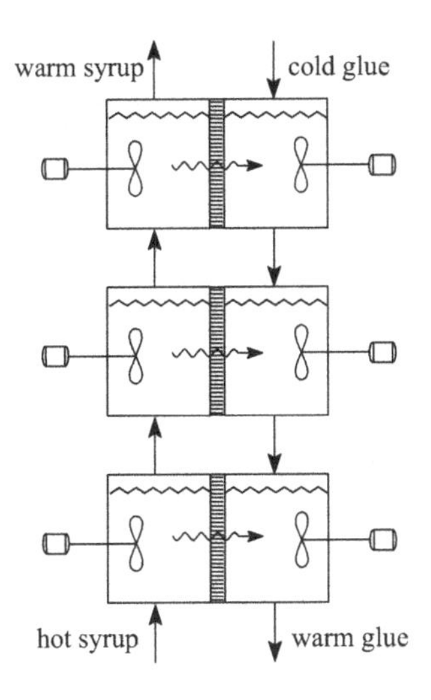

We wish to calculate the number of stages needed.

An energy balance on the top n stages yields the equation below, which has the form $y = (\text{slope})x + \text{intercept}$.

$$(T_{\text{syrup}})_{\text{into } n\text{th tray}} = \left(\frac{F_{\text{glue}} C_{\text{P,glue}}}{F_{\text{syrup}} C_{\text{P,syrup}}}\right)(T_{\text{glue}})_{\text{out of } n\text{th tray}} + \left[(T_{\text{syrup}})_{\text{out of top ray}} - \left(\frac{F_{\text{glue}} C_{\text{P,glue}}}{F_{\text{syrup}} C_{\text{P,syrup}}}\right)(T_{\text{glue}})_{\text{into top tray}}\right].$$

Determine the number of trays needed to achieve the following conditions: $T_{\text{syrup, out}} = 40°\text{C}$, $T_{\text{glue, in}} = 18°\text{C}$, $T_{\text{syrup, in}} = 89°\text{C}$, $F_{\text{syrup}} = 33.7$ kg/min, and $F_{\text{glue}} = 39.0$ kg/min.

(H) The multistage heat exchanger in part (G) is ideal; the syrup and glue leave a stage at the same temperature. Consider a non-ideal heat exchanger. The syrup leaving a stage is 4°C warmer than the glue leaving the stage.

Determine the number of trays needed with this non-ideal heat exchanger. Use the same conditions as for part (G): $T_{syrup,\ out} = 40°C$, $T_{glue,\ in} = 18°C$, $T_{syrup,\ in} = 89°C$, $F_{syrup} = 33.7$ kg/min, and $F_{glue} = 39.0$ kg/min.

Operating Lines for Multistage Countercurrent Separators: Distillation Columns

4.54 Use the data in Exercise 4.19 to prepare a liquid–vapor x–y diagram for methanol + water mixtures at 1 atm. Plot the data on the blank graph below. Label the axes and sketch a line through the data.

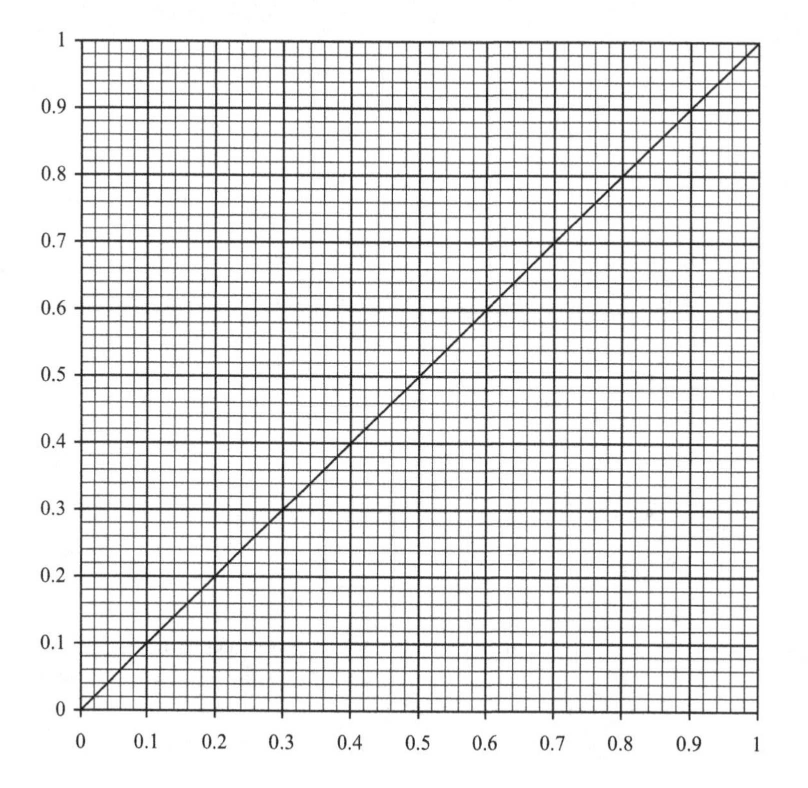

4.55 Tabulated below are vapor–liquid equilibrium data for nitric acid + water mixtures at 1.0 atm (these data may be downloaded from duncan.cbe.cornell.edu/Graphs). Use these data to plot a temperature–composition phase diagram (suitable for analyzing a flash drum) and a liquid–vapor x–y diagram (suitable for McCabe–Thiele analysis); use the blank graph for Exercise 4.54 or download a blank graph from duncan.cbe.cornell.edu/Graphs. Label the axes and label the regions of the plots (liquid, vapor, equilibrium line, too much H_2O in the vapor, etc.).

T (°C)	$x_{\text{nitric acid}}$	$y_{\text{nitric acid}}$
100.	0.0	0.0
104.	0.067	0.003
104.5	0.072	0.003
106.5	0.102	0.010
107.0	0.110	0.012
108.5	0.135	0.020
109.5	0.141	0.023
110.5	0.162	0.035
111.5	0.181	0.042
112.0	0.181	0.042
114.5	0.217	0.082
115.5	0.233	0.096
117.5	0.282	0.165
119.2	0.348	0.297
119.4	0.341	0.259
120.0	0.383	0.375
119.9	0.374	0.375
118.5	0.450	0.564
117.0	0.474	0.651
115.0	0.515	0.762
113.0	0.530	0.764
112.6	0.540	0.768
111.5	0.557	0.857
108.8	0.574	0.864
106.0	0.606	0.936
102.9	0.651	0.942
97.5	0.700	0.960
96.1	0.719	0.972
95.8	0.723	0.984
95.5	0.738	0.986
92.0	0.755	0.983
91.0	0.802	0.983
87.2	0.853	0.984
86.9	0.878	0.988
82.8	0.991	0.997
83.0	1.0	1.0

Experimental data from Gmehling, J., Onken, U., and Rarey-Nies, J. R. (1988). *Vapor-Liquid Equilibrium Data Collection - Aqueous Systems - Supplement 2*, Dechema, Frankfurt, Germany, vol. 1, part 1B, p. 384.

4.56 Consider the vapor–liquid equilibria of six binary mixtures represented by the temperature–composition diagrams, pressure–composition diagrams, and x–y diagrams below. Use these diagrams and the list of compounds to complete the following table. For example, binary mixture A corresponds to pressure–composition diagram 1. Which temperature–composition diagram corresponds to binary mixture *A*? What is the volatile component in binary mixture *A*? You may use a compound from the list more than once, or not at all.

	Binary mixture					
	A	*B*	*C*	*D*	*E*	*F*
pressure–composition diagram	1	2	3	4	5	6
temperature–composition diagram						
liquid–vapor x–y diagram						
volatile component						
second component						

Compound	Boiling point at 1 atm (°C)
ethylene oxide	14
2-methylbutane	28
methylene chloride	40
acetone	56
chloroform	61
hexane	69
2-butanone	80
trichloroethylene	87
3-pentanone	102
4-methyl-2-pentanone	116
3-hexanone	123
chlorobenzene	132
4-methylpyridine	145
furfural	162
phenol	182
acetophenone	203
octanoic acid	240

Pressure–composition diagrams for Exercise 4.56.

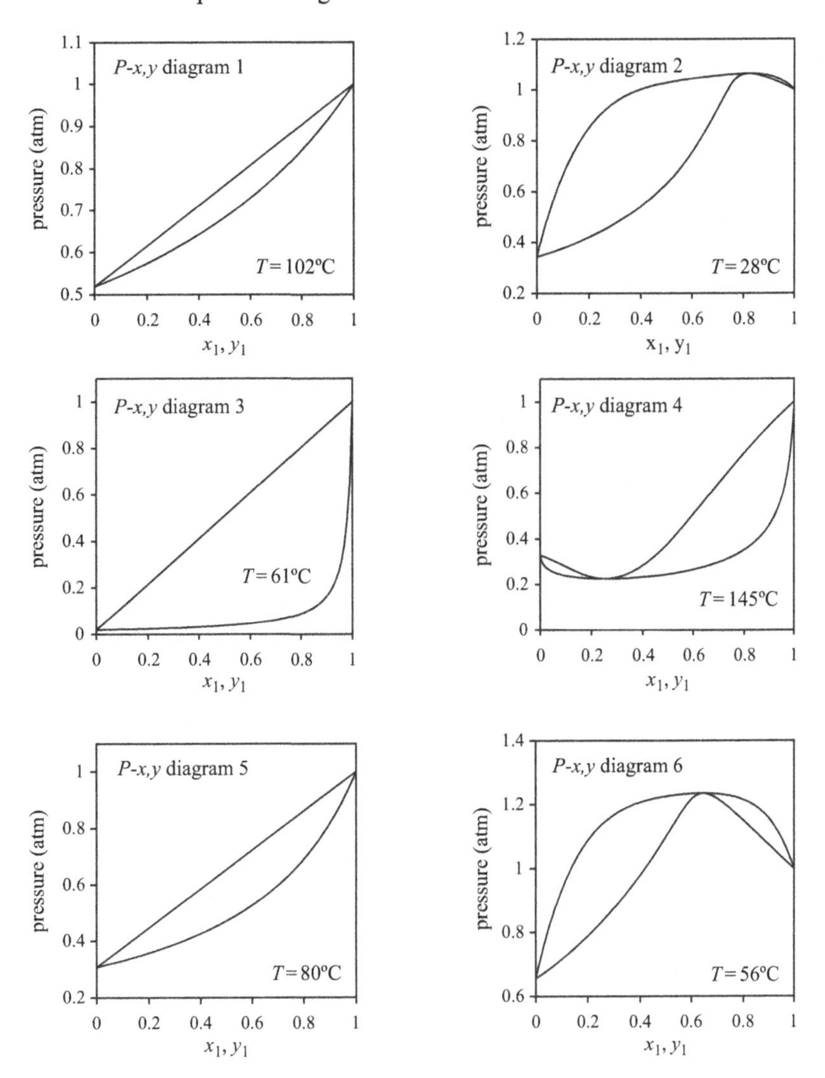

Temperature–composition diagrams for Exercise 4.56 (all diagrams represent mixtures at 1 atm).

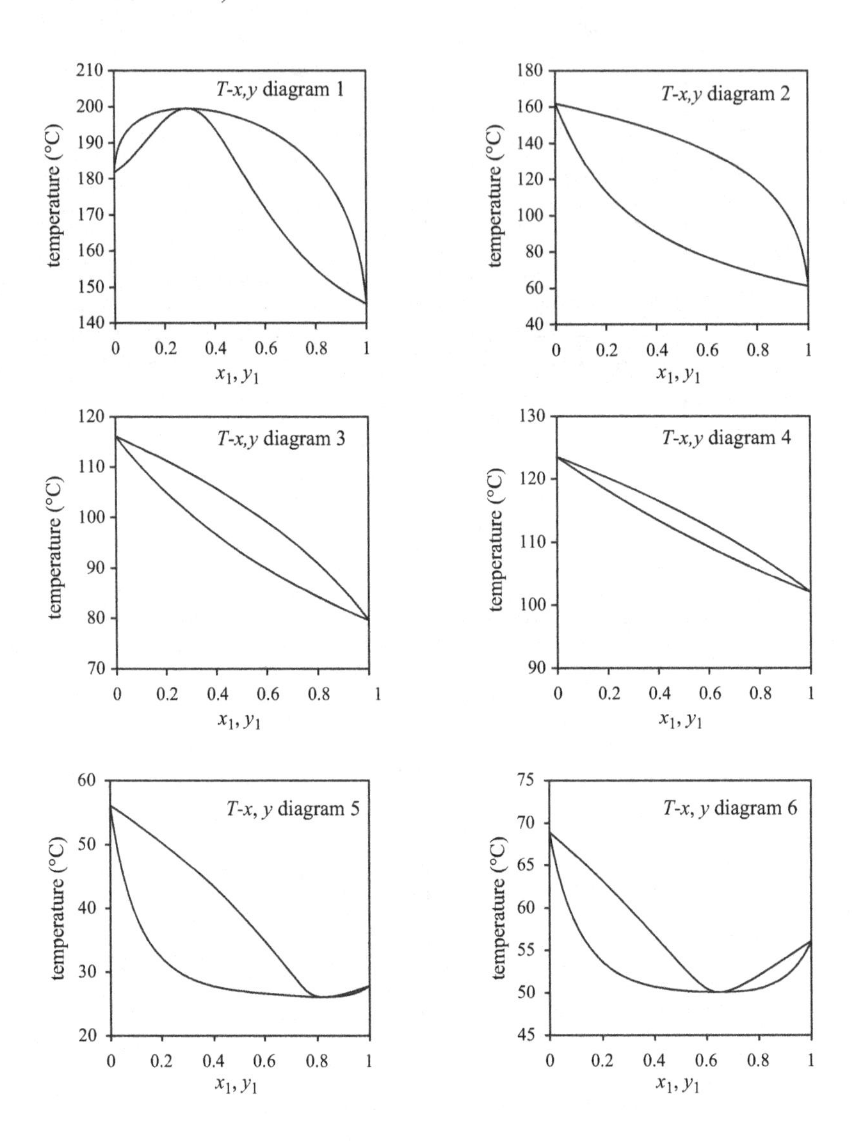

Liquid–vapor x–y diagrams for Exercise 4.56 (all diagrams represent mixtures at 1 atm).

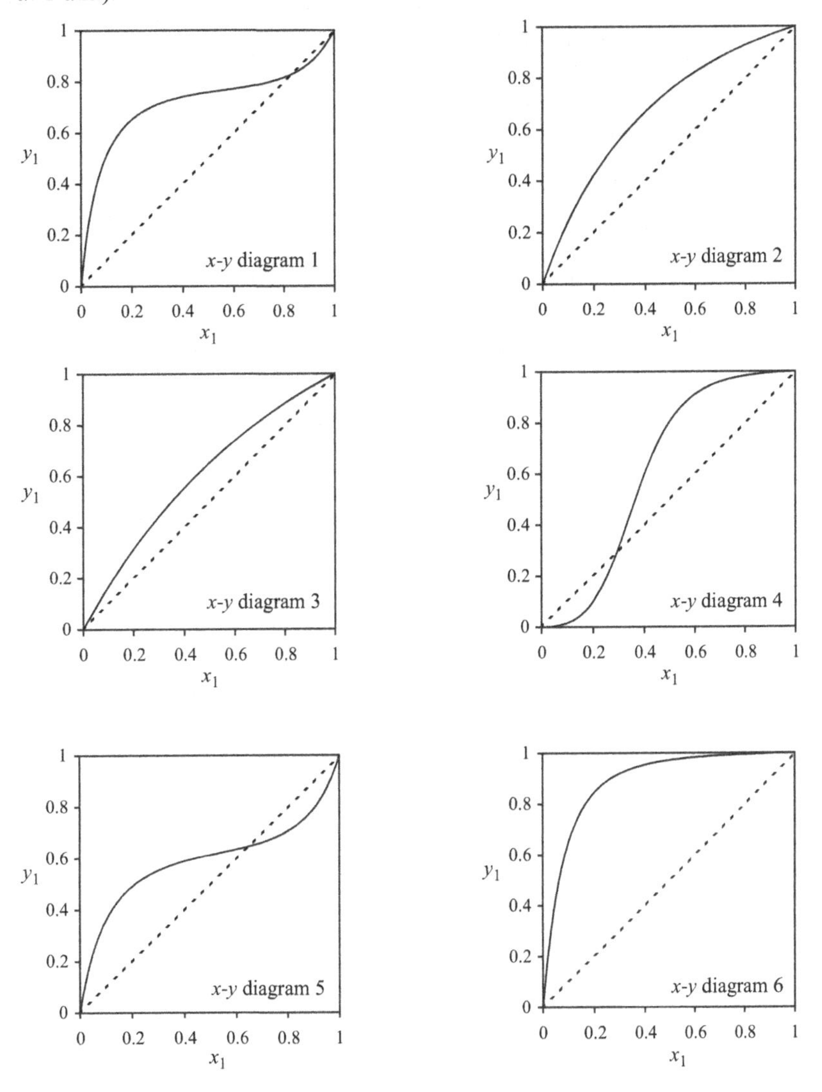

4.57 The temperature–composition diagram for $A + B$ mixtures at 1 atm is shown below.

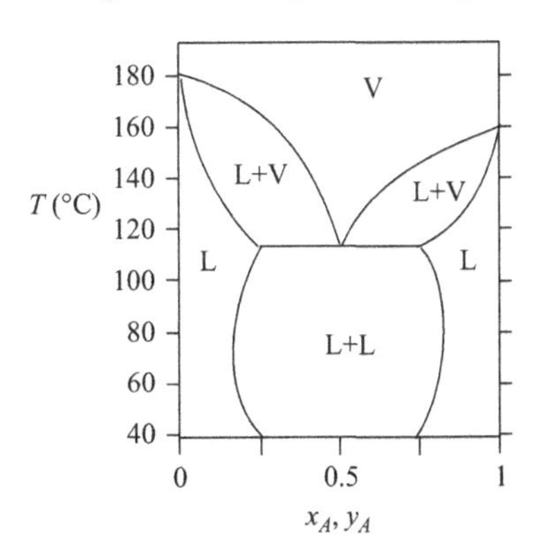

(A) Which substance in the table below is most likely component A?

	Melting point (°C)	Boiling point (°C)
2-methyl tetrahydrofuran	−65	80
isoamyl chloride	−104	100
tetrachloroethylene	−22	120
acetylacetone	−23	140
isoamyl propionate	17	160
ethyl acetoacetate	−45	180

(B) Which liquid–vapor x–y diagram below most likely corresponds to the temperature–composition diagram above?

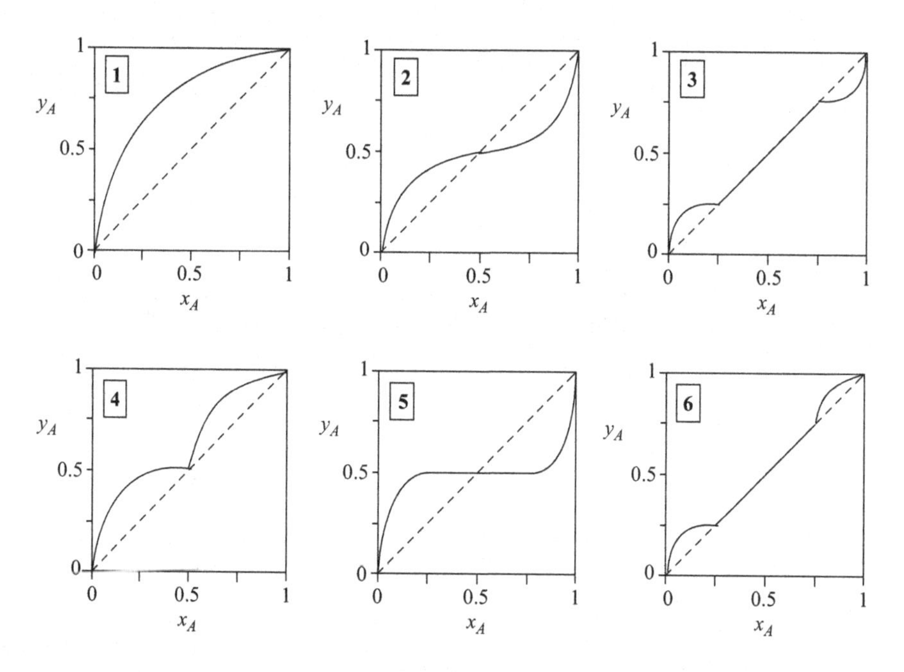

4.58 The temperature–composition diagram for $A + B$ mixtures at 1 atm is shown below (not the same A or B as in Exercise 4.57, or any other exercise).

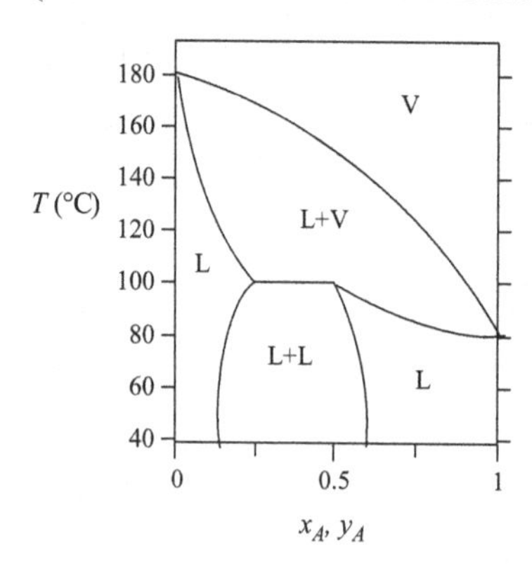

(A) Which substance in the table in Exercise 4.57(A) is most likely component A?

(B) Which liquid–vapor x–y diagram below most likely corresponds to the temperature–composition diagram above?

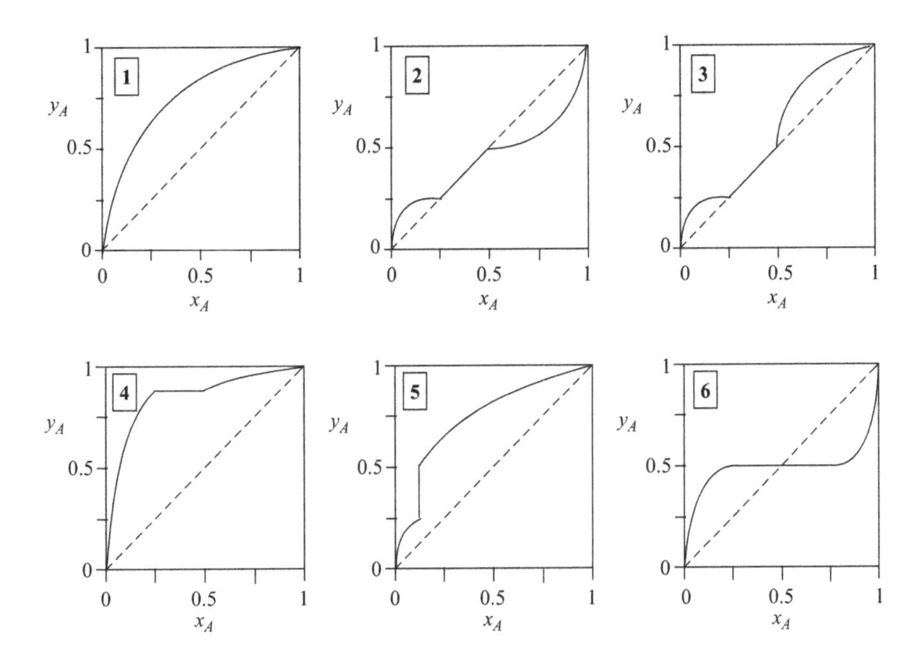

4.59 The temperature–composition diagram for $A + B$ mixtures at 1 atm is shown below (not the same A or B as in Exercise 4.57 or 4.58, or any other exercise).

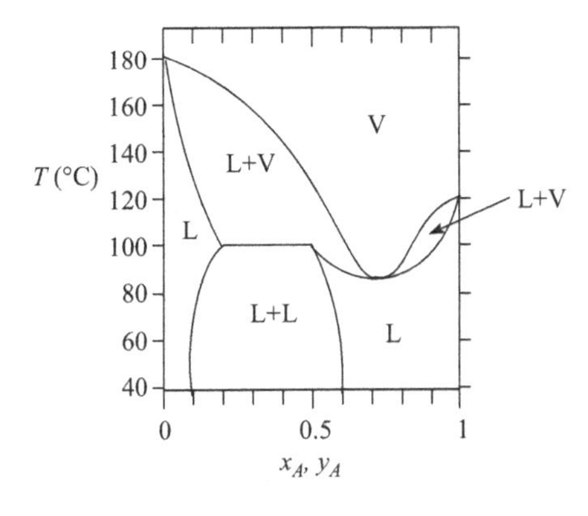

(A) Which substance in the table in Exercise 4.57(A) is most likely component A?
(B) Which liquid–vapor x–y diagram below most likely corresponds to the temperature–composition diagram above?

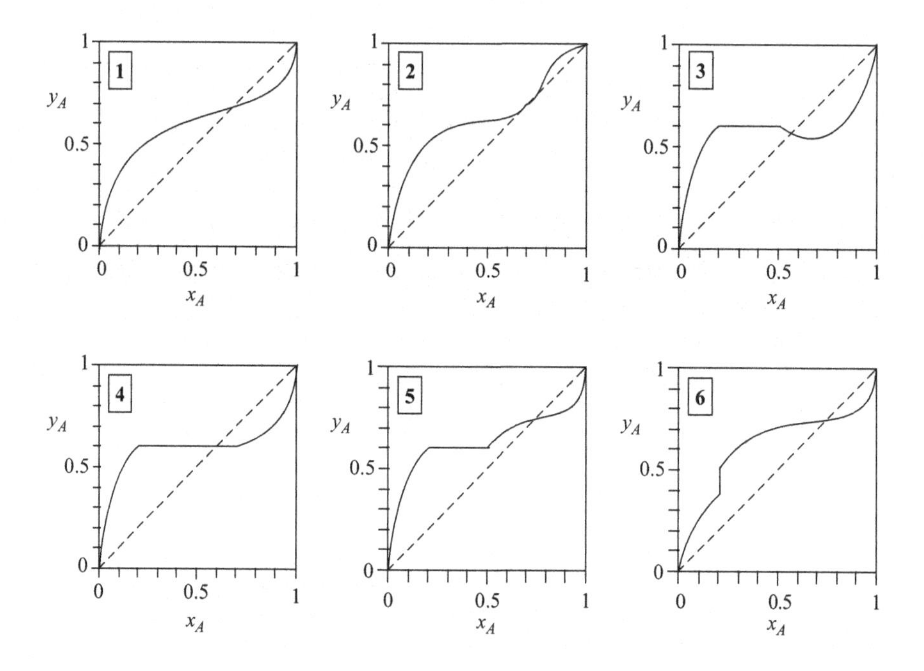

4.60 Which McCabe–Thiele constructions below are plausible analyses of a distillation column, which constructions cannot possibly represent a distillation column, and which constructions lack sufficient information to decide? Assume the equilibrium line in each diagram is correct. Decide if the operating lines and the steps are consistent with a distillation column with one feed stage and finite output from both the top and the bottom. *Disregard minor imperfections in the drawings.*

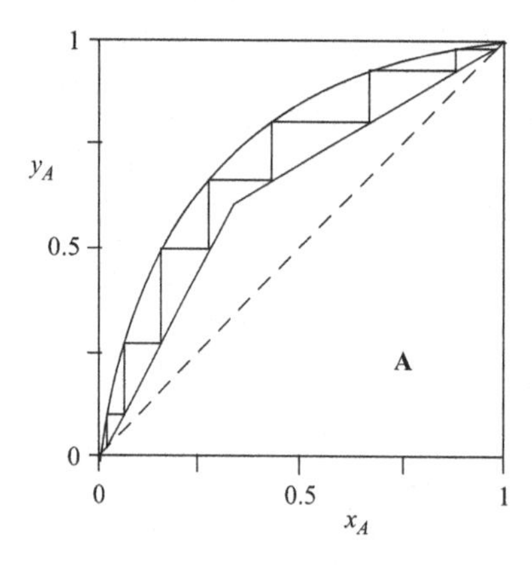

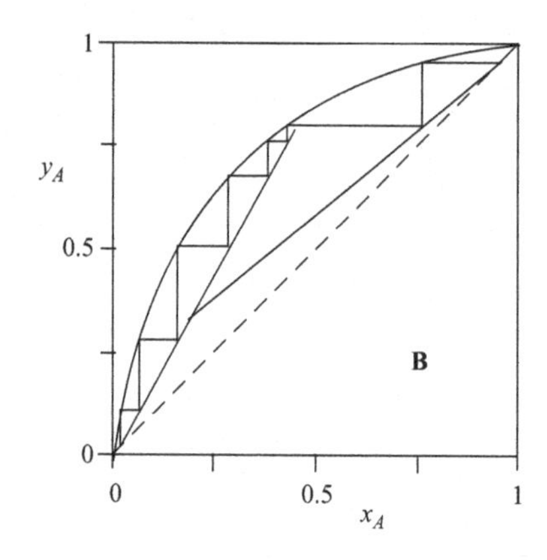

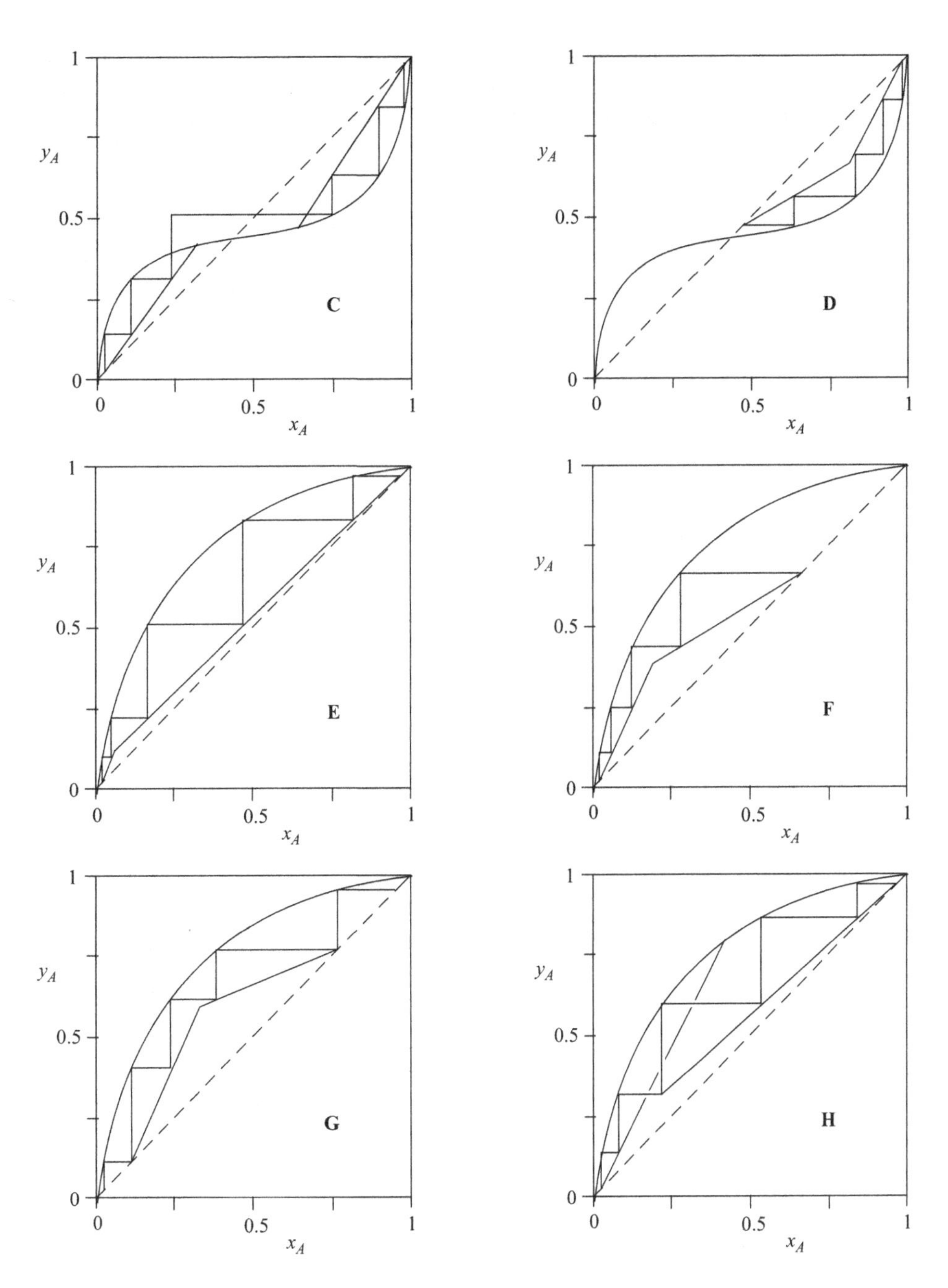

4.61 Which McCabe–Thiele constructions below are plausible analyses of a distillation column, which constructions cannot possibly represent a distillation column, and which constructions lack sufficient information to decide? Assume the equilibrium line in each diagram is correct. Decide if the operating lines and the steps are consistent with a distillation column with one feed stage and finite output from both the top and the bottom. *Disregard minor imperfections in the drawings.*

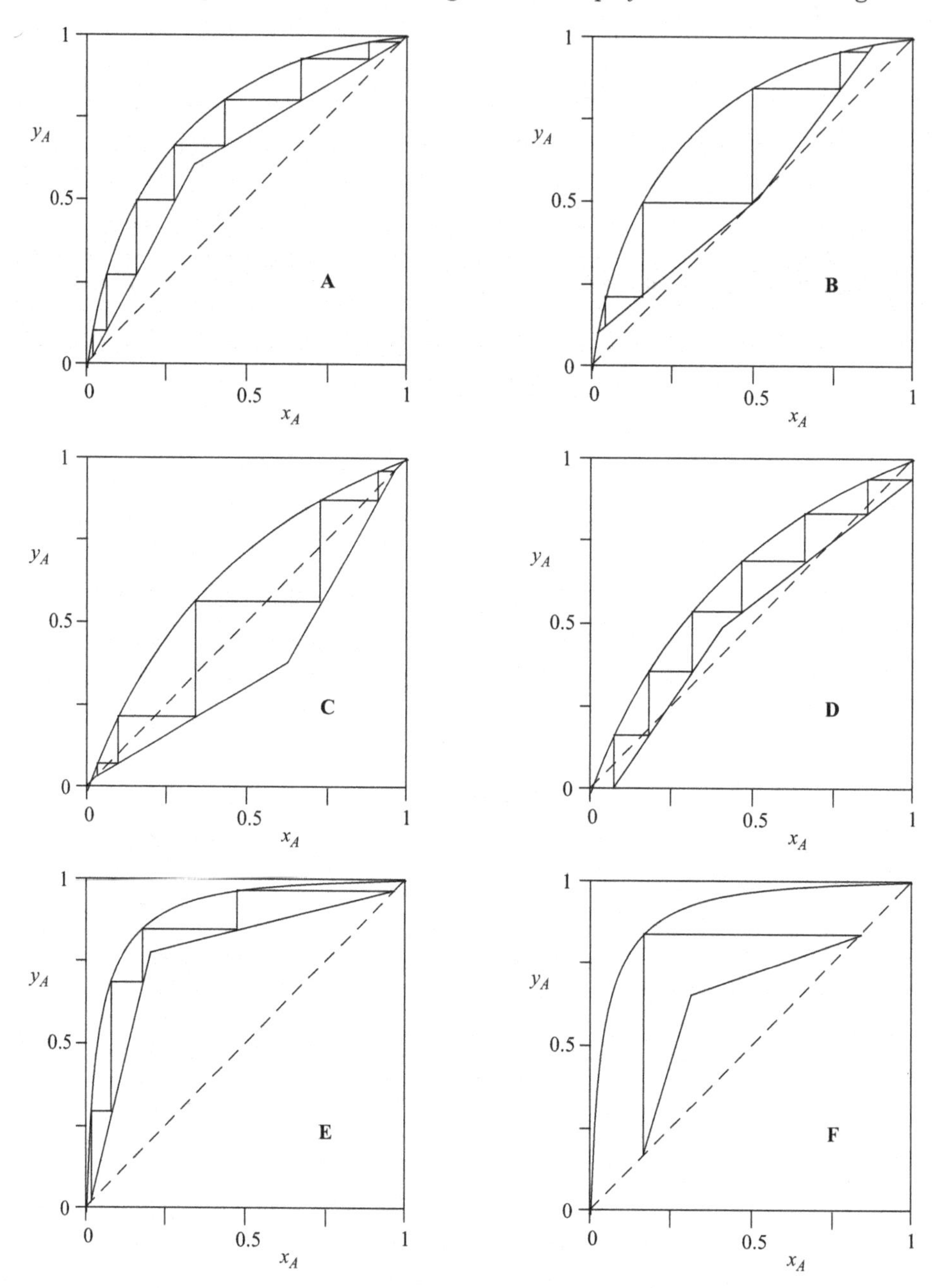

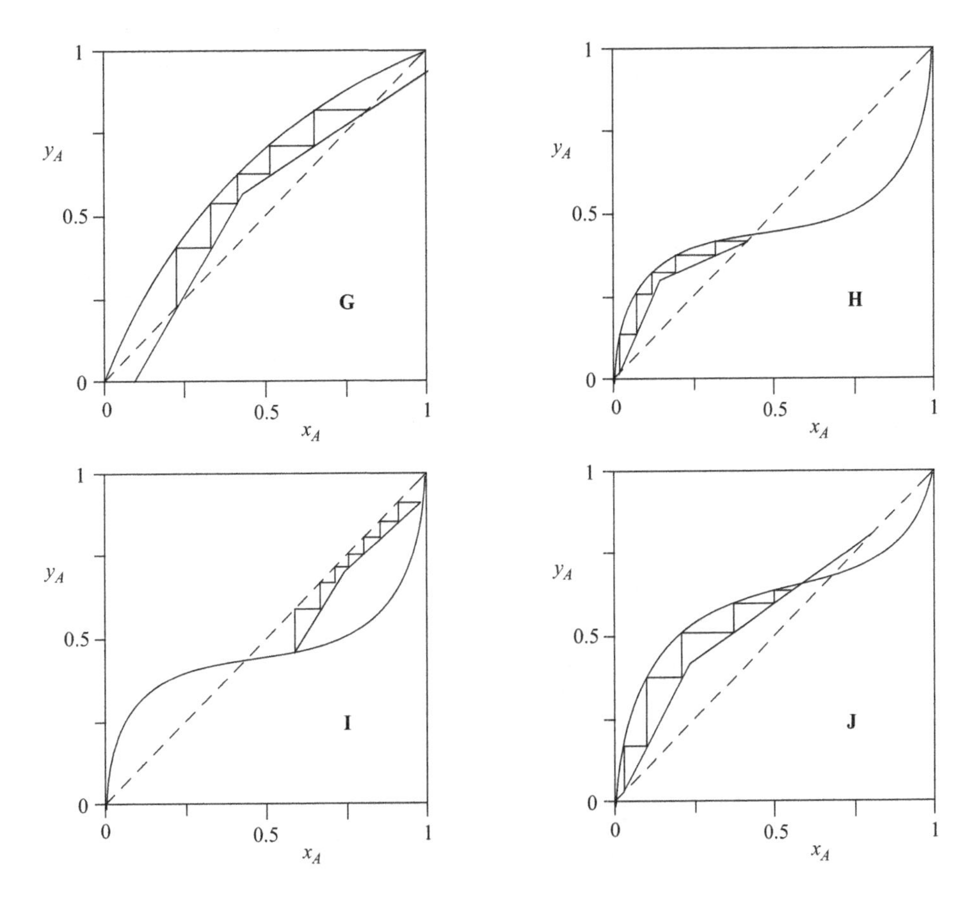

4.62 From which mixture would it be easiest to produce 99.9% A by distillation (not the same A as in any other exercise)? Assume the feed is a liquid + vapor mixture at equilibrium, with 50 mol% A in the vapor phase. Justify your choice, either with words or a graphical analysis.

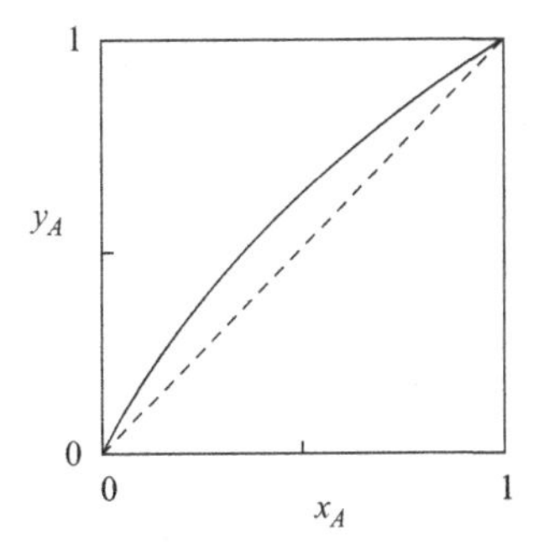

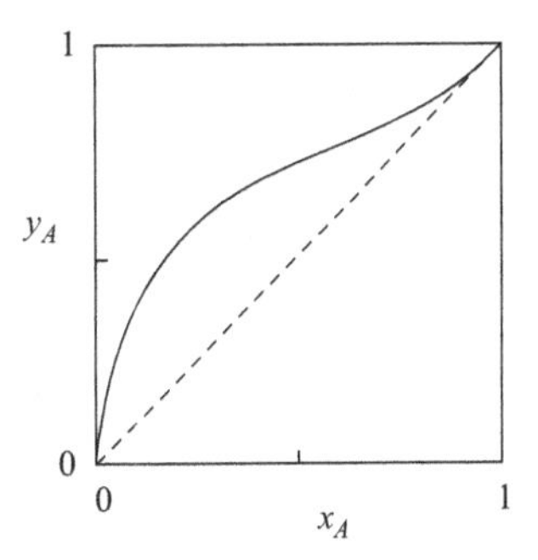

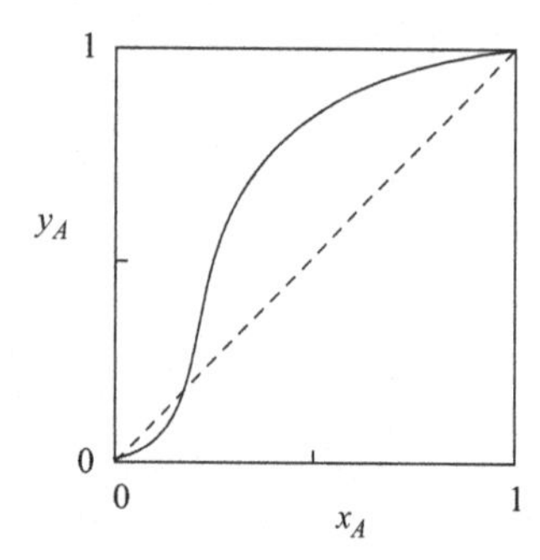

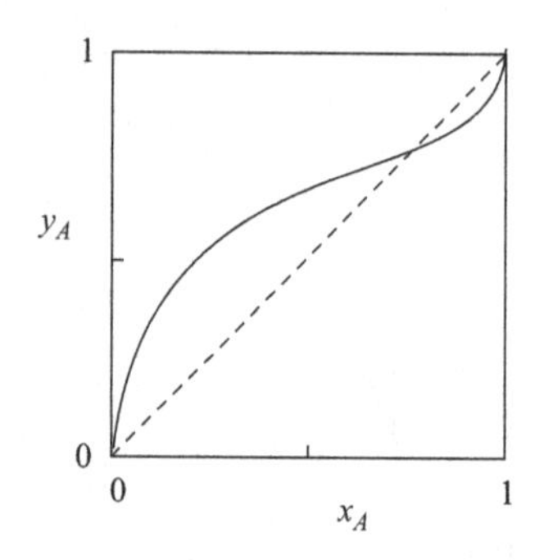

4.63 A distillation column separates acetonitrile ($CH_3C{\equiv}N$) and water at 1 atm with parameters in the McCabe–Thiele diagram shown below.

x–y diagram for liquid-vapor systems of acetonitrile + water mixtures at 1 atm.

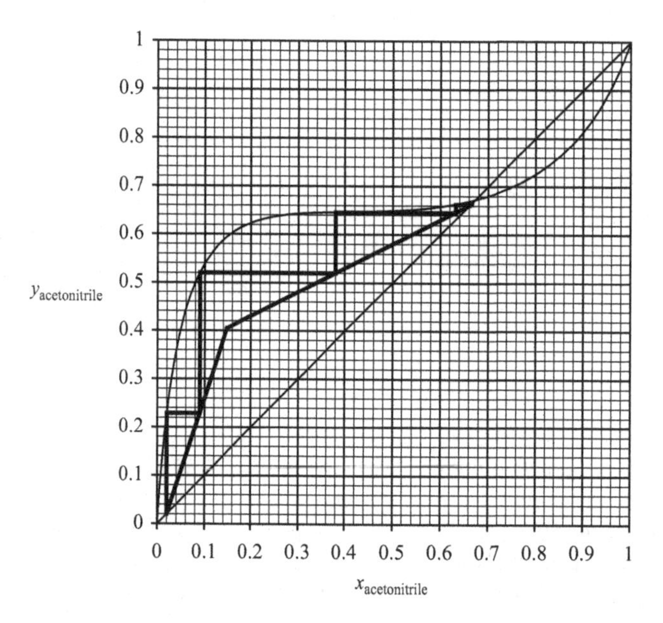

(A) How many equilibrium stages does this distillation column have?
(B) What is the mol% of acetonitrile in the acetonitrile-rich product?
(C) What is the mol% of acetonitrile in the water-rich product?
(D) What is the mol% of acetonitrile in the liquid and vapor streams that leave the feed stage?
(E) What is the ratio of liquid to vapor flow rates in the rectifying section (the upper section) of the column?
(F) Shown below is a phase diagram for acetonitrile + water mixtures at 1 atm as a function of temperature and composition. Label the four regions of the diagram.

Temperature–composition phase diagram for acetonitrile + water mixtures at 1 atm.

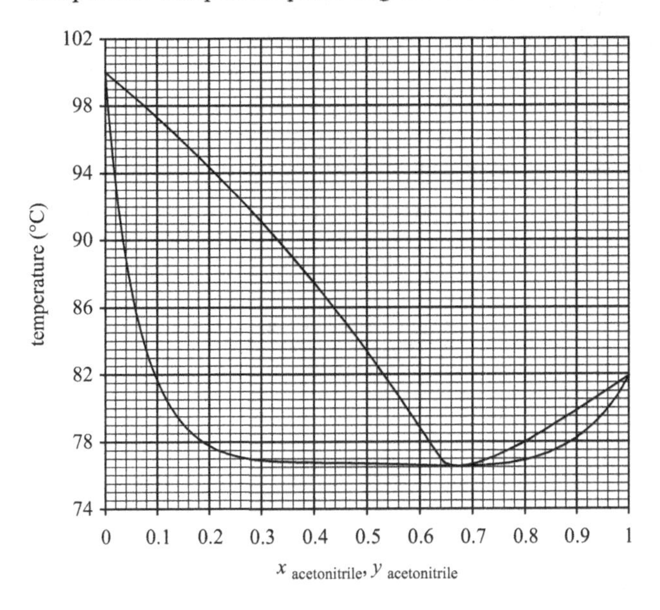

(G) The distillation column described above is preceded by a heater to bring the acetonitrile + water mixture (20 mol% acetonitrile) to match the conditions of the feed stage. Use the phase diagram of part (F) to determine the temperature in the preheater.

(H) What is the ratio of liquid to vapor in the stream leaving the preheater?

Diagrams generated from data compiled by Gmehling and Onken (1977), vol. 1, p. 78.

4.64 The temperature–composition phase diagram for acetonitrile (ACN) + 2-methylpyridine (MP) mixtures in Exercise 4.15 can be plotted as the x–y diagram below.

x–y diagram for liquid–vapor systems of acetonitrile + 2-methylpyridine mixtures at 1 atm.

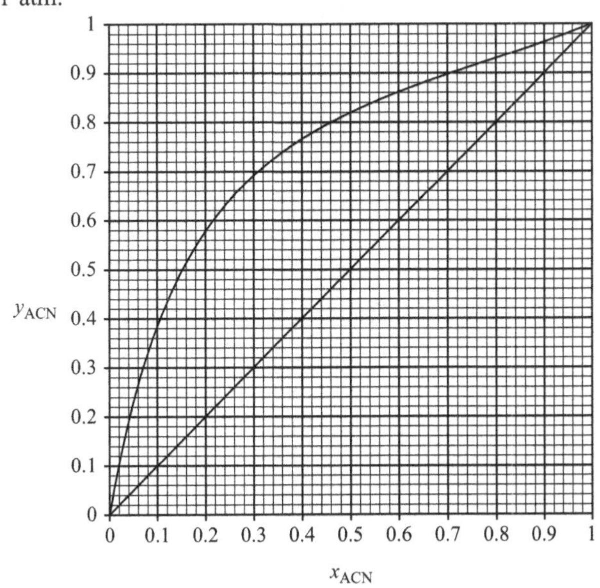

(A) Mark the point that corresponds to pure ACN.
(B) Mark the point that corresponds to a vapor + liquid mixture of ACN and MP at equilibrium at 98°C.
(C) Mark the point that corresponds to a vapor + liquid mixture *not* at equilibrium: vapor with 80% ACN above liquid with 70% ACN.
(D) After the system in (C) reaches equilibrium, will the *vapor* contain less than 80% ACN or more than 80% ACN? Justify your reasoning on the diagram.

4.65 The distillation column shown below distills a water + ethanol mixture.

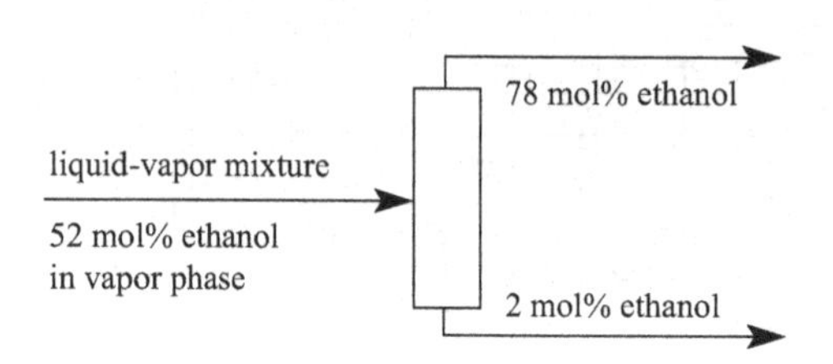

(A) What is the minimum number of stages this column could have?
(B) What is the maximum ratio of liquid/vapor flow for the stripping portion (the stages below the feed)?

x–y diagram for liquid–vapor systems of ethanol + water mixtures at 1 atm.

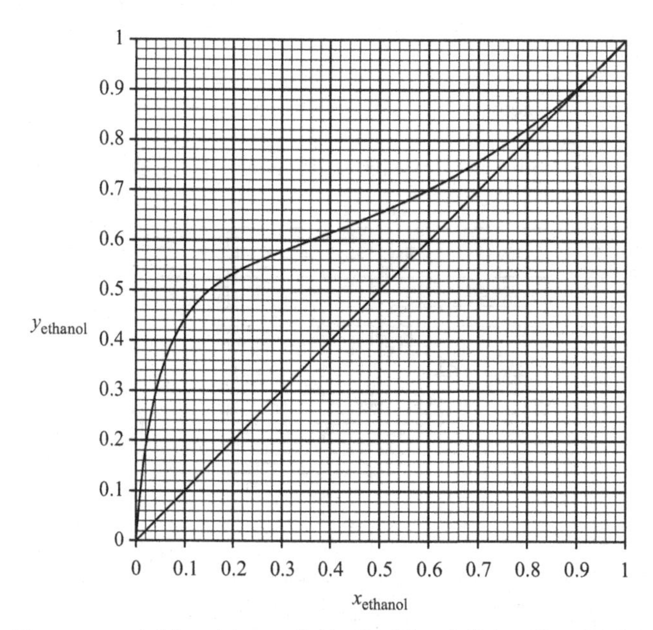

Diagram generated from data compiled by Gmehling, J., Onken, U., and Arlt, W. (1982). *Vapor-Liquid Equilibrium Data Collection – Organic Hydroxy Compounds: Alcohols*, Dechema, Frankfurt, Germany, vol. 1, p. 165.

4.66 Use the experimental data given in Exercise 4.19 (or the diagram you prepared in Exercise 4.54) to specify the operating conditions in the distillation column below. The column has nine trays. Determine operating conditions that will maximize L/V in the lower trays (stripping section) and minimize L/V in the upper trays (rectifying section). Also, specify the temperature of stream 1 for which the vapor–liquid composition of the input matches that of the feed tray.

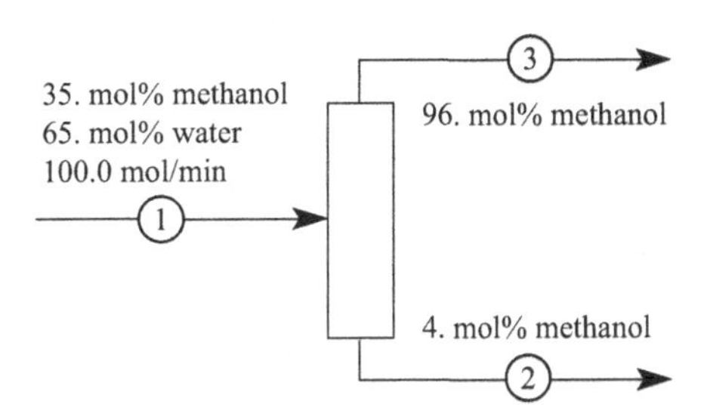

4.67 (A) The multistage distillation column shown below separates a mixture of F and Z. How many stages are needed?

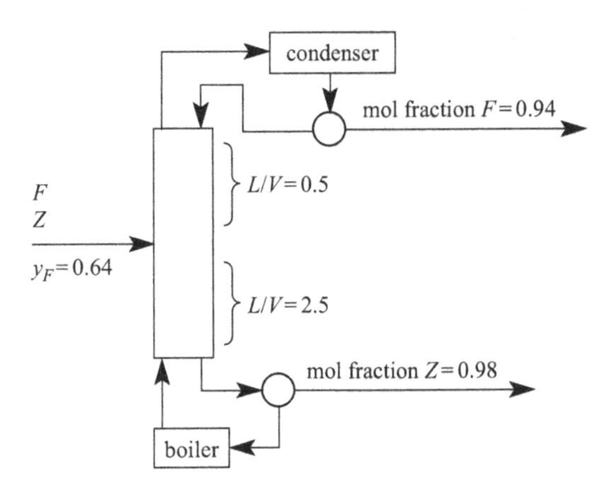

(B) The distillation column shown below provides an intermediate output stream. How many stages are needed? At what stage does the feed enter? Above what stage does the intermediate stream leave? **Hint:** Contrary to usual practice, begin stepping off stages at the top.

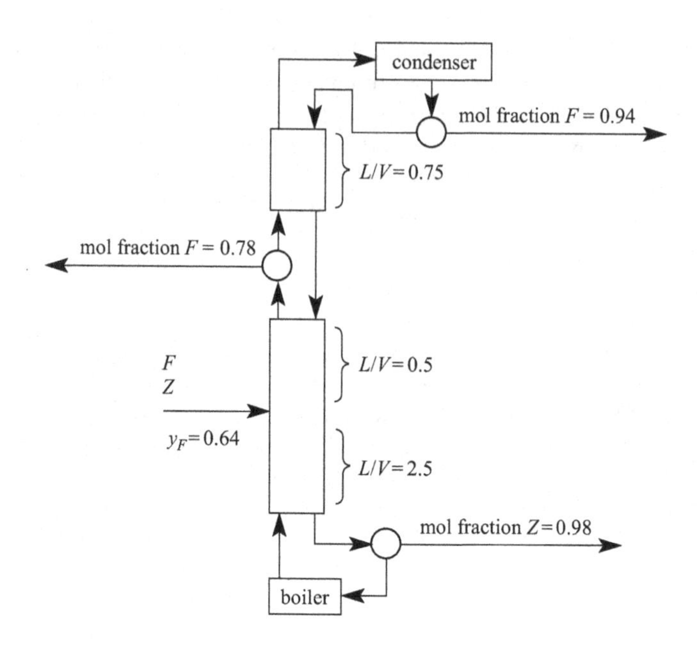

x–y diagram for liquid–vapor systems of $F + Z$ mixtures at 1 atm.

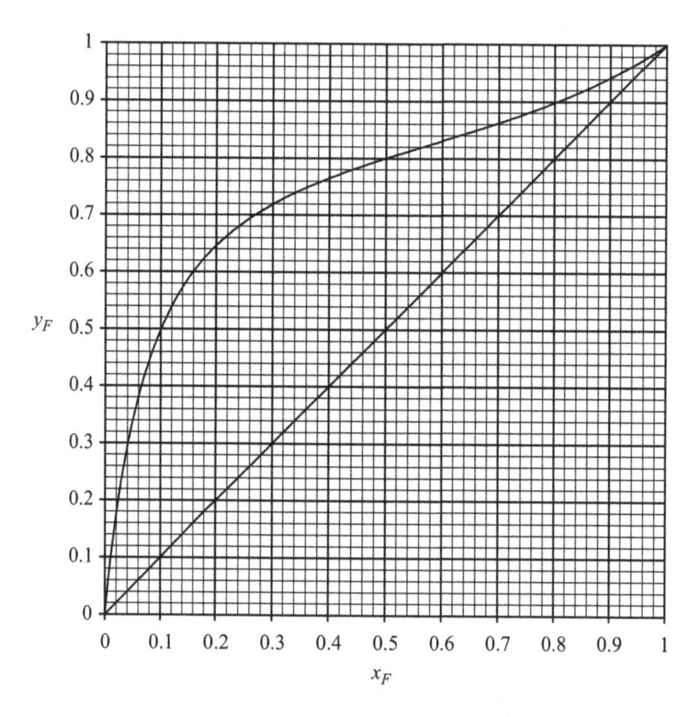

4.68 The distillation column shown below differs from traditional columns: the reboiler is an additional equilibrium stage. The liquid from the bottom stage of the column (stage 8) is *partially* reboiled. The liquid and vapor streams that leave the reboiler are in equilibrium.

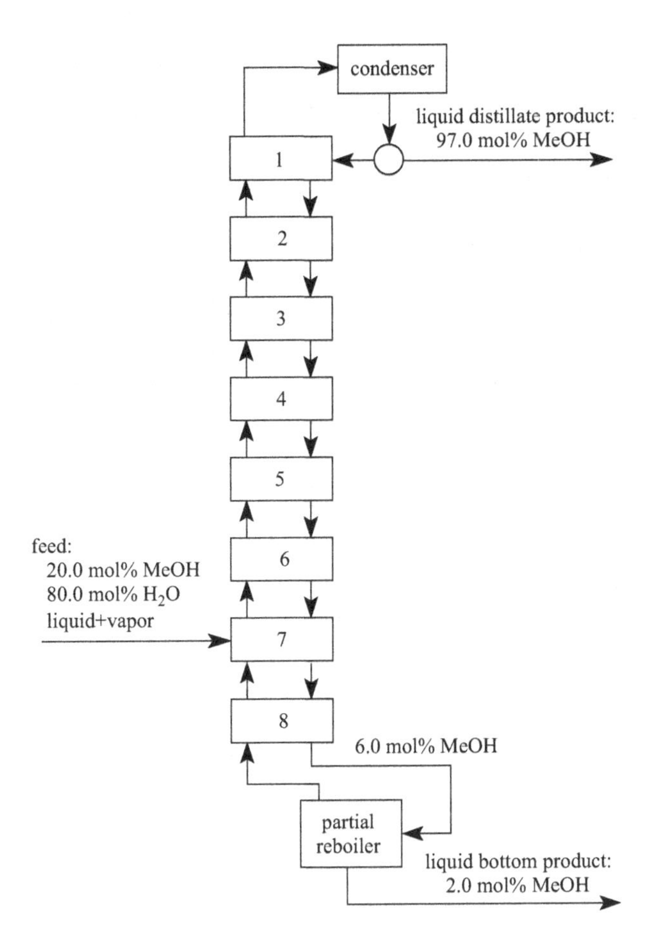

(A) Given the specifications on the column below, what is the temperature in the reboiler (in °C)?

Temperature–composition phase diagram for methanol + H_2O mixtures at 1 atm.

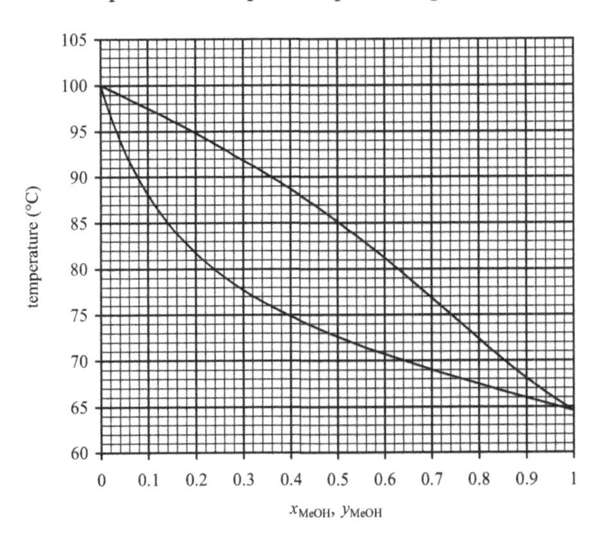

x–y diagram for liquid–vapor systems of methanol + H_2O mixtures at 1 atm.

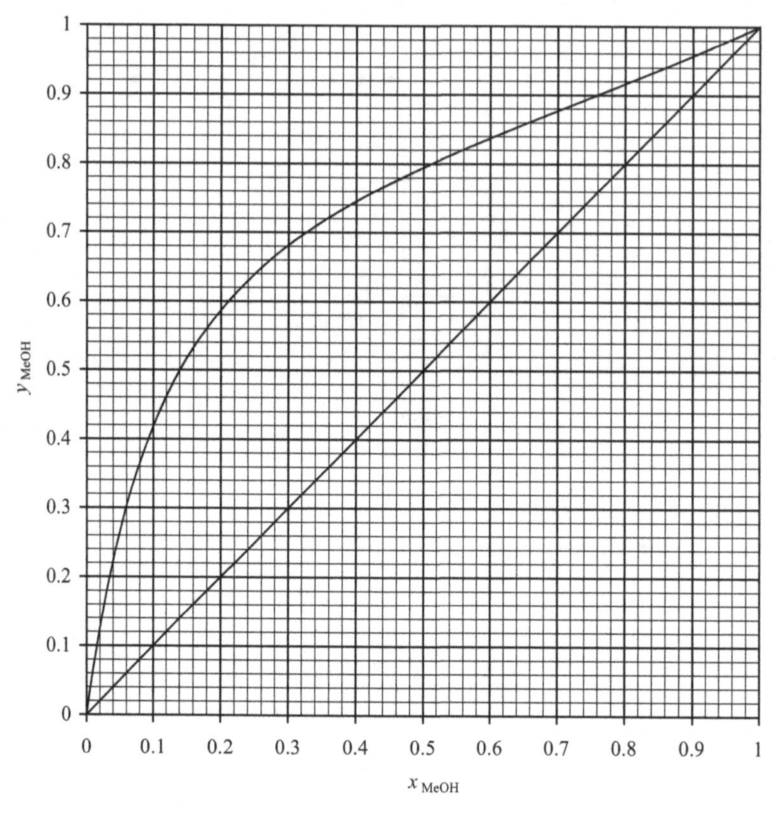

(B) What is the temperature in stage 8 (in °C)?
(C) What is the maximum temperature in the condenser (in °C)?
(D) What is the slope of the operating line for the stripping section (the stages below the feed stage)?
(E) Is stage 7 an appropriate feed stage for the specifications shown below? If not, which stage is appropriate?

4.69 The feed to the distillation column below is split: the two streams are heated and fed to a distillation column as shown in the following diagram. The McCabe–Thiele analysis is also shown below.

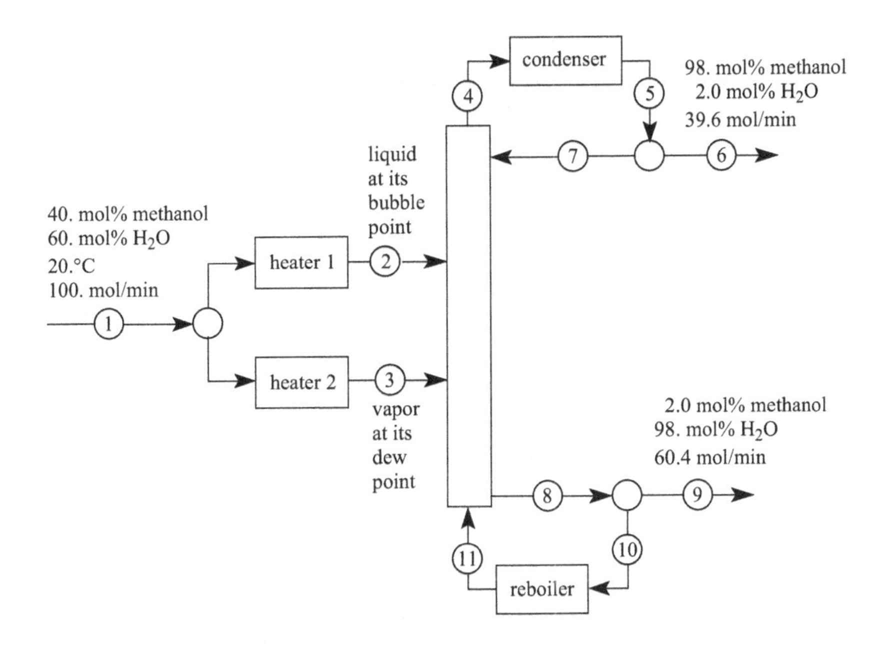

(A) Describe the design of the distillation column:
How many stages does the column have?
If the bottom is stage 1, stream 3 enters into what stage?

(B) Describe the operation of the distillation column:
What is the ratio of liquid flow to vapor flow in the stripping (bottom) section?
What is the ratio of liquid flow to vapor flow in the rectifying (top) section?
Stream 2 is the same temperature as its feed stage. What is the temperature of stream 2? Use the temperature–composition diagram with Exercise 4.68.

(C) What is the flow rate of stream 8, the liquid stream that leaves the bottom of the column, in mol/min?

(D) What is the flow rate of stream 3, in mol/min?

(E) Assume instead that stream 1 is heated and fed to the distillation column as a single stream. Further assume that the L/V ratios of the stripping and rectifying sections are kept the same.

How many stages will the new column require?
Stream 1 will be the same temperature as the feed stage. To what temperature should stream 1 be heated?
You may use the x–y diagram below, or use the blank diagram with Exercise 4.68.

x–y diagram for liquid–vapor systems of methanol + H_2O mixtures at 1 atm.

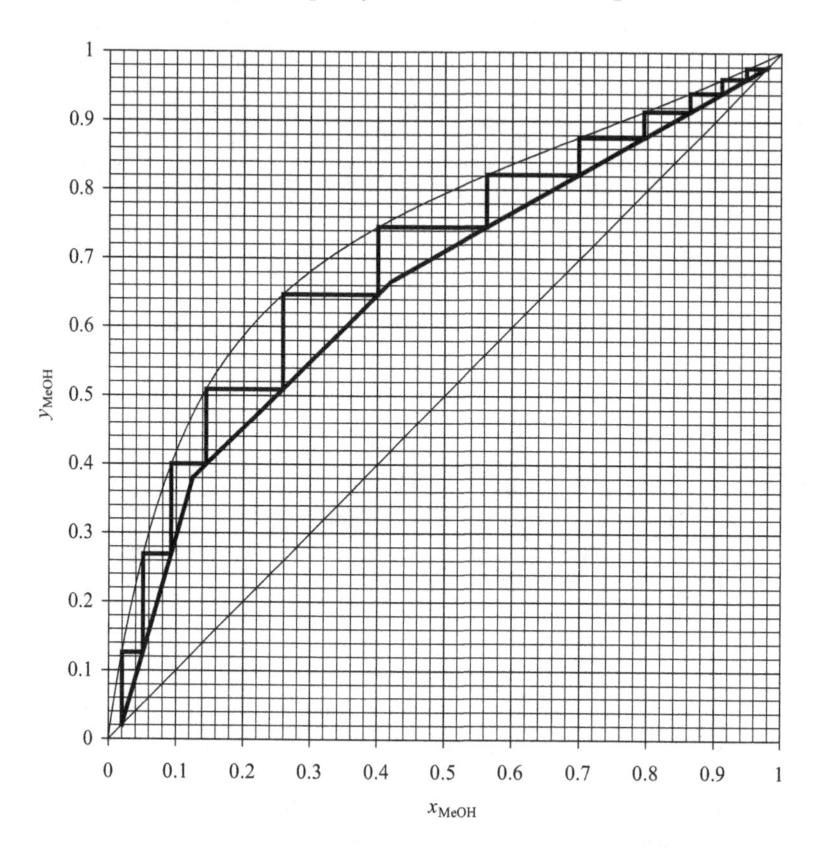

4.70 Match each distillation column from the choices below with one of the graphical models that follow them.

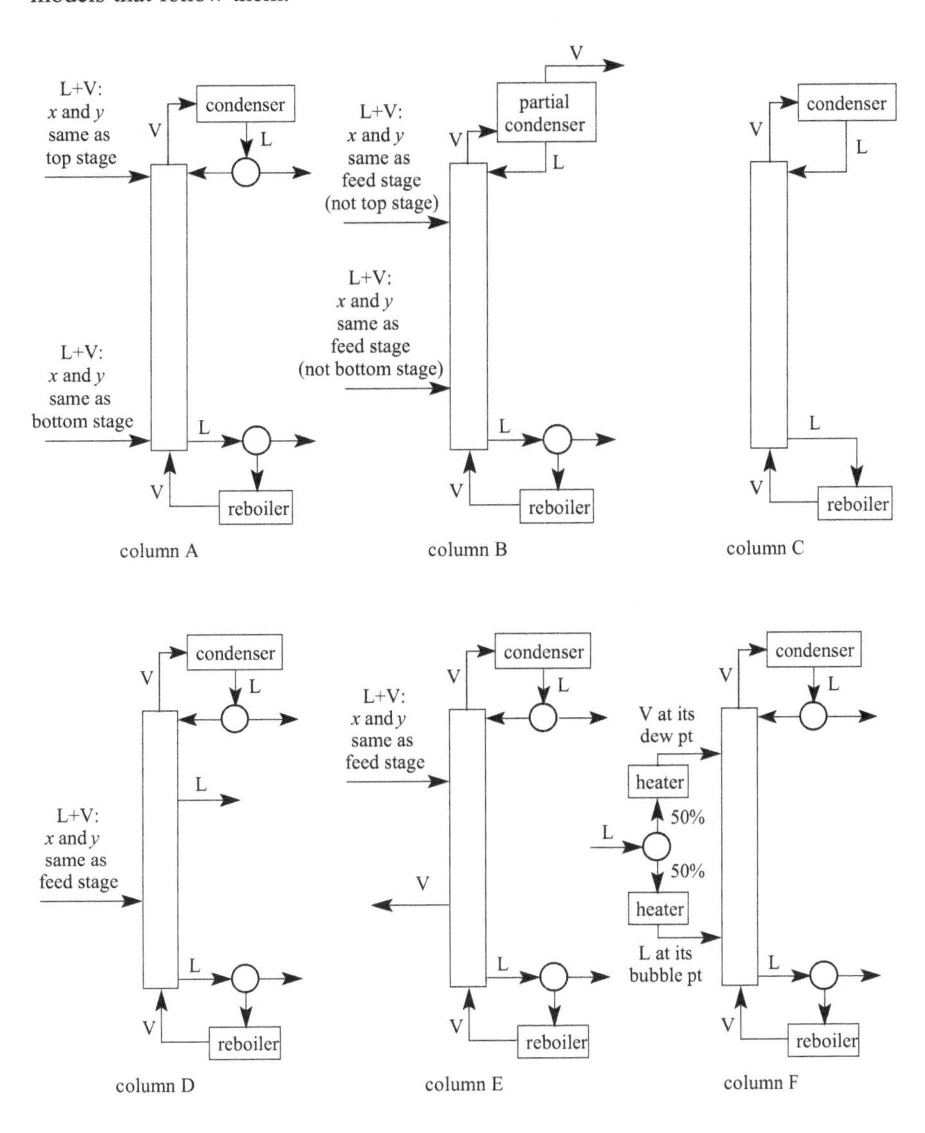

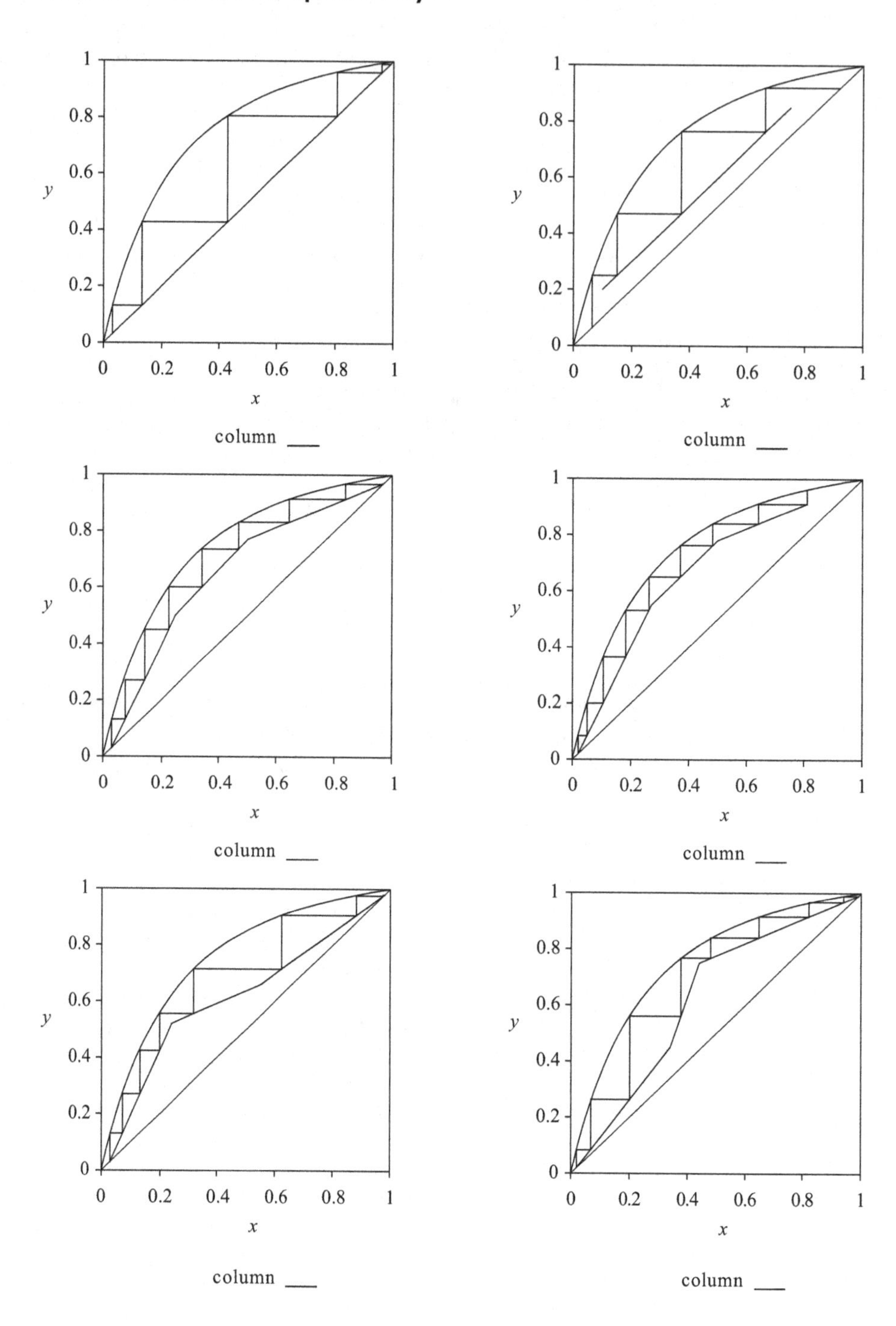

1
0.8
0.6
0.4
0.2
0
y
0
0.2
0.4
0.6
0.8
1
x
column ___
column ___
column ___
column ___
column ___
column ___

4.71 A typical flash drum has one liquid–vapor equilibrium stage. The effectiveness of a flash drum is improved by adding stages, as shown below. The liquid feed is split into two streams: a liquid stream is delivered to the top stage and the other stream is converted entirely to vapor and bubbled through the bottom stage.

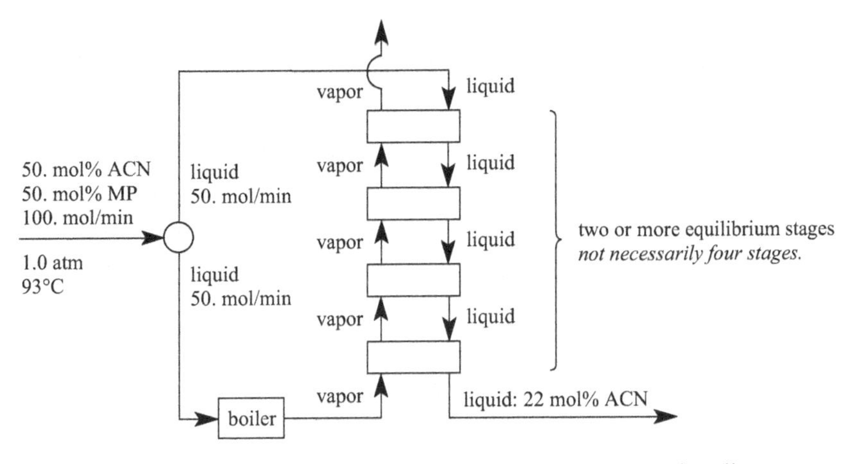

A multistage flash drum may be analyzed with an operating line on a x–y diagram, similar to the McCabe–Thiele analysis of a distillation column. But there are two differences.

(A) Unlike a distillation column, the liquid leaving the bottom stage has a different composition from the vapor entering the bottom stage. Locate the lower point of the operating line on the x–y diagram for ACN + MP mixtures given in Exercise 4.64, or download the x–y diagram from duncan.cbe.cornell.edu/Graphs.

(B) Unlike a distillation column, all stages have the same L/V ratio, the ratio of liquid flow rate to vapor flow rate. Draw the operating line on the x–y diagram.

(C) Use the operating line to determine the number of equilibrium stages required.

(D) Use the operating line to determine the composition of the vapor output.

(E) Given a feed with 50.% ACN, what is the maximum concentration of ACN in the vapor output? You may use any number of stages; you may use any L/V ratio. **Hint:** You must sketch a different operating line.

4.72 The process shown below converts E to M, then separates the unreacted E from the product M and recycles the E (with some M) to the reactor. The McCabe–Thiele analysis for the distillation column is also shown below.

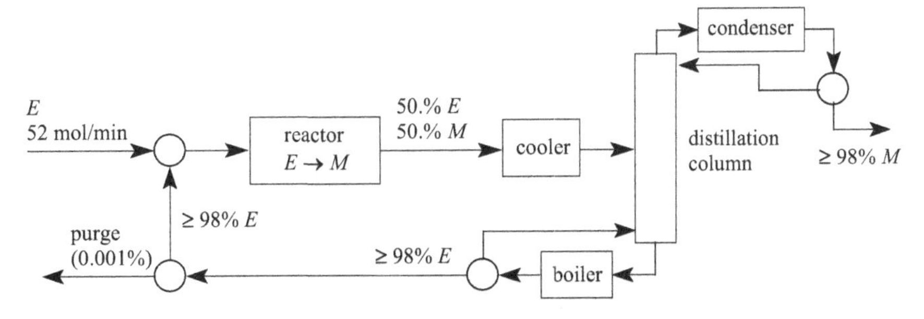

(A) Describe the design of the distillation column.
How many stages does the column have?
If the bottom is stage 1, what stage is the feed stage?

(B) Describe the operation of the distillation column.
What is the ratio of liquid flow to vapor flow in the stripping (bottom) section?
What is the ratio of liquid flow to vapor flow in the rectifying (upper) section?
If the feed enters as liquid + vapor mixture with the same composition as the feed stage, what is the temperature in the cooler?

x–y diagram for liquid–vapor systems of $E + M$ mixtures at 1 atm.

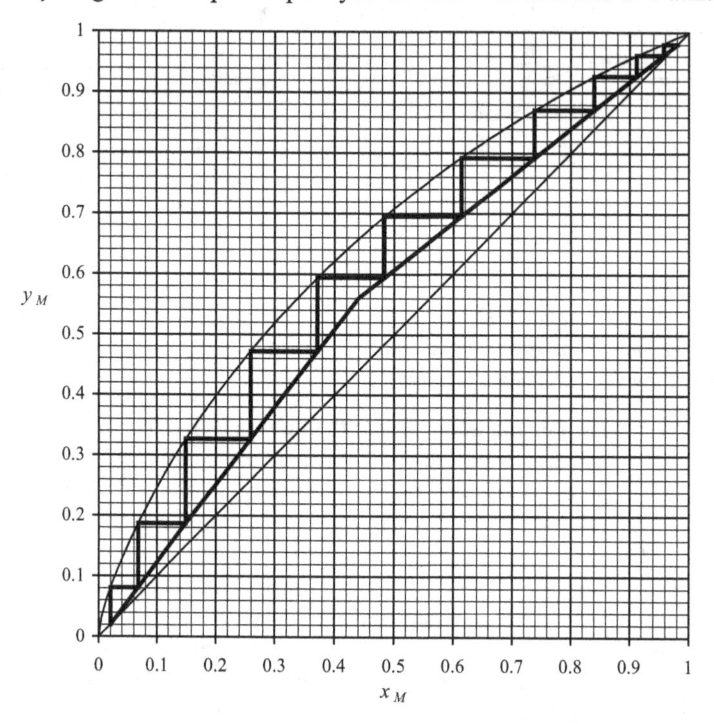

(C) What is the flow rate (mol/min) of the product stream?

(D) What is the flow rate of vapor into the condenser (mol/min)?

(E) The conversion in the reactor decreases; the reactor output changes to 40.% M. What operating parameters, if any, must be changed in the distillation column and the cooler to maintain $\geq$98% M in the distillate and $\geq$98% E in the bottoms? Operating parameters include reflux ratios and the cooler temperature. You may not change the location of the feed stage. If any operating parameters change, state the new values.

(F) The conversion in the reactor decreases further; the reactor output changes to 30.% M. What operating parameters, if any, must be changed in the distillation column and the cooler to maintain $\geq$98% M in the distillate and $\geq$98% E in the bottoms? Again you may not change the location of the feed stage. If any operating parameters change, state the new values.

(G) What is the minimum amount of M in the reactor output (in mol%) that can be fed to the distillation column and still provide some product with $\geq$98% M?

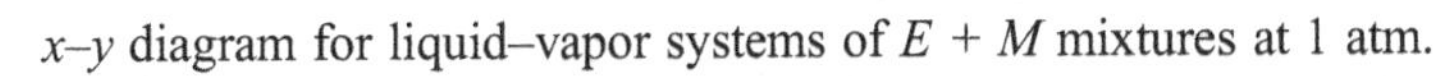

x–y diagram for liquid–vapor systems of $E + M$ mixtures at 1 atm.

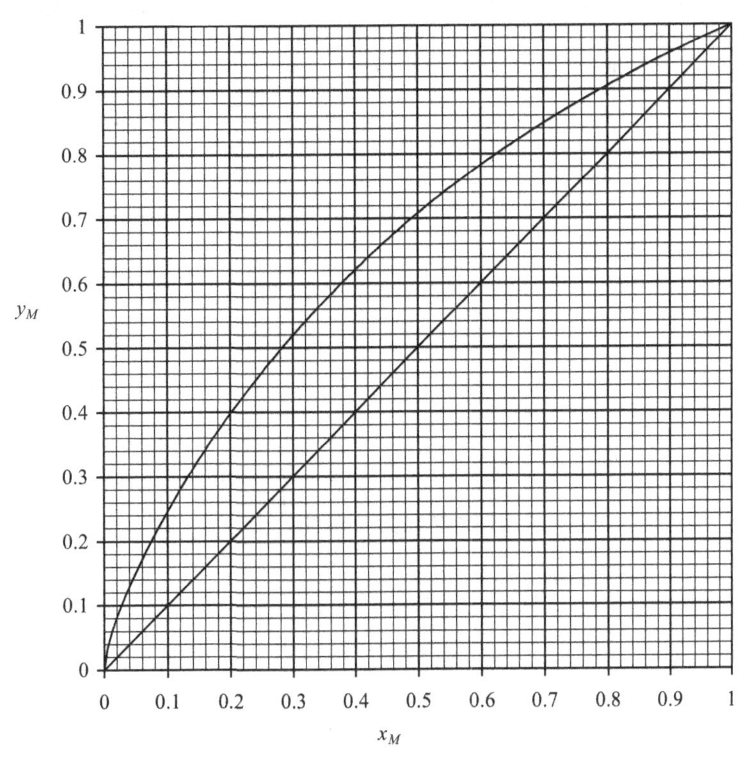

Temperature–composition phase diagram for $E + M$ mixtures at 1 atm.

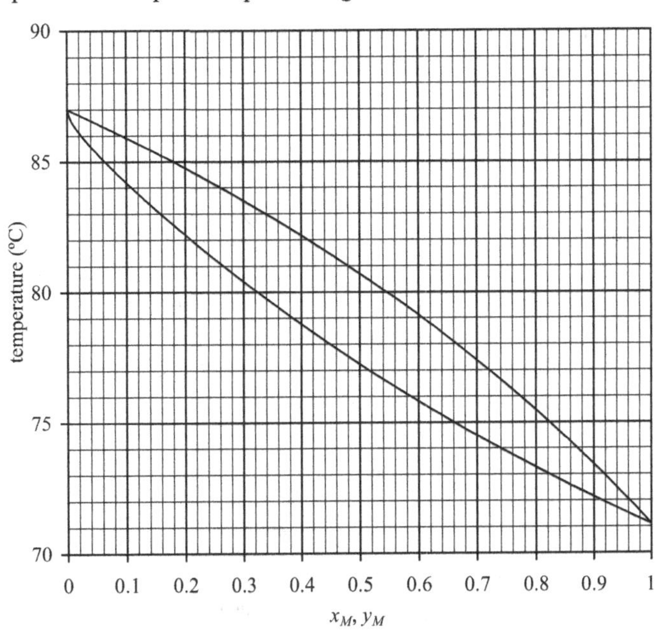

Phase Diagrams of Pure Substances

4.73 An air stream is contaminated by 100 ppm (on a molar basis) of compound *X*. We wish to remove compound *X* from the air by condensation. 100 ppm = 0.01 mol% = 10^{-4} mol fraction.

(A) If the pressure of the air is 1 atm, what is the partial pressure of compound *X*?
(B) If the temperature of the air is 300 K, what is the temperature of compound *X*?
(C) Consider the phase diagram of compound *X* shown below. Locate the point on the graph that corresponds to 100 molar ppm of compound *X* in air at 300 K and 1 atm.

Pressure–temperature phase diagram for compound *X*.

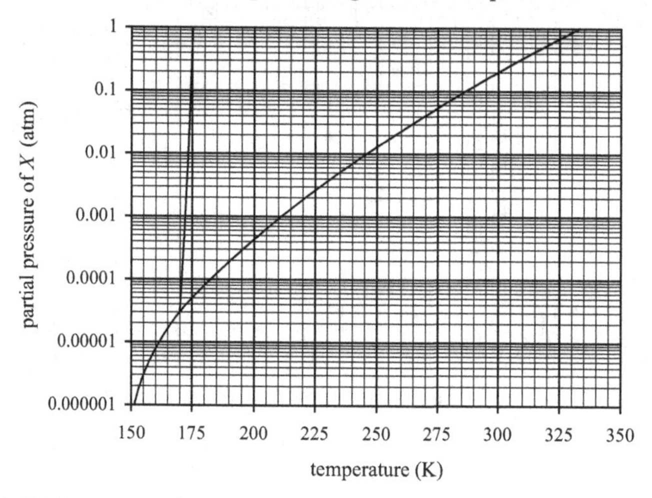

(D) If the contaminated air is cooled at constant pressure (1 atm), at what temperature will compound *X* start to condense?
(E) To what temperature must one cool the air (at total pressure of 1 atm) to reduce the contamination level to 10 ppm?
(F) If the contaminated air is compressed at constant temperature (300 K), at what pressure will compound *X* start to condense?
(G) To what pressure must one compress the air (at 300 K) to reduce the compound *X* level to 10 ppm?

4.74 The air conditioner below produces air at 20°C (68°F) and 43% humidity. That is, the partial pressure of H_2O vapor in stream 5 is 0.01 atm. Air with 100% humidity has the maximum partial pressure of H_2O vapor at a given temperature. The partial pressure of H_2O vapor at 100% humidity increases with temperature. For example, 100% humidity at 20°C is 0.023 atm of H_2O vapor and 100% humidity at 40°C is 0.073 atm of H_2O vapor.

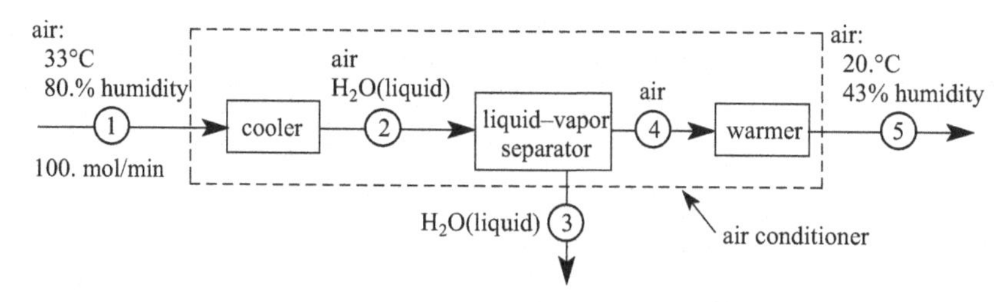

In this exercise, "air" is oxygen, nitrogen, and H_2O vapor. Streams 2 and 4 may contain H_2O vapor. The temperature in the liquid–vapor separator is the same as the temperature in the cooler. The total pressure is 1 atm throughout the process.

(A) What is the temperature in the cooler?

(B) What is the flow rate of H_2O liquid in stream 3 (in mol/min)?

Pressure–temperature phase diagram for H_2O.

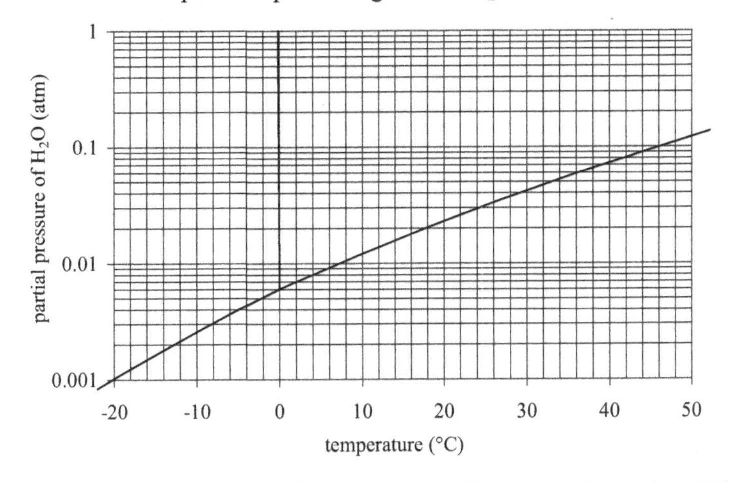

(C) In hotter, drier climates *a different air conditioner* produces air at 20°C (68°F). Again the temperature in the liquid–vapor separator is the same as the temperature in the cooler, the total pressure is 1 atm throughout the process, and streams 2 and 4 may contain H_2O vapor. What is the temperature in the cooler?

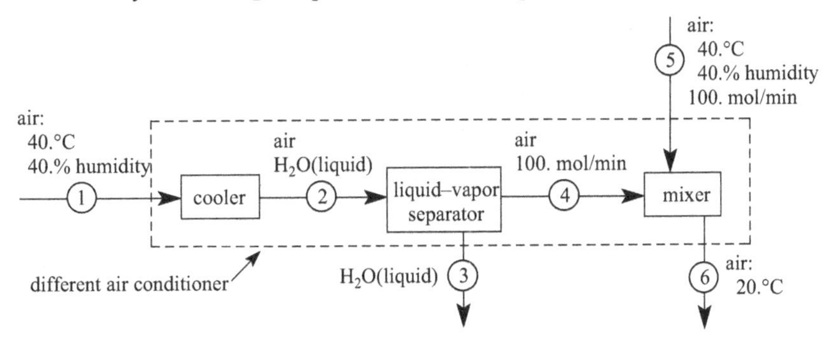

(D) What is the relative humidity in stream 6 (in %)? You may assume the heat capacity of air is independent of temperature and independent of H_2O vapor content.

4.75 When cold outside air is drawn into a home and heated, the warm air is too dry. The process below treats air at −8°C and 65% humidity to produce air at 24°C and 30% humidity.

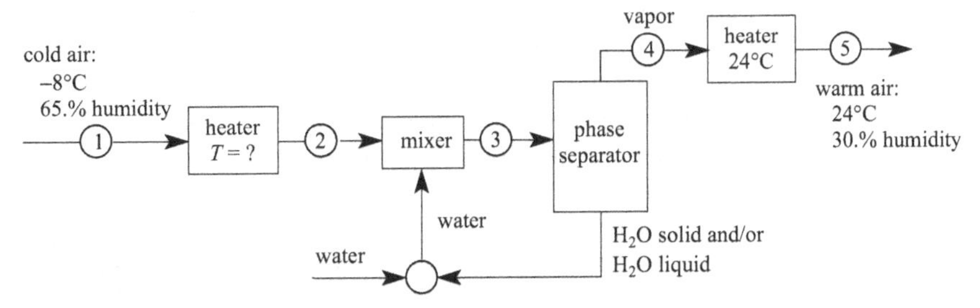

The temperature in the mixer and the separator is the same as the temperature in the first heater.

(A) Locate and label streams 1 to 5 on the phase map below.

Pressure–temperature phase diagram for H_2O.

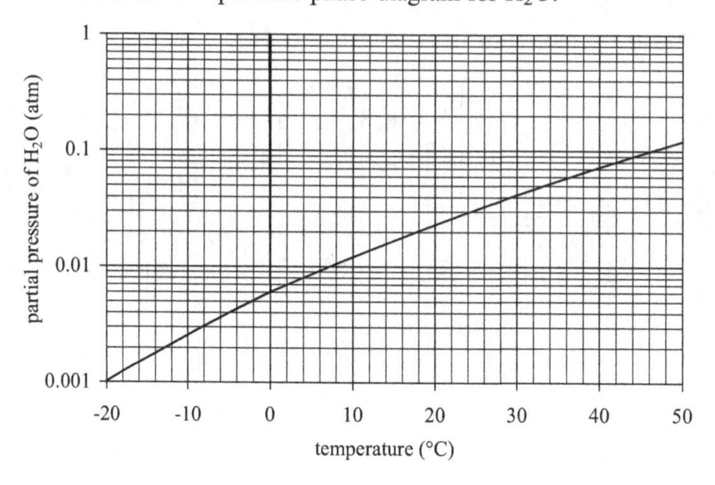

(B) What is the temperature in the first heater?

(C) If no water is added to the cold air (−8°C and 65% humidity), what is the humidity at 24°C?

4.76 Air with 100% humidity has the maximum partial pressure of H_2O at a given temperature. Air with 100% humidity is saturated with H_2O vapor.

(A) Indicate air at 29°C (84°F) with 25% humidity with an "*A* " on the pressure–temperature phase diagram for H_2O in Exercise 4.75, or download from duncan.cbe.cornell.edu/Graphs.

(B) Clouds are two-phase mixtures of H_2O vapor and H_2O liquid suspended in air. Indicate a cloud at 20°C (68°F) with a "*B* " on the pressure–temperature phase diagram.

(C) A cloud at ground level is fog. Early-morning fog usually disappears as the Sun rises and warms the air. Assume the fog is 0.02 mol fraction H_2O (liquid plus vapor) in air at 10°C (50°F) and 1 atm. Show the transition from fog to no fog on the diagram. At what temperature does the fog disappear?

(D) On a certain day in November the air over Ithaca, New York, is clear, 4°C (39°F), with 50% humidity. Later that night, this same air cools to −4°C (25°F). Do you predict snow overnight? Unlike TV weather reporters, *you* will get no credit for guessing "yes" or "no." You must justify your prediction in words or by drawing on the pressure–temperature phase diagram.

(E) A cold, dry mass of polar air moving south mixes with a warm, wet mass of gulf air moving north. The polar air has a temperature of −10°C and 0.% humidity. The gulf air has a temperature of 22°C and 95% humidity. Assume equal portions of polar air and gulf air are mixed. Calculate the temperature of the mixture and the mol fraction of H_2O in the mixture.

(F) How much rain will fall when the polar air and gulf air in (E) mix? Calculate the rainfall as the depth of water that will accumulate on the ground (in inches). Assume there are 14.7 lbs of air above each square inch of ground.

4.77 We need to compress air to 500. atm. However, we must not condense any liquid H_2O in the compressor; liquid droplets can ablate the turbine blades. What temperature is needed in the liquid–gas separator in the process below?

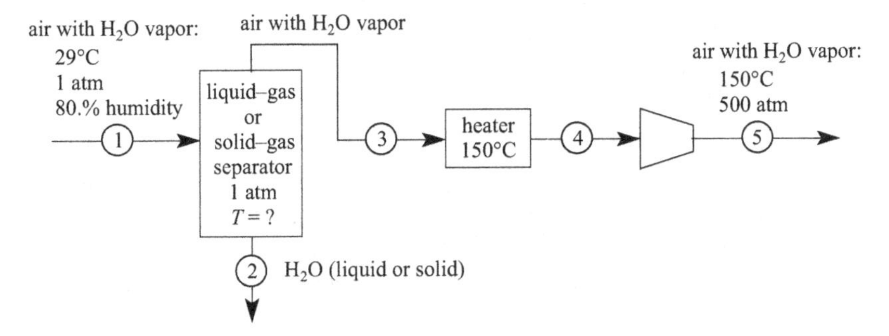

Pressure–temperature phase diagram for H_2O.

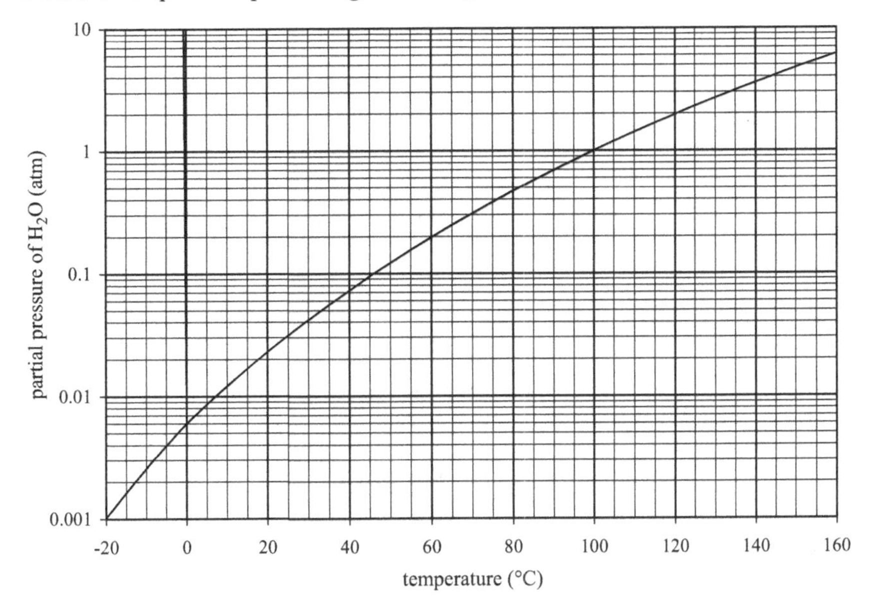

4.78 The following process removes water vapor from air by condensation.

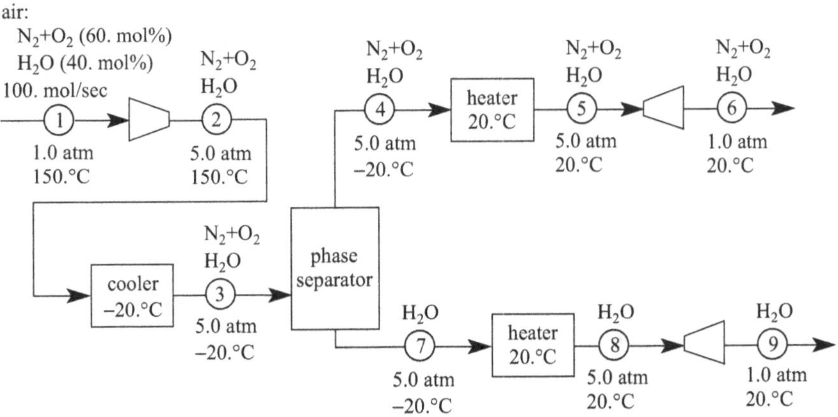

(A) Show the path of this process on the pressure–temperature phase diagram for H_2O in Exercise 4.77, or download from duncan.cbe.cornell.edu/Graphs. Represent each stream with a point on the map.
(B) What phase or phases are present in stream 2?
(C) What phase or phases are present in stream 3?
(D) What is the mol% of H_2O in stream 6?
(E) If the compressor fails, the pressure in stream 2 and all subsequent streams is 1 atm. What temperature in the cooler (and phase separator) will maintain the same composition in stream 6 when the compressor fails?

4.79 The process below transforms hot, wet air into comfortable air at 18°C (64°F). The process reduces the humidity without cooling the air below 18°C. Instead of using refrigeration, the process uses high pressure to condense H_2O vapor. Streams 2, 3, and 4 are at the same pressure.

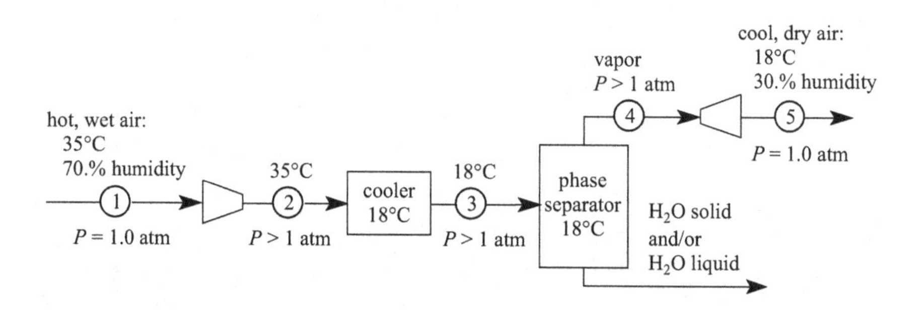

(A) Locate and label streams 1 to 5 on the phase map below.
(B) What is the pressure in stream 2?

Pressure–temperature phase diagram for H_2O.

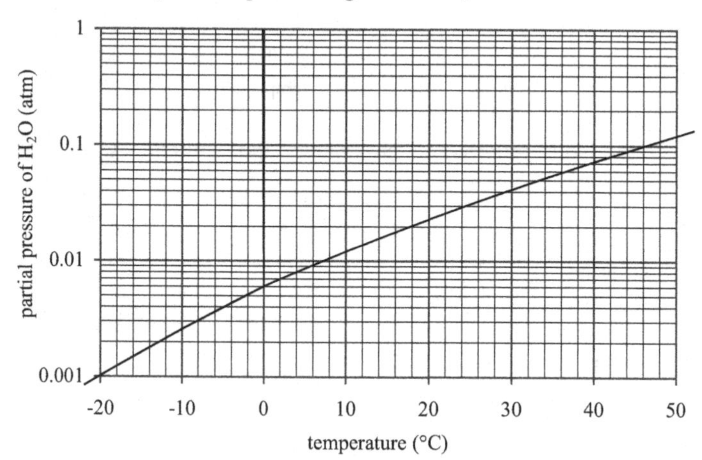

4.80(A) The process below produces liquid CO_2 from a mixture of 60. mol% He and 40. mol% CO_2. Calculate the flow rate of stream 4. You may assume that no He liquid condenses at the conditions in the liquid–gas separator.

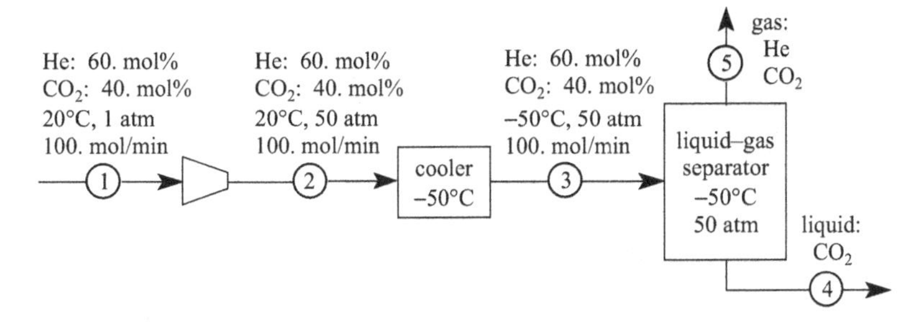

Pressure–temperature phase diagram for CO_2.

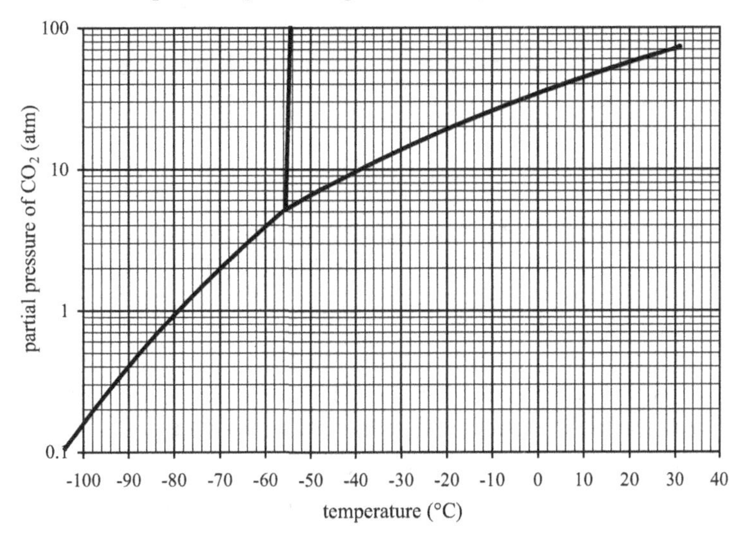

(B) Stream 4 (pure CO_2) is converted to solid pure CO_2 at 1 atm. What is the temperature of the solid CO_2?

(C) Given that the temperature in the liquid–gas separator is $-50°C$, what pressure is necessary to recover 90. mol% of the CO_2 in stream 1?

4.81 The process below uses helium gas to partially dry wet solids at low pressure.

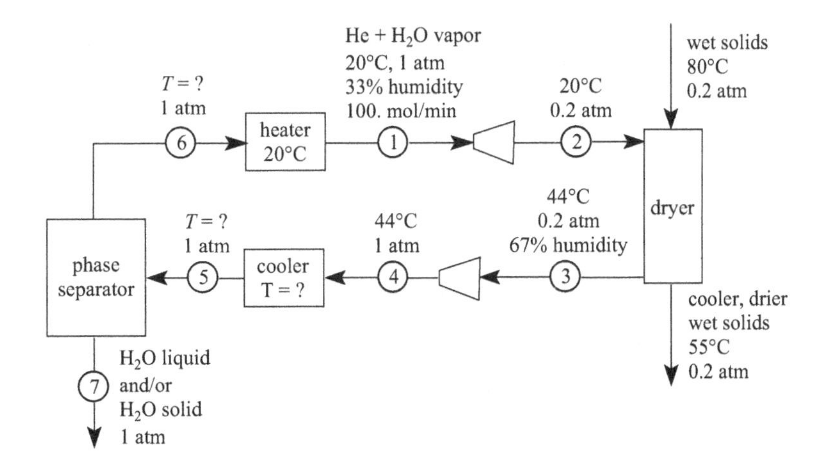

You may assume that:

- no He liquid condenses at any of the conditions in this process
- no He leaves with the "cooler, drier wet solids" stream
- streams 5, 6, and 7 are the same temperature.

(A) Show the locations of streams 1 through 7 on the phase diagram for H_2O below. *Be precise.*

Pressure–temperature phase diagram for H_2O.

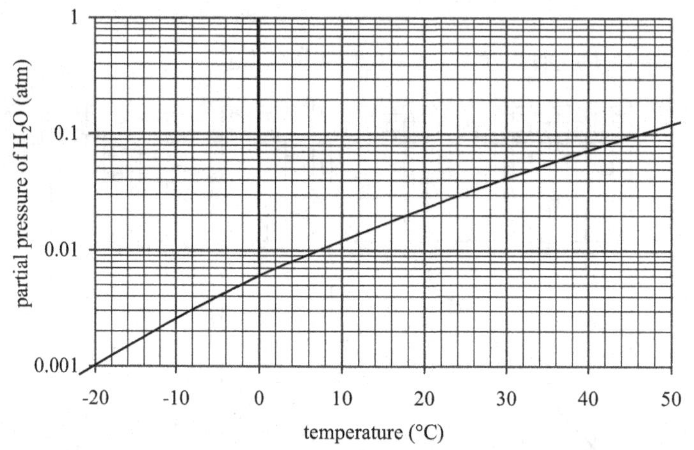

(B) Calculate the flow rate of condensed H_2O (liquid and/or solid) in stream 7. You may use mass balances and mathematical modeling.

4.82 Many homes use propane for heating and cooking. A storage tank on the homeowner's property supplies the propane. The tank contains propane liquid and propane vapor. "Propane" fuel need contain only 90 mol% propane. The remainder is chiefly propene, *n*-butane, *i*-butane, and methane with a trace of ethyl mercaptan (or thiophane) for odor. For this exercise, we will assume that propane fuel is pure propane.

A pressure–temperature phase diagram for propane is shown below. Liquid propane freezes at $-187°C$ at 1 atm. Propane's critical temperature and pressure are 96.7°C and 41.9 atm.

Pressure–temperature phase diagram for propane.

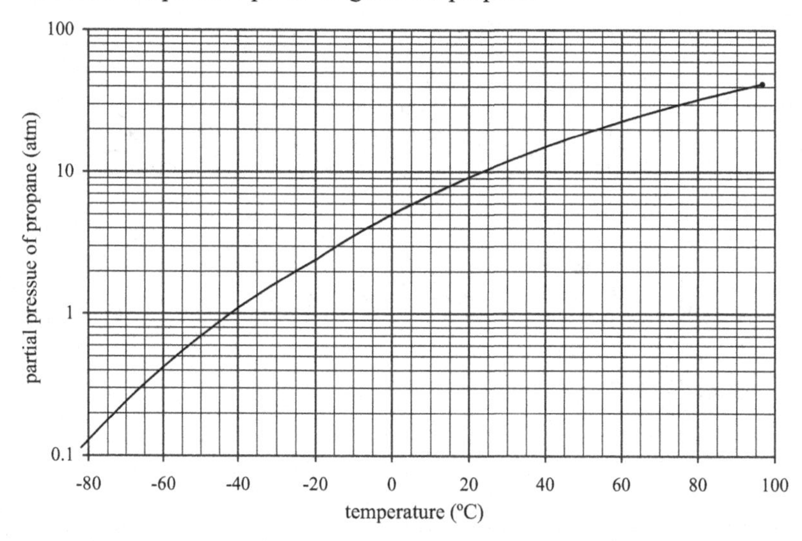

(A) No pump is needed to deliver the propane from the tank to the home if the pressure in the tank is greater than 1 atm. What is the tank pressure on a very hot day (38°C = 100°F)? What is the tank pressure on a very cold night (–29°C = –20°F)?

(B) At what temperature is the tank pressure 1 atm?

(C) A homeowner happens to read the tank pressure just before the propane supplier's truck arrives to recharge the tank. The truck driver connects the truck's propane tank to the homeowner's tank and appears to recharge the homeowner's tank. While recharging, the tank is vented; gas is released. The driver finishes and presents the homeowner with a bill for 460 kg (= 1010 lb) of propane. Later that day, the homeowner reads the tank pressure and finds that the tank pressure is the same as before the tank was recharged.

The homeowner calls you, the company engineer, to complain that he has been cheated. He has observed that the pressure in his oxygen tank increases as the oxygen tank is recharged. (Like many ex-smokers, he now needs an oxygen supplement to breathe.) Because the propane pressure is unchanged, the homeowner concludes no propane was delivered. Has the homeowner been cheated? Write an explanation for truck drivers to give homeowners, and thus prevent further interruptions of their more pressing duties.

(D) Consider a 1000 L tank with 999 L of propane vapor and 1 L of propane liquid. The tank is recharged without venting – nothing leaves the tank as propane liquid is pumped into the tank. The recharged tank contains 40 L of vapor and 960 L of liquid at 20°C. What is the pressure in the recharged tank?

(E) A new propane tank is charged with propane *without venting*. The new tank initially contains 1000 L of air at 20°C and 1 atm. After filling, the tank contains 900 L of propane liquid at 20°C. What is the pressure in the tank after equilibrium is reached? You may assume that liquid propane does not absorb air. **Hint:** Consider the total mols in the vapor phase: air plus propane vapor (if any).

(F) As in (E), a new tank is charged with propane, but the tank is vented during charging. Assume that, immediately after the tank is charged, the tank contains 100 L of air and 900 L of propane liquid and the tank pressure is 1 atm at 20°C, and the vent is closed. Assume there is no propane vapor immediately after the tank is charged. What is the pressure in the tank later?

(G) Mixtures of propane vapor and air will explode when ignited if the mixture is between 2.5 and 9.5 mol% propane. Which mixture (if any) is explosive – the vapor in part (E) or the vapor in part (F)?

4.83 Locate on *both* phase diagrams below the following systems of pure H_2O; *no air*. Note that some systems may be represented by a point, some may be represented by a line, and some may be represented by a region. Where appropriate, draw a tie line to represent a mixture and show the endpoints and fulcrum position of the tie line.

Indicate on *both* maps :

(A) all mixtures that contain both H_2O vapor and H_2O liquid

(B) H_2O liquid at its bubble point at 20 atm

(C) H_2O vapor at its dew point at 4 m^3/kg

(D) a mixture of 50 kg H_2O liquid and 50 kg H_2O vapor at 1 atm

(E) a mixture of 90 kg H_2O solid and 10 kg H_2O vapor at –13°C.

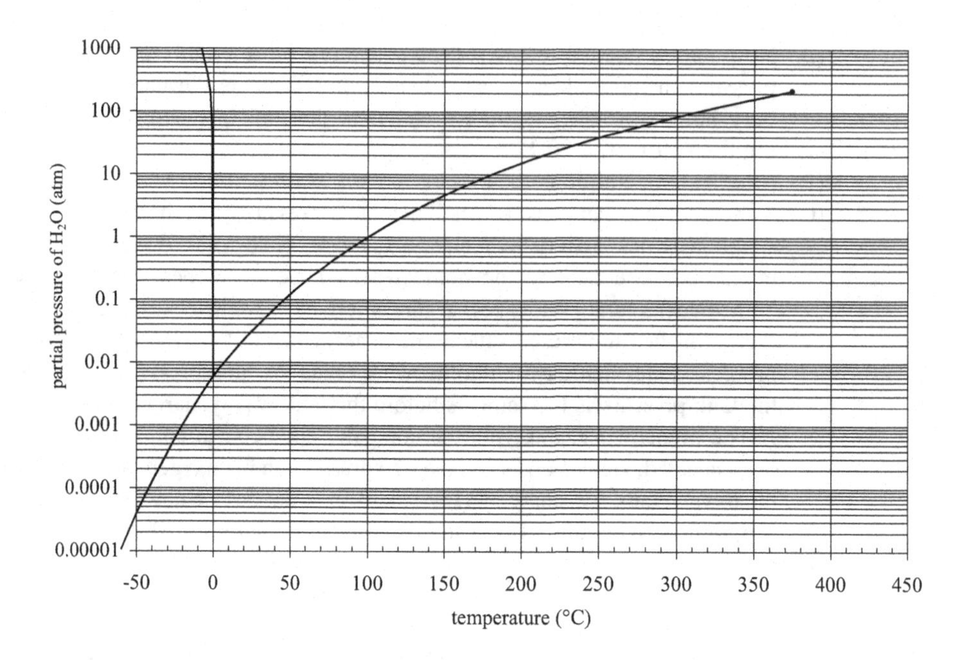

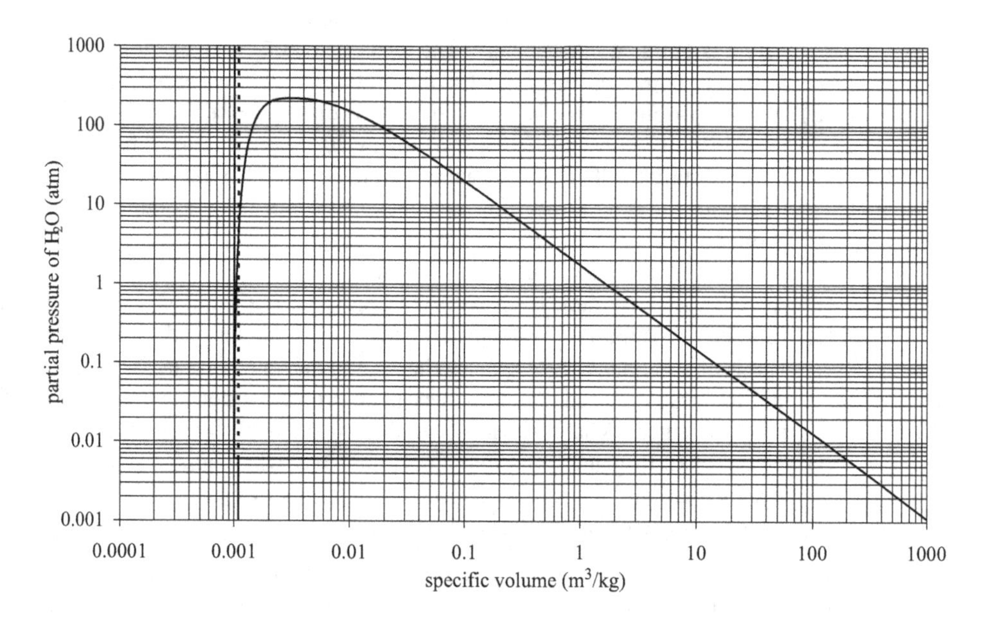

4.84 Shown below are the P–T and P–$\bar{V}$ phase diagrams for a pure compound Z. Assume that when Z is a gas, it is an ideal gas.

(A) Draw on the P–$\bar{V}$ diagram the path that corresponds to path 1 on the P–T diagram. What is the **liquid** specific volume (m^3/kg) at point A? What is the **liquid** specific volume (m^3/kg) at point B?

(B) Draw on the P–$\bar{V}$ diagram the path that corresponds to path 2 on the P–T diagram. What is the **vapor** specific volume (m^3/kg) at point C? What is the **vapor** specific volume (m^3/kg) at point D?

Pressure–temperature phase diagram for compound Z.

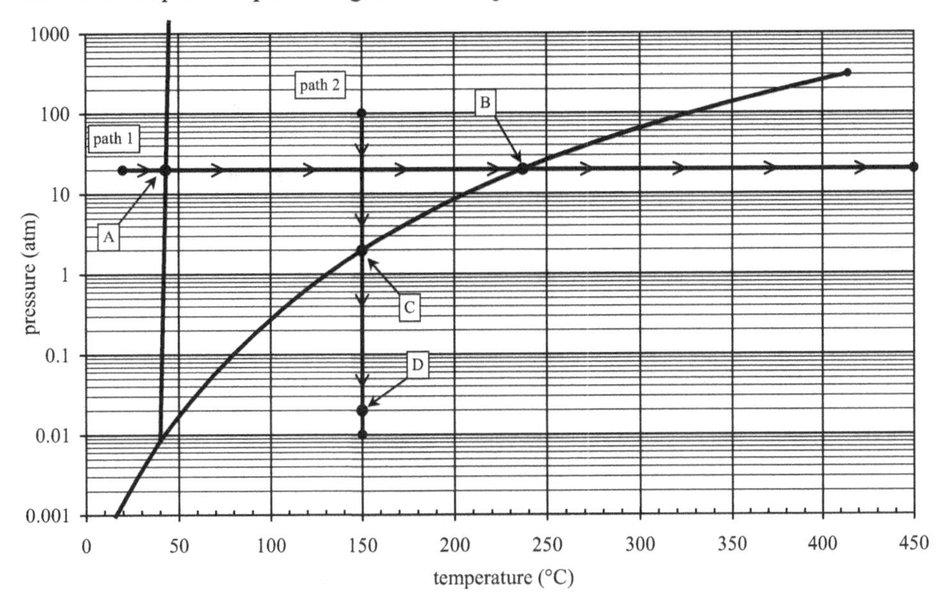

Pressure–volume phase diagram for compound Z.

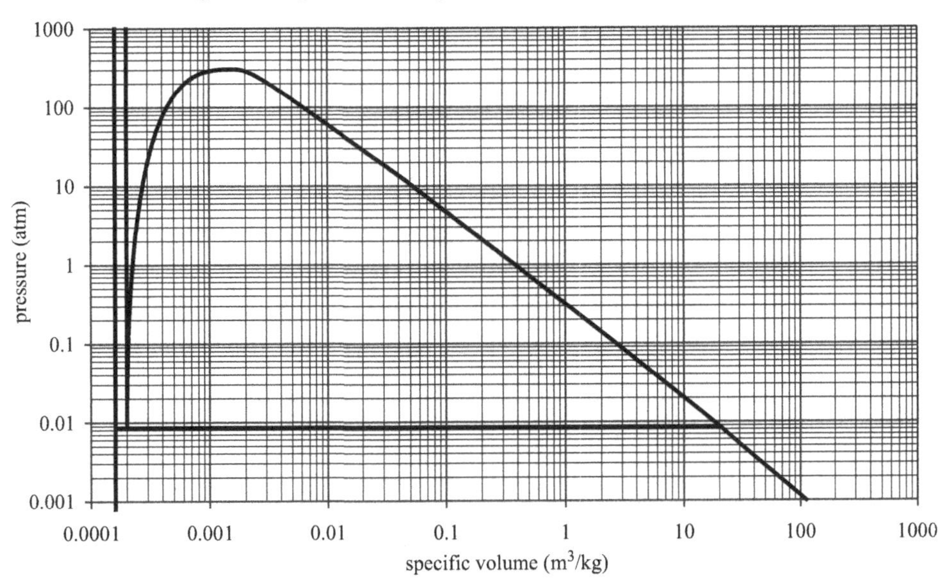

4.85 1,2-dichloroethane (1,2-DCE) is a common solvent and degreasing agent. We wish to reduce the partial pressure of 1,2-DCE in air to 0.01 atm from 0.1 atm. The phase map below shows the path for cooling the contaminated air at constant total pressure, as discussed for benzene in Section 4.3.2.

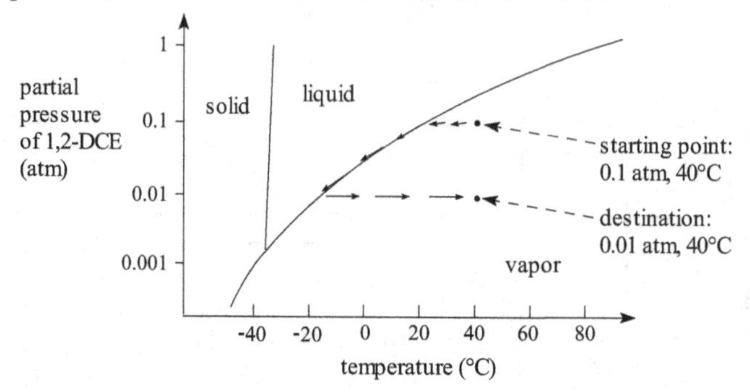

Use the phase map below to plot the same process as plotted above – cool the contaminated air at constant total pressure to reduce the 1,2-DCE in the vapor phase by 90 mol%.

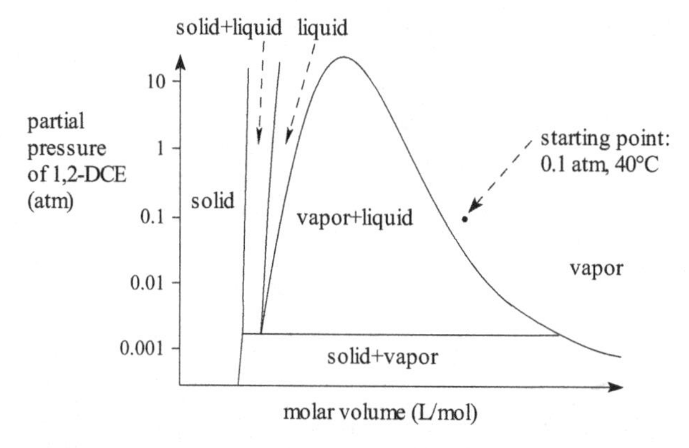

The x axis, molar volume, is a logarithmic scale. For reference, at 1 atm the vapor + liquid region spans from 0.09 L/mol (liquid molar volume) to 22.4 L/mol (vapor molar volume). A point within the two-phase region corresponds to the average molar volume, calculated as

$$\text{average molar volume} = (\text{liquid mol fract})(\text{liquid molar vol}) + (\text{vapor mol fract})(\text{vapor molar vol}).$$

A point midway between the two borders is a two-phase mixture with about 5% vapor and 95% liquid.

4.86 In Section 4.3.2 we analyzed two processes to reduce the partial pressure of benzene in air from 0.02 atm to 0.002 atm. One process reduced the temperature at constant pressure and another increased the pressure at constant temperature.

Consider here a process to reduce the partial pressure of 1,2-dichloroethane (1,2-DCE) by reducing the temperature and total pressure at constant total volume. Plot a process to reduce the partial pressure of 1,2-DCE in air from 0.1 atm to 0.01 atm on the two-phase maps below. See the previous exercise for a note on average molar volumes.

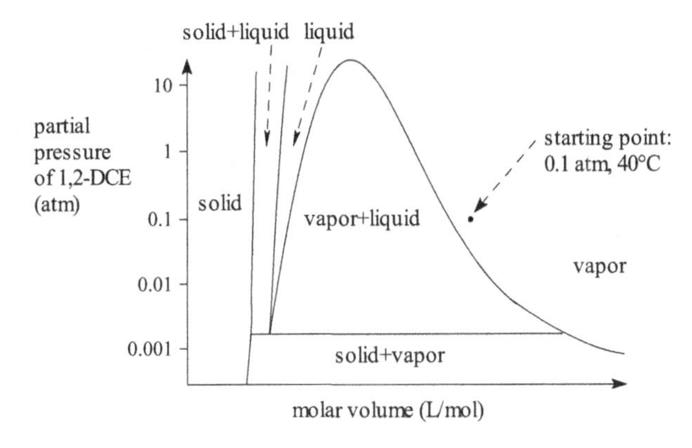

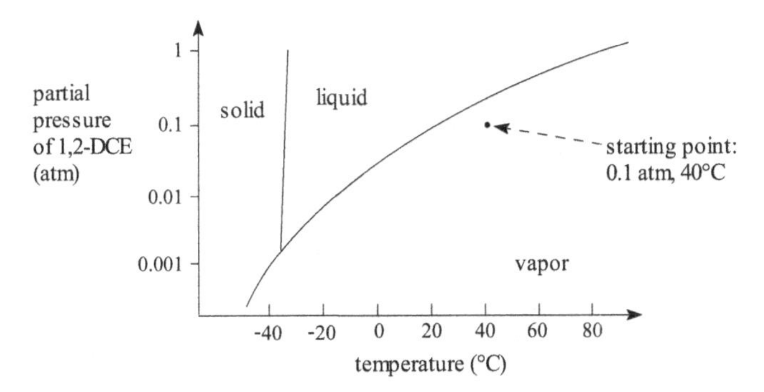

4.87 (A) The process described by steps 1 to 6 below condenses 99% of the H_2O from a mixture of H_2O vapor and air. Draw the process path on the *P–T* diagram for H_2O given in Exercise 4.83 (or download a figure from duncan.cbe.cornell.edu/Graphs). Assume air is a non-condensible gas and H_2O vapor is an ideal gas.

1. The initial mixture contains 90 mol air and 10 mol H_2O vapor at 200°C and a total pressure of 10 atm. Label this point 1 on the *P–T* diagram for H_2O (see Exercise 4.83).
2. The mixture is cooled at constant total pressure to its dew point. Label this point 2 and draw the path from point 1 to point 2.
3. The temperature is held constant after reaching point 2 while the total pressure is increased until 99% of the H_2O vapor condenses. Label this point 3 and draw the path from point 2 to point 3.

4. While maintaining the temperature and the total pressure of point 3, the H_2O liquid is drained from the system. Label this point 4 and draw the path from point 3 to point 4.
5. While maintaining the temperature constant, the total pressure is decreased to 10 atm. Label this point 5 and draw the path from point 4 to point 5.
6. While maintaining the total pressure at 10 atm, the system is heated to 200°C. Label this point 6 and draw the path from point 5 to point 6.

(B) Draw the path for the *same* process (steps 1 to 6 in part (A)) on the $P–\bar{V}$ diagram for H_2O given in Exercise 4.83. Again assume air is a non-condensible gas and H_2O vapor is an ideal gas.

(C) The $T–\bar{V}$ phase diagram for H_2O is shown below. This diagram complements the $P–T$ phase diagram and the $P–\bar{V}$ phase diagram. The $T–\bar{V}$ phase diagram is obtained from the 3D phase diagram in Figure 4.109 by viewing it from above. Draw the path for the *same* process (steps 1 to 6 in part (A)) on the $T–\bar{V}$ diagram below. Again assume air is a non-condensible gas and H_2O vapor is an ideal gas.

Temperature–volume phase diagram for H_2O.

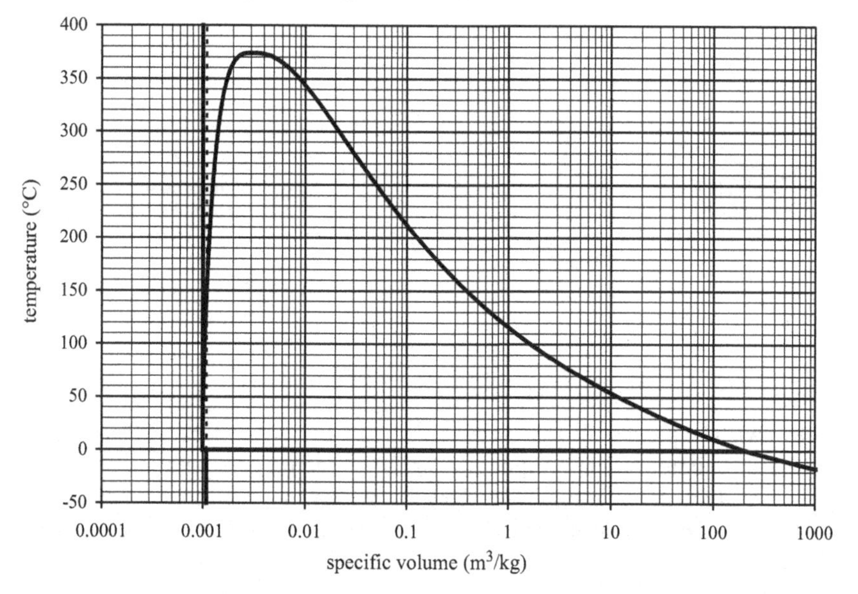

1. The initial mixture contains 90 mol air and 10 mol H_2O vapor at 200°C and a total pressure of 10 atm. Label this point 1 on the $T–\bar{V}$ diagram.
2. The mixture is cooled at constant total pressure to its dew point. Label this point 2 and draw the path from point 1 to point 2.
3. The temperature is held constant after reaching point 2 while the total pressure is increased until 99% of the H_2O vapor condenses. Label this point 3 and draw the path from point 2 to point 3.
4. While maintaining the temperature and the total pressure of point 3, the H_2O liquid is drained from the system. Label this point 4 and draw the path from point 3 to point 4.
5. While maintaining the temperature constant, the total pressure is decreased to 10 atm. Label this point 5 and draw the path from point 4 to point 5.

6. While maintaining the total pressure at 10 atm, the system is heated to 200°C. Label this point 6 and draw the path from point 5 to point 6.

4.88 In Section 4.2 we used a linear energy scale for energy balances on a pure substance, such as H_2O. A second dimension may be added, for example, for energy balances on systems in which the pressure changes. Shown below is a pressure–enthalpy diagram for ethylene.

Pressure–enthalpy phase diagram for ethylene.

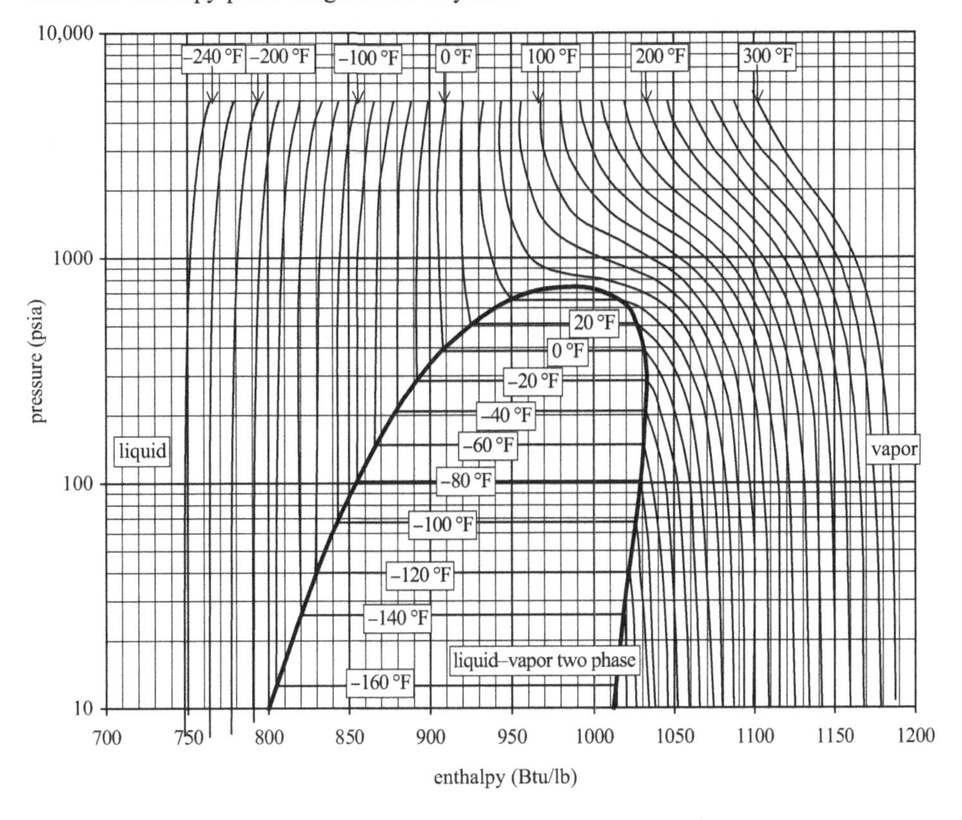

(A) What is the heat of vaporization of ethylene at 300 psia (in Btu/lb)?
(B) What is the heat of vaporization of ethylene at −120°F (in Btu/lb)?
(C) What is the boiling point (in °F) at 500 psia?
(D) What is the critical pressure (in psia) of ethylene?
(E) What is the critical temperature (in °F) of ethylene?
(F) What is the enthalpy change (in Btu) for 17 lb of ethylene, initially at 60 psia and −120°F, compressed to 1000 psia and heated to 200°F?
(G) What is the approximate heat capacity of ethylene at 200 psia and −120°F (in Btu/(lb·°F))? **Hint:** To estimate the energy needed to increase the temperature of 1 lb by 1°F, calculate the energy needed to increase the temperature of 1 lb by 20°F and divide by 20.
(H) Calculate the temperature and composition (liquid and vapor wt fractions) of stream 3 below.

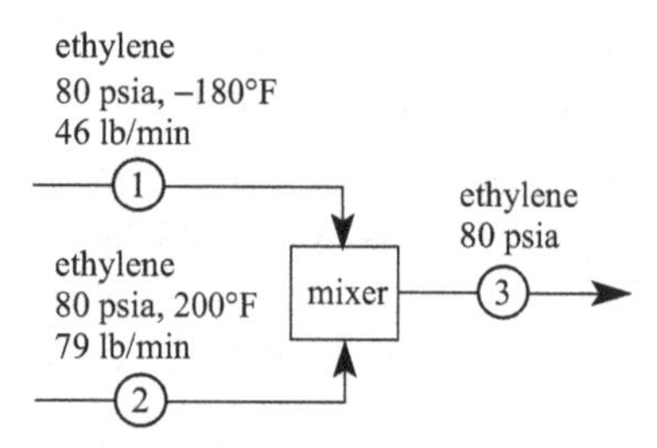

(I) Sketch the liquid–vapor border of the ethylene phase diagram on the graph below. Include as much of the border as you can determine from the pressure–enthalpy diagram. Label the critical point.

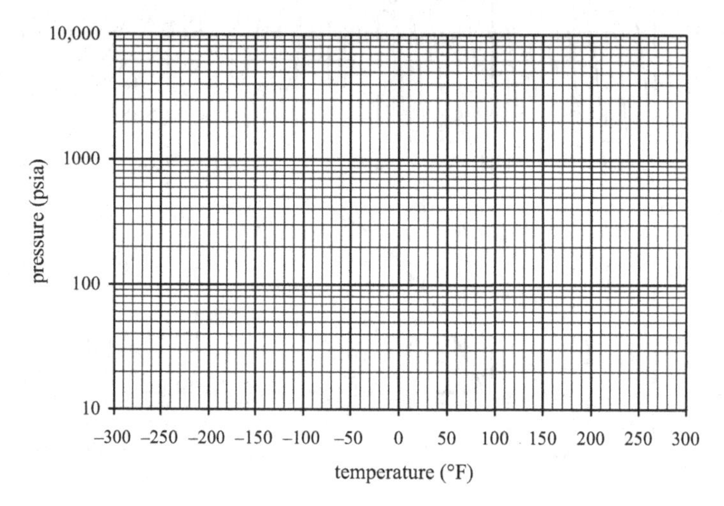

Process Design with Graphical Modeling

General Guidelines for Chapter 4 Design Exercises

- Design a steady-state continuous process.
- Unless instructed otherwise, assume solid–liquid separations produce wet solids.
- Solid + solid mixtures cannot be separated.
- Unless instructed otherwise, you may neglect heat exchangers, heaters, coolers, and pumps.
- Many designs are workable. Explore different options and choose your best design.

General Flowsheet Practice

- Use a discrete unit for each operation. For example, a reactor should not do double duty as a mixer and/or a separator. A reactor should have exactly one input and one output.
- Label each unit. For separators, indicate the physical basis of the separation, such as "liquid–gas separator." Where appropriate, indicate the temperature in the unit. You need not label combiners and splitters. You need not label mixers and heat exchangers if you use descriptive symbols (a stirring paddle or a zigzag stream, respectively).
- For most exercises, you need not write compositions or temperatures on your flowsheet streams. Instead, indicate the compositions and temperatures of process streams by labeling points with the stream number on a phase diagram.

- Where appropriate, list the physical state of the substances – gas, solid, liquid, or aq (in aqueous solution).
- Indicate the destination of any stream that leaves the process (for example, "vent to air," "to landfill," or "to market").
- Use conventions to improve the readability of your flowsheet. For a liquid–gas separator the gas should leave the top and the liquid should leave the bottom. If convenient, the process should start in the upper left-hand corner. It is helpful if streams entering the process start at the periphery and streams leaving the process terminate at the periphery.

Design Based on Single-Stage Separations

4.89 *Design based on Section 4.2.1 – Flash drums.* Using only flash drums, design a process to yield two products: (1) a mixture with $\geq$94. mol% M and $\leq$6. mol% K; and (2) a mixture with $\leq$6. mol% M and $\geq$94. mol% K. Your process should start with two mixtures of M and K: (1) 100 mol/min of 80. mol% M and 20. mol% K at 20°C; and (2) 100 mol/min of 20. mol% M and 80. mol% K at 20°C. See Exercise 4.16 for the temperature–composition phase diagram for $M + K$ mixtures at 1 atm.

Design Goals (in decreasing importance)

- Maximize product output.
- Minimize the number of flash drums.
- Minimize the size of each flash drum.

Design Rule

- Use only flash drums; use no distillation columns.

4.90 *Design based on Section 4.2.1 – Flash drums and liquid–liquid single-stage separators.* Design a process to separate a mixture of 50 mol% butanol and 50 mol% water at 60°C into two products: (1) a water-rich stream with at most 5 mol% butanol, and (2) a butanol-rich stream with at least 95 mol% butanol. See Exercise 4.26 for the temperature–composition phase diagram for butanol + water mixtures at 1 atm. **Hint:** Consider liquid–liquid separation as well as liquid–vapor separation.

Design Goals (in decreasing importance)

- Minimize waste.
- Minimize the number of units.
- Minimize the size of each of unit.

4.91 *Design based on Section 4.2.1 – Liquid–solid single-stage separators.* Design a process to separate a solid composed of 50 mol% C and 50 mol% D into two products: a stream with $\geq$98 mol% D, and a stream with $\leq$2 mol% D. Your process should minimize waste and the number of units. You may assume that liquid + solid mixtures can be separated perfectly; the liquid product contains no solid and the solid product contains no liquid. However, solid + solid mixtures cannot be separated.

The input stream is labeled by a circled "1" on the $C + D$ phase diagram below.

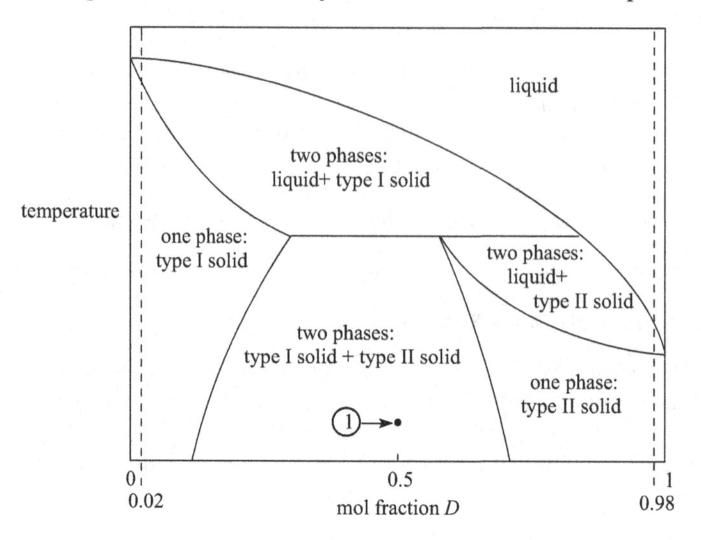

4.92 *Design based on Section 4.2.1 – Liquid–solid single-stage separators.* Design a process to produce pure Na from a liquid mixture of 65 wt% Na and 35 wt% K. Your process should maximize the amount of Na produced and minimize the number of units needed. You may assume that liquid + solid mixtures can be separated perfectly; the liquid product contains no solid and the solid product contains no liquid. However, solid + solid mixtures cannot be separated.

The input stream is labeled (1) on the Na + K phase diagram below.

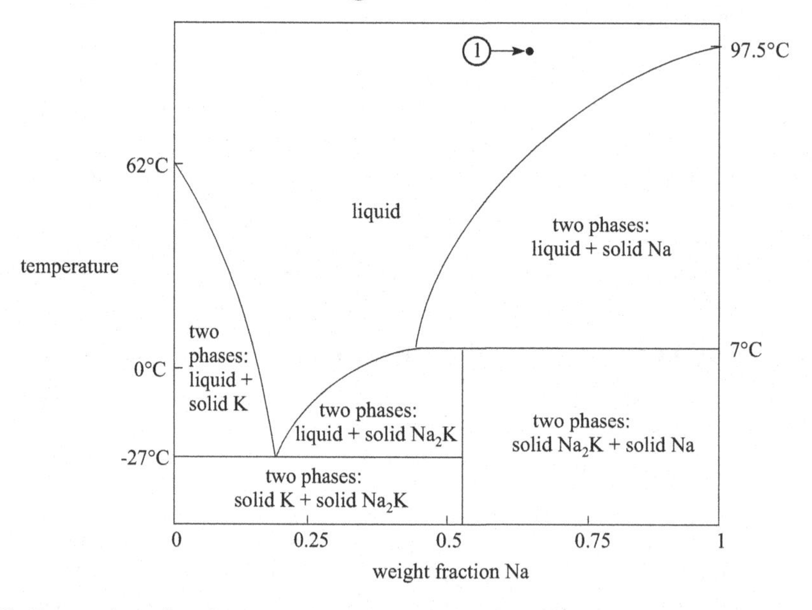

4.93 *Design based on Section 4.2.1 – Liquid–solid single-stage separators.* In Exercise 4.92 you designed a process to produce pure Na from a mixture of 65 wt% Na and 35 wt% K. The process discarded a waste stream with about 20 wt% Na.

A new engineer under your supervision proposes the process below to produce pure K from the waste stream. The process mixes pure K with the waste stream to shift the composition from the "two phases: liquid + solid Na_2K" region into the "two phases: liquid + solid K" region.

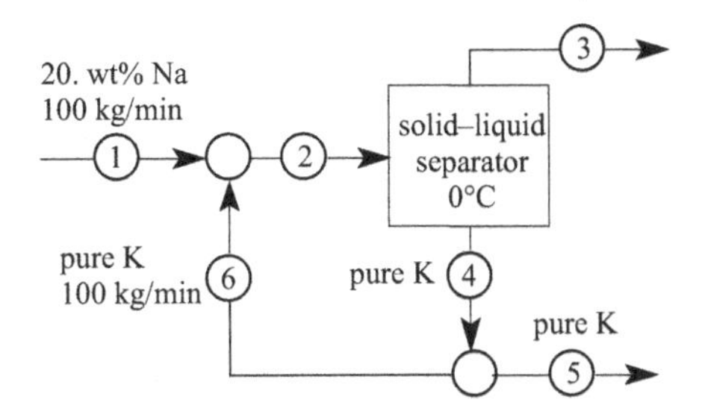

(A) What is the composition of stream 3?
(B) This process is not viable. Explain why. You may assume that the liquid–solid separator provides a perfect separation; the solid K is free of any liquid solution.
(C) Can you make the process viable by changing operating conditions, such as the temperature in the separator or the flow rate of stream 6? Do not add units and do not add feed streams. Explain your conclusion. Remember that one cannot separate solid + solid two-phase mixtures.

4.94 *Design based on Section 4.2.1 – Liquid–solid single-stage separators.*

(A) Design a process to produce solid Na_2SO_4 from a mixture of 60. wt% Na_2SO_4 and 40. wt% H_2O at 20°C. The phase diagram for mixtures of water and Na_2SO_4 has six regions, as shown below. Only the region labeled "liquid solution" is a single phase. All other regions are liquid + solid two-phase or solid + solid two-phase. A eutectic is a solid alloy with the lowest melting point. In this system, the eutectic composition is about 5 wt% Na_2SO_4.

Design Goals (in decreasing importance)

- Maximize the output of solid Na_2SO_4.
- Minimize the number of separators.

Design Guideline

- You may assume that solid + liquid mixtures can be separated perfectly. However, solid + solid mixtures cannot be separated.

(B) Assume that the flow rate of the input stream is 100. kg/min. Apply mathematical mass balances to calculate the flow rate of solid Na_2SO_4 produced by your process.
(C) Apply a graphical mass balance to calculate the flow rate of solid Na_2SO_4 by using a non-standard tie line. Draw a line segment connecting the Na_2SO_4 stream and the other stream leaving your process. Because the tie line connecting these two streams does not represent two streams in equilibrium, the tie line need not be horizontal and the tie line may cross borders between phase regions. Place the fulcrum at the composition of the feed stream (60 wt % Na_2SO_4). Then use the lever rule to calculate the flow rate of Na_2SO_4.

Temperature–composition phase diagram for Na_2SO_4 + water mixtures.

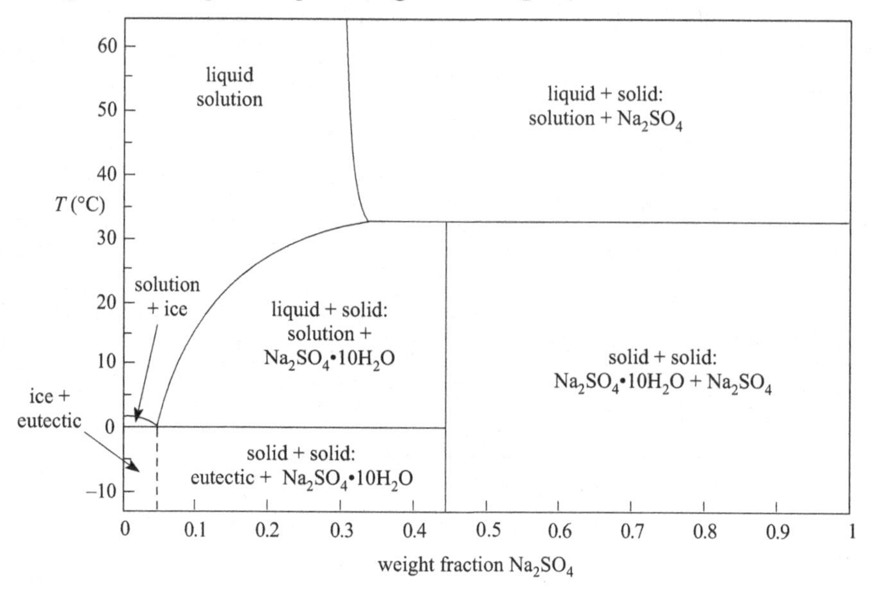

Diagram adapted from Hougen, O. A., Watson, K. W. and Ragatz, R. A. (1954). *Chemical Process Principles. Part I*, 2nd ed., Wiley, NY.

4.95 *Design based on Section 4.2.1 – Liquid–solid single-stage separators.*
Magnesium sulfate crystals exist in five compositions:

$MgSO_4 \cdot 12H_2O$	(0.36 wt fraction $MgSO_4$)
$MgSO_4 \cdot 7H_2O$	(0.48 wt fraction $MgSO_4$)
$MgSO_4 \cdot 6H_2O$	(0.53 wt fraction $MgSO_4$)
$MgSO_4 \cdot H_2O$	(0.87 wt fraction $MgSO_4$)
$MgSO_4$	(1.0 wt fraction $MgSO_4$)

The phases at equilibrium in a mixture of $MgSO_4$ and H_2O are shown on the phase diagram below. The region labeled "liquid solution" is a single phase. All other regions are two-phase; either liquid + solid or solid + solid.

Design a process to produce $MgSO_4 \cdot 6H_2O$ (0.53 wt fraction $MgSO_4$) from a liquid solution with 0.40 wt fraction $MgSO_4$ at 200°F. The input stream is indicated by a circled "1" on the phase diagram.

Design Goals (in decreasing importance)

- Maximize the output of solid $MgSO_4 \cdot 6H_2O$.
- Minimize the number of units.

Design Guideline

- You may assume that solid + liquid mixtures can be separated perfectly. However, solid + solid mixtures cannot be separated.

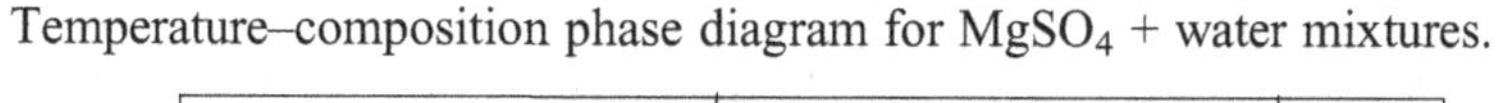

Temperature–composition phase diagram for $MgSO_4$ + water mixtures.

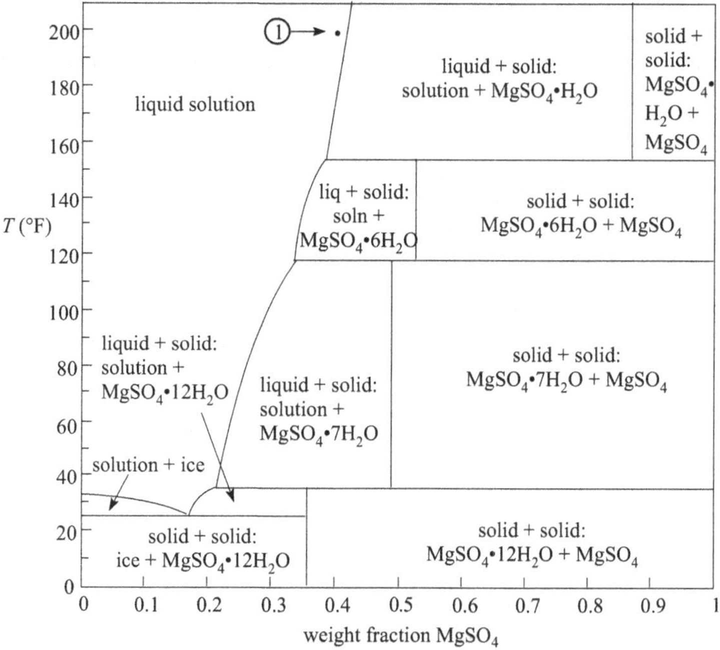

Exercise adapted from Felder, R. M., and Rousseau, R. W. (1986). *Elementary Principles of Chemical Processes*, 2nd ed, Wiley, NY, p. 259. Figure adapted from Perry, R. H. and Green, D. W. 1984. *Perry's Chemical Engineers' Handbook*, 6th ed., McGraw-Hill, NY, pp. 19–26. Copyright McGraw-Hill, Inc. Reproduced by permission.

4.96 *Design based on Section 4.2.1 – Liquid–liquid single-stage separators and ternary diagrams.* Design a process to treat 100 mol/min of a mixture of 50 mol% hexane and 50 mol% methylcyclopentane (MCP) to yield a product with *at least* 75 mol% hexane. The product's flow rate must be *at least* 35 mol/min.

Design Goals (in decreasing importance)

- Minimize the number of units.
- Minimize the size of each unit.

Design Rules

- Your process must use liquid–liquid separation, not distillation.
- You may use as much 100 mol% aniline as you wish.
- Specify the *approximate* flow rates of each stream. You may indicate compositions by labeling the phase diagram.
- You may ignore the fate of all by-product streams.
- The ternary phase diagram for MCP + hexane + aniline mixtures is given in Exercise 4.28 and may be downloaded from duncan.cbe.cornell.edu/Graphs.

Hint: Begin your process as follows:

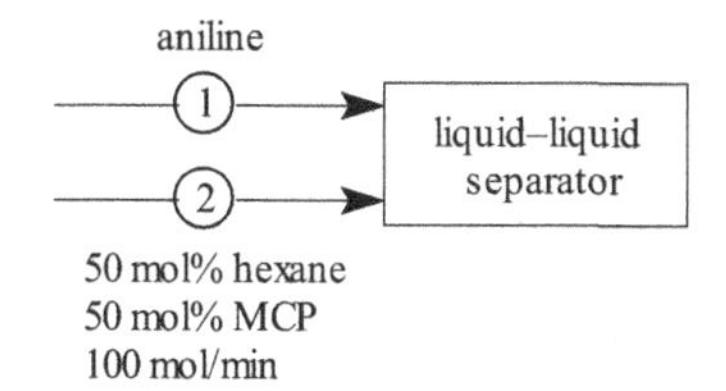

4.97 *Design based on Section 4.2.1 – Liquid–liquid single-stage separators and ternary diagrams.* The process below treats a mixture of methylcyclohexane (MCH) and aniline (A) by liquid–liquid extraction with *n*-heptane (H), and then distillation to remove the H and produce A with less than 2% MCH. The liquid–liquid separator has one equilibrium stage.

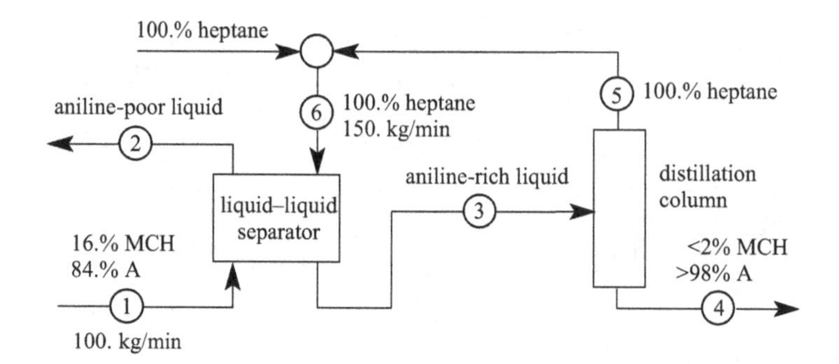

(A) Use mass balances and the ternary diagram for heptane + methylcyclohexane + aniline mixtures (see Exercise 4.31 or duncan.cbe.cornell.edu/Graphs) to determine the composition *and* flow rate of stream 4.

(B) Design a process to recover useful components from stream 2. That is, add units to the process above to minimize waste. You are not permitted to separate MCH and A by distillation. Stream 4 must have less than 2% MCH. **Hint:** Draw a border around your entire process to analyze your design.

4.98 *Design based on Section 4.2.1 – Liquid–liquid single-stage separators and ternary diagrams.* We wish to separate *A* (aromatic) and *P* (polar) from a mixture of 30. mol% CO_2, 40. mol% *A*, and 30. mol% *P*. These are the same *A* and *P* as in Exercise 4.29. Distillation is impractical because *A* and *P* have similar boiling points. Design a process to separate *A* and *P* using supercritical CO_2. Pure CO_2 is available at 20°C and 1 atm.

Design Goals (in decreasing importance)

- Produce two products: one with at least 98 mol% *A* and one with at least 98 mol% *P*.
- Maximize product flow rates.
- Minimize consumption of pure CO_2.
- Minimize pressure deviations from 1 atm.
- Minimize temperature deviations from 20°C.
- Minimize the number of units.
- Minimize the size of each unit.

Design Rules

- The mixture is initially at 20°C and 1 atm.
- Use only single-stage separators: liquid–liquid separators, liquid–vapor separators (flash drums), solid–liquid separators, or solid–vapor separators. Use no distillation columns and no multistage absorbers or strippers.

- Number each stream and indicate each stream on a phase diagram.
- Show all pumps, pressure reducers, heaters, and coolers. You may neglect heat exchangers.
- See Exercise 4.29 or duncan.cbe.cornell.edu/Graphs for a ternary diagram of CO_2 + A + P mixtures.

Temperature–composition phase diagram for CO_2 + A mixtures at 1 atm.

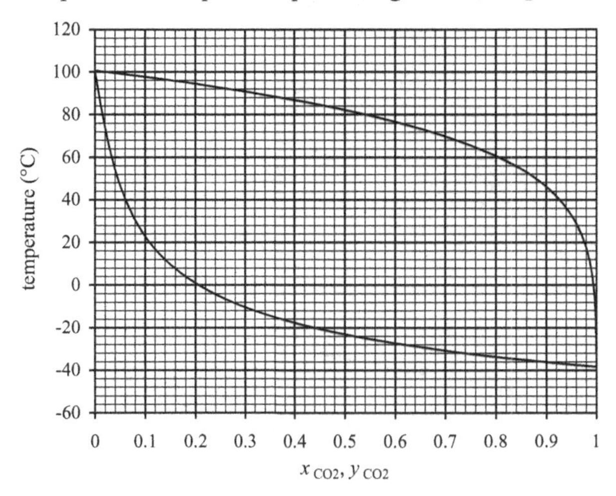

Temperature–composition phase diagram for CO_2 + P mixtures at 1 atm.

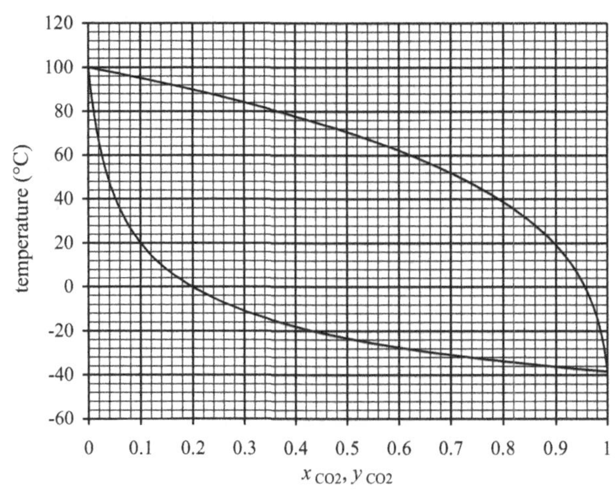

4.99 *Design based on Section 4.2.1 – Liquid–vapor single-stage separators and ternary diagrams*. Design a process to separate CO_2 and CH_4 by liquid–vapor separation at 40 atm and 220 K. Use N_2 to enhance the separation. The liquid–vapor phase diagram for CO_2 + CH_4 + N_2 mixtures at 40 atm and 220 K is shown below.

Ternary diagram for $CO_2 + CH_4 + N_2$ mixtures at 40 atm and 220 K.

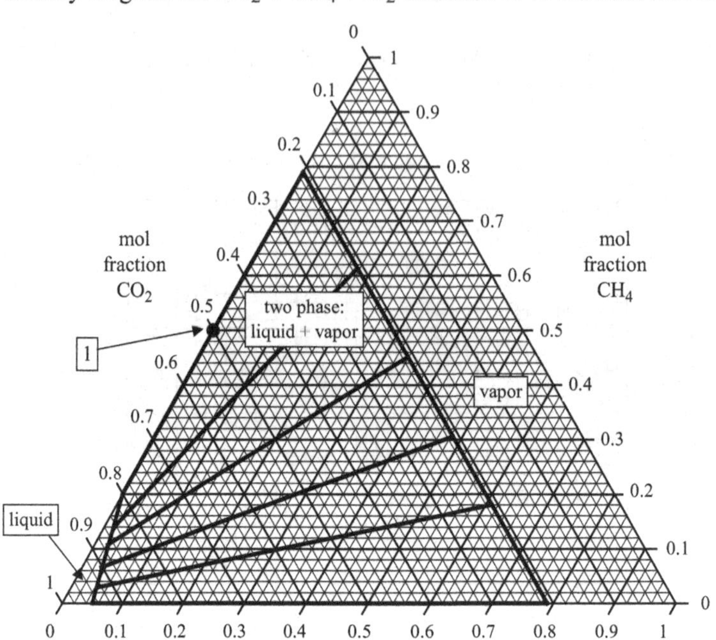

Design Goals (in decreasing importance)

- Produce a stream with at least 90. mol% CO_2.
- Maximize the flow rate of product and produce at least 35 mol/min.
- Minimize the number of units.
- Minimize the size of each unit.

Design Rules

- Start with 100. mol/min of 50. mol% CO_2 and 50. mol% CH_4.
- You may use as much N_2 as you wish.
- Use only single-stage liquid–vapor separators (flash drums) at 40 atm and 220 K.
- Number each stream and indicate each stream on the ternary phase diagram. Stream 1 (50. mol% CO_2 and 50. mol% CH_4) is already labeled on the ternary diagram.
- You may discard all streams other than the product stream.
- You need not specify flow rates.

4.100 *Design based on Section 4.2.1 – Solid–liquid single-stage separators and ternary diagrams*. Consider a solid–liquid separator, such as the one shown below.

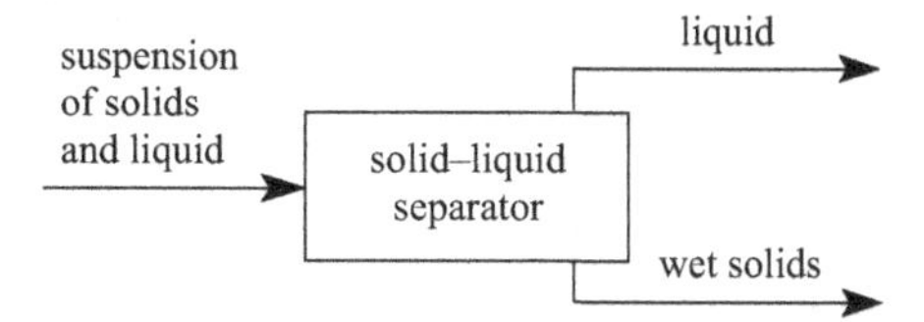

Like liquid–liquid separators and liquid–gas separators, a solid–liquid separator involves two phases at equilibrium.

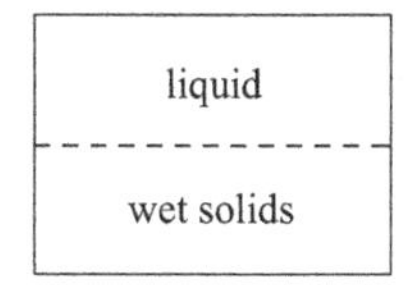

For liquids composed of two miscible liquids, such as oil and ethanol, one can represent the two phases on a ternary diagram, shown qualitatively below.

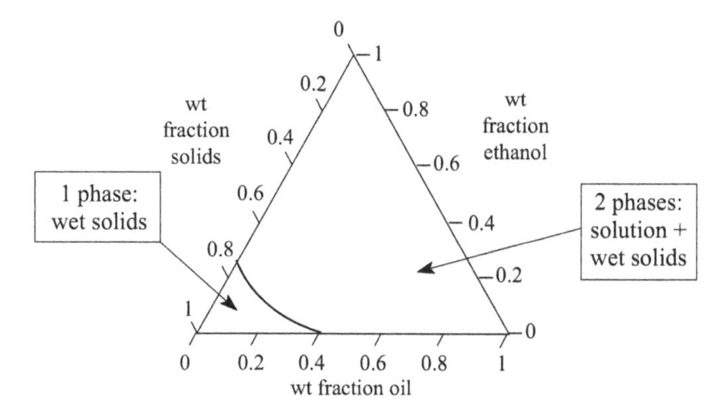

The dashed lines below show some representative tie lines. All tie lines pass through the pure solids point if extrapolated.

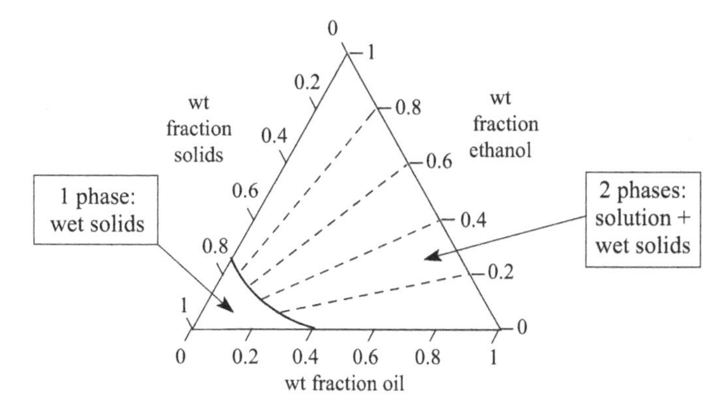

(A) Use the phase diagram for ternary mixtures of solids, oil, and ethanol to calculate the flow rates and compositions of the streams leaving the separator below.

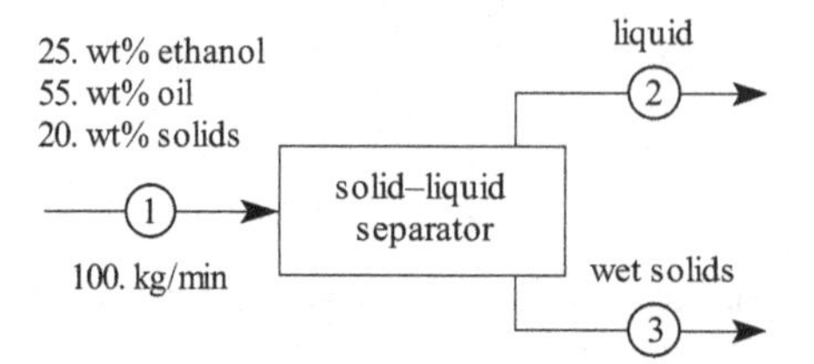

Ternary phase diagram for solids + oil + ethanol mixtures at 20°C and 1 atm.

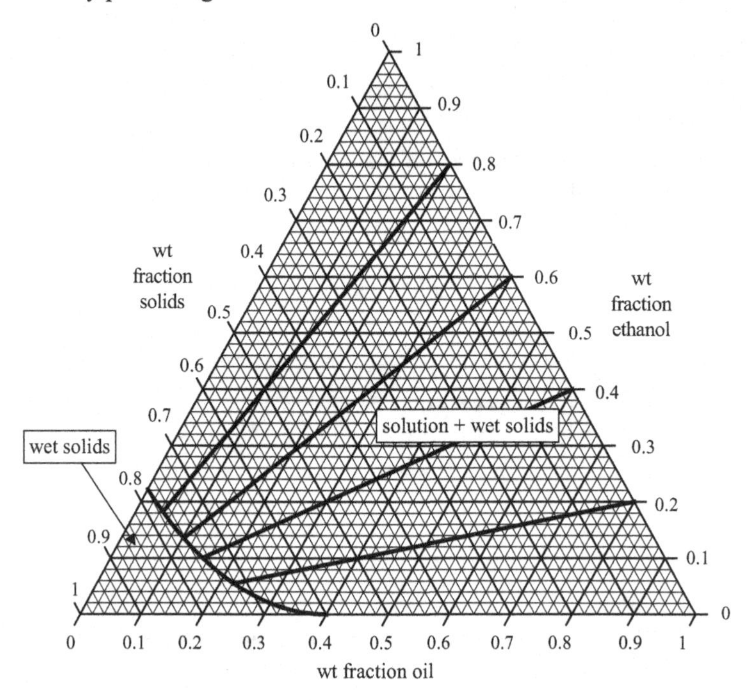

(B) Design a process to extract oil from a mixture of 60. wt% solids and 40. wt% oil. The oil cannot be evaporated from the solids because the oil is heat-sensitive. However, the oil is miscible with ethanol, and ethanol (boiling point = 78.5°C) is easily separated from the oil.

Design Goals (in decreasing importance)

- Produce a product with at least 98 wt% oil.
- Maximize product output; minimize waste.
- Minimize the number of units.
- Minimize the size of each unit.

Design Rules

- Start with 100. kg/min of wet solids: 60. wt% solids and 40. wt% oil.
- Do not heat the oil above 300°C.
- You may discard wet solids with $\leq$2 wt% oil.

- Use only single-stage separators: you may use liquid–liquid separators, liquid–vapor separators (flash drums), solid–liquid separators, or solid–vapor separators. You may not use distillation columns or multistage absorbers or strippers.
- Number each stream and indicate the stream on a phase diagram.
- You need not specify flow rates.

Temperature–composition phase diagram for ethanol + oil mixtures at 1 atm.

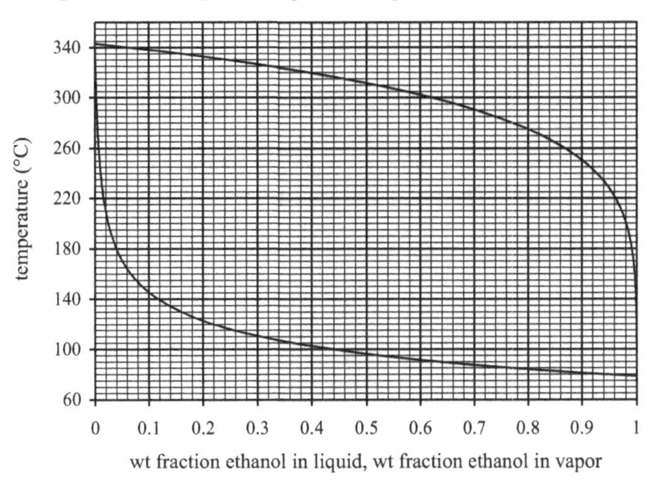

4.101 *Design based on Section 4.2.1 – Solid–liquid single-stage separators and ternary diagrams.* A chemical process produces a valuable liquid, P, with a solid by-product, S. Because P is heat-sensitive, we cannot separate P from the solid by evaporation. (The boiling point of P is 357°C.)

A P + S suspension may be partially separated in a solid–liquid separator.

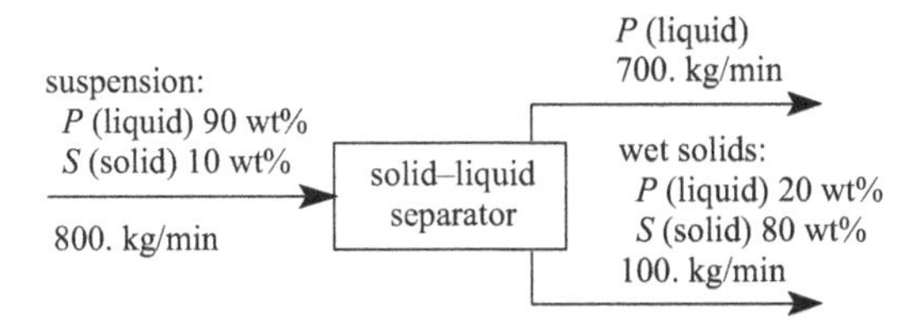

We want to recover P from the wet solids. Because P is miscible with water, we mix it with water and use a second solid–liquid separator.

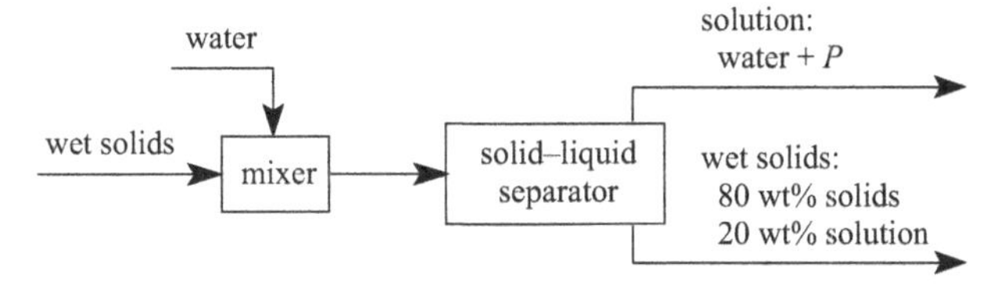

The solid–liquid separator has two phases in equilibrium: a liquid solution and wet solids. We can therefore construct a phase diagram on a ternary plot with three axes: solids, P, and water.

We prepare six compositions, mix thoroughly, and then allow each system to reach equilibrium. The compositions of the two phases in equilibrium are tabulated below.

	Solution phase		Wet solids phase		
Experiment	Wt fraction water	Wt fraction P	Wt fraction solids	Wt fraction water	Wt fraction P
1	1.0	0.0	0.80	0.20	0.0
2	0.0	1.0	0.80	0.0	0.20
3	0.5	0.5	0.80	0.10	0.10
4	0.75	0.25	0.80	0.15	0.05
5	0.25	0.75	0.80	0.05	0.15
6	0.90	0.10	0.80	0.18	0.02

(A) Plot these data on the ternary diagram below to create a phase diagram. Include tie lines. (The point labeled "1" refers to part (B).)

Ternary phase diagram for solids + P(liquid) + water mixtures at 20°C and 1 atm.

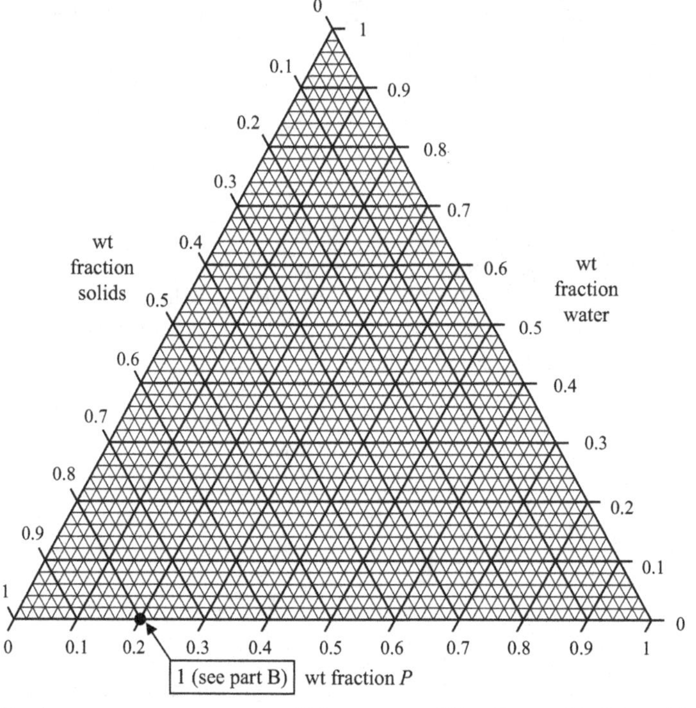

(B) Design a process to recover P from wet solids. Start with 20. wt% P and 80. wt% solids.

Design Goals (in decreasing importance)

- Produce a produce with at least 98 wt% P.
- Discard wet solids with less than 2.0 wt% P.
- Maximize product output; minimize waste.
- Minimize the number of units.
- Minimize the size of each unit.

Design Rules

- Do not heat P above 300°C.
- Use only single-stage separators: liquid–liquid separators, liquid–vapor separators (flash drums), solid–liquid separators, or solid–vapor

separators. Use no distillation columns and use no multistage absorbers/strippers.

- Number each stream and indicate the stream on a phase diagram. Stream 1 (20. wt% *P* and 80. wt% solids) is labeled on the ternary diagram.

Temperature–composition phase diagram for water + *P* mixtures at 1 atm.

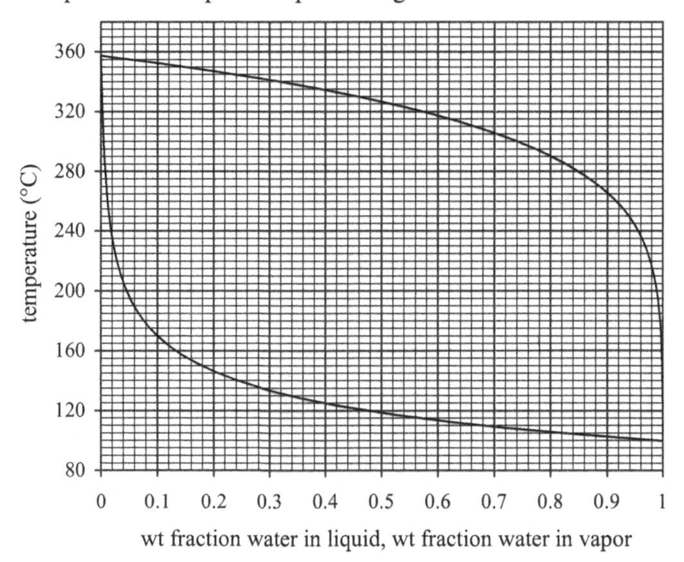

4.102 *Design based on Section 4.2.1 – Solid–liquid single-stage separators and ternary diagrams*. Design a process to separate a mixture of 50. wt% solid *A* and 50. wt% solid *B* into pure solid *A* and pure solid *B*. *A* and *B* have a eutectic point at 30. wt% *B*. *A* and *B* are soluble in water; *B* is more soluble than *A*. **Hint:** Use the difference in solubility to get past the eutectic point.

Design Goals (in decreasing importance)

- Produce two products: pure solid *A* and pure solid *B*.
- Maximize product output.
- Minimize consumption of water.
- Minimize the number of units.
- Minimize the size of each unit.

Design Rules

- Use only two inputs: (1) a mixture of 50. wt% *A* and 50. wt% *B*; and (2) water.
- List substances and approximate wt%'s for every stream.
- Use any number of single-stage separators: liquid–liquid separators, liquid–solid separators, solid–gas separators, and liquid–gas separators (flash drums). Indicate the temperature in each separator.
- Number all streams that enter and leave a separator and indicate these streams on a phase diagram.

Allowances

- You need not calculate flow rates.
- You may neglect minor purge streams.
- You may assume ideal solid–liquid separations for mixtures of *A* and *B* (no water).

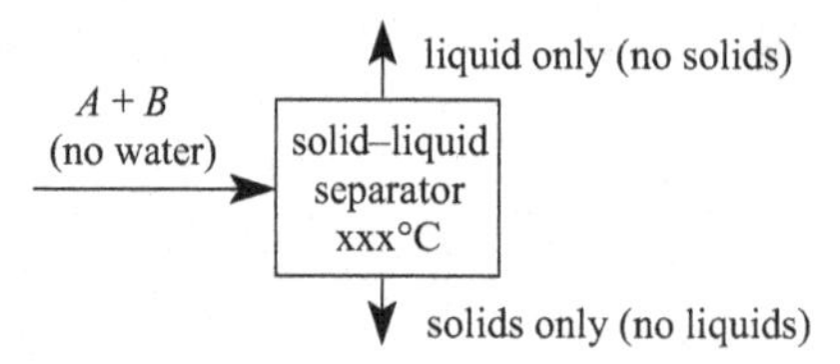

You may assume ideal solid–gas separations for mixtures of H_2O with A and/or B.

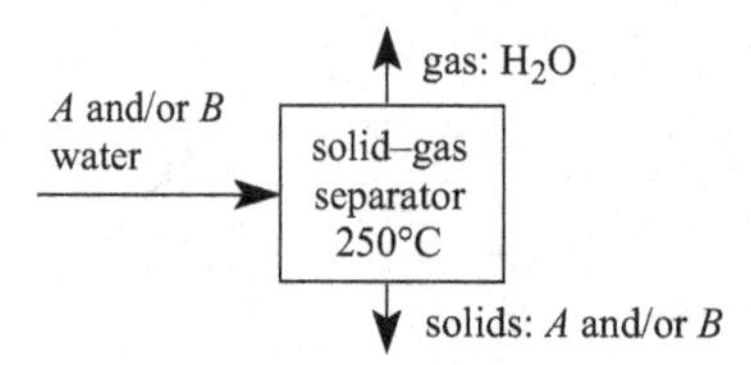

Temperature–composition phase diagram for mixtures of A and B at 1 atm.

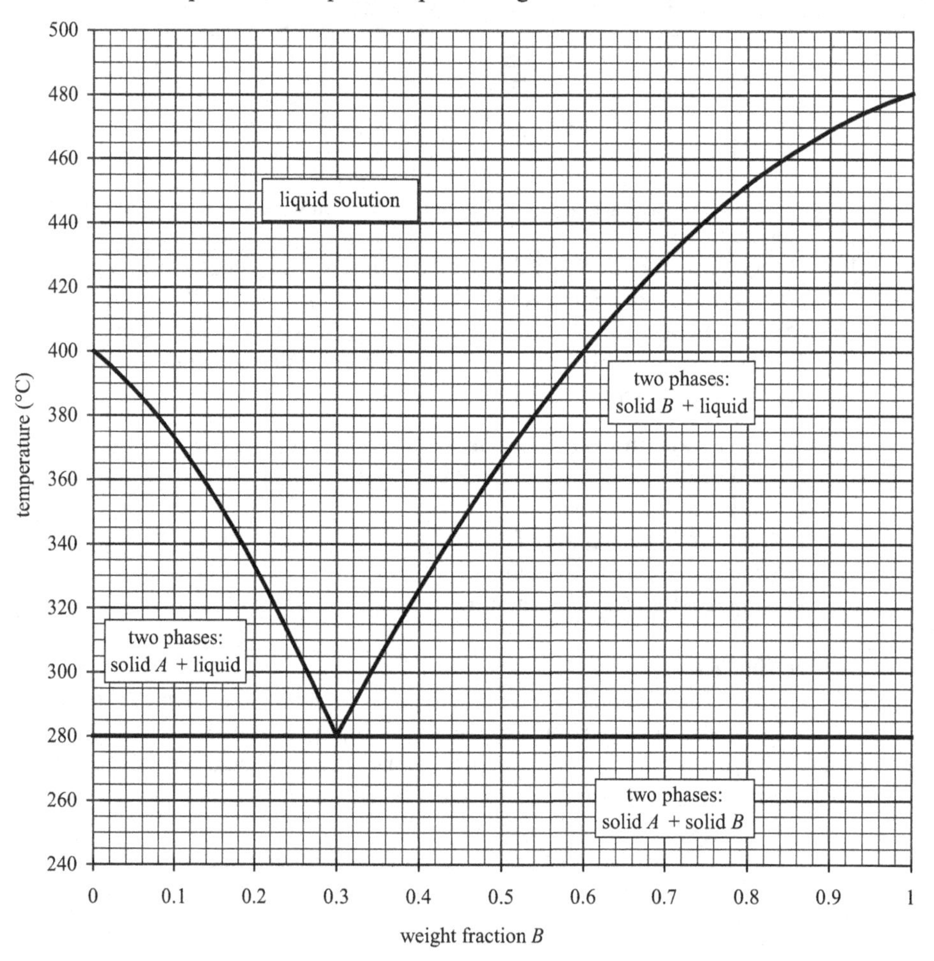

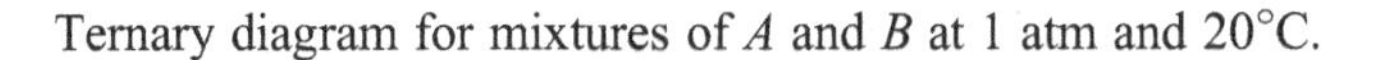
Ternary diagram for mixtures of A and B at 1 atm and 20°C.

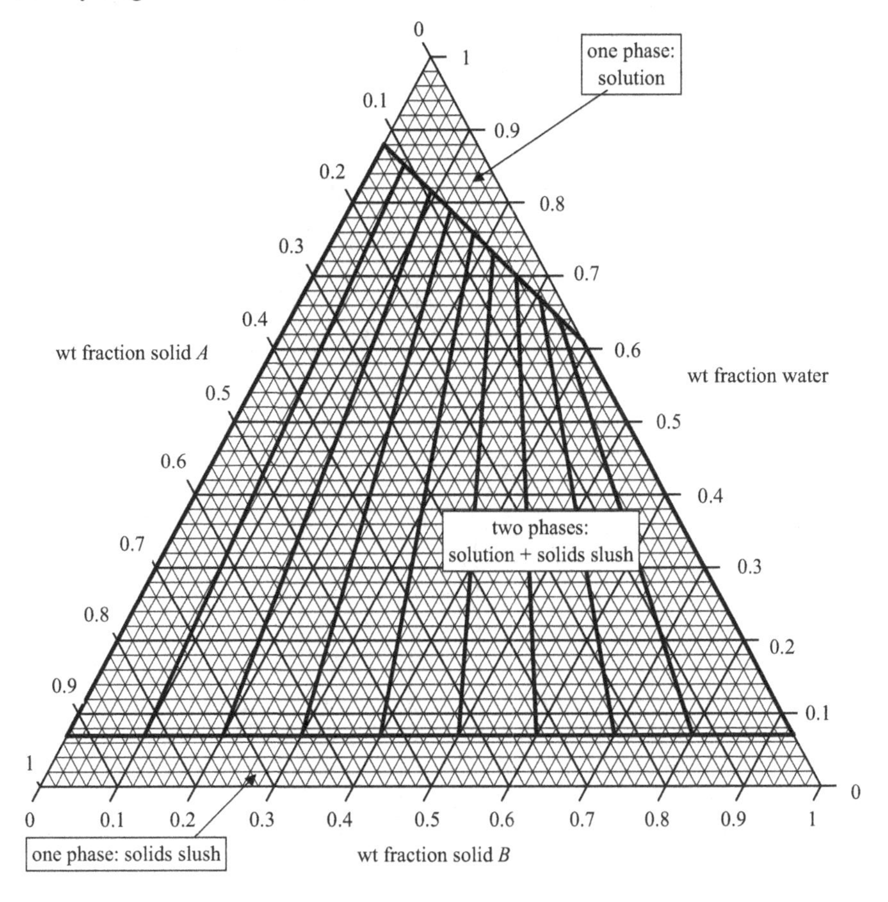

4.103 *Design based on Section 4.2.1 – Solid–liquid single-stage separators and ternary diagrams*. Both sugar and solid impurity I are soluble in water, but the solubilities differ. The phase diagram for mixtures of sugar, I, and water is shown below.

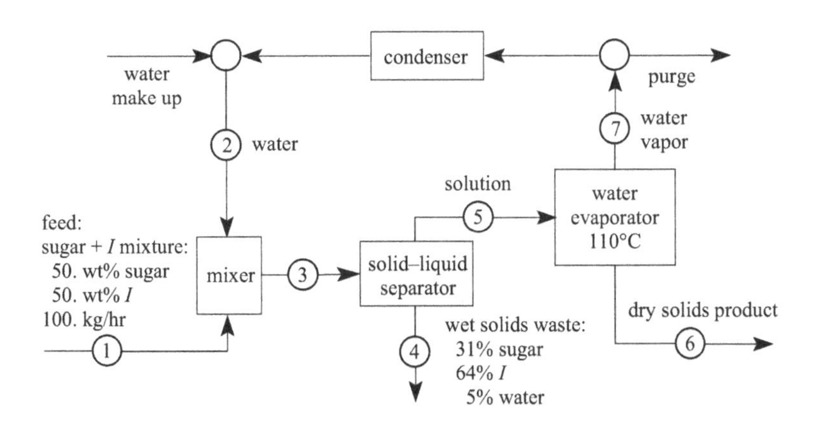

We wish to decrease the amount of impurity I in a mixture of sugar and I. The process below capitalizes on the solubility differences to decrease the wt fraction of I in the mixture.

Ternary phase diagram for sugar + I (soluble solid) + water mixtures at 1 atm.

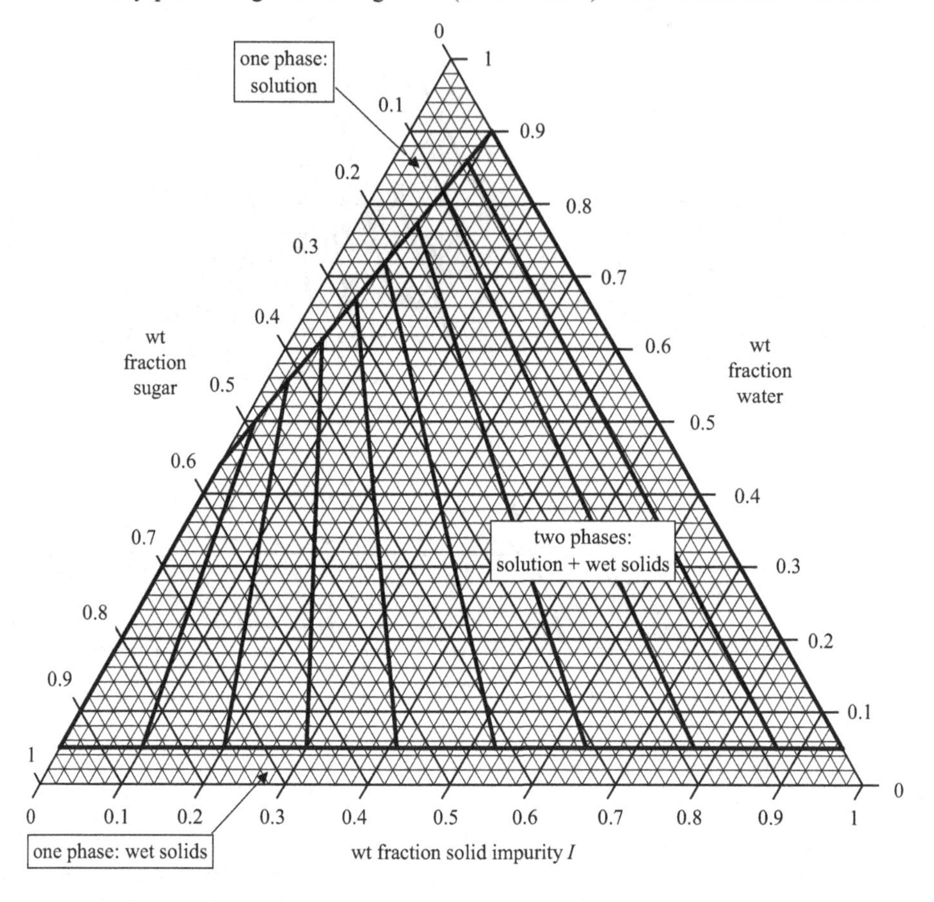

(A) Use graphical methods to calculate the flow rate of stream 2.

(B) Use graphical methods to calculate the *composition* of stream 6. You may use mass balances to check your answer, but you must show a graphical method for full credit.

(C) How would you change the operating parameters in the process above to maximize the wt fraction of sugar in stream 6? What is the maximum wt fraction of sugar in stream 6 that can be attained with the process above?

(D) Design a process to produce dry solids with less than 15. wt% impurity I. You may start with the process above and add units. Or you may start with the feed (stream 1) and design your own process. Number every stream that leaves a separator or an evaporator and indicate the stream on the ternary phase diagram.

Design Goals (in decreasing importance)

- Produce a solid product with less than 15. wt% impurity *I* (greater than 85. wt % sugar).
- Maximize the flow rate of product.
- Minimize the *number* of units.

Design Rules

- Use 100. kg/hr of a mixture of 50. wt% sugar and 50. wt% solid impurity *I*, and as much pure water as you need.
- No process unit may have a capacity larger than 1000. kg/hr.
- It is not necessary to calculate flow rates.

Hint: To maximize product flow rate, discard only one waste stream, with minimal sugar in the waste stream.

This exercise was created by Nicholas Hoh, Cornell University Chemical Engineering Class of 2008.

Design Based on Multistage Absorbers and Strippers

4.104 *Design based on Section 4.2.2 – Liquid–vapor and liquid–liquid absorbers and strippers.* Design a process to extract *E* and *X* from a dilute water solution and then separate the mixture of *E* and *X*. The water solution is 3 wt% *E*, 2 wt% *X*, and 95 wt% H_2O.

Design Goals (in decreasing importance)

- Produce an *E*-rich product with less than 10 wt% *X* and no H_2O.
- Produce an *X*-rich product with less than 10 wt% *E* and no H_2O.
- Minimize the number of units.
- Minimize waste.

Design Rules

- Label each stream with qualitative compositions.
- Label each separator with the temperature and the physical basis for the separation.
- You need *not* specify the operating parameters (number of stages, L/V ratios) for the separators.

You may assume water does not dissolve in the oil, the oil does not dissolve in water, and the oil does not evaporate into N_2 at 20°C. However, water evaporates into N_2 (1.4 wt% H_2O vapor in N_2 at 20°C).

Boiling points at 1 atm (°C).

E	*X*	H_2O	Oil	N_2
100	100	100	320	−196

Thermodynamic data at 20°C and 1 atm.

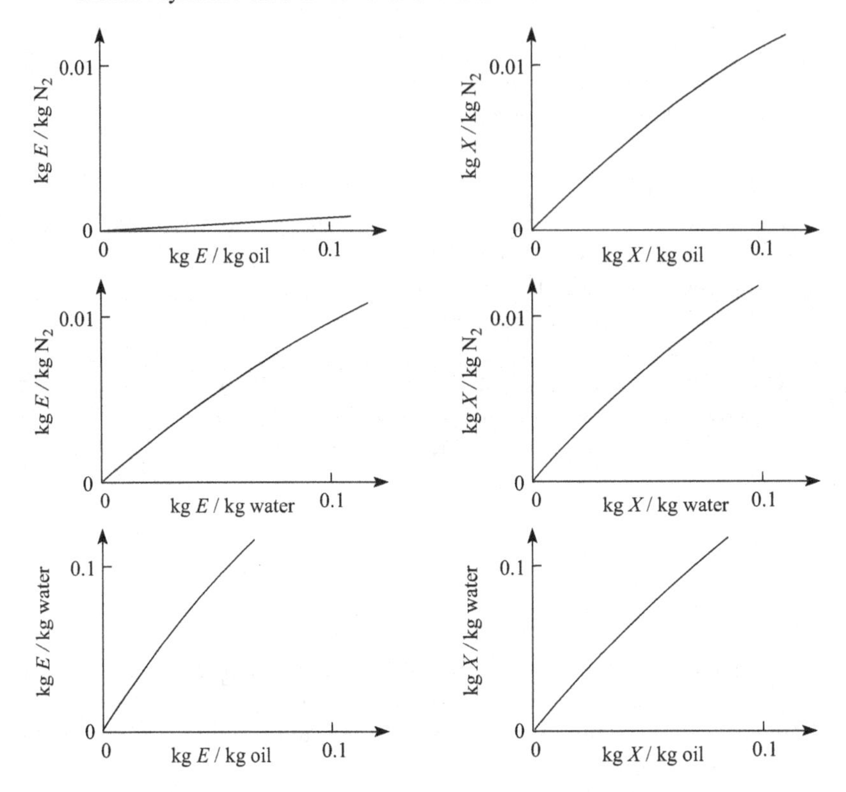

Design Based on Distillation Columns

4.105 *Design based on Section 4.2.4 – Distillation columns.* Design a distillation column to separate a mixture of ethanol and water into a water-rich stream (98 mol% water) and an ethanol-rich stream (80 mol% ethanol). The feed to the distillation column is a liquid + vapor mixture equilibrated such that $y_{\text{ethanol}} = 0.443$ and $x_{\text{ethanol}} = 0.10$. Use the ethanol–water x–y diagram from Exercise 4.65 (or download from duncan.cbe.cornell.edu/Graphs). State the following details of your design:

- the number of stages
- the reflux ratio (L/V) for the stripping section
- the reflux ratio (L/V) for the rectifying section
- the stage into which the feed enters
- whether the feed needs to be heated or cooled before it enters the feed stage.

For uniformity, number your stages from the bottom of the distillation column.

Although there are a multitude of viable designs for the column, avoid the absurd. For example, a column with reflux ratio of 1 (no output) or a column with maximum throughput (requiring an infinite number of stages) would be considered absurd.

4.106 *Design based on Section 4.2.4 – Distillation columns*. Design a distillation column to separate a mixture of DCE (1,2-dichloroethane, CH_2ClCH_2Cl) and 1-propanol ($CH_3CH_2CH_2OH$) into a DCE-rich stream (74 mol% DCE) and a propanol-rich stream (99 mol% propanol). The input to the distillation column is a liquid + vapor mixture equilibrated such that $x_{DCE} = 0.30$. Use the DCE-propanol vapor–liquid diagram below to determine operating conditions and the number of stages. You need not find the absolute optimal conditions; just find a workable solution. Specify:

- the number of stages
- the reflux ratio (L/V) for the stripping section
- the reflux ratio (L/V) for the rectifying section
- the stage into which the feed enters
- whether the feed needs to be heated or cooled before it enters the feed stage.

x–y diagram for liquid–vapor systems of 1,2-dichloroethane + 1-propanol mixtures at 1 atm.

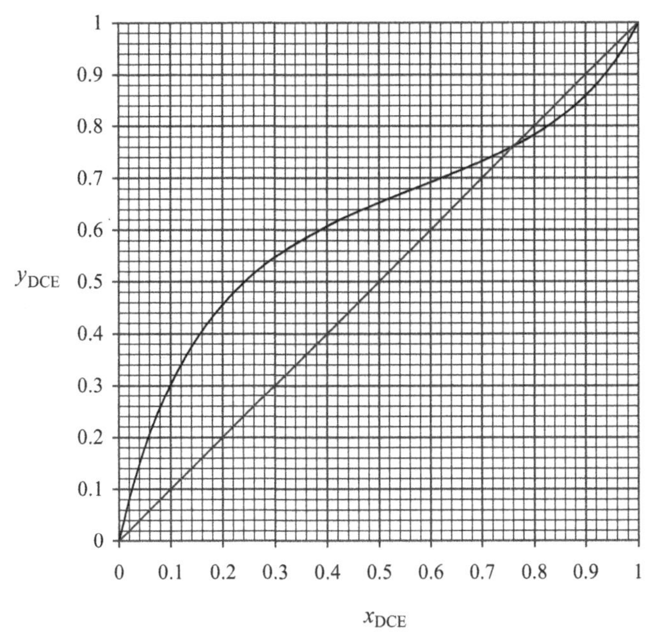

Diagram generated from data compiled by Gmehling, J., Onken, U., and Arlt, W. (1982). *Vapor-Liquid Equilibrium Data Collection - Organic Hydroxy Compounds: Alcohols*, Dechema, Frankfurt, Germany, vol. 1, part 2c, p. 477.

4.107 *Design based on Section 4.2.4 – Distillation columns*. Design a process to separate a mixture of 20. mol% acetonitrile and 80. mol% water into two streams: one with ≥98 mol% acetonitrile and the other with ≤2 mol% acetonitrile. Indicate the compositions of all streams in your process and the number of stages in the distillation column(s). Use the liquid–vapor equilibrium data for acetonitrile + water mixtures at 1.0 atm (Exercise 4.63), as well as the data at 0.2 atm shown below.

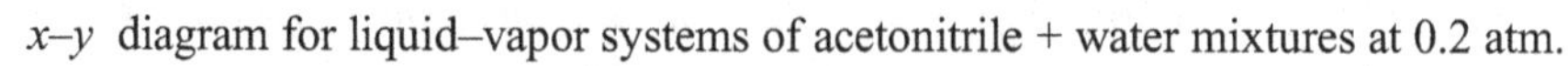
x–y diagram for liquid–vapor systems of acetonitrile + water mixtures at 0.2 atm.

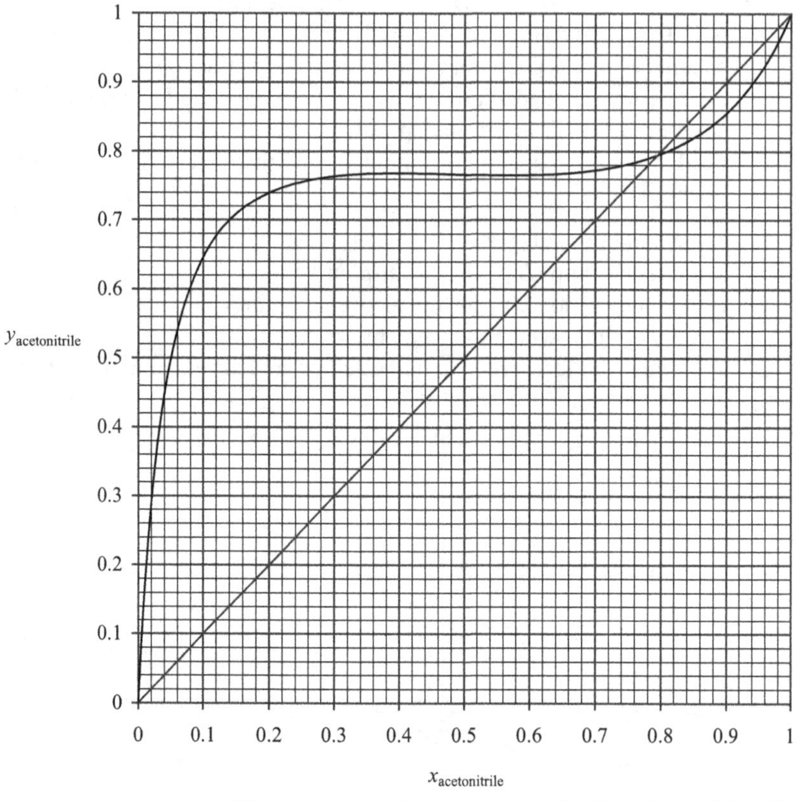

Diagram generated from data given in Gmehling, J. and Onken, U. (1977). *Vapor-Liquid Equilibrium Data Collection - Organic Hydroxy Compounds: Alcohols*, Dechema, Frankfurt, Germany, vol. 1, p. 79.

4.108 *Design based on Sections 4.1.4 and 4.2.4 – Liquid–liquid single-stage separators and distillation columns.*

The process below separates W and R.

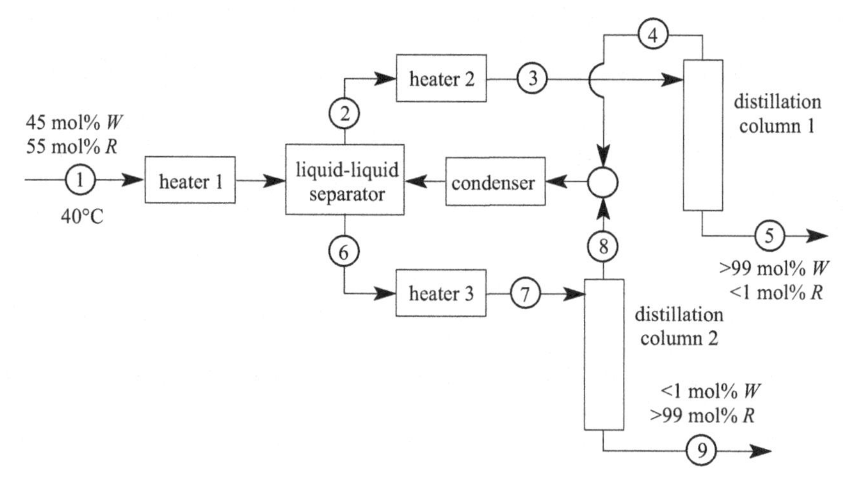

Use the W + R phase diagram and the x_W–y_W liquid–vapor diagram below to complete parts (A)–(D).

Temperature–composition phase diagram for $W + R$ mixtures at 1 atm.

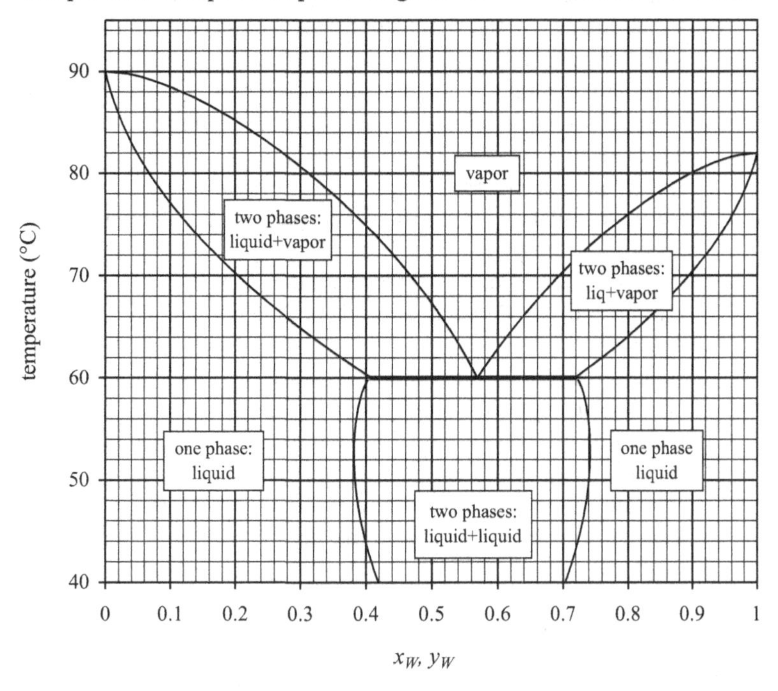

x–y diagram for liquid–vapor systems of $W + R$ mixtures at 1 atm.

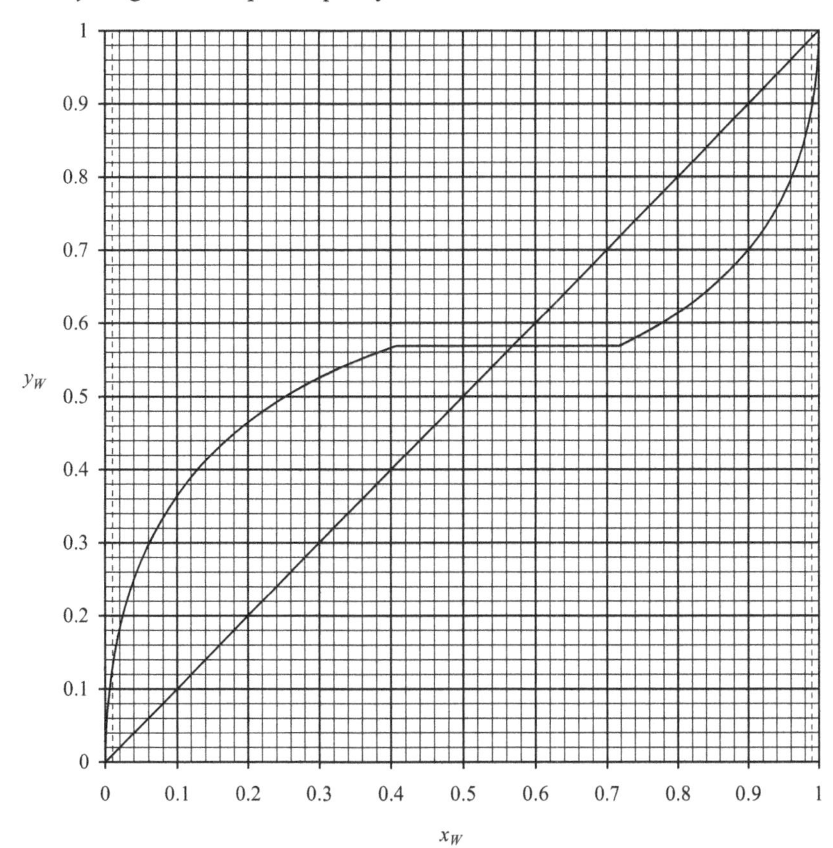

(A) What temperature in heater 1 will optimize the liquid–liquid separation?

(B) Label streams 2 and 6 on the temperature–composition phase diagram for $W + R$ mixtures.

In both distillation columns the feed enters at the top stage. The distillate stream from each column is thus vapor in equilibrium at the temperature of the preheater. It is not necessary to optimize the distillation columns – just specify parameters so each column will produce the output streams 5 and 9.

(C) Design distillation column 1 by completing the following steps:
Specify the temperature for heater 2.
Indicate stream 4 on the W–R phase diagram.
Indicate stream 3 (feed stream) on the x_W–y_W liquid–vapor equilibrium diagram.
Specify the number of equilibrium stages.
Specify the *approximate* L/V ratio.

(D) Design distillation column 2 by completing the following steps:
Specify the temperature for heater 3.
Indicate the composition of stream 8 on the W–R phase diagram.
Indicate stream 7 (feed stream) on the x_W–y_W liquid–vapor equilibrium diagram.
Specify the number of equilibrium stages.
Specify the *approximate* L/V ratio.

4.109 *Design based on Sections 4.1.4 and 4.2.4 – Liquid–liquid and liquid–vapor single-stage separators and distillation columns.* Design a process to separate a mixture of 50 mol% W and 50 mol% S.

Design Goals (in decreasing importance)

- Produce a W-rich product with at least 97 mol% W.
- Produce a S-rich product with at least 97 mol% S.
- Minimize waste.
- Minimize the number of units.

Design Rules

- Label each separator with the temperature and the physical basis for the separation.
- For liquid–vapor separators and liquid–liquid separators, the flow rate of the waste stream must be no more than five times the flow of the product stream.
- Use exactly one distillation column and specify the following for the distillation column: the total number of stages, the stage number at the feed stage, the temperature of the feed stream, the mol fraction vapor in the feed stream, the L/V for rectifying section (the section that produces the distillate), and the L/V for stripping section (the section that produces the bottoms).

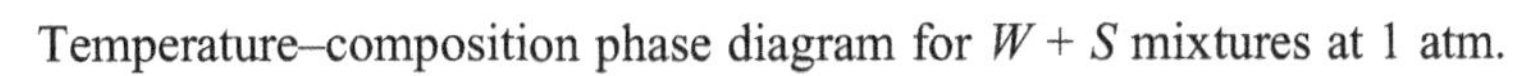

Temperature–composition phase diagram for $W + S$ mixtures at 1 atm.

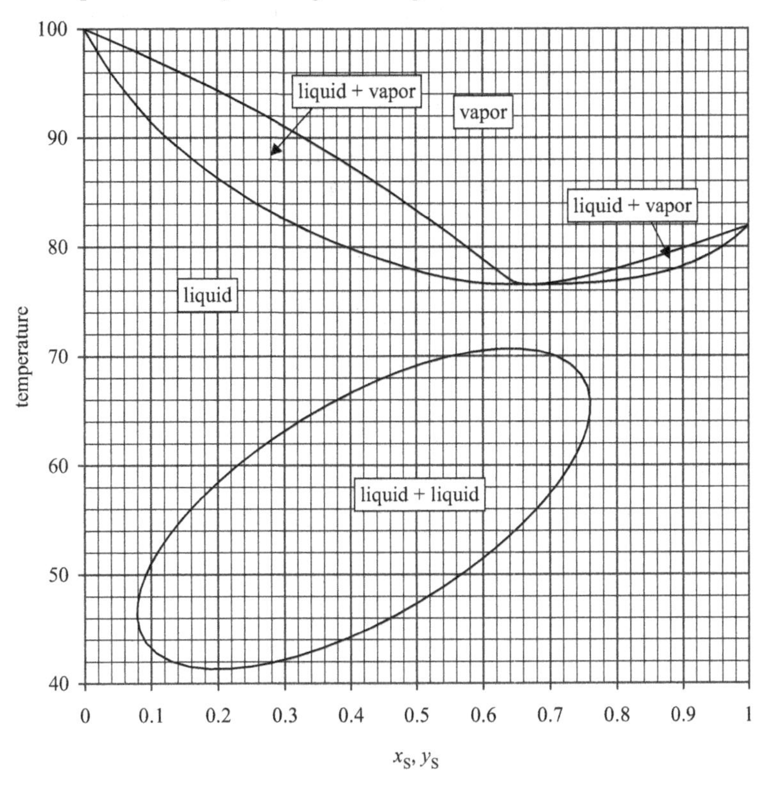

x–y diagram for liquid–vapor systems of $W + S$ mixtures at 1 atm.

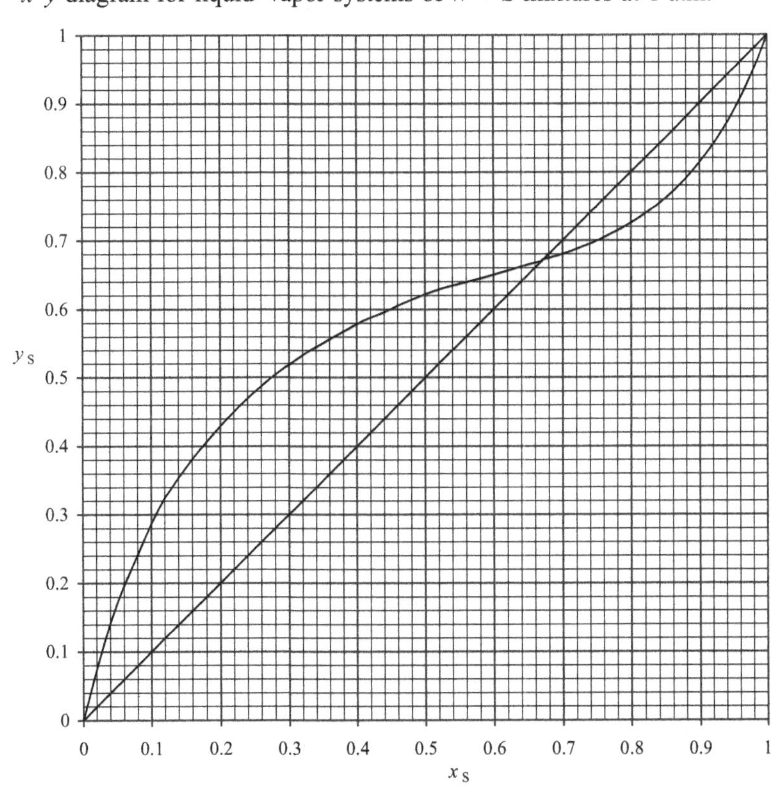

4.110 *Design based on Sections 4.1.4 and 4.2.4 – Liquid–liquid and liquid–vapor single-stage separators and distillation columns.* Design a process to produce 98 mol% *A* and 98 mol% *B* from a mixture of 20. mol% *A* and 80. mol% *B*. Because *A* and *B* form an azeotrope, your design should use extracting agent *Z*. Liquid *Z* is immiscible with *B*-rich mixtures of *A* and *B*, and liquid *Z* selectively absorbs *B*.

Design Goals (in decreasing importance)

- Produce two products: ≥98 mol% *A* and ≥98 mol% *B*.
- Maximize product output.
- Minimize consumption of *Z*.
- Minimize the number of units.
- Minimize the size of each unit.

Design Rules

- The process has only two inputs: (1) a mixture of 20. mol% *A* and 80. mol% *B* at 100. mol/min; and (2) pure *Z* at unlimited flow rate.
- Use no more than *two* distillation columns.

Ternary diagram for *A* + *B* + *Z* mixtures at 1 atm and 20°C.

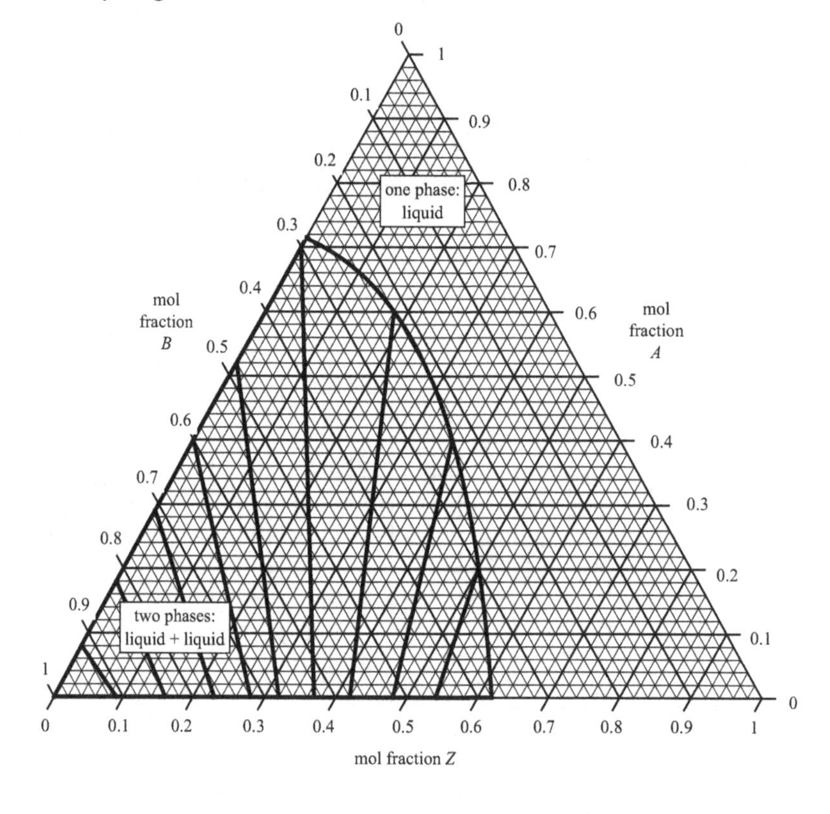

- Use any number of single-stage separators: liquid–liquid separators and liquid–vapor separators (flash drums).
- Number each stream and indicate each stream on a phase diagram.
- You need not calculate flow rates.
- If your process has a distillation column, specify the total number of stages, the stage number at the feed stage, and the temperature of the feed stream.

Temperature–composition phase diagram of $A + Z$ mixtures at 1 atm.

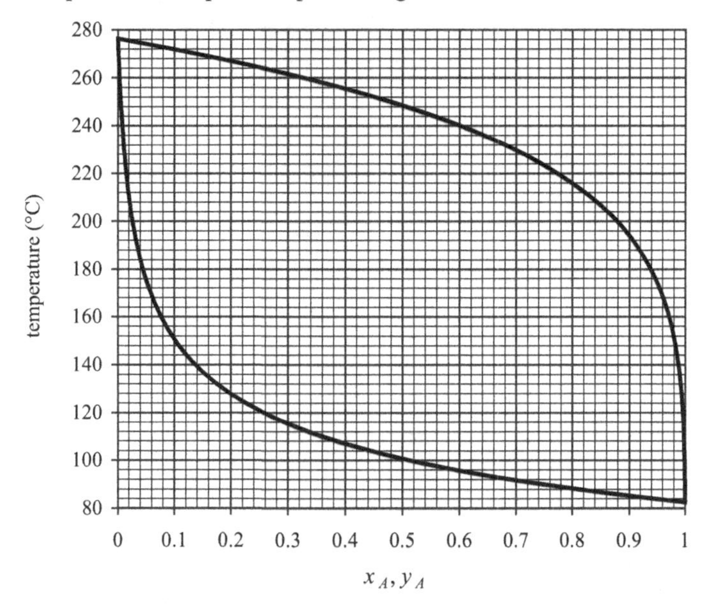

Temperature–composition phase diagram of $B + Z$ mixtures at 1 atm.

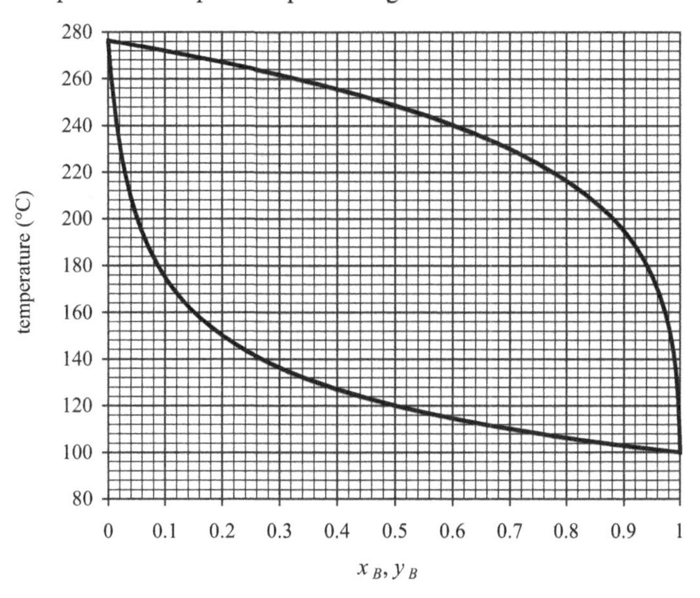

Temperature–composition phase diagram of $A + B$ mixtures at 1 atm.

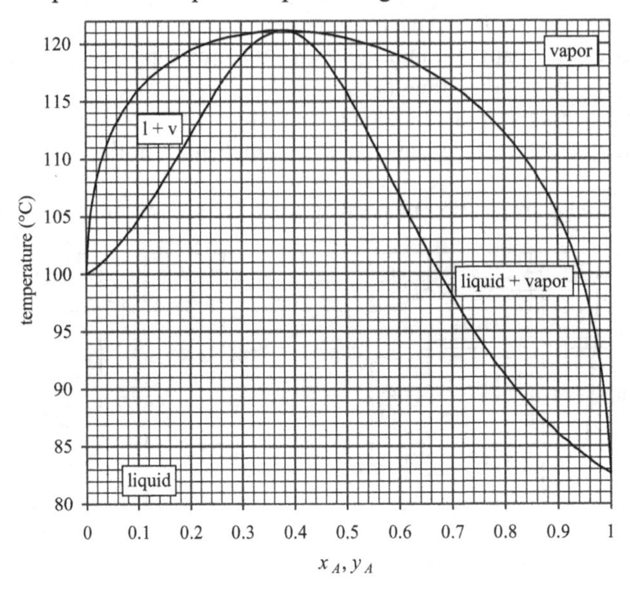

x–y diagram for liquid–vapor systems of $A + B$ mixtures at 1 atm.

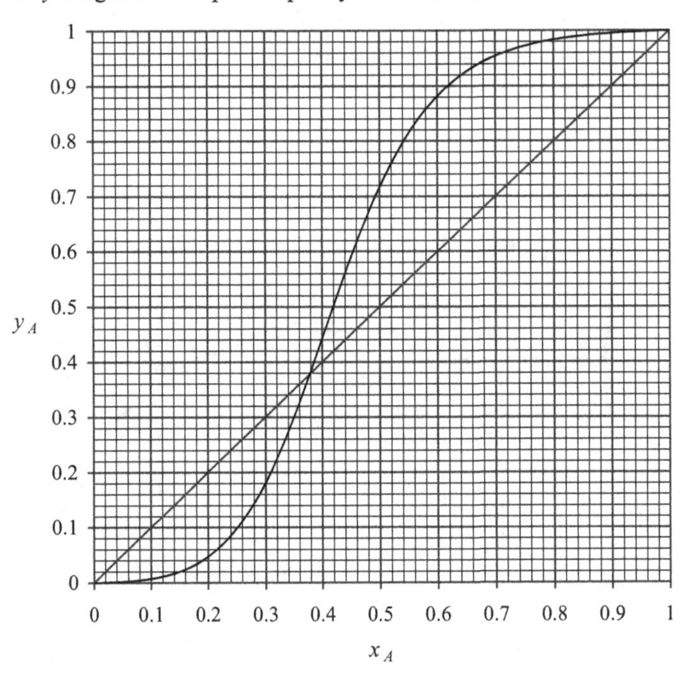

Design Based on Condensation from a Non-condensible Gas

4.111 *Design based on Section 4.3.2 – Condensation from a mixture with a non-condensible gas.* Design a process to transform gaseous chemical A at 300°C and 2 atm to solid chemical A at 0°C and 1 atm without forming liquid at any point in the process. Label the function of each unit and specify the temperature and pressure of each stream.

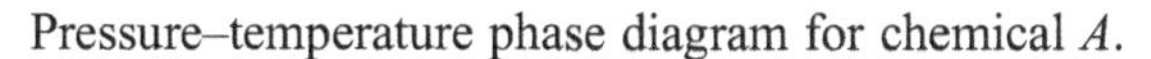

Pressure–temperature phase diagram for chemical A.

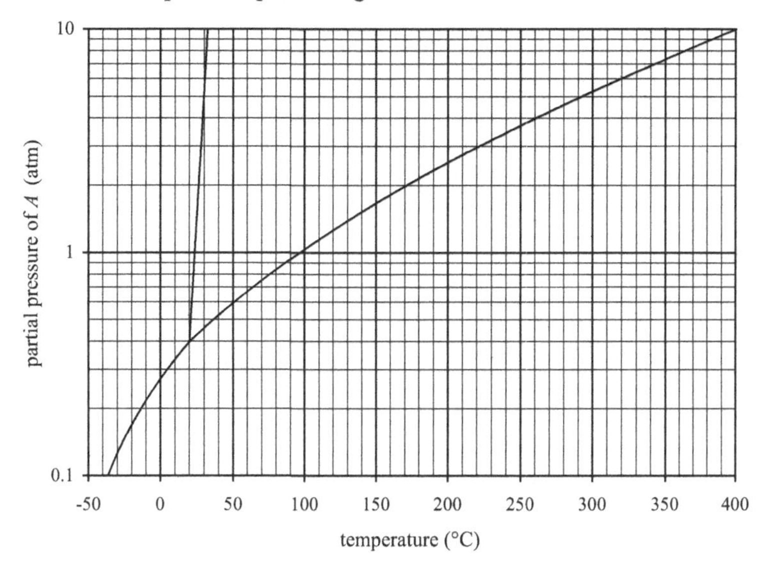

4.112 *Design based on Section 4.3.2 – Condensation from a mixture with a non-condensible gas.* Design a process to produce air at 1 atm and 18°C with 10.% humidity.

(A) Start with air at 1 atm and 29°C with 75% humidity. Show all separators, heaters, coolers, pumps, compressors, and expanders (pressure reducers); but you may neglect heat exchangers. Assume a liquid–solid separator produces solid-free liquid, but wet solids. Finally, the pressure in your process must not exceed 5 atm. Use the pressure–temperature phase diagram for H_2O in Exercise 4.74 or download from duncan.cbe.cornell.edu/Graphs.

(B) Design a different process to produce air at 1 atm and 18°C with 10.% humidity. Again start with air at 1 atm and 29°C with 75% humidity, but design a process without using solid–gas or solid–liquid separators. Use the same design guidelines and rules as in (A). Remember, the pressure must not exceed 5 atm.

4.113 *Design based on Sections 4.2.1, 4.2.2, 4.2.4, and 4.3.2* – Parts (A)–(D) concern portions of the process below. Each part can be worked independently.

The process below converts reactant A to product P. Because the reaction is highly exothermic, the reaction is conducted at 20°C in a liquid solvent. A mixture of A and P has an azeotrope at 60 mol% P, so we need to increase the P-to-A ratio to distill to 98 mol% P. The reactor effluent is first stripped with He gas; then A and P are condensed from the He gas; and finally the $A + P$ mixture is distilled.

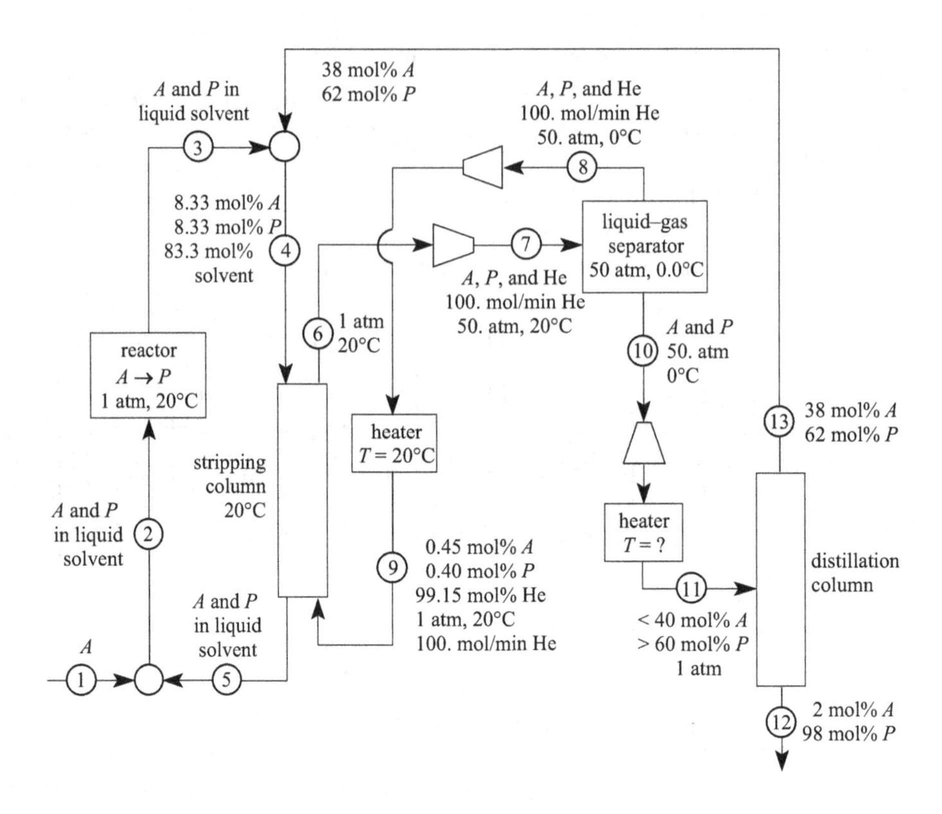

Boiling points at 1 atm (°C).

A	*P*	Solvent	He
72	65	357	−269

(A) The design and operation of the stripping column are shown graphically below. Calculate the following quantities:

1. The number of equilibrium stages.
2. The flow rate of liquid solvent, in mol/min.
3. The ratio (mol *P* / mol liquid solvent) in stream 5.
4. The ratio (mol *P* / mol He) in stream 6.
5. The ratio (mol *A* / mol liquid solvent) in stream 5.
6. The ratio (mol *A* / mol He) in stream 6.

P/solvent–P/helium solubility diagram.

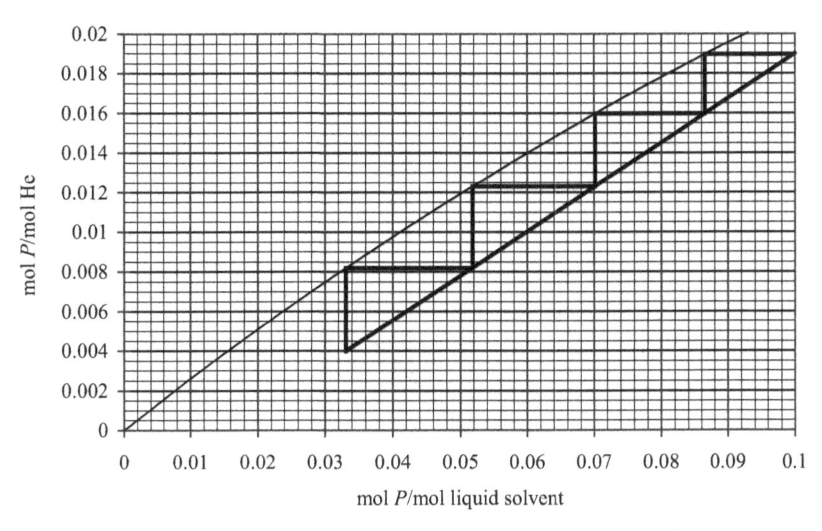

A/solvent–A/helium solubility diagram.

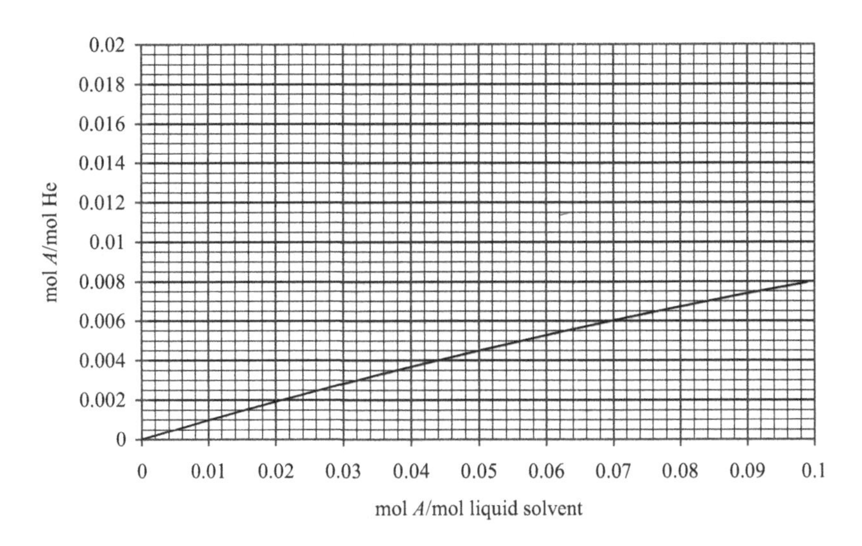

(B) Analyze the liquid–gas separator. Assume the partial pressure of P in stream 6 is 0.019 atm.

1. Show the locations of streams 6, 7, 8, and 9 on the diagram below.
2. Calculate the flow rate of P in stream 10, in mol/min. You may assume that no He condenses at 0°C and 50 atm.

Pressure–temperature phase diagram for chemical *P*.

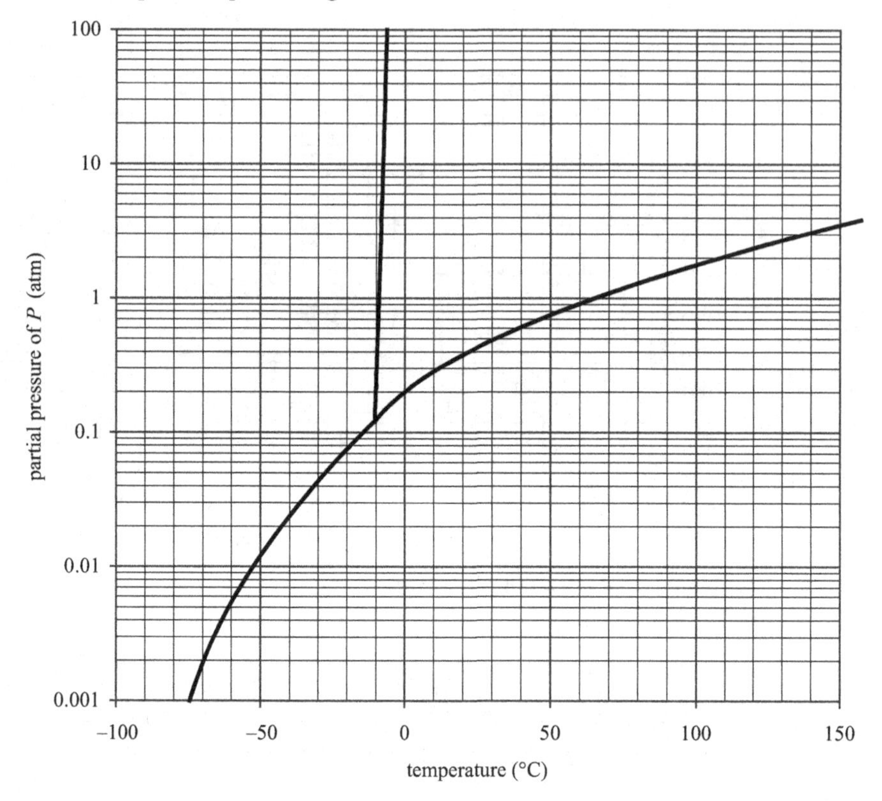

(C) Specify a design and operation for the distillation column. Assume stream 11 flows at 2.0 mol/min and contains 80. mol% *P*. Design a distillation column to produce two streams: 98 mol% *P* and 62 mol% *P*. List six specifications for your design:
1. the total number of stages
2. the stage number at the feed stage
3. the temperature of the feed stream
4. L/V for the stripping section
5. L/V for the rectifying section
6. flow rate of product stream 12, in mol/min.

x–y diagram for liquid–vapor systems of $A + P$ mixtures at 1 atm.

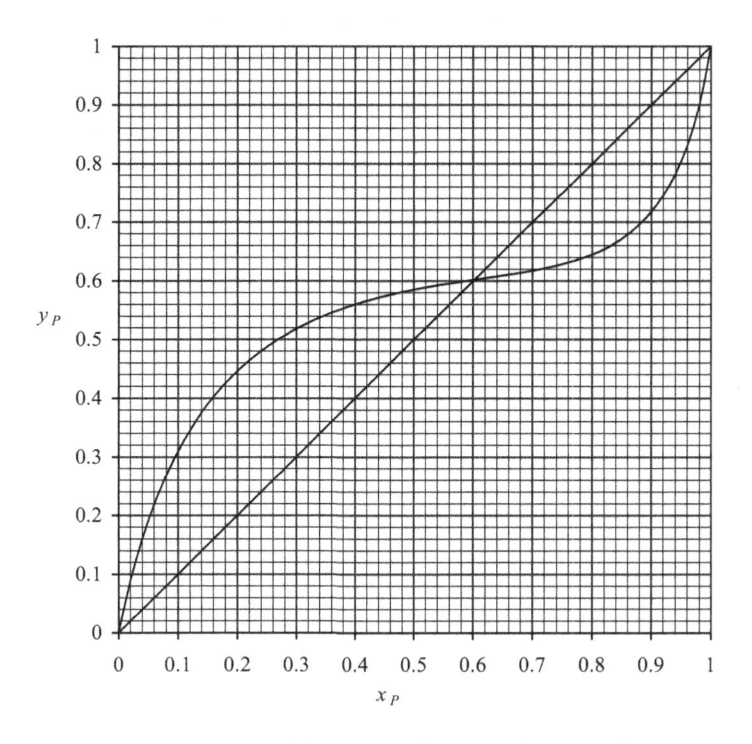

Temperature–composition phase diagram of $A + P$ mixtures at 1 atm.

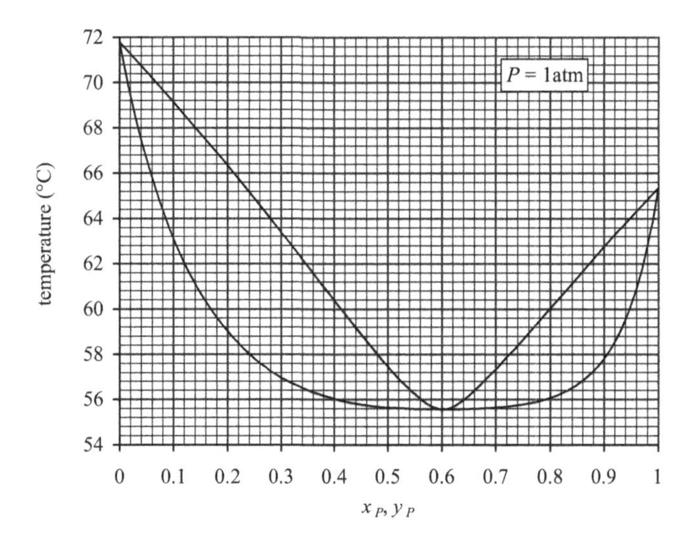

(D) Your engineer claims the stripping column and liquid–gas separator are not necessary to get past the azeotrope at 60 mol% *P*. Your engineer proposes the process below to produce 98 mol% *P*.

Provide an analysis that demonstrates the process is viable, or provide an analysis that demonstrates the process cannot work.

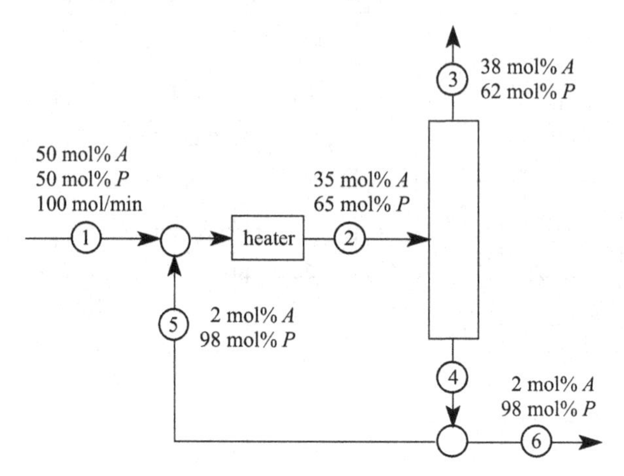

5

Dimensional Analysis and Dynamic Similarity

We have now considered two methods for modeling and analyzing processes: mathematical modeling based on laws and graphical analysis based on empirical measurements. In mathematical modeling we started with a general principle and substituted the specifics of the process. The general principle might be universal, such as the conservation of mass, (*rate of mass in*) = (*rate of mass out*) at steady state, or constrained, such as a constitutive equation like the ideal gas law, $PV = nRT$. For empirical analysis we first measured data, such as the solid–liquid–vapor phase diagram of a pure substance or a vapor–liquid composition diagram for a mixture of two components. We did not concern ourselves with why these data were the way they were. We did not dwell on why some systems are non-ideal, for example, why some systems have an azeotrope. We just used the data. We never extrapolated from the measured data. That is, if we measured vapor–liquid composition data for mixtures of benzene + toluene at 80°C as a function of pressure, we used these data only to analyze units that separated benzene + toluene mixtures at 80°C.

What if we had wanted to operate the unit at 60°C? We would have had to return to the lab and remeasure the data at 60°C. We had no means of extrapolating the data to different conditions. Thermodynamics provides the tools to extrapolate equilibrium data. But lacking knowledge of thermodynamics, is there a way to extrapolate? More generally, how does one extrapolate in systems too complex to yield an analytical solution? How does one design a large unit too complex to be analyzed mathematically? Should one build the unit and hope for the best? That approach would cost too much in time and money. Instead, we use a third means of modeling: we will study a convenient working model and extrapolate to a large industrial size or to a microscopic analytical size. As was stated eloquently by Leo Baekeland (1863–1944), inventor of Bakelite, the first synthetic plastic (Zlokarnik 1991, p. 5),

> Commit your blunders on a small scale and make your profits on a large scale.

A change in size or rate is called *scaling*. Extrapolation of a small unit to a large unit or extrapolation of a slow process to a fast process is called *scale-up*. Similarly, extrapolation of a small process to a microscopic process or extrapolation of a fast process to a slow process is called *scale-down*.

So how does one extrapolate from the small to the large? Are there formal rules? Let's look at some examples. We divide the examples into two types: static systems and dynamic systems.

Static systems – systems with no motion – include maps and molecular models. We scale up a map such as Figure 5.1 to calculate distances between cities, for example. Similarly, we scale down the dimensions of an atomic model such as Figure 5.2 to calculate interatomic distances. The rule of scaling in static systems is obvious: the size scales linearly. This is geometric scaling.

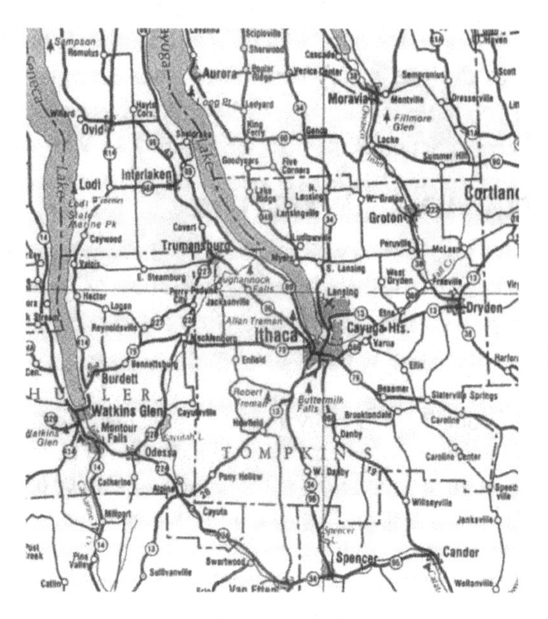

Figure 5.1 A two-dimensional static model, a map. Copyright Cornell University. Reproduced by permission.

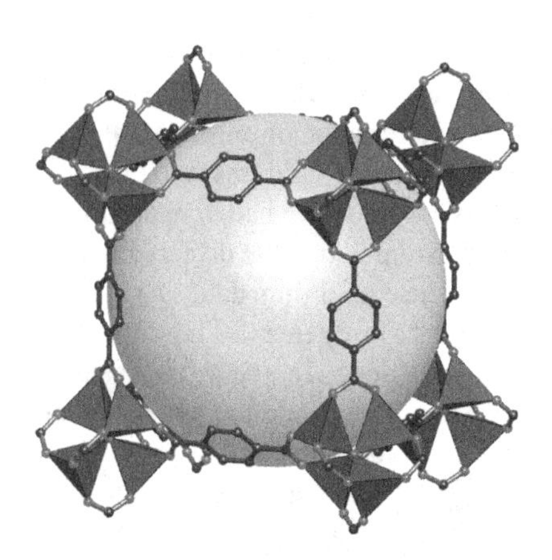

Figure 5.2 A drawing of a three-dimensional static model for the metal-organic framework MOF-5, having the formula Zn_4O $(BDC)_3$, where BDC ≡ 1,4-benzodicarboxylate.

Compare the scaling of maps and molecular models to the scaling of dynamic systems. Consider lizards, for example. Figure 5.3 is a picture of a small lizard, shown close to its actual size.

Figure 5.3 A small lizard, about 10 cm long. Copyright Deborah Duncan. Reproduced by permission.

This design works quite well for a small lizard. What if you wanted to scale-up this design to create a very large lizard – for example a lizard the size of a house? Should you scale the dimensions linearly? Linear scaling was used in most of the 1950s science fiction B-movies.

Figure 5.4 A common lizard scaled linearly to gigantic size. Scene from *The Giant Gila Monster* (1959). Movie Still Archives.

Why does the large lizard in Figure 5.4 look unrealistic? Perhaps because we know large lizards are proportioned differently than small lizards. Large lizards tend to be shaped like dinosaurs, a design that worked well for about 100 million years.

Figure 5.5 A large reptile, triceratops, about 10 m long. Copyright Deborah Duncan. Reproduced by permission.

Why do small lizards and large lizards have different proportions? Consider how various features of a lizard scale with size. The mass of a lizard is proportional to the volume of a lizard, which is proportional to its height cubed:

$$\text{body mass} \propto (\text{height})^3. \tag{5.1}$$

However, the compressive strength of a structural element, such as a bone, is proportional to its cross-sectional area, which is proportional to the height squared:

$$\text{bone strength} \propto (\text{height})^2. \tag{5.2}$$

If all the dimensions of a lizard were scaled linearly, the mass of a large lizard would exceed the capacity of its skeletal system. As dynamic systems get larger, the length dimensions cannot scale linearly. Scaling in dynamic systems is based on dimensional analysis, the subject of this chapter. But before we take up dimensional analysis we need to know what a dimension is.

5.1 UNITS AND DIMENSIONS

Context: The Système International (SI) and English systems of units.

Concepts: Units vs. dimensions, base units, multiple units, derived units, and dimensional consistency.

Defining Question: Why do we need multiple units and derived units?

Any physical quantity has dimensions – not just the usual dimensions of length and time, but others too. Base dimensions, listed in Table 5.1, are denoted by uppercase letters. For temperature we use Θ because we already used T for time (Θ is the Greek letter for Q, but "theta" starts with a "t"). The last two items in Table 5.1 will not be encountered again in this text.

Table 5.1 Base dimensions.

Base dimension	Symbol
length	L
mass	M
time	T
temperature	Θ
amount	N
electric charge	Q
luminous intensity	I

Each base dimension has a *base unit.* The base unit for each dimension depends on the system of units. Two systems dominate: Système International (SI) and English. The base units are listed in Table 5.2. Any physical quantity can be expressed as the product of a number and a combination of base units.

Table 5.2 The SI and English systems of base units.

Base dimension	SI (mks)	English
length, L	meter (m)	foot (ft)
mass, M	kilogram (kg)	pound-mass (lb_m)
time, T	second (s)	second (s)
temperature, Θ	degrees Celsius (°C) or kelvin (K)	degrees Fahrenheit (°F) or degrees Rankine (°R)
amount, N	mole	mole

The existence of two systems of units causes frustration, tedium, and error. It is important to develop a rigorous method of converting between systems. Two guidelines are:

- multiply by factors of one: for example, 2.54 cm/1 inch = 1
- keep meticulous accounting.

From base units we create *multiple units*. Multiple units are, for example, milligram and kilometer. Why multiple units? Because people like to use numbers between 0.1 and

100 and SI units are sometimes awkward. For example, consider length. Runners compete in a 10-kilometer race (aka a 10K race), rather than a 10,000-meter race. Similarly, when describing the distance between two atoms, it is convenient to say a C–H bond length is 0.11 nanometers, rather than 1.1×10^{-10} m. One describes a feature on a silicon wafer as approximately 1 μm, rather than 1×10^{-6} m. The common prefixes are listed in Table 5.3. For the dimension of time, multiple units include hour, day, year, and century. English units of length include inch, foot, yard, and mile. The English system is fraught with special units for each dimension.

Table 5.3 Prefixes for multiple units.

Prefix	Symbol	Multiplier
mega	M	10^6
kilo	k	10^3
centi	c	10^{-2}
milli	m	10^{-3}
micro	μ	10^{-6}
nano	n	10^{-9}
pico	p	10^{-12}

The third and final class of units is *derived units*. Derived units are the products of base units and have special names. Some important derived units are listed in Table 5.4.

Table 5.4 Derived units and dimensions.

Quantity	Dimensions	Units in SI
volume	L^3	m^3
velocity	L/T	m/s
acceleration	L/T^2	m/s^2
momentum (= mass × velocity)	ML/T	kg·m/s
force (= mass × acceleration)	ML/T^2	$kg{\cdot}m/s^2$ ≡ newton (N)
pressure (= force/area)	M/LT^2	$kg/(m{\cdot}s^2) = N/m^2$ ≡ pascal (Pa)
energy (= ½mv^2 or force × distance)	ML^2/T^2	$kg{\cdot}m^2/s^2$ = N·m ≡ joule (J)
power (= energy/time)	ML^2/T^3	$kg{\cdot}m^2/s^3$ = J/s ≡ watt (W)

We conclude this topic with two rules. The first rule is:

Dimensional consistency: all quantities in a sum must have the same dimensions.

Example: the total energy in a system is the sum of the kinetic and potential energy:

$$\text{total energy} = \text{kinetic energy} + \text{potential energy} \tag{5.3}$$

$$E = \frac{1}{2}mv^2 + mgh. \tag{5.4}$$

We introduce the convention that a quantity in square brackets, [quantity], means "dimensions of the quantity." Thus,

$$[E] = \frac{\mathrm{ML}^2}{\mathrm{T}^2}, \quad [m] = \mathrm{M}, \; [v] = \frac{\mathrm{L}}{\mathrm{T}}, \quad [g] = \frac{\mathrm{L}}{\mathrm{T}^2}, \quad [h] = \mathrm{L}. \tag{5.5}$$

Substitute these dimensions into equation 5.4:

$$[E] = \left[\frac{1}{2}mv^2 + mgh\right]$$
$$\frac{\mathrm{ML}^2}{\mathrm{T}^2} = \mathrm{M}\left(\frac{\mathrm{L}}{\mathrm{T}}\right)^2 + \mathrm{M}\left(\frac{\mathrm{L}}{\mathrm{T}^2}\right)\mathrm{L} \tag{5.6}$$

$$\frac{\mathrm{ML}^2}{\mathrm{T}^2} = \frac{\mathrm{ML}^2}{\mathrm{T}^2} + \frac{\mathrm{ML}^2}{\mathrm{T}^2}. \tag{5.7}$$

The second rule is:

All quantities in a sum must have the same units:

$$\text{distance} = 1\ \text{km} + 1\ \text{foot} + 1\ \text{mm} \neq 3 \tag{5.8}$$

$$= 1\ \text{km} + 1\ \text{foot}\left(\frac{0.3048\ \text{m}}{1\ \text{foot}}\right)\left(\frac{1\ \text{km}}{1000\ \text{m}}\right) + 1\ \text{mm}\left(\frac{1\ \text{km}}{10^6\ \text{mm}}\right) = 1.000306\ \text{km}. \tag{5.9}$$

5.2 DIMENSIONAL ANALYSIS

5.2.1 Dimensional Analysis of a Pendulum Swinging

Context: Predicting the period of a large pendulum.
Concepts: Dimensional consistency to derive a universal relation.
Defining Question: How does a pendulum's period scale with size? With mass?

To illustrate the principles of dynamic similarity, we consider the motion of a pendulum. Can we extrapolate the behavior of a small pendulum to a large pendulum, such as a person swinging on the end of a trapeze? What determines the period of a pendulum? How does the period depend on mass? If a child and an adult are side by side on identical swings, will they have the same period? Must the child and the adult swing at the same amplitude to have the same period? Our everyday experience with swings and trapezes suggests how pendulums scale with size and mass. Can we describe the scaling mathematically?

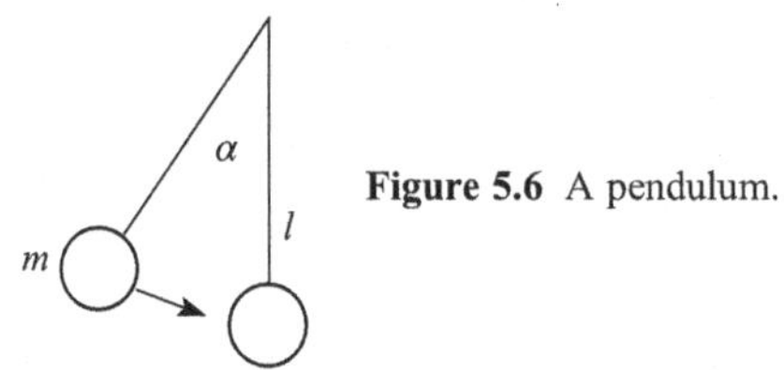

Figure 5.6 A pendulum.

We begin by listing in Table 5.5 the physical quantities that describe a pendulum. We also assign a mnemonic symbol to each quantity and list each quantity's dimensions.

Why does amplitude have no dimensions? Recall that amplitude is an angle measured in radians. The amplitude of an angle is the ratio of the length of a radius divided by the arc swept out by the radius. The dimensions of amplitude are thus L/L, which is dimensionless.

Table 5.5 The parameters of a pendulum.

Physical quantity	Symbol	Dimensions
period of oscillation	t_p	T
length of pendulum	ℓ	L
mass of pendulum	m	M
gravitational acceleration	g	L/T^2
amplitude	α	(none)

Have we neglected anything? How about the viscosity of the medium through which the pendulum moves? This would be important, for example, in designing an underwater pendulum. We neglect it here.

We wish to use a small model of a pendulum to predict the period of oscillation, t_p, from the physical parameters. We seek an equation of the form

$$t_p = \text{some function of } \ell, m, g, \text{ and } \alpha, \text{ or} \tag{5.10}$$

$$t_p = f(\ell, m, g, \alpha). \tag{5.11}$$

This function must be dimensionally consistent. Whatever combination of parameters we derive for the right side of equation 5.11, each term must have dimensions of time, T. That is, the function we seek will have some general form

$$t_p = (\text{term } 1) + (\text{term } 2) + (\text{term } 3) + \cdots, \tag{5.12}$$

and each term must have the dimensions of time. Consequently no term may contain the parameter m. Why? Because there is no complementary term to cancel its dimension of mass. Thus, the period of oscillation must be independent of mass. Dimensional analysis simplifies the function we seek to

$$t_p = f(\ell, g, \alpha). \tag{5.13}$$

Of course, it is possible we have neglected a parameter whose base dimensions include M. We discount that possibility and continue.

What about the length, ℓ? Does the pendulum's behavior depend on length? Yes, we may include ℓ because the constant g also contains the dimension of length:

$$[\ell] = \mathrm{L} \quad [g] = \mathrm{L}/\mathrm{T}^2 . \tag{5.14}$$

So, we must divide ℓ by g:

$$\left[\frac{\ell}{g}\right] = \frac{\mathrm{L}}{\mathrm{L}/\mathrm{T}^2} = \mathrm{T}^2. \tag{5.15}$$

Thus, *if* length is to appear in our function, it must appear in ratio with g. The function we seek thus has two variables:

$$t_p = f\left(\frac{\ell}{g}, \alpha\right). \tag{5.16}$$

Furthermore, the function is determined in part by dimensional consistency. Each term on the right side must have dimensions of time:

$$[t_p] = \mathrm{T}, \quad \left[\frac{\ell}{g}\right] = \mathrm{T}^2, \quad \left[\left(\frac{\ell}{g}\right)^{1/2}\right] = \mathrm{T}. \tag{5.17}$$

Therefore,

$$t_p = \left(\frac{\ell}{g}\right)^{1/2} f(\alpha). \tag{5.18}$$

Dimensional analysis has revealed an important relation: the pendulum period scales as the square root of the pendulum length. So if a small pendulum has a period of 1 s, a pendulum 100 times larger would have a period of 10 s. The relative scaling of parameters is key to designing models.

Because α has no dimensions, dimensional consistency cannot provide any information on $f(\alpha)$. Possibilities include

$$\begin{aligned} f(\alpha) &= \alpha^3, \quad \text{or} \\ &= \sin\alpha, \quad \text{or} \\ &= e^{\alpha}, \quad \text{or} \\ &= \text{a constant}. \end{aligned} \tag{5.19}$$

To proceed further, one must measure the behavior of a pendulum for different values of α and ℓ. Note that our experimental agenda has been shortened by dimensional analysis; we need not waste time measuring the effect of the pendulum mass. Furthermore, our analysis of the data is simplified. Dimensional analysis has assured us we may plot t_p vs. $(\ell/g)^{1/2}$ for constant amplitude, and the data *for all pendulums* will lie on a single curve.

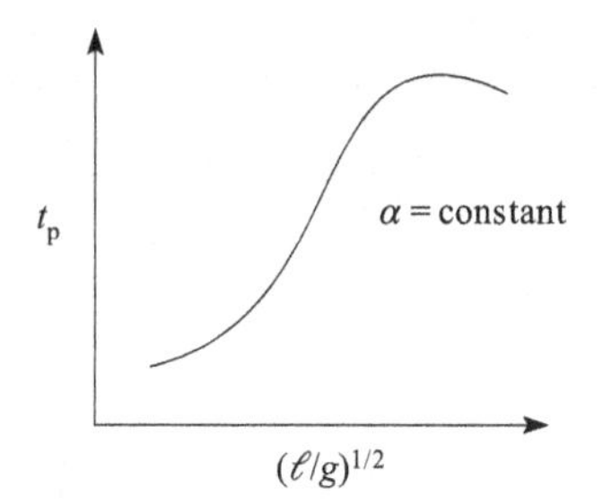

Figure 5.7a A possible plot of pendulum period as a function of $(\ell/g)^{1/2}$.

Plotting data measured with various amplitudes could generate a family of curves.

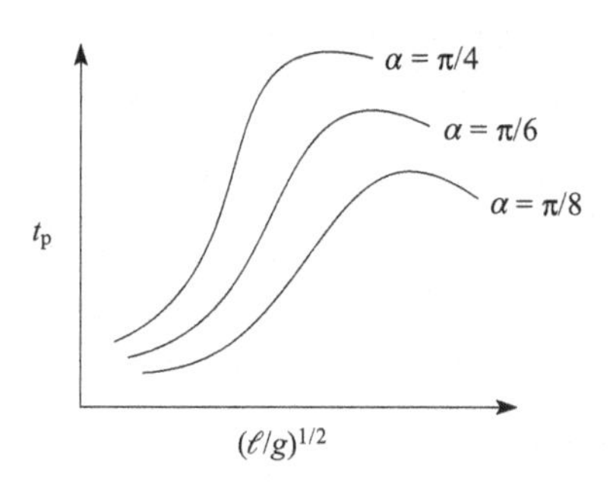

Figure 5.7b A possible plot of pendulum period as a function of $(\ell/g)^{1/2}$ for various amplitudes.

Upon measuring the pendulum behavior as a function of α and ℓ, one finds experimentally that the period of oscillation is independent of the amplitude, for small amplitudes (α less than $\pi/2$ radians, or 90°). That is, the curves for different amplitudes are all the same. Furthermore, the relationship between t_p and $(\ell/g)^{1/2}$ is linear, with a slope of 2π, as shown in Figure 5.7c.

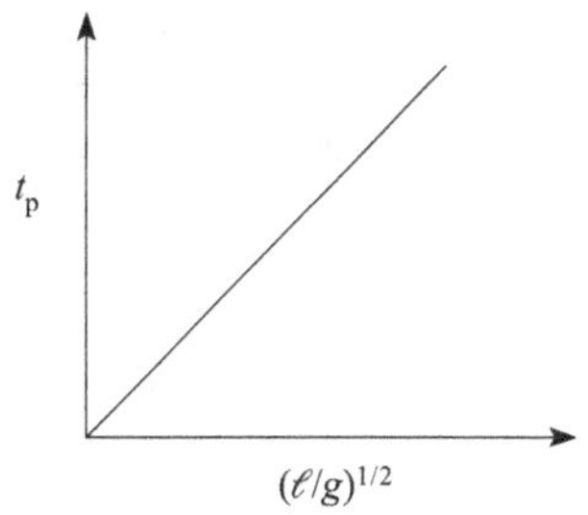

Figure 5.7c The actual plot of pendulum period as a function of $(\ell/g)^{1/2}$.

Any pendulum is thus described by a simple function:

$$t_p = 2\pi\left(\frac{\ell}{g}\right)^{1/2}. \tag{5.20}$$

Let's summarize what we found for the pendulum. Dimensional analysis reveals that the pendulum period is independent of mass and is proportional to the square root of the length. But dimensional analysis is limited; eventually we must conduct experiments. However, our experimental agenda is shortened by dimensional analysis. In addition, our analysis of the data will be simplified because we know how parameters are grouped.

It is worth noting that our analysis of the pendulum was aided by some key experiments that preceded ours. Specifically, we benefited from experiments that determined physical constants. In 1604, Galileo Galilei sought a formula for the time a mass m falls a distance h:

$$t_{fall} = f(m, h, \ldots). \tag{5.21}$$

Galileo discovered that t_{fall} was independent of mass and proposed a constant to relate t_{fall} to h. The constant is g. Other key experiments have provided the universal gravitational constant, the Planck constant, the Boltzmann constant, and the speed of light.

5.2.2 Dimensional Analysis of a Person Walking and Running

Context: The relation of walking velocity to leg length, stride, and mass.

Concepts: Π groups, a formal method for dimensional analysis, and the Froude number.

Defining Question: What is a core variable?

Let's apply dimensional analysis to another phenomenon – walking and running. Because all humans are constructed similarly, we should be able to predict a person's walking velocity from physical parameters. Clearly this system would be difficult to analyze at a fundamental level.

We begin by listing the possible parameters of walking. The first parameter is what we seek – velocity. What else might play a role? Tall people seem to walk fast, at least compared to children. So we might expect height to play a role. However, not everyone is proportioned the same. More precisely, the length of a person's leg is the pertinent parameter. We might also expect a person's mass to affect walking; an unladed person walks differently than a heavily laden person. Walking involves lifting one's mass during the step. Thus one would expect gravity is involved. That is, a person walking on the Moon has a distinctly different gait than on Earth, even in the same space suit. Finally, similar people have different styles of walking; some take small steps and some take long steps. And although stride length tends to scale with leg length, it is truly an independent parameter. People with identical leg lengths have different stride lengths. If the stride length is redundant with the leg length, this will be revealed in the experimental data. For example, a dimensionless group such as (leg length)/(stride length) will be found to be superfluous.

Table 5.6 The parameters of walking.

Parameter	Symbol	Dimensions
velocity	v	L/T
leg length	ℓ	L
mass	m	M
gravity	g	L/T^2
stride length	s	L

We list these five parameters of walking in Table 5.6, assign each a logical symbol, and list the dimensions of each.

We now apply dimensional analysis to guide us in measuring and analyzing data on walking. We could proceed in the haphazard fashion that yielded a functional relationship for the pendulum. Like the pendulum, dimensional analysis of walking is simple enough to guess the answer. But let's use this opportunity to introduce a more rigorous method, a method we will find useful for more complex systems.

Recall we started the pendulum analysis by looking for a formula for the period, t_p, in terms of four physical parameters:

$$t_\mathrm{p} = f(\ell, m, g, \alpha). \tag{5.11}$$

We applied dimensional analysis to simplify equation 5.11 to the form in equation 5.18:

$$t_\mathrm{p} = \left(\frac{\ell}{g}\right)^{1/2} f(\alpha). \tag{5.18}$$

From experimental data we determined the form of the function:

$$t_\mathrm{p} = 2\pi\left(\frac{\ell}{g}\right)^{1/2}. \tag{5.20}$$

We can rearrange equation 5.20 into a *dimensionless* form to obtain

$$2\pi = \frac{g^{1/2} t_\mathrm{p}}{\ell^{1/2}}. \tag{5.22}$$

We obtained equation 5.18 by finding a combination of ℓ, m, g, and α with dimensions of time. Alternatively, we could have obtained a dimensionless equation by finding a combination of t_p, ℓ, m, g, and α that was dimensionless. Although adding the parameter t_p may seem to complicate the analysis, this approach lends itself well to a general algorithm. We will use this dimensionless approach to analyze walking.

A dimensionless term is called a Π group (of parameters). In the vernacular of dimensional analysis, the parenthetical phrase is omitted. We seek the Π group(s) that describe walking. A Π group for walking will have the general form

$$\Pi = v^a \ell^b m^c g^d s^e, \tag{5.23}$$

where the exponents a, b, c, d, and e are to be determined. We express Π in terms of its dimensions, as listed in Table 5.6:

$$[\Pi] = \left[v^a \ell^b m^c g^d s^e\right] = \left(\frac{\mathrm{L}}{\mathrm{T}}\right)^a \mathrm{L}^b \mathrm{M}^c \left(\frac{\mathrm{L}}{\mathrm{T}^2}\right)^d \mathrm{L}^e \tag{5.24}$$

$$= \mathrm{L}^{a+b+d+e}\mathrm{T}^{-a-2d}\mathrm{M}^c. \tag{5.25}$$

Because Π must be dimensionless, the exponent on each term in equation 5.25 must be zero. This requirement yields an equation for each dimension:

$$\text{Length, L}: \quad a + b + d + e = 0 \tag{5.26}$$

$$\text{Time, T}: \quad -a - 2d = 0 \tag{5.27}$$

$$\text{Mass, M}: \quad c = 0. \tag{5.28}$$

Because equation 5.28 demands that c is zero, and c is the exponent of m in equation 5.23, there is no contribution from mass. The velocity of walking – like the period of a pendulum – is independent of mass. This is not too surprising when one considers the similarity of motion in the two systems. What remains are two equations and four unknowns. Because the number of unknowns exceeds the number of independent equations, there are an infinite number of combinations of a, b, d, and e that will satisfy these two equations. We need to specify something. We introduce the Buckingham Pi theorem, which states that

$$\text{number of dimensionless groups} = \text{number of parameters} - \text{number of dimensions,} \tag{5.29}$$

or, in other words,

$$\text{number of dimensionless groups} = \text{number of unknowns} - \text{number of equations.} \tag{5.30}$$

Applying equation 5.30 to our analysis of walking yields

$$N_{\Pi} = 4 - 2 = 2. \tag{5.31}$$

Fine. So how does one determine these two dimensionless groups? We need another rule:

For each Π group one must choose a "core variable."

A core variable is a parameter that forms the core of a dimensionless group, and appears in no other dimensionless groups. Because our goal is to find a function in terms of dimensionless groups, such as

$$\Pi_1 = f(\Pi_2, \ \Pi_3, \ \ldots, \ \Pi_n), \tag{5.32}$$

one core variable should be the parameter one wishes to predict. For walking, we wish to predict the velocity. The other core variable should be an independent parameter – a parameter one would vary in the experiments. For walking, stride length is a suitable choice for the second core variable.

We now derive the two Π groups, which we will designate Π_1 and Π_2. Π_1 will have velocity as its core variable and Π_2 will have stride length as its core variable.

Π_1: This dimensionless group contains v and not s. Thus the exponent on s must be zero; $e = 0$. We are free to choose any non-zero number for the exponent on v. It is usually convenient to choose 1; we set $a = 1$. Substituting $a = 1$ and $e = 0$ into equations 5.26 and 5.27, we obtain

$$1 + b + d + 0 = 0 \tag{5.33}$$

$$-1 - 2d = 0. \tag{5.34}$$

As stated in Chapter 3, methods for solving systems of equations are beyond the intended scope of this text. But this system is trivial. From equation 5.34 we see that $2d = -1$, or $d = -1/2$. Substituting into equation 5.33, we have $1 + b - 1/2 = 0$, or $b = -1/2$. Thus

$$\Pi_1 = v^1 \ell^{-1/2} m^0 g^{-1/2} s^0 = \frac{v}{(g\ell)^{1/2}}. \tag{5.35}$$

This dimensionless group appears in the analysis of many physical systems. In general, this group appears when an object's motion causes something to be lifted in a gravitational field. Walkers lift their bodies with every step. Ships create bow waves. This dimensionless group first appeared in William Froude's analysis of a bow wave's contribution to a ship's resistance in the 1870s. This Π group (actually its square) is named for Froude (rhymes with "food," not "cloud"). We adopt the conventional form for this first Π group:

$$\Pi_1 = \frac{v^2}{g\ell} \equiv \text{the Froude number} \equiv \text{Fr}. \tag{5.36}$$

Using the square of this dimensionless group will not affect the results of our analysis. However, adopting a conventional form will allow us to compare our analysis of walking to the analyses of similar systems.

In general, fractional exponents are eliminated by squaring, or cubing, etc., the Π group. Before electronic calculators, it was arduous to calculate square roots, cube roots, etc. But it was easy to calculate a square, a cube, etc.

Π_2: This dimensionless group contains s and not v. Thus the exponent on v must be zero; $a = 0$. Again we are free to choose any non-zero number for the exponent on the core variable, s. Again we choose 1; we set $e = 1$. When we substitute $a = 0$ and $e = 1$ into equations 5.26 and 5.27, we obtain

$$0 + b + d + 1 = 0 \tag{5.37}$$

$$-0 - 2d = 0. \tag{5.38}$$

Again the solution is trivial. From equation 5.38 we have $d = 0$. Substituting $d = 0$ into equation 5.37, we have $b = -1$. Thus,

$$\Pi_2 = v^0 \ell^{-1} m^0 g^0 s^1 = \frac{s}{\ell}. \tag{5.39}$$

We have determined that a dimensionless group containing stride length is a function of a dimensionless group containing velocity:

$$\Pi_2 = f(\Pi_1) \tag{5.40}$$

$$\frac{s}{\ell} = f\left(\frac{v^2}{g\ell}\right). \tag{5.41}$$

It is conventional to say that the *reduced* stride length is a function of the *reduced* velocity. That is, the square of the velocity has been scaled by $g\ell$. Similarly, stride length has been scaled by the leg length. With these scalings, data for all walkers should lie on a universal correlation, shown graphically in Figure 5.8. That is, a person walking with a stride length of 8 ft may be dynamically similar to a person walking with a stride length of 6 ft if their strides are scaled similarly. Similarly scaled strides would have the same reduced stride length s/ℓ and thus would have the same reduced velocity, $v^2/g\ell$. But the person with the longer stride would have a higher velocity.

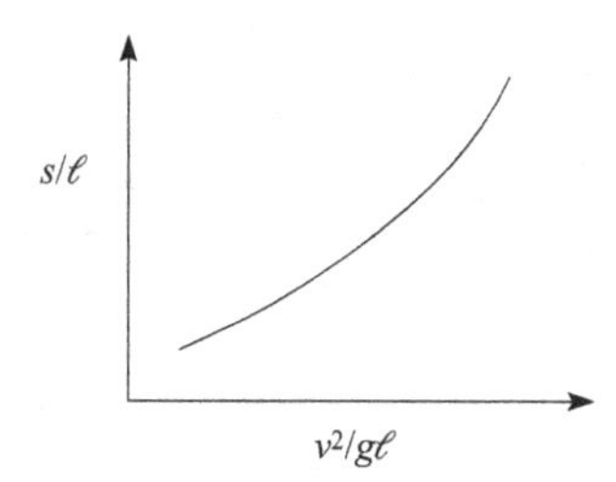

Figure 5.8 A possible universal correlation between reduced velocity and reduced stride length.

What is the correlation between reduced stride length and reduced velocity? We cannot determine that from dimensional analysis. To determine the correlation we must perform experiments.

Data obtained by first-year engineering students at Cornell University, shown in Figure 5.9, have an upward trend. Data for fast-walking students lie in the upper right portion of the plot, $s/\ell > 2$ and $v^2/g\ell > 0.8$. Slow-paced walkers correspond to the data in the lower left portion of the plot, $s/\ell < 1.5$ and $v^2/g\ell < 0.3$. The data in the middle were measured on students walking at their "natural" pace, including (surprisingly) people walking backwards at their natural pace. There is probably another domain at high s/ℓ corresponding to running. The curve might shift because the phenomenon of running is different from walking; at times both feet are off the ground.

The dimensional analysis of walking demonstrates an important concept in dynamic similarity. The magnitude of a dimensionless group indicates the "character" of the phenomenon. That is, if you are told a walker has a Froude number of 2.0, you could be reasonably confident the walker is "speed walking." Morticia Addams's style of walking (as portrayed by Carolyn Jones in the 1960s TV series *The Addams Family* and by Angelica Huston in the 1990s movies of the same title) is "slow walking." At the other extreme, the style of walking known as "trucking" (introduced in the late 1960s) would be typified by high values of s/ℓ, approaching the upper limit of 4 for walking.

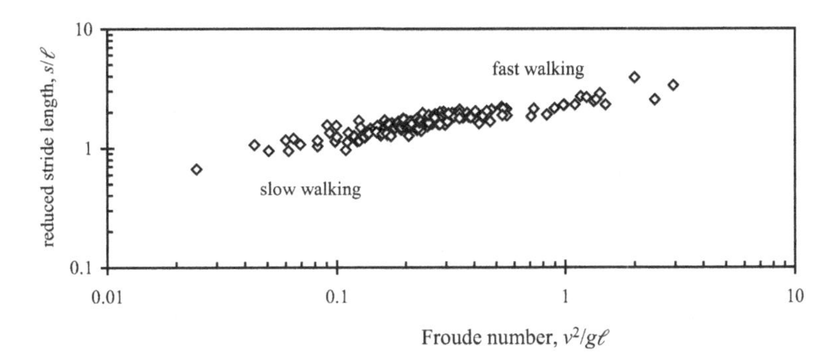

Figure 5.9 Experimental data for reduced stride vs. Froude number for first-year engineers at Cornell University (1993–1996).

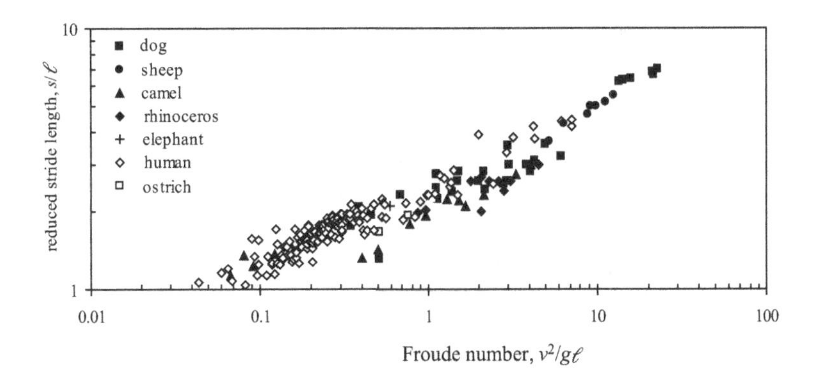

Figure 5.10 Experimental data for reduced stride as a function of Froude number for quadrupeds and bipeds. Adapted from Alexander (1989), figure 3.10.

The analysis of walking has been extended to include another biped – the ostrich – as well as quadrupeds such as dogs and elephants. As shown in Figure 5.10, the data for a diverse range of walkers lie on one general curve.

It has been noted (Alexander, 1989; 1992) that quadrupeds – from ferrets to rhinoceroses – change gait from trotting to galloping at a Froude number of about 2.6, although the absolute velocities of each are different. Galloping is more strenuous than trotting. Elephants don't gallop, but buffalo do. The dimensional analysis of bone strength suggests the types of gaits an animal can attain. A very large dinosaur, apatosaurus (34 metric tons), had a reduced bone strength similar to an elephant, suggesting that its fastest gait was a trot. Triceratops (6–9 metric tons), however, had a reduced bone strength more similar to an African buffalo, suggesting it was capable of galloping. If triceratops did gallop, the correlation in Figure 5.10 suggests that it could gallop at 9 m/s. (The best human sprinters can achieve 11 m/s.) The analysis of dinosaur running is an excellent example of the use of dynamic similarity to predict results for experimentally inaccessible systems. With dynamic similarity, for example, one may predict if triceratops could outrun a predator such as an allosaurus.

In this section we developed a general method for dimensional analysis of a system:

1. List the variables that describe the system, assign a variable to each, and list the dimensions of each.
2. Write a general equation for a dimensionless group of variables, substitute the dimensions of each variable, and collect terms of each dimension.
3. Use the exponents of each dimension to write an algebraic equation of the variables' exponents.
4. Apply the Buckingham Pi theorem to determine the number of dimensionless groups.
5. Choose a core variable for each dimensionless group. Solve the algebraic equations such that each dimensionless group contains exactly one core variable. For each dimensionless group, set the exponent of the core variable to any non-zero number and set the exponents of the other core variables to zero.

5.2.3 Dimensional Analysis of a Solid Sphere Moving Through a Fluid

Context: The terminal velocity of a solid sphere falling (or rising) through a fluid.

Concepts: The Reynolds number and transitions in dynamic systems.

Defining Question: How does terminal velocity scale with sphere diameter for laminar flow? For turbulent flow?

Let's apply dimensional analysis to another phenomenon – the terminal velocity of a sphere falling under the force of gravity.

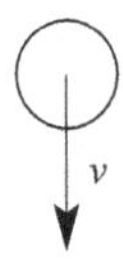

Figure 5.11 A solid sphere falling through a fluid.

This phenomenon is common in chemical engineering processes; for instance:

- Silt settling in water. How long does it take to clarify muddy water?
- Crystals of pharmaceuticals settling in a growth broth. Therapeutic proteins are often isolated by crystallization.
- Dust in air. Will ash from your company's smokestack fall on your company's property, in the nearby town, or in the ocean? How long will microscopic dust

persist in the upper atmosphere owing to a large meteor impact, a large volcanic eruption, or global thermonuclear war?
- What force is needed to tow a sphere through a fluid?

In these examples it is impractical to measure the actual process. The particles are too small (approximately 1 μm for some dust) and/or the terminal velocity is too slow (nuclear winter is predicted to last 1–10 years). We would much rather take measurements on more convenient systems, such as a plastic sphere of diameter 1 cm falling through water. We will apply dimensional analysis to reveal how to scale the results of our experiments to experimentally inaccessible real processes.

To aid in identifying the parameters of this system, let's examine the physics of the phenomenon. Why does a particle fall? Well, gravity has something to do with this. But why does a particle fall and not rise? A sphere of polyethylene will fall in air but rise in water. The density of the particle compared to the density of the fluid is important. Why does a falling particle reach a terminal velocity? Why doesn't the acceleration due to gravity increase the velocity without limit (with due respect to relativity)? The terminal velocity suggests there is an equal and opposing force: friction owing to the surrounding fluid.

Again we begin by constructing a table. Again the first choice is obvious – the terminal velocity of the sphere. The parameters that affect the terminal velocity can be divided into two groups: parameters that describe the sphere and parameters that describe the fluid. First we consider the sphere. A key parameter is the sphere's diameter. Another is the buoyancy of the sphere, given by the difference between the sphere density and the fluid density. Is there anything else about the sphere? How about its mass? Yes, the mass has an effect, but the mass is redundant with the buoyancy. How about the "smoothness" of the sphere's surface? We will ignore that effect for now and assume the sphere is perfectly smooth.

Next we consider the fluid. The frictional force exerted by the fluid is related to the viscosity of the fluid. The next fluid parameter is subtle. When the fluid passes the sphere, the inertia of the fluid changes. Imagine a stationary sphere in a moving fluid. The fluid must pass around the particle, which changes the fluid's inertia, as sketched in Figure 5.12. The fluid's inertia is proportional to the mass of the fluid and the change in velocity of the fluid. The velocity of the sphere already accounts for the change in velocity of the fluid. However, we need to account for the mass of the fluid. Again we use the fluid density.

Figure 5.12 Fluid moving around a solid sphere.

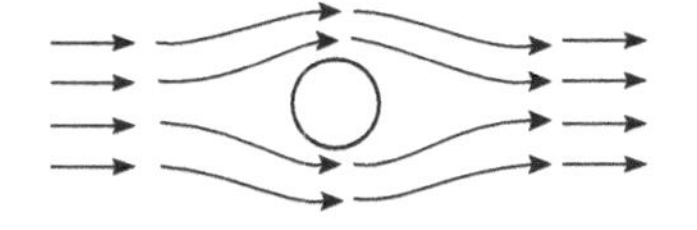

Finally, we add physical constants. In this case, we add the constant for gravitational acceleration. We list these six parameters of a sphere moving through a fluid in Table 5.7 on the next page, assign a logical symbol, and list the dimensions of each.

To obtain the dimensions of viscosity, one might search for the viscosity of water and use the units to convert to dimensions. But a typical viscosity has units of "poise." This doesn't help. Sometimes it is useful to use a fundamental law to determine the units of a quantity. For example, to remember the units of force one can use $F = ma$. For energy one can use $E = ½mv^2$. For viscosity, a law dear to the hearts of chemical engineers is Newton's law of viscosity:

$$-\text{viscosity} \times \frac{d\,(\text{velocity})}{d\,(\text{distance})} = \frac{\text{shear force}}{\text{area}} \tag{5.42}$$

Table 5.7 The parameters of a sphere moving through a fluid.

Parameter	Symbol	Dimensions
terminal velocity	v	L/T
sphere diameter	d	L
buoyancy	$\rho_{\text{sphere}} - \rho_{\text{fluid}}$	M/L^3
fluid viscosity	μ	M/LT
fluid density	ρ_{fluid}	M/L^3
gravitational acceleration	g	L/T^2

$$\left[-\text{viscosity} \times \frac{d\,(\text{velocity})}{d\,(\text{distance})}\right] = \left[\frac{\text{shear force}}{\text{area}}\right] \tag{5.43}$$

$$[\text{viscosity}]\frac{\text{L/T}}{\text{L}} = \frac{\text{ML/T}^2}{\text{L}^2} \tag{5.44}$$

$$[\text{viscosity}] = \frac{\text{M}}{\text{LT}}. \tag{5.45}$$

It may seem that Table 5.7 was constructed in an arbitrary manner. How did we know to include the density of the fluid? What would have happened if we used the density of the sphere instead of the buoyancy? The answer is that we would still be able to calculate dimensionless groups for this system. However, when we conducted experiments we would find that the dimensionless groups were not related by a single function – a plot of the results would yield considerable scatter instead of a single line. Why did we use specific parameters, such as diameter instead of radius? Either parameter would have been acceptable. We chose diameter so we could compare our results with those of previous workers.

We now determine the dimensionless groups. Each will have the general form

$$\Pi = v^a d^b \left(\rho_{\text{sphere}} - \rho_{\text{fluid}}\right)^c \mu^d \rho_{\text{fluid}}^e g^f \tag{5.46}$$

$$[\Pi] = \left(\frac{\text{L}}{\text{T}}\right)^a \text{L}^b \left(\frac{\text{M}}{\text{L}^3}\right)^c \left(\frac{\text{M}}{\text{LT}}\right)^d \left(\frac{\text{M}}{\text{L}^3}\right)^e \left(\frac{\text{L}}{\text{T}^2}\right)^f \tag{5.47}$$

$$[\Pi] = \text{M}^{c+d+e}\text{T}^{-a-d-2f}\text{L}^{a+b-3c-d-3e+f}. \tag{5.48}$$

Again, for Π to be dimensionless, each exponent in equation 5.48 must be zero. This requirement yields three equations:

$$\text{Mass, M}: \quad c + d + e = 0 \tag{5.49}$$

$$\text{Time, T}: \quad -a - d - 2f = 0 \tag{5.50}$$

$$\text{Length, L}: \quad a + b - 3c - d - 3e + f = 0. \tag{5.51}$$

We have six unknowns (a to f) and three equations. The Buckingham Pi theorem thus dictates we have 6 – 3 = 3 Π groups. To solve equations 5.49 to 5.51, we must fix three of the unknowns, which we do by selecting three core variables. What do the guidelines suggest for core variables?

- What do we want to know? The dependent variable is *velocity*.
- Which parameters will we vary in our experiments?

We will eventually go to the lab and drop spheres of different diameters and densities through fluids with different densities and viscosities. Candidates for the independent parameters are the sphere diameter, the buoyancy, the fluid viscosity, and the fluid density. However, to compare our analysis with previous work, we should consider some alternative choices. Consider the following rationalization. We wish to describe the terminal velocity of a sphere falling (or rising) through a fluid. But we can also analyze a more general system. That is, we are concerned only with the resistive force on a sphere moving through a fluid. It does not matter how the sphere is propelled through the fluid. For terminal velocity, buoyancy propels the sphere. However, the system applies also to a mutually buoyant sphere ($\rho_{sphere} = \rho_{fluid}$) being towed through a fluid, or a sphere fixed in place with a fluid flowing past it. We can describe systems with broader range if we do not regard velocity as a *dependent* variable. We may wish to vary the velocity, for example, by towing a sphere through the fluid. The choice of core variables is chiefly for convenience. The Π groups derived with one valid set of core variables can be related to the Π groups derived with other valid sets of core variables.

The traditional core variables for the terminal velocity of a sphere are $(\rho_{sphere} - \rho_{fluid})$, g, and μ. We now use these to derive the Π groups.

Π_1 contains $(\rho_{sphere} - \rho_{fluid})$, but not g and not μ. Thus, the exponents on the parameters g and μ in equation 5.46 must be zero; set $f = 0$ and set $d = 0$. The exponent of the core variable $(\rho_{sphere} - \rho_{fluid})$ may be any non-zero number; $c = 1$ is convenient. We substitute these exponents into equations 5.49, 5.50, and 5.51:

$$\text{Mass, M}: \quad 1 + 0 + e = 0 \quad \Rightarrow \quad e = -1 \tag{5.52}$$

$$\text{Time, T}: \quad -a - 0 - 2(0) = 0 \quad \Rightarrow \quad a = 0 \tag{5.53}$$

$$\text{Length, L}: \quad 0 + b - 3(1) - 0 - 3(-1) + 0 = 0 \quad \Rightarrow \quad b = 0\,. \tag{5.54}$$

In summary, for Π_1 we have

$$\left.\begin{array}{l} a = 0 \\ b = 0 \\ c = 1 \\ d = 0 \\ e = -1 \\ f = 0 \end{array}\right\} \quad \Pi_1 = \left(\rho_{sphere} - \rho_{fluid}\right)^1 \left(\rho_{fluid}\right)^{-1} = \frac{\rho_{sphere} - \rho_{fluid}}{\rho_{fluid}}\,. \tag{5.55}$$

Thus the buoyancy is scaled by the density of the fluid. This dimensionless group has no name. We will call it reduced buoyancy.

Π_2 contains g, but not $(\rho_{sphere} - \rho_{fluid})$ and not μ. Thus, we set the exponents of $(\rho_{sphere} - \rho_{fluid})$ and μ to zero; from equation 5.46, $c = 0$ and $d = 0$. The exponent on the core variable g can be any non-zero number. Again, $f = 1$ is convenient. We substitute these exponents into equations 5.49, 5.50, and 5.51:

$$\text{Mass, M}: \quad 0 + 0 + e = 0 \quad \Rightarrow \quad e = 0 \tag{5.56}$$

$$\text{Time, T}: \quad -a - 0 - 2(1) = 0 \quad \Rightarrow \quad a = -2 \tag{5.57}$$

$$\text{Length, L}: \quad -2 + b - 3(0) - 0 - 3(0) + 1 = 0 \quad \Rightarrow \quad b = 1\,. \tag{5.58}$$

In summary, for Π_2 we have

$$\left.\begin{aligned} a &= -2 \\ b &= 1 \\ c &= 0 \\ d &= 0 \\ e &= 0 \\ f &= 1 \end{aligned}\right\} \Pi_2 = v^{-2} d^1 g^1 = \frac{dg}{v^2}. \tag{5.59}$$

The reciprocal of this dimensionless group is v^2/dg, the Froude number. Similar to the analysis of walking where we arbitrarily squared the Π group, we will arbitrarily invert Π_2 to obtain the Froude number as our second dimensionless group. We would have obtained the Froude number if we had set the exponent of g to -1. Inverting Π_2 will not affect our analysis, but it will allow comparison with analyses of similar systems.

Because this is our second encounter with the Froude number, it warrants further inspection. Like the other two dozen or so common dimensionless groups in chemical engineering, the Froude number can be expressed as a ratio of forces. Specifically, the Froude number can be expressed as the ratio of inertial energy to potential energy in a gravitational field:

$$\Pi_2 = \mathrm{Fr} = \frac{v^2}{dg} = \frac{(\text{mass})v^2}{(\text{mass})\,gd} \propto \frac{\text{inertial (kinetic) force}}{\text{potential energy in a gravitational field}}. \tag{5.60}$$

Π_3 contains μ, but not $(\rho_{\text{sphere}} - \rho_{\text{fluid}})$ and not g. Thus, from equation 5.46, set $c = 0$, set $f = 0$, and set d to any non-zero number. Again, $d = 1$ is convenient. Substituting into equations 5.49, 5.50, and 5.51 yields

$$\text{Mass, M}: \quad 0 + 1 + e = 0 \quad \Rightarrow \quad e = -1 \tag{5.61}$$

$$\text{Time, T}: \quad -a - 1 - 2(0) = 0 \quad \Rightarrow \quad a = -1 \tag{5.62}$$

$$\text{Length, L}: \quad -1 + b - 3(0) - 1 - 3(-1) + 0 = 0 \quad \Rightarrow \quad b = -1. \tag{5.63}$$

In summary, for Π_3 we have

$$\left.\begin{aligned} a &= -1 \\ b &= -1 \\ c &= 0 \\ d &= 1 \\ e &= -1 \\ f &= 0 \end{aligned}\right\} \Pi_3 = v^{-1} d^{-1} \mu^1 \rho_{\text{fluid}}^{-1} = \frac{\mu}{dv\rho_{\text{fluid}}}. \tag{5.64}$$

The reciprocal of Π_3 is also a famous dimensionless group – perhaps *the* most famous dimensionless group – the Reynolds number, Re. As with the Froude number, the Reynolds number can be expressed as a ratio of forces:

$$\mathrm{Re} = \frac{dv\rho_{\text{fluid}}}{\mu} \propto \frac{\text{fluid inertia}}{\text{viscous effects}}. \tag{5.65}$$

One encounters the Reynolds number throughout chemical engineering.

Let's summarize to this point our analysis of a sphere moving through a fluid. We combined six parameters to form three dimensionless groups. What is the functional relationship between these groups? That is, what is the function f such that $\Pi_2 = f(\Pi_1, \Pi_3)$, or

$$\frac{v^2}{dg} = f\left(\frac{\rho_{\text{sphere}} - \rho_{\text{fluid}}}{\rho_{\text{fluid}}}, \frac{dv\rho_{\text{fluid}}}{\mu}\right)? \tag{5.66}$$

Dimensional analysis can take us no further. To determine the function f we must appeal to theory and/or experiment. The theory lies outside the realm of this text. Moreover, there is no theory that encompasses the functional form of f for all values of Fr, reduced buoyancy, and Re. To proceed further we must measure terminal velocities of spheres falling through fluids.

Before proceeding, let's consider the effect of choosing different core variables. What would be the result of choosing the obvious candidates for core variables: v, $(\rho_{\text{sphere}} - \rho_{\text{fluid}})$, and μ? If we use an exponent of 1 for each core variable, one obtains for Π_1

$$\Pi_1 = \frac{v}{\sqrt{dg}}. \tag{5.67}$$

This is the square root of the Froude number. For Π_2 one obtains

$$\Pi_2 = \frac{\rho_{\text{sphere}} - \rho_{\text{fluid}}}{\rho_{\text{fluid}}}. \tag{5.68}$$

This is the reduced buoyancy, the same result as before. Finally, for Π_3 one obtains

$$\Pi_3 = \frac{\mu}{d^{3/2}\rho_{\text{fluid}}g^{1/2}}. \tag{5.69}$$

The inverse square of this group is the Galileo number:

$$\Pi_3^{-2} = \frac{d^3\rho_{\text{fluid}}^2 g}{\mu^2} = \text{Ga}. \tag{5.70}$$

The Galileo number is the ratio of gravitational effects to viscous effects. The Galileo number is also the square of the Reynolds number divided by the Froude number:

$$\text{Ga} = \frac{\text{Re}^2}{\text{Fr}}. \tag{5.71}$$

To determine experimentally the function in equation 5.66, first-year engineers at Cornell University observed spheres of various materials (steel, brass, aluminum, glass, lucite, nylon, teflon, and polypropylene) falling through various fluids (water, vegetable oil, and glycerol). The students measured each sphere's mass and diameter, the distance each sphere fell, and the time elapsed. The students then used spreadsheets to calculate the parameters of Table 5.7 in SI units, and finally they combined these parameters to calculate the three dimensionless groups: Fr, Re, and the reduced buoyancy.

To determine the relation in equation 5.66, it is illustrative to plot the data. Plotting the walking data was straightforward because there were only two dimensionless groups. Because spheres moving through fluids are described by three dimensionless groups, plotting is more difficult. We have at least two options. We could prepare a three-dimensional plot with axes Re, Fr, and reduced buoyancy. The correlation would then be a surface in this three-dimensional space. We tried this and it gets ugly – trust us. Another option is to plot, for example, reduced buoyancy vs. Re for values of constant Fr. We would then have a family of curves for different values of Fr. This is shown in Figure 5.13 on the next page.

The trends in the data sets in Figure 5.13 suggest the correlation can be simplified. Note that when the Froude number increases by a factor of 10, the reduced buoyancy increases by a factor of 10. This suggests that one can divide the reduced buoyancy by the Froude number and all the data will collapse onto one universal curve. The data in Figure 5.13 are replotted as reduced buoyancy/Fr vs. Re in Figure 5.14. The data seem to lie on a single correlation.

Figures 5.13 and 5.14 contain only a portion of the experimental data. Let's see if the correlation in Figure 5.14 holds for data with arbitrary values of Fr. Figure 5.15 shows all

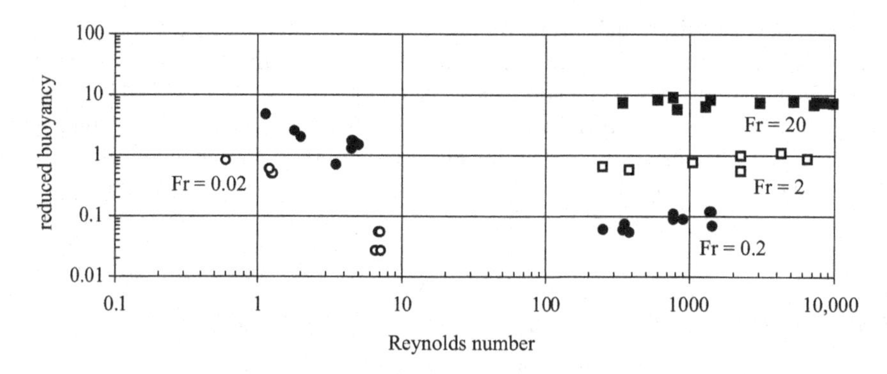

Figure 5.13 Data for a sphere moving through a fluid, plotted for Froude numbers of 0.02, 0.2, 2, and 20 (±10%).

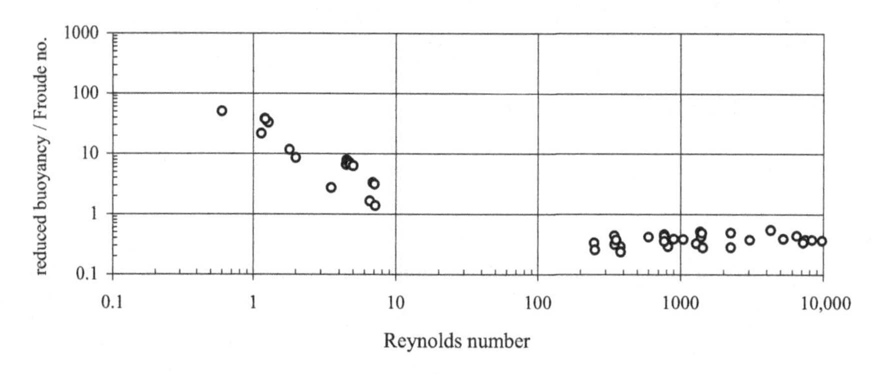

Figure 5.14 Data for a sphere moving through a fluid.

the data plotted in the manner of Figure 5.14, but with the reduced buoyancy/Fr multiplied by 4/3.The factor of 4/3 appears when one derives the relation with fluid mechanics; 4/3 comes from the volume of a sphere, $(4/3)\pi r^3$. The reduced buoyancy/Fr multiplied by 4/3 is known as the friction factor, as given in equation 5.72:

$$\text{friction factor} = \frac{4}{3}\frac{\rho_{\text{sphere}} - \rho_{\text{fluid}}}{\rho_{\text{fluid}}}\frac{1}{\text{Fr}}. \tag{5.72}$$

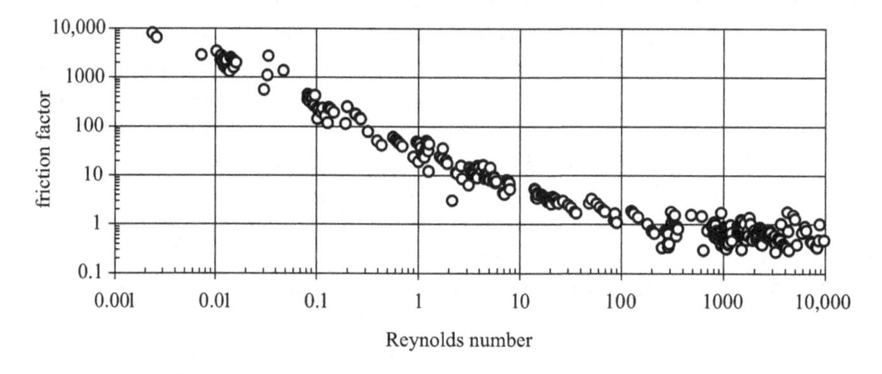

Figure 5.15 Data for a sphere moving through a fluid.

It should come as no surprise that the correlation in Figure 5.15 has been studied before. The data measured by the first-year engineers agree well with the correlation determined previously, shown in Figure 5.16.

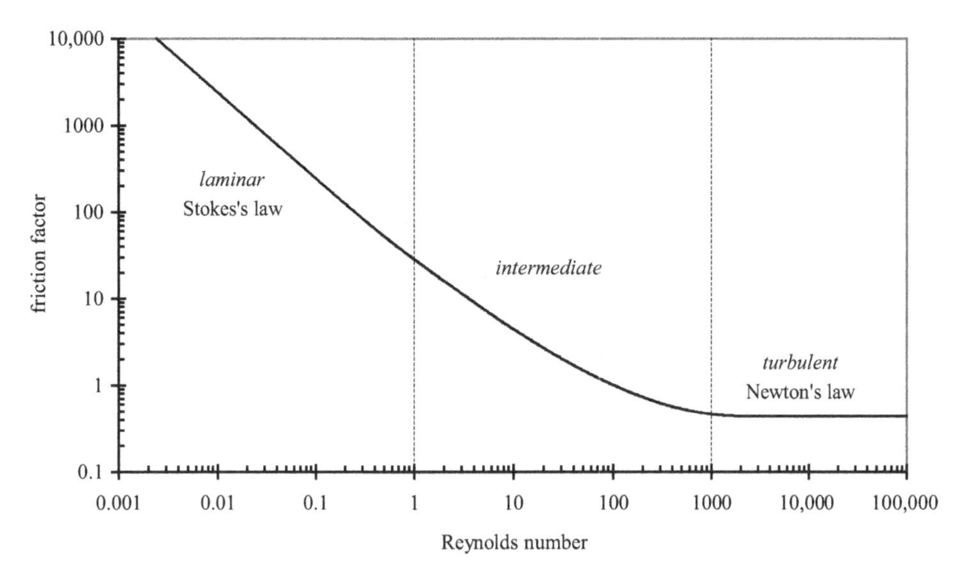

Figure 5.16 Correlation for a sphere moving through a fluid. Adapted from Bird, Stewart, and Lightfoot (1960), figure 6.3–1.

The correlation between friction factor and Reynolds number in Figure 5.16 has three distinct regions. There is a linear region for Re < 1, a curved region for $1 <$ Re < 1000, and a flat region for Re > 1000.

For Re < 1, the plot is linear with a negative slope. Recall the definition of the Reynolds number:

$$\mathrm{Re} = \frac{d v \rho_{\mathrm{fluid}}}{\mu}. \tag{5.65}$$

A low Reynolds number occurs for low velocity, low fluid density, a small sphere, or high fluid viscosity. Let's derive the equation that corresponds to this regime. Because the line is straight we know the functional form is

$$y = mx + b. \tag{5.73}$$

We choose two convenient points on the line to calculate the slope: (Re, F) = (0.5, 50) and (0.005, 5000). Because this is a log–log plot, one must be careful calculating the slope. Specifically, we calculate the distance between the logarithms of these coordinates, not the distances between the coordinates:

$$\begin{aligned} m = \text{slope} &= \frac{\text{rise}}{\text{run}} = \frac{\log_{10}5000 - \log_{10}50}{\log_{10}0.005 - \log_{10}0.5} \\ &= \frac{\left(\log_{10}5 + \log_{10}10^{3}\right) - \left(\log_{10}5 + \log_{10}10^{1}\right)}{\left(\log_{10}5 + \log_{10}10^{-3}\right) - \left(\log_{10}5 + \log_{10}10^{-1}\right)} = \frac{3-1}{-3-(-1)} = -1. \end{aligned} \tag{5.74}$$

Substitute the slope into equation 5.73:

$$y = (-1)x + b \tag{5.75}$$

$$\log_{10}\mathrm{F} = (-1)\log_{10}\mathrm{Re} \; + b \tag{5.76}$$

$$\log_{10}\mathrm{F} = \log_{10}\mathrm{Re}^{-1} + b. \tag{5.77}$$

Raise both sides of the equation to the power of 10:

$$10^{\log_{10}\mathrm{F}} = 10^{\left[\log_{10}\mathrm{Re}^{-1} + b\right]} \tag{5.78}$$

$$10^{\log_{10}\mathrm{F}} = 10^{\log_{10}\mathrm{Re}^{-1}}10^{b} \tag{5.79}$$

$$\mathrm{F} = \mathrm{Re}^{-1}10^{b}. \tag{5.80}$$

The constant b may be calculated by inserting a point on the correlation into equation 5.80, such as (Re, F) = (0.5, 50). Doing so, one calculates $10^b = 25$. Thus,

$$\text{for Re} < 1, \quad \frac{4}{3}\frac{\rho_{\text{sphere}} - \rho_{\text{fluid}}}{\rho_{\text{fluid}}}\frac{1}{\mathrm{Fr}} = \frac{25}{\mathrm{Re}}. \tag{5.81}$$

Equation 5.81 is a form of Stokes's law. It is not a universal law, as the conservation of mass is, because it is restricted to flows with Re $<$ 1. Stokes's law is a constrained law.

For Re $>$ 1000, the plot is essentially flat. Thus the friction factor is independent of the Reynolds number:

$$\text{For Re} > 1000, \quad F \approx 0.44 = \frac{4}{3}\frac{\rho_{\text{sphere}} - \rho_{\text{fluid}}}{\rho_{\text{fluid}}}\frac{v^2}{dg} \tag{5.82}$$

$$v \approx 0.57\left(dg\frac{\rho_{\text{fluid}}}{\rho_{\text{sphere}} - \rho_{\text{fluid}}}\right)^{1/2}. \tag{5.83}$$

Equation 5.83 is known as Newton's law of resistance, not to be confused with Newton's laws of motion, or Newton's law of viscosity, or Newton's law of cooling. For turbulent fluid flow, terminal velocity scales as the square root of the sphere diameter.

In the intermediate region, 1 $<$ Re $<$ 1000, the system is in transition. We won't attempt to extract an equation from our data.

The fluid streamlines around the sphere are distinctly different for each region. Figure 5.17 shows fluid streamlines for flow around a cylinder, which is similar to flow around a sphere. (It is difficult to photograph streamlines around a sphere.) At low Reynolds number, the fluid passes smoothly past the object; the streamlines gently warp around the cylinder. This type of flow is called laminar, and Stokes's law applies. At higher Reynolds number, eddies develop behind the object. At even higher Reynolds number, the eddies break up and no longer form steady streamlines; the pattern in Figure 5.17f changes with time. For Re $>$ 1000, the flow is turbulent and the friction factor becomes independent of Re.

Let's use our universal correlation to extrapolate outside the range we measured, for example to microscopic dust settling in air. This is an important consideration in a "clean room" for the manufacture of integrated circuits. If a person who has recently inhaled tobacco smoke exhales into the room, how soon do the smoke particles settle from the air? Or, imagine a large dust cloud injected into the upper atmosphere, for example, after a volcanic eruption, a large asteroid impact, or global thermonuclear war. If the dust cloud is above the cleansing action of rain, how long before the dust particles settle to Earth?

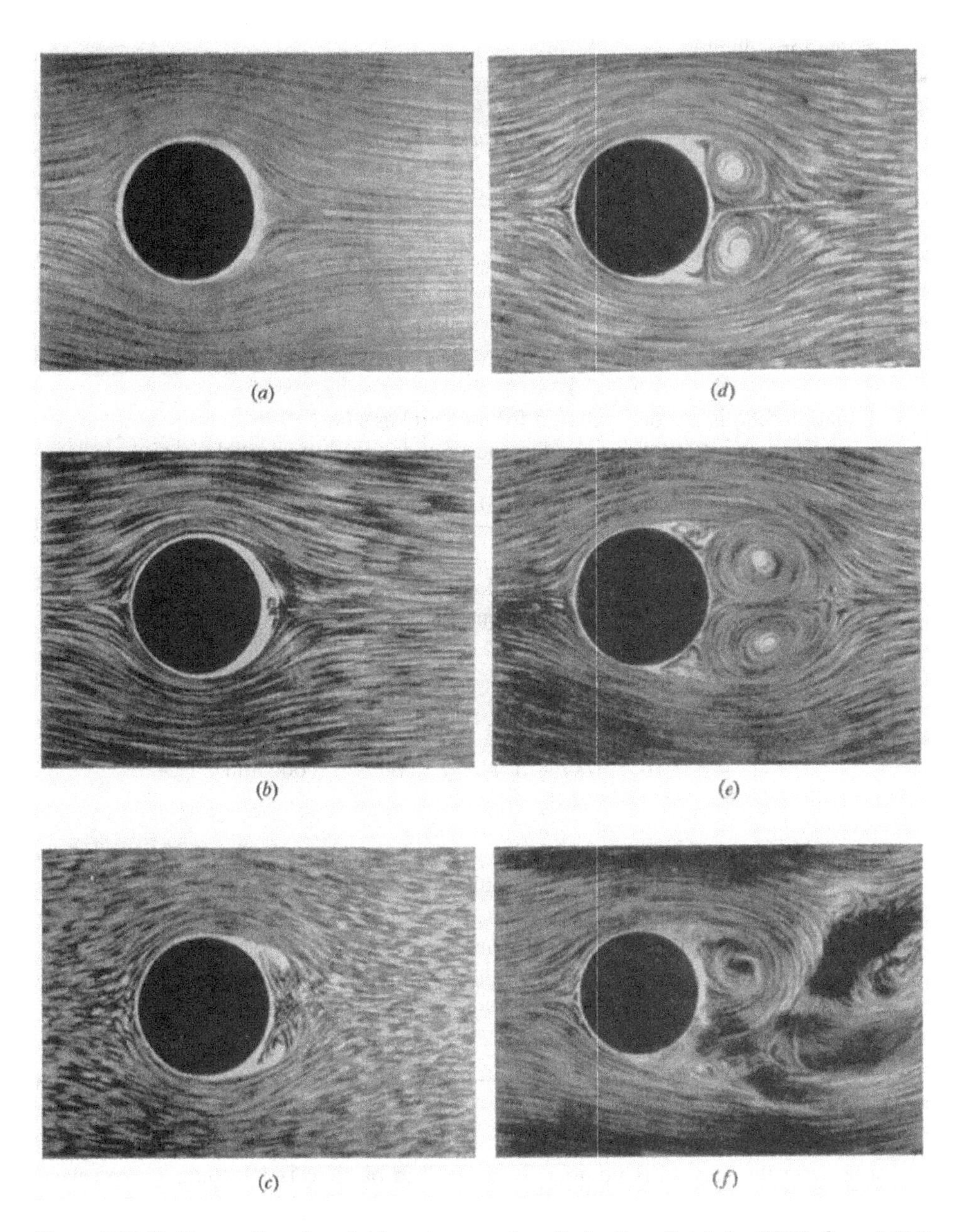

Figure 5.17 Fluid streamlines for a fluid moving around a cylinder. From Batchelor (1967), figure 5.11.3, on plate 10. Copyright Cambridge University Press. Reproduced by permission.

Assume microscopic dust settling in air is in the low Reynolds number regime. We start with Stokes's law, equation 5.82, to derive an equation for the terminal velocity, and substitute the definitions for the Froude number and Reynolds number:

$$\frac{4}{3}\left(\frac{\rho_{\text{sphere}}-\rho_{\text{fluid}}}{\rho_{\text{fluid}}}\right)\left(\frac{dg}{v^2}\right)=\frac{25}{\left(\frac{dv\rho_{\text{fluid}}}{\mu}\right)}. \tag{5.84}$$

Solve for velocity:

$$v = \frac{4}{75}\frac{gd^2}{\mu}\left(\rho_{\text{sphere}} - \rho_{\text{fluid}}\right). \tag{5.85}$$

We will approximate the prefactor of 4/75 by 1/18, which is the result obtained by mathematical modeling of a sphere at low Reynolds number. Equation 5.85 can be simplified further for solids falling through air. Because $\rho_{\text{sphere}} \approx 2000\ \text{kg/m}^3$ and $\rho_{\text{fluid}} \approx 1\ \text{kg/m}^3$, $\rho_{\text{sphere}} - \rho_{\text{fluid}} \approx \rho_{\text{sphere}}$, which yields

$$v = \frac{1}{18}\frac{gd^2\rho_{\text{sphere}}}{\mu}. \tag{5.86}$$

Equation 5.86 is a more common form of Stokes's law.

Substitute into Stokes's law the constants for a 10 μm particle in air:

$$v = \frac{1}{18}\frac{(9.8\ \text{m/s}^2)(10^{-5}\text{m})^2(2000\ \text{kg/m}^3)}{1.8\times10^{-5}\ \text{Pa·s}} = 6\times10^{-3}\text{m/s} = 0.6\ \text{cm/s}. \tag{5.87}$$

The dust settles quite slowly. How long does it take a 10 μm particle to settle from a height of 50 km? Because distance equals velocity × time,

$$\text{time} = \text{distance}\times\frac{1}{\text{velocity}} \tag{5.88}$$

$$= 50\ \text{km}\left(\frac{1}{6\times10^{-3}\text{m/s}}\right)\left(\frac{1000\ \text{m}}{1\ \text{km}}\right)\left(\frac{1\ \text{min}}{60\ \text{s}}\right)\left(\frac{1\ \text{hr}}{60\ \text{min}}\right)\left(\frac{1\ \text{day}}{24\ \text{hr}}\right) = 96\ \text{days}. \tag{5.89}$$

Now, without performing a calculation similar to equations 5.87 to 5.89, estimate the time for a 100 μm (=0.1 mm) particle to settle from a height of 50 km. Because equation 5.86 reveals that the velocity scales as d^2, the velocity of a 100 μm particle is 100 times greater than that of a 10 μm particle. A 100 μm particle will settle in about 1 day.

Before we conclude, we must check our initial assumption that the terminal velocity of a 10 μm particle is in the laminar regime. Calculate the Reynolds number:

$$\text{Re} = \frac{dv\rho_{\text{fluid}}}{\mu} = \frac{(1\times10^{-5}\ \text{m})(6\times10^{-3}\ \text{m/s})(1.3\ \text{kg/m}^3)}{1.8\times10^{-5}\ \text{Pa·s}} = 4\times10^{-3}. \tag{5.90}$$

The Reynolds number is less than 1, so it was legitimate to apply Stokes's law. Likewise the Reynolds number for a 100 μm particle is less than 1. However, for a particle of diameter 1 mm, the Reynolds number using v predicted by Stokes's law is 400, which lies outside the valid range of Stokes's law. So, will a 1 mm particle fall faster or slower than predicted by Stokes's law?

The universal correlation in Figure 5.16 is awkward for obtaining the terminal velocity of a sphere because velocity is a parameter in the dimensionless groups on both the abscissa and the ordinate. That is, velocity is a parameter in the Reynolds number (x axis) and the Froude number (y axis). Exercise 5.29 describes methods to obtain the value for a parameter that occurs on both the abscissa and the ordinate. Another method is to identify the character of the fluid flow (laminar or turbulent), and use Stokes's law (equation 5.84) or Newton's law of resistance (equation 5.82), respectively. Yet another method is to use a set of dimensionless groups with velocity as one of the core variables. For the terminal velocity of a sphere, such a set is the Froude number, the reduced density, and the Galileo number (equations 5.67, 5.68, and 5.72). To obtain a universal correlation similar to Figure 5.16, express the Reynolds number in terms of the new dimensionless groups. Translate the abscissa (Re) into this new set of dimensionless groups. Specifically, multiply Re by (reduced buoyancy)Re/Fr to obtain (reduced buoyancy) × Ga. A universal correlation with velocity in the ordinate only is shown in Figure 5.18. The lines through the data are Stokes's law (laminar region) and Newton's law of resistance (turbulent region).

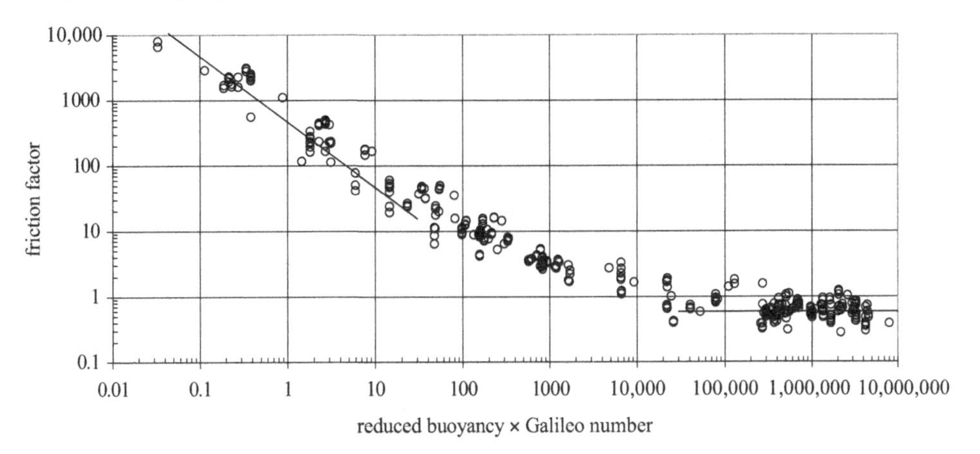

Figure 5.18 Data for a sphere moving through a fluid.

Correlations between dimensionless groups are *design tools*. Just as we derived operating equations to analyze the desalinator in Chapter 3 and studied the McCabe–Thiele method to analyze distillation columns in Chapter 4, the correlation between the Reynolds number and the friction factor in Figure 5.16 can be used to analyze spheres moving through fluids. Similarly, the correlation between reduced stride and Fr in Figure 5.10 is a design tool for devices that walk.

There are myriad correlations between dimensionless groups. We will examine a few more examples in this chapter. Again, it is not necessary to memorize each correlation. Rather, it is important to know how to apply the principles of dynamic similarity to create new design tools.

5.3 DYNAMIC SIMILARITY

5.3.1 Dynamic Similarity of Fluid Flow in a Smooth Pipe

Context: The Alaskan pipeline.

Concepts: Design of a dynamically similar model.

Defining Question: Is it necessary to obtain a functional relation between dimensionless groups?

In the three previous studies – the pendulum, walking, and spheres falling through fluids – we followed the same steps: dimensional analysis, then experiment, and finally analysis of the data. In the analysis of walking and the terminal velocity of spheres we measured many data to obtain functional relationships. We can use these relations to extrapolate to dimensionally similar phenomena such as dinosaurs walking or dust settling in air. What if we don't care to derive the functional relation? That is, assume we know the numerical value of the dimensionless groups of the system of interest. If we are to perform a single experiment on a model system to predict the behavior of the real system at these specific conditions, what should be the specific conditions in the model? That is, how do we design a model? Dynamic similarity tells us the numerical values of the dimensionless groups must be the same for the model system and the real system.

We wish to specify pumps to send oil through the Alaskan pipeline. Assume pumps are stationed every 1000 m.

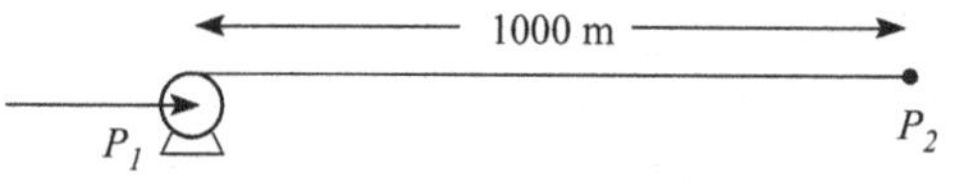

1000 m

P_1 P_2

Figure 5.19 The Alaskan pipeline.

What pressure must each pump deliver?

$$P_1 - P_2 = \Delta P = ? \tag{5.91}$$

It would be impractical to build a full-scale system to measure the pressure drop; the pipe diameter is 1 m and the flow is 425,000 barrels/day. Convert the Alaskan pipeline flow rate from English units to SI units:

$$\left(\frac{425{,}000\ \text{bbl}}{\text{day}}\right)\left(\frac{0.159\ \text{m}^3}{1\ \text{bbl}}\right)\left(\frac{1\ \text{day}}{24\ \text{hr}}\right)\left(\frac{1\ \text{hr}}{60\ \text{min}}\right)\left(\frac{1\ \text{min}}{60\ \text{s}}\right) = 0.8\ \text{m}^3/\text{s}. \tag{5.92}$$

Convert the volumetric flow rate to an average fluid velocity. Dimensional analysis suggests how. Volumetric flow rate has dimensions of L^3/T. Fluid velocity has dimensions of L/T. We must account for a difference in dimensions of L^2, the dimensions of area. It is thus dimensionally consistent to write

$$\text{volumetric flow rate} = (\text{average fluid velocity}) \times (\text{cross-sectional area of pipe}). \tag{5.93}$$

Note that dimensional consistency cannot guarantee that you are correct, but it can reveal when you are wrong. We calculate

$$\text{flow rate} = v\pi\left(\frac{d}{2}\right)^2 \tag{5.94}$$

$$v = \frac{4\,(\text{flow rate})}{\pi d^2} = \frac{4(0.8\ \text{m}^3/\text{s})}{\pi(1\ \text{m})^2} = 1\ \text{m/s}. \tag{5.95}$$

Let's build a model for the Alaskan pipeline. But rather than measure a lot of data with our model to find a functional relation, we will measure exactly one datum with our model. The numerical values of the dimensionless groups for the model will be set equal to the numerical values of the dimensionless groups for the Alaskan pipeline. The model and the Alaskan pipeline are thus dynamically similar. To do this we must (1) derive the dimensionless groups for flow through a pipe and (2) equate the dimensionless groups for the actual system and the model.

Step 1. Prepare a table of the parameters.

The first parameter is what we seek, the pressure drop of the fluid. Two physical parameters of the pipe are relevant. The pipe diameter is important; it is harder to force a given flow of fluid through a small pipe than through a large pipe. Similarly, for a given diameter of pipe, a fluid flows faster through a short pipe than a long pipe. Next, we consider properties of the fluid, starting with the dynamic property, the fluid velocity, v. We expect viscosity to be important. Also, the inertia of the fluid, ρv, will play a role, so we include the fluid density. We list these six parameters of fluid flow through a pipe in Table 5.8, assign a logical symbol, list the dimensions of each, and list the specifics of the Alaskan pipeline.

Table 5.8 The parameters of fluid flow through a pipe, with specifics for crude oil in the Alaskan pipeline.

Parameter	Symbol	Dimensions	Alaskan pipeline
pressure drop	ΔP	M/LT^2	?
pipe length	ℓ	L	1,000 m
pipe diameter	d	L	1 m
velocity	v	L/T	1 m/s
viscosity	μ	M/LT	10 Pa·s
density	ρ_{fluid}	M/L^3	800 kg/m^3

Step 2. Write a general expression for each dimensionless group, Π:

$$\Pi = (\Delta P)^a \ell^b d^c v^d \mu^e \rho_{\text{fluid}}^f \tag{5.96}$$

$$[\Pi] = \left(\frac{\mathrm{M}}{\mathrm{LT}^2}\right)^a \mathrm{L}^b \mathrm{L}^c \left(\frac{\mathrm{L}}{\mathrm{T}}\right)^d \left(\frac{\mathrm{M}}{\mathrm{LT}}\right)^e \left(\frac{\mathrm{M}}{\mathrm{L}^3}\right)^f \tag{5.97}$$

$$\Pi = \mathrm{M}^{a+e+f}\mathrm{T}^{-2a-d-e}\mathrm{L}^{-a+b+c+d-e-3f}. \tag{5.98}$$

Step 3. Set all exponents equal to zero to satisfy the requirement that Π is dimensionless:

$$\text{Mass, M}: \quad a + e + f = 0 \tag{5.99}$$

$$\text{Time, T}: \quad -2a - d - e = 0 \tag{5.100}$$

$$\text{Length, L}: \quad -a + b + c + d - e - 3f = 0. \tag{5.101}$$

Step 4. How many Π groups? Apply the Buckingham Pi theorem:

$$6 \text{ variables} - 3 \text{ equations} = 3 \ \Pi \ \text{groups}.$$

Step 5. Choose three core variables, one for each Π group:

- ΔP – This is the parameter whose value we seek; it would be convenient if ΔP appears only once, so we derive an explicit expression for ΔP.
- ℓ – This is what we will vary if ΔP is too high or too low, so it would be convenient if ℓ appears only once.
- μ – This is a fluid property.

As foretold, there are rules for a valid set of core variables.

Rule 1. All dimensions of the system must be represented in the core variables.

The dimensions of flow through a pipe are M, L, and T. M is represented by ΔP and μ, L is represented by all three, and T is represented by ΔP and μ.

Rule 2. The core variables must not form a dimensionless group.

Rule 2 requires more effort to check. The process of calculating the Π groups will reveal when Rule 2 is violated; one will be unable to find a set of exponents for every Π group.

Step 6. Derive the Π groups.

Π_1 contains ΔP and not ℓ or μ. Set $a = 1$, $b = 0$, and $e = 0$.

$$\text{Mass, M}: \quad 1+0+f=0 \quad \Rightarrow \quad f=-1 \tag{5.102}$$

$$\text{Time, T}: \quad -2(1)-d-0=0 \quad \Rightarrow \quad d=-2 \tag{5.103}$$

$$\text{Length, L}: \quad -1+0+c+(-2)-0-3(-1)=0 \quad \Rightarrow \quad c=0\,. \tag{5.104}$$

In summary, for, Π_1,

$$\left.\begin{aligned} a&=1\\ b&=0\\ c&=0\\ d&=-2\\ e&=0\\ f&=-1 \end{aligned}\right\} \Pi_1 = (\Delta P)^1 v^{-2}(\rho_{\text{fluid}})^{-1} = \frac{\Delta P}{v^2\rho_{\text{fluid}}}. \tag{5.105}$$

This dimensionless group is the Euler (pronounced "oiler") number, Eu, another useful group in chemical engineering:

$$\text{Eu} = \frac{\Delta P}{v^2\rho_{\text{fluid}}} = \frac{\left(\frac{\Delta P}{v}\right)d}{v\rho_{\text{fluid}}d} = \frac{\text{friction force}}{\text{inertial force}}. \tag{5.106}$$

Π_2 contains ℓ and not ΔP or μ. Set $b = 1$, $a = 0$, and $e = 0$.

$$\text{Mass, M}: \quad 0+0+f=0 \quad \Rightarrow \quad f=0 \tag{5.107}$$

$$\text{Time, T}: \quad -2(0)-d-0=0 \quad \Rightarrow \quad d=0 \tag{5.108}$$

$$\text{Length, L}: \quad -0+1+c+0-0-3(0)=0 \quad \Rightarrow \quad c=-1\,. \tag{5.109}$$

In summary, for Π_2,

$$\left.\begin{aligned} a&=0\\ b&=1\\ c&=-1\\ d&=0\\ e&=0\\ f&=0 \end{aligned}\right\} \Pi_2 = \ell^1 d^{-1} = \frac{\ell}{d}. \tag{5.110}$$

Π_3 contains μ and not ΔP or ℓ. Set $e = 1$, $a = 0$, and $b = 0$.

$$\text{Mass, M}: \quad 0+1+f=0 \quad \Rightarrow \quad f=-1 \tag{5.111}$$

$$\text{Time, T}: \quad -2(0)-d-1=0 \quad \Rightarrow \quad d=-1 \tag{5.112}$$

$$\text{Length, L}: \quad -0+0+c+(-1)-1-3(-1)=0 \quad \Rightarrow \quad c=-1\,. \tag{5.113}$$

In summary, for Π_3,

$$\left.\begin{aligned} a&=0\\ b&=0\\ c&=-1\\ d&=-1\\ e&=1\\ f&=-1 \end{aligned}\right\} \Pi_3 = d^{-1}v^{-1}\mu^1(\rho_{\text{fluid}})^{-1} = \frac{\mu}{d\,v\rho_{\text{fluid}}}. \tag{5.114}$$

The inverse of Π_3,

$$\Pi_3^{-1} = \frac{d\, v \rho_{\text{fluid}}}{\mu}, \tag{5.115}$$

should look familiar. It is the Reynolds number, with a slight twist; the pertinent length is the diameter of the pipe. In our first encounter with the Reynolds number, the length scale was represented by the diameter of the sphere. Recall that the Reynolds number is the ratio

$$\text{Re} = \frac{dv\rho_{\text{fluid}}}{\mu} \propto \frac{\text{fluid inertia}}{\text{viscous effects}}. \tag{5.65}$$

Note that the Reynolds number is just as appropriate for the third dimensionless group as is the particular form we derived. We arrived at the particular group in equation 5.115 because we set $e = 1$ for convenience. It is equally appropriate to set $e = -1$ and thus arrive at the Reynolds number.

We now design a model system to predict the characteristics of the real system. To do so, the numerical value of each dimensionless number must be the same for the model system and the real system:

$$\text{Eu}_{\text{model}} = \text{Eu}_{\text{real}} \tag{5.116}$$

$$\Pi_{2,\text{model}} = \Pi_{2,\text{real}} \tag{5.117}$$

$$\text{Re}_{\text{model}} = \text{Re}_{\text{real}}. \tag{5.118}$$

This is *dynamic similarity.*

Let's design a model one could build in one's kitchen. In modeling fluid systems, the fluid in the model must be different from the fluid in the actual system. Because our model pipe diameter will be smaller than the Alaskan pipeline diameter, dynamic similarity of the Reynolds number dictates that the model fluid viscosity must be less than the crude oil viscosity. Water is a convenient fluid, with a viscosity 10,000 times less than crude oil.

Substitute the specifics of the Alaskan pipeline and our kitchen model into equations 5.116 to 5.118. Use the Euler number to derive an equation:

$$\begin{aligned} \left(\frac{\Delta P}{v^2\rho}\right)_{\text{model}} &= \left(\frac{\Delta P}{v^2\rho}\right)_{\text{Alaska}} \\ \frac{\Delta P_{\text{model}}}{v_{\text{model}}^2 (1000\ \text{kg/m}^3)} &= \frac{\Delta P_{\text{Alaska}}}{(1\ \text{m/s})^2 (800\ \text{kg/m}^3)} \\ 0.8 \frac{\Delta P_{\text{model}}}{v_{\text{model}}^2} &= \Delta P_{\text{Alaska}}. \end{aligned} \tag{5.119}$$

Use Π_2 to derive a second equation:

$$\begin{aligned} \left(\frac{\ell}{d}\right)_{\text{model}} &= \left(\frac{\ell}{d}\right)_{\text{Alaska}} \\ \frac{\ell_{\text{model}}}{d_{\text{model}}} &= \frac{1000\ \text{m}}{1\ \text{m}} \\ \ell_{\text{model}} &= 1000 d_{\text{model}} \end{aligned} \tag{5.120}$$

Use the Reynolds number to derive a third equation.

$$\left(\frac{d\,v\rho_{\text{fluid}}}{\mu}\right)_{\text{model}} = \left(\frac{d\,v\rho_{\text{fluid}}}{\mu}\right)_{\text{Alaska}}$$
$$\frac{d_{\text{model}}v_{\text{model}}(1000\ \text{kg/m}^3)}{10^{-3}\ \text{Pa·s}} = \frac{(1\ \text{m})(1\ \text{m/s})(800\ \text{kg/m}^3)}{10\ \text{Pa·s}} \tag{5.121}$$
$$v_{\text{model}} = \frac{8\times10^{-5}\text{m/s}}{d_{\text{model}}}.$$

The dynamic similarity equations reveal that we can choose the pipe diameter in our model and then use equations 5.120 and 5.121 to specify the pipe length and fluid velocity. We arbitrarily choose $d_{\text{model}} = 0.003$ m (= 3 mm), and thus $\ell_{\text{model}} = 3$ m and $v_{\text{model}} = 2.7 \times 10^{-2}$ m/s (= 2.7 cm/s). Both v_{model} and ℓ_{model} are reasonable. If either parameter were not reasonable (a length too long or a velocity too fast) we could choose a different fluid and/or a different pipe diameter.

We now measure ΔP with our model system. What is the volumetric flow rate in our model?

$$\text{flow rate} = v \times \text{area} = \left(2.7\times10^{-2}\ \text{m/s}\right)\times\pi\left(\frac{0.003\ \text{m}}{2}\right)^2\left(\frac{100\ \text{cm}}{1\ \text{m}}\right)^3\left(\frac{60\ \text{s}}{1\ \text{min}}\right)$$
$$= 11\ \text{cm}^3/\text{min}. \tag{5.122}$$

We adjust the kitchen faucet to deliver the desired flow rate. How does one measure the pressure drop? Drill two holes in the tube, separated by a distance of 3.0 m, and attach vertical glass tubes. Measure the difference in the water height in the vertical tubes. We observe in our model system a difference of 2.9 cm water, which we convert to SI units:

$$2.9\ \text{cm water}\left(\frac{98\ \text{Pa}}{1\ \text{cm water}}\right) = 290\ \text{Pa}. \tag{5.123}$$

Use this experimental result from our model to predict the pressure drop in the real system; use equation 5.119:

$$\Delta P_{\text{Alaska}} = 0.8\frac{\Delta P_{\text{model}}}{v^2_{\text{model}}} = 0.8\frac{290\ \text{Pa}}{\left(2.7\ \times\ 10^{-2}\ \text{m/s}\right)^2} = 3.2\times10^5\ \text{Pa}. \tag{5.124}$$

Although SI units are convenient for calculation, we have no sense for a pressure of 3.2×10^5 Pa. So we convert to more convenient metric units and English units:

$$3.2\times10^5\ \text{Pa}\left(\frac{1.01\times10^5\ \text{atm}}{1\ \text{Pa}}\right) = 3.2\ \text{atm}\left(\frac{14.7\ \text{psi}}{1\ \text{atm}}\right) = 46\ \text{psi}. \tag{5.125}$$

Thus, as promised, one experiment performed on a model has yielded the pressure drop for the actual system. However, we did not obtain a functional relationship. If any parameters of the real system change, we must perform another experiment with the model to predict the new pressure drop.

5.3.2 Dynamic Similarity of Fluid Flow in a Rough Pipe

Context: A universal relation for flow through a pipe.

Concepts: Dimensional reduction of pipe roughness.

Defining Question: How does pipe roughness affect friction for laminar flow? For turbulent flow?

In the previous section we derived three Π groups to describe fluid flow in pipes:

$$\frac{\Delta P}{v^2 \rho} \equiv \text{Eu} \propto \frac{\text{friction force}}{\text{inertial force}} \tag{5.106}$$

$$\frac{\ell}{d} \equiv \text{reduced length} \tag{5.110}$$

$$\frac{dv\rho}{\mu} \equiv \text{Re} \propto \frac{\text{fluid inertia}}{\text{viscous force}}. \tag{5.115}$$

For real pipes, we need an additional parameter to describe the flow – the tube roughness. Roughness is an inherent property of the pipe material; glass tubes are smoother than concrete drains. Roughness is also caused by deposits that accumulate in a pipe, known as scaling. Because roughness varies considerably, there is no precise way to characterize it. A sensible descriptor of roughness is the average height of features along the tube wall, k. Some typical values of k are given in Table 5.9.

Table 5.9 Roughness factors for flow through a pipe.

Material	k (mm)
glass	0.002
drawn copper	0.002
cast iron	0.3
concrete	2.0

From Perry and Green, 1984, Figure **5**–6, p. **5**–24.

The addition of another parameter requires a fourth dimensionless group. To find this fourth group we could return to the formal method used in the previous section. But because the dimensions of k are simple, we can guess the new Π group. How does one compensate for k's dimension of L? Divide by a parameter with dimension of L. What is the logical parameter to combine with k? The two candidates are ℓ and d. Because ℓ is a core variable, it can appear in only one Π group. So we divide k by the tube diameter to form k/d. And this makes sense. The roughness should be normalized by the diameter of the tube. Whereas 1 mm of roughness may be insignificant in a pipe of diameter 1 m ($k/d = 0.001$), it would be significant in a pipe of diameter 1 cm ($k/d = 0.1$).

In our analysis of the terminal velocity of a sphere we found we could combine two of the Π groups to form a composite dimensionless group, the friction factor:

$$\text{friction factor} = \frac{4}{3}\frac{\rho_{\text{sphere}} - \rho_{\text{fluid}}}{\rho_{\text{fluid}}}\frac{1}{\text{Fr}}. \tag{5.72}$$

A similar analysis of flow through pipes reveals that the Euler number and the reduced length can be combined into one dimensionless group, also a called a friction factor. Again, a seemingly arbitrary factor appears, again as a result of theoretical modeling. Beware: because the factor of 1/8 is arbitrary it is omitted from some plots. The Fanning friction factor is given by

$$\text{Fanning friction factor} \equiv \frac{1}{8}\frac{d}{\ell}\text{Eu}. \tag{5.126}$$

As shown in Figure 5.20, roughness increases the friction factor, but only in the region of high Reynolds number. What type of fluid flow is typical of this region? Turbulence. Why should roughness have the largest effect here?

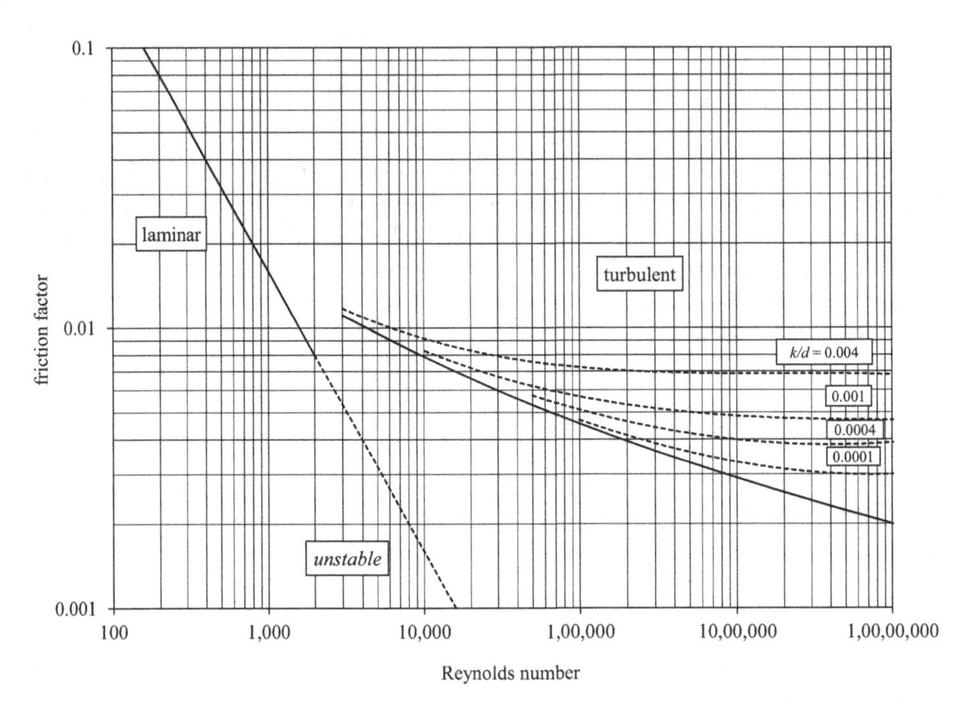

Figure 5.20 The friction factor (equation 5.125) as a function of Re, with effect of pipe roughness, *k*. Adapted from Bird, Stewart, and Lightfoot (1960), figure 6.2–2.

The region intermediate to laminar and turbulent is labeled "unstable." The flow pattern in this region depends on the history of the flow. If a laminar flow increases in velocity gently, it can remain laminar for Re well above 1000. However, once perturbed to turbulence, the flow remains turbulent.

5.3.3 Dynamic Similarity of Heat Transfer from a Fluid Flowing in a Tube

Context: A heat exchanger.

Concepts: The Stanton and Prandtl numbers.

Defining Question: What dimensionless groups describe the dynamics of heat flow? What dimensionless group describes a fluid's thermal properties?

We have incorporated heat exchangers in many of our process designs, which we represented by the generic unit shown in Figure 5.21. Heat exchangers are commonly of the "shell and tube" design. One liquid passes through a bank of parallel tubes. The other liquid flows around the tubes, countercurrent to the flow in the tubes contained by the outer shell, as diagrammed in Figure 5.22.

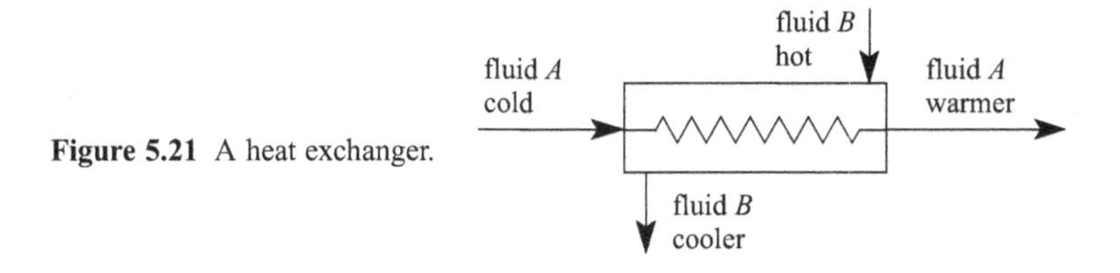

Figure 5.21 A heat exchanger.

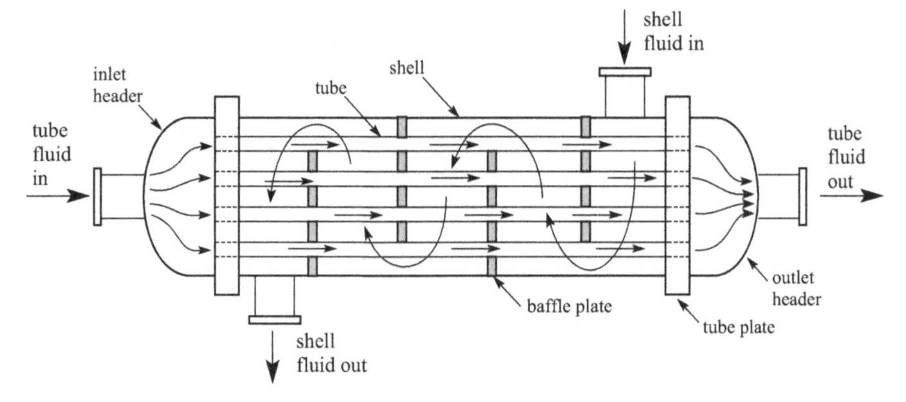

Figure 5.22 Schematic cross-section of a cylindrical shell and tube heat exchanger. The tube fluid enters the inlet header and is distributed into tubes. Typical heat exchangers have hundreds of tubes. The shell fluid flows inside a cylindrical shell, around the outside of the tubes

An important consideration in designing heat exchangers is the rate of heat transfer to the fluid inside the tubes. Clearly this will depend on the flow rate of the fluid. If the fluid passes through too quickly, it will not absorb much heat. Therefore we expect that dimensional analysis of a heat exchanger includes the parameters of flow in a tube, summarized in Table 5.8. In addition, we have new parameters associated with heat transfer, listed in Table 5.10. The dimensions of the heat-transfer parameters are less obvious than those in Table 5.8. It is useful to include an intermediate column for the units of these parameters.

We would not expect a novice engineer to be able to devise the list in Table 5.10. However, we hope a novice engineer can appreciate the relevance of these parameters. The temperature difference is the driving force for heat transfer, just as the density difference was the driving force for a sphere to fall (or rise) in a fluid. The heat capacity is literally the fluid's capacity to absorb heat; the absorption of energy increases the fluid temperature. The thermal conductivity is the transmittance of energy, or the inverse of the resistance to heat transfer; a high conductivity means a low resistance to heat transfer. Analogously, the resistance to a sphere moving through a fluid is viscosity.

Table 5.10 The parameters of heat transfer for fluid flow through a pipe.

Parameter	Symbol	Units (SI)	Dimensions
rate of heat transfer	q	J/s	$\mathrm{ML^2/T^3}$
temperature difference	$T_{\text{fluid}} - T_{\text{pipe}}$	°C	Θ
heat capacity	C_{P}	J/(kg·°C)	$\mathrm{L^2/T^2\Theta}$
thermal conductivity	k	J/(s·m·°C)	$\mathrm{ML/T^3\Theta}$

The rate of heat transfer, q, is usually combined with the temperature difference and the surface area of the tube (calculated from the pipe diameter and length) to yield a *heat-transfer coefficient*, h:

$$h = \frac{q}{\left(T_{\text{fluid}} - T_{\text{pipe}}\right)(\pi d \ell)}, \quad [h] = \frac{\mathrm{ML^2/T^3}}{\Theta \mathrm{L}^2} = \frac{\mathrm{M}}{\Theta \mathrm{T}^3}. \tag{5.127}$$

With these parameters one can form four dimensionless groups, three of which are

$$\frac{dv\rho}{\mu} = \mathrm{Re} = \text{Reynolds number} = \frac{\text{fluid inertia}}{\text{viscous force}} \tag{5.128}$$

$$\frac{h}{C_P v \rho} = \mathrm{St} = \text{Stanton number} = \frac{\text{heat transferred}}{\text{thermal heat capacity}} \tag{5.129}$$

$$\frac{C_P \mu}{k} = \mathrm{Pr} = \text{Prandtl number} = \frac{\text{momentum diffusivity}}{\text{heat diffusivity}}. \tag{5.130}$$

A fourth dimensionless group includes the pressure drop, ΔP. However, experiments would reveal that this dimensionless group was irrelevant, just as the pendulum angle was revealed to be irrelevant.

The Reynolds number characterizes the dynamics of the fluid flow. The Stanton number characterizes the dynamics of the heat flow. The Prandtl number characterizes the substance being heated as it flows. The Prandtl number varies from substance to substance and varies with temperature; for water, Pr = 7.7 at 15°C and Pr = 1.5 at 100°C.

Given three dimensionless groups, one can plot, for example, St (heat-transfer dynamics) vs. Re (fluid-flow dynamics) as a function of Pr (fluid type), as shown in Figure 5.23. The lines through the data sets in Figure 5.23 are fits to equations of the form $\mathrm{St} = c \cdot \mathrm{Re}^{-0.2}$. The proportionality c varies with fluid and with Pr. Note that as Pr increases, the data lie lower on the plot. This suggests that c varies systematically with Pr. Indeed, Figure 5.24 shows that a plot of $\mathrm{St(Re)}^{0.2}$ (= c) as a function of Pr is a straight line. The line drawn in Figure 5.24 is $\mathrm{St(Re)}^{0.2} = 0.023 \cdot \mathrm{Pr}^{-0.6}$.

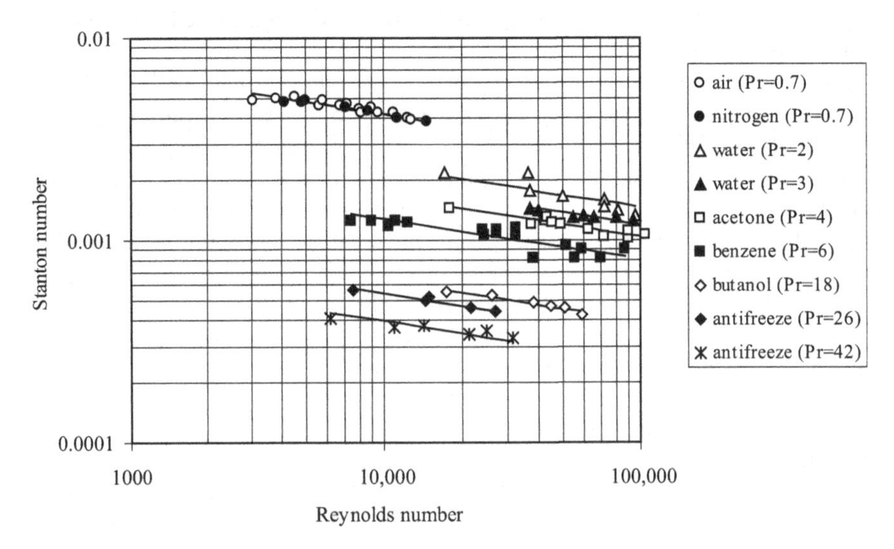

Figure 5.23 St vs. Re for various fluids, each of which corresponds to a different value of Pr. Adapted from Brown, *et al.* (1950), chapter 29.

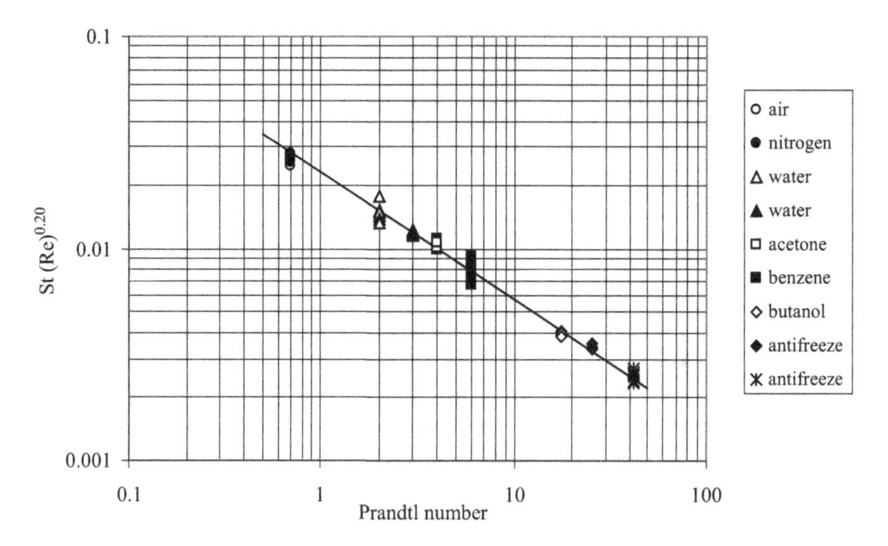

Figure 5.24 $St(Re)^{0.2}$ as a function of Pr. Adapted from Brown, *et al.* (1950), chapter 29.

Figure 5.24 illustrates the universal relation between a dimensionless measure of a fluid's thermal properties (heat capacity and thermal conductivity) and the dynamics of heat transfer from flowing fluids to a solid surface.

There is substantial scatter in Figures 5.23 and 5.24. This is common in correlations between dimensionless numbers that describe complex phenomena. Some scatter can be attributed to errors in the measurements, but most scatter indicates that one or more parameters have been neglected. Most importantly, note that predictions from such correlations will not be precise. Whereas the mass balances in Chapter 3 were quite accurate, a prediction from a correlation such as in Figure 5.24 has an accuracy of about ±50%.

5.3.4 Dynamic Similarity of Vapor–Liquid Equilibrium Stages

Context: Absorbers and distillation columns

Concepts: Blow-through and dumping, flooding, and weeping.

Defining Question: For a given reflux ratio, how should the column diameter change to reduce flooding? To reduce weeping?

A powerful application of dynamic similarity is the analysis of complex systems, systems impractical to model analytically. Examples include two process units we encountered in Chapter 4: absorbers and distillation columns. As you might expect, engineers have developed many design tools for absorbers and distillation columns based on correlations of dimensionless groups. In this section we delve into the inner workings of multistage countercurrent units and examine typical design tools for predicting their behavior.

In the previous chapter we found that a series of equilibrium stages is an effective design for absorbers and distillation columns, as shown in Figures 4.60 and 4.78. Liquid cascades down the column and vapor permeates up the column. At each stage the liquid and vapor mix and components transfer from one phase to the other to approach equilibrium. For efficient performance, we want equilibrium to be approached rapidly at each stage. To improve the rate of mass transfer between the liquid and vapor phases, one can increase the amount of interface between the liquid and vapor phases. A common design uses a series of sieve plates, linked by downcomers, as shown in Figure 5.25. Liquid passes down a slot and spills onto a sieve plate. Vapor bubbles through holes in the sieve plate and creates a froth, which increases the amount of liquid–vapor interface.

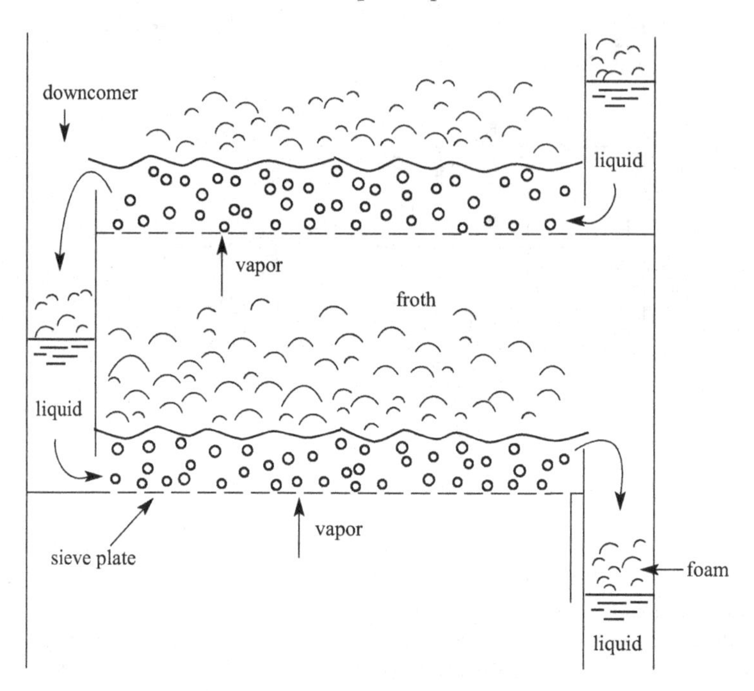

Figure 5.25 Schematic cross-section of a cylindrical distillation column with sieve plates. Liquid flows from a downcomer, across the sieve plate, and into the next downcomer. Vapor flows through the holes in the sieve plate, creating froth above the liquid.

When we designed distillation columns with the graphical McCabe–Thiele method, we specified the *relative* flow rates of liquid and vapor to obtain the operating lines. What diameter of column is needed to accommodate the *absolute* flow rates? If the column is too narrow (i.e. the sieve plate area is too small), the liquid will pass over the sieve plate too quickly and not equilibrate with the vapor. If the column is too wide, the liquid will not cover the tray completely and the vapor will blow through without equilibrating.

Figure 5.26 on the next page shows a map of the possible behaviors for a distillation column. The abscissa is a dimensionless group formed from the product of the reflux ratio (L/V) and the square root of the vapor–liquid density ratio. The square root of the vapor–liquid density ratio is about 0.04 at 1 atm. The ratio decreases as the pressure in the column is decreased. The ordinate is the vapor flow rate (mass/time) divided by the area of the tray.

With the map in Figure 5.26 one can predict the effect of different tray diameters with constant reflux ratio (a vertical path on the map) or of changing the reflux ratio at constant vapor flow rate (a horizontal path). *Weeping* occurs when there is insufficient vapor flow through the sieve holes; the liquid "weeps" through the sieve holes. At lower reflux ratios liquid *dumps* onto the lower tray at low vapor flow and vapor *blows through* the liquid with insufficient contact at high vapor flow. At intermediate vapor flow and low L/V, liquid is entrained by the vapor and carried up to the next tray. Of course, the borders on the map are fuzzy; changes from one behavior to another are gradual, not abrupt.

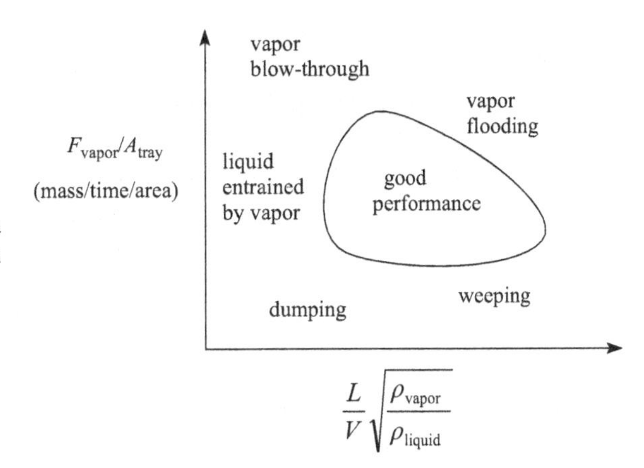

Figure 5.26 Operating regimes for a distillation column. Adapted from King (1971), figure 12–8.

Many more dimensionless design tools for absorbers and distillation columns may be found in King's text *Separation Processes* (1971) and in *Perry's Chemical Engineers' Handbook* (1984).

5.4 APPLICATIONS OF DIMENSIONAL ANALYSIS

5.4.1 Dimensional Analysis of Gases

Context: A universal relation for the molar volume of a non-ideal gas.

Concepts: Compressibility, reduced temperature, and reduced pressure.

Defining Question: What gas property best describes non-ideal behavior?

We have applied dimensional analysis to two process units: heat exchangers and distillation columns. Dimensional analysis can also be applied to fundamental systems, such as a gas of a pure substance. Given the properties of a gas, we would like to predict the molar volume at a given temperature and pressure.

In Chapter 3 we cited the ideal gas law as an example of a constrained law. If one is modeling a gas at high temperature and low pressure, the ideal gas law is valid. To extend beyond the limits of the ideal gas law, one could measure the properties of a specific gas and develop methods of graphical analysis similar to those introduced in Chapter 4. A qualitative plot of the behavior outside the ideal limits was shown in Figure 4.1. What if one wants to model a gas outside the ideal gas region but does not have access to experimental data? Can the behavior of all gases be scaled to a universal curve?

Dimensional analysis provides the basis for a universal curve for gases. As always, we begin with the parameters of a gas: pressure P, molar volume $\bar{V} = V/n$, and temperature T. These parameters are related in the ideal gas equation by the proportionality R, the gas constant.

We need additional parameters to characterize a non-ideal gas. Let's examine two causes of non-ideality. The equation for an ideal gas predicts that the volume of a gas approaches zero as the temperature approaches zero and as the pressure approaches infinity. But a gas is composed of molecules that have a finite (albeit small) volume. At the limits of low temperature and high pressure, the molar volume approaches the molar volume of the liquid (or solid), and not zero. Thus at low temperature and high pressure: $\bar{V}_{real} > \bar{V}_{ideal}$.

The equation for an ideal gas predicts that the volume becomes very large at low temperature, as the pressure goes to zero. But because molecules mutually attract, the molar volume is less than predicted: $\bar{V}_{real} < \bar{V}_{ideal}$.

So we could add parameters for the molar volume of the liquid and the attractive force between the gas molecules. The molar volume of the condensed phase is a convenient parameter, but the intermolecular force is not. Consider a different approach. Rather than use the *causes* of non-ideality, let's examine the *effects* of non-ideality. For example, rather than try to quantify intermolecular forces, one might use the boiling point of the liquid. Given two gases with equal molecular weight, the gas with the stronger attractive forces will boil at a higher temperature. However, experiments would reveal that the boiling point at a standard pressure is not a good predictor for the non-ideal behavior of a gas.

What other properties characterize a gas? Recall the generic phase map for a pure substance, shown in Figure 5.27. What landmarks do all substances have in common? Every substance has a triple point – the temperature and pressure at which solid, liquid, and vapor coexist. We could use the pressure and temperature of the triple point as the distinguishing features of our gas. However, like the boiling point at a standard pressure, the triple point is not a good predictor for the non-ideal behavior of a gas.

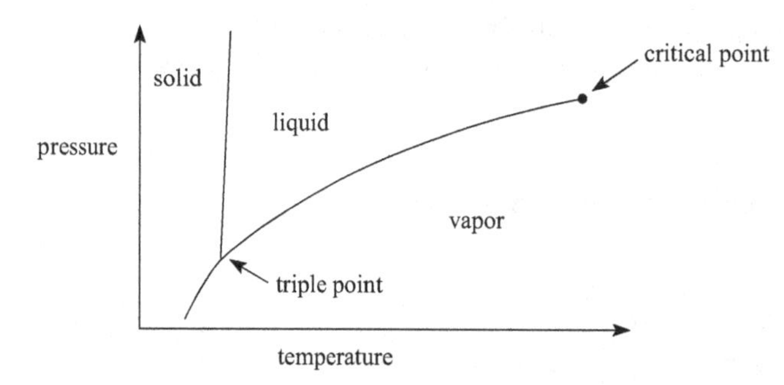

Figure 5.27 Phase diagram for a pure substance.

The other landmark common to all substances is a critical point – the point at which liquid and gas become a single phase. Experimental studies reveal that the critical temperature and pressure are good indicators of the properties of a gas. We add the critical temperature and pressure to our table of gas parameters.

Table 5.11 The parameters of a gas.

Parameter	Symbol	Dimensions
pressure	P	M/LT^2
volume	$\bar{V}$	L^3
temperature	T	Θ
gas constant	R	$ML^2/T^2\Theta$
critical temperature	T_c	Θ
critical pressure	P_c	M/LT^2

You would not have been able to predict that critical temperature and pressure were appropriate for scaling the non-ideal behavior of a gas. Until experiments were performed (and you had completed a course in molecular thermodynamics), other distinguishing features of a gas – boiling point at 1 atm, the triple point, molar volume, etc. – are reasonable choices. And how can we know the list of parameters in Table 5.11 is sufficient? Again, we would not know until we measured and analyzed data for many gases. If all the data lie on the same line, the list is sufficient.

We now derive the dimensionless groups. As before we begin by writing a general expression for the dimensionless group, Π:

$$\Pi = P^a V^b T^c R^d T_c^e P_c^f \tag{5.131}$$

$$[\Pi] = \left(\frac{\mathrm{M}}{\mathrm{LT}^2}\right)^a \left(\mathrm{L}^3\right)^b \Theta^c \left(\frac{\mathrm{ML}^2}{\mathrm{T}^2\,\Theta}\right)^d \Theta^e \left(\frac{\mathrm{M}}{\mathrm{LT}^2}\right)^f \tag{5.132}$$

$$[\Pi] = \mathrm{M}^{a+d+f}\mathrm{L}^{-a+3b+2d-f}\mathrm{T}^{-2a-2d-2f}\Theta^{c-d+e}. \tag{5.133}$$

We then set all exponents equal to zero to satisfy the requirement that Π is dimensionless:

$$\text{Mass, M}: \; a + d + f = 0 \tag{5.134}$$

$$\text{Length, L}: \; -a + 3b + 2d - f = 0 \tag{5.135}$$

$$\text{Time, T}: \; -2a - 2d - 2f = 0 \tag{5.136}$$

$$\text{Temperature, } \Theta: \; c - d + e = 0. \tag{5.137}$$

How many Π groups describe a gas? The Buckingham Pi theorem tells us 6 variables – 4 equations = 2 Π groups. But there is a problem with the four equations. This problem would become apparent if you attempted to solve the equations by successive substitution, for example. The equations for mass and time are not independent; multiply equation 5.134 by −2 and it becomes identical to equation 5.136. This is because M appears only in ratio with T^2. So there are actually only three *independent* equations, and thus there are 6 – 3 = 3 dimensionless groups.

There are many valid choices for the three core variables. However, the standard dimensionless groups used to describe a gas are

$$\frac{T}{T_c} \equiv \text{reduced temperature} \equiv T_r \tag{5.138}$$

$$\frac{P}{P_c} \equiv \text{reduced pressure} \equiv P_r \tag{5.139}$$

$$\frac{P\bar{V}}{RT} \equiv \text{compressibility} \equiv Z. \tag{5.140}$$

Compressibility is a useful indicator of the ideality of a gas. It is conventional to plot Z vs. P_r as a function of T_r. Let's predict qualitatively the general appearance of the family of curves on such a plot.

If a gas is ideal, $P\bar{V} = RT$. We can rewrite the ideal gas law in a dimensionless form,

$$\frac{P\bar{V}}{RT} = 1 = \text{ideal compressibility}. \tag{5.141}$$

Thus at very high temperature, we expect $Z = 1$, as shown in Figure 5.28.

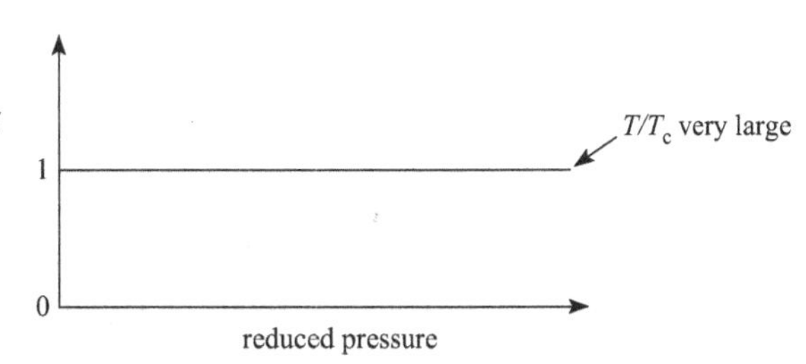

Figure 5.28 Compressibility as a function of pressure at a high temperature.

As discussed above, we expect non-ideal behavior at low temperature. At high pressures, because of the finite molar volume, $\bar{V}_{\text{real}} > \bar{V}_{\text{ideal}}$ and thus $Z > 1$. At low pressures, intermolecular attractions cause $\bar{V}_{\text{real}} < \bar{V}_{\text{ideal}}$ and thus $Z < 1$. The predicted behavior of Z is shown on the phase maps in Figures 5.29 and 5.30.

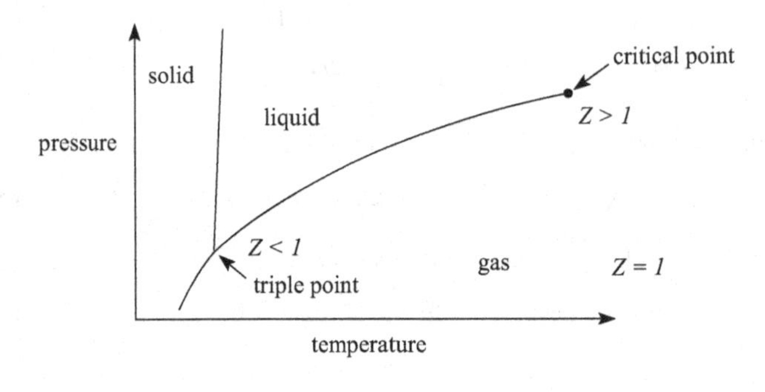

Figure 5.29 Compressibility as a function of pressure and temperature.

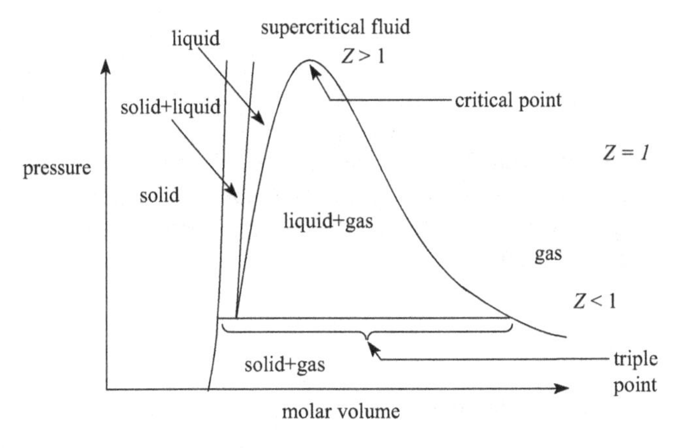

Figure 5.30 Compressibility as a function of pressure and molar volume.

Thus we would expect the isotherm for low temperature to have the general shape given in Figure 5.31.

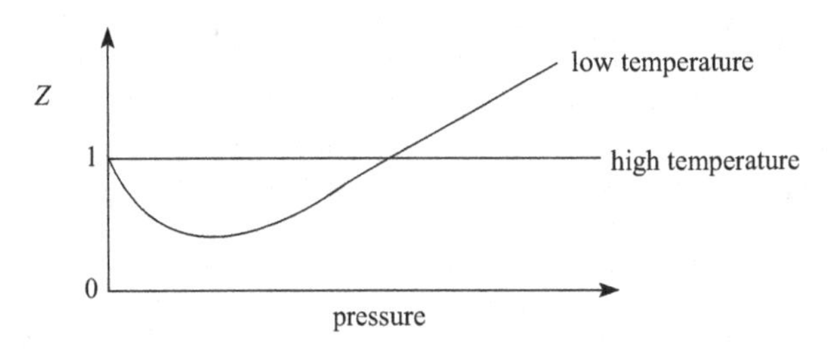

Figure 5.31 Compressibility as a function of pressure at two temperatures.

The universal plots of Z vs. P_r as a function of T_r are shown for moderate pressures in Figure 5.32a and for high pressures in Figure 5.32b on the next page.

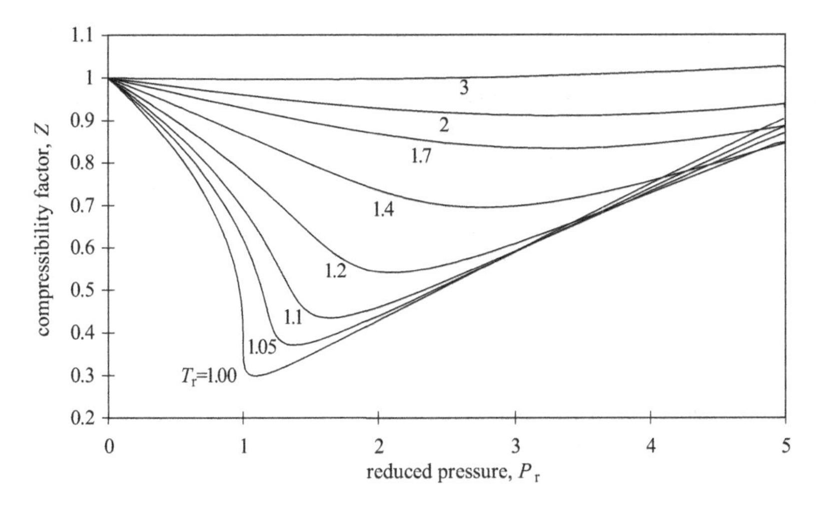

Figure 5.32a Z vs. P_r as a function of T_r for moderate pressures and low temperatures. Curves generated with the van der Waals equation.

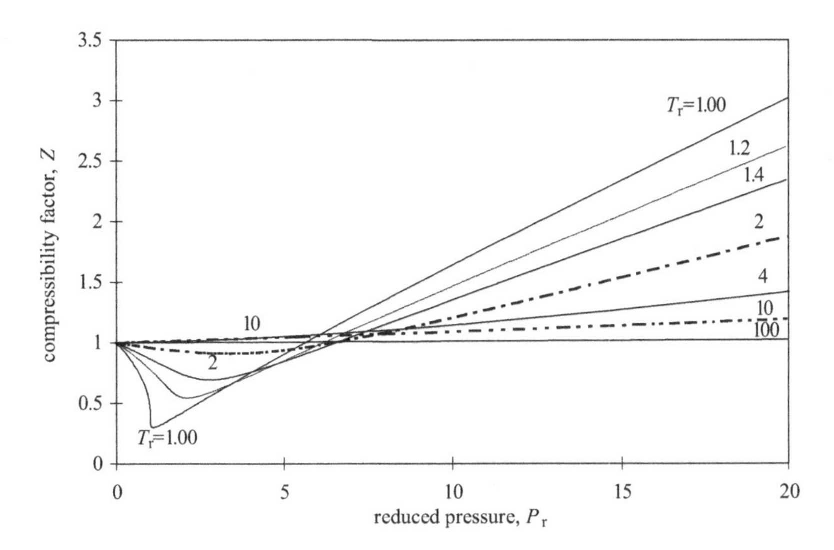

Figure 5.32b Z vs. P_r as a function of T_r for high pressures. Curves generated with the van der Waals equation.

5.4.2 Dimensional Analysis of Biological Systems

Context: The scaling of animal behavior with animal size.

Concepts: Projectile trajectory.

Defining Question: Why don't ants throw stones?

As discussed in the introduction to this chapter, an animal's size dictates an animal's proportion and dynamics. Wing area increases as the square of body length, whereas body mass increases as the cube of body length; a condor's wings are proportionally larger than a sparrow's wings. A quetzalcoatl's wings are yet larger proportionally, but not large enough to allow flight, as convincingly argued with dimensional analysis by Octave

Levenspiel (Levenspiel, 2000; Levenspiel, Fitzgerald, and Pettit, 2000), a scholar of chemical reactor design and analysis at Oregon State University.

Species survival depends on offensive and defensive skills. Early man developed hunting with projectiles, likely rocks thrown, then rocks flung with slings, spears chucked, and arrows launched with bows. Insects lack intelligence, but the sheer numbers of insects over millions of years have evolved diverse innovations such as poisons for offense and chemical sprays for defense. So why have no insects evolved to attack or defend by throwing stones? Ants are proportionally much stronger than humans, so ants could throw proportionally large stones.

The answer is the scaling of the stone's inertia to the stone's frictional losses. To demonstrate, we design a dynamically similar model: an ant throwing a stone in air is like a human throwing a ... in ... ?

We must consider the Reynolds number of the projectile – a dimensionless ratio of the projectile's inertia to the projectile's frictional losses:

$$\text{Re} = \frac{\text{projectile inertia}}{\text{viscous force}} = \frac{(\rho v d)_{\text{projectile}}}{\mu_{\text{fluid}}}. \tag{5.142}$$

A typical ant is 5 mm long. Assume ambitiously that the diameter of the ant's projectile is 1 mm; remember, ants are proportionally stronger. It is difficult to predict an ant's pitch velocity. Human athletes have a pitch velocity of 40 m/s (100 mph). We assume an average human has a pitch velocity of 20 m/s. An ant can move its appendage faster, but its appendage is shorter. Let's estimate that an ant's pitch velocity is at most 2 m/s. Assume the projectile is a grain of sand; ρ_{sand} = 2600 kg/m^3. Calculate the Reynolds number for the ant's projectile in air:

$$\text{Re for sand grain in air} = \frac{(2600 \text{ kg/m}^3)(2 \text{ m/sec})(0.001 \text{ m})}{1.8 \times 10^{-5} \text{ Pa·sec}} = 3 \times 10^5. \tag{5.143}$$

Similarly, we calculate Re = 3 × 10^8 for a human throwing a 10-cm stone in air. Experience tells us the human's stone travels far enough to be an effective weapon. How far does a human-sized projectile travel with a Reynolds number that is a factor of 1000 smaller, for Re = 3 × 10^5? What is a dynamically similar human-sized projectile? We must decrease the projectile's inertia and/or increase the projectile's frictional losses.

Imagine throwing a 10-cm stone underwater (μ_{water} = 0.001 Pa·s). The water's viscosity slows our throwing arm; we assume the pitch velocity is ten times slower, v = 2 m/s:

$$\text{Re for stone in water} = \frac{(2600 \text{ kg/m}^3)(2 \text{ m/sec})(0.1 \text{ m})}{0.001 \text{ Pa·sec}} = 5 \times 10^5. \tag{5.144}$$

A stone travels only a few feet underwater before friction stops the stone. A human cannot fend off a shark by throwing stones.

An equivalent, convenient model is a balloon in air. We ignore that a balloon is not an effective weapon; we are interested only in the distance the balloon travels. Assume the balloon is slightly more dense than air: ρ_{balloon} = 1.4 kg/m^3. For a 20-cm balloon,

$$\text{Re for balloon in air} = \frac{(1.4 \text{ kg/m}^3)(20 \text{ m/sec})(0.2 \text{ m})}{1.8 \times 10^{-5} \text{ Pa·sec}} = 3 \times 10^5. \tag{5.145}$$

Throw a 20-cm balloon as fast as you can. The balloon loses momentum and falls to the ground a few meters away. The trajectory is about 10 balloon diameters. By dynamic similarity, the ant's 1-mm sand grain will travel no more than 1 cm. Flinging a projectile is an ineffective defense or offense at the scale of an ant.

Firearm ballistics scale similarly. With comparable muzzle velocities, birdshot from a shotgun travels ~100 m, a bullet from a rifle travels ~1000 m, and a 20-inch projectile from a battleship travels ~10,000 m. The projectile range scales by the ratio of the projectile's inertia to its frictional losses.

Consider the scaling of other phenomena from the insect scale to the human scale.

- Why do ants survive falls 100 times their height, whereas humans do not survive falls 100 times their height?
- How can some insects alight on vertical surfaces?
- How can some insects walk on water?
- Why don't fleas need buckets to carry water?
- Why does a paramecium propel itself by cilia, whereas a dolphin propels itself by flexing its body?

5.4.3 Dimensional Analysis of Microchemical Systems

Context: The length scale of fluid flow phenomena.

Concepts: Reynolds, Bond, and Péclet numbers for microfluidics.

Defining Question: Why is the piping in a chemical plant different from the fluid channels etched on a microchip?

Microchemical systems are ideal for chemical analysis, pharmaceutical testing, combinatorial screening, and fuel cells, for example. Microchemical blood analysis requires only a microliter sample. Whereas testing on small mammals requires grams of a pharmaceutical, testing with the "human on a chip" (Demirci, *et al.*, 2012) that mimics a mammalian system of cells from key organs – lungs, stomach, intestine, liver, and kidney – requires only a microgram of a pharmaceutical. Microchemical arrays for combinatorial chemistry, in which key parameters such as promoters in a catalyst are varied systematically, afford rapid screening of hundreds of variations. Although technology for battery storage of electrical energy progresses steadily, the energy density of the best battery is far below that of a hydrocarbon. A microchemical electrical cell fed by a reservoir of butanol comparable to a pocket lighter could power a laptop for months.

The scaling for microchemical processes presents challenges and opportunities, chiefly owing to the scaling of fluid flow. Table 5.12 lists four key energies in fluid flow and their scaling with the characteristic length of the system, such as the width of the fluid channel or the diameter of a sphere or bubble moving through a fluid.

Table 5.12 The length scaling of mass and system energies.

Property	Expression	Dimensions	Scaling with respect to length
mass	$\rho\ell^3$	M	ℓ^3
inertial (kinetic) energy	$\frac{1}{2}mv^2 \propto (\rho\ell^3)v^2$	ML^2/T^2	ℓ^3
viscous energy loss	$\propto \mu v\ell^2$	ML^2/T^2	ℓ^2
potential energy in a gravitational field	$mg\ell \propto (\rho\ell^3)\ell$	ML^2/T^2	ℓ^4
surface (interface) energy	$\propto \gamma\ell^2$	ML^2/T^2	ℓ^2

Consider the ratio of the inertial (kinetic) energy to the viscous energy loss. This ratio is represented by the dimensionless Reynolds number and scales as $\ell^3/\ell^2 = \ell$. Thus we expect kinetic energy to dominate in large systems (Re > 100) and viscous energy loss to dominate in microfluidics (Re < 1). Consider swimming, for example.

When one stops paddling, one glides a few body lengths. This is important for safely swimming to the edge of a pool. An ocean vessel, such as a cruise ship, glides for tens of ship lengths. A cruise ship reduces propulsion miles from port. At the microfluidic level, a single-cell organism stops moving immediately after its flagellum stops rotating. Macroscopic flow must consider massive momentum changes of a 90° turn. Neglect of this momentum caused the Flixborough disaster. On the contrary, momentum may be neglected in microfluidic flow. A sharp right-angle turn in a microchannel causes negligible loss of fluid momentum relative to the viscous losses.

Consider the ratio of the gravitational potential energy to surface (interface) energy. This ratio is represented by the dimensionless Bond number, Bo, and scales as $\ell^4/\ell^2 = \ell^2$. Thus we expect potential energy to dominate in large systems (Bo > 10) and surface energy loss to dominate in microfluidics (Bo < 1). Consider a concurrent flow of air and water. In a large pipe such as a city's sewer system, the flow is stratified: a continuous flow of air above a continuous flow of water. But in microfluidics the flow is interspaced plugs of air and water, as illustrated in Exercise 5.22. The potential energy lost by water filling the tube from top to bottom is small compared to the surface energy gained by reducing the total water–air interface area.

Fluid streams mix slowly at the microscopic scale. The rate of convective mass transfer is proportional to $(v/\ell)(dC/dx)$. The rate of diffusive mass transfer is proportional to $(D/\ell^2)(d^2C/dx^2)$, where D is the diffusion coefficient (in units of m^2/s) and C is the concentration. The ratio of diffusive to convective mass transfer is represented by the dimensionless Péclet number, as described in Exercise 5.31. In macroscopic systems, convection dominates mixing. Diffusive mass transfer would require hours for a scent to move across a lecture hall, whereas experience tells us the scent is detected within minutes or seconds on the other side of a room. Because convective mass transfer is small relative to diffusive mass transfer in microfluidic flow, special provisions must be made to mix two parallel flows. One method is to etch a herringbone pattern into the microchannel to stretch and twist the two parallel flows, like pulling taffy to mix red and white (Stroock, *et al.*, 2002). But the lack of convective mixing can also be an advantage. A macroscopic liquid–liquid absorber uses two immiscible liquid phases, such as oil and water. At the microscopic level, two water streams are essentially immiscible owing to the lack of convective mixing. Consequently, molecules may be selectively absorbed from one water phase to another water phase. Such a device is called an H filter, and is described in Exercise 5.39.

SUMMARY

Dimensional analysis has five advantages:

- *To reveal the relative scaling of parameters.* If the length of the pendulum arm doubles, the period decreases by a factor of the square root of 2. If the diameter of a sphere doubles, its terminal velocity increases by a factor of 4, for low Reynolds number.
- *To optimize an experimental study.* In some cases, dimensional analysis reveals that certain parameters are not pertinent. Mass is irrelevant to the period of a pendulum and to the velocity of walking.
- *To optimize analysis.* For each system we studied – the pendulum, walking, and spheres moving through fluids – a complex phenomenon described by several variables was reduced to a function of a few dimensionless groupings of the variables. In the analysis of walking, the parameters h, ℓ, g, and m reduced to $s/\ell = f(v^2/g\ell)$. In the analysis of the terminal velocity of spheres, the six parameters v, d, g, buoyancy, ρ_{fluid}, and μ reduced to a relation between three Π groups:

$$\frac{v^2}{dg} = f\left(\frac{\rho_{\text{sphere}} - \rho_{\text{fluid}}}{\rho_{\text{fluid}}}, \frac{\rho_{\text{fluid}} v d}{\mu}\right). \tag{5.66}$$

- *To reveal the different characteristics of a phenomenon.* For quadrupeds, the style of running changes from a trot to a gallop at a Froude number of about 2.6. The fluid flow pattern around a sphere has two distinct regions – laminar and turbulent – characterized by the Reynolds number.
- *Expedience.* There are hundreds of phenomena in chemical engineering. When the equation describing one of these phenomena is expressed in a dimensionless form, the equation reduces to one of about eight differential equations. The advantages of this are obvious: the equation has already been solved and dimensional similarity reveals physical similarity. For example, heat transfer by conduction is mathematically similar to mass transfer by diffusion.

The dimensionless groups that describe a system can be expressed as the ratio of two phenomena, such as a ratio of forces (the Reynolds number is the ratio of inertial forces to viscous forces) or a ratio of energies (the Froude number is the ratio of kinetic energy to potential energy in a gravitational field). Dynamically similar models are designed by setting equal the dimensionless groups of the model and the actual system. If the numerical magnitude of a dimensionless group is very large or very small, one of the effects in the ratio is negligible and that dimensionless group need not be precisely equal in the model and in the actual system. For example, inertia is negligible in creeping flows ($\text{Re} < 10^{-6}$) and viscous forces are negligible in highly turbulent flows ($\text{Re} > 10^6$). The model and the actual system must both have creeping flow, for example, but need not have precisely the same Reynolds number.

Finally, using a convenient model to predict the behavior of an inconvenient system is not extrapolation; it is dynamic similarity.

REFERENCES

Alexander, R. McN. 1983. "Posture and gait of dinosaurs," *Zoological Journal of the Linnean Society*, **83**, 1–25.

Alexander, R. McN. 1989. *Dynamics of Dinosaurs and Other Extinct Giants*, Columbia University Press, New York, NY.

Alexander, R. McN. 1992. "How dinosaurs ran," *Scientific American*, April, 4–10.

Anderson, J. F., Hall-Martin, A., and Russell, D. A. 1985. "Long-bone circumference and weight in mammals, birds, and dinosaurs," *Journal of Zoology, London (A)*, **207**, 53–61.

Batchelor, G. K. 1967. *Introduction to Fluid Mechanics*, Cambridge University Press, London.

Bird, R. B., Stewart, W. E., and Lightfoot, E. N. 1960. *Transport Phenomena*, John Wiley & Sons, New York, NY.

Brown, G. G., Foust, A. S., Katz, D. L., *et al.* 1950. *Unit Operations*, John Wiley & Sons, New York, NY.

Colbert, E. H. 1962. "The weights of dinosaurs," *American Museum Novitates*, **2076**, 1–16.

Demirci, U., Khademhossein, A., Langer, R., and Blander, J. 2012 *Microfluidic Technologies for Human Health*, World Scientific Publishing Company, Singapore.

King, C. J. 1971. *Separation Processes*, McGraw-Hill, New York, NY.

Levenspiel, O. 2000. "Earth's early atmosphere," *Chemical Innovation*, **30**, 47–50.

Levenspiel, O., Fitzgerald, T.J., and Pettit, D. 2000. "Earth's atmosphere before the age of dinosaurs," *Chemical Innovation*, **30**, 50–55.

Perry, R. H. and Green, D. 1984. *Perry's Chemical Engineers' Handbook*, 6th edn., McGraw-Hill, New York, NY.

Schaefer, C. A. and Thodos, G. 1959. "Thermal conductivity of diatomic gases: Liquid and gaseous states," *AIChE Journal*, **5**, 367–372.

Schmidt, R. and Housen, K. 1995. "Problem solving with dimensional analysis," *The Industrial Physicist*, **1**, 21–24.

Sherwood, T. K., Pigford, R. L., and Wilke, C. R. 1975. *Mass Transfer*, McGraw-Hill, New York, NY.

Skorobogatiy, M. and Mahadevan, L. 2000. "Folding of viscous sheets and filaments," *Europhysics Letters*, **52**, 532–538.

Stroock, A. D., Dertinger, S. K. W., Ajdari, A., *et al.* 2002. "Chaotic mixer for microchannels." *Science*, **295**, 647–651.

Zlokarnik, M. 1991. *Dimensional Analysis and Scale-Up in Chemical Engineering*, Springer-Verlag, New York, NY.

EXERCISES

Units and Dimensions

5.1 Consider the following table.

Unit	Base dimensions	Type of unit	Converted to SI units
second	T	base	1 s
gram	M	multiple	1×10^{-3} kg
liter	L^3	derived	1×10^{-3} m^3

Construct a similar table for the following units: acre, Btu, carat, parsec, kilobar, and centipoise.

5.2 Show that the following equations are dimensionally consistent.

(A) Einstein's mass–energy equation, $E = mc^2$, where E is energy, m is mass, and c is the velocity of light.

(B) Newton's law of gravitation,

$$F = G\frac{m_1 m_2}{d^2},$$

where F is the force between two masses m_1 and m_2 separated by a distance d, and G is the universal gravitational constant.

(C) Einstein's equation for the molar heat capacity of a solid, $\bar{C}_V$ (in J/(mol·K)),

$$\bar{C}_V = 3R\left(\frac{hv}{kT}\right)^2 \frac{\exp(hv/kT)}{(\exp(hv/kT) - 1)^2},$$

where R is the gas constant, h is the Planck constant, v is a frequency, k is the Boltzmann constant, and T is the temperature of the solid.

(D) The Clausius–Clapeyron equation,

$$\frac{\Delta\bar{H}_{vap}}{T(\bar{V}_{gas} - \bar{V}_{liquid})} = \frac{dP}{dT},$$

which predicts the change in vapor pressure, P, caused by a change in temperature, T. $\Delta\bar{H}_{vap}$ is the molar heat of vaporization (in J/mol) and $\bar{V}_{gas}$

and $\bar{V}_{liquid}$ are the molar volumes of the gas and liquid phases, respectively. **Hint:** Use the definition of a derivative,

$$\frac{dy}{dx} = \lim_{x_1 \to x_2} \frac{y_2 - y_1}{x_2 - x_1},$$

to determine the dimensions of a derivative.

5.3 Show that the following equations are dimensionally consistent.

(A) The ideal gas law, $PV = nRT$, where P is pressure, V is volume, n is gas mols, R is the gas constant, and T is temperature.

(B) The Hagen–Poiseuille (pronounced *pwah-zø-yah*) law for the volumetric flow rate, Q, of a fluid with viscosity μ through a pipe of radius r and length ℓ, with a pressure drop ΔP:

$$Q = \frac{\pi(\Delta P)r^4}{8\mu\ell}.$$

(C) The energy of an ideal gas, given by the equation

$$\rho\bar{C}_V\left(\frac{dT}{dt}\right) = k\left(\frac{d^2T}{dx^2}\right) - P\left(\frac{dv}{dx}\right),$$

where ρ is the molar density (in mol/m^3), $\bar{C}_V$ is the molar heat capacity at constant volume (in J/(mol·K)), dT/dt is the derivative of temperature with respect to time, k is the thermal conductivity, d^2T/dx^2 is the second derivative of temperature with respect to distance, P is pressure, and dv/dx is the derivative of velocity with respect to distance. See the Hint in Exercise 5.2(D).

5.4 Condensation is an important phenomenon in liquid–vapor separations. In a common arrangement, liquid condenses on chilled vertical tubes. In 1916, William Nusselt (see Exercise 5.28) derived the following relation for the heat-transfer coefficient between a vapor of a pure substance at its dew point and the chilled surface of a vertical tube:

$$h = 0.943\left(\frac{k^3\rho^2 g\Delta\bar{H}_{vap}}{\ell\mu(\Delta T)}\right)^{1/4},$$

where h is the mean heat-transfer coefficient (in kJ/(m^2·s·°C)), k is the thermal conductivity of the liquid (in kJ/(m·s·°C)), ρ is the liquid density (in kg/m^3), g is gravitational acceleration (9.8 m/s^2), $\Delta\bar{H}_{vap}$ is the specific heat of vaporization (in kJ/kg), μ is the liquid viscosity (in Pa·s), ℓ is the tube length (in m), and ΔT is the temperature difference between the tube and the fluid (in °C). What are the units of the constant, 0.943?

Deriving Dimensionless Groups

5.5 The mean velocity v of a molecule in a gas is determined by the mass m of the molecule, the temperature T of the gas, and the Boltzmann constant, $k = 1.38 \times 10^{-23}$ J/K.

(A) List the dimensions of each of these four quantities.

(B) Combine the four parameters v, m, T, and k to form a dimensionless group.

5.6 The molecules in a fluid move constantly. On average, a molecule will diffuse a distance d in a time period t. Assume molecular diffusion may be described by the following parameters: distance (d), time (t), mass (m), and diffusion coefficient (D, with $[D] = L^2/T$).

(A) Derive the dimensionless group(s) of molecular diffusion.
(B) The distance diffused is proportional to what power of t? That is, for $d \propto t^x$, what is x?

5.7 A rock of mass m falls in a vacuum under the acceleration of gravity, g. Express the distance s that it falls in terms of m, g, and t. Compare your relation to the relation obtained in elementary physics.

5.8 An instantaneous explosion (from high explosives or a nuclear device) causes a blast wave, a shock wave that expands supersonically from the source. For atmospheric explosions far above the ground, the radius of a blast wave r depends on the energy E of the detonation, the time t since the detonation, and the air density, ρ. Express the blast wave radius r in terms of E, t, and ρ.

5.9 When a viscous liquid is poured from a sufficient height, the liquid stream coils and/or folds as it impinges on a flat surface. Viscous coiling is observed for common fluids such as syrup on pancakes and vegetable oil on a cold skillet. The onset of this phenomenon is described by five parameters: h, the height of the liquid source above the surface; r, the characteristic radius of the liquid stream; μ, the liquid viscosity; ρ, the liquid density; and g, gravitational acceleration.

Derive the dimensionless group or groups that describe viscous coiling of the stream of an impinging liquid. For the first core variable, use r. If there is a second dimensionless group, use μ as the core variable. If there is a third dimensionless group, use ρ.

Skorobogatiy and Mahadevan (2000).

5.10 Waves on a liquid surface below a gas (such as air) are described by the following parameters:

Parameter	Symbol	Units
wave velocity	v	m/s
wavelength	λ	m
liquid density	ρ	kg/m^3
liquid viscosity	μ	Pa·s
liquid surface tension	γ	N/m
gravitational acceleration	g	m/s^2

Derive a set of dimensionless groups that describe waves on a liquid surface. Use the wave velocity v as the core variable for the first dimensionless group. If there is a second dimensionless group, use μ as its core variable. If there is a third dimensionless group, use γ as its core variable. If there is a fourth dimensionless group, use g as its core variable.

5.11 In Section 5.3.1, three dimensionless groups were derived to describe fluid flow through a pipe. Core variables ΔP, ℓ, and μ were chosen and we obtained the Euler number, the reduced length, and the Reynolds number.

(A) Derive the dimensionless groups that correspond to the core variables v, ρ, and ℓ.
(B) Express the Euler number and the Reynolds number in terms of the groups derived in (A).

5.12 We wish to determine the rate at which a liquid drains from a large tank with a small hole in the bottom. This system is described by the following parameters:

Parameter	Symbol	Units
height of fluid	h	m
diameter of hole	d	m
volumetric flow rate	Q	m^3/s
fluid density	ρ	kg/m^3
fluid viscosity	μ	Pa·s
gravitational acceleration	g	m/s^2

Derive a dimensionless group or set of dimensionless groups that characterize this system.

5.13 Heat transfer by forced convection is characterized by the following parameters:

Parameter	Symbol	Units
pipe diameter	d	m
fluid velocity	v	m/s
fluid density	ρ	kg/m^3
fluid specific heat capacity	$\bar{C}_P$	J/(kg·°C)
fluid viscosity	μ	Pa·s
fluid thermal conductivity	k	J/(m·s·°C)

Derive a dimensionless group or set of dimensionless groups that characterize this system.

5.14 The unsteady-state rate of heat conduction into a solid sphere is characterized by the following parameters:

Parameter	Symbol	Units
sphere diameter	d	m
time elapsed	t	s
sphere density	ρ	kg/m^3
sphere specific heat capacity	$\bar{C}_P$	J/(kg·°C)
sphere thermal conductivity	k	J/(m·s·°C)

Derive a dimensionless group or set of dimensionless groups that characterize this system.

5.15 Fan-assisted ovens, used by serious chefs, circulate hot air around the food being cooked. We wish to perform experiments to determine the rate q (in units kJ/s) at which heat transfers from the hot air to the food. We approximate our food, such as a potato or a Thanksgiving turkey, as a sphere of diameter d. We hypothesize that the heat-transfer rate will also depend on the temperature difference between the air and the food (ΔT), the air's specific heat capacity ($\bar{C}_P$ in units kJ/(kg·°C)), the air density (ρ), the air viscosity (μ), and the air velocity (v).

(A) Derive dimensionless groups for cooking with a fan-assisted oven. Choose core variables in the following order: q, ΔT, μ, ρ, $\bar{C}_P$, d, and v.

(B) Design experiments to determine the equation for the rate of heat transfer in a fan-assisted oven. That is, describe what parameters to vary to determine the constants a, b, c, etc. in an equation of the form

$$\Pi_1 = a(\Pi_2)^b(\Pi_3)^c(\Pi_4)^d \cdots.$$

The density, viscosity, and heat capacity of air as a function of temperature are available. You may assume you have a heat flux sensor to measure q. The fan speed on your oven has only one setting; v is fixed.

5.16 A gas is bubbled through a fluid by forcing the gas through an orifice plate. The bubble diameter depends on the orifice diameter and the fluid properties listed below.

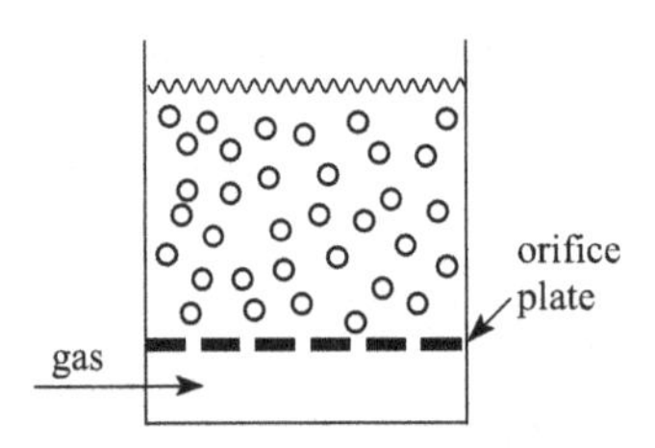

Parameter	Symbol	Units
bubble diameter	d_b	m
orifice diameter	d_o	m
fluid density	ρ	kg/m^3
fluid viscosity	μ	Pa·s
fluid surface tension	γ	N/m
gravitational acceleration	g	m/s^2

Derive dimensionless groups for this system. Use d_b as the core variable for the first dimensionless group. If there is a second dimensionless group, use μ as its core variable. If there is a third dimensionless group, use γ as its core variable. If there is a fourth dimensionless group, use g as its core variable.

5.17 The Prandtl number, Pr, is a dimensionless number used to characterize heat transfer in fluids. The Prandtl number comprises three parameters: the fluid specific heat capacity ($\bar{C}_P$, in J/(kg·°C)), the fluid viscosity (μ, in Pa·s), and the fluid thermal conductivity (k, in J/(m·s·°C)). Given that the exponent of $\bar{C}_P$ is 1, derive the exponents of μ and k.

5.18 The Péclet number is a dimensionless group that appears in the analysis of heat transfer via forced convection. The Péclet number comprises five parameters: a characteristic length (ℓ), fluid velocity (v), specific heat capacity ($\bar{C}_P$), density (ρ), and thermal conductivity (k). Given that the exponent of ρ is 1, derive the Péclet number.

5.19 The Lewis number is a dimensionless group composed of four parameters: fluid density (ρ), fluid specific heat capacity ($\bar{C}_P$), fluid thermal conductivity (k), and molecular diffusivity (D). Diffusivity D appears in Fick's second law,

$$\frac{dC}{dt} = D\frac{d^2C}{dx^2},$$

where t is time, x is distance, and C is concentration (in units of mol/volume). Use k as the core variable and derive an expression for the Lewis number.

5.20 For steady laminar flow in a cylindrical tube (Re < 1000), the fluid velocity profile is a paraboloid. A cross-section of the velocity profile is parabolic, as shown below. The fluid velocity is zero at the tube wall and is highest at the tube center, as shown by the velocity vectors.

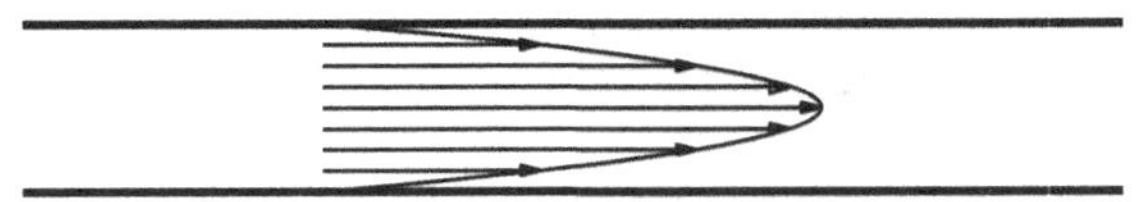

For pulsed laminar flow, such as the heart's rhythmic pumping of blood, the fluid velocity profile can change, as shown below.

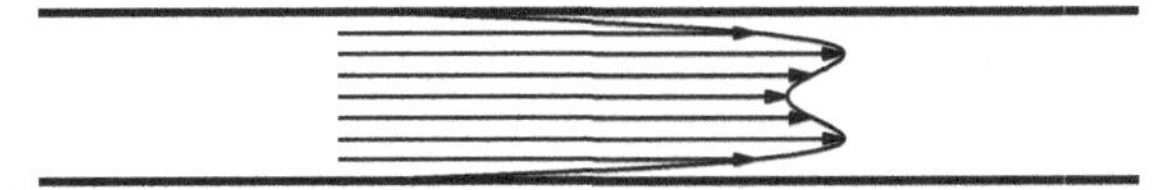

The change in velocity profile affects, for example, cholesterol deposition on arterial walls and the radial concentration distribution of blood cells. A dimensionless group called the Womersley number, α, characterizes the deviation from a parabolic profile in pulsed flow. The parameters are the tube radius (r), the fluid density (ρ), the fluid viscosity (μ), and the pulse frequency (ω, units of radians/s).

(A) Given that r is the core variable, derive the Womersley number.

(B) For Re = 1000, the parabolic velocity profile distorts for $\alpha > 10$. Calculate the radius of an artery for which $\alpha = 10$. For blood at 20°C, $\mu = 3.5 \times 10^{-3}$ Pa·s and $\rho = 1060$ kg/m^3. A typical heart rate of 72 beats/minute (1.2 cycles/s) is a pulse frequency $\omega = 2\pi \times 1.2$ radians/s = 7.5 radians/s.

5.21 In Section 5.4 we derived a set of dimensionless groups to describe the terminal velocity of a sphere moving through a fluid. Bubbles rising through liquids, however, are not necessarily spheres. Large bubbles with low surface tension deform into flattened spheres and even convex shapes. The shape of a rising bubble is determined by six parameters: the bubble diameter (d), the bubble velocity (v), the liquid density (ρ), the liquid viscosity (μ, units of Pa·s), the gas–liquid surface tension (γ, units of N/m), and gravitational acceleration (g).

(A) Derive the dimensionless group or groups that describe a gas bubble rising in a liquid. For the first core variable, use μ. If there is a second dimensionless group, use γ as the core variable. If there is a third dimensionless group, use g. If there is a fourth dimensionless group, use ρ.

(B) Some dimensionless correlations for rising bubbles use the Eötvös group (Eo) and the Morton group (Mo):

$$\text{Eo} = \frac{\rho d^2 g}{\gamma}, \quad \text{Mo} = \frac{g\mu^4}{\rho\gamma^3}.$$

Eo is the ratio of gravitational to surface (capillary) forces and Mo contains only fluid properties. What are the core variables for Eo and Mo?

(C) Express Eo and Mo in terms of the dimensionless groups you derived in (A). For example, Eo can be expressed by the general equation

$$\text{Eo} = (\Pi_1)^a(\Pi_2)^b \cdots (\Pi_n)^n.$$

Find the exponents a, b, etc.

(D) If Eo and Mo are the first two dimensionless groups in an alternative representation of the rising bubble system, are there additional dimensionless groups? If so, what is (are) the core variable(s) of the other dimensionless group(s) in this representation?

(E) Derive the additional dimensionless group(s) predicted in part (D), if any. If you find fractional exponents, multiply each exponent by the least common denominator to obtain integral exponents.

5.22 As the size of a system decreases, the ratio of volume to surface decreases as the characteristic length of the system. For example, the ratio of a sphere's volume to its surface area is $(4/3)\pi r^3/(4\pi r^2) \propto r$.

(A) In multiphase systems, the ratio of gravitational potential energy (a volume property) to interface energy (a surface property) decreases with the size of the system. As a system becomes smaller, surface tension becomes important. For example, in macroscopic channels, immiscible liquids combined at a T-junction will flow in parallel. The potential energy benefit of the higher density fluid (A) below the lower density fluid (B) outweighs the surface energy cost of the large interface between the two fluids.

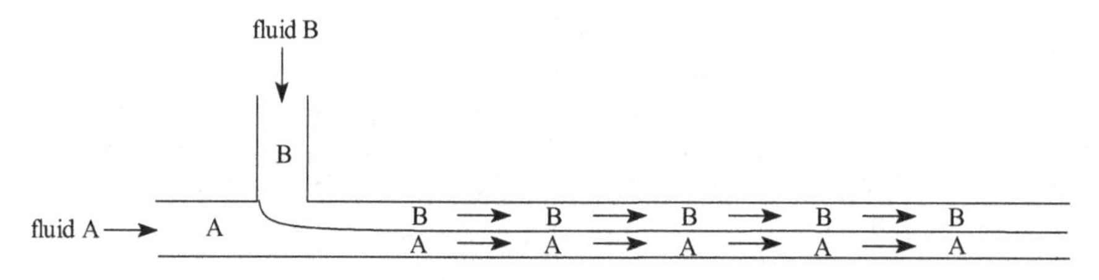

But in microscopic channels, surface effects dominate and the flow breaks into droplets, called slug flow.

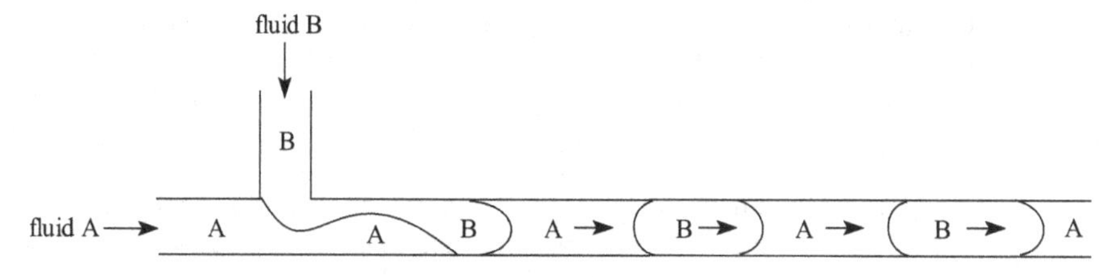

The relative importance of gravitational and surface forces is characterized by a dimensionless group known as the Bond number, Bo. Surface forces are important when Bo < 10. This group comprises the density difference between fluids ($\Delta\rho$), the surface tension between the two fluids

(γ, units of N/m), the characteristic size for the system (d), and gravitational acceleration (g). Set the exponent on g equal to 1 and derive the Bond number.

(B) The ratio of inertial energy (a volume property) to interface energy (a surface property) affects flow patterns in multiphase systems. Fluid velocity also affects the transition from parallel flow to slug flow; at higher velocity the droplets shorten from plugs to spheres. The relative importance of inertial and surface forces is characterized by a dimensionless group known as the Capillary number, Ca. Surface forces become important when $Ca < 0.001$, which occurs in flow through porous media, such as oil in porous rock and coffee in a dunked cookie. The Capillary group comprises the fluid velocity (v), the fluid viscosity (μ), and the surface tension between the two fluids (γ). Set the exponent on v equal to 1 and derive the Capillary number.

5.23 The dynamics of drop formation from a tube, as shown by the sequence of sketches below, is important in chemical engineering applications such as spray-drying, inkjet printers, paint sprayers, micro-dispensing DNA solutions, and leaky faucets.

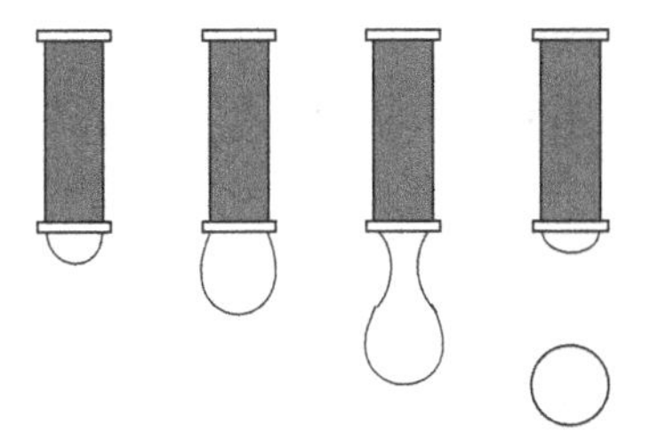

Drop formation is characterized by the following parameters:

Parameter	Symbol	Units
tube diameter	d	m
fluid velocity	v	m/s
fluid density	ρ	kg/m^3
fluid viscosity	μ	Pa·s
fluid surface tension	γ	N/m
gravitational acceleration	g	m/s^2

These parameters may be assembled into three dimensionless groups: the Weber number (We), the Ohnesorge number (Oh), and the gravity number (G). Professor Osman Basaran, of the School of Chemical Engineering at Purdue University, developed dimensionless correlations for drop characteristics. For example, he has identified a regime of We and Oh in which the drop's neck pinches off to form a "satellite" drop behind the primary drop, as shown below.

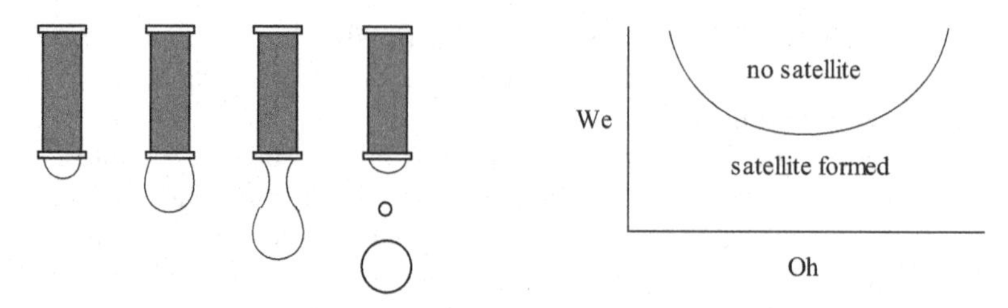

(A) Derive the dimensionless groups We, Oh, and G. Use the core variables v, μ, and g, respectively. Set the exponent on v equal to 2 for the Weber number.

(B) Derive an alternative set of dimensionless groups to describe drop formation. Use the Reynolds number as the first group, $\mathrm{Re} = \rho v d/\mu$. The core variable for Re is μ. Choose core variables for the two other dimensionless groups and derive the other two dimensionless groups, Π_2 and Π_3. If you derive a dimensionless group that is similar to a named group, such as the Euler number, state your result in terms of the named group.

(C) Show that each dimensionless group in the original set (We, Oh, and G) can be expressed as a product (or ratio) of the dimensionless groups in part (B). For example, express We as a product,

$$\mathrm{We} = \mathrm{Re}^a \Pi_2^b \Pi_3^c,$$

and find the exponents a, b, and c. Repeat for Oh and G.

Analyzing Graphical Data

5.24 A fluid of viscosity μ and density ρ flows at velocity v through a rough pipe of diameter d when pressure drop ΔP is applied over a length ℓ. With all other parameters held constant, ΔP is varied and v is measured.

ΔP (10^5 Pa)	v (m/s)
1	1
4	2
9	3
16	4
100	10

(A) Derive a (set of) dimensionless group(s) that characterizes this system.

(B) Refer to Figure 5.20. Is the flow in the pipe laminar, unstable, or turbulent? Explain your conclusion.

(C) How does ΔP depend on μ in this regime? That is, if we write $\Delta P \propto \mu^n$, what is the value of n?

5.25 A fluid is pumped through a pipe that contains an orifice plate. The diameter of the orifice is half the diameter of the pipe, D_1.

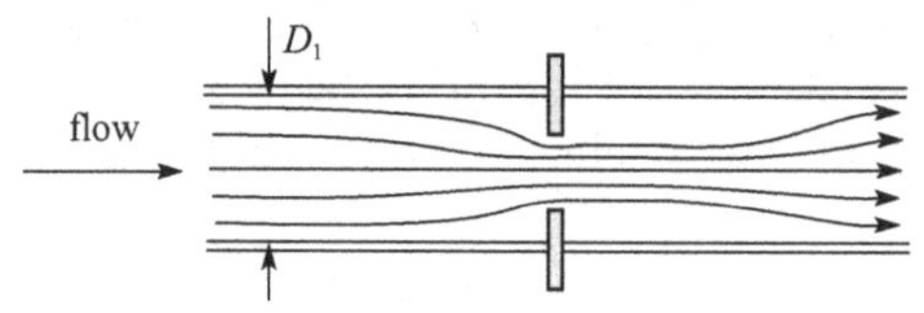

The flow rate $Q_{orifice}$ through the pipe with the orifice is measured as a function of the Reynolds number. $Q_{orifice}$ is divided by the flow rate Q_0 through the pipe without the orifice, and the result is plotted below. Note that the data are plotted on a semilog coordinate system, not a log–log system. Obtain an expression for the flow rate through the orifice (i.e. $Q_{orifice} = ?$) for Re < 100. Check your expression by substituting at least two points in the range Re < 100.

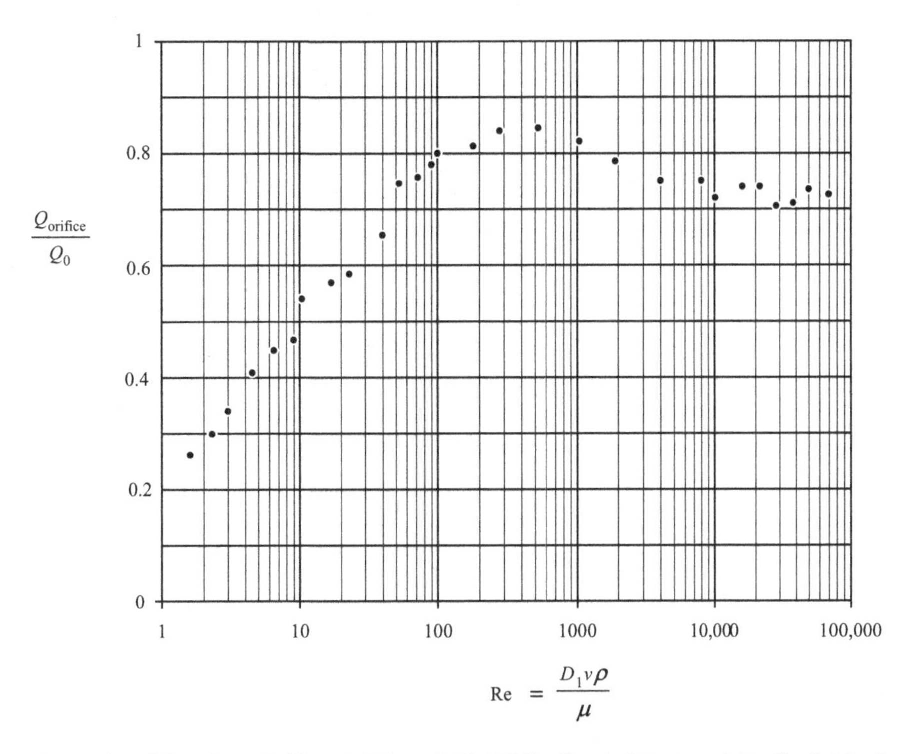

Figure adapted from Perry, R. H., and Chilton, C. H. (1973). *Chemical Engineers' Handbook*, 5th ed. McGraw-Hill, New York, figure **5**-18, p. **5**-13.

5.26 Use the dimensionless correlation and the table of critical constants to find the thermal conductivity k for nitrogen (N_2) at 500 K and 2000 atm.

Substance	T_c (K)	P_c (atm)	k_c (J/(s·m·K))
Br_2	584	102	0.0284
CO	133	34.5	0.0360
Cl_2	417	76.1	0.0402
F_2	144	55	0.0399
HBr	363	84	0.0344
HCl	325	82	0.0469
HF	503	95	0.1088
I_2	785	116	0.0265
NO	180	64	0.0495
N_2	126	33.5	0.0358
O_2	155	50	0.0435

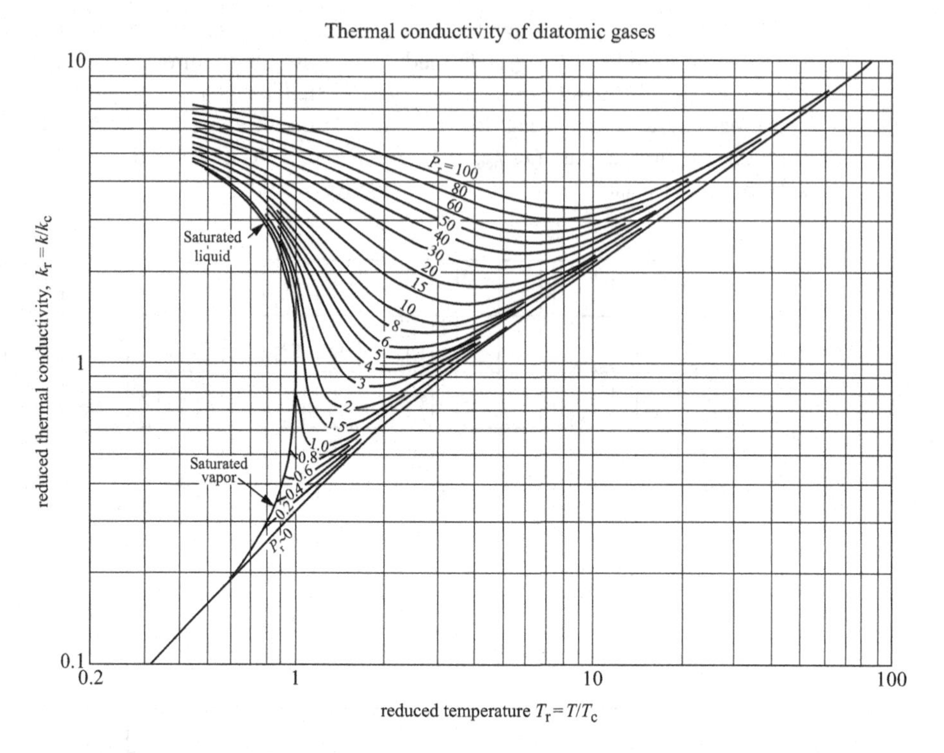

Figure from Schaefer and Thodos, AIChE Journal, **20** 367–362 (1959), reprinted in *Chemical Engineers' Handbook*, 5th edition, R. H. Perry and C. H. Chilton, McGraw-Hill **5**-12, (1973).

5.27 As discussed in Section 5.3.3, heat transfer to a fluid flowing through a tube may be described by three dimensionless groups: the Reynolds number (Re), the Prandtl number (Pr), and the Stanton number (St). Use the line drawn through the data in Figure 5.24 to derive an equation that relates Re, Pr, and St. Specifically, derive an equation of the form St = . . .

5.28 Heat transfer from a solid sphere moving through a fluid is characterized by three dimensionless numbers: the Nusselt number (Nu), the Reynolds number (Re), and the Prandtl number (Pr), respectively defined as

$$\mathrm{Nu} = \frac{hd}{k}, \quad \mathrm{Re} = \frac{dv\rho}{\mu}, \quad \text{and} \quad \mathrm{Pr} = \frac{\bar{C}_P\mu}{k},$$

where h is the heat-transfer coefficient, ρ is the fluid density, v is the sphere velocity, k is the fluid thermal conductivity, μ is the fluid viscosity, d is the sphere diameter, and $\bar{C}_P$ is the fluid specific heat capacity. Use the graph below to calculate h for a solid sphere of diameter d = 3.7 mm moving through water at v = 0.29 m/s.

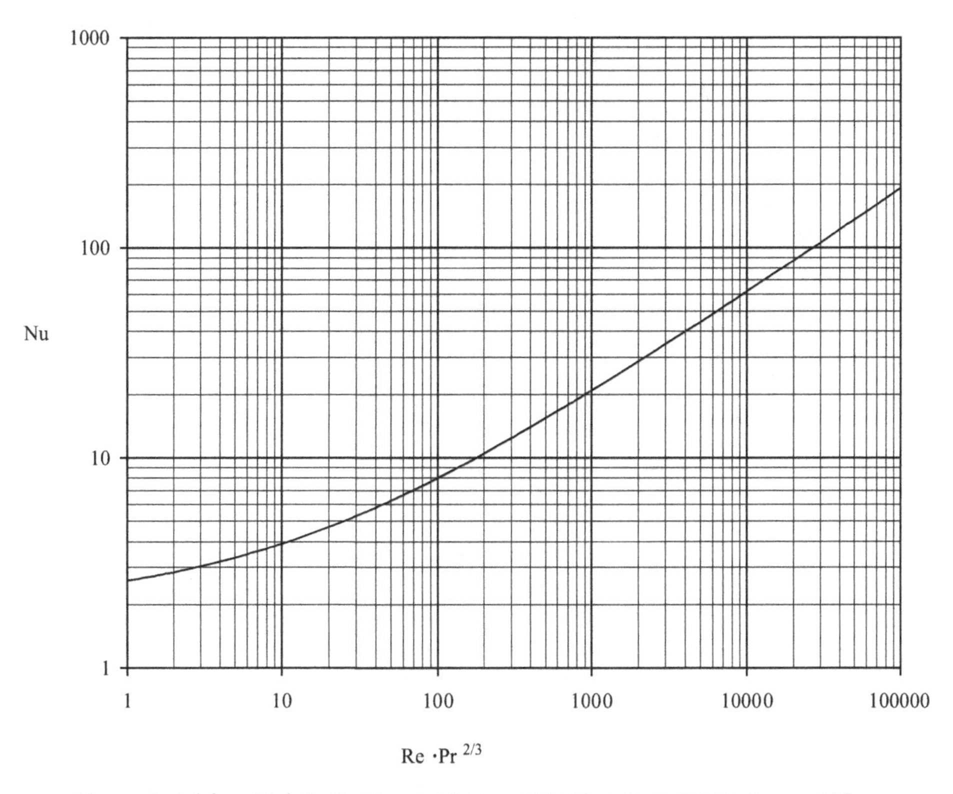

Figure adapted from Bird, R. B., Stewart, W. E., and Lightfoot, E. N. (1960). *Transport Phenomena*, Wiley, New York, figure 13.3-2, p. 409.

5.29 A cold metal sphere warms as it passes through a hot fluid. As stated in the preceding exercise, this phenomenon is described by three dimensionless groups: the Reynolds number (Re), the Nusselt number (Nu), and the Prandtl number (Pr). The desired values for the parameters in this exercise are h = 2500 J/(m^2·s·K), v = 0.17 m/s, k = 0.59 J/(m·s·K), $\mu = 1.0 \times 10^{-3}$ Pa·s, ρ = 1000. kg/m^3, and $\bar{C}_P$ = 4170 J/(kg·K).

Use the graph in the preceding exercise to determine the diameter of the sphere needed to obtain these parameters. Note that because d appears in both Nu and Re, this is not a straightforward task. There are several approaches; one is outlined in steps (A), (B), and (C) below. You are not obliged to follow this approach; you may devise your own.

(A) Start with the definition for Nu and derive an equation of the form $d = f_1(\text{Nu})$.

(B) Start with the definitions for Re and Pr and derive an equation of the form $d = f_2(\text{Re}\cdot\text{Pr}^{2/3})$.

(C) Plot the function $f_1(\text{Nu}) = f_2(\text{Re}\cdot\text{Pr}^{2/3})$ on the graph in the preceding exercise to find d.

5.30 Hot air is used to pop popcorn and separate popped corn from corn kernels, as follows. An upward flow of hot air levitates a corn kernel; the kernel's terminal velocity equals the velocity of the hot air.

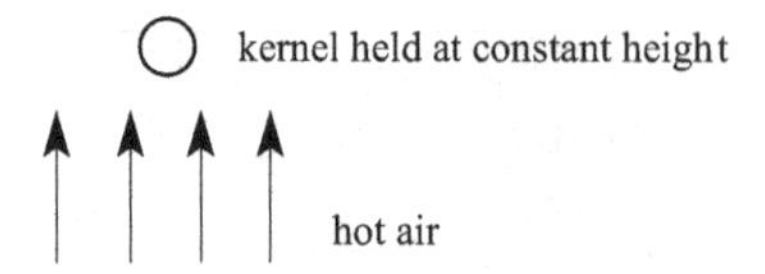

After the kernel pops, the popped corn rises with the hot air.

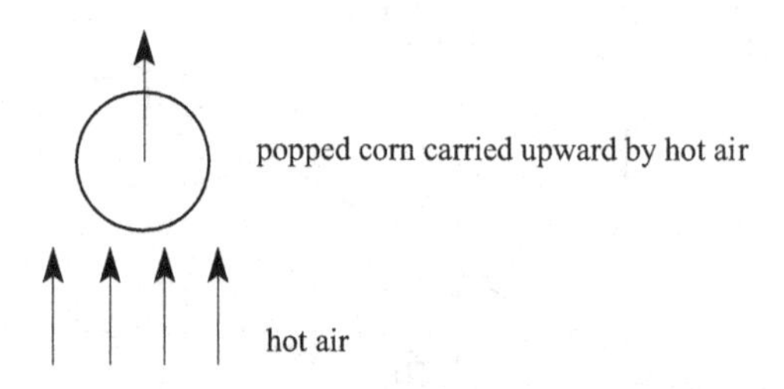

(A) Use the plot below to calculate the velocity of the hot air. Note that velocity appears in both Re and the friction factor. The kernel is approximately spherical of diameter 5 mm and mass 0.07 g. The popped corn is also approximately spherical and has diameter 3 cm and mass 0.05 g. The hot air has density 0.9 kg/m^3 and viscosity 3×10^{-5} Pa·s.

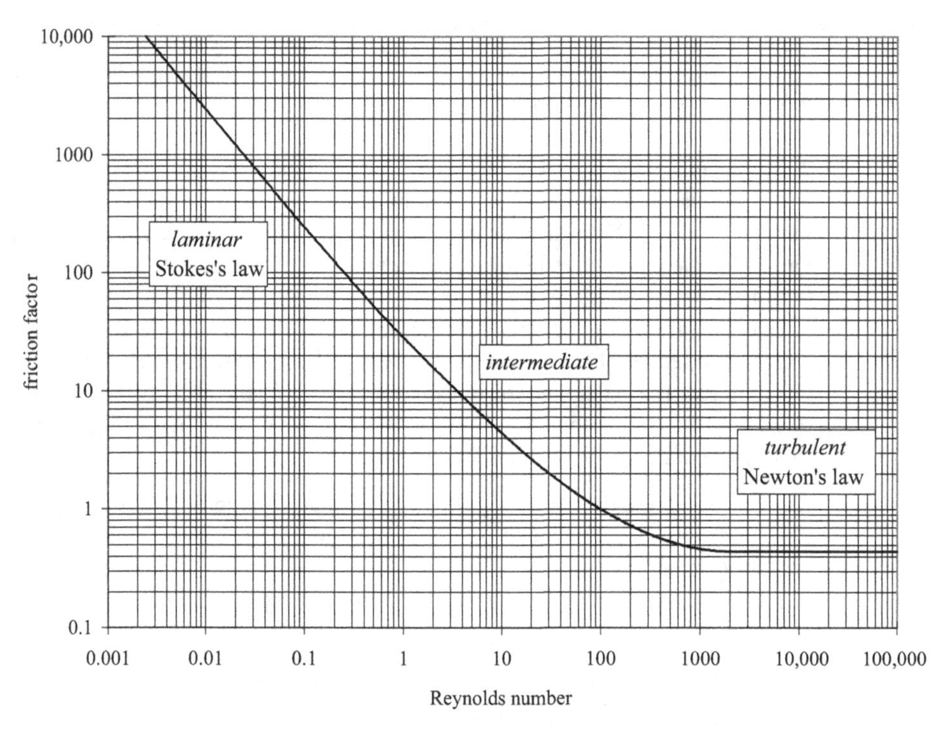

Figure adapted from Bird, R. B., Stewart, W. E., and Lightfoot, E. N. (1960). *Transport Phenomena*, figure 6.3-1.

(B) Calculate the upward velocity of popped corn.

5.31 Mass transfer between a fluid and a sphere moving through the fluid is described by the following four parameters:

Parameter	Symbol	Units
sphere diameter	d	m
sphere velocity	v	m/s
mass transfer coefficient	k	m/s
diffusion coefficient	D	m^2/s

These parameters may be assembled into two dimensionless groups, the Sherwood number (Sh) and the Péclet number (Pe):

$$\mathrm{Sh} = \frac{kd}{D}, \quad \mathrm{Pe} = \frac{vd}{D}.$$

Use the dimensionless correlation below to find the diffusion coefficient D for a sphere of diameter 0.0034 m moving with velocity 0.0025 m/s through a fluid. The mass transfer coefficient is 3.2×10^{-5} m/s.

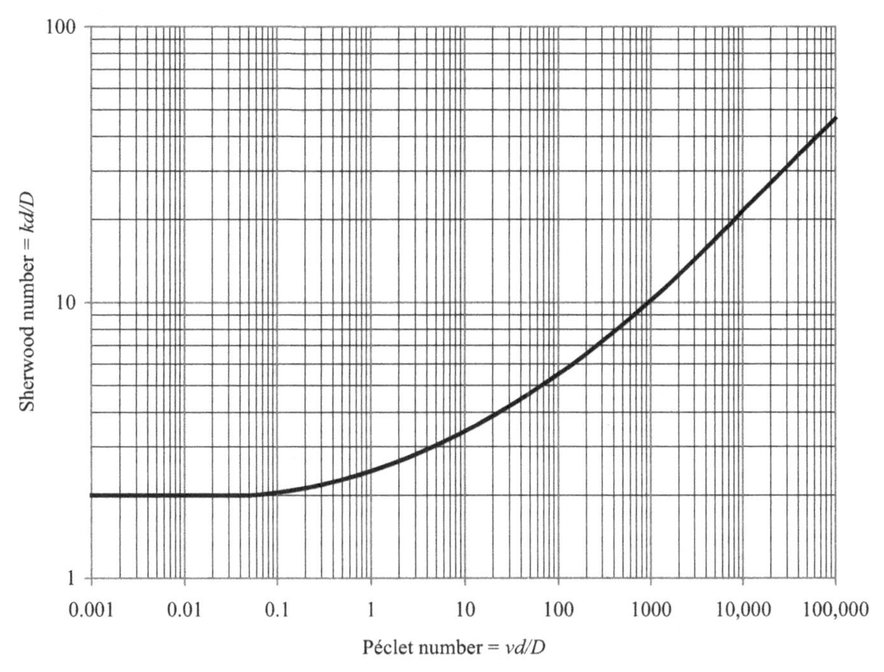

Figure adapted from T. K. Sherwood, R. L. Pigford, and C. R. Wilke, *Mass Transfer*, McGraw-Hill (1975); figure 6.8, p. 216.

Design of Dynamically Similar Models

5.32 Waves formed behind a vessel traveling on the surface of a fluid account for much of the frictional loss. Two dimensionless groups describe this phenomenon: the Froude number and the Reynolds number. The quantity D in both dimensionless groups is the depth of the vessel below the surface of the fluid. We wish to study the waves behind a real battleship by using a model battleship. The real battleship's depth is 10. m and its velocity is 12 m/s.

(A) The model battleship is scaled such that its depth below the fluid surface is 0.62 m. At what velocity should we move the model through the fluid?

(B) Calculate the desired viscosity of the fluid in the model system. Assume the model fluid will have a typical density of 1.0×10^3 kg/m^3.

5.33 A meteor of mass m and density ρ_{meteor} strikes the Earth with velocity v and creates a crater of volume V. Additional parameters are the soil density ρ_{soil} and gravitational acceleration g. A set of three dimensionless groups that describe this event are

$$\frac{\rho_{meteor}}{\rho_{soil}}, \quad \frac{\rho_{soil} V}{m}, \quad \text{and} \quad \frac{g}{v^2}\left(\frac{m}{\rho_{meteor}}\right)^{1/3}.$$

We wish to model the impact of an enormous meteor ($m = 5.2 \times 10^8$ kg), a round rock of diameter 100. m with velocity 1.1 km/s (2500 mph). Our model will use a metal ball with twice the density of the meteor and with velocity 0.26 km/s. We will perform our experiment in a large centrifuge, so gravitational acceleration is replaced by centrifugal acceleration, which is $500. \times g_{earth} = 4.9$ km/s^2. The density of the soil in our model is double the density of the soil at the predicted point of impact. What is the mass of the "meteor" in our model?

Adapted from Schmidt and Housen (1995).

5.34 Dimensional analysis can be applied to study Earth's plate tectonics. The goal is to design a small model whose behavior over a period of minutes will predict the movement of geological structures that develop over millions of years.

The Earth's crust is solid rock over molten rock. Professor Eberhard Bodenschatz and his research group model the Earth's tectonics with a solid wax crust above melted wax. They observe the effects of stretching, compressing, and shearing of the wax crust. A detailed description and pictures of wax tectonics are posted at http://foalab.earth.ox.ac.uk/publications/Katz_etal_NJP05.pdf.

Spreading tectonic plates are characterized by the parameters in the table below.

Parameter	Symbol	Units	Earth	Wax model
crust thickness	h	m	4.0×10^4	
velocity of moving plate	v	m/s	1.0×10^{-9}	
crust density	ρ	kg/m^3	3,300	790
thermal conductivity	k	J/(m·s·°C)	2.5	0.087
specific heat capacity	$\bar{C}_P$	J/(kg·°C)	1,360	2,260
specific heat of melting	$\Delta\bar{H}_{melt}$	J/kg	3.67×10^5	2.2×10^5
difference between melting point and surface temperature	$T_m - T_s$	°C	1200	

(A) How many dimensionless groups are needed to describe the tectonics of spreading plates?

(B) Derive the dimensionless groups. Use $\Delta\bar{H}_{melt}$ as the core variable of the first dimensionless group. If there is a second group, use k as the core variable. If there are third, fourth, and fifth groups use v, $\bar{C}_P$, and ρ, respectively, for the core variables.

(C) A wax model used to simulate spreading plate tectonics typically has a crust thickness of a few centimeters, and the difference between the wax melting point and the surface temperature is about 50°C. Typically the wax is pulled apart at about 1 mm/s. Is this model dynamically similar to the plate tectonics of the Earth?

5.35 Three dimensionless groups describe flow through an artificial kidney: the Reynolds number Re, the Euler number Eu, and the reduced length. Use these groups to design a model for flow through an artificial kidney. In this hypothetical artificial kidney, blood (density = 1000. kg/m^3 and viscosity = 1.0×10^{-2} Pa·s) is pumped through 10. m of tubing of diameter 0.0010 m. The average velocity of the blood is 0.010 m/s. To determine the pressure drop in the artificial kidney, we want to measure the pressure drop in a model using glycerin (density = 1260. kg/m^3, viscosity = 1.49 Pa·s) flowing through a tube of diameter 0.0050 m.

(A) What length of tube do we need for the model system?
(B) What should be the average velocity of the glycerin in our model system?
(C) Assume we measure a pressure drop of 2.25×10^7 Pa in our model system. What pressure drop do we expect for blood flowing through 10. m of tube in our artificial kidney?
(D) The walls of the tubing in the artificial kidney are very thin to enhance the removal of wastes from the blood. Consequently, the maximum pressure drop must be less than 1.0×10^4 Pa. Given this limit on the pressure drop, clearly our artificial kidney cannot be a single tube 10 m long. Rather, it must be many short tubes in parallel such that the total length is 10 m. You may assume the pressure changes linearly with tube length. What is the maximum length of one of the many short tubes in parallel?

Warning: blood flow in a kidney is more complex than described here. Do not use these parameters to design an artificial kidney.

5.36 We need a convenient, dynamically similar model for heat transfer from an explosive, toxic fluid flowing in a large pipe. The proposed model uses water in a small pipe and is described in the table below. Heat transfer from the flowing fluid is completely described by the seven parameters below.

Parameter	Symbol	Units	Actual system	Proposed model
pipe diameter	d	m	1.0	0.020
fluid density	ρ	kg/m^3	770	1,000.
fluid viscosity	μ	Pa·s	1.0	0.0010
fluid specific heat capacity	$\bar{C}_P$	J/(kg·°C)	1,470	4170
fluid thermal conductivity	k	J/(m·s·°C)	0.21	0.59
fluid velocity	v	m/s	2.6	0.10
heat-transfer coefficient	h	J/(m^2·s·°C)	to be predicted from the model	to be measured with the model

Is the model valid for predicting the heat-transfer coefficient in the actual system? You must justify your answer for full credit.

5.37 A cold metal sphere is warmed as it passes through a hot fluid. This phenomenon is described by three dimensionless groups: the Reynolds number (Re), the Nusselt number (Nu), and the Prandtl number (Pr), defined in Exercise 5.28.

We wish to analyze a microscopic process with a macroscopic model. In the microscopic process a copper sphere of diameter 1.0×10^{-4} m moves through water at a velocity of 1.2×10^{-3} m/s. In our model we will use a copper sphere of diameter 0.10 m in water.

(A) At what velocity should our model sphere move through the water?
(B) We measure a heat-transfer coefficient of 2500 J/(m^2·s·K) with our model. What is the heat-transfer coefficient in the microscopic system?

5.38 When an impeller stirs a fluid in a cylindrical tank, a vortex forms; the liquid surface falls at the center and rises at the wall. The parameters that describe this system are given below.

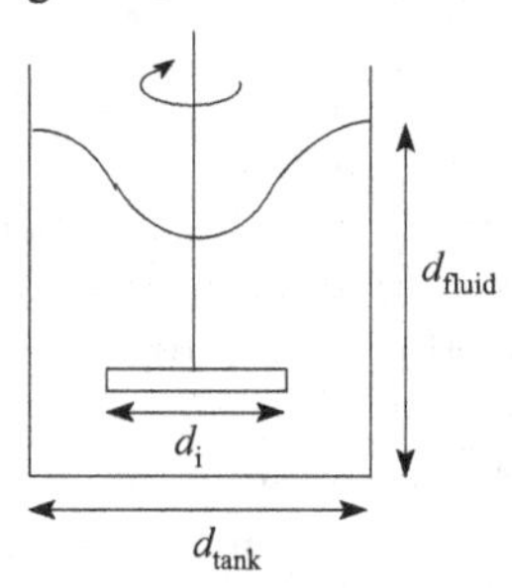

Parameter	Symbol
impeller rotation rate	n
impeller diameter	d_i
tank diameter	d_{tank}
fluid depth at wall	d_{fluid}
fluid density	ρ
fluid viscosity	μ
gravitational acceleration	g

A vortex in a stirred tank is described by four dimensionless groups:

$$\frac{d_{fluid}}{d_i}, \quad \frac{d_{tank}}{d_i}, \quad \frac{d_i^2 n\rho}{\mu}, \quad \text{and} \quad \frac{d_i n^2}{g}.$$

As part of a microanalytical device, blood is stirred in a microtank (diameter = 0.001 m, impeller diameter = 0.0007 m, impeller rotation rate = 0.57/s). Design a conveniently-sized model to predict the fluid depth in the microtank. Choose a fluid for your model (blood, water, vegetable oil, or glycerin) and specify parameter values for your model. Gravitational acceleration is a constant; you cannot change g in your model.

Fluid	Density (kg/m^3)	Viscosity (Pa·s)
water	1,000.	0.0010
blood	1,025.	0.0040
vegetable oil	930.	0.050
glycerin	1250.	1.2

5.39 In the macroscopic everyday world, fluids mix by convection. However, in the microscopic world, temperature gradients and concentration variations do not cause convection. Rather, fluids mix by molecular diffusion. The Péclet number (Pe), which characterizes the ratio of convective heat transfer to conductive heat transfer (see Exercise 5.18), also characterizes the ratio of convective mass transfer to diffusive mass transfer. For mass transfer (mixing), the form of the Péclet number is

$$\text{Pe} = \frac{vd}{D} = \frac{v}{D/d} = \frac{\text{flow velocity}}{\text{diffusion velocity}},$$

where v is the fluid velocity, d is the diffusion distance, and D is the diffusion coefficient. Diffusivity is a property of the dissolved or suspended substance, the solvent, and the temperature. The diffusion coefficient decreases with particle size, as shown by typical values of D for particles in water at 25°C.

Particle	Size	Diffusion coefficient D
glycine (75 g/mol)	0.5 nm	1×10^{-9} m²/s
ribonuclease (13,700 g/mol)	2 nm	1×10^{-10} m²/s
myosin (4.4×10^{5} g/mol)	20 nm	1×10^{-11} m²/s
virus	0.1 μm	2×10^{-12} m²/s
bacterium	1 μm	2×10^{-13} m²/s
human cell	10 μm	2×10^{-14} m²/s

(A) When diffusion dominates, differences in diffusivity can be exploited for separations. For example, a device called an H filter extracts a protein from a mixture of water, protein, and cells. Two streams meet and flow side by side. The adjacent flows are laminar; there is no convective mixing.

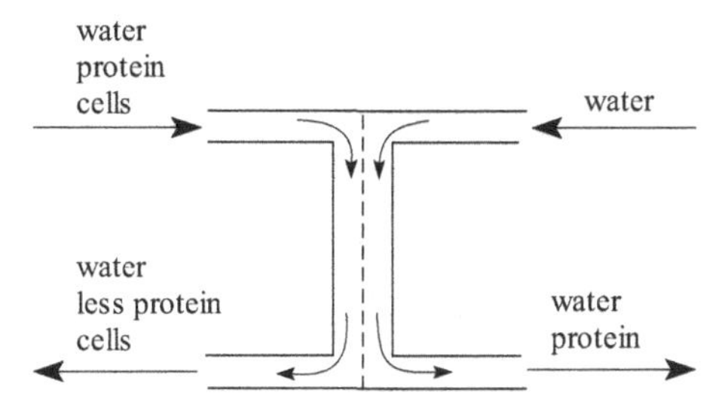

Protein diffuses from the water + protein + cells stream into the pure water stream. The cells also diffuse, but 1000 times more slowly. For a substance to diffuse to a uniform concentration across the interdiffusion zone as it travels through the H filter, the Péclet number must be less than or equal to ℓ/w:

$$\text{Pe} = \frac{v}{D/w} \leq \frac{\ell}{w},$$

where ℓ is the interdiffusion zone length and w is the interdiffusion zone width.[1] Consider an H filter with ℓ = 1.0 cm and w = 20 μm. What fluid velocity is needed to separate a therapeutic protein ($D = 10^{-11}$ m²/s) from cell debris ($D = 10^{-13}$ m²/s)?

(B) How much time is required for the fluid to travel the length of the interdiffusion zone?

(C) If the protein diffuses to a uniform concentration across the interdiffusion zone, what fraction of the protein is extracted?

[1] Squires, M., and Quake, S. R. 2005. "Microfluidics: Fluid physics at the nanoliter scale," *Reviews of Modern Physics*, **77**, 977–1026.

(D) How might one redesign the cell to extract a greater percentage of protein, assuming the protein diffuses to a uniform concentration across the interdiffusion zone?

5.40 Polymer molecules are long chains. For example, polyethylene molecules are typically 10^4 to 10^6 $-CH_2-$ units. A fully extended polyethylene molecule with 10^5 $-CH_2-$ units is about 0.015 mm (15 μm) long. However, polymer molecules in solution curl and fold into spheres. As one will derive in physical chemistry, a polymer molecule composed of N freely jointed units of length ℓ exists as spheres with effective radii given by the radius of gyration,

$$R_g = \left(\frac{N}{6}\right)^{1/2} \ell.$$

Assuming $\ell = 1.5 \times 10^{-10}$ m (1.5 Å) for a $-CH_2-$ unit in a polyethylene molecule, a chain with 105 units forms a sphere of radius 0.02 μm (200 Å).

In microfluidics, the channel diameters are comparable to the length of the polymer molecule, and the flow can elongate the molecular spheres into ellipsoids. Flow in a microchannel stretches the polymer molecules, and thus the flow of a polymer solution (such as a DNA solution) in a microanalytical device will be different than what is observed in a lab-scale device.

The degree of polymer extension in a flow is characterized by the elasticity number (El), a dimensionless group composed of the molecular elasticity force constant k_H, the hydrodynamic resistance of the molecular spheres ξ (the Greek symbol xi, pronounced "*kuh-see*"), the fluid viscosity μ, the fluid density ρ, and the channel width w:

$$\mathrm{El} = \frac{\xi\mu}{k_H \rho w^2}.$$

Use El to estimate the channel width w for which elasticity becomes important for a solution of polyethylene molecules with 10^6 units. That is, for what value of w is the elasticity number approximately 1?

k_H is the proportionality between the force F applied to stretch an elastic material (such as a polymer chain) and the distance x the material stretches:

$$F = k_H x \quad \text{(Hooke's law)}.$$

For a polymer chain, the spring constant is

$$k_H = 3\frac{k_B T}{N\ell^2},$$

where k_B is the Boltzmann constant 1.38×10^{-23} J/K and T is the temperature in K. The hydrodynamic resistance of a sphere of radius R_g in laminar flow (Re < 1) is

$$\xi = 6\pi\mu R_g.$$

5.41 Engineers at your chemical company have proposed a process to produce a new chemical. The key unit in the process is unprecedented. It is also large and expensive, and the phenomena inside the unit are complex. You need to know how this unit will perform under a given set of operating conditions, such as the composition of the input and the temperature.

List a sequence of steps to model this unique unit. The first step is given below. *Do not* execute the steps; just list the steps.

Step 1. Apply dimensional analysis to obtain the Π group(s).

Data Analysis on Spreadsheets

Dimensional analysis allows one to perform experiments with a convenient, dynamically similar model. An engineer should pose two fundamental questions when analyzing the experimental data. First, an engineer should assess the *certainty* that the data exhibit a specific relationship. For example, experiments on walking will yield a collection of Froude numbers and reduced strides. What is the certainty that data for reduced stride vs. Froude number lie on a straight line on a log–log plot? That is, what is the certainty that $s/\ell \propto \mathrm{Fr}^m$? If the certainty is low, perhaps a different functional form is needed.

An engineer's second question is similar, but there is a subtle, important difference. Given a specific functional relationship, what is the *certainty* of the fitted parameters in the equation? For example, *given* that data for walking are represented by the functional form $s/\ell \propto \mathrm{Fr}^m$, what is the certainty that m is a specific value? Are the confidence limits on m ± 0.01, or ± 0.1, or ± 1, for example?

This tutorial illustrates the value of spreadsheets in reducing experimental data, plotting the reduced data, and analyzing trends. Like the tutorial on Mass Balances on Spreadsheets in the Chapter 3 exercises, this tutorial assumes a basic knowledge of spreadsheets, such as how to use a mouse to select a cell and how to copy and paste. However, the tutorial on Mass Balances on Spreadsheets is not a prerequisite.

Spreadsheet Example 1

Experimental data for the terminal velocities of spheres in water, vegetable oil, and glycerin are presented in three tables below. The data are a subset of the experimental data measured by Cornell University first-year engineers to obtain a general correlation between the Reynolds number,

$$\mathrm{Re} = \frac{D v \rho_{\mathrm{fluid}}}{\mu}, \tag{5.63}$$

and the friction factor,

$$\text{friction factor} = \frac{4}{3}\frac{\rho_{\mathrm{sphere}} - \rho_{\mathrm{fluid}}}{\rho_{\mathrm{fluid}}}\frac{1}{\mathrm{Fr}}. \tag{5.72}$$

Use these data to prepare a plot of the friction factor vs. the Reynolds number.

Spheres moving through water.[a]

Sphere	Diameter (mm)	Mass (g)	Height (m)	Time (s)
nylon	3.18	0.019	0.50	5.53
nylon	6.35	0.150	0.50	3.90
nylon	6.35	0.150	0.51	3.80
lucite	3.18	0.021	0.51	4.60
teflon	12.70	2.480	0.51	0.75
glass	2.80	0.038	0.50	1.27
aluminum	6.35	0.361	0.51	0.80
steel	3.18	0.133	0.50	0.76
steel	6.35	1.064	0.51	0.45
steel	9.53	3.572	0.51	0.40
steel	9.53	3.572	0.51	0.35
steel	12.70	8.486	0.51	0.40

[a] Viscosity = 0.0010 Pa·s, density = 1000 kg/m^3.

Spheres moving through vegetable oil.[a]

Sphere	Diameter (mm)	Mass (g)	Height (m)	Time (s)
nylon	3.18	0.019	0.64	34.72
nylon	3.18	0.019	0.64	32.80
lucite	6.35	0.172	0.64	9.49
teflon	3.18	0.038	0.64	7.47
aluminum	6.35	0.361	0.51	2.23
steel	1.59	0.016	0.64	5.05
steel	1.59	0.016	0.64	4.99
steel	1.59	0.016	0.64	4.99
steel	3.18	0.133	0.64	1.94
steel	3.18	0.133	0.51	1.79
steel	6.35	1.064	0.51	1.17
steel	9.53	3.572	0.51	0.70

[a] Viscosity = 0.05 Pa·s, density = 930 kg/m^3.

Spheres moving through glycerin.[a]

Sphere	Diameter (mm)	Mass (g)	Height (m)	Time (s)
polypropylene	3.18	0.014	0.09	41.76
polypropylene	6.35	0.113	0.223	45
nylon	3.18	0.019	0.05	58.79
nylon	3.18	0.019	0.05	64.94
glass	2.80	0.038	0.11	22.56
glass	2.80	0.038	0.11	21.30
aluminum	6.35	0.361	0.23	7.91
steel	1.59	0.016	0.23	33.78
steel	1.59	0.016	0.223	30
steel	3.18	0.133	0.11	3.47
steel	3.18	0.133	0.18	4.98
steel	6.35	1.064	0.23	1.69
steel	6.35	1.064	0.28	1.82

[a] Viscosity = 1.2 Pa·s (absorbed moisture from air), density = 1250 kg/m^3.

Solution to Spreadsheet Example 1

A spreadsheet solution to this exercise is shown in Figures 5.33a and 5.33b. The steps below will lead you through the process of creating this spreadsheet. The method used here is not unique, nor is it the most efficient. If you are skilled at spreadsheets we encourage you to pursue a different approach to produce a column of Reynolds numbers and a column of friction factors.

1. *Enter* the title. Select cell A1, the cell in column A and row 1; use the mouse to position the cursor in the cell and click the mouse. The border around the cell A1 should be highlighted. Cell A1 is now the *active cell.* Type "Example 1. Terminal velocity of a sphere in a fluid" into cell A1. Although only a portion of the title will be visible in cell A1, the entire text is displayed in the *formula bar*, above the worksheet and below the tool bar. After you hit *enter*, the entire text appears in the spreadsheet, because the adjacent cells are empty.
2. Make a table of the three fluids and their properties. Type "fluid" into cell A3, then type the names of the three fluids – "water," "vegetable oil," and "glycerin" – into cells A4 through A6. You may adjust the width of column A by positioning the cursor at the line between the column headings A and B, at the top of the worksheet. The cursor will change to a vertical bar with a horizontal double arrow. Drag the cursor to change the column width.

Add the properties of the fluids in columns B and C. In cell B2 type "viscosity" and in cell B3 type its units, "Pa s". Type the fluid viscosities below this column heading. Similarly, type "density" into cell C2, the units "kg/m^3" into cell C3, and the fluid densities below the heading. The in-line caret (^) is the symbol for exponentiation in most spreadsheets.

3. Enter the data for spheres in water. Begin by entering column headings, starting in cell A8 for "sphere" and ending with "t (s)", into cell E8. Now enter the data into rows 9 through 20. The tedium and error associated with entering data can be reduced by copying and pasting. For example, after typing "nylon" into cell A9, *copy* cell A9, *select* cells A10 through A11, and then *paste.* To select a group of cells, such as A10 through A11, first *select* cell A10, then *drag* down to cell A11 while depressing the clicker. Or, first *select* cell A10, then *select* the final cell in the group by holding down the *shift* key when clicking. Later, copy "steel" from cell A16 and paste into cells A17 through A20.
4. It will be useful to list the fluid in each row. Enter the column heading "fluid" into cell F8. Type "water" into cell F9, then *copy* and *paste* into cells F10 through F20.
5. Enter the data for spheres in vegetable oil and glycerin. Follow steps (3) and (4) to complete rows 21 through 45, columns A through F.
6. Our goal is to reduce the data to the Reynolds number and the friction factor. We will begin by calculating the values of two parameters that constitute these dimensionless groups – the terminal velocity and the buoyancy. We must take care to calculate all parameters *in mks units.*

   Velocity is (distance traveled)/(time elapsed), assuming a uniform velocity. Distance and time are each in mks units, so we don't need any conversion factors. Enter the formula for the first velocity by typing "= D9/E9" into cell G9. The number 0.0904... should appear in cell G9 after you hit *enter.* The number of significant figures will depend on the format and width of the cell.

	A	B	C	D	E	F	G	H	I	J	K	L
1	*Example 1. Terminal velocity of a sphere in a fluid*											
2		viscosity	density									
3	fluid	Pa s	kg/m^3		constants							
4	water	0.0010	1000		pi	3.1416						
5	vegetable oil	0.05	930		g	9.8						
6	glycerin	1.2	1250									
7							*velocity*	*sphere*	*buoyancy*	*Reynolds*	*Froude*	*friction*
8	*sphere*	*dia (mm)*	*m (g)*	*ht (m)*	*t (s)*	*fluid*	*m/s*	*density*	*kg/m^3*	*number*	*number*	*factor*
9	nylon	3.18	0.019	0.5	5.53	water	0.0904	1128	128	288	0.26	0.65
10	nylon	6.35	0.15	0.5	3.9	water	0.1282	1119	119	814	0.26	0.60
11	nylon	6.35	0.15	0.51	3.8	water	0.1342	1119	119	852	0.29	0.55
12	lucite	3.18	0.021	0.51	4.6	water	0.1109	1247	247	353	0.39	0.84
13	teflon	12.7	2.48	0.51	0.75	water	0.6800	2312	1312	8636	3.72	0.47
14	glass	2.8	0.038	0.5	1.27	water	0.3937	3306	2306	1102	5.65	0.54
15	aluminum	6.35	0.361	0.51	0.8	water	0.6375	2693	1693	4048	6.53	0.35
16	steel	3.18	0.133	0.5	0.76	water	0.6579	7899	6899	2092	13.89	0.66
17	steel	6.35	1.064	0.51	0.45	water	1.1333	7936	6936	7197	20.64	0.45
18	steel	9.53	3.572	0.51	0.4	water	1.2750	7882	6882	12151	17.41	0.53
19	steel	9.53	3.572	0.51	0.35	water	1.4571	7882	6882	13887	22.73	0.40
20	steel	12.7	8.486	0.51	0.4	water	1.2750	7912	6912	16193	13.06	0.71
21	nylon	3.18	0.019	0.64	34.72	oil	0.0184	1128	198	1.09	0.01	26.09
22	nylon	3.18	0.019	0.64	32.8	oil	0.0195	1128	198	1.15	0.01	23.29
23	lucite	6.35	0.172	0.64	9.49	oil	0.0674	1283	353	7.97	0.07	6.92
24	teflon	3.18	0.038	0.64	7.47	oil	0.0857	2257	1327	5.07	0.24	8.08
25	aluminum	6.35	0.361	0.51	2.23	oil	0.2287	2693	1763	27.01	0.84	3.01
26	steel	1.59	0.016	0.64	5.05	oil	0.1267	7602	6672	3.75	1.03	9.28
27	steel	1.59	0.016	0.64	4.99	oil	0.1283	7602	6672	3.79	1.06	9.06
28	steel	1.59	0.016	0.64	4.99	oil	0.1283	7602	6672	3.79	1.06	9.06
29	steel	3.18	0.133	0.64	1.94	oil	0.3299	7899	6969	19.51	3.49	2.86
30	steel	3.18	0.133	0.51	1.79	oil	0.2849	7899	6969	16.85	2.60	3.84
31	steel	6.35	1.064	0.51	1.17	oil	0.4359	7936	7006	51.48	3.05	3.29
32	steel	9.53	3.572	0.51	0.7	oil	0.7286	7882	6952	129.1	5.68	1.75

Figure 5.33a Spreadsheet solution to Example 1.

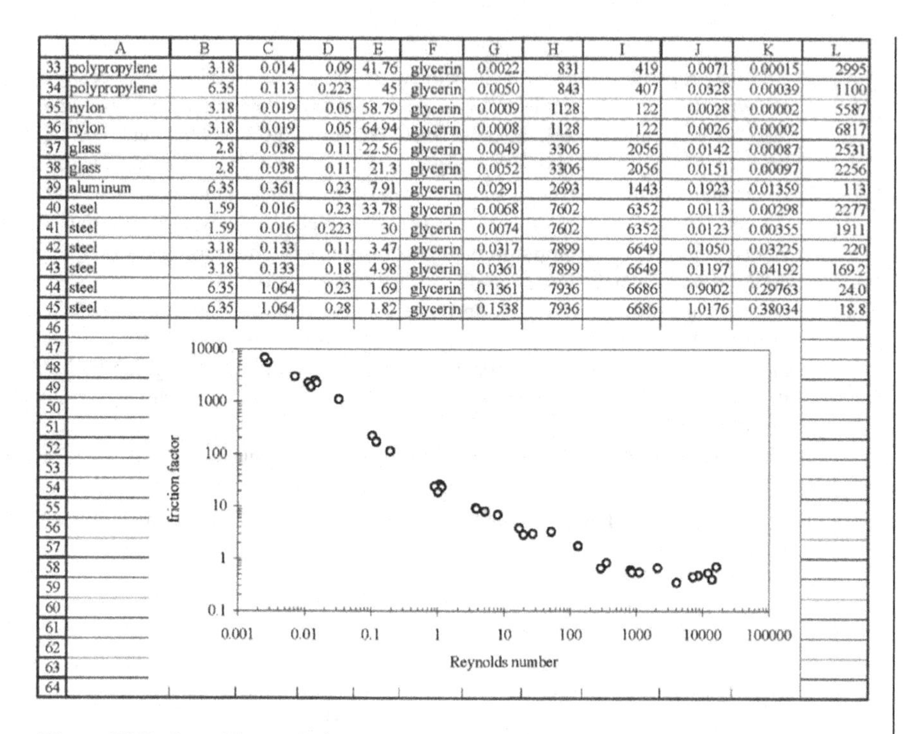

	A	B	C	D	E	F	G	H	I	J	K	L
33	polypropylene	3.18	0.014	0.09	41.76	glycerin	0.0022	831	419	0.0071	0.00015	2995
34	polypropylene	6.35	0.113	0.223	45	glycerin	0.0050	843	407	0.0328	0.00039	1100
35	nylon	3.18	0.019	0.05	58.79	glycerin	0.0009	1128	122	0.0028	0.00002	5587
36	nylon	3.18	0.019	0.05	64.94	glycerin	0.0008	1128	122	0.0026	0.00002	6817
37	glass	2.8	0.038	0.11	22.56	glycerin	0.0049	3306	2056	0.0142	0.00087	2531
38	glass	2.8	0.038	0.11	21.3	glycerin	0.0052	3306	2056	0.0151	0.00097	2256
39	aluminum	6.35	0.361	0.23	7.91	glycerin	0.0291	2693	1443	0.1923	0.01359	113
40	steel	1.59	0.016	0.23	33.78	glycerin	0.0068	7602	6352	0.0113	0.00298	2277
41	steel	1.59	0.016	0.223	30	glycerin	0.0074	7602	6352	0.0123	0.00355	1911
42	steel	3.18	0.133	0.11	3.47	glycerin	0.0317	7899	6649	0.1050	0.03225	220
43	steel	3.18	0.133	0.18	4.98	glycerin	0.0361	7899	6649	0.1197	0.04192	169.2
44	steel	6.35	1.064	0.23	1.69	glycerin	0.1361	7936	6686	0.9002	0.29763	24.0
45	steel	6.35	1.064	0.28	1.82	glycerin	0.1538	7936	6686	1.0176	0.38034	18.8

Figure 5.33b Spreadsheet solution to Example 1, continued.

As was illustrated in the tutorial on Mass Balances on Spreadsheets in the Chapter 3 exercises, it is easier to create a formula by selecting cells rather than typing cell addresses. For example, the formula for the first velocity can be created by first typing "=" into cell G9. This notifies the spreadsheet that you are entering a formula. Then select cell D9 with the mouse. The characters "D9" will appear in cell G9 and the formula bar. Type the symbol for division, "/". Now select cell E9. This completes the formula. Hit *enter*.

Calculate all the velocities. Copy the contents of cell G9 and paste into cells G10 through G45. Note that the cell addresses in the formula are incremented automatically. For example, the formula in G10 is "= D10/E10" and so on down the column.

7. To calculate the buoyancy, it is prudent to first calculate the sphere density, in mks units. Type the headings "sphere" into cell H7 and "density" into cell H8. Density is mass/volume, and the volume of a sphere is $\pi d^3/6$. We could simplify this formula by calculating the constant $\pi/6$. However, this formula provides an excuse to introduce absolute cell references. Start a list of constants. Type the heading "constants" into cell E3 and type "pi" into cell E4. Now enter a value for π in cell F4. Five significant figures is more than enough, given the accuracy of the data; type "3.1416" into cell F4. Now create the formula for density in cell H9. Enter "= (C9/1000)/(F4*(B9/1000)^3/6)" into cell H9. Note that both mass and diameter must be converted to mks units: grams to kilograms and millimeters to meters. The first pair of parentheses are only for clarity. The same result obtains without the first pair of parentheses.

   Paste the formula for sphere density into the cells below. But wait – the reference to the cell that contains π, cell F4, is incremented to F5, F6, and so on. We want the reference to cell F4 to remain unchanged down the column. We must indicate an *absolute cell reference* by replacing F4 with F4. Now you can paste the formula from cell H9 into cells H10 through H45. Because only the row is incremented as we paste down a column, we only needed to indicate that the row reference was absolute. Replacing F4 with F$4 is sufficient.

Computing the sphere density as a separate quantity allows one to scan for errors – errors in the formula for density and errors in the data. Check the nylon spheres: all should have about the same density, about 1120 kg/m^3. Teflon should be about 2300 kg/m^3, steel about 7700 kg/m^3, polypropylene about 840 kg/m^3, and so on.

8. Calculate the buoyancy – the difference between the sphere density and the fluid density – in column I. Create formulas that use the densities for water, oil, and glycerin in cells C4, C5, and C6, respectively. If you used the absolute references correctly, the buoyancy of a given material in a given fluid should be constant. Also, a given material, such as steel, should be least buoyant in oil and most buoyant in glycerin.

   You probably noticed something suspicious about some of the buoyancies – some are negative. This was also evident to the students who measured the data. Nylon and polypropylene spheres float on glycerin. Negative buoyancies represent spheres *rising* in glycerin. Spheres rising at their terminal velocity lie on the same correlation as spheres falling at their terminal velocity.

   However, negative buoyancies will wreak havoc with the friction factor, which must be a positive number. Let's guarantee all the buoyancies are positive. A simple solution is to calculate the absolute value. Enter the formula "= ABS(H9-C4)" into cell I9 and paste down the column, taking care to reset the absolute reference to the densities of oil and water.
9. We can now calculate the dimensionless groups. Start with the Reynolds number. Apply your spreadsheet skills to create a formula for Re in cell J9, taking care that all parameters are in mks units. Paste the formula down column J. Remember to use absolute references for ρ_{fluid} and μ, and remember to change the absolute references when the fluid changes. Compare your results with the spreadsheet printed in Figures 5.33a and 5.33b.
10. We now calculate the Froude number,

$$\text{Fr} = \frac{v^2}{gD}, \tag{5.36}$$

    as an intermediate to the friction factor. Gravitational acceleration on Earth is 9.8 m^2/s. We add g to the list of constants. Type "g" into cell E5 and "9.8" into cell F5. Calculate Fr in column K and check your results.
11. Finally, we calculate the friction factor. Calculate the friction factor in column L and check your results. Did you remember to use an absolute reference for ρ_{fluid} and to change the absolute reference when the fluid changed?
12. Prepare a log–log plot of the data. First, select the data to be plotted. The x values are in column J, rows 9 through 45. The y values are in column L, rows 9 through 45. Because the columns are not adjacent, selecting both at once requires special attention. Dragging from cell J9 to cell L45 would select column K as well. First, select cells J9 through J45. Then hold the *Ctrl* key while selecting cells L9 through L45.

The details of plotting vary with software type. The procedure for Microsoft Excel begins with clicking on the ChartWizard icon in the *tool bar*, which changes the cursor from an arrow to cross-hairs. Position the cross-hairs where you would like one corner of your plot, drag the cross-hairs to the position of the opposite corner, and release. This sets the size of your graph. The size is easily changed later, by dragging any corner or border.

ChartWizard now leads you through five steps, which are self-explanatory. At the second step, choose a *XY (Scatter) chart* and choose the log–log option, designated by non-uniform spacing of the gridlines in both the x and y directions. After step 5 a plot appears, albeit an esthetically unpleasant plot. Explore various formatting options by clicking on various features of the plot – the axes, the body, the gridlines – to produce a plot of publication quality, as shown in the example.

Spreadsheet Example 2

Determine the equation for a straight line through the data in the preceding example for Re < 2. Evaluate the certainty that the data lie on a straight line. Given that a straight line will be fit to the data, evaluate the certainty of the fitted parameters, the slope, and the intercept.

Solution to Spreadsheet Example 2

Begin by culling the data for Re < 2. One could do this point by point since there are only 37 data points to inspect. But this would be tedious for larger data sets. Plus, selecting point by point lacks style. Instead, we will arrange the data in ascending Re, and then truncate the data for Re > 2.

1. Open a new worksheet. Select cell A1 and enter the title "Example 2. Terminal velocity of a sphere in Stokes regime". Now, copy the column of Reynolds numbers from the previous worksheet into column A of the new worksheet. Copy the heading as well, but do not copy the formulas in each cell, as follows. Change windows to the first worksheet, select cells J7 through J45, and *copy*. Change windows to the new worksheet and select cell A2. Pull down the *edit* menu and choose *paste special*; specify that only the *values* are pasted. Check that the formulas were not pasted, by selecting a cell in column A. Only its value should appear in the *formula bar*, no formula.

   Paste the friction factors from the first worksheet to column B of the new worksheet. You may now close the first worksheet.
2. Sort the data. Select the data in columns A and B. Pull down the *Data* menu and choose *Sort* to open a dialog box. Specify an *ascending* sort on column A. Because you selected both columns A and B, the Re and f pairs are retained when the data are sorted.
3. Delete the data for Re > 2. If all is well to this point, you should find that the data in row 19 and below have Re > 2. Select these data (both columns A and B) and delete, by pulling down the *Edit* menu and choosing *Clear* (specify *All* in the dialog box). The first two columns of your spreadsheet should resemble the example in Figure 5.34, give or take a few insignificant figures.

	A	B	C	D	E	F	G	H	I	J
1	*Example 2. Terminal velocity of a sphere in Stokes' regime*									
2	*Reynolds*	*friction*					*log(Re)**			
3	*number*	*factor*	*log(Re)*	*log(f)*	*[log(Re)]^2*	*[log(f)]^2*	*log(f)*	*f (calc'd)*	*deviat'n^2*	
4	0.0026	6817.3	-2.593	3.834	6.726	14.697	-9.942	9071.1	0.015	
5	0.0028	5587.2	-2.550	3.747	6.503	14.041	-9.556	8221.6	0.028	*large deviation*
6	0.0071	2995.3	-2.146	3.476	4.607	12.086	-7.462	3280.1	0.002	
7	0.0113	2277.3	-1.948	3.357	3.794	11.272	-6.540	2087.6	0.001	
8	0.0123	1910.7	-1.910	3.281	3.647	10.766	-6.266	1914.2	0.000	
9	0.0142	2531.3	-1.847	3.403	3.412	11.583	-6.286	1659.9	0.034	*large deviation*
10	0.0151	2256.4	-1.822	3.353	3.320	11.245	-6.110	1568.3	0.025	*large deviation*
11	0.0328	1100.5	-1.484	3.042	2.203	9.251	-4.515	727.3	0.032	*large deviation*
12	0.1050	219.9	-0.979	2.342	0.958	5.486	-2.293	230.1	0.000	
13	0.1197	169.2	-0.922	2.228	0.850	4.966	-2.054	202.2	0.006	
14	0.1923	113.3	-0.716	2.054	0.513	4.219	-1.471	126.5	0.002	
15	0.9002	24.0	-0.046	1.380	0.002	1.903	-0.063	27.5	0.004	
16	1.0176	18.8	0.008	1.273	0.000	1.621	0.010	24.4	0.013	
17	1.0903	26.1	0.038	1.417	0.001	2.006	0.053	22.8	0.003	
18	1.1541	23.3	0.062	1.367	0.004	1.869	0.085	21.5	0.001	
19		**sum:**	**-18.86**	**39.56**	**36.54**	**117.01**	**-62.41**		**0.17**	
20										
21	points:	n	15							
22	slope:	m	-0.988							
23	intercept:	b	1.395							
24	uncertainty in y		0.113							
25	uncertainty in slope		0.032							
26	uncert'nty in intercept		0.049							
27										
28										
29										
30										
31										
32										

Figure 5.34 Spreadsheet solution to Example 2.

At this point one could plot the data and choose a canned routine to fit a straight line to the data. Admittedly, this approach has style. However, canned procedures often omit details of the fit. Is there a subset of data that consistently lies above the curve? Perhaps all the data for spheres in oil? Maybe our value for the viscosity for oil is too low? More important, not all canned routines provide the information necessary to evaluate the certainty of the model and the certainty of the fitted parameters.

In the tutorial that follows, we take up the method of least squares for fitting a straight line to a set of data.

Linear Regression and the Method of Least Squares

We seek the best fit of a straight line,

$$y = mx + b,$$

to a set of n data points. The slope and intercept that minimize the sum of the squared deviations between each data point and the line are

$$\text{slope} = m = \frac{n\sum xy - \sum x \sum y}{n\sum x^2 - \left(\sum x\right)^2}$$

$$\text{intercept} = b = \frac{\sum y - m\sum x}{n}.$$

The certainty that the data are represented by a straight line can be evaluated by calculating the uncertainty in the values of y, the sum of the squared deviations between the experimental data and the calculated data:

$$\text{uncertainty in } y \equiv \sigma_y \approx \left[\frac{\sum(y - mx - b)^2}{n-2}\right]^{1/2}.$$

Given that a straight line will be used, we can estimate the uncertainty in the slope and intercept with the following equations:

$$\text{uncertainty in the slope} \equiv \sigma_m \approx \left[\frac{n\sigma_y^2}{n\sum x^2 - \left(\sum x\right)^2}\right]^{1/2}$$

$$\text{uncertainty in the intercept} \equiv \sigma_b \approx \left[\frac{\sigma_y^2 \sum x^2}{n\sum x^2 - \left(\sum x\right)^2}\right]^{1/2}.$$

We now use a spreadsheet to calculate the least-squares fit and evaluate the uncertainty of the calculated values and the uncertainty of the fitted parameters.

4. First we must remember that because the data are linear on log–log paper, x is $\log_{10}$Re, not Re, and y is $\log_{10}$f, not f. We create two new columns: $\log_{10}$Re and $\log_{10}$f. Type the label "log(Re)" into cell C3 and enter the formula "= log10(A4)" into cell C4. Paste the formula from cell C4 into cells C5 through C18. Type "log f" into cell D3 and *copy* the formula from cell C4 into cell D4. The formula in cell D4 should read "= log10(B4)". Complete column D.
5. Compute various summations, Σx, Σx^2, Σy, Σy^2, and Σxy. Spreadsheets provide functions to calculate summations. Begin with Σx. In cell C19 enter the formula "= SUM(C4:C18)". To calculate Σy *copy* cell C19 and *paste* into cell D19. To remind us that row 19 contains sums, type "sum:" into cell B19.

In columns E, F, and G, compute Σx^2, Σy^2, and Σxy. Finally, calculate the sums of these columns by *copying* cell C19 and *pasting* into cells E19, F19, and G19. Compare your spreadsheet through column G with the spreadsheet example in Figure 5.34.

6. Compute the slope and intercept. Type the labels "n", "m", and "b" into cells B21, B22, and B23. Enter the number of data points (15) into cell C21. Use the formulas above to calculate the slope and intercept in cells C22 and C23. You should obtain −0.988 for the slope and 1.395 for the intercept.
7. Use the slope and intercept to calculate values for the friction factor, given the values for the Reynolds numbers in column A. As was discussed in the text, the equation

$$\log_{10} f = m\,[\log_{10}\mathrm{Re}] + b$$

can be rearranged to

$$f = 10^b \mathrm{Re}^m.$$

Use the above formula and calculate friction factors in cells H4 through H18. Did you remember to use absolute references for the slope and intercept?

8. Calculate squared deviations between the experimental values for y and the calculated values for y. In cell I4 enter the formula "= (D4 – log10(H4))^2". Remember – we must calculate the differences between base-10 logarithms of friction factors, not differences between friction factors. Paste the formula from cell I4 down the column. In cell I19 calculate the sum of cells I4 through I18.
9. Calculate the uncertainty in y, the logarithm of the friction factor. In cell C24 enter the formula "= SQRT(I19/(C21−2))". We obtain an uncertainty of 0.113, which means the typical deviation between the logarithm of the experimental friction factor and the logarithm of the calculated friction factor is 0.113. The difference between logarithms of two numbers is equivalent to the logarithm of the ratio of the two numbers. In other words, the ratio of two numbers is 10 raised to the difference in logarithms. In this case, the ratio is $10^{0.113} \approx 1.30$. Thus a typical calculated friction factor will deviate by as much as a factor of 1.30; the calculated friction factor will typically be between 30% larger and 30% smaller than the experimental friction factor.

   Does an uncertainty in y of 0.113 justify a fit to a straight line? The uncertainty *alone* is insufficient to judge. We need to compare the uncertainty of ±30% to the confidence limit of the experimental data. If the experimental uncertainty is ±40%, a straight line is justified. If the experimental uncertainty is ±10%, a straight line is questionable. We would try different functional forms until the calculated uncertainty was comparable to the experimental uncertainty. Similarly, if the experimental uncertainty is ±40%, there is no justification for replacing our straight line with, for example, a quadratic function, even though the calculated uncertainty might be improved to ±20%.

   Estimating the experimental uncertainty lies beyond the intended range of this book. It would require details about the experiment not pertinent to this tutorial and would require an explanation of the method of propagation of errors.
10. Calculate the uncertainty in the fitted parameters – the slope and the intercept. In cell C25 enter the formula "=SQRT(C21*C24*C24 / (C21*E19−C19*C19))" This should yield a value of 0.032. The slope is thus −0.988 ± 0.032. When reporting uncertainties, only one significant figure is meaningful, unless the first digit is 1. Similarly, it is not meaningful to report the fitted parameters to decimal places beyond the uncertainty. Thus the slope is −0.99 ± 0.03. We calculate the uncertainty of the intercept to be 0.049. Thus the intercept is 1.40 ± 0.05. Stokes's law predicts a slope of −1 and an intercept of 1.398 ($10^{1.398} \approx 25$). Our fitted parameters and their uncertainties are consistent with Stokes's law.

11. Plot the experimental data (columns A and B) and the calculated data (columns A and I). Plot the data as points and the calculated data as a line (no points). You are invited to prepare a plot larger than presented here (we endeavored to fit the example onto one page of this textbook).
12. Examine the results. The largest deviations are for the points with Re = 0.0028, 0.0142, 0.0151, and 0.0328. Reopen the spreadsheet from the first example and locate these data points. The points correspond to the data in rows 35, 37, 38, and 34, respectively. All data are from experiments with glycerin, which is a difficult fluid to work with because it is sticky and gooey. But the problem is not the value for the viscosity of glycerin, because some glycerin points lie above the line and some lie below. Two of the data points are for glass spheres – that's suspicious. The densities of these glass spheres are 3300 kg/m^3, which is abnormally high. Silica glass typically has a density of 2200 kg/m^3. The mass or the diameter might be wrong. The mass was measured on a calibrated electronic balance, so the mass is probably correct. Sphere diameters (2.8 mm) were measured with calipers with an accuracy of $\pm$0.1 mm. What diameter would yield the typical glass density of 2200 kg/m^3? 3.2 mm. An error of 0.4 mm is conceivable.

We can use the spreadsheets to investigate further. Start with the first spreadsheet and save a copy under a new filename. Now change the diameters of the glass spheres in cells B37 and B38 to 3.2 mm. Go to the second spreadsheet and save a copy under a new filename. Paste the "adjusted" Reynolds numbers and friction factors for the glass spheres into the second spreadsheet. Note that uncertainty in y decreases to 0.086 from 0.113. However, the fitted slope and intercept deviate further from the Stokes's law parameters: the slope changes to -0.97 ± 0.02 from -0.99 ± 0.03 (the theoretical value is 1) and the intercept changes to 1.39 ± 0.04 from 1.40 ± 0.05 (the theoretical value is 1.398).

Do we leave the glass sphere diameters at 3.2 mm because the data fit the correlation better? Absolutely not! Delete both files with the "adjusted" data. "Adjusting" data is fraud. Such scientific misconduct would warrant professional sanctions. "Adjusting" data goes against the fundamental goal of science – to seek the truth. Can we delete the data for the glass spheres because we suspect they are faulty? Absolutely not! It is also fraud to discard data because it doesn't agree with your pet theory or (as has been known to happen) because it doesn't agree with your mentor's pet theory.

Searching for the source of experimental deviation can suggest new measurements. In this case, we suspect it is difficult to accurately measure the diameter of small spheres. What do you think *you* would find if *you* remeasured the glass spheres? Of course, it would be hard to avoid finding 3.2 mm, because we are prejudiced. Rather you should ask someone else to measure the glass spheres, someone not biased toward any particular answer. And this person should remeasure the diameters of *all* the small spheres. If the method of measuring small spheres is suspect, it is suspect for all small spheres, not just those with the largest deviations from the correlation. Or perhaps another method of determining a sphere's diameter is needed? Perhaps one could place several identical small spheres in a graduated cylinder with water, measure the volume displaced, and then calculate the diameter?

It is important to recognize potential sources of scientific fraud, especially self-deception. These topics are discussed in the engaging book *Betrayers of the Truth*, by William Broad and Nicholas Wade.[2]

Now, use the skills you acquired in the preceding tutorials to complete the following exercise.

5.42 Analyze the relationship between a quadruped's body mass and the circumferences of its major leg bones, given by the data below.

(A) Prepare a plot of body mass (y axis) vs. the sum of the humerus and femur bone circumferences (x axis). Simply adding the bone circumferences may seem

[2] Broad, W., and Wade, N. 1982. *Betrayers of the Truth*, Simon & Schuster, New York, NY.

naïve. If we knew the dominant stress on the bones – compressive, bending, or torsional – we would combine the bone circumferences differently. For example, if compressive forces dominated, because the strain caused by compressive forces goes as the cross-sectional area, we would add the squares of the circumferences, then take the square root. However, because the humerus and femur circumferences are comparable, the method of combining the two has little effect. The larger effect is the differences in body structures.

(B) Determine the *proportionality constant* and exponent m in a relation of the form

$$\text{(body mass)} = \text{(proportionality constant)} \times \text{(humerus + femur bone circumference)}^m$$

Bone circumferences and body masses of quadrupedal mammals.

	Leg bone circumference (mm)		
Mammal	Humerus	Femur	Mass (kg)
meadow mouse	4.9	5.5	0.047
guinea pig	10	15	0.385
gray squirrel	10	13	0.399
opossum	27	23	3.92
gray fox	28	26	4.20
raccoon	30	28	4.82
nutria	21	28	4.84
bobcat	31	32	5.82
porcupine	30	34	7.20
otter	32	28	9.68
coyote	36	36	12.7
cloud leopard	45	41	13.5
duiker	31	46	13.9
yellow baboon	55	57	28.6
cheetah	67	69	38.0
cougar	62	60	44.0
wolf	62	62	48.1
bushbuck	56	62	50.9
impala	65	69	60.5
warthog	83	72	90.5
nyala	99	97	135.
lion	104	94	144.
black bear	98	94	218.
grizzly bear	124	107	256.
blue wildebeest	115	100	257.
Cape Mountain zebra	132	143	262.
kudu	140	135	301.
Burchell's zebra	129	147	378.
polar bear	158	135	448.
giraffe	192	173	710.
bison	192	168	1,179.
hippopotamus	209	208	1,950.
elephant ("Jumbo")	459	413	5,897.

Data from Anderson, J. F., Hall-Martin, A., and Russell, D. A. (1985). "Long-Bone Circumference and Weight in Mammals, Birds, and Dinosaurs," *Journal of Zoology*, London (A) **207**: 53–61.

from a least-squares fit to the data. Calculate uncertainties for the proportionality constant and the exponent m.

You should find m is surprisingly close to 3. This suggests that animals *do* scale linearly, contrary to the argument at the beginning of this chapter. What gives? The key issue is the *stress* on the bone, not the *strength* of the bone. Strain doesn't necessarily scale with body mass. The stress, or force, on a bone is proportional to body mass *times acceleration*. Earth's gravitational field causes acceleration, whether an animal is standing, walking, galloping, or jumping. So if bears hopped around as squirrels do, bears would surely break bones. Similarly, if elephants jumped about as terriers do, elephants would surely break bones. Again, one must consider the *dynamics* when scaling animate systems.

(C) Use the relation you determined in (B) to draw a line through the data and calculate the individual deviations between the actual body masses and the predicted body masses. Critically analyze the fit. Is this a universal correlation? Are there any types of mammals that always lie above or below the curve? Catlike mammals? Deerlike mammals? Bearlike mammals? Are small mammals consistently above or below the curve? Large mammals? Medium mammals? Do the data deviate from a straight line, suggesting a different functional form is needed?

(D) Use the correlation developed in (B) to estimate the masses of the dinosaurs in the table below. Check one of your estimates with ours: we estimate the apatosaurus mass to be 39,000 kg, to two significant figures.

Bone circumferences and body masses of quadrupedal dinosaurs.

Dinosaur	Leg bone circumference (mm)		Estimated mass (kg)[a]
	Humerus	Femur	
styracosaurus	288	370	4,080
diplodocus	320	405	10,560
brachiosaurus	654	730	78,260
apatosaurus	629	845	27,870
apatosaurus	629	845	32,420

[a] Masses were estimated from the volumes of models, assuming a density of 900 kg/m^3. More recent studies assume densities comparable to water, 1000 kg/m^3 (see Alexander, 1983), which would increase the masses in this table by 11%.

Data from Alexander et al. (1985) and Colbert, E. H. (1962). "The Weights of Dinosaurs," *American Museum Novitates* **2076**: 1–16. Weights were estimated from the volumes of models, assuming a density of 900 kg/m3. More recent studies assume densities of comparable to water, 1000 kg/m3, (see Alexander, R. McN. 1983 "Posture and Gait of Dinosaurs," *Zoological Journal of the Linnean Society*, **83**: 1–25) which would increase the masses in the table above by 11%.

(E) As opposed to the mammal masses, the dinosaur masses tabulated above are not experimental data. There are no data for the masses of live dinosaurs. Colbert molded dinosaur models, measured the models' volumes, and estimated dinosaur masses by assuming a dinosaur density and lung volume. Based on your predictions from dynamic similarity, which of Colbert's models was overly stocky or overly thin? That is, which of your estimates deviate excessively from Colbert's estimates? (An excessive deviation is one that exceeds the typical deviations you observed for mammals.)

(F) *Optional*: Increase the range of the correlation by adding data for insects. Insects have an exoskeleton, so the leg "bone" circumference is just the leg circumference. True, insect legs are fleshy inside, but so are mammal bones. And as you will learn in statics, the strength of a cylindrical member depends little on the material in the center.

6

Transient-State Processes

In Chapter 3 we analyzed processes by developing models based on fundamental laws, such as conservation principles. We restricted our analyses to systems at steady state, namely where none of the variables changed with time. This simplified the mathematical modeling since we dealt only with algebraic equations; such models, as we saw, are extremely useful for analyzing steady-state systems.

Many chemical processes, however, do not operate at steady state. Chemical reactions performed one batch at a time, such as the experiments you conducted in your secondary school and first-year chemistry classes, are examples of unsteady processes. We need design tools for inherently time-dependent processes such as these. Furthermore, even in designing a steady-state process, time-dependent problems occur: for example, how does one *start* one of the elegant designs we accomplished in Chapter 2? A member of the faculty at a leading US university tells an interesting anecdote. When he was a design engineer employed at a major oil company, he designed a continuous reactor that was constructed but could not be started. Furthermore, this happened twice! The hard-learned lesson is that *a thorough understanding of transients is required to start, stop, and control a continuous process.*

A non-steady process, or transient-state process, is one where at least one process variable changes with time. Modeling a transient-state process usually involves a differential equation. Since your calculus course has prepared you to deal with simple differential equations, you can now appreciate the engineering aspects of transient phenomena.

In this chapter we apply the analytical methods presented in the previous chapters – mathematical modeling, graphical analysis, and dimensional analysis – to transient-state systems. We will begin with mathematical modeling based upon fundamental and constitutive laws such as those described in Chapter 3. Graphical methods will be used to assess stability. Dimensional analysis will be used to provide supporting equations. And as before, the choice of system boundaries can greatly simplify an analysis.

We begin with mathematical modeling and a new formal statement of a conservation principle:

The rate of change of a quantity within a system equals the rate the quantity enters that system plus the rate that quantity is generated within the system minus the rate that quantity leaves the system and minus the rate that quantity is consumed within the system.

A convenient distillation of the above statement is

The rate of accumulation equals input plus generation minus output minus consumption, or

$$\text{accumulation} = \text{input} + \text{generation} - \text{output} - \text{consumption}.$$

The previous chapters on mass conservation dealt with systems at steady state, namely when accumulation equaled zero. We now generalize mass conservation to include accumulation.

6.1 TRANSIENT-STATE MASS BALANCES: A SURGE TANK

Context: A surge tank.

Concepts: Integrating a transient-state process unit into a steady-state system.

Defining Question: How can we use a batch fermentor to continuously produce citric acid at steady state?

A surge tank is a simple process unit, yet rich in transient phenomena and important engineering lessons. In its typical application, a surge tank may be used to average transients in a flow stream; the flow into the tank may surge and ebb, but the flow out of the tank will be stable. We will use the production of citric acid as a case study in the use of surge tanks, and in the suppression of transients in process design. The process flowsheet in Figure 6.1 shows the overall design for citric acid production from the fungus *Aspergillus niger*. A fermentor is followed by an absorber to transfer citric acid from the aqueous phase into the organic phase, thereby separating the product from the reaction mixture.

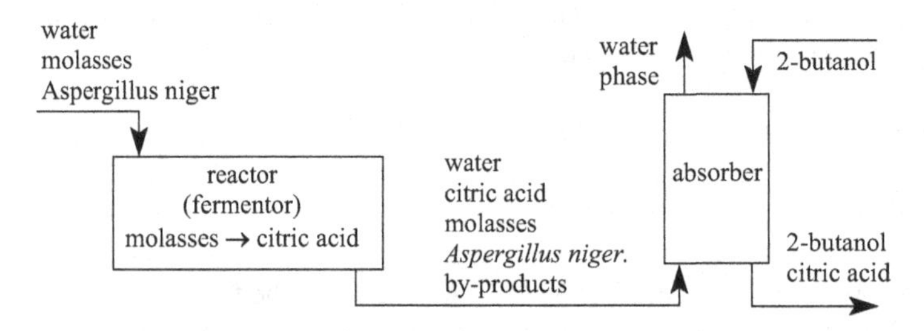

Figure 6.1 Citric acid produced in a fermentor and extracted in an absorber.

The process is more complicated than shown in Figure 6.1, because the flow out of the fermentor is incompatible with the flow required by the absorber. The fermentor produces citric acid in batches, whereas the absorber operates best as a continuous device. A surge tank allows us to connect batch fermentation with continuous separations, as shown below.

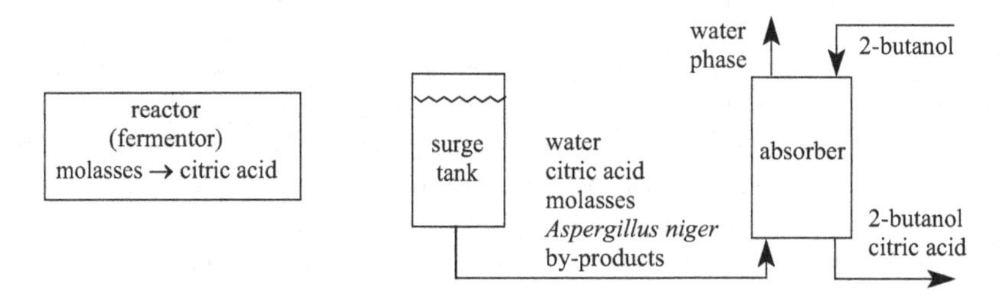

Figure 6.2 Batch fermentation takes place while the surge tank provides a steady flow to the absorber.

Upon completion of the fermentation process, the fermentor will empty its contents into the surge tank. The fermentor will then be refilled, and the process begins again, as shown in Figure 6.3 on the next page.

Figure 6.4 shows three graphs depicting the stream flow rate out of the fermentor and into the surge tank, the stream flow rate into the absorber, and the level of fluid in the surge tank. Fermentation occurs during time A to time B; there is no flow out of the fermentor and the fluid level in the surge tank falls. The fermentor is emptied into the surge tank during time B to time C, and the level in the surge tank rises. The process repeats, refilling the fermentor and

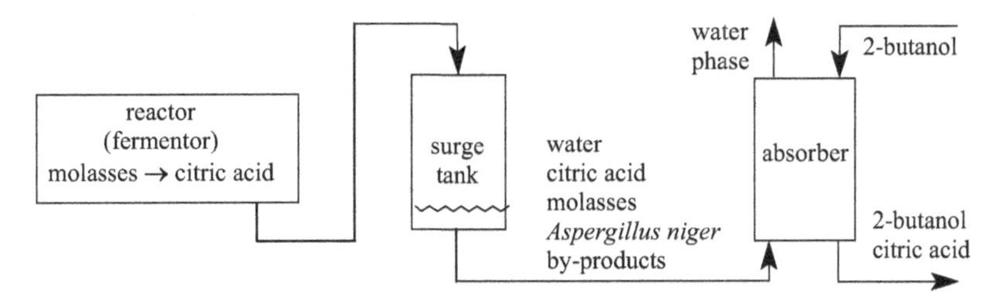

Figure 6.3 Fermentor empties its contents into the surge tank.

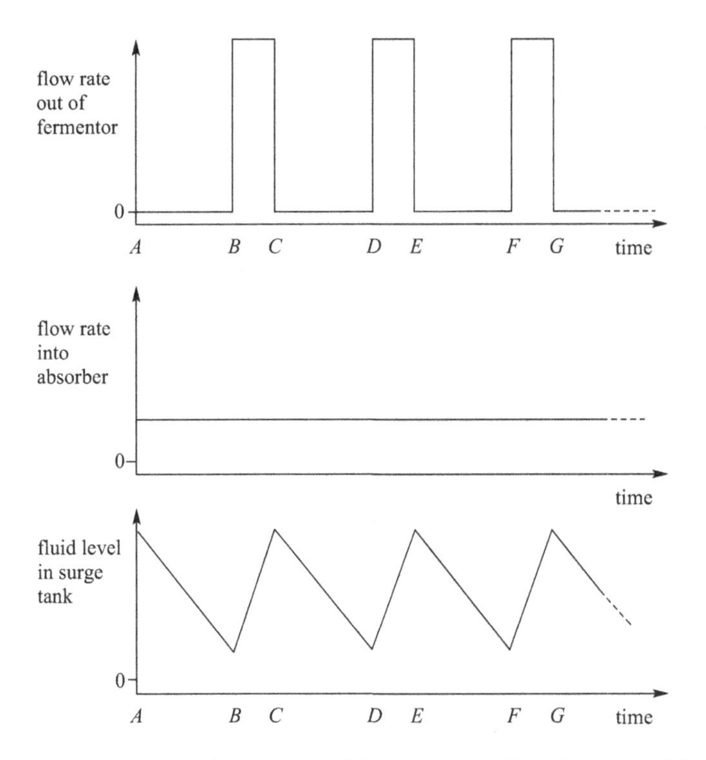

Figure 6.4 Stream flow rates out of the fermentor and into the surge tank (top), stream flow rates into the absorber (middle), and fluid level in the surge tank (bottom).

conducting fermentation from time C to time D, and draining the fermentor from D to E. The surge tank accepts the large transients arising from batch production and provides the steady flow required by the absorber. Of course, the surge tank must never overflow or empty completely. We also see the timescales of the transients that the surge tank must suppress. We now have a design problem: what specifications does the surge tank need to perform this function?

Let's consider just the surge tank, and apply some of the mathematical analysis tools we developed in Chapter 3. Consider a surge tank with volumetric flow rate Q_{in} in and volumetric flow rate Q_{out} out, as shown in Figure 6.5 on the next page.

We first analyze the simple scenario in which Q_{in} and Q_{out} are pumped at constant, but not necessarily equal, rates. Further assume that the cross-sectional area of the surge tank, A_{tank}, is constant along the height of the tank, typical for cylindrical tanks. We apply

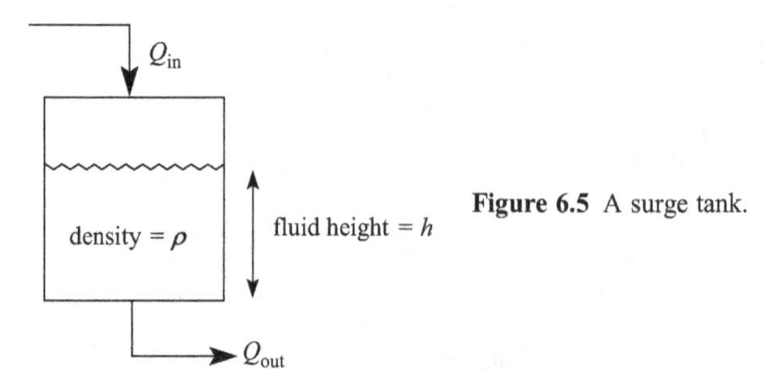

Figure 6.5 A surge tank.

conservation principles to this surge tank to model its behavior. Specifically, we wish to model the height of the fluid as a function of time. What is conserved? Mass is conserved. We must express mass in terms of height. To introduce the dimension of length, one can express the total mass in terms of the density ρ and the volume V:

$$\text{mass} = \left(\frac{\text{mass}}{\text{volume}}\right)\text{volume} = (\text{density})(\text{volume}) \tag{6.1}$$

$$M_{\text{fluid}} = \rho_{\text{fluid}} V_{\text{fluid}}. \tag{6.2}$$

Equation 6.2 requires uniform density throughout the tank, which is valid for many liquid systems; we could add a stirring mechanism if we were concerned about density gradients. Because the cross-sectional area of the tank is constant, the volume of the fluid is the product of the fluid height h and the area A_{tank}:

$$V = A_{\text{tank}} h \tag{6.3}$$

$$M_{\text{fluid}} = \rho_{\text{fluid}} A_{\text{tank}} h. \tag{6.4}$$

We can use equation 6.4 to translate an expression in terms of mass into an expression in terms of height.

We begin with the conservation of mass – *the rate of accumulation equals input plus generation minus output minus consumption* – and translate into mathematics. What are the appropriate system boundaries? The tank. Convention and good sense suggests we associate the generation and consumption terms of our formal conservation statement with the reactants and products of chemical reactions. In the absence of any reactions, therefore, both the generation and consumption terms are zero. As with the total mass in the system, we express the mass flow rates F_{in} and F_{out} in terms of density and volumetric flow rates:

$$\begin{aligned} \text{rate of mass flow in} &= \frac{\text{mass}}{\text{time}} = \left(\frac{\text{mass}}{\text{volume}}\right)\left(\frac{\text{volume}}{\text{time}}\right) \\ &= (\text{density of incoming fluid}) \times (\text{volumetric flow rate in}) \end{aligned} \tag{6.5}$$

$$F_{\text{in}} = \rho_{\text{in}} Q_{\text{in}}. \tag{6.6}$$

Similarly, we have

$$F_{\text{out}} = \rho_{\text{out}} Q_{\text{out}}. \tag{6.7}$$

The accumulation is the rate at which total mass in the system changes with time:

$$\text{rate of accumulation} = \frac{dM_{\text{tank}}}{dt}, \tag{6.8}$$

which can be rewritten in terms of h using equation 6.4:

$$\frac{dM_{\text{tank}}}{dt} = \frac{d}{dt}(\rho_{\text{fluid}} A_{\text{tank}} h) = \rho_{\text{fluid}} A_{\text{tank}} \frac{dh}{dt}. \tag{6.9}$$

We can combine the terms in the conservation statement

$$\frac{dM_{\text{tank}}}{dt} = F_{\text{in}} - F_{\text{out}} = \rho_{\text{in}} Q_{\text{in}} - \rho_{\text{out}} Q_{\text{out}} \tag{6.10}$$

and arrive at the equation

$$\rho_{\text{fluid}} A_{\text{tank}} \frac{dh}{dt} = \rho_{\text{in}} Q_{\text{in}} - \rho_{\text{out}} Q_{\text{out}}. \tag{6.11}$$

We assume further that the density of the fluid entering the tank is equal to that exiting the tank, and equal to that in the tank:

$$\rho_{\text{fluid}} = \rho_{\text{in}} = \rho_{\text{out}}, \tag{6.12}$$

which simplifies equation 6.11 to

$$A_{\text{tank}} \frac{dh}{dt} = Q_{\text{in}} - Q_{\text{out}}. \tag{6.13}$$

We now solve equation 6.13. Its solution is simplified because Q_{out} is constant and Q_{in} is piecewise constant, as shown in Figure 6.4. This allows us to move all variables dependent on h to one side of the equation and all variables dependent on t to the other side (the constants may reside on either side). Thus,

$$A_{\text{tank}} dh = (Q_{\text{in}} - Q_{\text{out}}) dt. \tag{6.14}$$

The differential equation 6.13 is thus *separable* and as such is easily solved by integration:

$$A_{\text{tank}} \int dh = (Q_{\text{in}} - Q_{\text{out}}) \int dt \tag{6.15}$$

$$A_{\text{tank}} h + (\text{a constant}) = (Q_{\text{in}} - Q_{\text{out}}) t + (\text{another constant}). \tag{6.16}$$

Combining the constants and solving for h yields

$$h = \frac{1}{A_{\text{tank}}} [(Q_{\text{in}} - Q_{\text{out}}) t + (\text{a constant})]. \tag{6.17}$$

This formula for the fluid height as a function of time contains an unknown constant; can we eliminate this from our analysis? Let's exploit the fact that our analysis tells us height is linearly proportional to time; equation 6.17 defines a straight line for which we know the slope, but not the intercept. To uniquely determine the line on a plot of h vs. t we must specify one point on the line, and thus we need to know h at a given time to determine the constant in equation 6.17. This datum is an example of an *initial condition*. The solution of differential equations containing a derivative with respect to time may be solved completely only when an initial condition is specified. Similarly, differential equations that contain a derivative with respect to position may be solved completely only if the conditions of the system are specified at a geometric boundary, hence the term *boundary condition*. For the surge tank we will specify a condition at a specific time; we arbitrarily designate time $t = 0$ such that the height of the fluid h is h_0. Substituting into equation 6.17 we have

$$h_0 = \frac{1}{A_{\text{tank}}}[(Q_{\text{in}} - Q_{\text{out}}) \times 0 + (\text{a constant})] \tag{6.18}$$

$$A_{\text{tank}} h_0 = \text{a constant}. \tag{6.18}$$

Thus,

$$h = \frac{1}{A_{\text{tank}}}(Q_{\text{in}} - Q_{\text{out}})t + h_0. \tag{6.20}$$

A more efficient juncture at which to introduce the initial condition(s) for equation 6.15 is before integration. We create two definite integrals using the initial condition $h = h_0$ at $t = 0$ for the lower limits of each integral,

$$A_{\text{tank}} \int_{h_0} dh = (Q_{\text{in}} - Q_{\text{out}}) \int_0 dt, \tag{6.21}$$

and the unknown h at some later t as the upper limits,

$$A_{\text{tank}} \int_{h_0}^{h} dh = (Q_{\text{in}} - Q_{\text{out}}) \int_0^{t} dt. \tag{6.22}$$

Equation 6.22 represents an evolution from a certain height h_0 at time = 0 to a different h at a later time. Paying careful attention to the limits of integration, we have

$$A_{\text{tank}}(h - h_0) = (Q_{\text{in}} - Q_{\text{out}})(t - 0), \tag{6.23}$$

which yields the same result as equation 6.20. Thus the height of fluid in the tank increases linearly with time if $Q_{\text{in}} - Q_{\text{out}} > 0$, decreases linearly with time if $Q_{\text{in}} - Q_{\text{out}} < 0$, and remains constant if $Q_{\text{in}} = Q_{\text{out}}$. These three mathematical conditions make intuitive sense: when $Q_{\text{in}} - Q_{\text{out}} > 0$ the rate of accumulation is positive and the surge tank fills with fluid. When $Q_{\text{in}} - Q_{\text{out}} < 0$ the rate of accumulation is negative and the surge tank empties. The situation when $Q_{\text{in}} = Q_{\text{out}}$ is exactly at steady state, and the fluid height in the surge tank remains constant.

A surge tank drained by gravity poses a slightly more challenging problem. The flow rate of fluid through an orifice at the bottom is a function of the height of water in the tank, as shown in Figure 6.6.

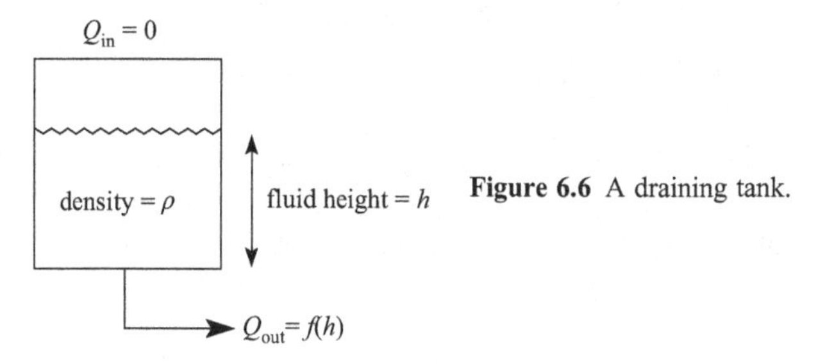

Figure 6.6 A draining tank.

Qualitatively, one expects the flow rate to decrease as the fluid height decreases. Thus a model for a draining surge tank will include the relationship between the height of fluid in the tank and the flow rate through the hole in the bottom. We also expect the flow rate to depend on the geometry of the hole.

Dimensional analysis may be used to obtain the relationship between h and Q_{out}. Using the techniques presented in Chapter 5, we first list the relevant parameters in Table 6.1 on the next page.

Table 6.1 The parameters of a draining tank.

Parameter	Variable	Dimensions
fluid height	h	L
volumetric flow rate	Q_{out}	L^3/T
orifice cross-sectional area	$A_{orifice}$	L^2
gravitational acceleration	g	L/T^2

Upon arranging the exponents of the three key dimensions (mass, time, and length) into algebraic equations and choosing h and Q_{out} as core variables, we derive two dimensionless groups:

$$\Pi_1 = \frac{h^2}{A_{orifice}} \quad \text{and} \quad \Pi_2 = \frac{Q^2_{out}}{g A^{5/2}_{orifice}}. \tag{6.24}$$

Thus we know there exists a functional relationship of the form

$$\Pi_1 = f(\Pi_2) \tag{6.25}$$

$$\frac{h^2}{A_{orifice}} = f\left(\frac{Q^2_{out}}{g A^{5/2}_{orifice}}\right). \tag{6.26}$$

One then goes to the laboratory and measures Q_{out} as a function of $A_{orifice}$ and h. A plot of the data reveals a simple relationship:

$$Q_{out} = 0.6 A_{orifice}(gh)^{1/2}. \tag{6.27}$$

This is the so-called *orifice equation*, a constitutive law that relates h and Q_{out}.

Using this expression in our mass balance equation, along with the fact that $Q_{in} = 0$, for a freely draining tank we have

$$\frac{dh}{dt} = \frac{1}{A_{tank}}(Q_{in} - Q_{out}), \quad \text{with} \quad Q_{in} = 0 \tag{6.13}$$

$$= -\frac{1}{A_{tank}}\left(0.6 A_{orifice}(gh)^{1/2}\right). \tag{6.28}$$

Although this differential equation is more complicated than equation 6.13, it is still separable. Upon separation of variables we have

$$h^{-1/2} dh = -0.6 \frac{A_{orifice}}{A_{tank}} g^{1/2} dt, \tag{6.29}$$

and we can integrate both sides with appropriate initial conditions:

$$\int_{h_0}^{h} h^{-1/2} dh = -0.6 \frac{A_{orifice}}{A_{tank}} g^{1/2} \int_0^t dt. \tag{6.30}$$

Equation 6.30 uses the same initial conditions as before. We have moved the constants out of the integral over time. Evaluating the integrals yields

$$2\left(h^{1/2} - h_0^{1/2}\right) = -0.6 \frac{A_{orifice}}{A_{tank}} g^{1/2}(t - 0) \tag{6.31}$$

$$h = \left[h_0^{1/2} - 0.3 \frac{A_{orifice}}{A_{tank}} g^{1/2} t\right]^2. \tag{6.32}$$

Equation 6.32 may be generalized using dimensional consistency and scaling. Rather than posing a problem specific to this tank, one may pose a problem applicable to all draining tanks. How does one eliminate the information specific to this problem to solve a more general problem?

As we discussed in Chapter 5, systems may be rendered dynamically similar by converting to a dimensionless form. A step toward solving the more general problem is to recognize that the ratio h/h_0 is dimensionless and represents the fraction of water remaining in the cylindrical tank. We define

$$x \equiv \text{fraction of water remaining in tank} = \frac{h}{h_0}, \tag{6.33}$$

and it follows that

$$\frac{dx}{dh} = \frac{d}{dh}\left(\frac{h}{h_0}\right) = \frac{1}{h_0}. \tag{6.34}$$

Thus,

$$dh = h_0 dx. \tag{6.35}$$

We also introduce a dimensionless time. This is not as straightforward as the dimensionless height because no other variable has dimensions of time. The gravitational constant g has dimensions of L/T^2, and thus $(h_0/g)^{1/2}$ has dimensions of time; its reciprocal has dimensions of inverse time. We thus define a dimensionless time τ:

$$\tau \equiv \text{reduced time} = \left(\frac{g}{h_0}\right)^{1/2} t, \tag{6.36}$$

and it follows that

$$\frac{d\tau}{dt} = \frac{d}{dt}\left(\frac{g}{h_0}\right)^{1/2} t = \left(\frac{g}{h_0}\right)^{1/2}. \tag{6.37}$$

Thus,

$$dt = \left(\frac{h_0}{g}\right)^{1/2} d\tau. \tag{6.38}$$

Finally, we define a dimensionless ratio of the relative cross-sectional areas of the tank and the orifice:

$$\alpha \equiv \text{relative areas of tank and orifice} = \frac{A_{\text{orifice}}}{A_{\text{tank}}}. \tag{6.39}$$

Returning to our mass balance expression,

$$\frac{dh}{dt} = -\frac{1}{A_{\text{tank}}}\left(0.6A_{\text{orifice}}(gh)^{1/2}\right), \tag{6.28}$$

we substitute the dimensionless variables from equations 6.33, 6.35, 6.38, and 6.39 to yield

$$\frac{h_0 dx}{\left(\frac{h_0}{g}\right)^{1/2} d\tau} = -0.6\,\alpha(gh_0 x)^{1/2}, \tag{6.40}$$

which simplifies to

$$\frac{dx}{d\tau} = -0.6\alpha x^{1/2}. \tag{6.41}$$

Again we separate variables, which gives

$$x^{-1/2}dx = -0.6\alpha\, d\tau. \tag{6.42}$$

We now integrate, and apply the initial condition $h = h_0$ at $t = 0$, which in dimensionless quantities corresponds to $x = 1$ at $\tau = 0$:

$$\int_1^x x^{-1/2}dx = -0.6\alpha \int_0^\tau d\tau. \tag{6.43}$$

Solving the integrals and substituting the limits gives

$$2\left(x^{1/2} - 1^{1/2}\right) = -0.6\alpha(\tau - 0), \tag{6.44}$$

which can be rearranged to obtain

$$x = (1 - 0.3\alpha\tau)^2. \tag{6.45}$$

Equation 6.45 predicts that the fraction of water remaining in *any similar cylindrical tank will vary as* (*time*)^2. Notice also that, when posed in a dimensionless form, the equation is simpler and thus is easier to manipulate mathematically.

We return to the design problem illustrated in Figure 6.4. If we choose a gravity-drained surge tank before our absorber, the flow rate into the absorber will not be constant: equation 6.27 tells us it will vary as the square root of fluid level. Thus we must use a pump, or some other controlling device, after the surge tank to guarantee constant input to the absorber.

Any mathematical model is subject to the limits imposed by its founding assumptions. For example, when we invoked the orifice equation we implicitly assumed the fluid height in the tank preserved the physics of the draining process. When the tank is nearly empty, however, the physics will be different – a whirlpool of air will form in the outlet, for example. Thus at long times the orifice equation, and subsequently equation 6.45, will not be valid. Similarly, a tank with a closed top, or a tank with variable cross-sectional area (such as a funnel), or a tank with density gradients in the fluid, will require different mathematical models.

A surge tank can allow an intermittent process, such as fermentation, to be coupled to a continuous process (e.g. absorption). The design for such a system proceeds by using a *transient* mass conservation statement and solving *initial value* problems. This example of the draining tank was adapted from the textbook by Russell and Denn (1972). You are encouraged to study the other analyses of dynamic systems in their textbook.

6.2 RESIDENCE TIMES AND SEWAGE TREATMENT

Context: Sewage treatment.

Concepts: Component mass balances and residence time.

Defining Question: How long do we have to fix a problem after something breaks?

Consider a mathematical model for a more complex physical situation where the fluid streams contain solutes. This situation is common in industrial practice, for example in the treatment of residential sewage.

A common treatment is to inculcate bacteria into residential sewage to digest the organic material into CO_2 and water. Consider the system diagrammed in Figure 6.7. In this simplified process, sewage is pumped into a well-mixed aeration tank, and $[\text{bacteria}]_{\text{tank}} = [\text{bacteria}]_{\text{treated}}$ = 8 g bacteria/gal sewage. Treated sewage is then sent to a settling tank where bacteria, whose density is greater than that of water, settle to the bottom. The settling tank recovers bacteria to be recycled to the aeration tank. In this example the *design* was based on a steady-state analysis; the *operation*, however, will be plagued with transients.

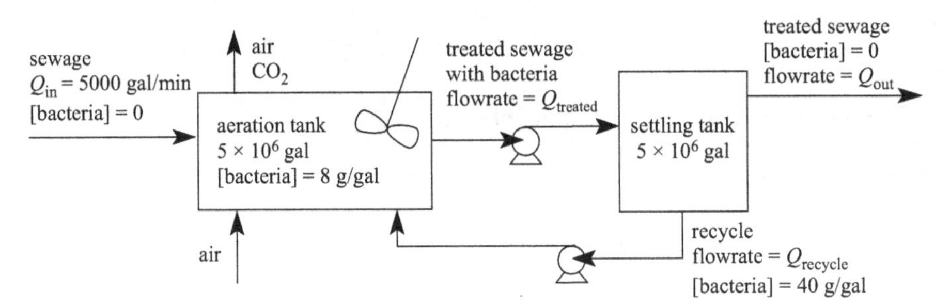

Figure 6.7 A process for treating residential sewage.

Let's analyze the steady-state operation to determine the volumetric flow rates Q_{treated} and Q_{recycle}. We will assume (1) the sewage density is independent of the bacteria concentration or organic matter concentration, (2) the process is at steady state, (3) the aeration tank is well mixed, (4) the flow rate of the inlet air is equal to the flow rate of the outlet air plus CO_2, and (5) the growth rate of the bacteria is zero (this is controlled by the flow rate of air to the aeration tank).

Given that there are two unknowns, Q_{treated} and Q_{recycle}, you should suspect you need two independent equations. A mass balance on the entire system is trivial, namely 5000 gallons per minute enter and leave the treatment facility. We recognize, however, that in addition to the overall mass balance, we can apply the conservation of mass principle to the *components* of the fluid streams. Thus we can write a mass balance for the bacteria as well as for the total mass.

Let's choose the system boundaries to be the aeration tank, because we have the most information about this unit. We continue to use the symbol F for the rate of total mass input and Q for volumetric flow rates; applying the conservation of mass to the aeration tank at steady state yields

$$F_{\text{total, in}} + F_{\text{total, recycle}} = F_{\text{total, treated}} \tag{6.46}$$

$$Q_{\text{in}}\rho_{\text{in}} + Q_{\text{recycle}}\rho_{\text{recycle}} = Q_{\text{treated}}\rho_{\text{treated}}. \tag{6.47}$$

Because we assumed the densities are equal and we are given that Q_{in} = 5000 gal/min,

$$5000 + Q_{\text{recycle}} = Q_{\text{treated}}. \tag{6.48}$$

A steady-state bacteria balance around the aeration tank yields

$$F_{\text{bacteria, in}} + F_{\text{bacteria, recycle}} = F_{\text{bacteria, treated}} \tag{6.49}$$

$$Q_{\text{in}}[\text{bacteria}]_{\text{in}} + Q_{\text{recycle}}[\text{bacteria}]_{\text{recycle}} = Q_{\text{treated}}[\text{bacteria}]_{\text{treated}} \tag{6.50}$$

$$0 + Q_{\text{recycle}}(40) = Q_{\text{treated}}(8) \tag{6.51}$$

$$5Q_{\text{recycle}} = Q_{\text{treated}}. \tag{6.52}$$

Equations 6.48 and 6.52 can be solved by substitution to yield

$$Q_{\text{recycle}} = 1250 \text{ gal/min} \qquad \text{and} \qquad Q_{\text{treated}} = 6250 \text{ gal/min}. \tag{6.53}$$

Imagine that the pump on the recycle stream fails. As the managing engineer for this process, you must take action. For example, you must predict how long before the sewage treated by the process is unsafe to release. That is, when will the bacteria concentration in the aeration tank decrease to a level such that the organic matter is not adequately digested?

Your first priority is to control the transients. What problem should be addressed first? Without the recycle, the level in the aeration tank will fall and the level in the settling tank will rise. Your first objective should be to maintain the levels in each tank. How does one drain the aeration tank more slowly and fill the settling tank more slowly? The best solution is to throttle the pump for the Q_{treated} stream down to 5000 gallons per minute from 6250 gallons per minute.

We expect the bacteria concentration in the aeration tank to fall with a concomitant rise in the level of sewage in the discharge, as sketched in Figure 6.8. Of course, neither change will necessarily be linear with time. Figure 6.8 is only a qualitative sketch.

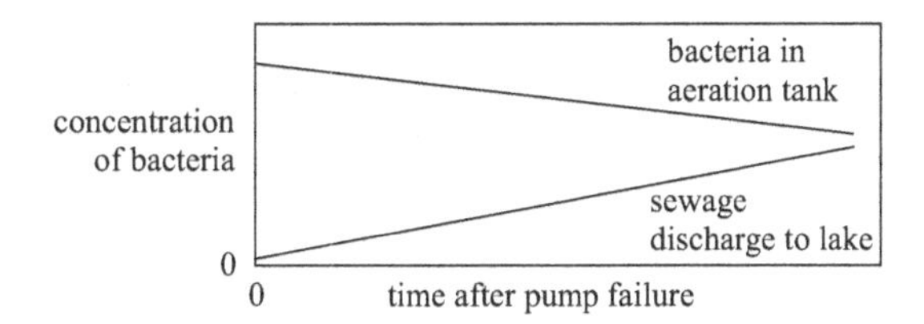

Figure 6.8 Qualitative expectations for the concentrations of sewage and bacteria after failure of the recycling pump.

We need to know how long we can continue to discharge treated sewage. Thus we need to know the transient behavior of the bacteria concentration in the aeration tank and the sewage concentration in the settling tank. After consulting with various bacteriologists and treatment experts, we learn that a minimum level of 4 g bacteria per gallon in the aeration tank is necessary to ensure safe levels of sewage in the discharge.

Now we have a transient problem with regards to the bacteria concentration in the treatment tank. We recall our general statement of mass conservation:

The rate of accumulation equals input plus generation minus output minus consumption, or

$$\text{accumulation} = \text{input} + \text{generation} - \text{output} - \text{consumption}.$$

Clearly we need to rewrite our mass conservation expression for bacteria, this time including the accumulation term as a derivative with respect to time of the mass of bacteria in the tank. Bacteria are neither generated nor consumed (remember we set the air flow rate to suppress bacterial growth), but bacteria do enter and exit the control volume. We thus have

$$\frac{d}{dt}\left([\text{bacteria}]_{\text{tank}} V\right) = [\text{bacteria}]_{\text{in}} Q_{\text{in}} - [\text{bacteria}]_{\text{treated}} Q_{\text{treated}}, \tag{6.54}$$

where V is the volume of the aeration tank. Recall that the aeration tank is well mixed, so $[\text{bacteria}]_{\text{tank}} = [\text{bacteria}]_{\text{treated}}$. Because density is constant and $[\text{bacteria}]_{\text{in}} = 0$, and because the volume is stabilized by throttling the Q_{treated} pump, we have

$$\frac{d}{dt}[\text{bacteria}] = -\left(\frac{Q_{\text{treated}}}{V}\right)[\text{bacteria}], \tag{6.55}$$

and, upon separating the variables dependent on concentration from those dependent on time,

$$\frac{d[\text{bacteria}]}{[\text{bacteria}]} = -\left(\frac{Q_{\text{treated}}}{V}\right)dt. \tag{6.56}$$

Just as with the case of the draining tank, we now must incorporate the *initial conditions.* Define $t = 0$ as the time the pump failed. At $t = 0$ the bacteria concentration in the tank was 8 g bacteria/gal sewage. We are interested in calculating the time t_{bad} when the bacteria concentration falls to 4 g bacteria/gal sewage. We can now set the limits and integrate equation 6.56:

$$\int_8^4 \frac{d[\text{bacteria}]}{[\text{bacteria}]} = -\left(\frac{Q_{\text{treated}}}{V}\right)\int_0^{t_{\text{bad}}} dt. \tag{6.57}$$

Upon integrating and substituting the limits, we obtain

$$\ln(4) - \ln(8) = -\frac{Q_{\text{treated}}}{V}(t_{\text{bad}} - 0) \tag{6.58}$$

$$t_{\text{bad}} = -\frac{V}{Q_{\text{treated}}}\ln\frac{4}{8} = 700 \text{ min}. \tag{6.59}$$

Analysis of this sewage treatment process is interesting for three reasons. First, the choice of the system boundaries requires some judgement (in this case, the unit for which you have the most data). Second, it represents an actual problem a chemical engineering alumnus related to us. Third, it demonstrates that safely operating a process requires an understanding of both transient-state and steady-state modeling.

Equations 6.55 to 6.59 contain an interesting ratio of quantities, the flow rate out of the tank divided by the fluid volume in the tank, Q_{treated}/V. This has dimensions of reciprocal time:

$$\left[\frac{Q_{\text{treated}}}{V}\right] = \frac{\text{volumetric flow rate}}{\text{fluid volume}} = \frac{\text{L}^3/\text{T}}{\text{L}^3} = \frac{1}{\text{T}}. \tag{6.60}$$

The inverse of this ratio indicates the average period of time a molecule resides in the tank and is called the *residence time* of water in the tank:

$$\frac{\text{fluid volume}}{\text{volumetric flow rate}} \equiv \text{residence time} \equiv \tau. \tag{6.61}$$

Returning to equation 6.58, we can write a general equation that relates the concentration C to the initial concentration C_0:

$$\ln\left(\frac{C}{C_0}\right) = -\frac{1}{\tau}t. \tag{6.62}$$

Solving for the concentration,

$$C = C_0 \exp(-t/\tau). \tag{6.63}$$

Equation 6.63 tells us that in all systems of this type the concentration of solute will fall to zero *exponentially.* The time scale of the exponential decrease is governed by the residence time of the system. Figure 6.9 on the next page shows the bacteria concentration in the aeration tank as a function of time.

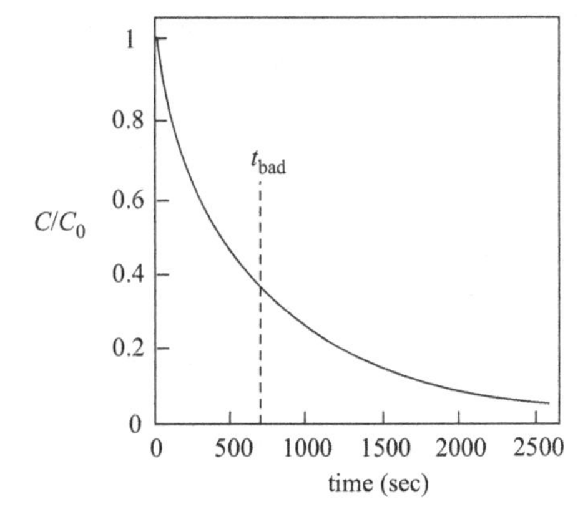

Figure 6.9 The decay of bacteria concentration as a function of time.

The exponential term on the right-hand side of equation 6.63 is a number that multiplies the initial concentration C_0. Consider the magnitude of the exponential for various increments of the ratio t/τ. The data in Table 6.2 suggest that, to an accuracy of about 5%, a period of about three residence times must pass for the bacteria concentration to decay to zero.

Table 6.2 Exponential decay.

t/τ	$\exp(-t/\tau)$
0.1	0.90
0.5	0.61
1	0.37
2	0.14
3	0.05
5	0.007

It is useful to compare the residence time to the dimensionless time introduced in the draining-tank analysis. For the draining tank, the quantity $(h_0/g)^{½}$ has dimensions of time and also forms a characteristic residence time: the time for the tank to drain. Indeed, if you use equation 6.45 and set x equal to zero, one finds that the tank empties in a time $(h_0/g)^{½}/(0.3\alpha)$.

Let us generalize from our study of the sewage treatment plant, the surge tank, and the draining tank. Characteristic timescales describe the dynamics of chemical engineering systems. In the case of well-stirred tanks at steady state, analysis of *component* mass balances reveals the *residence* time, given by the ratio of the tank volume to the characteristic total output (or input) flow rate. The effects of transients on tanks can be estimated by comparing the timescale of the transient perturbation to the characteristic residence time; input fluctuations that are rapid with respect to the characteristic residence time have little effect. Input fluctuations that are slow with respect to the characteristic time affect the output of the tank; thus the tank serves no purpose. A surge tank's residence time must be longer than timescales of the anticipated fluctuations. This is a key design criterion for a transient-state process with multiple components.

The relationship among what goes in, what comes out, and their fluctuations is called *process control*. Chemical engineering students are required to study this subject in some detail, usually as a course and laboratory in the latter half of the undergraduate curriculum. Prerequisite is a thorough understanding of chemical processes, as well as second-year college calculus and differential equations.

6.3 RATE CONSTANTS: MODELING ATMOSPHERIC CHEMISTRY

Context: Thermal decomposition of ozone-depleting N_2O gas.

Concepts: Chemical reaction rates in batch reactors.

Defining Question: How do we measure chemical reaction rates?

Anthropogenic activities have prompted many scientists and engineers to look at the effects of chemicals on the composition of the Earth's upper atmosphere. Of particular concern is the role certain chemicals play in creating or destroying stratospheric ozone, O_3. Ozone in the stratosphere is important to life on Earth because it absorbs solar radiation in the *ultraviolet* (*uv*) portion of the electromagnetic spectrum.[1] This high-energy portion of the solar spectrum is capable of indiscriminately breaking chemical bonds and thus can be harmful to plant and animal life on Earth; degradation of this layer is of international concern. Use your favorite search engine and look up "ozone hole."

One occasionally hears that man-made ozone should be used to replace the ozone destroyed by chemical reaction with man-made pollutants. This "geoengineering" viewpoint (using chemical engineering on a global scale to remedy a pollution problem) is naïve. Approximately 10^{13} W of solar radiation are used in the production of Earth's natural ozone layer; this is three times the total power produced by humans. In other words, replacing 10% of the stratospheric ozone would take about one-third of all the power plants on Earth! Assuming one could solve the power requirement, one must also engineer a means of delivering the ozone to the stratosphere, with the constraint that ozone spontaneously explodes at low concentrations. The present international strategy is to ban production of the chemicals that react with ozone in the upper atmosphere.

It is difficult to conduct experiments in the upper atmosphere, and thus our understanding of transport and chemistry in the stratosphere relies on laboratory measurements informed by modeling. These models use sophisticated computer algorithms that include hundreds of chemical reactions, diffusion of different species throughout the atmosphere, fluctuating energy input from the solar radiation, the presence of particles, and many other factors. The process is diagrammed in Figure 6.10. To use these models we need to know the rate at which chemicals are generated and consumed in the atmosphere.

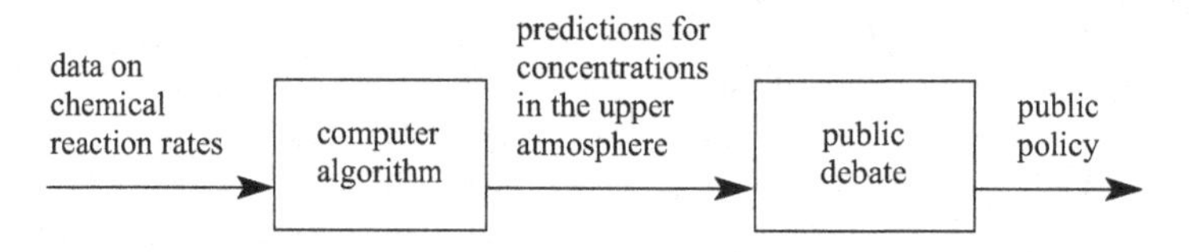

Figure 6.10 A process to convert technical data into public policy.

Nitrous oxide (N_2O) contributes to ozone depletion through photochemical reactions in the stratosphere. It reacts with oxygen atoms to form NO; N_2O is the chief source of NO in the stratosphere through the reaction

$$N_2O + O \rightarrow 2NO. \qquad \text{rxn (6.1)}$$

[1] Ozone in the troposphere, down where we live, is harmful to human health and is to be avoided.

NO catalyzes the removal of ozone in the upper stratosphere through the cycle in reactions 6.2 and 6.3, the net result of which is reaction 6.4:

$$NO + O_3 \rightarrow NO_2 + O_2 \qquad \text{rxn (6.2)}$$

$$NO_2 + O \rightarrow NO + O_2 \qquad \text{rxn (6.3)}$$

$$net: O + O_3 \rightarrow 2O_2. \qquad \text{rxn (6.4)}$$

Nitrous oxides are formed naturally by decomposition of nitrogen compounds in the soil. N_2O is also a man-made chemical used as a propellant for food spray cans and as an anesthetic. It can also be a by-product of pollution control devices installed on smokestacks at power plants. In any case, man-made nitrous oxides formed in the lower part of the atmosphere diffuse to the upper atmosphere where they contribute to ozone depletion. Nitrous oxides can also be introduced into the upper atmosphere directly from the exhaust of jet engines.

To model the concentration of N_2O in the stratosphere we need to measure the rate of one of the chemical reactions that dominates its decomposition into oxygen and nitrogen:

$$2N_2O \rightarrow 2N_2 + O_2. \qquad \text{rxn (6.5)}$$

Our goal is to design a chemical reactor that measures the rate of reaction 6.5. Our chemical reactor will be a simple "batch" reactor (i.e. one that is ostensibly simple because it has no input or output). For our purposes, it is a constant-volume, constant-temperature vessel, as illustrated in Figure 6.11. We surmise that if we measure the concentration of chemicals in the reactor as a function of time we will be able to write a characteristic rate equation for the rate of N_2O decomposition. Usually one would measure the pressure in the reactor to calculate the N_2O concentration in the reactor. However, we will simplify our analysis by assuming we have a spectrometer that can measure the N_2O concentration in the reactor.

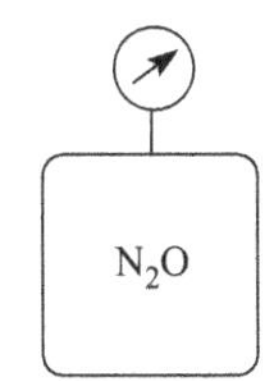

Figure 6.11 A batch reactor for studying the decomposition of N_2O.

For this reactor we will need model equations that describe the concentration of nitrous oxide with time. How will this system evolve with time? Clearly the N_2O concentration will fall and the concentrations of both N_2 and O_2 will rise such that the concentration of N_2 is double the concentration of O_2. (This latter fact comes from the stoichiometry of the reaction.) Because we have assumed reaction 6.5 is irreversible, ultimately the concentration of N_2O will fall to zero; we expect, then, something like Figure 6.12.

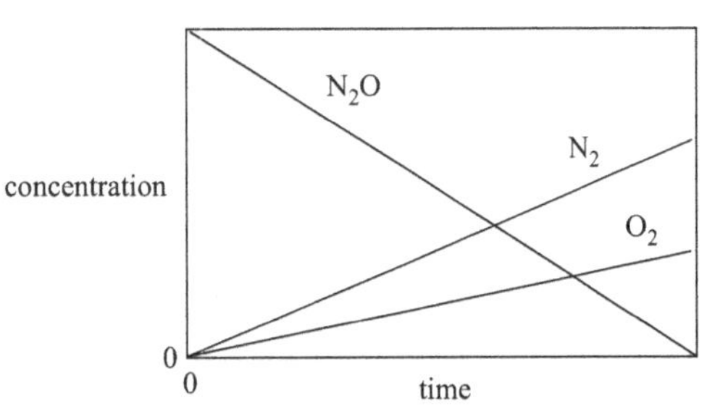

Figure 6.12 Qualitative expectations for the concentrations of N_2O, N_2, and O_2 as a function of time.

Our model equations will come from the principle of conservation of mass. We will apply the statement

rate of accumulation equals input plus generation minus output minus consumption

to this particular physical situation.

The total mass balance is trivial. With no input, output, generation, or consumption there is no accumulation. What we seek, however, is a conservation statement for the nitrous oxide inside the reactor. Thus we choose the reactor as the system boundaries and consider only the mass of nitrous oxide. There is neither input nor output of N_2O. The chemical reaction given by reaction 6.5 tells us, however, that nitrous oxide is *consumed* by the reaction to produce nitrogen and oxygen. Thus we have

$$\begin{aligned} &\textit{input} &&= 0 \\ &\textit{output} &&= 0 \\ &\textit{generation} &&= 0 \\ &\textit{consumption} &&= \textit{an expression derived using reaction } 6.5 \end{aligned}$$

What is this expression for consumption? In other words, how do we mathematically articulate consumption by chemical reaction? We first define a variable for the rate at which a substance is consumed by chemical reaction. We seek an intensive variable – that is, a variable that does not depend on the size of the reactor. This will allow us to use the rate in systems of different scale. We write

$$\begin{aligned} \text{rate at which } N_2O \text{ is consumed by chemical reaction} &= (\text{reaction rate}) \times (\text{reactor volume}) \\ &= rV. \end{aligned} \tag{6.64}$$

$$\text{Because } [rV] = \frac{\text{moles}}{\text{time}}, \text{ the units of } r \text{ are } [r] = \frac{\text{moles}}{(\text{volume})(\text{time})}. \tag{6.65}$$

The rate of accumulation of nitrous oxide is the rate at which the amount of N_2O changes with time, which in mathematical terms is

$$\text{rate of accumulation of } N_2O = \frac{d}{dt}(\text{moles of } N_2O \text{ in the reactor}) \tag{6.66}$$

$$= \frac{d}{dt}([N_2O]\, V) \tag{6.67}$$

$$= V\frac{d}{dt}[N_2O]. \tag{6.68}$$

Our mass conservation expression for N_2O in the reactor may now be written

$$V\frac{d}{dt}[N_2O] = 0 + 0 - 0 - rV \tag{6.69}$$

$$\frac{d}{dt}[N_2O] = -r. \tag{6.70}$$

The differential equation 6.70 appears trivial until we realize that the reaction rate r depends on the N_2O concentration. In other words, we expect the reaction rate of N_2O will be higher at higher N_2O concentrations, as shown qualitatively in Figure 6.13 on the next page.

Figure 6.13 Qualitative expectation for the amount of N_2O as a function of time for different initial concentrations. The rate of decay, which is the slope of the line, is faster for higher initial concentrations.

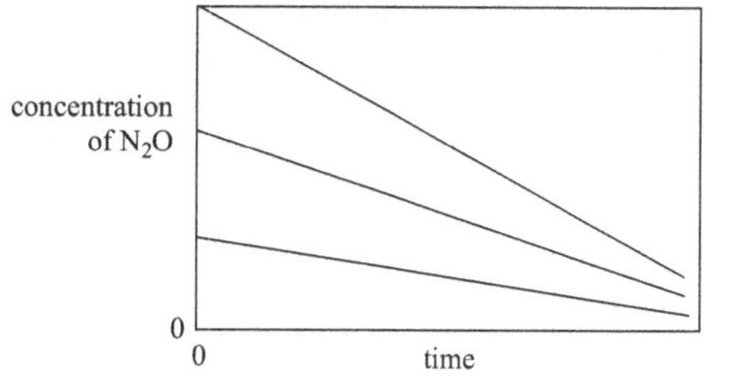

We must now determine how the reaction rate depends on the N_2O concentration.

Modern chemical physics uses principles of quantum mechanics and statistical mechanics to model reaction rates in terms of concentrations and temperature. Reactions in gases are caused by collisions between molecules. One can use statistics to argue that the reaction rate of N_2O is proportional to the rate of N_2O–N_2O collisions. The rate of N_2O–N_2O collisions is proportional to the square of the N_2O concentration. We *assume* a nitrous oxide molecule must collide with another nitrous oxide molecule to react, so

$$r = k[N_2O]^2, \tag{6.71}$$

where k is the *rate constant*. Although k is called a "constant" it is not a universal constant. It will almost certainly depend on temperature.

We use equation 6.71 to substitute for r in equation 6.70 and obtain

$$\frac{d}{dt}[N_2O] = -k[N_2O]^2. \tag{6.72}$$

Again we find that the differential equation may be solved by separation of variables:

$$\frac{d[N_2O]}{[N_2O]^2} = -k\,dt. \tag{6.73}$$

Now we implement an initial condition. At time $t = 0$ we designate the concentration of nitrous oxide to be $[N_2O]_0$, and at a time t later the concentration is $[N_2O]$. We add these limits to the integrals formed from equation 6.73:

$$\int_{[N_2O]_0}^{[N_2O]} \frac{d[N_2O]}{[N_2O]^2} = -k\int_0^t dt. \tag{6.74}$$

Evaluating the integrals, we obtain

$$-\left(\frac{1}{[N_2O]} - \frac{1}{[N_2O]_0}\right) = -k(t-0), \tag{6.75}$$

which we solve for $[N_2O]$:

$$[N_2O] = \frac{[N_2O]_0}{1 + kt[N_2O]_0}. \tag{6.76}$$

In a typical experiment, we would charge the batch reactor with N_2O and then measure the time required for the N_2O concentration to drop by 50%, i.e. the half-life, $t_{1/2}$. Table 6.3 shows half-life data for N_2O at 1015 K.

Table 6.3 Half-lives for the decomposition of N_2O at 1015 K.

$[N_2O]_0$ (mol/L)	$t_{1/2}$ (s)
0.135	1,060
0.286	500
0.416	344
0.683	209

Equation 6.76 may be used to derive a formula for $t_{1/2}$, the time for 50% conversion of nitrous oxide. When 50% of the nitrous oxide is reacted, then $[N_2O] = 0.5\ [N_2O]_0$, and equation 6.76 yields

$$t_{1/2} = \frac{1}{k[N_2O]_0}. \tag{6.77}$$

Thus a plot of $t_{1/2}$ vs. $1/[N_2O]_0$ should yield a straight line with slope of $1/k$. The data from Table 6.3 are plotted in Figure 6.14.

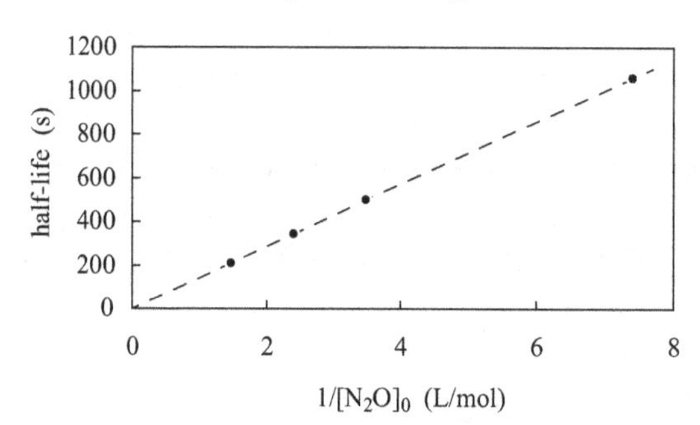

Figure 6.14 The half-life for the decomposition of N_2O as a function of the reciprocal of the concentration.

The straight line is obtained from a best fit of a straight line to the four data points. Because the data fit well to a straight line, our assumptions for the rate equation, equation 6.71, are justified. Using the slope of the line in Figure 6.14 (143 mol·s/L), and equation 6.77, we calculate the rate constant for reaction 6.5:

$$k = \frac{1}{\text{slope}} = \frac{1}{143\ \text{mol·s/L}} = 0.007\ \text{L/(mol·s)}. \tag{6.78}$$

Our batch reactor may be used to generate rate constants for a variety of temperatures and pressures. These laboratory-derived rate constants can subsequently be used in quantitative models for atmospheric chemistry. In other words, if we choose the upper atmosphere for the system boundary, then we may write a conservation of mass statement for nitrous oxide that looks like:

rate of accumulation equals input plus generation minus output minus consumption,

where at least one of the consumption terms is $rV = k[N_2O]^2$ and $k = 0.007$ L/(mol·s) at 1015 K.

Chemical engineers are renowned for their ability to use data from batch reactors to devise mathematical models containing chemical rates. From these comparisons come fundamental rate constants that inform science and technology. The example above is straightforward; imagine the perspicacity needed to model a system that contains substances with non-trivial rate expressions, as well as hundreds of other substances that participate in side reactions. The modeling of chemical reactions in the atmosphere, and the lower atmosphere in particular, is one of the great successes of chemical engineering.

6.4 OPTIMIZATION: BATCH REACTORS

Context: Optimal synthesis of a chemical transiently produced in a reactor.

Concepts: Graphical approximations to differential equations. Ordinary differential equations in series.

Defining Question: How long do we conduct a reaction in a batch reactor?

Batch reactors are rarely used in the chemical industry to produce bulk commodity chemicals. In specialty chemical applications, however, batch reactors may be used because it is relatively easy and inexpensive to conduct a synthesis on an occasional basis. The citric acid production described earlier is another example: fermentation is inherently a batch process. Whereas the N_2O decomposition illustrates how batch reactors may be used to measure reaction rates, commercial applications of batch reactors often involve more complicated chemical systems, such as enzymatic reactions or reactions taking place in the presence of multiple phases. We now consider operational issues associated with producing a chemical in a batch reactor where the chemical is an intermediate in a series of reactions.

Suppose we wish to dehydrogenate *n*-butane, $CH_3CH_2CH_2CH_3$, to produce 1-butene, $CH_2{=}CHCH_2CH_3$, in a batch reactor:

$$CH_3CH_2CH_2CH_3 \rightarrow CH_2{=}CHCH_2CH_3 + H_2. \qquad \text{rxn (6.6)}$$

Suppose further that we want to avoid dehydrogenation of 1-butene to butadiene:

$$CH_2{=}CHCH_2CH_3 \rightarrow CH_2{=}CHCH{=}CH_3 + H_2. \qquad \text{rxn (6.7)}$$

Reactions 6.6 and 6.7 may each be assigned rates, in units of moles/(volume × time):

$$r_1 \equiv \text{rate at which } CH_3CH_2CH_2CH_3 \text{ is consumed} \tag{6.79}$$

$$r_2 \equiv \text{rate at which } CH_3{=}CHCH_2CH_3 \text{ is consumed.} \tag{6.80}$$

The principle of conservation of mass may now be applied to the batch reactor; we apply

rate of accumulation equals input plus generation minus output minus consumption

to each component. For $CH_3CH_2CH_2CH_3$, the five terms in the above expression are

$$\begin{aligned} \text{accumulation} &= \frac{d}{dt}[CH_3CH_2CH_2CH_3]V \\ \text{generation} &= 0 \\ \text{input} &= 0 \\ \text{output} &= 0 \\ \text{consumption} &= -r_1 V. \end{aligned} \tag{6.81}$$

Because volume is constant in our batch reactor, we combine the terms in equation 6.81 to arrive at

$$\frac{d}{dt}[CH_3CH_2CH_2CH_3] = -r_1. \tag{6.82}$$

Applying mass conservation to each component, we obtain three additional equations:

$$\frac{d}{dt}[CH_2{=}CHCH_2CH_3] = r_1 - r_2 \tag{6.83}$$

$$\frac{d}{dt}[H_2] = r_1 + r_2 \tag{6.84}$$

$$\frac{d}{dt}[CH_2{=}CHCH{=}CH_2] = r_2. \tag{6.85}$$

We now postulate rate expressions for r_1 and r_2. In the N_2O example we assumed that the rate expression varied as the concentration of reactant squared; this is a *second-order* rate expression because the exponent 2 appears in the rate expression. For this case, however, we note that chemists have previously found that each decomposition is proportional to the amount of reactant. That is, reactions 6.6 and 6.7 are *first-order.* The expressions for the rates r_1 and r_2 are thus given in equations 6.86 and 687:

$$r_1 = k_1[CH_3CH_2CH_2CH_3] \tag{6.86}$$

$$r_2 = k_2[CH_2{=}CHCH_2CH_3]. \tag{6.87}$$

From equations 6.82 to 6.85 we construct the differential equations governing the compositions in our reactor:

$$\frac{d}{dt}[CH_3CH_2CH_2CH_3] = -k_1[CH_3CH_2CH_2CH_3] \tag{6.88}$$

$$\frac{d}{dt}[CH_2{=}CHCH_2CH_3] = k_1[CH_3CH_2CH_2CH_3] - k_2[CH_2{=}CHCH_2CH_3] \tag{6.89}$$

$$\frac{d}{dt}[H_2] = k_1[CH_3CH_2CH_2CH_3] + k_2[CH_2{=}CHCH_2CH_3] \tag{6.90}$$

$$\frac{d}{dt}[CH_2{=}CHCH{=}CH_2] = k_2[CH_2{=}CHCH_2CH_3]. \tag{6.91}$$

The chemical composition in our reactor at any time may be determined by solving the system of coupled differential equations 6.88 to 6.91. The solutions are straightforward with the appropriate mathematics training. We will solve equations 6.88 to 6.91 intuitively.

Equation 6.88 may be solved by separation and integration. Separating into terms that depend on $[CH_3CH_2CH_2CH_3]$ on the left and terms that depend on time on the right, integrating, and adding limits gives us

$$\int_{[CH_3CH_2CH_2CH_3]_0}^{[CH_3CH_2CH_2CH_3]} \frac{d[CH_3CH_2CH_2CH_3]}{[CH_3CH_2CH_2CH_3]} = -\int_0^t k_1\, dt \tag{6.92}$$

$$\ln\,[CH_3CH_2CH_2CH_3] - \ln\,[CH_3CH_2CH_2CH_3]_0 = -k_1 t \tag{6.93}$$

$$[CH_3CH_2CH_2CH_3] = [CH_3CH_2CH_2CH_3]_0 \exp(-k_1 t). \tag{6.94}$$

The concentration of $CH_3CH_2CH_2CH_3$ thus decreases exponentially with a time constant of $1/k_1$, as shown in Figure 6.15 on the next page.

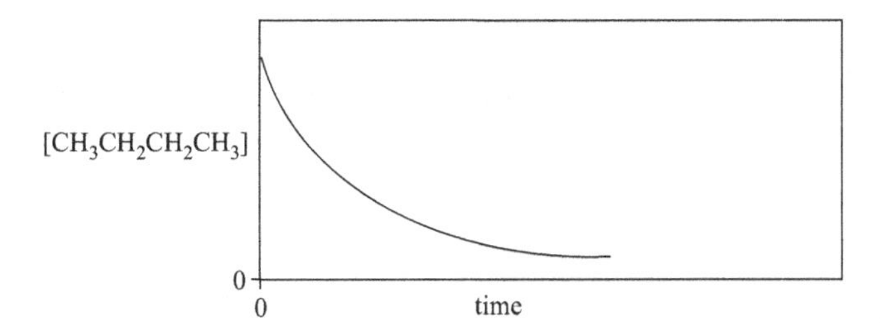

Figure 6.15 A sketch of the concentration of $CH_3CH_2CH_2CH_3$ as a function of time.

Recall that this is the same behavior we predicted for the bacteria in the sewage tank, as shown in Figure 6.9. We substitute the expression for $[CH_3CH_2CH_2CH_3]$ from equation 6.94 into equation 6.89 to obtain

$$\frac{d}{dt}[CH_2{=}CHCH_2CH_3] = k_1[CH_3CH_2CH_2CH_3]_0 \exp(-k_1 t) - k_2[CH_2{=}CHCH_2CH_3]. \quad (6.95)$$

We need to solve equation 6.95 to optimize the production of $CH_2{=}CHCH_2CH_3$. However, equation 6.95 cannot be separated. The procedure to solve equation 6.95 may not be in your repertoire. Throughout your career as an engineer you will encounter equations that are beyond your ability to solve analytically, even after four semesters of calculus. Indeed, some of the equations you will encounter will not have analytical solutions. Let's explore a method of approximating the time dependence of the concentration of $CH_2{=}CHCH_2CH_3$. We will follow the incremental change in $[CH_2{=}CHCH_2CH_3]$.

Consider the rate expression for $[CH_2{=}CHCH_2CH_3]$. Assuming there is initially no $CH_2{=}CHCH_2CH_3$ in the reactor, at *short times* equation 6.95 is approximately

$$\frac{d}{dt}[CH_2{=}CHCH_2CH_3] \approx k_1[CH_3CH_2CH_2CH_3]_0 \exp(-k_1 t), \quad (6.96)$$

where we see that the right-hand side of the equation is greater than zero. A derivative greater than zero means the function grows with time; thus at short times $[CH_2{=}CHCH_2CH_3]$ increases, as sketched in Figure 6.16.

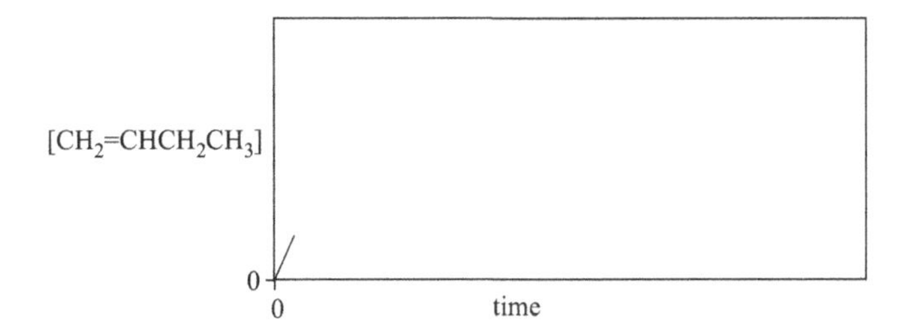

Figure 6.16 A sketch of the concentration of $CH_2 = CHCH_2CH_3$ as a function of time for short times.

The rate of increase of $[CH_2{=}CHCH_2CH_3]$ decreases because $\exp(-k_1/t)$ decreases with increasing time and the second term in equation 6.95 comes into play. This trend continues, as shown in Figure 6.17 on the next page. Clearly the accuracy of these incremental estimates can be improved by decreasing the time increment.

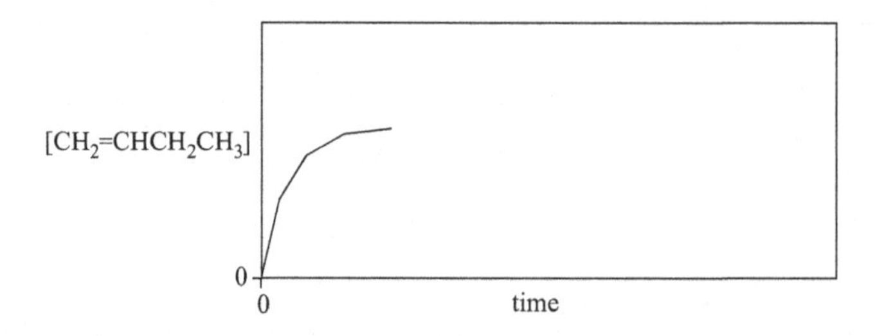

Figure 6.17 A sketch of the concentration of $CH_2{=}CHCH_2CH_3$ as a function of time for short times.

At long times, $CH_3CH_2CH_2CH_3$ is depleted, thus $[CH_3CH_2CH_2CH_3] \approx 0$; the rate expression for $CH_2{=}CHCH_2CH_3$ becomes

$$\frac{d}{dt}[CH_2{=}CHCH_2CH_3] \approx -k_2[CH_2{=}CHCH_2CH_3], \tag{6.97}$$

where we see that the right-hand side is less than zero. A derivative less than zero means the function decreases with time; thus the rate of decrease of $[CH_2{=}CHCH_2CH_3]$ decreases for longer times, because $[CH_2{=}CHCH_2CH_3]$ decreases. We add the long-time behavior for three increments to yield Figure 6.18. The behavior for intermediate times can be estimated by sketching a smooth curve between the estimates at short and long times.

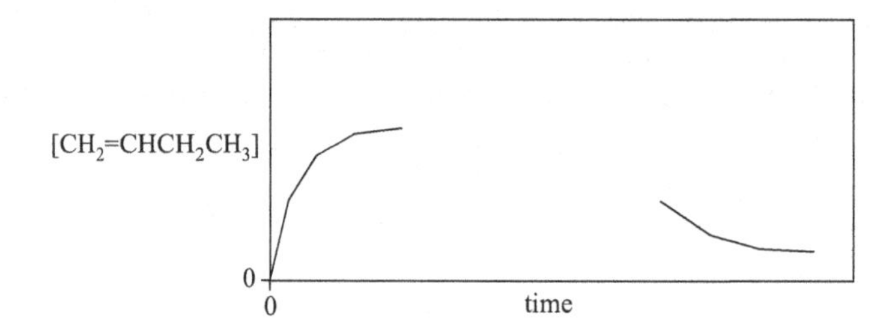

Figure 6.18 A sketch of the concentration of $CH_2{=}CHCH_2CH_3$ as a function of time for short and long times.

These limiting cases clearly suggest that $[CH_2{=}CHCH_2CH_3]$, the concentration of desired product, passes through a maximum. Recall from entry-level calculus that a function is at a maximum when its derivative equals zero. Thus the maximum $[CH_2{=}CHCH_2CH_3]$ is obtained by setting equation 6.95 equal to zero:

$$0 = k_1[CH_3CH_2CH_2CH_3]_0 \exp(-k_1 t_{max}) - k_2[CH_2{=}CHCH_2CH_3]_{max} \tag{6.98}$$

$$[CH_2{=}CHCH_2CH_3]_{max} = \frac{k_1}{k_2}[CH_3CH_2CH_2CH_3]_0 \exp(-k_1 t_{max}). \tag{6.99}$$

From equation 6.99 one obtains the time at which the maximum of $[CH_2{=}CHCH_2CH_3]$ occurs:

$$t_{max} = \frac{1}{k_1} \ln\left(\frac{k_1}{k_2}\frac{[CH_3CH_2CH_2CH_3]_0}{[CH_2{=}CHCH_2CH_3]_{max}}\right). \tag{6.100}$$

Thus if k_1 is large (the desired reaction 6.6 is fast) and k_2 is small (the undesired reaction 6.7 is slow), the maximum concentration of $CH_2{=}CHCH_2CH_3$ is comparable to the initial concentration of CH_2=$CHCH_2CH_3$, and the maximum is reached in a short time. If the converse is true – if k_1 is small and k_2 is large – the maximum concentration of CH_2=$CHCH_2CH_3$ is small.

Finally, we sketch the time behavior of the undesired product, CH_2=CHCH=CH_2. From equation 6.91, we see we can estimate [CH_2=CHCH=CH_2] as a function of time from Figure 6.14. First, we see that the rate of increase of [CH_2=CHCH=CH_2] is never negative. At short times, there is little CH_2=$CHCH_2CH_3$ in the reactor and thus the rate of increase of [CH_2=CHCH=CH_2] is slow. At longer times, the amount of CH_2=$CHCH_2CH_3$ increases and the rate of increase of [CH_2=CHCH=CH_2] increases.

At much longer times, [CH_2=$CHCH_2CH_3$] falls to zero and the rate of increase of [CH_2=CHCH=CH_2] falls to zero. In other words, at long times, [CH_2=CHCH=CH_2] asymptotically reaches a maximum. This is sketched in Figure 6.19.

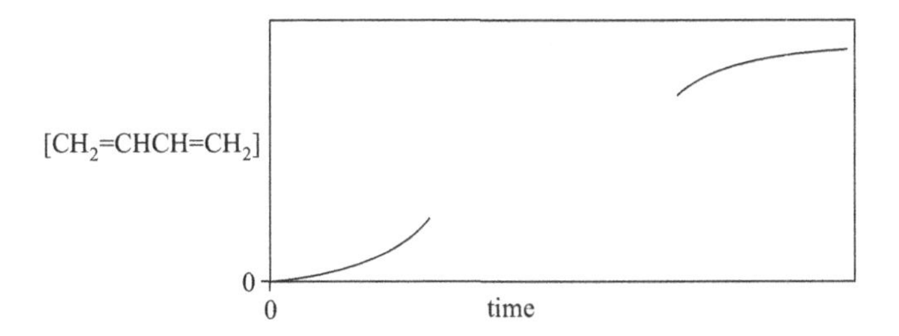

Figure 6.19 A sketch of the concentration of CH_2=CHCH=CH_2 as a function of time for short and long times.

Without solving differential equations 6.88 to 6.91, we have predicted the qualitative concentrations of the reactant, intermediate, and undesired product in our batch reactor. Quantitative estimation using time steps done with computers is called *numerical integration*. An introduction to the subject, Numerical Integration of Differential Equations, is given in the exercises at the end of this chapter. The discussion above has large time steps set by our intuition of the behavior of each differential equation.

The species concentrations as a function of time for a typical reaction $A \rightarrow B \rightarrow C$, obtained either by numerical or by analytical methods, yields a plot similar to Figure 6.20.

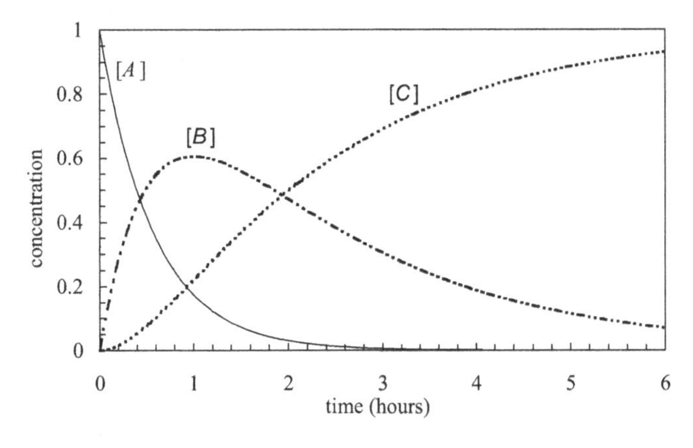

Figure 6.20 Concentrations of reactant *A*, product *C*, and intermediate *B* as a function of time for a series reaction conducted in a batch reactor.

There is clearly a time at which the amount of compound B reaches a maximum – about one hour in the example in Figure 6.20. A chemical plant for the continuous production of compound B, then, would contain a large reactor that was charged with reactants, run for one hour, and then emptied into a surge tank, which feeds a separator. The reactor is subsequently recharged, and the process repeated.

Batch production is not compatible with the *steady-state* processes we designed in Chapter 2. We could insert a surge tank after a batch reactor, as was shown for citric acid production in Section 6.1. Instead, we could use a *plug flow reactor* (PFR) to produce compound B continuously and produce the same maximum concentration of B as the batch reactor. The plug flow reactor is so named because a *plug* of fluid moves along the reactor *without mixing with the "plug of fluid" in front of it or behind it*: it is a pipe. We will use our understanding of the batch reactor to describe how a plug flow reactor works.

Imagine a batch reactor on a conveyor belt, as shown in Figure 6.21. The reactor is filled with reactants and placed on the conveyor. The conveyor's speed is adjusted so the batch reactor reaches the end just as the concentration of B reaches its maximum.

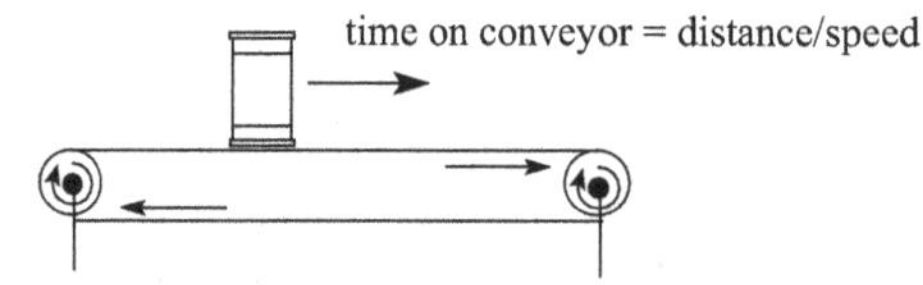

Figure 6.21 A batch reactor on a conveyor belt.

As the batch reactor falls off the end of the conveyor, we catch it and empty its contents into our separation process. The reactor is then recharged and placed at the beginning of the conveyor.

The flow from this conveyor reactor can be smoothed by adding more reactors, as shown in Figure 6.22. If the conveyor is loaded with many tiny reactors, the output is nearly continuous.

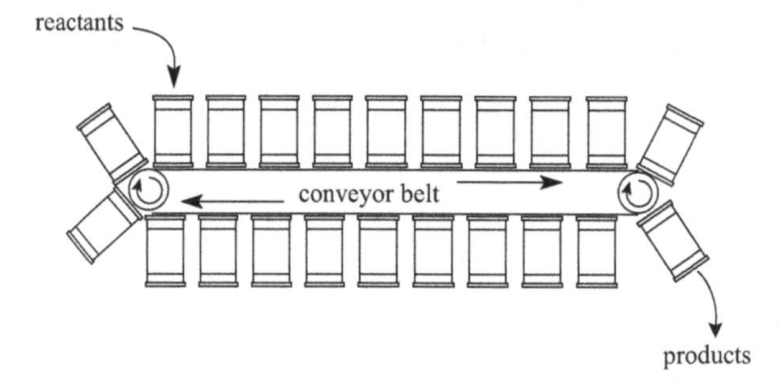

Figure 6.22 A continuous series of batch reactors on a conveyor belt.

The last step toward visualizing a plug flow reactor is to imagine that the batch reactors become infinitesimally small and infinitely numerous. The conveyor is replaced by a pipe, and the conveyor belt is replaced by a solvent that contains the reactants. Each "plug" of liquid moving down the pipe is the equivalent of a batch reactor on the conveyor. The plugs are infinitesimally thin and thus infinitely numerous. Such a plug flow reactor is shown in Figure 6.23. The time each plug spends in the pipe, or residence

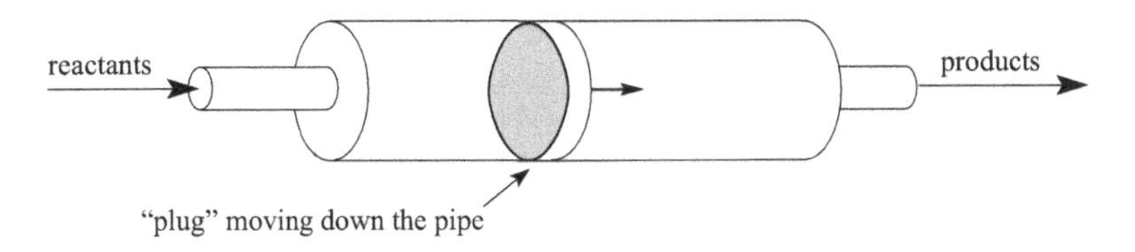

Figure 6.23 The plug flow reactor (PFR).

time, is the pipe volume divided by the volumetric flow rate. The volumetric flow rate is adjusted so the time in the pipe maximizes the amount of intermediate B. Or, given a desired volumetric flow rate, the volume of the plug flow reactor is designed to achieve the desired residence time.

In this section we considered a batch reactor in which multiple chemicals were being created and consumed at the same time. A mathematical model for such a system led to coupled ordinary differential equations, the solution of which may be visualized prior to a proper mathematical solution. The notion of an optimal time to produce one of the intermediary chemicals in such a reactor suggests reactions conducted in a pipe of specified length. Chemical engineering students take a course entitled Reaction Engineering during their junior year; this course teaches both analytical and graphical methods for designing plug flow reactors for complex reaction kinetics and heterogeneous physical systems.

6.5 MULTIPLE STEADY STATES: CATALYTIC CONVERTERS

Context: Automotive catalytic converters.

Concepts: Multiple steady states in a continuous, stirred-tank reactor (CSTR).

Defining Question: How do we design a catalytic converter for an automobile?

An automobile catalytic converter is also a *continuous reactor* that expedites three different chemical reactions. First, it catalyzes the reduction of NO (with the concomitant oxidation of CO):

$$NO + CO \rightarrow \frac{1}{2}N_2 + CO_2. \qquad \text{rxn (6.8)}$$

The CO in excess of reaction 6.8 is oxidized via

$$CO + \frac{1}{2}O_2 \rightarrow CO_2, \qquad \text{rxn (6.9)}$$

as are hydrocarbons via

$$\text{hydrocarbons} + O_2 \rightarrow CO_2 + H_2O. \qquad \text{rxn (6.10)}$$

The three-way catalyst for these reactions is a formulation of metal particles supported on a metal-oxide surface. In this section we will study the design and operation of a reactor to conduct reaction 6.8.

Reactions at catalytic surfaces are characterized by unusual rate expressions. Unlike simple reactions in liquids, where the reaction rate increases with reactant concentration, the rate for a surface reaction often depends on reactant concentration in the manner shown in Figure 6.24. The features in Figure 6.24 arise from the mechanism of the surface-catalyzed reaction; for CO to react, both CO and the co-reactant NO must be adsorbed on the surface of the solid catalyst. At low [CO], increasing [CO] increases the

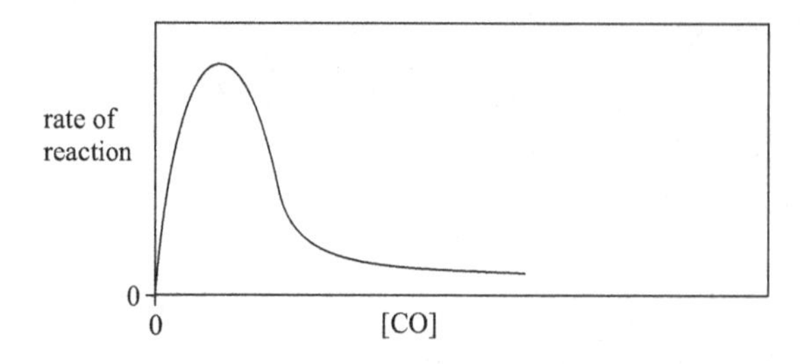

Figure 6.24 Rate of a surface-catalyzed reaction as a function of reactant concentration.

amount of CO on the surface and thus increases the rate of reaction of CO. But above a critical concentration of CO, too much CO adsorbs on the surface and excludes the adsorption of NO. Thus the rate of reaction decreases at high [CO]. The reaction mechanism comprises a series of elementary steps involving CO, NO, and adsorption sites, designated S:

$$\mathrm{CO + S \leftrightarrow CO{-}S} \qquad \text{rxn (6.11)}$$

$$\mathrm{NO + S \leftrightarrow NO{-}S} \qquad \text{rxn (6.12)}$$

$$\mathrm{CO{-}S + NO{-}S \rightarrow \frac{1}{2}N_2 + CO_2 + 2S.} \qquad \text{rxn (6.13)}$$

The double-headed arrows indicate reversible reactions. Because these reversible adsorption/desorption reactions 6.11 and 6.12 are much faster than the surface reaction 6.13, the reversible reactions are at equilibrium. The mechanism of reactions 6.11 to 6.13 leads to the rate expression

$$\text{rate of reaction} = \frac{k_1[\mathrm{CO}][\mathrm{NO}]}{(1 + K_1[\mathrm{CO}] + K_2[\mathrm{NO}])^2}, \qquad (6.101)$$

where k_1 is the rate constant of the surface reaction, and K_1 and K_2 are the equilibrium constants for the adsorption/desorption reactions. A qualitative plot of equation 6.101 is given in Figure 6.24.

As noted in the previous section, most chemical reactors are *continuous*, unlike the batch reactor. There are several types of continuous reactors, including the PFR. Another common industrial reactor is the continuous, stirred-tank reactor (CSTR), diagrammed in Figure 6.25.

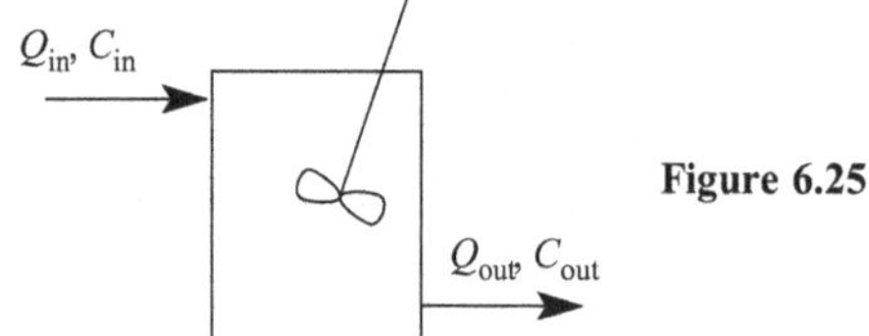

Figure 6.25 A continuous, stirred-tank reactor (CSTR).

Because the contents of a CSTR are *well stirred*, the composition of the effluent is the same as the composition at every point inside the reactor. Uniform composition is a key assumption in the analysis of a CSTR.

We will use a CSTR to catalytically reduce NO with CO, reaction 6.8. The catalytic converter in your car is *not* a CSTR; it is a PFR. Our analysis will suggest one reason *why* it is not a CSTR.

We apply the mass conservation principle to the CO in the reactor. First list the five terms in the time-dependent mass balance on CO:

$$\begin{aligned} \text{input} &= [\text{CO}]_{\text{in}} Q_{\text{in}} \\ \text{output} &= [\text{CO}]_{\text{out}} Q_{\text{out}} \\ \text{generation} &= 0 \\ \text{consumption} &= rV \\ \text{accumulation} &= \frac{d}{dt}[\text{CO}]_{\text{reactor}} V. \end{aligned} \tag{6.102}$$

Combining these terms, our mass balance for CO is given by

$$\frac{d}{dt}[\text{CO}]_{\text{reactor}} V = [\text{CO}]_{\text{in}} Q_{\text{in}} - [\text{CO}]_{\text{out}} Q_{\text{out}} - rV. \tag{6.103}$$

At *steady state* the derivative with respect to time equals zero and $Q_{\text{in}} = Q_{\text{out}} \equiv Q$, which yields

$$[\text{CO}]_{\text{in}} - [\text{CO}]_{\text{out}} = r\frac{V}{Q}. \tag{6.104}$$

Recall that the ratio V/Q is the residence time for a molecule in the reactor; thus we write

$$[\text{CO}]_{\text{in}} - [\text{CO}]_{\text{out}} = r\tau. \tag{6.105}$$

Thus the difference in concentration between inlet and outlet equals the rate of reaction times the residence time. This makes sense. If the rate of reaction is increased, for example, by increasing the temperature, equation 6.105 predicts that the difference between inlet and outlet concentrations should increase. If the reactor is enlarged, the residence time increases and the difference between inlet and outlet concentrations should increase.

Equation 6.101 provides a mathematical expression for the rate of reaction r in terms of the concentration of CO. But rather than solve this non-linear algebraic equation analytically (i.e. derive an expression $[\text{CO}]_{\text{out}} = \ldots$) we will solve equation 6.105 graphically, as follows. First, we plot the left side of equation 6.105 as a function of $[\text{CO}]_{\text{reactor}}$ in Figure 6.26. Recall that $[\text{CO}]_{\text{reactor}} = [\text{CO}]_{\text{out}}$ for a CSTR; thus the functional form of the left-hand side of equation 6.105 can be plotted as $(1-x)$ as a function of x.

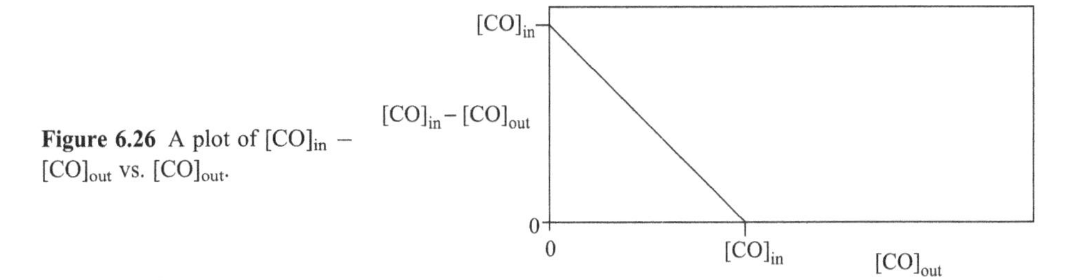

Figure 6.26 A plot of $[\text{CO}]_{\text{in}} - [\text{CO}]_{\text{out}}$ vs. $[\text{CO}]_{\text{out}}$.

Now we superimpose on Figure 6.26 the right side of equation 6.105 as a function of [CO]. This curve is obtained by multiplying the rate shown in Figure 6.24 by the residence time.

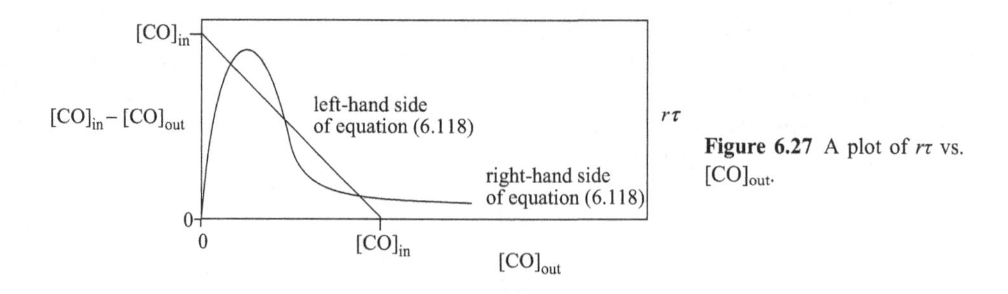

Figure 6.27 A plot of $r\tau$ vs. $[CO]_{out}$.

The superposition of the two curves in Figure 6.27 generates a *locus* of points that satisfy equation 6.105. Figure 6.28 reveals three such points, known as the steady-state operating points for our CSTR. Note that we can change these operating points by, for example, changing the size of the reactor (design) or the flow rate through the reactor (operation). Increasing the reactor size increases τ, which moves the rate curve up, which changes the intersections. Similarly, decreasing the flow rate increases τ, which moves the rate curve up, which changes the intersections.

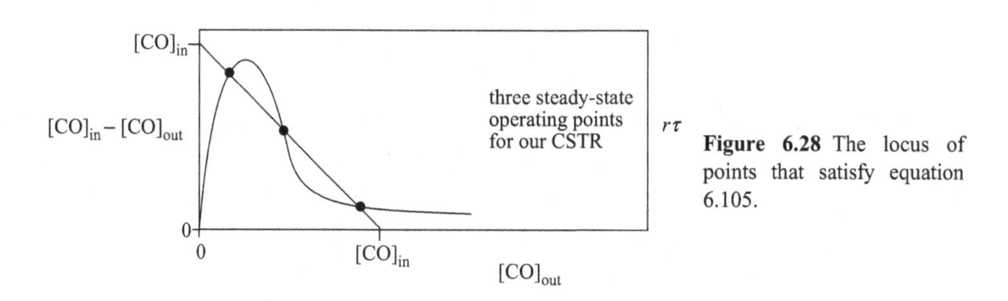

Figure 6.28 The locus of points that satisfy equation 6.105.

Not all the steady-state operating points derived from Figure 6.27 can be obtained in practice, however. Some of the steady-state points are stable and some are unstable. If the reactor is perturbed slightly from steady state and it returns to the same steady state, the operating point is *stable*. Similarly, if the reactor is perturbed slightly from steady state and it then migrates further from steady state, the operating point is *unstable*. To examine the stability of a steady-state point, we must examine the time dependence of the system, as given in equation 6.103:

$$\frac{d}{dt}[CO]_{reactor}V = [CO]_{in}Q_{in} - [CO]_{out}Q_{out} - rV. \tag{6.103}$$

Assume the reactor volume is constant and the flow rate in equals the flow rate out. Divide each side of equation 6.103 by the flow rate, to obtain

$$\tau\frac{d}{dt}[CO]_{reactor} = \left([CO]_{in} - [CO]_{out}\right) - r\tau. \tag{6.106}$$

Equation 6.106 is interesting: it tells us that the time dependence of [CO] is given by the difference between the two curves in Figure 6.28. To best demonstrate how the graph in Figure 6.28 illustrates the stability of our steady-state operating conditions, consider perturbations that change [CO] in the reactor. Such changes are illustrated in Figure 6.29 with a blow-up of the left-hand operating point from Figure 6.28.

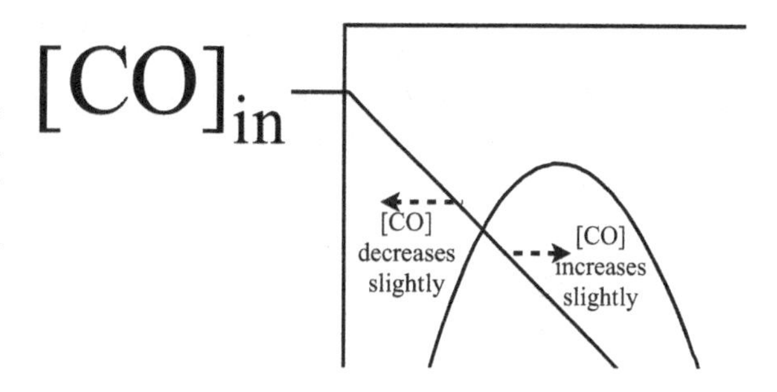

Figure 6.29 A blow-up of the left-hand stable operating point in Figure 6.28. The dashed arrows indicate perturbations that drive [CO] slightly lower (left) and slightly higher (right).

A slight decrease in [CO] occurs in a region of Figure 6.29 where the line for $[CO]_{in} - [CO]_{out}$ is above the line for $r\tau$. We see from equation 6.106 that $d[CO]/dt$ is positive and thus [CO] increases back to its steady-state operating point. Similarly, a slight increase in [CO] occurs in a region where the derivative is negative (the difference between the line and the curve). Thus $d[CO]/dt$ is negative and [CO] decreases back to its steady state operating point. The left-hand steady-state operating point is stable with respect to perturbations in [CO].

The same analysis reveals that the second steady-state operating point is unstable. If for some reason [CO] falls below the second steady-state point, [CO] continues to fall until it reaches the first steady-state point. Why? Because the difference between the line for $[CO]_{in} - [CO]_{out}$ and the line for $r\tau$ is negative. Similarly, if [CO] momentarily rises above the second steady-state point, [CO] continues to rise until it reaches the third steady-state operating point. An analysis of the third operating point reveals that it is stable.

We conclude with an interesting question on the operation of our CSTR. Recall that we wish to oxidize as much CO as possible to mitigate emissions from our automobile. When we start our reactor, $[CO]_{out}$ is equal to $[CO]_{in}$, the very right-hand side of the graph in Figure 6.28. The reaction starts and [CO] decreases and then reaches the third operating point. Whenever [CO] decreases below the third steady-state point, our stability analysis predicts [CO] will return to the third steady-state point. How does one reach the desirable operating point at low [CO]? Vexing.

The steady-state operation of this CSTR predicted acceptable operation. However, start-up will be tricky, if not impossible. Perhaps a CSTR is a poor choice for this reaction; indeed your automobile has a PFR for this reaction. Note that this CSTR design problem would not be detected by analyzing the steady-state operation alone. The transient behavior must also be analyzed.

Highly non-linear chemical kinetics can result in unusual conditions for the operation of a continuous, stirred-tank reactor. In this section we examined the unusual kinetics associated with chemical reactions taking place with multiple species on a catalytic surface, such as the situation within a catalytic converter in an automobile. The result is stable operation of the reactor under three separate steady-state operating conditions. The start-up and shut-down of such a reactor, and its control, might be very complicated. The analysis of multiple steady states in reactors is described in more detail in the text by Denn (1986), from which this example was adapted. We encourage you to investigate this textbook on process modeling.

6.6 MASS TRANSFER: CITRIC ACID PRODUCTION

Context: Transfer of substances between immiscible liquid phases.
Concepts: Overall mass transfer coefficient.
Defining Question: How do we design the absorber for citric acid production?

In this and the next section we briefly introduce two other transient-state processes that dominate engineering design: heat and mass transfer. These rate processes embody a legacy that is in the heart of all chemical engineers: every rate process is defined by a driving force, a famous number, and a geometry.

The movement of mass from one place to another is labeled *mass transfer* and is the subject of at least one course in the chemical engineering curriculum. We assumed rates of mass transfer were infinitely fast when, in Chapter 4, we considered separation processes involving vapor–liquid equilibria. Liquid–liquid absorbers operate on the same principle, as seen in Figure 6.30. Two liquid phases (usually oil and water) partition citric acid between them. The froth appearing in the absorber (see Figure 5.25) comprises small oil and water droplets; we must ensure equilibrium is reached for the transfer of citric acid from water to butanol within each stage of the absorber. Figure 6.30 schematically shows a typical absorber stage. How do we account for the transfer of mass from the water to the oil phase? How do we know when equilibrium is reached?

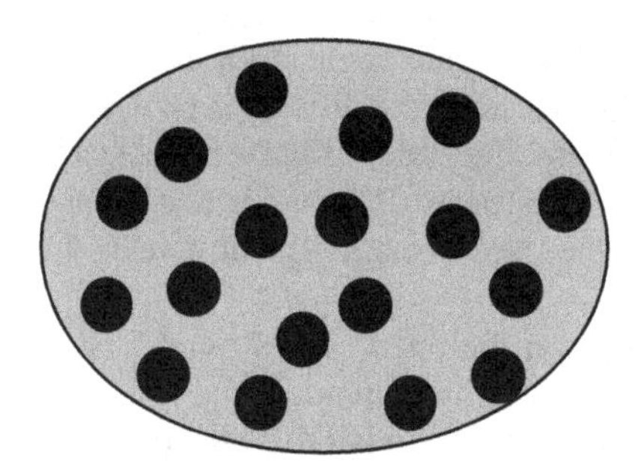

Figure 6.30 Water (gray) and oil (black) phases within the stage of a liquid–liquid absorber used for producing citric acid. The solute molecule citric acid undergoes mass transfer from the water to the oil phase.

We start with a mass balance for citric acid (CA) for both the water and oil phases as control volumes:

$$\textit{accumulation of citric acid in water} = \textit{input} + \textit{generation} - \textit{output} - \textit{consumption}$$
$$\textit{accumulation of citric acid in oil} = \textit{input} + \textit{generation} - \textit{output} - \textit{consumption}$$

The accumulation will be the familiar derivative with respect to time, $d([\text{CA}]_{\text{water}} V_{\text{water}})/dt$ and $d([\text{CA}]_{\text{oil}} V_{\text{oil}})/dt$. The volumes of oil and water are given by the ratio of water and oil mass flow rates in the absorber (assuming oil and water are immiscible). No citric acid is generated or consumed in either phase. The input and output terms are given by *mass transfer* from one phase to the other.

We will guess an equation for the rate of mass transfer based upon analogy to chemical reaction rates, r. In the case of chemical reactions, we found the rate of chemical reaction was given by

$$[r_{\text{rxn}}] = \text{mass}/(\text{volume} \times \text{time}), \tag{6.107}$$

and r_{rxn} was a rate constant times the concentrations of reactants to some power. Chemical reactions occur in a volume so the dimensions of $r_{\text{rxn}}V$ are mass per time.

Mass transfer occurs across a surface. We surmise an overall rate of mass transfer r_{mass} such that

$$[r_{\text{mass}}] = \text{mass}/(\text{area} \times \text{time}), \tag{6.108}$$

so $r_{\text{mass}}A$ has dimensions of mass/time, where A is the total interfacial area between the oil and water phases. We further define this rate to be the net rate of citric acid transfer into the oil phase. We have mass conservation:

$$\frac{d}{dt}[\text{CA}]_{\text{water}}V_{\text{water}} = -Ar_{\text{mass}} \tag{6.109}$$

$$\frac{d}{dt}[\text{CA}]_{\text{oil}}V_{\text{oil}} = Ar_{\text{mass}}. \tag{6.110}$$

What is the functional dependence of r_{mass} on the concentrations? We recognize a fundamental law due to Adolf Fick in the middle of the nineteenth century, namely that the *driving force* for mass transfer is a gradient in concentration. In the absence of a gradient there is no mass transfer. The rate of mass transfer is proportional to the concentration gradient, as expressed in equation 6.111:

$$r_{\text{mass}} \propto \Delta(\text{concentration}). \tag{6.111}$$

The proportionality converts to an equality by adding a factor known as the overall mass transfer coefficient, K_{mass}:

$$r_{\text{mass}} = K_{\text{mass}}\Delta(\text{concentration}). \tag{6.112}$$

Because r_{mass} has units of mass/(area $\times$ time), dimensional consistency demands that K_{mass} have units of length/time. A similar expression can be written for the oil phase, with a sign change.

What is the form of Δ(concentration)? We surmise it to be a gradient between the oil and water phase concentrations: let's write this function as the difference $[\text{CA}]_{\text{oil}} - f([\text{CA}]_{\text{water}})$. Why a difference? When $[\text{CA}]_{\text{oil}}$ equals $f([\text{CA}]_{\text{water}})$ no mass transfers, consistent with Fick's law. We find $f([\text{CA}]_{\text{water}})$ through laboratory measurement. Conveniently, the function is typically just a number times $[\text{CA}]_{\text{water}}$. The number is called the *distribution coefficient* or *partition coefficient.* We label it D and look it up in tables. Equation 6.112 becomes

$$r_{\text{mass}} = K_{\text{mass}}\left([\text{CA}]_{\text{oil}} - D[\text{CA}]_{\text{water}}\right). \tag{6.113}$$

We substitute equation 6.113 into equations 6.109 and 6.110 to obtain our transient-state mass balances:

$$\frac{d}{dt}[\text{CA}]_{\text{water}}V_{\text{water}} = -AK_{\text{mass}}\left([\text{CA}]_{\text{oil}} - D[\text{CA}]_{\text{water}}\right) \tag{6.114}$$

$$\frac{d}{dt}[\text{CA}]_{\text{oil}}V_{\text{oil}} = AK_{\text{mass}}\left([\text{CA}]_{\text{oil}} - D[\text{CA}]_{\text{water}}\right). \tag{6.115}$$

Citric acid leaves one phase and enters the other phase. If we define the system boundaries as the entire volume of Figure 6.30, there is no accumulation of citric acid. Mathematically, we add equations 6.114 and 6.115 to get

$$\frac{d}{dt}[\text{CA}]_{\text{water}}V_{\text{water}} + \frac{d}{dt}[\text{CA}]_{\text{oil}}V_{\text{oil}} = 0. \tag{6.116}$$

For simplicity, let's set the volumes of oil and water identical in our absorber, so V_{oil} and V_{water} are equal to a number, V. Equation 6.116 simplifies to

$$V\frac{d}{dt}[\text{CA}]_{\text{water}} + V\frac{d}{dt}[\text{CA}]_{\text{oil}} = 0 \quad (6.117)$$

$$\frac{d}{dt}[\text{CA}]_{\text{oil}} = -\frac{d}{dt}[\text{CA}]_{\text{water}}. \quad (6.118)$$

We integrate equation 6.118 and get equation 6.119: the amount of citric acid that leaves the water phase is equal to the amount of citric acid that enters into the oil phase:

$$[\text{CA}]_{\text{oil,0}} - [\text{CA}]_{\text{oil}} = -\left([\text{CA}]_{\text{water,0}} - [\text{CA}]_{\text{water}}\right). \quad (6.119)$$

We solve equation 6.119 for $[\text{CA}]_{\text{oil}}$ and insert it into equation 6.114 to yield the following differential equation:

$$\frac{d}{dt}[\text{CA}]_{\text{water}}V_{\text{water}} = -AK_{\text{mass}}\left([\text{CA}]_{\text{water,0}} - [\text{CA}]_{\text{water}} + [\text{CA}]_{\text{oil,0}} - D[\text{CA}]_{\text{water}}\right) \quad (6.120)$$

$$\frac{d}{dt}[\text{CA}]_{\text{water}} = \frac{AK_{\text{mass}}(1+D)}{V}[\text{CA}]_{\text{water}} - \frac{AK_{\text{mass}}D}{V}\left([\text{CA}]_{\text{water,0}} + [\text{CA}]_{\text{oil,0}}\right). \quad (6.121)$$

The differential equation 6.121 has the form $dx/dt = ax + b$, which can be separated and integrated to yield equation 6.122:

$$[\text{CA}]_{\text{water}} = \frac{1}{1+D}\left([\text{CA}]_{\text{water,0}} + [\text{CA}]_{\text{oil,0}}\right) + \frac{1}{1+D}\left([\text{CA}]_{\text{water,0}} - D[\text{CA}]_{\text{oil,0}}\right)e^{-\frac{AK_{\text{mass}}(1+D)}{V}t}. \quad (6.122)$$

It is always prudent to check an equation at known limits. Equation 6.122 simplifies to $[\text{CA}]_{\text{water}} = [\text{CA}]_{\text{water, 0}}$ at $t = 0$ and $[\text{CA}]_{\text{water}}$ equals the first term in the limit $t \to \infty$. The limits are correct.

Note the time dependence of the exponential in equation 6.122. Time appears only in the argument of the exponent, suggesting comparison to the *residence* time τ described in equation 6.63:

$$\tau = \frac{V}{AK_{\text{mass}}(1+D)}. \quad (6.123)$$

Equation 6.123 reveals that the residence time for mass transfer is governed by the overall mass transfer coefficient and the size of the droplets dispersed into the absorber, represented by the interfacial area A. As with residence time (Section 6.2), we associate the approach to equilibrium as a decrease of the time dependence in equation 6.123 to approximately 5% of its initial value; that is, the argument of the exponent is about -3. We have

$$\frac{AK_{\text{mass}}(1+D)}{V}t_{\text{equil}} \approx 3, \quad (6.124)$$

and with typical numbers for $K_{\text{mass}} \approx 3 \times 10^{-3}$ cm/s and $D \approx 2$, we arrive at

$$t_{\text{equil}} \approx \frac{3 \times 10^{3}\ \text{sec}}{\text{cm}}\left(\frac{V}{A}\right). \quad (6.125)$$

The ratio V/A for a spherical bubble is easily estimated from geometric arguments, and we calculate the equilibrium timescales in Table 6.4 on the next page.

Table 6.4 Timescales for mass transfer equilibrium.

Droplet radius (cm)	V/A (cm)	t_{equil} (s)
0.01	⅓ × 10^{-2}	10
0.1	⅓ × 10^{-1}	100
1	⅓	1,000

One can use the equilibrium timescale with the dimensionless relation in Chapter 5 to calculate the terminal velocity of an oil droplet rising through water and thus the required depth of water on an equilibrium stage in an absorber column.

Let's review the dimensions of our mass transfer timescale:

$$\tau = \frac{V}{AK_{\text{mass}}(1+D)} = \frac{\text{volume}}{(\text{area})(\text{mass transfer coefficient})(1+D)} \tag{6.126}$$

$$\tau \propto \frac{\text{volume/area}}{\text{mass transfer coefficient}} = \frac{\text{sphere charateristic length}}{\text{mass transfer coefficient}}. \tag{6.127}$$

Equation 6.127 reveals that τ, the characteristic time for mass transfer, is the characteristic length of the sphere (= V/A) divided by the velocity of mass transfer. We see from our numerical estimates that mass transfer is rapid compared to hydraulic residence times for this absorber. This is good news: a countercurrent multistage absorber can nearly attain (within 5%) equilibrium at each stage while the fluids traverse the apparatus. The bad news: droplet sizes must be small, necessitating mechanical mixers to disperse the droplets (think kitchen blender as a single-stage liquid–liquid absorber!).

We have identified a new rate process relevant to chemical engineering: the transfer of mass between two immiscible phases. The engineering reaction rate is governed by an overall rate which has dimensions of mass per time. Mass transfer is also governed by a fundamental quantity, the overall mass transfer coefficient (also dimensions of mass per time). The overall mass transfer coefficient connects the rate of master transfer to the driving forces for mass transfer (often, but not always, given by a concentration gradient) and the geometry of the system (in this case, spherical droplets). Aspects of modeling mass transfer can be found in the texts by Russell and Denn (1972) and Cussler (2009).

6.7 HEAT TRANSFER: CHEMICAL REACTOR RUNAWAY

Context: Temperature control in an adiabatic reactor.

Concepts: Overall heat-transfer coefficient.

Defining Question: When will the conditions in my reactor become dangerous?

The CSTR depicted in Figure 6.25 is broadly used in the chemical industry. The analysis conducted in Section 6.5 was for a particularly simple system in which the temperature of the reactor was assumed to be held constant. How does one do that? With a heat exchanger in the form of a jacket, as shown in Figure 6.31 on the next page. The cooling jacket uses water as a heat-transfer fluid because water is cheap and readily available (with a nod to sustainability concerns).

Endothermic and exothermic reactions were discussed in Chapter 3. The excess energy produced by exothermic reactions is delivered as heat that must be transferred to the cooling jacket to operate the reactor safely. Without the cooling jacket the energy

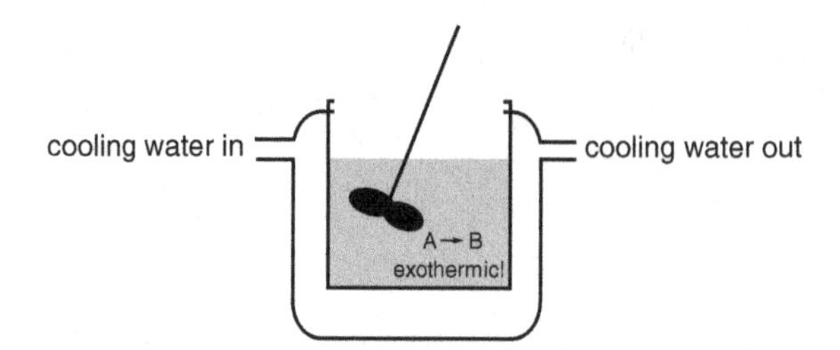

Figure 6.31 The jacketed CSTR. Cooling water carries away the excess energy (in the form of heat) from the chemical reaction.

released in the chemical reaction appears as heat that accelerates the reaction, which in turn produces more energy released as heat, and the reactor becomes a "runaway."

The question: how to prevent the apparatus depicted in Figure 6.31 from becoming a runaway? We require an analysis for safety whereby we consider the effects of energy released from the chemical reaction balanced against the ability to transfer heat from the reactor to the cooling water.

We begin with an energy balance on the reactor. Steady-state energy balances were described in Section 5.3.3. We must now include the rates of energy generated, consumed, input, and output:

$$\textit{accumulation of energy} = \textit{input} + \textit{generation} - \textit{output} - \textit{consumption}. \tag{6.128}$$

We recognize that the accumulation of energy will be the change in internal energy of our system with respect to time; see equation 3.111 for the relationship between internal energy and temperature. We have

$$\textit{accumulation of energy} = \frac{dU}{dt} = M\bar{C}_P\frac{dT}{dt} \quad \text{and} \tag{6.129}$$

$$\left[M\bar{C}_P\frac{dT}{dt}\right] = (\text{mass})\frac{\text{energy}}{(\text{mass})(\text{temperature})}\frac{\text{temperature}}{\text{time}} = \frac{\text{energy}}{\text{time}}, \tag{6.130}$$

with M as the total mass in the reactor and $\bar{C}_P$ its average heat capacity per mass, taken as a constant. Energy is generated by the exothermic reaction as described in Section 3.6.1, so we write

$$\textit{generation} = (-\Delta\bar{H}_{\text{rxn}})(rV_{\text{reactor}}) \quad \text{and} \tag{6.131}$$

$$[(-\Delta\bar{H}_{\text{rxn}})(rV_{\text{reactor}})] = \frac{\text{energy}}{\text{mass}}\frac{(\text{mass})}{(\text{volume})(\text{time})}(\text{volume}) = \frac{\text{energy}}{\text{time}}. \tag{6.132}$$

Why is there a negative sign? You will derive this result in more detail in a course on chemical reactor design, but you can recognize for now that an exothermic reaction has a negative $\Delta\bar{H}_{\text{rxn}}$, so the generation term becomes positive for an exothermic reaction. Check; equation 6.131 makes sense.

Consumption of energy is represented as heat transfer from the reactor to the jacket. Section 5.3.3 discussed heat transfer to a tube (Figure 5.21). At that time, we used dimensional arguments to posit that the rate of heat transfer, dq/dt, is usually combined with the temperature difference between the fluids, the surface area of the heat-transfer device, and a heat-transfer coefficient h.

$$\frac{dq}{dt} = hA(T - T_{\text{cool}}) \quad \text{and} \tag{6.133}$$

$$[hA(T - T_{\text{cool}})] = \frac{\text{energy}}{(\text{area})(\text{temperature})(\text{time})}(\text{area})(\text{temperature}) = \frac{\text{energy}}{\text{time}}. \tag{6.134}$$

Equation 6.128 becomes

$$M\bar{C}_P\frac{dT}{dt} = (-\Delta\bar{H}_{rxn})(rV_{reactor}) - \frac{dq}{dt} = (-\Delta\bar{H}_{rxn})(rV_{reactor}) - hA(T - T_{cool}). \tag{6.135}$$

Equation 6.135 is often called the basic design equation for a jacketed reactor. It is a transient-state energy balance.

The expression for the chemical reaction rate r must be included in equation 6.135. Our chemistry colleagues tell us this particular reaction is independent of concentration of the reactant, or, following Section 6.3,

$$r = k. \tag{6.136}$$

How do chemical reactions depend on temperature? All chemical rate constants k depend strongly on temperature, as given by the Arrhenius expression for a chemical reaction rate constant k,

$$k = k_0 e^{-E_a/RT}. \tag{6.137}$$

You will discover in your physical chemistry courses the origin and physical underpinnings of this profound equation. For now, we just use it. E_a is known as the *activation energy* and is tabulated (or can be calculated) for many reactions. Equation 6.135 becomes

$$M\bar{C}_P\frac{dT}{dt} = (-\Delta\bar{H}_{rxn})\left(k_0 e^{-E_a/RT}V_{reactor}\right) - hA(T - T_{cool}). \tag{6.138}$$

At steady state our reactor temperature is $T_{steady\text{-}state}$, and this time derivative will be zero. Thus at steady state equation 6.138 is

$$(-\Delta\bar{H}_{rxn})\left(k_0 e^{-E_a/RT_{steady\text{-}state}}V_{reactor}\right) - hA\left(T_{steady\text{-}state} - T_{cool}\right) = 0. \tag{6.139}$$

One can solve equation 6.139 for $T_{steady\text{-}state}$. To determine if the operating temperature $T_{steady\text{-}state}$ is a stable steady state, one might plot the two terms in equation 6.139 and ascertain if the reactor is stable to both positive and negative perturbations in the temperature, as in Section 6.5. We will use calculus instead. We take the derivative of equation 6.139 with respect to temperature and set it equal to zero:

$$\frac{d}{dT}\left((-\Delta\bar{H}_{rxn})\left(k_0 e^{-E_a/RT_{steady\text{-}state}}V_{reactor}\right) - hA\left(T_{steady\text{-}state} - T_{cool}\right)\right) =$$
$$(-\Delta\bar{H}_{rxn})\frac{E_a}{RT^2_{steady\text{-}state}}\left(k_0 e^{-E_a/RT_{steady\text{-}state}}V_{reactor}\right) - hA = 0. \tag{6.140}$$

Using equation 6.139 to eliminate the exponential term in equation 6.140 (it is essential to use equation 6.139 because this defines $T_{steady\text{-}state}$), we find a quadratic equation for $T_{steady\text{-}state}$ that depends only on the activation energy of the reaction and the cooling water temperature:

$$hA\left(T_{steady\text{-}state} - T_{cool}\right)\left(\frac{E_a}{RT^2_{steady\text{-}state}}\right) - hA = 0. \tag{6.141}$$

This yields

$$T_{steady\text{-}state} = \frac{E_a \pm \sqrt{E_a^2 - 4RT_{cool}E_a}}{2R}. \tag{6.142}$$

For a cooling jacket at 25°C and a typical activation energy of 120 kJ/mol, the maximum operating temperature is 25°C! We see that this reactor operates on a knife edge: if the reactor temperature rises slightly above the jacket temperature – runaway to boom! We would never build such a reactor.

This section provides a third and final rate process: a rate by which energy flows from one region of space to another: heat transfer. Like reaction rates and mass transfer rates, the heat-transfer rate has a driving force (typically a temperature difference), a geometry (in this case a cylindrical jacket), and a famous number: the overall heat-transfer coefficient. We see that failure to properly cool a reactor can lead to a system that is unsafe.

Heat transfer is essential for operating chemical reactors with exo- and endothermic reactions. For endothermic reactions, a failure of the heat-transfer apparatus quenches the chemical reaction because energy in the form of heat transfer is no longer supplied to the reactor. For exothermic reactions, the cooling jacket tempers the heat released by the chemical reaction. Care must be taken in working with such systems.

This example was adapted from a case study published by the Center for Chemical Process Safety of the AIChE.[2] We recommend other case studies in transient-state processes described there.

SUMMARY

While steady-state analyses provide the necessary information to design a process, transient-state mass and energy balances are needed to start, shut down, and control the process in a safe and effective manner. These balances necessitate the use of differential equations, often in combination with graphical and dimensional analyses. Underpinning these differential equations one often finds combinations of quantities that yield a characteristic timescale for these rate processes. These timescales can be used to broadly categorize how much time an engineer has to deal with a process interruption.

REFERENCES

Cussler, E. L. 2009. *Diffusion: Mass Transfer in Fluid Systems*, 3rd edn., Cambridge University Press, Cambridge, UK.

Denn, M. M. 1986. *Process Modeling*, Pitman Publishing, Marshfield, MA.

Russell, T. W. F. and Denn, M. M. 1972. *Introduction to Chemical Engineering Analysis*, John Wiley & Sons, New York, NY.

EXERCISES

Transient-State Processes

6.1 A round lake of diameter *d* is fed by a river and drained by seepage. The flow rate of the river is equal to the rate of seepage and thus the water level in the lake is constant. Coincidentally, the level in the lake matches the height of a wide spillway at its north end. Should the water level increase, water flows over the spillway into a flood plain.

[2] Center for Chemical Process Safety, Welker, J. R., and Springer, C. 1990. *Safety, Health, and Loss Prevention in Chemical Processes: Problems for Undergraduate Engineering Curricula*, American Institute of Chemical Engineers, New York, NY.

Suddenly the flow rate of the river increases to 1.5 times its normal rate. The extra water flows over the spillway at a rate determined by the height of the lake above the spillway,

$$Q_{\text{spillway}} = \alpha h^{3/2} (\text{in gal/ min}),$$

such that α is a constant determined in part by the width of the spillway.

(A) If the flow rate of the river persists indefinitely at 1.5 times the normal rate, to what level will the lake rise, relative to the height of the spillway? You may assume that the surface area of the lake remains approximately constant as the level rises.
(B) After the lake rises to the steady-state flood level calculated in (A), the flow rate of the river suddenly returns to normal. Derive a formula for the height of the lake as a function of time after the river flow returns to normal.

6.2 A different lake is fed by a river and is depleted by evaporation (no seepage). The rate of evaporation equals the rate of flow from the river. Suddenly a pollutant (compound X) is continuously dumped into the river. The pollutant does not evaporate and thus accumulates in the lake.

(A) Calculate the concentration of pollutant in the lake as a function of time (in kg pollutant/gal water):

$$\text{volume of water in lake} \equiv V_{\text{lake}} = 1.8 \times 10^{10} \text{gal}$$
$$\text{flow rate of river} \equiv Q_{\text{river}} = 3.1 \times 10^{8} \text{gal/day}$$
$$\text{concentration of } X \text{ in river} \equiv [X] = 1.3 \times 10^{-6} \text{kg/gal water.}$$

(B) The pollutant decomposes to inert substances at a rate proportional to its concentration in water:

$$\text{rate of decomposition of pollutant} = k[X].$$

$[X]$ has units of (kg X)/(gal water) and k is a constant with units of (gal water)/day. Calculate the concentration of pollutant in the lake as a function of time, in (kg pollutant)/(gal water), when decomposition is included.
(C) Calculate the steady-state concentration of pollutant in the lake.

6.3 A meteor strikes the Earth and forms a conical crater. The cone is shaped such that, at a vertical distance x from the bottom of the crater, the diameter of the crater is $4x$. Thus the volume when filled to a depth x is

$$\text{volume} = \frac{4}{3}\pi x^3.$$

Much later a river with volumetric flow rate Q_{river} begins to fill the crater.

(A) Assume that no water leaves the crater by evaporation or by seepage and derive an equation for the depth of water as a function of time.
(B) Assume now that water escapes from the crater lake by evaporation. The rate of evaporation (in gal/min) is proportional to the surface area of the lake:

$$\text{rate of evaporation} = kx^2,$$

where k is a constant determined by the temperature, humidity, and rate of solar heating. What will be the level of the lake at steady state?

(C) After the lake reaches steady state, the flow to the crater suddenly stops (because the river is diverted to Los Angeles). The water level in the lake drops, owing to evaporation. Derive a formula for the level of the lake as a function of time, starting from the time the river is diverted to Los Angeles.

6.4 A round lake of diameter d is fed by a river and drained by seepage. The flow rate of the river is equal to the rate of seepage and thus the water level in the lake is constant. Should the lake level rise, the water from the lake flows through a triangular notch in a retaining wall and into a flood plain. Normally the water level matches the bottom of the triangular notch.

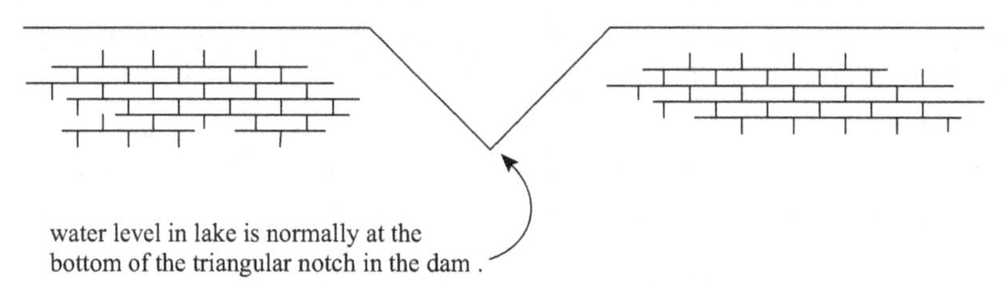
view of dam from flood plain
water level in lake is normally at the
bottom of the triangular notch in the dam .

Suddenly the flow rate of the river increases to 1.5 times the normal rate. The extra water flows through the triangular notch at a rate determined by the height of the lake above the bottom of the notch:

$$Q_{\text{notch}} = \alpha h^{5/2} \text{ (in gal/min)},$$

where α is a constant determined by the specifics of the notch.

(A) If the flow rate persists indefinitely at 1.5 times the normal rate, to what level will the lake level rise, relative to the height above the bottom of the notch? You may assume that the surface area of the lake remains approximately constant as the level rises.
(B) After the lake level has risen to the steady-state flood level calculated in (A), the flow rate of the river suddenly returns to normal. Derive a formula for the height of the lake as a function of time after the river flow returns to normal.

6.5 Spherical tanks are commonly used to store liquids and gases. For a sphere of radius R, a liquid at height h occupies a volume $V = \pi(Rh^2 - ⅓h^3)$.

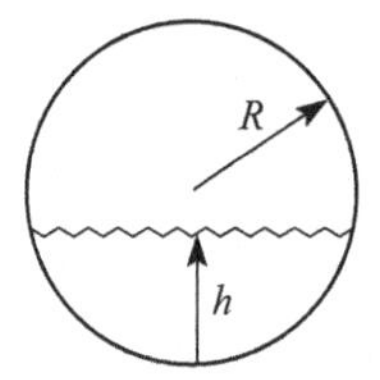
R
h

(A) A liquid is delivered to an empty spherical tank at a constant flow rate F_{in} (in kg/min). Derive an equation for the time the liquid rises to height h. Assume the liquid has a constant density ρ (in kg/m^3).
(B) Calculate the time the liquid reaches height $h = (3/2)R$.
(C) The liquid drains through a hole at the bottom of the spherical tank. For a finite liquid height h, the volumetric flow rate is $Q_{\text{out}} = kh^{1/2}$ (in m^3/min), where k is a proportionality constant determined by the diameter of the drain hole. Consider a

spherical tank initially filled with liquid to a height $h = (3/2)R$ with no flow into the tank. Derive an equation for the time for the liquid height to decrease to h.

(D) Calculate the times at which the draining liquid is at heights $h = (3/2)R$ (this will check your equation), $h = R$, and $h = ½R$.

6.6 Methanol is withdrawn from a 5.00 m^3 storage tank at a rate that increases linearly with time. Initially the tank contains 750. kg of methanol and the withdrawal rate is 750. kg/hr. Five hours later the withdrawal rate has increased to 1000. kg/hr. Methanol is continuously added to the tank at a rate of 1200. kg/hr to replenish the supply. The density of methanol is 0.792 g/cm^3.

(A) Derive a formula for the methanol withdrawal rate, $F_{out}(t)$, in kg/hr.
(B) Perform a mass balance to obtain a differential equation for the mass of methanol in the tank as a function of time.
(C) Calculate how much methanol will be in the tank at $t = 5.00$ hr.
(D) Calculate the time at which the methanol level in the tank is a maximum. What percentage of the tank volume is filled at this time?
(E) Calculate the time at which the tank is empty.

6.7 A well-mixed tank initially contains 741 L of B. At $t = 0$, A is pumped into the tank at 56 L/min and liquid is pumped from the tank at 56 L/min.

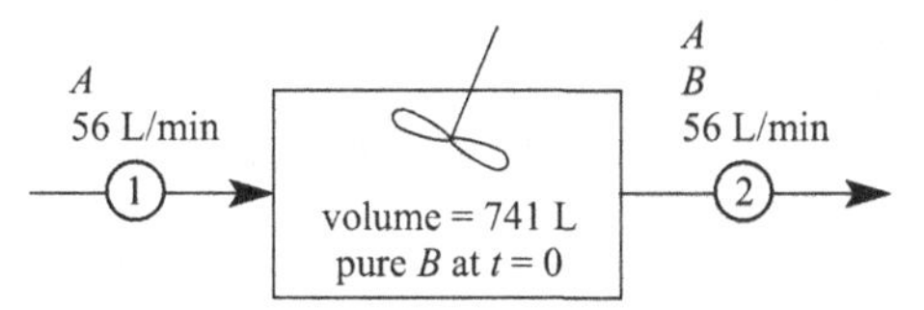

(A) Calculate the concentration of A (in kg/L) in stream 2 as a function of time. Data: density of A = density of B = density of A–B mixtures = 0.78 kg/L.
(B) Check your answer at three times: $t = 0$, $t = \infty$, and at t = residence time (characteristic time) for this process.
(C) In another system, this well-mixed tank is connected to different piping, as shown below. The tank initially contains 741 L of B. At $t = 0$, the flow rate of stream 1 is 56 L/min of A and stream 5 is 56 L/min of liquid. Initially the flow bypasses the tank; there is no flow in streams 2 and 3. $Q_4 = 56$ L/min and $Q_2 = Q_3 = 0.0$ L/min at $t = 0$.

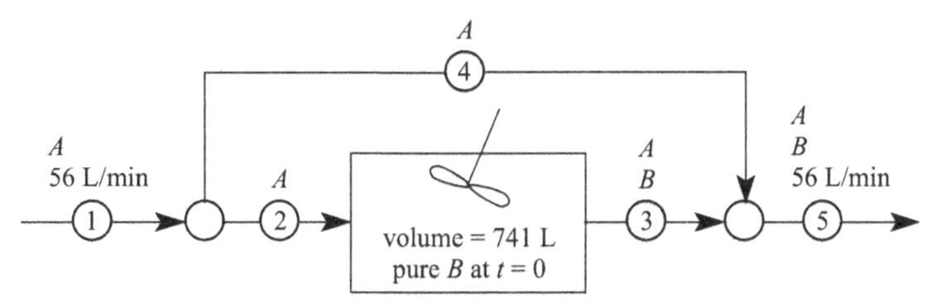

For $t \geq 0$ the flow in streams 2 and 3 increases linearly and the flow in stream 4 decreases linearly, as follows:

$$Q_2 = Q_3 = 2.0t\,\text{L/min for } 0 \leq t \leq 28\,\text{minutes}$$

$$Q_4 = 56 - 2.0t\,\text{L/min for } 0 \leq t \leq 28\,\text{minutes.}$$

For $t \geq 28$ minutes, $Q_4 = 0.0$ L/min and $Q_2 = Q_3 = 56$ L/min. Calculate the concentration of A (in kg/L) in stream 5 as a function of time for $0 \leq t \leq 28$ minutes.

6.8 Your company mixes three fruit juices – orange juice, cranberry juice, and guava juice – to produce two beverages – Red Sunshine Refresher and Best Morning Start – in the series of well-mixed tanks shown below. The splitter sends 40% of stream 3 to tank 2. Both tanks 1 and 2 are full when the process operates at steady state.

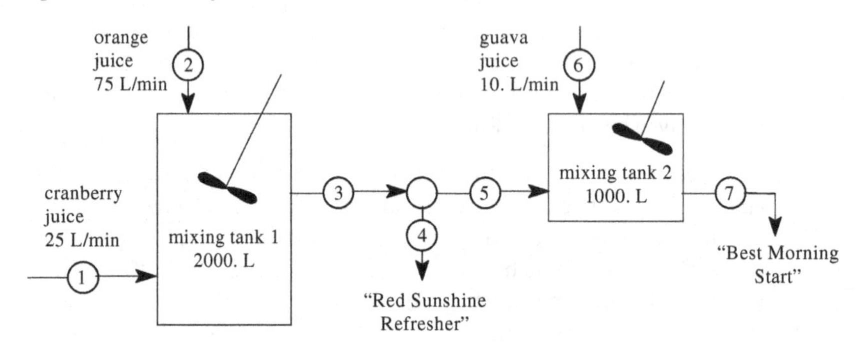

The pump that delivers cranberry juice to tank 1 fails. The orange juice pump continues at 75 L/min and the pump on stream 3 continues at 100 L/min.

(A) At what time does the volume of juice in tank 1 decrease to 1000. L? You may assume all juice densities are approximately equal to that of water (1.0 kg/L).

(B) What is the concentration of cranberry juice in tank 1 (in kg/L), 10 minutes after the pump fails? **Hint:** The composition in stream 3 is the same as the composition in tank 1.

(C) Propose a simple response to prevent either tank from emptying after the cranberry juice pump fails. You may decrease the flow rate of any stream but you may not increase the flow of any input streams. You may change the ratio of the splitter from the steady-state ratio (40% to tank 1 and 60% to Red Sunshine Refresher). The best response is the response with the fewest changes.

6.9 Consider the process shown below, previously described in Exercise 3.11. The pump providing the cranberry juice to the first tank fails. The orange juice pump continues at 75 L/min and the pump on the output of tank 1 continues at 100 L/min.

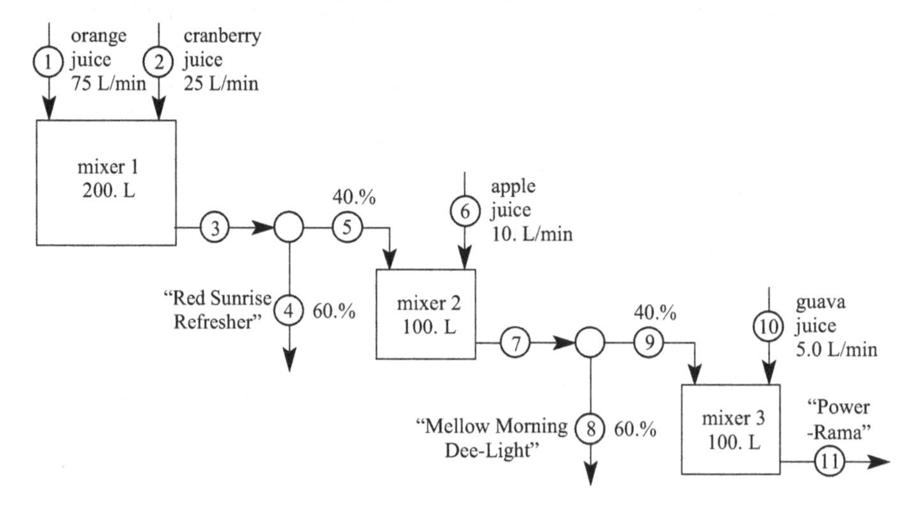

(A) At what time after the pump fails is mixer 1 empty?
(B) What is the concentration of cranberry juice in the mixer 1 (in kg/L), 2 minutes after the pump fails?
(C) Propose a simple response to prevent any of the mixers from emptying after the cranberry juice pump fails. You can decrease the flow rate of any stream but you cannot increase the flow of any input streams. You can change the ratio of either splitter from the normal ratio of 40% to 60%.

6.10 A portion of a simplified freezer–desalinator, discussed in Chapter 3, is shown below.

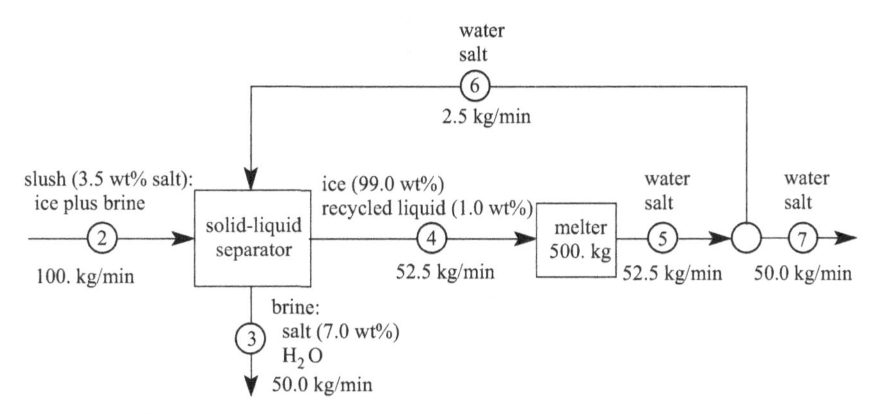

When the desalinator is started, the melter contains 500. kg of salt water (3.5 wt% salt). As the desalinator runs, the salt concentration in the melter decreases to a potable level.

(A) Calculate the wt fraction of salt in the melter as a function of time. The melter is well mixed, so the salt concentration in stream 5 is the same as the salt concentration in the melter. Ice is pure H_2O. The recycled liquid in stream 4 has the same composition as the liquid in stream 6.
(B) At what time does the process produce potable water (≤ 0.0005 wt fraction salt in stream 5)?

6.11 Compose an emergency response plan for a community near a chemical plant based on the following scenario.

Your chemical plant uses acrolein (acrylaldehyde) as an intermediate. Because acrolein is immediately lethal at a concentration of 6 ppm (0.0006%) in air, you need to advise the local community how to respond to an acrolein spill. The spill will be liquid acrolein on a non-porous surface. The liquid will slowly evaporate such that the air above the spill will contain a constant 10 ppm acrolein as long as the liquid remains. A steady wind will carry the acrolein-laden air to the local community, which is contiguous to the chemical plant.

An emergency response team will don equipment, arrive at the spill, and neutralize the acrolein. Assume that the air at the spill contains 10 ppm acrolein immediately after the spill and that the air contains no acrolein immediately after the spill is neutralized. Also assume that the acrolein does not diffuse laterally in the air, but travels only because of the wind. A detector in the local community would thus measure the acrolein concentration shown below. The spill occurs at time = 0. At a time t_1 the wind has traveled from the spill to the detector. After the acrolein has been neutralized, the air passing over the spill site will contain no acrolein and this air reaches the detector at time t_2.

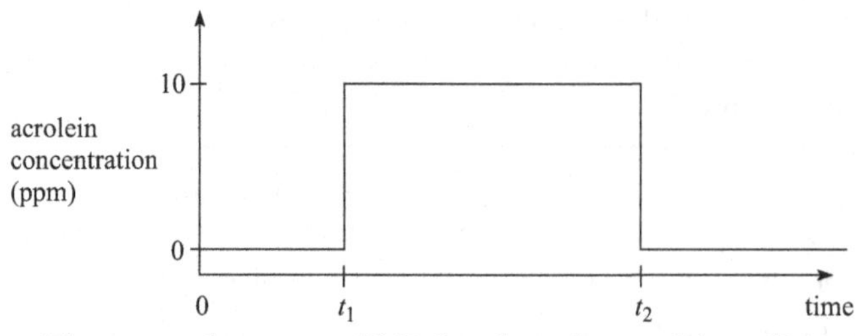

The nearest homes are 3000 feet from the acrolein spill. Assume the wind travels from the spill toward the nearest homes at 2.2 miles per hour. Assume the emergency response team neutralizes the acrolein 15 minutes after the spill occurs.

(A) How soon after the spill must the nearest residents react? That is, what is time t_1?
(B) How long before the danger passes? That is, what is time t_2?
(C) How should the nearest residents react? Should they flee or should they go inside and close all the doors and windows? Assume the residents live in small, single-family homes. The air in each home is replaced by outside air three times an hour and the air in each home is well mixed.
To justify your emergency response plan you should show the concentration of acrolein in a home nearest the spill, as well as outside the home, as a function of time after the spill.
(D) What if the response team needs 30 minutes to neutralize the acrolein? Will this change your emergency response plan?

6.12 In Section 6.3, the concentration of N_2O in the batch reactor was measured by spectroscopy. Assume that we don't have a spectrometer and instead must use the pressure in the reactor to calculate the concentration of N_2O. The reactor initially contains only N_2O at a concentration $[N_2O]_0$, and the initial pressure is P_0. Derive an expression for $[N_2O]$ in terms of P, P_0, and $[N_2O]_0$.

6.13 Consider the first-order reaction $A \rightarrow P$. We wish to produce 4752 moles of P during a 10-hour period and convert 99% of the A to P. We will use a batch reactor. For each reactor batch, one must charge the reactor, conduct the reaction, and empty the reactor. Charging the reactor with A and heating to the reaction temperature takes 0.26 hours. Discharging the reactor and preparing the reactor for the next batch takes 0.90 hours.

The reaction rate is $k[A]$ in units of mol/(L·min) and the rate constant k is 0.0573/min at the reaction temperature. The reaction is conducted in a solvent. A, P, and the solvent are all organic liquids and have the same density, 0.85 g/mL. The initial concentration of A is 8.0 mol/L.

(A) Calculate the time required to convert 99% of the A to P.
(B) How many batches can the reactor produce in 10 hours?
(C) Calculate the reactor volume, in L.
(D) Calculate the reactor volume if one neglects the time required to charge and discharge the reactor.
(E) Based upon your calculations in parts (C) and (D), how would you categorize the costs of charging and discharging of the batch rector – operating costs, or capital costs, both, or neither? That is, if you decrease the time required to charge and discharge the reactor, will you decrease operating costs, or capital costs, both, or neither?

6.14 "Spirit vinegar" is vinegar fortified with residual ethanol. Spirit vinegar is traditionally produced by the aerobic fermentation of cider, malt, or wine in which the bacteria *Acetobacter aceti* converts ethanol to acetic acid by the following reaction:

$$2C_2H_5OH + 2O_2 \rightarrow 2CH_3COOH + 2H_2O.$$

We will produce spirit vinegar by a "fed-batch" process: oxygen is fed to a batch of cider. Air is bubbled through the cider such that the concentration of dissolved oxygen is constant during the reaction.

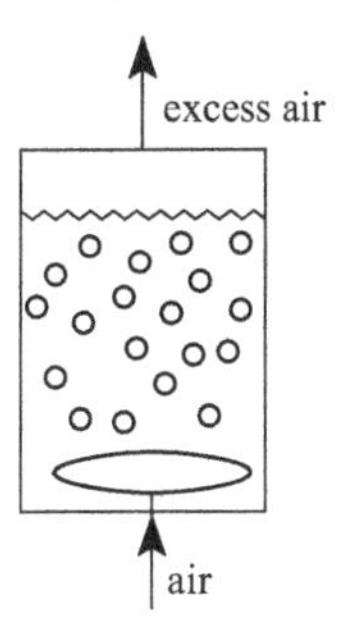

The reaction rate for this process is second-order in the ethanol concentration,

$$\text{reaction rate} = k[C_2H_5OH]^2 \text{ (in mol/(day·L))},$$

and the concentration of ethanol has units of mol/L.

(A) What are the units of the reaction rate constant k?

(B) Calculate the time required to convert 85% of the ethanol to acetic acid at 20°C. The reactor contains 2500 L of cider. You may assume the cider volume is constant. The initial concentration of ethanol is (1.09 mol ethanol)/(1 L cider). The numerical value of k is 0.01156 at 20°C.

6.15 Consider the reversible reaction

$$A \leftrightarrow B.$$

A is consumed in the forward reaction at a rate = k_f [A] (in moles/(time × volume)). A is generated in the backward reaction at a rate = k_b[B].

(A) Write a mass balance on A for a batch reactor that contains both A and B.

(B) Consider a batch reactor that initially contains only A at a concentration $[A]_0$, in moles/liter. Express [B] in terms of $[A]_0$ and [A].

(C) Substitute your expression for [B] into the mass balance and solve for [A]. Your expression for [A] should contain only k_f, k_b, $[A]_0$, and t. (You may wish to consult the table of indefinite integrals in Appendix E.)

(D) Check your expression in the limits $t = 0$ and $t = \infty$. At equilibrium, the ratio of concentrations in the batch reactor will be $[A]/[B] = k_b/k_f$.

6.16 A simultaneously reacts to form B and C:

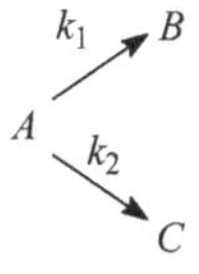

The rate of the reaction $A \rightarrow B$ is $r_1 = k_1[A]$ and the rate of the reaction $A \rightarrow C$ is $r_2 = k_2[A]$. Consider a batch reactor that initially contains only A.

(A) Derive an equation for $[A]$ as a function of time. In other words, derive an equation of the form $[A] = \ldots$, where the right side contains only constants, t, and $[A]_0$.
(B) Derive an equation for $[B]$ as a function of time.

6.17 An electric coil is used to heat 10.0 kg of water in an insulated vessel. The coil delivers a constant 1.0 kW to the water. The water is initially at 20°C. The pressure in the vessel remains constant at 1 atm.

(A) Apply an energy balance to obtain a differential equation for the water temperature. Assume the heat capacity of water at 1 atm is constant and equal to 4.18 kJ/(kg·°C). You may neglect the energy required to heat the coil.
Recall that heat capacity is defined as the change in internal energy, dU, with change in temperature, at constant pressure. For a unit of mass,

$$\left(\frac{dU}{dT}\right)_P = C_P$$

$$\Delta U = \int_{U_1}^{U_2} dU = \int_{T_1}^{T_2} C_P dT.$$

If heat capacity is independent of temperature, then $\Delta U = C_P \Delta T$, as in Chapter 3.
(B) Calculate the time at which the water reaches 100°C.
(C) The heat capacity for water is not constant; it varies with temperature. A good approximation for C_P over the range 20°C to 100°C is

$$C_P = 4.19 - 6.8 \times 10^{-4} T + 9.4 \times 10^{-6} T^2.$$

Apply an energy balance and calculate the time at which the water reaches 100°C.

6.18 The "heat capacity" of my house is 3300 kJ/C. That is, 3300 kJ raises the temperature of my house by 1°C. The heater in my house can supply heat at a maximum rate of 5.2×10^4 kJ/hr.

(A) I return from vacation to a cold house. The inside temperature is 5°C (41°F) and the outside temperature is −15°C (5°F). I set the heater at its maximum rate at 8:00 pm. At what time will the temperature in my house be 20°C (68°F)?
(B) Heat escapes from my house by conduction through the walls and roof. The rate q_{loss} of heat loss (in kJ/hr) is proportional to the difference between the inside temperature and the outside temperature:

$$q_{loss} = k(T_{inside} - T_{outside}),$$

where $k = 740$ kJ/(°C hour). Repeat the calculation in (A), but include heat loss by conduction through the walls and roof.

Numerical Integration of Differential Equations

For simple rate equations, the integrated form can be obtained analytically. For example, the rate equation

$$\frac{d[A]}{dt} = -k[A] \tag{1}$$

can be separated and integrated to yield

$$[A] = [A]_0 \exp(-kt) \; . \tag{2}$$

Some complex rate laws cannot be integrated analytically (or are too difficult to integrate analytically), and one must appeal to numerical methods. Numerical methods yield $[A]$ at time increments of Δt, beginning with $[A]_0$:

$$[A]_{\Delta t} = [A]_0 + \left(\frac{d[A]}{dt}\right)_0 (\Delta t). \tag{3}$$

That is, equation (1) is used to calculate $d[A]/dt$ at $t = 0$; $(d[A]/dt)_0 = -k[A]_0$ and equation (3) becomes

$$[A]_{\Delta t} = [A]_0 - k[A]_0(\Delta t). \tag{4}$$

We continue in the same manner to calculate $[A]$ at $t = 2\Delta t$, using $(d[A]/dt)_{\Delta t} = -k[A]_{\Delta t}$ and

$$[A]_{2\Delta t} = [A]_{\Delta t} - k[A]_{\Delta t}(\Delta t). \tag{5}$$

In general, $[A]$ at any time t can be used to calculate $[A]$ at a time $t + \Delta t$ as follows:

$$[A]_{t + \Delta t} = [A]_t + \left(\frac{d[A]}{dt}\right)_t (\Delta t). \tag{6}$$

This algorithm of numerical integration is called the Euler method. The error is proportional to Δt. A more detailed discussion of the Euler method, as well as methods with better accuracy (and concomitant higher complexity), can be found in any text on numerical methods. A fine text with many examples in the context of chemical engineering is *Applied Numerical Methods* by B. Carnahan, H. A. Luther, and J. O. Wilkes.[3]

6.19 Integrate numerically rate equation 1 given above, with $k = 1/\text{s}$. Use a spreadsheet (or write a computer program) to produce a table with at least the following six columns:

Column 1. n, the iteration step. This column should begin with 0 and increase in increments of 1.

Column 2. t, the time. This column should begin with 0 and increase in increments of Δt.

Column 3. $[A]_t$, the numerically integrated value for $[A]$. The column should begin with an arbitrary $[A]_0$ of your choosing, and subsequent values should be computed using equation 6 above.

Column 4. $(d[A]/dt)_t$. This is computed with the formula

$$\left(\frac{d[A]}{dt}\right)_t = -k[A]_t.$$

Column 5. $[A]_t$ obtained from the analytical solution, equation 2 above.

Column 6. the relative error, defined as the difference between the numerical and analytical values for $[A]_t$, divided by the analytical value for $[A]_t$.

[3] Carnahan, B, Luther, H. A., and Wilkes, J. O. 1990. *Applied Numerical Methods*, Krieger Publishing, Melbourne, FL.

Use your spreadsheet (or program) to supply the following information. (You are encouraged to add additional rows to the table.)

$[A]_0$:______

Analytical value for $[A]_t$ at 4 s:______

Δt	Number of steps	$[A]_t$ at 4 s (numerical)	Relative error (%)
0.333	12		
0.1	40		
0.0333	120		
0.01	400		

This exercise appeared on an exam. It was estimated that it could be completed in 30 minutes.

6.20 Use numerical integration to obtain $[A]$, $[B]$, and $[C]$ as functions of time for the following reaction:

$$A \overset{k_a}{\rightarrow} B \overset{k_b}{\rightarrow} C,$$

such that $k_a = 0.8/\text{s}$ and $k_b = 0.7/\text{s}$ and for initial concentrations of $[A]_0 = 1.0$, $[B]_0 = [C]_0 = 0.0$. Plot $[A]$, $[B]$, and $[C]$ as functions of time for $t = 0$ to $t = 10$ s. Assume that each reaction is first-order. That is, $r_a = k_a[A]$ and $r_b = k_b[B]$.

Compare the results of your numerical integration to the analytical results:

$$[A] = [A]_0 e^{-k_a t}$$

$$[B] = [A]_0 \frac{k_a}{k_a - k_b} \left(e^{-k_b t} - e^{-k_a t}\right)$$

$$[C] = [A]_0 \frac{1}{k_a - k_b} \left[k_a\left(1 - e^{-k_b t}\right) - k_b\left(1 - e^{-k_a t}\right)\right].$$

6.21 Use numerical integration to obtain $[A]$, $[B]$, and $[C]$ as functions of time for the following reaction:

$$A \underset{k_{-a}}{\overset{k_a}{\leftrightarrow}} B \overset{k_b}{\rightarrow} C,$$

where $k_a = 0.8/\text{s}$, $k_{-a} = 1/\text{s}$, and $k_b = 0.7/\text{s}$, and for the initial concentrations of $[A]_0 = 1.0$ and $[B]_0 = [C]_0 = 0.0$. Plot $[A]$, $[B]$, and $[C]$ as functions of time for $t = 0$ to $t = 10$ s. Assume that each reaction is first-order. That is, $r_a = k_a[A]$, $r_{-a} = k_{-a}[B]$, and $r_b = k_b[B]$.

Compare the results of your numerical integration to the analytical results,[4]

$$[A] = [A]_0 \left[\frac{k_a(\lambda_2 - k_b)}{\lambda_2(\lambda_2 - \lambda_3)} e^{-\lambda_2 t} + \frac{k_a(k_b - \lambda_3)}{\lambda_3(\lambda_2 - \lambda_3)} e^{-\lambda_3 t}\right]$$

[4] Moore, J. W., and Pearson, R. G. 1981. *Kinetics and Mechanism*, 3rd edn., John Wiley & Sons, New York, NY, p. 313.

$$[B] = [A]_0 \left[\frac{-k_a}{\lambda_2 - \lambda_3} e^{-\lambda_2 t} + \frac{k_a}{\lambda_2 - \lambda_3} e^{-\lambda_3 t} \right]$$

$$[C] = [A]_0 \left[\frac{k_a k_b}{\lambda_2 \lambda_3} + \frac{k_a k_b}{\lambda_2 (\lambda_2 - \lambda_3)} e^{-\lambda_2 t} - \frac{k_a k_b}{\lambda_3 (\lambda_2 - \lambda_3)} e^{-\lambda_3 t} \right],$$

where

$$\lambda_2 = \frac{1}{2} \left(k_a + k_{-a} + k_b + \left[(k_a + k_{-a} + k_b)^2 - 4 k_a k_b \right]^{1/2} \right)$$

$$\lambda_3 = \frac{1}{2} \left(k_a + k_{-a} + k_b - \left[(k_a + k_{-a} + k_b)^2 - 4 k_a k_b \right]^{1/2} \right).$$

Note that when the reaction progressed from $A \rightarrow B$ to $A \leftrightarrow B \rightarrow C$ in the preceding three exercises, the complexity of the analytical solutions increased dramatically. However, the complexity of the numerical calculation increased only modestly.

While working the preceding exercises, you may encounter integrals that may not be in your repertoire. Even after studying the calculus for a year you will encounter integrals you cannot solve. Yes, it's true. When training and/or memory fail you, it is useful to consult a table of integrals. Appendix E contains a brief list of indefinite integrals germane to the preceding exercises. The *CRC Handbook of Chemistry and Physics* has a short table and there are several books with larger compilations. Our favorites are *Tables of Integrals and Other Mathematical Data* by H. B. Dwight (300+ pages) and *Table of Integrals, Series, and Products* by I. S. Gradshteyn and I. M. Ryzhik (1100+ pages).

Appendix A

List of Symbols

A	area (m^2)
C_P	molar heat capacity at constant pressure (J/(kg·K) or J/(mol·K))
C_V	molar heat capacity at constant volume (J/(mol·K)
f	a function
$F_{A,i}$	mass flow rate of component A in stream i (kg/s)
g	gravitational constant (9.8 m/s^2)
k	Boltzmann constant, or rate constant for a chemical reaction, or thermal conductivity (J/(s·m·°C)), or roughness factor for flow in pipes
L	liquid flow rate in a separation unit (mol/s)
$M_{A,i}$	mass of component A in phase i (kg)
N_A	Avogadro constant (6.02×10^{23}/mol)
P	pressure (Pa, although atm is also used)
q_i	rate of heat flow of stream i (J/s)
Q_i	volumetric flow rate of stream i (m^3/s)
R	gas constant (8.314 m^3·Pa/(mol·K))
T	temperature (°C or K)
t	time (s)
v	velocity (m/s)
V	volume (m^3) or vapor flow rate in a separation unit (mol/s)
x_i	mol fraction of i in the liquid phase
y_i	mol fraction of i in the vapor phase

GREEK SYMBOLS

μ	viscosity (Pa·s)
ρ	density (kg/m^3)
γ	surface tension (N/m)
ω	frequency (1/s)
ξ	hydrodynamic resistance (kg/s)

DIMENSIONS

L	length
M	mass
N	amount

T time
Θ temperature

SYMBOLS

≡	is defined as
[physical quantity]	dimensions of physical quantity
[*A*]	molar concentration of compound *A*

DIMENSIONLESS GROUPS

Bo	Bond number
Ca	Capillary number
El	elasticity number
Eo	Eötvös number
Fr	Froude number
G	gravity number
Le	Lewis number
Mo	Morton number
Nu	Nusselt number
Oh	Ohnesorge number
Pe	Péclet number
Pr	Prandtl number
Re	Reynolds number
Sh	Sherwood number
St	Stanton number
We	Weber number
α	Womersley number

Appendix B

Units, Conversion Factors, and Physical Constants

B.1 UNITS

We use the mks (meter-kilogram-second) SI (Système International) system of units in (most) calculations in this text. The base units (see Chapter 5) of the mks system are given in Table B.1.

Table B.1 Base units in the mks system.

Dimension	Unit	Symbol
length	meter	m
mass	kilogram	kg
time	second	s
temperature	kelvin, or degrees Celsius	K, or °C
amount	mole	mol

The base units are combined to form derived units, some of which are shown in Table B.2.

Table B.2 Some derived units in the mks system.

Quantity	Units	Name	Symbol
area	m^2	–	–
volume	m^3	–	–
velocity	m/s	–	–
acceleration	m/s^2	–	–
momentum	m·kg/s	–	–
force	$m \cdot kg/s^2$	newton	N
energy	$m^2 \cdot kg/s^2$	joule	J
power	$m^2 \cdot kg/s^3$	watt	W
pressure	$kg/(m \cdot s^2)$	pascal	Pa

B.2 CONVERSION FACTORS

Area		
1 acre	=	$4047\ m^2$
Force		
1 dyne	≡	1×10^{-5} N
1 pound-force (lbf)	=	4.448 N
Energy		
1 Btu	=	1054.4 J
1 calorie	=	4.184 J
1 erg	≡	1×10^{-7} J
Length		
1 angstrom (Å)	≡	1×10^{-10} m
1 fathom	=	1.829 m
1 foot (ft)	=	0.3048 m
1 inch (in)	=	0.0254 m
1 light year	=	9.46×10^{15} m
1 mile	=	1609 m
1 yard	=	0.9144 m
Mass		
1 ounce (avoirdupois)	=	2.835×10^{-2} kg
1 ounce (troy)	=	3.110×10^{-2} kg
1 pound (lb, 16 oz avoirdupois)	=	0.4536 kg
1 pound (12 oz troy)	=	0.3732 kg
1 ton (short, 2000 lb)	=	907.2 kg
1 amu (atomic mass unit, ^{12}C scale)	=	1.661×10^{-27} kg
Power		
1 horsepower (hp)	=	745.7 W
Pressure		
1 atmosphere (atm)	=	1.013×10^5 Pa
1 bar	≡	1×10^5 Pa
1 inch of Hg	=	3386 Pa
1 foot of water	=	2989 Pa
1 pound-force/in^2 (psi)	=	6895 Pa
1 torr (1 mm Hg)	=	133.32 Pa
Temperature		
kelvin (K)	=	degrees Celsius + 273.15
degrees Fahrenheit (°F)	=	1.8 × (degrees Celsius) + 32
Time		
1 day	=	86,400 s
1 hour (hr)	=	3600 s
1 minute (min)	=	60 s
1 year (yr)	=	3.154×10^7 s
Velocity		
1 knot	=	0.5144 m/s
1 mile/hour (mph)	=	0.4470 m/s
Viscosity		
1 centipoise (cp)	=	1×10^{-3} Pa·s

Volume

1 barrel (oil, 42 gal)	=	0.1590 m^3
1 barrel (bbl, 31.5 gal)	=	0.1192 m^3
1 fluid ounce (US)	=	2.957×10^{-5} m^3
1 gallon (US, liquid)	=	3.785×10^{-3} m^3
1 liter (L)	≡	1×10^{-3} m^3
1 quart (US, liquid)	=	9.464×10^{-4} m^3

B.3 PHYSICAL CONSTANTS

Avogadro constant	N_A	6.022×10^{23}/mol
Boltzmann constant	$k = R/N_A$	1.381×10^{-23} J/K
gas constant	R	8.314 J/(mol·K)
		8.314 m^3·Pa/(mol·K)
		0.08206 L·atm/(mol·K)
gravitational acceleration at sea level	g	9.8 m/s^2
Planck constant	h	6.626×10^{-34} J·s
speed of light in vacuum	c	3.0×10^8 m/s

Appendix C

Significant Figures

Quantities are composed of two parts: a number (e.g. 5.31) and a unit (e.g. grams). The number must have the proper number of significant figures, defined as follows:

Significant figures – the digits from the first non-zero digit on the left to the last non-zero digit on the right.

The number of significant figures in a number is perhaps determined most easily from the scientific notation for the number. Some examples are listed in Table C.1.

Table C.1 Significant figures.

Number	Scientific notation	Number of significant figures
3.0	3.0×10^0	2
23	2.3×10^1	2
0.0353	3.53×10^{-2}	3
1000	1×10^3	1
1000.	1.000×10^3	4
1000.0	1.0000×10^3	5

The numbers "1000" and "1000." both have only one significant figure if one applies the definition above. This is wrong: "1000" has one significant figure and "1000." has four. We need to augment the definition, as follows:

Significant figures – the digits from the first non-zero digit on the left to the last non-zero digit on the right, or to the last digit if there is a decimal point.

How does one determine the number of significant figures in a number obtained from an arithmetic operation? Follow this algorithm for addition and subtraction:

1. Note the position of the last significant figure in each number being added or subtracted.
2. Note the position of the left-most last significant figure of all the numbers to be added.
3. The last significant figure of the sum is given by the left-most last significant figure.

Consider two examples.

$$\begin{array}{r} 6750 \\ +\ 10.3 \\ \hline 6760.3 \end{array}$$

The correct sum is 6760 after truncating to three significant figures.

$$\begin{array}{r} 1.0000 \\ +\,0.22 \\ \hline 1.2200 \end{array}$$

The correct sum is 1.22 after truncating to three significant figures.

Here is a more relevant example. In the 2004–2005 academic year, the top private universities spent an estimated average \$27,000 per undergraduate per semester. At Cornell University, the 2004–2005 tuition was \$15,083.50 per semester. The average grant awarded was \$4700 per undergraduate per semester (Ehrenberg, 2000; averages estimated from the 1994–1995 data, pages 8–10). Calculate the total subsidy a university gives each student:

$$\text{total subsidy} = \$27{,}000 - \$15{,}083.50 + \$4700 = \$16{,}616.50 \text{ per student per semester.}$$

Because the estimated costs were accurate to the "thousands," the total subsidy is \$17,000 per student per semester. How can universities afford to lose \$17,000 per student per semester? The subsidy is paid chiefly by the revenue from endowments and donations, which are chiefly due to the benevolence of generous alumni.

Follow this algorithm for multiplication and division:

1. Determine the number of significant figures in each multiplicand or divisor.
2. The number of significant figures in the product is equal to the number of significant figures in the multiplicand or divisor which has the smallest number of significant figures.

Example: convert 87.0 kg water/min to units of gal/hr:

$$\left(\frac{87.0\,\text{kg water}}{\text{min}}\right)\left(\frac{60\ \text{min}}{1\ \text{hour}}\right)\left(\frac{1.000\,\text{g}}{1\ \text{kg}}\right)\left(\frac{1.000\,\text{mL water}}{1.000\,\text{g water}}\right)\left(\frac{1\ \text{gal}}{4.405\times 10^{-3}\ \text{mL}}\right)$$
$$= 1185.0\,\text{gal/hour}.$$

The units are converted by multiplying by factors of 1. For example, because 60 min = 1 hr,

$$\frac{60\ \text{min}}{1\ \text{hour}} = 1,$$

by definition. And each such quantity has an infinite number of significant figures, by definition. Another factor of 1 is formed from a physical property specific to this problem, the density of water, for which 1.000 mL water = 1.000 g water, and

$$\frac{1.000\ \text{mL water}}{1.000\ \text{g water}} = 1.$$

It is assumed here that the density of water is known to four significant figures. The multiplicand with the smallest number of significant figures is 87.0 kg water/min, which has three. Thus the proper converted quantity is 1180 gal/hr, or 1.18×10^3 gal/hr.

The last example brings us to the convention for rounding off numbers that end in a 5:

If the digit after the last significant figure is a 5 and the last significant figure is even, round down. If the digit after the last significant figure is a 5 and the last significant figure is odd, round up.

Thus 1185 rounds off to 1180 and 1175 rounds off to 1180.

Consider another example. Convert Cornell University's 2018–2019 tuition to units of \$/(in-class hour). Assume that the average load of a first-semester first-year student at Cornell is 14 credits, as recommended by the enlightened policies of the College of Engineering. Further assume that 14 credits entails about 16 in-class hours per week. Therefore,

$$\left(\frac{\$27{,}409.00}{1\text{ semester}}\right)\left(\frac{1\text{ semester}}{14\text{ weeks}}\right)\left(\frac{1\text{ week}}{16\text{ in-class hours}}\right)=\$122.36/\text{in-class hour}.$$

Although 27,409.00 has seven significant figures, our assumptions have only two. The answer is thus \$120/(in-class hour). A bargain at any price (Ehrenberg, 2000; averages estimated from the 1994–1995 data, pages 8–10).

Here are examples of conversions that use an incorrect number of significant figures. The August 26, 1991 issue of *Chemical and Engineering News* reported that "In Japan blue roses cost \$78." Why the arbitrary price of \$78? Why not \$75, or \$80? It seems that blue roses cost 10,000 yen, which is a nice round number with one significant figure. When one converts yen to dollars using the exchange rate at that time, one gets

$$\left(\frac{10{,}000\text{ yen}}{\text{blue rose}}\right)\left(\frac{\$1.000}{128\text{ yen}}\right)=\$78/\text{blue rose}.$$

Converting to one significant figure, the correct price is \$80 per blue rose.

On average, the temperature of the human body is 37°C. What is the average temperature in degrees Fahrenheit? We use the well-known formula to convert from degrees Celsius to degrees Fahrenheit,

$$37°\text{C}\left(\frac{9.00°\text{F}}{5.00°\text{C}}\right)+32.00°\text{F}=98.6°\text{F}.$$

Converting the answer to two significant figures, we get an average body temperature of 99°F.

Finally, retain extra digits during a calculation and truncate numbers to the proper number of significant figures only at the end. Do not truncate at an intermediate stage of your calculation.

REFERENCES

Ehrenberg, R. G. 2000. *Tuition Rising: Why College Costs So Much*, Harvard University Press, Cambridge, MA.

Appendix D

Log–Log Graph Paper

Experimental data often span many orders of magnitude. An example is the friction factor of a sphere moving through a fluid as a function of the Reynolds number, as discussed in Chapter 5. Some typical data for laminar flow are presented in Table D.1.

Table D.1 Data for a sphere moving through a fluid.

Reynolds number	Friction factor
0.9	27
0.33	73
0.12	200
0.074	324
0.023	300
0.0091	2,635
0.0065	3,700
0.0027	8,888
0.0011	21,900

It is useful to analyze experimental data by plotting. However, the plot shown in Figure D.1 is not very useful. The data are crowded near the origin or lie on either the x or y axes. Figure D.1 would not be useful for interpolating between points or scanning for suspicious data, for example.

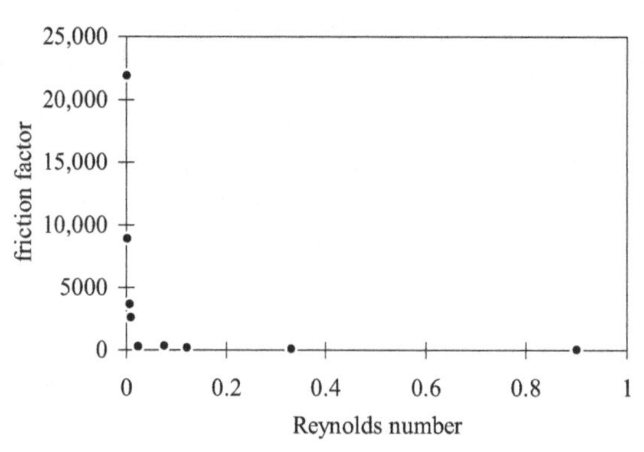

Figure D.1 An x–y plot of Reynolds number vs. friction factor.

The utility of the plot is improved by plotting the logarithm of the data, as shown in Figure D.2 on the next page. Figure D.2 clearly shows a linear correlation, as reinforced by the straight line through the data. Figure D.2 also reveals that the point measured for Re = 0.023 is suspect; it lies below the correlation of the other data.

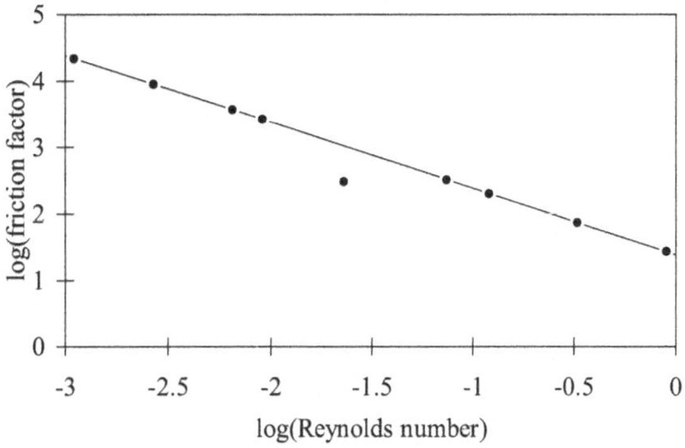

Figure D.2 An x–y plot of $\log_{10}$ (Reynolds number) vs. $\log_{10}$(friction factor).

Rather than compute the base-10 logarithm of each coordinate, one can plot directly onto *log–log graph paper*, as shown in Figure D.3. The gradations on both the abscissa and ordinate increase logarithmically.

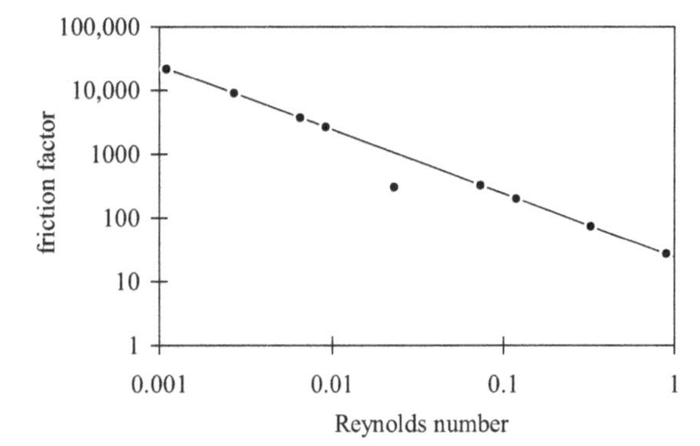

Figure D.3 A log–log plot of Reynolds number vs. friction factor.

For clarity, the gridlines have been omitted in Figure D.3. An example of log–log paper with gridlines is shown in Figure D.4 on the next page. Specifically, this is called 2-cycle × 2-cycle log–log graph paper because there are two powers of ten on each axis. (Figure D.3 is 3-cycle × 5-cycle log–log graph paper.) Note that the gridlines are not evenly spaced. The first gridline moving from the left is 2, not 1.1. The first gridline to the left of 10 is 9. The distance between 1 and 2 is greater than the distance between 9 and 10. Note that 10 is halfway between 1 and 100 because 10^1 is halfway between 10^0 and 10^2. That is, $\log_{10}10^1$ (= 1) is halfway between $\log_{10}10^0$ (= 0) and $\log_{10}10^2$ (= 2). Distances on a log scale are proportional to the exponent of ten. The midpoint between 1 and 10 is not 5.5. Rather it is the midpoint between 10^0 and 10^1, which is $10^{0.5} = 3.16$. To find the midpoint between two numbers, average their logarithms (base 10) and raise 10 to that power. What is the midpoint between 1 and 2 on the graph? It is not 1.5 but 10 raised to the power of the average of $\log_{10}(1)$ and $\log_{10}(2)$, which is $10^{0.15051} = 1.414 = 2^{0.5}$.

Graph paper with a logarithmic scale on the ordinate and a linear scale on the abscissa is called *semilog* graph paper. An example of 2-cycle semilog paper is shown in Figure D.5. The range on the abscissa in both the log–log graph paper in Figure D.4 and the

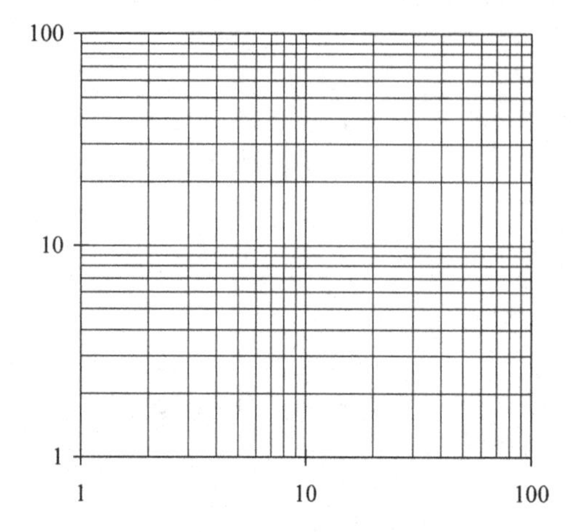

Figure D.4 2-cycle × 2-cycle log–log graph paper.

semilog graph paper in Figure D.5 is 1 to 100. What if your data are in a different range, for example, from 0.001 to 0.1? No problem. Just replace 1 with 0.001, replace 10 with 0.01, and replace 100 with 0.1.

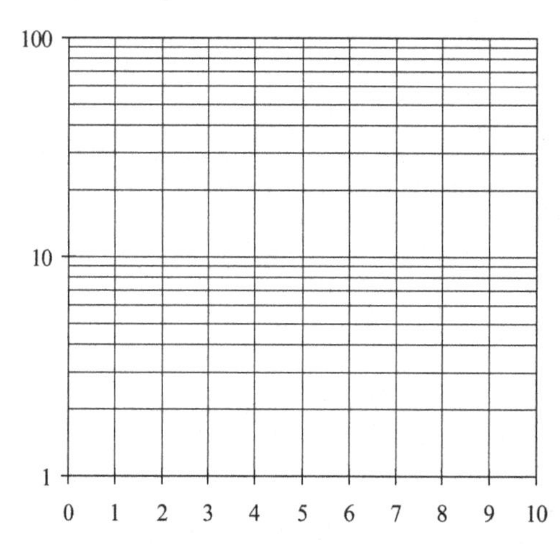

Figure D.5 2-cycle semilog graph paper.

Log–log plots and semilog plots are useful for determining the functional relationships of data. The human eye is adept at recognizing straight lines but less able to distinguish x^2 from e^x, especially if only a portion of the data is available. However, $y = x^2$ is a straight line on log–log graph paper, but it is not a straight line on x–y or semilog paper. The function $y = e^x$ is a straight line on semilog paper but not on x–y or log–log paper. How does one determine the mathematical equation of a straight line on log–log paper, such as the line shown in Figure D.6 (on the next page)?

A straight line means the functional form is

$$y = mx + b, \tag{D.1}$$

where m is the slope and b is the y-intercept. But be careful – Figure D.6 is not a plot of x vs. y, but a plot of $\log_{10}(x)$ vs. $\log_{10}(y)$. So a straight line on a log–log plot means the functional form is

$$\log_{10}(y) = m \log_{10}(x) + b. \tag{D.2}$$

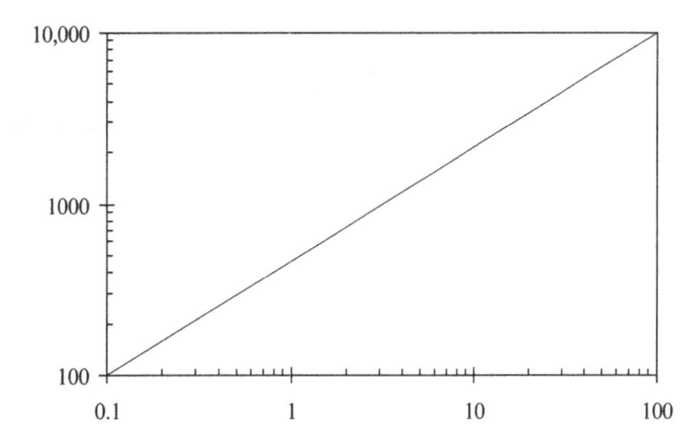

Figure D.6 A straight line on a log–log plot.

Let's calculate the slope m of the line in Figure D.6. We will compute the slope from the ratio of rise (the distance traveled vertically) to the run (the distance traveled horizontally):

$$\text{slope} = \frac{\text{rise}}{\text{run}} = \frac{10{,}000 - 100}{100 - 0.1} = 99.1. \tag{D.3}$$

Correct? No, we didn't use the proper coordinates to calculate the rise and run of the line. Equation D.3 uses x and y. We need to use $\log_{10}(x)$ and $\log_{10}(y)$. Therefore

$$\begin{aligned}\text{slope} &= \frac{\text{rise}}{\text{run}} = \frac{\log_{10}(10{,}000) - \log_{10}(100)}{\log_{10}(100) - \log_{10}(0.1)} = \frac{\log_{10}\left(10^4\right) - \log_{10}\left(10^2\right)}{\log_{10}\left(10^2\right) - \log_{10}\left(10^{-1}\right)} \\ &= \frac{4-2}{2-(-1)} = \frac{2}{3}. \end{aligned} \tag{D.4}$$

And what of b, the y-intercept? Again be careful – although the line in Figure D.6 crosses the y axis at $y = 100$, this is not the intercept. Note that the y axis is at $x = 0.1$, not $x = 0$. There is no intercept on a log–log graph because x is never equal to zero. To find b we substitute the slope into equation D.2:

$$\log_{10}(y) = \frac{2}{3}\log_{10}(x) + b \tag{D.5}$$

$$= \log_{10}\left(x^{2/3}\right) + b. \tag{D.6}$$

We now raise both sides of equation D.6 to the exponent of 10:

$$10^{\log_{10}(y)} = 10^{\left[\log_{10}\left(x^{2/3}\right) + b\right]} \tag{D.7}$$

$$y = 10^{\left[\log_{10}\left(x^{2/3}\right)\right]} \times 10^b \tag{D.8}$$

$$= x^{2/3} \times 10^b. \tag{D.9}$$

To determine b (or, actually, to determine 10^b), we substitute a point that lies on the line, such as $x = 0.1$, $y = 100$. This gives

$$10^b = \frac{y}{x^{2/3}} = \frac{100}{0.1^{2/3}} = 464. \tag{D.10}$$

Thus the equation of the straight line in Figure D.6 is

$$y = 464x^{2/3}. \tag{D.11}$$

It is always prudent to check one's solution. Does the line go through another known point, such as $x = 100$, $y = 10{,}000$? Does

$$10{,}000 = 464 \times 100^{2/3}? \tag{D.12}$$

This yields

$$10{,}000 \approx 9{,}997. \tag{D.13}$$

It checks.

Now, test your skills by determining the equation for the line in Figure D.7.

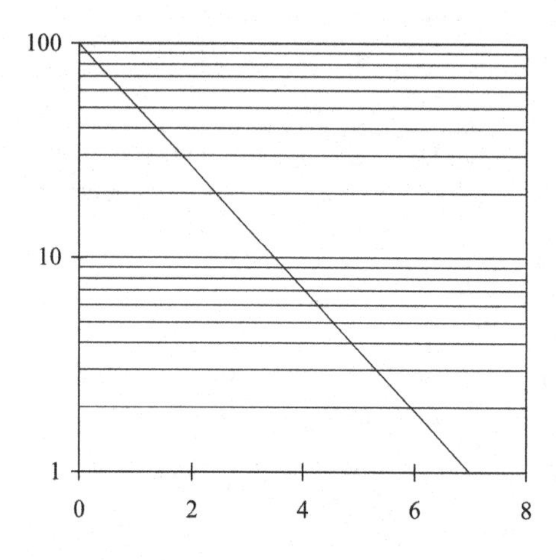

Figure D.7 A straight line on a semilog plot.

(We calculate that the line corresponds to $y = 100(10^{-2x/7})$, or $y = 100e^{-0.66x}$.)

Appendix E

Mathematics, Mechanics, and Thermodynamics

ALGEBRA

$$a^m a^n = a^{m+n} \qquad (ab)^m = a^m b^m$$
$$(a^m)^n = a^{mn} \qquad a^0 = 1$$
$$a^{-m} = \frac{1}{a^m}$$

$$\ln e^x = x \qquad \ln x^n = n \ln x$$
$$\ln a + \ln b = \ln (ab)$$

Quadratic Formula

The roots of

$$ax^2 + bx + c = 0 \text{ are}$$
$$x = \frac{-b \pm \sqrt{b^2 - 4ac}}{2a} \quad \text{for } a \neq 0.$$

GEOMETRY

A = area, C = circumference, V = volume,
r = radius, b = length of base, h = height

Planar Geometry

Circle: $A = \pi r^2, C = 2\pi r$
Triangle: $A = bh/2$

Solid Geometry

Sphere: $V = \frac{4}{3}\pi r^3,\ A = 4\pi r^2$

Cylinder: $V = \pi h r^2$

Cone: $V = \frac{1}{3}\pi r^2 h$

Theorem of Pythagoras

For a right triangle with hypotenuse c and legs a and b,

$$a^2 + b^2 = c^2.$$

TRIGONOMETRY

$$\sin^2\alpha + \cos^2\alpha = 1$$

$$\sin(\alpha+\beta) = \sin\alpha\cos\beta + \cos\alpha\sin\beta$$

$$\cos(\alpha+\beta) = \cos\alpha\cos\beta + \sin\alpha\sin\beta$$

$$e^{ix} = \cos x + i\sin x$$

THE CALCULUS

Differentiation

$$\frac{d}{dx}x^n = nx^{n-1}$$

$$\frac{d}{dx}e^{ax} = ae^{ax} \qquad \frac{d}{dx}\ln(x) = \frac{1}{x}$$

Product Rule

$$\frac{d}{dx}[f(x)g(x)] = f(x)\frac{d}{dx}g(x) + g(x)\frac{d}{dx}f(x)$$

Integration

$$\int x^n dx = \frac{x^{n+1}}{n+1}$$

$$\int \frac{1}{x}dx = \ln|x|$$

$$\int (a+bx)\,dx = \frac{1}{2b}(a+bx)^2$$

$$\int \frac{dx}{a+bx} = \frac{1}{b}\ln|a+bx|$$

$$\int \frac{dx}{x(a+bx)} = -\frac{1}{a}\ln\left|\frac{a+bx}{x}\right|$$

$$\int \frac{dx}{(a+bx)^2} = \frac{-1}{b(a+bx)}$$

$$\int \frac{dx}{(a+fx)(c+gx)} = -\frac{1}{ag-cf}\ln\left|\frac{c+gx}{a+fx}\right|$$

MECHANICS

Linear momentum: $p = mv$
Kinetic energy: K.E. $= \frac{1}{2}mv^2$
Potential energy at Earth's surface: P.E. $= mgh$

THERMODYNAMICS

Ideal Gas Law
Heat Capacity

$$PV = nRT$$

For an ideal gas,

$$C_P = C_V + R.$$

For a monatomic ideal gas (He, Ne, Hg),

$$C_P = \frac{5}{2}R.$$

For a diatomic ideal gas (O_2, N_2),

$$C_P = \frac{7}{2}R.$$

THE GREEK ALPHABET

alpha	α	Α
beta	β	Β
gamma	γ	Γ
delta	δ	Δ
epsilon	ε	Ε
zeta	ζ	Ζ
eta	η	Η
theta	θ	Θ
iota	ι	Ι
kappa	κ	Κ
lambda	λ	Λ
mu	μ	Μ
nu	ν	Ν
xi	ξ	Ξ
omicron	ο	Ο
pi	π	Π
rho	ρ	Ρ
sigma	σ	Σ
tau	τ	Τ
upsilon	υ	Υ
phi	ϕ,φ	Φ
chi	χ	Χ
psi	ψ	Ψ
omega	ω	Ω

Appendix F

Glossary of Chemical Engineering

absorption the assimilation of a chemical (usually a gas or liquid) *into* a liquid or porous solid.

adsorption the deposition of a chemical (usually a gas or liquid) *onto* the surface of a solid. Charcoal in an aquarium filter or in a tap water purifier *adsorbs* organic chemicals from water.

azeotrope a mixture of two or more components such that the composition of the liquid is the same as the vapor. Not all mixtures form an azeotrope. Water and ethanol form an azeotrope at about 95% water.

base units the units of the base dimensions, which are length, time, mass, temperature, amount, electric charge, and luminous intensity. In the mks SI system, the base units are meter, second, kilogram, kelvin (or degrees Celsius), mole, ampere, and candela.

batch reactor a vessel into which reactants are loaded, induced to react (by increasing the temperature, and/or increasing the pressure, and/or adding a catalyst), then discharged. A batch reactor is not at steady state, although its contents may approach equilibrium. A batch reactor is inherently non-continuous but can be integrated into a continuous process if it is preceded by a hold-up tank and followed by a surge tank. A crock pot is a batch reactor.

biochemical engineering the design and analysis of processes based on biological reactions, for example, to produce a drug or destroy a pollutant.

bioengineering the application of engineering design and analysis to biological systems. The two chief types of bioengineering are biochemical engineering and biotechnology.

biotechnology the application of science and engineering to organisms, cells, and biomolecular processes. Specific examples are the design and analysis of devices used in biological systems, such as artificial organs or diagnostic equipment.

bubble point the temperature at which the first bubble of vapor forms in a liquid at a given pressure, or the pressure at which the first bubble of vapor forms in a liquid at a given temperature. The bubble point of a pure liquid is the boiling point.

capital cost the expenditure to purchase and install equipment. The chief capital cost in a chemical process is usually the reactor.

catalyst a substance that accelerates a chemical reaction without being consumed or modified. A catalyst increases the rate that a system proceeds to equilibrium but does not change the equilibrium concentrations.

centrifugal pump a pump that increases the kinetic energy of a fluid by centrifugal force, then converts the kinetic energy to increase the fluid pressure and reduce the fluid velocity.

centrifugation separation by density quickened by spinning the mixture. The increase in separation rate is determined by the ratio of centripetal acceleration in the spinning mixture to gravitational acceleration. Centrifugation is common for biological separations, such as separating blood into plasma (the liquid portion) and the cells and platelets.

chemical kinetics study of the mechanism and rate of chemical reactions.

closed system an isolated system; a system with no mass and/or energy flow across its borders.

condensation the conversion from vapor to liquid.

conservation law a statement that the amount of some thing is invariant. Invariant things include energy, momentum, and electric charge. The amount of mass is invariant, unless mass is converted into energy.

constitutive equation a constrained law that relates a driving force to a flow of mass, energy, or momentum. For example, Fourier's law of heat conduction relates the driving force of a temperature gradient to a flux of energy.

constrained law A statement governing physical or chemical behavior restricted to certain conditions. The ideal gas law, $PV = nRT$, is a constrained law; it is valid only at low pressure and high temperature.

continuous, stirred-tank reactor (CSTR) a vessel into which reactants are introduced at a constant rate, and effluent is discharged at the same rate. The CSTR's contents are stirred to maintain uniform composition and temperature. To aid stirring, the vessel usually has comparable height and width, like a tank. The effluent has the same composition and temperature as the vessel's contents. It is more accurately called a continuous-flow, stirred-tank reactor, but this terminology is less common.

core variable a variable that appears in exactly one dimensionless group describing some phenomenon.

derived unit a unit created by the product of the base units. The joule (J), a unit of energy in the mks SI system, is the product of kg, m, and s: $1\ \text{J} = 1\ \text{kg}\cdot\text{m}^2/\text{s}^2$.

dew point the temperature at which the first drop of liquid condenses from a vapor at a given pressure, or the pressure at which the first drop of liquid condenses from a vapor at a given temperature. The dew point of a pure liquid is the boiling point.

dimension a physical quality. Any thing can be completely described by specifying its base dimensions: length, time, mass, temperature, amount, electric charge, and luminous intensity. Any parameter, such as energy or momentum, has dimensions. The dimensions of energy are $\text{mass} \cdot \text{length}^2/\text{time}^2$.

dimensional analysis the process of combining the parameters that describe a phenomenon (such as mass, velocity, or viscosity) into dimensionless groups (such as the Reynolds number).

displacement pump a pump that decreases the volume of a chamber that contains the fluid, which increases the pressure of the fluid, forcing it out of the chamber. The chamber is then expanded and refilled. Your heart is a displacement pump.

distillate the components with lower boiling points, and thus higher volatility. The distillate comes off the top of a distillation column. Also known as *light ends* or *tops*.

distillation separation of liquid mixtures based on differences in component volatilities.

dynamic similarity two systems are dynamically similar if the magnitudes of the dimensionless groups describing the systems are equal. A steel sphere falling through molasses may be dynamically similar to a weather balloon rising in air.

elastomer A polymeric material that returns to its initial shape after deformation. Elastomers are usually formed by crosslinking polymer molecules. Rubber bands, silicone rubbers, and automobile tires are elastomers.

electrolytic reactor a device for converting chemical energy into electrical energy. Some electrolytic reactors convert H_2 and O_2 into H_2O and generate electric current.

empirical analysis the study of a phenomenon by measuring behavior to find the correlation between parameters. Empirical analysis does not delve into underlying principles. Also known as parametric analysis.

enthalpy an energy used to characterize open systems. The change in enthalpy is the sum of (1) the change in internal energy upon moving through the system and (2) the work applied to move that substance through the system.

entrainment liquid droplets carried upward by a vapor bubbled through the liquid.

equation of state a mathematical expression that relates various thermodynamics properties. The ideal gas law, $PV = nRT$, is an equation of state for a gas.

equilibrium a system in which the conditions at any position do not change with time, and the conditions are uniform with position. The height of the water in a lake is at equilibrium; the water height is the same everywhere.

extensive property	a thermodynamic parameter that depends on the amount (extent) of a substance. Mass, volume, and internal energy are extensive properties. If 1 kg of water is added to 1 kg of water, the total mass is 2 kg.
extraction	a separation that extracts a component from a liquid mixture by dissolving the component into an immiscible solvent.
flash drum	A vessel in which a liquid is vaporized (*flashed*) by suddenly increasing the temperature and/or decreasing the pressure.
flowsheet	a diagram used to represent a chemical process. Process units are represented by simple geometric shapes, and interconnecting pipes are represented by lines and arrows.
fluid	a substance capable of flowing; a gas, a liquid, and in some cases, a finely particulate solid.
fluid dynamics	the study of the properties and dynamics of fluids. Fluid dynamics includes rheology, acoustics, plasma physics, and quantum fluids.
heat capacity	the ratio of change in enthalpy to the change in temperature of a substance. The heat capacity is a measure of a substance's capacity to absorb heat.
heat exchanger	a process unit that warms one fluid and cools another, by transferring heat from the fluid at the higher temperature. The fluids remain isolated from each other; inlet compositions are the same as outlet compositions.
heavy ends	the components with higher boiling points, and thus lower volatility. The heavy ends come off the bottom of a distillation column. Also known as *heavies* or *bottoms*.
hold-up tank	a process unit that accumulates material and discharges material discontinuously. A hold-up tank might precede a batch reactor in a continuous process, or it might be placed at the end of a process to fill shipping containers.
immiscible	two liquids that do not form a single phase when mixed. Oil and water are immiscible.
intensive property	a thermodynamic property that does not depend on the amount of a substance. Temperature and pressure are intensive parameters. If 1 kg of water at 20°C is added to 1 kg of water at 20°C, the temperature of the mixture is 20°C, not 40°C.
internal energy	an extensive property given by the sum of the electronic, vibrational, rotational, and translational energies of the molecules or atoms constituting the substance. The ratio of an incremental change in internal energy to an incremental change in temperature is the heat capacity at that temperature.
light ends	*see* distillate
mass balance	an accounting for the mass entering, leaving, and accumulating in (or depleting from) a system.

mathematical model an equation or set of equations that predicts the behavior of a phenomenon.

mole the amount of a substance containing 6.022×10^{23} units, usually atoms or molecules. 1 mole of H_2O has a mass of 18 g.

monomer from Greek, meaning one (*mono*) part (*meros*); a molecular building block that can be chemically linked to form polymers (thousands of units) or oligomers (tens to hundreds of units). Ethylene, $CH_2{=}CH_2$, is the monomer to the polymer polyethylene, $(-CH_2-CH_2-)_n$; tetrafluoroethylene, $CF_2{=}CF_2$, is the monomer to the polymer polytetrafluoroethylene (teflon) $(-CF_2-CF_2-)_n$.

oligomer from Greek, meaning a few (*oligo*) part (*meros*); a molecule with modest molecular weight (typically 100 to 1000 amu), composed of several chemical units (monomers). *See* polymer.

open system a system with flow of mass and/or energy across its borders.

operating cost The expense of producing a commodity or providing a service, after equipment has been purchased and installed. The chief operating cost in a chemical process comprises reactants and utilities.

operating line a straight line drawn on an equilibrium map of vapor composition vs. liquid composition, used in the graphical analysis of distillation columns and absorption columns. An operating line is usually devised from a point given by the compositions at the top or bottom of the column and a slope given by the ratio of flow rates moving up and down the column.

pi group Π group; a dimensionless product and ratio of parameters. A system may be characterized by the magnitude of its dimensionless group(s). For example, the magnitude of the Reynolds number indicates whether a fluid flow is laminar or turbulent.

plastic a moldable polymeric material. A thermoplastic polymer such as polyvinyl chloride(PVC) becomes pliable when heated, and can be reformed by reheating. Thermosetting polymers such as epoxies or polyphenol-formaldehydes (Bakelite) cannot be reformed by reheating.

plug flow reactor (PFR) a vessel into which reactants are introduced at a constant rate and effluent is discharged at the same rate. However, the composition and temperature in a PFR vary along the length of the reactor, which is usually long and narrow like a pipe.

polymer from Greek, meaning many (*poly*) parts (*meros*); a molecule with high molecular weight (typically 10^4 to 10^6 amu), composed of many chemical units (monomers). Common synthetic polymers are polyvinyl chloride (PVC, $(-CH_2-CHCl-)_n$), polypropylene ($((-CH_2-CH(CH_3)-)_n$), and polystyrene ($(-CH_2-CH(C_6H_5)-)_n$). Common natural polymers are cellulose, RNA, and DNA. The two chief types of polymers are plastics and elastomers (rubbers).

pump a process unit that increases the pressure of a fluid. Common types are centrifugal and displacement.

rate equation a mathematical expression for the rate of consumption of reactant in a chemical reaction. The rate equation usually includes reactant concentration(s), product concentration (s), and temperature.

reactor a vessel in which contents are transformed by chemical reaction. Common reactor types are batch, continuous stirred-tank, and plug flow.

rectifying section the trays (stages) in a distillation column above the feed tray, including the condenser.

recycle stream a flow of material taken from the principal flow in a process and returned to an earlier point in the process. Material is usually recycled from reactors with low conversions and from separators with poor separation.

reduced parameter a physical parameter, such as length, multiplied and divided by other parameters to form a dimensionless group. For flow through a pipe of length ℓ, the reduced length is ℓ/d, where d is the pipe diameter.

reflux ratio the ratio of the molar flow of liquid flowing down a distillation column to vapor flowing up; reflux ratio = L/V. The reflux ratio is less than 1 in the rectifying section and greater than 1 in the stripping section. Occasionally defined (elsewhere) as the ratio of liquid to distillate; reflux ratio = $L/D = L/(V-L)$.

residence time The mean time a molecule spends in a given vessel. It is most useful in reactors, where the probability that a molecule reacts increases with time spent in the reactor. For a continuous reactor, the residence time will increase if the flow rate into (and out of) the reactor is decreased.

return on investment (ROI) profit divided by capital cost.

rheology broadly defined as the study of the deformation and flow of fluids. In common usage, rheology refers to the study of non-classical and viscoelastic fluids.

rubber an elastomer.

stage a section of a distillation column or absorber in which two fluids contact, and components transfer between fluids to approach equilibrium compositions. In a distillation column, one fluid is liquid moving down the column and the other is vapor moving up the column. Also called a tray or plate.

static system a system that may be at steady state or at equilibrium.

steady state a system in which the conditions at any position do not change with time, although the conditions may vary with position. The height of the water in a river may be at steady state, although the water is higher upstream and lower downstream. A chemical process may be at steady state, although the chemical composition varies along the process; it varies from reactants at the inlet to products at the outlet.

Stokes's law predicts the terminal velocity of a sphere moved through a fluid by gravity, given the sphere's diameter (D) and density (ρ), the viscosity of the fluid (μ), and gravitational acceleration (g):

$$v = \frac{1}{18}\frac{gD^2\rho}{\mu}.$$

stripper A process unit that extracts (*strips*) a volatile component from a liquid mixture by contacting the liquid mixture with a gas stream. Bubbling air through a water/benzene mixture will extract (strip) the benzene.

stripping section the trays (stages) in a distillation column below the feed tray, including the evaporator. In the lower stages, the vapor *strips* the light ends from the mixture.

sublimation the transition from solid to vapor.

surge tank a process unit with an unsteady flow rate in but a steady flow rate out. The quantity of material in a surge tank varies with time.

system a subset of the universe demarcated by boundaries; the demarcation is usually made for the convenience of a mathematical model.

thermodynamics the study of empirical relations between the various forms of energy. The term derives from the initial emphasis of the field: the study of heat engines and the interconversion of heat (*thermo*) and mechanical motion (*dynamics*).

tie line a line on a phase diagram that crosses a two-phase region and connects coordinates on the two borders of the region, representing the two phases in equilibrium. On temperature–composition and pressure–composition phase diagrams, a tie line is always horizontal.

transient process a process in which the conditions change with time.

tray *see* stage.

turbine (or turboexpander) a process unit that converts the pressure energy of a fluid into mechanical energy. Steam generated at high pressure in power plants is fed to a turbine to convert the pressure energy to mechanical energy, which is subsequently converted to electrical energy in generators.

universal law a statement governing physical or chemical behavior valid at any conditions, at all times. The conservation of energy is a universal law.

viscoelastic fluid a fluid whose resistance to motion depends on how much the fluid has been deformed. Silly Putty and molasses are viscoelastic fluids.

Index

absorber, 33
  multistage, cascade, 272–277
absorbers, 32–35
  dimensional analysis, 457–459
  exercises, 53–54, 158, 166, 329–350
  mass transfer in, 528–531
  multistage, 272–277
  single stage, 267–272
absorption, 267, 562
acetanilide
  synthesis of, 72
acetone, thermodynamic properties, 86, 200, 498
acetonitrile–methylpyridine mixtures, 307, 363
acetonitrile–nitromethane mixtures, 328
acetonitrile–water mixtures, 362, 409
acetylene
  reaction with hydrogen chloride, 69, 174
  thermodynamic properties, 86
Addams, Morticia, 435
adsorbents, 300
adsorbers, 34, 300, 343
adsorption, 170, 300, 342, 524, 562
AIChE, 3
air
  cooling by water evaporation, 191
  thermodynamic properties, 125
air conditioning, 132, 161, 376
air drying, 160
air pollution, 4
Alaskan pipeline, 448
alkylation, 67, 161
aluminum
  impurity in silicon, 26
  reaction with hydrogen chloride, 27
  reaction with propanol, 22
  thermodynamic properties, 59, 86
ammonia
  adiabatic reactor for, 143
  as a refrigerant, 128
  as fuel, 195
  chemical equilibrium, 140
  from manure, 83
  mixtures with water, 266, 325–326
  reaction to aniline, 72
  reaction to hydrazine, 70
  reaction to hydrogen cyanide, 197
  reaction to nitric acid, 61
  reaction to urea, 197
  synthesis of, 8–16
  thermodynamic properties, 87, 200
aniline–heptane–methylcyclohexane mixtures, 322, 396
aniline–hexane–methylcyclopentane mixtures, 318, 395
aniline
  synthesis of, 72
antibiotics, 5, 72
artificial kidney, 157, 483
astronauts, protection of, 81
atmospheric chemistry, 512–517
automobile catalytic converter, 523
azeotrope, 562

bacteria
  for ethanol to acetic acid, 541
  in drinking water, 82
  in sewage treatment, 508–512
Baekeland, Leo, 423
Bakelite, 4
Basaran, Osman, 475
base dimensions, 426
base units, 426
batch process, 9, 517
batch reactor, 512
battleship waves, 481
beeswax, 192
benzene
  absorption from air, 158, 267
  adsorption from air, 170, 342
  condensation from air, 294
  isomerization of allyl to propenyl, 99
  mixtures with toluene, 253
  pressure–volume phase diagram, 291
  reaction to acetanilide, 73
  reaction to aniline, 72
  reaction to phenol, 55

benzene (cont.)
  reaction to styrene, 67, 161
  thermodynamic properties, 86
biotechnology, 6, 562
body mass
  relation to bone diameter, 495–496
  relation to wing area, 463
Bond number, 474
bottoms product, distillation, 283
brine, 91
bubble point
  mixture, 253
  pure substance, 292
bubbler, 267
bubbles, rising in liquids, 473
Buckingham Pi theorem, 433
buoyancy, 437
butane, dehydrogenation, 517
butanol
  as an absorber fluid, 528
  thermodynamic properties, 498
butanol–water mixtures, 315
butene, dehydrogenation, 517

calcium carbonate
  in the Solvay process, 60
  thermodynamic properties, 86
calcium chloride
  in the Solvay process, 60
  thermodynamic properties, 86
Capillary number, 475
capital costs, 147
carbon cycle for Earth, 165
carbon dioxide
  in Earth's carbon cycle, 165
  in fuel cell cycle, 30
  in hydrogen fuel cycle, 39
  in synthesis of urea, 197
  supercritical fluid for extractions, 319
  thermodynamic properties, 86
  transfer from lungs to air, 331
carbon monoxide
  from carbon dioxide, 198
  in automobile catalytic converters, 523
  in syngas, 135
  oxidation of, 523
  reaction with nitric oxide, 524
  thermodynamic properties, 86
carbon tetrachloride
  synthesis of, 54
  thermodynamic properties, 86
carbonate ions, in fuel cells, 30
catalytic cracking, 4
charcoal, as benzene adsorbent, 342
chemical engineering
  achievements, 3
  definition, 2
  opportunies, 5
chemical kinetics
  definition, 563
  in reactor design, 136, 527
chlorine
  in synthesis of acetanilide, 72
  reaction to carbon tetrachloride, 54
  reaction to hydrazine, 70
  reaction with ethylene, 68
  thermodynamic properties, 87
chlorobenzene
  in acetanilide synthesis, 72
  mixtures with hexane, 313
  mixtures with tetrachloromethane, 306
chlorofluorocarbons, 128, 190
chlorosilanes
  in silicon processing, 28
  removal from air, 84
citric acid, synthesis of, 500
Clausius–Clapeyron equation, 468
closed system, 91, 563
clothes, process to wash, 58
coal
  conversion to diesel fuel, 48
  conversion to methanol, 173
  in hydrogen synthesis, 35
  in silicon synthesis, 26
  sulfur emissions, 202
  underground gasification, 35
cocurrent flow, 126
coffee
  mug design, 192
  process for instant, 168
combiner, 10
compressibility, gas, 461
compressor, 18
condensation, 289
  for separating benzene from air, 294
condenser, in distillation columns, 282
conservation
  of assets, 145, 148
  of energy, 118–119, 126
  of mass, 91, 95, 104
conservation law, 155, 563
constitutive equation, 155
continuous process, 9
continuous reactor, 523
continuous, stirred-tank reactor (CSTR), 524
conversion factors, 549
cooling
  by evaporation, 191
  lake source, 190
core variable, 433
  guidelines for selecting, 438
  rules for a valid set of, 449
countercurrent flow, 126
  in multistage separators, 278
critical point, 293
  in dimensional analysis of gases, 460

Index

crude oil, in Alaskan pipeline, 449
CSTR, 524
cyclohexane
  partial oxidation to adipic acid, 74
  thermodynamic properties, 86, 200

Dalton's law, 140
dehydrogenation
  of butane to butene, 517
  of ethylbenzene to styrene, 67
depreciation, 148
derived units, 427
desalination of seawater, 90, 95–99
  analysis of energy flow, 118
  by freezing, 95
design tools
  for transient-state processes, 499
  from dimensional analysis, 447
  from graphical analysis, 301
  from mathematical modeling, 156
desulfurization of natural gas, 32
dew point, 563
  of a mixture, 258
  on enthalpy–composition diagrams, 264
  pure substance, 292
dichloroethane
  by-product in vinyl chloride synthesis, 174
  condensation from air, 386–387
  mixtures with propanol, 408
  thermodynamic properties, 86
diesel fuel, 48
differential equation, 499
  boundary conditions for, 503
  initial conditions for, 503
  numerical integration of, 521, 542
  separable, 503
dimensional analysis
  advantages of, 466
  design tools from, 447
  general method for, 436
dimensional consistency, 427
dimensionless groups, 547
  method for deriving, 448
  selecting core variables, 438, 449
dimensions, base, 426
dinosaurs, 425
  bone diameters and mass, 497
  speed of, 436
dioxin, 81
distillate, 564
distillate product, distillation, 283
distillation
  by flashing, 254
  multistage, 277
distillation columns, 283–289
  dimensional analysis of, 457
  exercises, 350–374
  schematic with sieve plates, 458
dough, recipe for, 58
dust in air, terminal velocity, 444
dynamic similarity, 448, 451, 467, 564

earthenware, cooling by evaporation, 191
Einstein, Albert, on problem solving, 23
Einstein's equation for the molar heat capacity of a solid, 468
Einstein's mass–energy equation, 468
elasticity number, 486
electric arc furnace, 26
elevators, delays of, 83
energy flow rates, rules of thumb, 122
engineering, description of, 2
English units, 427
enthalpy–composition diagrams, 262–267
  exercises, 324–328
equilibrium line
  for absorbers and strippers, 269
  for distillation columns, 280
ethanol
  in spirit vinegar, 541
  measured by a breathalyzer, 332
  mixtures with methanol, 158
  mixtures with water, 364, 408
  mixtures with water and methanol, 316
  reaction to acetic acid, 541
  synthesis from grain, 182
ethylene
  chlorination of, 68
  pressure–enthalpy diagram, 389
  reaction to styrene, 67, 161
  thermodynamic properties, 86
ethylene oxide
  synthesis from ethylene, 49
  thermodynamic properties, 86
Euler method of numerical integration, 543
Euler number, 450
explosion, dimensional analysis of, 470

Fanning friction factor, 453
feed tray, distillation, 283
fermentor, 500
fertilizer, 8
fibers, optical, 241
Fick's second law of diffusion, 473
flash drum, 254, 259
  mass balance on, 158
  mulitistage cascading, 277
flash drums
  exercises, 306–312
flooding in vapor–liquid columns, 457
flow
  from a draining tank, 504
  in a rough pipe, 452
  in a smooth pipe, 447
  over a spillway, 534
  through a notch, 536
  through an orifice plate, 476

flowsheets, 9
  conventions, 17, 23
  general practices, 57
  tips for analyzing, 40
fluid flow in a pipe, 447
  heat transfer, 454–457
  roughness, 453
fraud, scientific, 495
friction factor
  Fanning, 453
  for a sphere in a fluid, 554
Froude number
  for a sphere moving through a fluid, 440
  in walking, 434
  relation to the Galileo number, 441
fuel cells, 29

Galileo, 431
Galileo number for a sphere moving through a fluid, 441
galloping
  and bone strength, 497
  Froude number, 436
geometric scaling, 423
glucose, oxidation of, 172
graph paper, 554–558
groundwater pollution, 342

Hagen–Poiseuille law, 469
half-life, 515
heat exchanger, 15, 565
  energy balance on, 123–128
  in series with adiabatic reactors, 136
heat pump, 128
  coefficient of performance, 151
heat transfer
  by forced convection, 471
  fluid in a pipe, 455
  in fan-assisted ovens, 471
  Péclet number, 472, 484
  Prandtl number, Pr, 472
  solid sphere in a fluid, 478
  temperature control in a reactor, 531–534
heat-transfer coefficient, 456
  in a chemical reactor, 531
  vertical tube, 469
Henry's law, 155
heptane
  mixtures with aniline and methylcyclohexane, 322
  process to purify, 19–23
  process to purify, recycle, 115
hexane
  mixtures with aniline and methylcyclopentane, 318
  mixtures with chlorobenzene, 313
  mixtures with pentane and heptane, 212
home heating, energy analysis of, 192
humidity, 161, 376, 378

hydrazine
  synthesis of, 70
  thermodynamic properties, 87
hydrochlorination
  of acetylene, 69
  of ethylene, 68
hydrogen
  storage as a metal hydride, 44–47
  storage options, 42–44
  synthesis from coal, 35–36
  synthesis from methane, 29–32
  synthesis from natural gas, 32–35
  synthesis from water, 36–40
  thermodynamic properties, 87
hydrogen chloride
  reaction to chlorine, 69
  reaction with acetylene, 69, 174
  reaction with silicon, 27
  thermodynamic properties, 87
hydrogen cyanide
  in processing steel, 61
  synthesis by BMS process, 197
  synthesis from ammonia and methane, 53
  synthesis of, 197
  thermodynamic properties, 87
hydrogen peroxide
  in synthesis of hydrazine, 70
  synthesis of, 71
  thermodynamic properties, 87
hydrogen sulfide
  absorption from hydrogen, 164
  from manure digestion, 82
  in natural gas, 32
  thermodynamic properties, 87

ideal gas law, 155, 243, 423, 459–463, 469, 561
ideal solutions, 155, 295
immiscible liquids, phase diagrams, 318, 321, 474
instant coffee, 168
integrals, table of, 560
iron
  impurity in silicon, 26
  steel processing, 60
  thermodynamic properties, 27
iron oxide
  as catalyst, 32
  in cycle to convert carbon dioxide to methane, 73
  in steel processing, 61

just-in-time strategy for repairs, 214

Keynes, John Maynard, 145
kidney, artificial, 5, 157, 470

lake-source cooling, 190
laws, universal vs. constrained, 155
lead, contamination in water, 82
least squares, method of, 493

Leibniz, Gottfried, on conservation of energy, 118
Levenspiel, Octave, 463
lever rule, 244–247
  applied to tie lines, 259
  derivation of, 159
  for energy balances, 247–253
  for enthalpy–composition diagrams, 265
Lewis number, 473
limestone
  as sulfate scrubber, 202
  in Solvay process, 60
linear regression, 493
lions, strategy for, 80
log–log graph paper, 555
lungs, modeled as an absorber, 330

magnesium hydroxide, as sulfate scrubber, 62
magnesium sulfate
  mixtures with water, phase diagram, 394
make-up stream, 34
map, thermodynamic, 255
mass transfer
  across liquid–liquid interfaces, 528
  across vapor–liquid interfaces, 267, 457, 481
  Fick's law, 473, 529
  for concentration gradients, 466
  Péclet number, 484
McCabe–Thiele method, 283–289
membrane separation of air, 158
meteor crater, dimensional analysis of, 482
methane
  in natural gas, 32
  in synthesis of ammonia, 12
  in synthesis of hydrogen, 29–32, 64
  in synthesis of hydrogen cyanide, 197
  in synthesis of nitric acid, 62
  oxidation, adiabatic temperature rise, 134
  oxidation, heat of reaction, 134
  partial oxidation to methanol, 65
  synthesis from carbon dioxide and water, 73
  synthesis from syngas, 35, 135–139
  thermodynamic properties, 86
methane burner, 13
methanol
  from partial oxidation of methane, 65
  mixtures with ethanol, 158
  mixtures with water, 309, 358
  mixtures with water and ethanol, 316
  oxidation, adiabatic temperature rise, 196
  synthesis from coal, 173
  synthesis from methane, 66
  synthesis from syngas, 36, 136, 140, 198
  thermodynamic properties, 86
methylcyclohexane
  mixtures with aniline and heptane, 322, 396
methylcyclopentane
  mixtures with aniline and hexane, 318, 395
miscible liquids, ternary diagrams, 399
mixer, 17
mixing lines, 246–247, 272
  on enthalpy–composition diagrams, 266
  on ternary diagrams, 316
monoethanolamine, for hydrogen sulfide absorption, 34
myricyl palmitate
  in coffee mug design, 192
  thermodynamic properties, 200

natural gas
  in synthesis of hydrogen, 32
  removal of hydrogen sulfide, 32
Newton, Isaac, on mathematical modeling, 155
newton, unit of force, 427, 548
Newton's
  law of gravitation, 468
  law of resistance, 444
  law of viscosity, 437
  second law, 89, 155
nitric acid
  mixtures with water, 350
  synthesis of, 61
  thermodynamic properties, 87
nitric oxide
  reaction with carbon monoxide, 524
  thermodynamic properties, 87
nitrogen
  heat capacity, 125
  in synthesis of ammonia, 8–16
  thermodynamic properties, 87
nitrogen dioxide
  pollutant in air, 62
  thermodynamic properties, 87
nitrous oxides, pollutants in air, 513
numerical integration, 542
Nusselt number, 478

Ohnesorge number, 475
operating costs, 146
  depreciation, 148
  vs. capital costs, 147, 149, 151
operating lines
  for absorbers, 267–272
  for multistage cascade absorbers, 272–277
  for multistage cascade flash drums, 283–289
orange juice, concentration of, 159
orifice
  diameter, in a bubble plate, 472
  equation, 505
  in bottom of draining tank, 504
  plate in pipe, Reynolds number, 477
osmotic separation, 345
oxychlorination of ethylene, 69
oxygen
  absorption in lungs, 330
  atmospheric reactions, 512
  in methane burner, 14
  in synthesis of adipic acid, 74

oxygen (cont.)
  partial reduction to hydrogen peroxide, 71
  thermodynamic properties, 88
ozone, role in the stratosphere, 512

paintbrush, options for cleaning, 156
partial pressure
  and Dalton's law, 140
  and Henry's law, 155
  relation to thermodynamic activity, 140
Pauling, Linus, on good ideas, 89
Péclet number, 466, 472, 481, 485
pendulum, dimensional analysis of, 428
PFR, 522, 566
phase diagrams
  enthalpy–composition, 262–267
  pressure–composition, 253–262
  pure substances, 289–293
  temperature–composition, 253–262
  ternary systems, 318
phenol, thermodynamic properties, 303
photolithography, 50
physical constants, 550
Pi group, 432, 566
Pi theorem, 433
pinch point, 276, 287
pipes, cost of insulation, 203
pizza dough, recipe for, 58
plate tectonics, dimensional analysis of, 482
plug flow reactor (PFR), 522, 566
poise, base dimensions of, 437
popcorn, separation from corn kernels, 479
potassium–sodium mixtures, 392
Prandtl number, 456, 472, 478–479, 483
pressure–composition vapor + liquid phase diagrams, 253–262
pressure-swing processes, 288, 409
problem definition, 38, 47, 80
problem solving
  defining the real problem, 19
  McMaster heuristic, 23
  successive, for design evolution, 8
process control, 1, 7, 511
profit, 146, 149
  vs. return on investment, 150, 152
propane, phase diagram, 382
propanol
  mixtures with dichloroethane, 409
  mixtures with dimethylamine, 312
  removal from heptane, 19–23
  thermodynamic properties, 20
pump, 18
purge stream, 14, 22
pyrolysis reactions, 52, 68

rain
  and runoff pollution, 85, 342
  prediction of, 378

rate constant, in chemical reactions, 515, 533
reactor, 8
  and process viability, 17
  batch, 513, 517
  continuous, stirred-tank (CSTR), 524
  flowsheet guidelines for, 19
  plug flow, 523
  tips for flowsheet analysis, 40
rectifying section, distillation, 283, 567
recycle
  economic analysis of, 152
  mass balances on, 95–99, 110–114
  with purge, 14, 34, 104
  with separation, 21, 104
reduced parameters, 434, 439, 459, 567
refrigeration cycle, 130
repairs, just-in-time strategy, 214
residence time, 510–511, 523, 525, 530
return on investment (ROI), 149, 567
  vs. profit, 150, 152
ROI, *see* return on investment

sand
  as a projectile, 464
  in silicon synthesis, 25
scale-up, 4, 423
scrubbers, 202
seawater, composition, 90
semilog graph paper, 555
separator
  and phase lines for analysis, 41
  and process economics, 17
  flowsheet conventions, 18
  liquid–gas, 10
  solid–liquid, 20
sewage treatment, 508–512
Sherwood plot, 244
SI units, 427
Siemens process, 28
sieve plates, 458
significant figures, 551–553
silicon
  process for electronic-grade, 25–29
  thermodynamic properties, 27
silicon wafers
  process to coat for lithography, 50, 85, 241
snow
  mass of, 241
  prediction of, 378
soda ash, Solvay process, 60
sodium carbonate
  in Solvay process, 60
  thermodynamic properties, 87
sodium sulfate, mixtures with water, phase diagram, 393
sodium–potassium mixtures, 392
solid–liquid separators, 20, 215
Solvay process for soda ash, 60

soybeans, oil extracton from, 166
sphere, terminal velocity of, 437
splitter, 14
spreadsheets
  for data analysis, 487–493
  for mass balances, 178–186
  for numerical integration, 542–545
stages, equilibrium
  in absorbers, 275
  in distillation, 283
Stanton number, 456, 478
steady state
  definition, 92, 499
  multiple states, 523
  stability of, 526
Stokes's law, 155, 444, 568
stratosphere, reactions in, 512
stream, process, 9
stripper, 568
  model for lungs, 331–332
stripping section, distillation, 283
styrene
  synthesis of, 67
  thermodynamic properties, 87, 200
sulfur
  oxidation of, 51
  thermodynamic properties, 88
sulfur dioxide
  pollutant in air, 62
  thermodynamic properties, 88
sulfuric acid
  in a bubbler to dry air, 166
  in methanol synthesis, 67
  in steel processing, 60
  mixtures with water, 327
  thermodynamic properties, 87
sulfur–iodine cycle for hydrogen synthesis, 38
surge tank, 500
syngas, 135
  from coal, 35
  from methane and water, 31, 65
  in fuel cells, 29
  reaction to methanol, 198
synthesis gas, *see* syngas
system
  borders for mass balances, 92, 96, 114
  open vs. closed, 91
  steady-state, 92
  transient-state, 499

tank
  draining, 471, 504
  sewage, 508
  surge, 500
temperature–composition vapor + liquid phase diagrams, 253–262
ternary diagrams, 324–328
tetrachloroethylene, synthesis of, 52
tetrachloromethane
  mixtures with chlorobenzene, 306
tie line, 249
toluene
  mixtures with benzene, 253
trajectory line
  for reactors on temperature-conversion maps, 145
  for single-stage absorber, 272
trays, *see* stages
triple point, 293, 460
trucking, 435

uncertainty, in linear regression, 493
unit operations, 2, 17
units, base and derived, 426
units, process, 17
urea, synthesis of, 197

valve, flowsheet symbol, 18
vapor–liquid equilibrium diagram, 280
vapor–liquid phase diagram, 257
vinyl chloride
  synthesis of, 68
  thermodynamic properties, 86
viscosity, dimensions of, 437
viscous coiling, 470

walking
  analysis of, 431–436
  universal correlation for, 434
water
  and humidity in air, 378
  condensation from air, 387
  cooling by evaporation, 191
  energy–phase line, 250
  extraction by osmotic separation, 345
  mixtures with acetonitrile, 362
  mixtures with ammonia, 325
  mixtures with butanol, 315
  mixtures with ethanol, 364
  mixtures with methanol, 309
  mixtures with methanol and ethanol, 316
  mixtures with nitric acid, 350
  mixtures with sulfuric acid, 327
water cycle for Earth, 165
water–gas shift reaction, 32
waves
  blast, from explosions, 470
  bow, 434, 481
  dimensional analysis of, 470
weather, prediction of rain and snow, 378
Weber number, 475
Wellington, Arthur Mellen, and engineering, 33

zeolites, 170
zinc oxide, catalyst, 32

Printed in the United States
by Baker & Taylor Publisher Services